Array Signal Processing

PRENTICE HALL SIGNAL PROCESSING SERIES

Alan V. Oppenheim, Series Editor

ANDREWS AND HUNT *Digital Image Restoration*
BRIGHAM *The Fast Fourier Transform*
BRIGHAM *The Fast Fourier Transform and Its Applications*
BURDIC *Underwater Acoustic System Analysis, 2/E*
CASTLEMAN *Digital Image Processing*
COWAN AND GRANT *Adaptive Filters*
CROCHIERE AND RABINER *Multirate Digital Signal Processing*
DUDGEON AND MERSEREAU *Multidimensional Digital Signal Processing*
HAMMING *Digital Filters, 3/E*
HAYKIN, ED. *Advances in Spectrum Analysis and Array Processing, Vols. I & II*
HAYKIN, ED. *Array Signal Processing*
JAYANT AND NOLL *Digital Coding of Waveforms*
JOHNSON AND DUDGEON *Array Signal Processing: Concepts and Techniques*
KAY *Modern Spectral Estimation*
KINO *Acoustic Waves: Devices, Imaging, and Analog Signal Processing*
LEA, ED. *Trends in Speech Recognition*
LIM *Two-Dimensional Signal and Image Processing*
LIM, ED. *Speech Enhancement*
LIM AND OPPENHEIM, EDS. *Advanced Topics in Signal Processing*
MARPLE *Digital Spectral Analysis with Applications*
MCCLELLAN AND RADER *Number Theory in Digital Signal Processing*
MENDEL *Lessons in Digital Estimation Theory*
OPPENHEIM, ED. *Applications of Digital Signal Processing*
OPPENHEIM AND NAWAB, EDS. *Symbolic and Knowledge-Based Signal Processing*
OPPENHEIM, WILLSKY, WITH YOUNG *Signals and Systems*
OPPENHEIM AND SCHAFER *Digital Signal Processing*
OPPENHEIM AND SCHAFER *Discrete-Time Signal Processing*
QUACKENBUSH ET AL. *Objective Measures of Speech Quality*
RABINER AND GOLD *Theory and Applications of Digital Signal Processing*
RABINER AND SCHAFER *Digital Processing of Speech Signals*
ROBINSON AND TREITEL *Geophysical Signal Analysis*
STEARNS AND DAVID *Signal Processing Algorithms*
STEARNS AND HUSH *Digital Signal Analysis, 2/E*
TRIBOLET *Seismic Applications of Homomorphic Signal Processing*
VAIDYANATHAN *Multirate Systems and Filter Banks*
WIDROW AND STEARNS *Adaptive Signal Processing*

Array Signal Processing: Concepts and Techniques

Don H. Johnson
Rice University

Dan E. Dudgeon
Lincoln Laboratory,
Massachusetts Institute
of Technology

P T R Prentice Hall, Englewood Cliffs, NJ 07632

Library of Congress Cataloging-in-Publication Data

Johnson, Don H.
Array signal processing : Concepts and techniques / Don H. Johnson, Dan E. Dudgeon.
p. cm.
Includes bibliographical references and index.
ISBN 0-13-048513-6 :
1. Signal processing--Digital techniques. I. Dudgeon, Dan E. II. Title.
TK5102.5.J615 1993
621.382′2--dc20 92-10476
CIP

Editorial/production supervision: bookworks
Cover design: Bruce Kenselaar
Manufacturing buyer: Mary McCartney

A Division of Simon & Schuster
Englewood Cliffs, New Jersey 07632

The publisher offers discounts on this book when ordered in bulk quantities. For more information, contact:

Corporate Sales Department
PTR Prentice Hall
113 Sylvan Avenue
Englewood Cliffs, New Jersey 07632
Phone: 201-592-2863
Fax: 201-592-2249

Printed in the United States of America
10 9 8 7 6 5 4 3 2 1

ISBN 0-13-048513-6

Prentice-Hall International (UK) Limited, *London*
Prentice-Hall of Australia Pty. Limited, *Sydney*
Prentice-Hall Canada Inc., *Toronto*
Prentice-Hall Hispanoamericana, S.A., *Mexico*
Prentice-Hall of India Private Limited, *New Delhi*
Prentice-Hall of Japan, Inc., *Tokyo*
Simon & Schuster Asia Pte. Ltd., *Singapore*
Editora Prentice-Hall do Brasil, Ltda., *Rio de Janeiro*

To Katy, Alexa, and Ken;
and to Judy, Lindsay, and Jeffrey

Contents

Preface

This book discusses problems, algorithms, and solutions for processing signals received by arrays of sensors. The book incorporates the latest results from the field of digital signal processing into a coherent discussion of array processing applications. We wrote it primarily to serve as a text for a graduate course in advanced digital signal processing, but practicing engineers should also find it useful as a reference for array processing. We assume the reader is familiar with the fundamental concepts of digital signal processing, gained perhaps by studying a text such as *Discrete-Time Signal Processing* by Oppenheim and Schafer (Prentice Hall) or its predecessor, *Digital Signal Processing*, also by Oppenheim and Schafer (Prentice Hall). Our goal is to assemble in a coherent way a variety of theoretical and practical approaches to sensor array processing problems. Mastery of the concepts presented in this book will give the reader a strong foundation for approaching problems in applications areas such as acoustic signal processing, sonar, radar, geophysical processing, and, to a lesser extent, tomography, computed imaging, and ultrasonic imaging.

We envision several ways in which this text may be incorporated into a graduate curriculum in digital signal processing. This book contains more than sufficient material and depth of discussion to support a one-term course in sensor array processing. Some of the topics discussed, such as sampling in space and time and derivation of array patterns, are elaborations on fundamental concepts of digital signal processing, giving the student some enlightening variations on the themes learned in a first course on digital signal processing. Other topics, such as detection, parameter estimation, and tracking, may reinforce material covered in courses on detection and estimation theory. Alternatively, this book could be used in conjunction with other texts, for example *Multidimensional Digital Signal Processing* by Dudgeon and Mersereau (Prentice Hall), for a course on advanced topics in digital signal processing. Part of the motivation for writing this book is the realization that many of these topics are being taught in advanced courses using informal notes.

In addition to its use as a graduate text, this book should be helpful to the practicing engineer, particularly one who has recently graduated and has become immersed in an applications area, such as sonar, not covered explicitly during his or her formal education. It forms a convenient bridge for such a person, connecting basic understanding of digital signal processing concepts to sensor array processing concepts useful in applications. For the experienced engineer, the book draws together a variety of techniques—some familiar, some modern and perhaps less familiar—into a common framework and provides an analytical basis for reaching a deeper understanding of the processing techniques that have been developed in his/her particular specialty. Its range of topics and its references to the technical literature should make this book a useful summary of array processing techniques.

The book begins with a discussion of the objectives of sensor array processing and describes some typical applications, such as signal detection, estimation of propagation direction, and measurement of frequency content. We then introduce the physics of wave propagation along with an entry-level discussion of how media affect propagation. Those aspects of propagating waves that can be exploited by array processing algorithms are highlighted, foreshadowing what the remainder of the book discusses. One important concept that arises from the physics is the notion of decomposing a general space-time signal into a superposition of propagating plane waves using the Fourier Transform. From this viewpoint, the wavenumber-frequency spectrum is introduced as an alternative description of space-time signals. Since the use of modern signal processing algorithms implies the use of digital computers and sampled-data systems, we discuss the sampling of signals in space as well as time. This treatment includes multidimensional sampling patterns, the Sampling Theorem, and the multidimensional discrete Fourier Transform. In subsequent chapters, the concepts of beamforming, detection, and power spectrum estimation are developed in detail. The goal of these chapters is to show how these techniques can be applied to problems in sensor array processing rather than duplicating results found in existing texts. Results from estimation and optimization theory are derived in the context of sensor array processing to motivate the development of high-resolution, adaptive beamforming techniques including methods based on the eigenanalysis of covariance matrices. Finally, multiarray tracking is discussed, to give the engineer examples of systems built upon the foundation of sensor array processing.

This book is a technical book; it relies on mathematics to reach its many observations and conclusions. We have tried to write it, however, in a style that appeals to the reader's intuition, and we strive to develop and refine that intuition for the topics discussed. In some cases, mathematical rigor and lengthy derivations have been sacrificed to get to the point. We hope that this style does not offend purists unduly; we recognize the importance of mathematical precision in any analytical endeavor, but we feel the need for engineering judgment in solving real-world problems. We firmly believe that mathematics should be used to support and verify intuition, not substitute for it.

Many have helped us in our writing efforts. Rice University, through its sabbatical program, enabled us to work together at Lincoln Laboratory for a year during the initial stages. Throughout the project, both of our institutions have supplied generous background support, allowing unfettered access to word processing systems. Many reviewed manuscript drafts; we thank B. Aazhang, R. D. DeGroat, D. J. Edelblute, D. A. Linebarger, J. H. McClellan, and D. B. Williams for their efforts. The book's quality has been enhanced by their reviews. Finally, this book has been formatted within the LaTeX typesetting system; all in all, it helped more than hindered, but we did find a few bugs along the way.

A note about notation: We have exploited the referential capabilities of LaTeX in making our book useful as a reference. We use page-number references, enclosed in braces, in addition to equation-number and section-number references. For example, the wave equation, which enables array processing, is presented in §2.2 as Eq. 2.1 {11}. The index is extensive and indicates, when possible, where we define terms as well as where we give examples exploiting their underlying notions. We provide a list of symbols and where they are defined to mitigate the problem of symbol overload common in any lengthy technical document such as this one. Boxed equations are important, representing the ones the reader must understand to appreciate fully the concepts of array signal processing.

We sincerely hope that you find this book useful and learn as much by studying it as we did in writing it.

Don H. Johnson
Dan E. Dudgeon

Houston, Texas
Acton, Massachusetts

Array Signal Processing

Chapter 1

Introduction

We survive because we can sense the environment around us. Our eyes detect electromagnetic radiation in a band from roughly 450 THz to 750 THz.† This propagating radiation carries an immense amount of information that we use to great advantage. Our optical processing system tells us the intensity, the direction of propagation, and even the spectral content of the incoming light, allowing our brain to recognize features, objects, those we love, and the beauty and dangers of the world in which we live.

Similarly our ears detect, filter, and process acoustic radiation. Our two-sensor acoustic array is remarkably accurate at estimating the direction from which sound waves originate. We can process acoustic signals to extract information about distant events, such as an approaching thunderstorm or the squeal of a car's tires, that help us make intelligent decisions and take intelligent action. Even more remarkably, our acoustic processing system enables us to develop speech as a form of communication, representing complex and abstract ideas as a sequence of acoustic modulations. We can communicate acoustically with fellow humans at close range or, with electronic intermediaries, over extremely long distances.

Humankind strives to extend its senses. From the primitive cupping of one's hand to one's ear, we have evolved long-range acoustic detection systems (Fig. 1.1) and underwater arrays that can "hear" ships across a thousand miles of ocean. Human eyes were not good enough for Galileo, who developed a telescope and discovered moons around Jupiter. Today's "eyes" look to the edge of the universe and into crystals to see atoms oscillating in place. We "see" at a variety of wavelengths, from radio to X-ray.

The goal of signal processing is to extract as much information as possible from our environment. Array signal processing is a specialized branch of signal processing that focuses on signals conveyed by propagating waves. Here, an *array*—a group of sensors located at distinct spatial locations—is deployed to measure a propagating wavefield be it electromagnetic, acoustic, or seismic. As indicated in Fig. 1.2, an array samples the field $f(\vec{x}, t)$ at sensor locations $\{\vec{x}_m\}_{m=0}^{M-1}$ and time instants $\{t_n\}$. The sensors serve as transducers, converting field energy to electrical energy. In the transduction process,

† One THz equals 10^{12} Hertz.

Figure 1.1 An acoustic array was used by the French in World War I to detect enemy aircraft. The array consists of two "sensors," each of which consists of six subsensors arranged hexagonally; six hexagonally shaped continuous apertures comprise the subsensors. We calculate this array's directivity pattern in Prob. 4.5 {191}. This intricate listening device was developed by the then Sergeant Jean Perrin (right), who is demonstrating the array's use. Perrin later received the Nobel Prize in Physics for his pioneering work in atomic physics. *Reprinted by permission of the AIP Neils Bohr Library.*

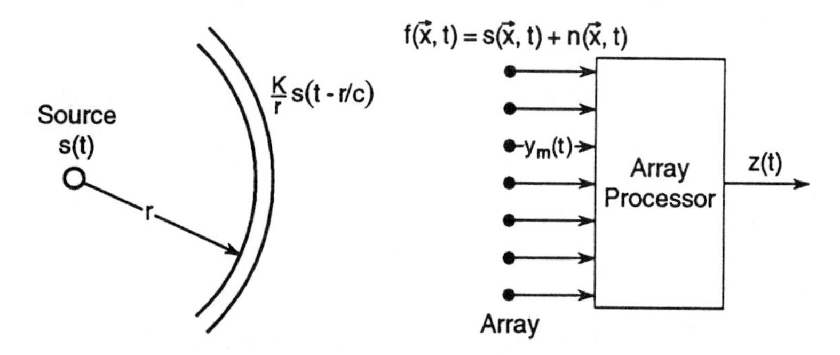

Figure 1.2 An array consists of a collection of sensors that spatiotemporally measure a wavefield $f(\vec{x}, t)$. These measurements are merged by the signal processing algorithm to accomplish some or all of the three goals described in the text. In some situations, the outputs of several arrays are merged for tracking the locations of the sources producing the propagating energy.

the mth sensor's output $y_m(t)$ is related to the field at the sensor's location by at least a conversion factor (gain), and possibly by temporal and spatial filtering. An array collects these waveforms to create the output signal $z(t)$; in the process of creating this output, information about the propagating signal(s) is extracted. Several arrays sampling a common field may be merged to produce more refined information about the signal(s). The goals of array processing are to combine the sensors' outputs cleverly and then the arrays' outputs so as

- to **enhance** the signal-to-noise ratio beyond that of a single sensor's output
- to **characterize** the field by determining the *number* of sources of propagating energy, the *locations* of these sources, and the *waveforms* they are emitting
- to **track** the energy sources as they move in space

1.1 Signal Detection and Enhancement

Often the information we want is carried by a signal degraded by noise or interference. Over several decades, signal processing techniques have been developed to detect the presence of a desired signal and to reduce the effects of unwanted noise. In many applications, this signal enhancement can be achieved by processing the waveform received by a single sensor, but often it is advantageous to use an array of sensors.

The *detection* problem is to determine whether a signal is present in a measurement contaminated by noise. The signal is distinguished from noise by having some special property. For example, we may know the signal waveform save for a few parameters (for example, we know it's a sinusoid, but the amplitude and phase may not be known). Given this information and observations of the field, how can the signal be (optimally) detected? This problem has a long history, being studied in the areas of radar, sonar, and pattern classification. In array processing, the signal's sole distinguishing property may be that it is propagating; that is what our ears and the system shown in Fig. 1.1 focus on. As we shall learn in Chap. 5, performance of detection systems depends heavily on how strong the signal is relative to the noise. Thus, signal enhancement and signal detection are intimately related.

Signal enhancement with an array can be illustrated by a simple example. A common assumption for some array processing problems decrees that the waveform $y_m(t)$ produced at each of M sensors consists of a signal, identical from sensor to sensor, plus random noise, statistically independent from sensor to sensor. We may write this model as

$$y_m(t) = s(t) + n_m(t)$$

where $s(t)$ is the signal and $n_m(t)$ is the noise at the mth sensor. In this simple case, we can reduce the degradation owing to noise by averaging over the waveforms received by the sensors.

$$z(t) = \frac{1}{M}\sum_{m=0}^{M-1} y_m(t) = s(t) + \frac{1}{M}\sum_{m=0}^{M-1} n_m(t)$$

The signal components at each sensor reinforce while the noise components tend to cancel because of their random, uncorrelated nature. Note that nowhere in this procedure is the signal assumed to be known; the signal was simply the component of the observations common to all sensor outputs. You can easily show (see Prob. 1.1) that this averaging increases the ratio of signal power to noise power [the signal-to-noise ratio (*SNR*)] by a factor of M compared with the *SNR* for a single sensor.† In Chap. 4, we shall discuss a filtering operation—beamforming—that reduces noise and interference; Chap. 5 describes how beamforming and detection are related.

1.2 Signal Characterization

In many applications, we want to describe or characterize in a quantitative way the signals received by an array. Often we can develop a model of the signal that results from an event of interest. By measuring various characteristics of received waveforms, we can draw conclusions about the events that caused them. For example, by analyzing the frequency content of a speech signal as it evolves in time, we deduce the sequence of phonemes produced by the speaker's vocal tract and ultimately the concepts intended by the speaker.

One characteristic of propagating signals that has fundamental importance is the *direction of propagation*. The direction of arrival of a beam of light shows us along which radius a star is located. The direction of arrival of sound waves in the ocean can be used to locate ships. Radio direction-finding equipment is used to determine position and to navigate safely. When we discuss beamforming in Chap. 4, we shall investigate methods that create a preference for signals propagating in certain directions over those propagating in other directions. Later, in Chaps. 7 and 8, we shall develop techniques to estimate the direction of propagation given an array of received waveforms.

The speed of propagation of a signal can also be an important characteristic. In seismic signal processing, for example, acoustic energy travels at different speeds through different materials. Deducing the speed of sound through some zone deep below the Earth's surface can give important clues to the type of material found in that zone. Knowing the structure of the subsurface zones gives the geophysicist information about possible locations for trapped oil and gas deposits. As described in Chaps. 3 and 4, the direction-finding problem and the propagation speed problem are duals: You need to know one to find the other unless some other constraint beyond the wave equation can be imposed on the propagating signal.

Signals can be characterized by their frequency content. As we shall see in Chap. 2, propagating signals have spatial as well as temporal spectra. The waveform received by a single sensor has a temporal spectrum containing varying amounts of energy as a function of frequency in cycles/s (Hz). The waveform can be characterized as a sum of many sinusoids having various frequencies, phases, and amplitudes. The waveform

†Because it is the ratio of signal power to noise power that increases by a factor of M, the effect of averaging the outputs of M sensors is qualitatively the same as reducing the noise amplitude in a single sensor by $\sqrt{M}$. Thus, it takes an array of 100 sensors whose outputs are averaged together to achieve the same effect as reducing the noise amplitude by a factor of 10 in a single sensor system.

appearing across an array at one instant of time can be similarly characterized as a sum of many sinusoids having various wavelengths, offsets, and amplitudes. Its spatial frequency content is measured as a function of cycles/m in the three spatial-frequency dimensions.

Characterizing a signal in terms of its frequency content is very similar to developing a model of the signal and estimating values for the model parameters that best match the model to the received waveform. A frequency-domain characterization is somewhat special because any reasonable signal can be modeled as a sum of sinusoids. Other models are possible, of course, and for applications where there is a relatively complete theory of signal production, such models offer important characterizations of received signals. For example, speech signals are often characterized by their pitch and by eight to twelve parameters characterizing the spectral shape of the vocal tract transfer function. We shall discuss parameter estimation in Chap. 6 and its application to array processing in Chap. 7.

1.3 Tracking

In many applications, array processing techniques are used as part of a system to monitor the positions of certain objects in the environment. In defense applications, these objects may pose threats to precious resources that must be defended for the benefit of society. In less hostile situations, the system may be used to keep objects safely apart as they navigate within some confined region. Such systems are often called tracking systems, or simply "trackers."

Tracking systems rely on the emission or reflection of propagating signals from remote objects to provide information about them. For example, aircraft emit acoustic and infrared radiation, and reflect electromagnetic radiation in the radio band (radar). Ships at sea emit and reflect underwater sound waves (sonar). Array processing techniques can be used to measure signal characteristics such as direction of propagation and frequency content. These measurements, perhaps from several arrays, combined with a model for the object's dynamics, form the fundamental data exploited by the tracking system to estimate the object's position and velocity.

Consider the case of a ship whose propulsion system emits a periodic, low-frequency tone that propagates across the ocean to several underwater acoustic arrays, as shown in Fig. 1.3. Array processing techniques applied to each sensor array can be used to detect the ship's signal and measure its direction of propagation. These measurements, often

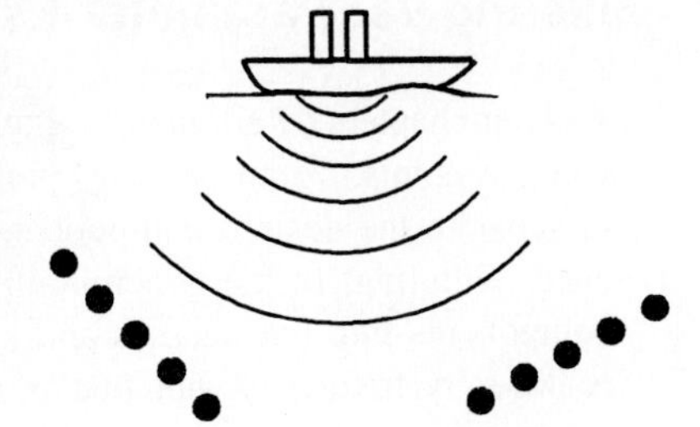

Figure 1.3 A ship in the ocean emits sounds that may be detected and characterized by several underwater acoustic arrays. As the ship moves, the bearing measurements slowly change.

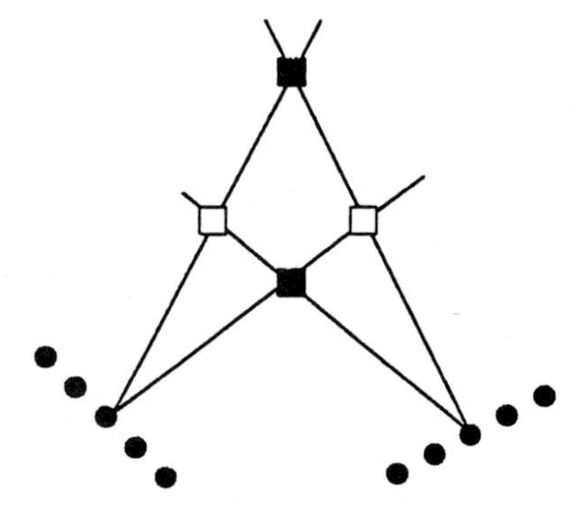

Figure 1.4 Two bearing measurements made at two arrays give ambiguous object positions. One possibility is two real (solid squares) positions and two "ghosts" (open squares). How many possible associations are there? The answer is two. Why?

called bearing measurements, can be used by a tracker to estimate the ship's position. In this simple static case, the calculation boils down to basic triangulation. From the size of the bearing measurement errors, we can compute the position error.

As the ship moves, the bearing measurements slowly change, and the position estimate is updated. Successive changes in position as a function of time allow the tracking system to estimate the ship's velocity (both speed and direction). Additional information may be available, however, to improve the velocity and position estimates. Because of the finite speed of sound, the ship's signal arrives at the arrays at different times. Furthermore, as the ship moves relative to the arrays, a Doppler shift is induced on the emitted signal. By measuring the differences in time of arrival and Doppler shift, we can infer the distances between the ship and the arrays as well as the relative velocities.

An interesting problem arises when several arrays are being used to track several objects. Let's assume that the tracking system is maintaining position and velocity estimates on two ships. At appropriate intervals the arrays report detected signals and corresponding bearing measurements. But which measurement should be associated with which ship? If an incorrect association is made, it leads to a new position estimate that corresponds to a fictitious "ghost" object, as shown in Fig. 1.4. Often this problem can be addressed by using the estimated position and velocity of a ship to predict what its bearing measurement should be from each array. If a bearing measurement is close to the predicted value for one particular ship, it is associated with that ship. In Chap. 8, we shall discuss tracking systems and their reliance on array processing measurements in much more detail.

1.4 Filtering vs. Parameter Estimation

As later chapters detail, many signal processing problems can be viewed from two distinct vantage points. From one viewpoint, a signal processing problem may call for filtering to separate the desired components of a received waveform from any noise or interference. Filtering, in the general sense of the word, simply means segregating waveform components into two classes and keeping one of the classes. Typically this segregation is done by frequency, but that is only one possibility. When a signal is low-pass filtered, for example, the low-frequency components are retained and the high-frequency components are discarded. An alternative view is to see a signal processing problem as estimating some parameters of a signal model from the observed waveform. We estimate these parameters by minimizing some error criterion such as the squared mean difference between the model signal and the received waveform.

Let's briefly contrast the two approaches for the case of determining the directions of propagations of several signals moving past an array. As we shall see in Chap. 4, we can build a beamformer, one type of spatiotemporal filter to pass signals propagating in a narrow cone of directions and attenuate those propagating in other directions. We can then measure the average power in the beamformer output to determine the strength of a signal, if any, propagating in the specified direction. If we build a beamformer for each possible direction of propagation, we can determine the direction of propagation of the strongest signals by examining the average power in the various beamformer outputs.

Alternatively we might assume that the received waveform can be adequately modeled as the sum of several plane waves plus white noise. The problem then becomes one of estimating the amplitude and direction of propagation of each plane wave from the waveforms received by the sensor array. Within this model, we can frame an optimization problem to find the parameter values that most closely match the model to the waveform. Of course, different signal models may be assumed. The choice of signal model depends on the extent of knowledge about the physical processes that generate the signal.

The filtering viewpoint and the parameter estimation viewpoint are not always distinct. Consider the analysis of the frequency content of a waveform received by a single sensor. We could build a bank of narrowband filters and measure the power in their outputs to determine the frequency content of the waveform. If the narrowband filters are chosen appropriately, we can obtain the same result by computing the squared magnitude of the waveform's Fourier Transform. Alternatively, we could model the waveform as a sum of sinusoids using Fourier Transform theory and estimate the amplitudes of the sinusoids. By doing so, we are led to the same computation as before. The Fourier Transform computation can be interpreted both as a filtering operation and as a parameter estimation operation.

Historically, filtering has been used in real-time signal processing systems, especially before the advent of digital signal processing, and parameter estimation has been used only as an off-line, signal analysis technique. As technology continues to improve the capabilities of digital systems, however, we begin to see parameter estimation techniques being applied to real-time problems. In addition to using an electronic analog or digital circuit to implement a filtering operation, we can use programmable resources to estimate the parameters of a received waveform.

Problems

1.1 Assume that M signals $y_m(t), m = 0, \ldots, M-1$ are observed. Each has a common deterministic component $s(t)$ and a stochastic component $n_m(t)$. The stochastic component has zero mean. We want to explore how beneficial averaging can be in enhancing the signal-to-noise ratio beyond that of a single observation. For this problem, we define signal-to-noise ratio as the energy in the signal component divided by the noise variance.

(a) If the noise components are statistically independent of each other, show that averaging M observations increases the signal-to-noise ratio by a factor of M over that in a single observation.

(b) Now assume that the noise components are correlated in such a way that $\mathcal{E}[n_{m_1}(t)n_{m_2}(t)] = \rho^{|m_1-m_2|}\sigma_n^2$. How is the signal-to-noise ratio gain affected by correlation?

1.2 Suppose two sensors are separated by a distance d. A wave consisting of a very narrow pulse propagates past the two sensors with a speed c. Define ϕ as the angle from the perpendicular bisector of the line segment connecting the two sensor locations. Thus $\phi = 0$ represents the perpendicular bisector itself, and $\phi = \pi/2$ represents the line drawn through the sensors. Depending on the angle of incidence of the wave, there is a relative delay between the pulses measured by the two sensors.

(a) Relate the angle of incidence ϕ to the relative delay Δ.

(b) What other angles of incidence cause the same relative delay? Consider both the two-dimensional problem (two sensors in a plane) and the three-dimensional problem (two sensors in space).

(c) Can a single sensor tell the direction of propagation of the wave? In other words, can you tell from what direction a sound is coming with one ear covered? Why?

Chapter 2

Signals in Space and Time

In many signal processing applications, we are concerned with waveforms that are functions of a single variable, which usually represents time. In array processing, however, propagating waves convey signals from the source to the array. These signals are thus functions of position as well as time and have properties governed by the laws of physics, in particular the wave equation. To become proficient in understanding and developing array signal processing algorithms, we need to understand the properties of spatiotemporal signals and noise, and of propagating signals in particular. Signal processing algorithms exploit the structure of propagating signals to fulfill the three goals of array processing {3}.

2.1 Coordinate Systems

In most situations, a three-dimensional Cartesian grid represents space, with time being the fourth dimension. A space-time signal is written as $s(x, y, z, t)$, for example, with x, y, and z being the three spatial variables in a right-handed orthogonal coordinate system as shown in Fig. 2.1. We represent the unit vectors in the three spatial directions as $\vec{i}_x$,

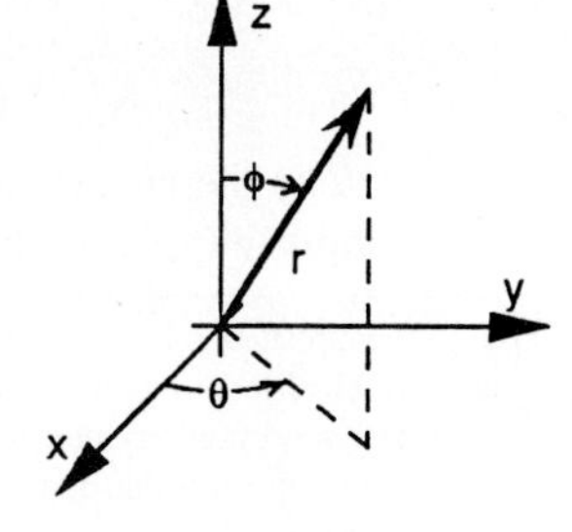

Figure 2.1 A right-handed orthogonal coordinate system showing the x, y, and z axes is depicted along with a spherical coordinate system. The angle θ is known as the azimuth and ϕ the elevation. The distance of a point from the spherical coordinate origin is denoted by r.

$\vec{\imath}_y$, and $\vec{\imath}_z$.

$$\vec{\imath}_x \cdot \vec{\imath}_x = \vec{\imath}_y \cdot \vec{\imath}_y = \vec{\imath}_z \cdot \vec{\imath}_z = 1$$
$$\vec{\imath}_x \cdot \vec{\imath}_y = \vec{\imath}_y \cdot \vec{\imath}_z = \vec{\imath}_z \cdot \vec{\imath}_x = 0$$
$$\vec{\imath}_x \times \vec{\imath}_y = \vec{\imath}_z$$

We shall use the *position vector* $\vec{x}$ to denote the triple of spatial variables (x, y, z). Using this notation we can write a spatiotemporal signal as $s(\vec{x}, t)$. Rather than combine the time variable with the three spatial variables to form a four-dimensional vector, convenience dictates keeping it separate. Of course, other coordinate systems may be defined. For certain problems, spherical coordinates represent space most appropriately. Here, a point is represented by its distance r from the origin, its azimuth θ within an equatorial plane containing the origin, and its angle ϕ down from the vertical axis (Fig. 2.1). The spherical coordinates of a point are related to the Cartesian coordinates by simple trigonometric formulas.

$$\begin{aligned} x &= r \sin\phi \cos\theta \\ y &= r \sin\phi \sin\theta \\ z &= r \cos\phi \end{aligned} \qquad \begin{aligned} r &= \sqrt{x^2 + y^2 + z^2} \\ \theta &= \cos^{-1}\left(\frac{x}{\sqrt{x^2+y^2}}\right) = \sin^{-1}\left(\frac{y}{\sqrt{x^2+y^2}}\right) \\ \phi &= \cos^{-1}\left(\frac{z}{\sqrt{x^2+y^2+z^2}}\right) \end{aligned}$$

Here, the (x, y) plane forms the equatorial plane, and azimuths are measured as angles counterclockwise from the x axis. Spherical coordinates are usually used to address problems with spherical symmetry. For example, an isotropically spreading spherical wave may be represented by $s(r, t)$ because it does not depend on the angular coordinates ϕ and θ.

2.2 Propagating Waves

Information about distant events is carried to our sensors by propagating waves. The physics of propagation is described by wave equation for the appropriate medium and boundary conditions. For example, electromagnetic fields obey Maxwell's equations, which for free space take the form of

$$\nabla \times \vec{E} = -\frac{\partial(\mu \vec{H})}{\partial t} \qquad \nabla \cdot (\epsilon \vec{E}) = 0$$
$$\nabla \times \vec{H} = \frac{\partial(\epsilon \vec{E})}{\partial t} \qquad \nabla \cdot (\mu \vec{H}) = 0$$

where $\vec{E}$ is the electric field intensity, $\vec{H}$ is the magnetic field intensity, ϵ is the dielectric permittivity, μ is the magnetic permeability,† and ∇ represents the gradient vector

†The constant ϵ is the proportionality constant between the electric field intensity expressed in volts/meter and the electric flux density in coulombs/square meter. As a result, ϵ has units of farads/meter. Similarly, μ is the proportionality constant between the magnetic field intensity in amperes/meter and the magnetic flux density in volt-seconds/square meter and has units of henrys/meter.

operator.

$$\nabla = \frac{\partial}{\partial x}\vec{\imath}_x + \frac{\partial}{\partial y}\vec{\imath}_y + \frac{\partial}{\partial z}\vec{\imath}_z$$

The wave equation can be easily derived directly from Maxwell's equations (see Prob. 2.1)

$$\boxed{\nabla^2 \vec{E} = \frac{1}{c^2}\frac{\partial^2 \vec{E}}{\partial t^2}}$$

where ∇^2 represents the Laplacian operator

$$\frac{\partial^2}{\partial x^2} + \frac{\partial^2}{\partial y^2} + \frac{\partial^2}{\partial z^2}$$

Using this definition and using $s(\vec{x}, t)$ to represent a general scalar field,† the wave equation is

$$\boxed{\frac{\partial^2 s}{\partial x^2} + \frac{\partial^2 s}{\partial y^2} + \frac{\partial^2 s}{\partial z^2} = \frac{1}{c^2}\frac{\partial^2 s}{\partial t^2}} \tag{2.1}$$

Later we shall see that the only parameter of the wave equation c can be interpreted as the speed of propagation. In electromagnetics, $c = 1/\sqrt{\epsilon\mu}$. In acoustics, the basic laws of physics and the conservation of mass lead to an equation governing the relation between pressure fluctuations and density fluctuations in a fluid. Remarkably, this equation takes the same form as the wave equation, with $s(\vec{x}, t)$ representing sound pressure at a point in space and time.‡ In this case, c is the speed of sound, which is related to the fluid parameters by

$$c^2 = \frac{dP}{d\rho}$$

where P is the pressure and ρ is the density. For a gas, the constant c^2 equals $\gamma RT_0/M$, where γ is the specific heat ratio (equaling 1.4 for a diatomic gas), R is the gas constant per mole, M is the molar mass, and T_0 is the ambient temperature. In a fluid, c^2 equals $\gamma B/\rho$, where B is the isothermal bulk modulus and ρ is density. In a solid, acoustic waves propagate very differently than in fluids. Solids can support deformations of the material that do not lead to changes in volume. Thus, in addition to *compressional* acoustic waves propagated by fluids, solids can propagate *transverse* waves that fluids cannot. Because of the linearity of the wave equation, each wave propagates separately and has its own wave equation constant. The constant for longitudinal motion (compressional waves) is $c_l^2 = (K + 4G/3)/\rho$ where K is the bulk modulus of the solid medium, G is the shear modulus, and ρ is the density. The constant for transverse motion (shear waves) is $c_t^2 = G/\rho$. The relations are summarized in Table 2.1 along with commonly used values for the material constants and the resulting propagation speeds.

†We do this to free us from notation of electromagnetism and to focus on the scalar wave equation rather than the more general vector wave equation. For a transverse electromagnetic wave any particular component of the electric or magnetic fields, say E_x, satisfies the scalar wave equation.

‡Deriving the wave equation for acoustics waves is more complicated than in the electromagnetic case because no single unified set of equations governs acoustics. See any good acoustics book, such as Kinsler et al. (Chap. 5), for a derivation.

Wave Type	Medium	Propagation Speed c	
		Formula	Value
Electromagnetic	Free space	$1/\sqrt{\epsilon\mu}$	3×10^8 m/s $\epsilon = 1/36\pi \times 10^{-9}\text{C}^2/\text{N-m}^2$, $\mu = 4\pi \times 10^{-7}$W/A-m
Electromagnetic	Glass	$1/\sqrt{\epsilon\mu}$	2×10^8 m/s[a]
Acoustic	Air	$\sqrt{\gamma R T_0/M}$	330.7 m/s $R = 8.3 \times 10^7$ erg/°K, $\gamma = 1.4$, $T_0 = 273$ °K, $M = 29$ g
Acoustic	Sea water	$\sqrt{\gamma B/\rho}$	1,498 m/s[b] $\gamma = 1.01$, $B = 2.28 \times 10^9\text{N/m}^2$, $\rho = 1.026 \times 10^3\text{kg/m}^3$
Water waves (Shallow water)		$\sqrt{gH}$	$3.13\sqrt{H}$ m/s $g = 9.8$ m/s^2, H = water depth (m)
Acoustic (longitudinal)	Granite	$\sqrt{\frac{K+4G/3}{\rho}}$	3,310 m/s[c]
Acoustic (transverse)	Granite	$\sqrt{G/\rho}$	5,770 m/s[c]

[a]This value is derived from the refractive index, which for glass equals about 1.5.

[b]Because of the complicated properties of most fluids, experimental, serieslike expressions for sound propagation speeds are common. For sea water, the expression is

$$c = 1,449.2 + 4.623T - 0.0546T^2 + (1.391 - 0.012T) * (S - 35) + \cdots$$

where T is temperature in °C, and S is salinity in parts per thousand.

[c]This value derived from direct measurements.

TABLE 2.1 Formulas and Values for Propagation Speed c of Various Types of Waves in Different Media.

The wave equation is *the* equation in array signal processing. As developed in succeeding sections of this chapter, its solution is indeed a propagating wave. Consequently, it governs how signals pass from a source radiating energy to an array. Array processing algorithms, which all attempt to extract information from propagating waves, rely on accurate characterization of how the medium affects propagation; in other words, through the wave equation's solution for a given source-medium-array situation, how can information conveyed by the propagating wave, source position and source waveform, for example, best be extracted?

2.2.1 Solutions of the Wave Equation in Cartesian Coordinates

Many texts on mathematical physics suggest hypothesizing a separable solution when one is confronted by a partial differential equation. In our case, this procedure means assuming a solution to the wave equation of the general form

$$s(x, y, z, t) = f(x)g(y)h(z)p(t)$$

For simplicity, let's initially assume that $s(x, y, z, t)$ has a complex exponential form

$$s(x, y, z, t) = A \exp\{j(\omega t - k_x x - k_y y - k_z z)\} \tag{2.2}$$

where A is a complex constant and k_x, k_y, k_z, and ω are real constants with $\omega \geq 0$. Substituting this form into the wave equation, we obtain

$$k_x^2 s(x, y, z, t) + k_y^2 s(x, y, z, t) + k_z^2 s(x, y, z, t) = \frac{\omega^2 s(x, y, z, t)}{c^2}$$

Canceling $s(x, y, z, t)$ gives us an equation constraining the real constants of the complex exponential form.

$$k_x^2 + k_y^2 + k_z^2 = \frac{\omega^2}{c^2}$$

As long as this constraint is satisfied, signals with the form of Eq. 2.2 satisfy the wave equation (Eq. 2.1).

The solution to the wave equation given by Eq. 2.2 may be interpreted as a *monochromatic* plane wave. The term "monochromatic," meaning "one color," refers to the temporal behavior of $s(x, y, z, t)$. If we place a sensor at some fixed position, say the origin $(x, y, z) = (0, 0, 0)$, and observe the signal, it has the form of a complex exponential with frequency ω:

$$s(0, 0, 0, t) = A \exp\{j\omega t\} = A \cos \omega t + jA \sin \omega t$$

The term "plane wave" arises because at any instant of time t_0, the value of $s(x, y, z, t_0)$ is the same at all points lying in a plane given by $k_x x + k_y y + k_z z = C$, where C is a constant. Using vector notation for position and for (k_x, k_y, k_z), we can write the monochromatic solution to the wave equation more concisely as

$$\boxed{s(\vec{x}, t) = A \exp\{j(\omega t - \vec{k} \cdot \vec{x})\}} \tag{2.3}$$

The planes of constant phase, the planes where $\vec{k} \cdot \vec{x}$ is constant, are perpendicular to the vector $\vec{k}$ (Prob. 2.2).

If the signal $s(\vec{x}, t)$ is truly a propagating wave, planes of constant phase move by an amount $\delta\vec{x}$ as time advances by an amount δt. This gives us $s(\vec{x} + \delta\vec{x}, t + \delta t) = s(\vec{x}, t)$, which in turn implies

$$\omega \delta t - \vec{k} \cdot \delta\vec{x} = 0 \tag{2.4}$$

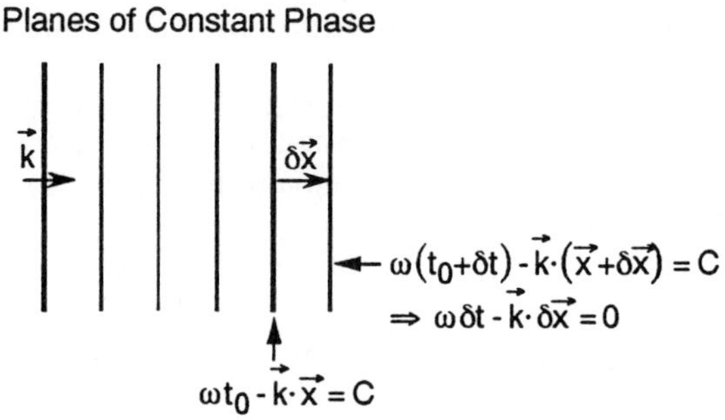

Planes of Constant Phase

$\vec{k}$

$\delta\vec{x}$

$\omega(t_0+\delta t)-\vec{k}\cdot(\vec{x}+\delta\vec{x})=C$

$\Rightarrow \omega\delta t-\vec{k}\cdot\delta\vec{x}=0$

$\omega t_0-\vec{k}\cdot\vec{x}=C$

Figure 2.2 We may take $\delta\vec{x}$ to have the same direction as $\vec{k}$ when we see how much a plane wave moves during δt seconds. This direction may be interpreted as the direction of propagation.

We may take the direction of $\delta\vec{x}$ to be the same as the direction of $\vec{k}$ because this choice gives us the minimum magnitude for $\delta\vec{x}$ (Fig. 2.2). With this interpretation, *the direction of propagation $\vec{\zeta}^o$ of the plane wave corresponds to the direction of $\vec{k}$*: $\vec{\zeta}^o = \vec{k}/k$, where $k = |\vec{k}|$ denotes the magnitude of the wavenumber vector. Because $\vec{k}$ and $\delta\vec{x}$ have the same direction, the dot product $\vec{k}\cdot\delta\vec{x}$ is simply $|\vec{k}||\delta\vec{x}|$, giving us $\omega\delta t = |\vec{k}||\delta\vec{x}|$ or $|\delta\vec{x}|/\delta t = \omega/|\vec{k}|$. The ratio $|\delta\vec{x}|/\delta t$ can be interpreted as the speed of propagation of the plane wave. Because $\vec{k}$ and ω are related by $|\vec{k}|^2 = \omega^2/c^2$, we conclude that

$$\frac{|\delta\vec{x}|}{\delta t} = c$$

assuming that c is positive. *The speed of propagation is indeed the parameter c that appears in the wave equation (Eq. 2.1).*

How far does a plane wave propagate during one temporal period $T = 2\pi/\omega$? After one period, the monochromatic plane wave appears (in space) as it did before, but moved one cycle forward. The distance propagated during one temporal period is the *wavelength* λ. Using the relation in Eq. 2.4 with $\delta t = 2\pi/\omega$, we conclude that $\delta\vec{x} = \lambda = 2\pi/|\vec{k}|$. The vector $\vec{k}$ is called the *wavenumber vector*; its magnitude expresses the number of cycles (in radians) per meter of length that the monochromatic plane wave exhibits in the direction of propagation. The wavenumber vector may therefore be considered as a *spatial frequency variable* just as ω is a temporal frequency variable. Because there are three spatial dimensions, the spatial frequency content of waves must be represented as a three-dimensional vector, a role that the wavenumber vector $\vec{k}$ fulfills nicely.

A spatiotemporal signal $s(\vec{x},t)$ that satisfies the wave equation can be expressed as a function of a single variable $s(\cdot)$. To see this simplification, rewrite Eq. 2.3 as

$$s(\vec{x},t) = A\exp\{j\omega(t-\vec{\alpha}\cdot\vec{x})\}$$

where $\vec{\alpha} = \vec{k}/\omega$. Then, $s(\vec{x},t) = s(t-\vec{\alpha}\cdot\vec{x})$, where $s(u) = A\exp\{j\omega u\}$. The vector $\vec{\alpha}$ has a magnitude equal to the reciprocal of the propagation speed because of the relationship between $\vec{k}$ and ω. For this reason, it is often called the *slowness vector*. It points in

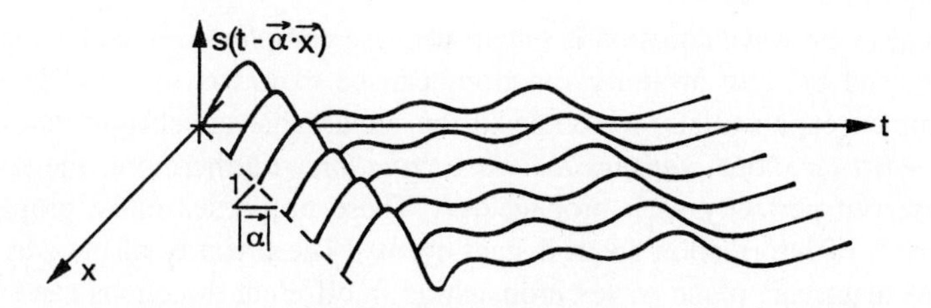

Figure 2.3 A propagating wave $s(t - \vec{\alpha} \cdot \vec{x})$ with an arbitrary wave shape satisfies the wave equation.

the same direction as $\vec{\zeta}$, the direction of propagation of the plane wave, and has units of reciprocal velocity.

The wave equation is a linear equation: If $s_1(\vec{x}, t)$ and $s_2(\vec{x}, t)$ are two solutions to the wave equation, then the linear combination $as_1(\vec{x}, t) + bs_2(\vec{x}, t)$, where a and b are scalars, is also a solution. Because $A \exp\{j\omega(t - \vec{\alpha} \cdot \vec{x})\}$ is a solution to the wave equation, we can build up more complicated solutions by expressing them as sums or integrals of complex exponentials. For example, the waveform

$$s(\vec{x}, t) = s(t - \vec{\alpha} \cdot \vec{x}) = \sum_{-\infty}^{\infty} S_n \exp\{jn\omega_0(t - \vec{\alpha} \cdot \vec{x})\}$$

which has the form of a harmonic series with a fundamental frequency of ω_0, is a solution. Because of Fourier's Theorem, any arbitrary periodic waveform $s(u)$ with period $T = 2\pi/\omega_0$ can be represented by such a series. The coefficients S_n are given by

$$S_n = \frac{1}{T} \int_0^T s(u) e^{-jn\omega_0 u} \, du$$

In this case, $s(\vec{x}, t) = s(t - \vec{\alpha} \cdot \vec{x})$ represents a propagating periodic wave with an arbitrary wave shape, as shown in Fig. 2.3. The wave propagates in the direction specified by the slowness vector $\vec{\alpha}$ with a speed $c = 1/|\vec{\alpha}|$. The various components of the wave have different frequencies $\omega = n\omega_0$ and wavenumber vectors $\vec{k}$, but the frequencies and wavenumber vectors must satisfy the constraint $\vec{k}/\omega = \vec{\alpha}$.

More generally, we can use Fourier Theory to form an integral of complex exponentials to represent an arbitrary (nonperiodic) wave shape. Let

$$s(\vec{x}, t) = s(t - \vec{\alpha} \cdot \vec{x}) = \frac{1}{2\pi} \int_{-\infty}^{\infty} S(\omega) \exp\{j\omega(t - \vec{\alpha} \cdot \vec{x})\} \, d\omega$$

Because $s(\vec{x}, t)$ is a superposition of solutions to the wave equation, it too is a solution. The function $s(\cdot)$ is arbitrary,[†] and its frequency representation $S(\omega)$ is given by the Fourier Transform

$$S(\omega) = \int_{-\infty}^{\infty} s(u) e^{-j\omega u} \, du$$

[†]Well, not quite arbitrary. The function must have a defined and convergent Fourier integral.

Because the wave equation is linear, because complex exponentials solve the wave equation, and because arbitrary functions can be expressed as a weighted superposition of complex exponentials, we come to the remarkable conclusion that *any signal, no matter what its shape, satisfies the wave equation.* Furthermore, the shape of the wave is preserved perfectly as it propagates.‡ These properties make propagating waves ideal carriers of information about distant events. The linearity of the wave equation also implies that many plane waves propagating in different directions can exist simultaneously. Linear wave theory leads us to conclude that the plane waves pass through each other unperturbed: the Superposition Principle applies.

In summary, the propagation of plane waves is governed by the equations

Propagating plane wave	$s(t - \vec{\alpha} \cdot \vec{x})$	
Propagating sinusoidal plane wave	$\sin(\omega t - \vec{k} \cdot \vec{x})$	
Slowness vector	$\vec{\alpha} = \vec{k}/\omega, \quad \lvert\vec{\alpha}\rvert = 1/c$	(2.5)
Wavenumber vector	$\vec{k} = \omega\vec{\alpha}, \quad \lvert\vec{k}\rvert = 2\pi/\lambda$	
Frequency and wavelength	$c = \lambda \cdot \omega/2\pi$	

2.2.2 Solution to the Wave Equation in Spherical Coordinates

The wave equation (Eq. 2.1 {11}) can be reexpressed in the spherical coordinates (r, ϕ, θ). The calculations, which are straightforward but intricate, result in the general spherical wave equation

$$\frac{1}{r^2}\frac{\partial}{\partial r}\left(r^2\frac{\partial s}{\partial r}\right) + \frac{1}{r^2 \sin\phi}\frac{\partial}{\partial\phi}\left(\sin\phi\frac{\partial s}{\partial\phi}\right) + \frac{1}{r^2\sin^2\phi}\frac{\partial^2 s}{\partial\phi^2} = \frac{1}{c^2}\frac{\partial^2 s}{\partial t^2}$$

Complicated though it is, this equation can be solved by the method of separation of variables. General solutions involve Bessel functions and associated Legendre polynomials, but they are not of interest here. Rather, the spherical wave equation is generally used in situations in which the problem suggests that the solution exhibits spherical symmetry. In these cases, $s(r, \phi, \theta, t)$ does not depend on ϕ or θ, and simpler solutions emerge. For spherically symmetric problems, the general spherical wave equation becomes

$$\frac{1}{r^2}\frac{\partial}{\partial r}\left(r^2\frac{\partial s}{\partial r}\right) = \frac{1}{c^2}\frac{\partial^2 s}{\partial t^2} \tag{2.6}$$

With a little manipulation, this equation can be cast into the form of a one-dimensional wave equation.

$$\frac{\partial^2(rs)}{\partial r^2} = \frac{1}{c^2}\frac{\partial^2(rs)}{\partial t^2}$$

‡In §2.3 {21}, we shall discover conditions under which the wave's shape is not preserved.

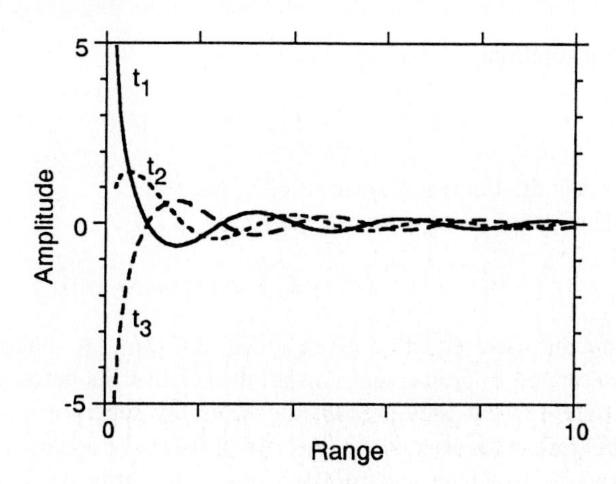

Figure 2.4 A plot of $s(r,t) = [\cos(\omega t - kr)]/r$ as a function of r for three time instants $t_1 < t_2 < t_3$.

One solution to this wave equation is the monochromatic one: $rs(r,t) = A\exp\{j(\omega t - kr)\}$ or

$$\boxed{s(r,t) = \frac{A}{r}\exp\{j(\omega t - kr)\}}$$

This solution can be interpreted as a spherical wave propagating outward from the origin. At any point in space,[†] the wave has a temporal frequency of ω. The parameter k can be interpreted as the spatial frequency (or wavenumber) corresponding to a wavelength of $\lambda = 2\pi/k$. Substituting $s(r,t)$ back into the simplified spherical wave equation, we see that k and ω are related by $k^2 = \omega^2/c^2$. Using this relationship, we can write $s(r,t)$ without explicit dependence on k as

$$s(r,t) = \frac{A}{r}\exp\{j\omega(t - r/c)\}$$

Fig. 2.4 shows a plot of the related spherical wave given by $\cos(\omega t - kr)/r$. The distance between consecutive zero-crossings, the points where $s(r,t) = 0$, is given by $kr = 2\pi$ or $r = 2\pi/k = \lambda$. It is also instructive to compute the distance between consecutive maxima of this function (Prob. 2.4).

Just as in the Cartesian case, we can build up more complicated solutions to the spherical wave equation by superposing complex exponentials. In this case, however, the wave shape is altered as the wave propagates outward from the origin because of the $1/r$ factor. For example, suppose we have a well-behaved function $s(\cdot)$. Then it is straightforward to verify that the wave $s(r,t) = s(t - r/c)/r$ satisfies the spherically symmetric wave equation. We may think of this solution as a superposition of complex

[†]We neglect the singularity at the origin.

exponential solutions

$$s(r,t) = \frac{1}{2\pi r}\int_{-\infty}^{\infty} S(\omega)\exp\{j\omega(t - r/c)\}\,d\omega$$

where $S(\omega)$ is the Fourier Transform of $s(\cdot)$.

Propagating spherical waves of the form

$$s(r,t) = \frac{B}{r}\exp\{j(\omega t + kr)\}$$

also satisfy the wave equation given in Eq. 2.6 {16}. A spherical wave with this form can be interpreted as propagating toward the origin.† As before, it is possible to build up more complicated inwardly propagating waves by superposing many complex exponentials of this form. Because of the linearity of the wave equation, solutions that consist of the superposition of both inwardly and outwardly propagating waves are possible. Furthermore, we can have waves radiating outward from (or inward toward) different points simultaneously. Although the spherical wave solutions were derived from the spherical wave equation, they also satisfy the Cartesian wave equation (Eq. 2.1).‡

2.2.3 Doppler Effect

When either a radiating source or a sensor move, the propagating wave's measured frequency content differs from that produced by the source. This phenomenon has been known since the early nineteenth century for both acoustic and light waves. The physics underlying the *Doppler effect* differs in these two cases: In acoustics, the propagation speed differs in moving frames of reference, whereas in electromagnetics the propagation speed remains constant. We consider first the more difficult case in which wave propagation speed is not an invariant. Consider a stationary source with a sensor moving away from it parallel to the direction of propagation with velocity v_{sensor}. At $t = 0$, assume the distance between source and sensor equals ℓ. Waves emitted by the source at this time must travel a distance $ct = \ell + v_{\text{sensor}}t$ to reach the moving sensor. Waves emitted δt later must travel $c(t - \delta t) = \ell + v_{\text{sensor}}\delta t + v_{\text{sensor}}(t - \delta t)$ to reach the sensor. The number of "waves" emitted during δt by the source must equal the number the sensor receives within the interval given by the difference of the times t that solve these two equations. Assuming a source frequency of ω, the frequency measured by the sensor thus equals the number of emitted waves divided by the time taken to receive them.

$$\omega' = \frac{\dfrac{\omega\delta t}{2\pi}}{\dfrac{\delta t}{2\pi}\dfrac{c}{c - v_{\text{sensor}}}} = \omega\left(1 - \frac{v_{\text{sensor}}}{c}\right)$$

---

†Here, we assume that ω and k are positive.

‡If you are interested in demonstrating or testing your mastery of partial differentiation, you can verify this statement by substituting the Cartesian form for the spherical wave

$$s(x,y,z,t) = \frac{A}{\sqrt{x^2+y^2+z^2}}\exp\left\{j\omega\left(t - \frac{1}{c}\sqrt{x^2+y^2+z^2}\right)\right\}$$

into Eq. 2.1.

In the second situation, the source moves toward the sensor at velocity v_{source} parallel to the direction of propagation of the waves reaching the sensor. Waves emitted at $t = 0$ when the distance equals ℓ must travel $ct = \ell$ to reach the stationary sensor. Waves emitted δt later must travel $c(t - \delta t) = \ell - v_{\text{source}}\delta t$ to reach the sensor. Similar calculations for the sensor's measured frequency yield

$$\omega' = \frac{\frac{\omega\delta t}{2\pi}}{\frac{\delta t}{2\pi}\left(1 - \frac{v_{\text{source}}}{c}\right)} = \omega\left(\frac{1}{1 - \frac{v_{\text{source}}}{c}}\right)$$

Clearly, this formula does not equal the first with the sign of the velocity reversed. Only the first applies to electromagnetic waves (when $v \ll c$).† Taking into account the general situation when the direction of propagation is not parallel to the motion's velocity vector, the frequency received by the sensor is Doppler shifted according to

Nonrelativistic	Relativistic ($v \ll c$)
$\omega\left(\frac{1 - \vec{\alpha}\cdot\vec{v}_{\text{sensor}}}{1 - \vec{\alpha}\cdot\vec{v}_{\text{source}}}\right)$	$\omega\big(1 - \vec{\alpha}\cdot(\vec{v}_{\text{sensor}} - \vec{v}_{\text{source}})\big)$

(2.7)

2.2.4 Array Processing Implications

Whenever the wave equation accurately describes the propagation of energy through space and time, the propagating signals have the following properties that can be exploited by array processing algorithms.

- *Propagating signals are functions of a single variable, $s(\cdot)$, with space and time linked by the relation $t - \vec{\alpha}\cdot\vec{x}$.* Assuming for the moment that our signal is bandlimited, by temporally sampling the propagating signal at one location *or* by spatially sampling the signal at one instant, we can reconstruct the signal over all space and time: Based on "local" (either in space or time) information, we can infer the signal over all space and time. Consequently, we can determine the signal radiated by the source.
- *The speed of propagation depends on physical parameters of the medium.* Knowing this speed and determining propagation direction go hand in hand. If the characteristics of medium are known, the speed can be calculated and direction of propagation inferred. Conversely, if the direction of propagation is known, the speed of propagation can be measured, allowing at least a partial characterization of the medium through the speed's dependence on physical parameters.

---

†The exact formula for the Doppler shift in the relativistic case is

$$\omega\sqrt{\frac{1 - \vec{\alpha}\cdot(\vec{v}_{\text{sensor}} - \vec{v}_{\text{source}})}{1 + \vec{\alpha}\cdot(\vec{v}_{\text{sensor}} - \vec{v}_{\text{source}})}}$$

- *Signals propagate in a specific direction $\vec{\zeta}^o$ represented equivalently by either $\vec{\alpha}$ or $\vec{k}$.* By knowing the signal waveform and by comparing signal values measured at several locations, this direction can be inferred *if* the locations are properly chosen. At a given time instant, the spatial samples provide a set of values $\{s(t_0-\vec{\alpha}\cdot\vec{x}_m)\}$ that can be used to determine $\vec{\alpha}$ if these samples occur on unambiguous portions of the waveform. Subsequently, conditions for unambiguous determination of propagation direction emerge. Assuming we can find this direction, we have the opportunity to trace the signal back to its source.

- *Spherical waves describe the radiation pattern of most sources (at least near their locations).* Most signal sources (as opposed to noise sources) are situated at discrete locations and radiate energy in no preferred direction. The medium may, however, distort the wave pattern as it propagates, causing it to cease being a spherical wave. Far removed from the source, however, the wave resembles a plane wave as the wavefront's curvature decreases. Thus, we use the plane wave model in such *far-field* conditions, but we must revert to some other model in *near-field* conditions.

- *The Superposition Principle applies, allowing several propagating waves to occur simultaneously without interaction.* Consequently, distinguishing multiple sources radiating to a common field rests with the array processing algorithm's ability to separate—spatiotemporally filter—received waveforms. Interestingly, we shall find that *nonlinear* algorithms can be used to advantage.

When does the wave equation apply, allowing us to extract information from propagating signals in the ways just outlined?

- The medium must be homogeneous, having constant propagation speed throughout space and time. If not, *refractive* effects distort the propagation.

- Regardless of the wave's amplitude and its frequency content, the wave must interact with the medium in the same way. Otherwise, nonlinearities and frequency-dependent propagation, *dispersion*, occur.

- The medium is lossless. The propagating wave may still *attenuate* as it propagates, but only as a consequence of the ideal wave equation's properties (as with the spherical wave solution).

Note that these conditions apply to the medium. Unfortunately, ideal, uncomplicated conditions apply only in restricted situations. For example, dispersion may not be significant below some temporal frequency. It behooves the wise signal processor to understand well the physical characteristics of the propagation medium before developing algorithms. Consequently, the remainder of this chapter summarizes the "bad" situations, with attention paid to how significant departures can be from the solution to the ideal wave equation.

2.3 Dispersion and Attenuation

Dispersive waves have frequency-dependent propagation characteristics and, unfortunately, occur frequently in nature. The nature and variety of circumstances under which dispersion and attenuation occur can be understood by a linear circuits/systems analogy. Consider the waveform radiated by a source to be the input to a circuit or system with the output equaling the result of propagation through some medium. Nondispersive, lossless, and homogeneous media, propagation through which is governed by the wave equation of the previous section, can be likened to a system exhibiting a pure delay: What goes in comes out once the propagation delay has passed. Lossy media correspond to resistive circuits, which pass a signal unaltered save for an attenuation. Dispersive media are like systems that contain capacitors and inductors, imposing a frequency-dependent phase shift on its inputs. Now the signal is altered by the phase shift, with various frequency components delayed by an amount dependent on the phase shift. Just as the phase shift imposed by the system can not be categorized, a medium's dispersive effects cannot be traced to a single physical phenomenon. Consequently, media can be lossy or dispersive or both and, just as *RLC* circuits are ubiquitous, so are dispersive, lossy media.

Dispersive media are characterized by different propagation speeds for components with different wavelengths. For example, light traveling through a prism suffers dispersion. Different colors (wavelengths) travel at slightly different speeds and are bent—dispersed—at interfaces by slightly different amounts. In the ocean, waves of different lengths also travel at different speeds. There are many causes of dispersion. It has been said that the physics of *all* material media leads to dispersive propagation at some frequency. In each case, the physics leads to a wave equation having an augmented form whose solutions exhibit different propagation speeds for different wavelengths. Nonmonochromatic signals—transients that have a significant bandwidth, for example—thereby disperse, having longer durations as they propagate through space. A propagating signal being dispersed by the medium does not lose energy: The signal energy measured at any moment equals a constant. Many media do attenuate a signal as it propagates; such media also usually disperse the signal. We must realize, however, that dispersion and attenuation are two separate phenomena having different physical bases.

2.3.1 Dispersion

Wave equations augmented for dispersion can often be solved by assuming a complex exponential (monochromatic) solution and deriving a relationship that must hold among the wavenumber vector $\vec{k}$, the temporal frequency ω, and c, the wave equation parameter. In the previous section, we saw that the solution of the classic wave equation led to the relation $\omega = ck$, where k is the magnitude of the wavenumber vector. Such relations between temporal frequency ω and wavenumber magnitude k are called *dispersion relations*. In this case, the relation is linear, indicating a nondispersive wave.† The linear dispersion relation is plotted in Fig. 2.5.

†Yes, despite the fact that the wave equation does not yield a dispersive solution, the relation it imposes between ω and k is *still* called a dispersion relation.

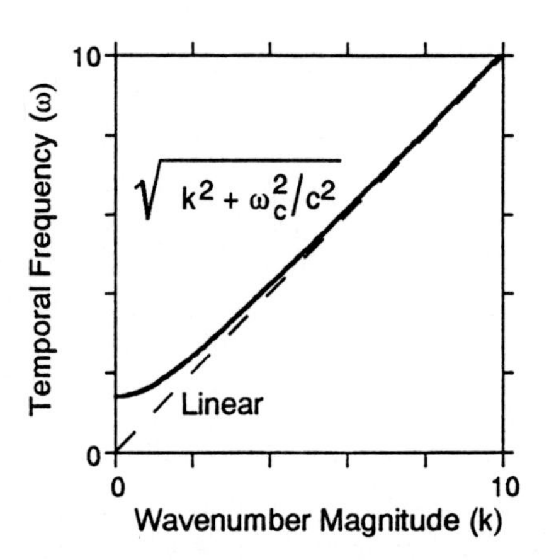

Figure 2.5 The dispersion relation for the wave equation (Eq. 2.1) is linear, indicating that frequency is proportional to wavenumber. In contrast, the dispersion relation for Eq. 2.8 is nonlinear, thereby indicating that different frequency components of a dispersive propagating wave have different propagation speeds. For the nonlinear relation, the quantity ω_0^2/c^2 equals two.

Example

One example of an augmented wave equation is

$$\nabla^2 s = \frac{1}{c^2}\frac{\partial^2 s}{\partial t^2} + \frac{\omega_c^2}{c^2} s$$

where ω_c and c are real constants. This equation arises when the medium has a stringlike stiffness and resists being deformed by the propagating wave. If, as before, we assume the complex exponential solution $s(\vec{x}, t) = A\exp\{j(\omega t - \vec{k}\cdot\vec{x})\}$ and substitute it into our augmented wave equation, we find that the assumed $s(\vec{x}, t)$ is indeed a solution as long as the relation

$$k^2 = \frac{1}{c^2}(\omega^2 - \omega_c^2) \quad \text{or} \quad \omega = c\sqrt{k^2 + \frac{\omega_c^2}{c^2}} \tag{2.8}$$

holds. This dispersion relation is also plotted in Fig. 2.5.

Group and phase velocity. One way of deriving the speed of propagation is expressed by Fig. 2.6. At any point in space, $s(\vec{x}, t)$ oscillates with a temporal frequency ω. During one period of oscillation ($T = 2\pi/\omega$), the wave propagates forward (in the direction of $\vec{k}$) by one wavelength $\lambda = 2\pi/k$. We can express the speed of propagation as the speed at which planes of constant phase, $\vec{k}\cdot\vec{x}$ = constant, move forward. This definition of propagation speed is often called the *phase velocity* and is denoted as the magnitude of a velocity vector $\vec{v}_p$.

$$\boxed{|\vec{v}_p| = \frac{\lambda}{T} = \frac{\omega}{k}}$$

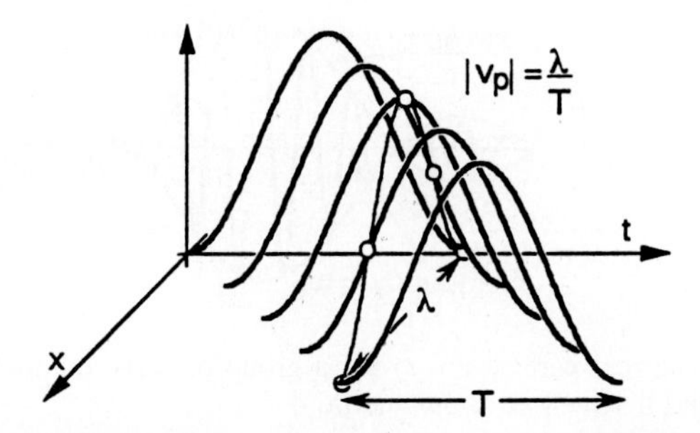

Figure 2.6 The speed of propagation of planes of constant phase is equal to the wavelength divided by temporal period of oscillation. Here, five successive snapshots of the sinusoidal propagating wave were taken at five spatial locations. The time at which the wave reached the last location equaled the period of the sinusoid. Sampling the waveforms at time T demonstrates that the spatial waveform is also sinusoidal. The spatial period of the wave equals its wavelength.

Taking $\vec{v}_p$ and $\vec{k}$ to point in the same direction, we see that the phase velocity is given by

$$\boxed{\vec{v}_p = \frac{\omega\vec{k}}{k^2}}$$

Because $|\vec{v}_p| = \omega/k$, we can interpret the phase velocity magnitude as the slope of the line connecting the origin to a point (k, ω) on the dispersion relation. For the linear dispersion relation, phase velocity magnitude equals c. From the dispersion relation in Eq. 2.8, we find the phase velocity to be

$$\vec{v}_p = \frac{c^2\omega}{\omega^2 - \omega_c^2}\vec{k}$$

With some algebraic manipulation we can write the magnitude of the phase velocity as

$$|\vec{v}_p| = \sqrt{c^2 + \lambda^2\left(\frac{\omega_c}{2\pi}\right)^2} \quad \text{or} \quad |\vec{v}_p| = c\sqrt{\frac{\omega^2}{\omega^2 - \omega_c^2}}$$

thereby showing that $|\vec{v}_p|$ depends on the wavelength λ or, equivalently, temporal frequency ω.

Another propagation velocity, the *group velocity*, can be associated with dispersive waves. It is the velocity at which a group of closely spaced (in frequency) complex exponential waves propagate. For example, let's suppose that $s(\vec{x}, t)$ consists of a superposition of complex exponentials propagating in the same direction with temporal frequencies lying in a narrow band $[\omega_0 - \delta\omega, \omega_0 + \delta\omega]$.

$$s(\vec{x}, t) = \int_{\omega_0-\delta\omega}^{\omega_0+\delta\omega} \exp\{j(\omega t - \vec{k}(\omega)\cdot\vec{x})\}\, d\omega$$

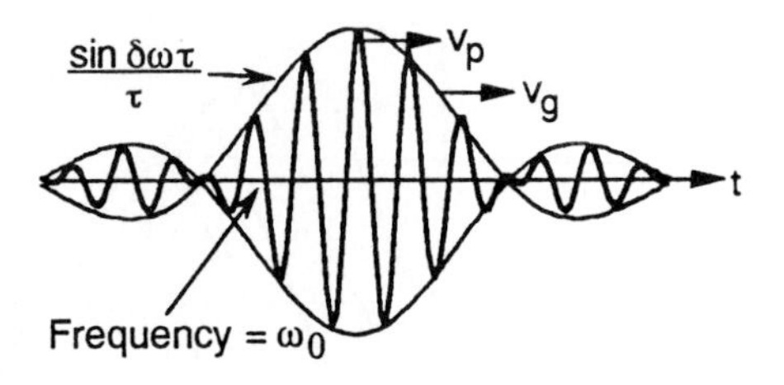

Figure 2.7 The real part of $s(\vec{x}, t)$ for a group of wave components with temporal frequencies in a narrow band between $\omega_0 - \delta\omega$ and $\omega_0 + \delta\omega$.

The wavenumber $\vec{k}$ depends on ω because of the dispersion relation. For sufficiently small $\delta\omega$, we may accurately approximate the wavenumber as

$$\vec{k}(\omega) \approx \vec{k}(\omega_0) + \tilde{\omega} \left. \frac{d\vec{k}}{d\omega} \right|_{\omega=\omega_0}$$

for $-\delta\omega \leq \tilde{\omega} \leq \delta\omega$. Using this approximation, we can write the expression for $s(\vec{x}, t)$ as

$$\begin{aligned} s(\vec{x}, t) &= \exp\{j(\omega_0 t - \vec{k}(\omega_0) \cdot \vec{x})\} \int_{-\delta\omega}^{\delta\omega} \exp\left\{ j\tilde{\omega} \left(t - \vec{x} \cdot \left. \frac{d\vec{k}}{d\omega} \right|_{\omega=\omega_0} \right) \right\} d\tilde{\omega} \\ &= 2\exp\{j(\omega_0 t - \vec{k}(\omega_0) \cdot \vec{x})\} \frac{\sin \delta\omega\tau}{\tau} \end{aligned}$$

where

$$\tau = t - \vec{x} \cdot \left. \frac{d\vec{k}}{d\omega} \right|_{\omega=\omega_0}$$

The real part of this function is sketched in Fig. 2.7 and consists of a $\sin \delta\omega\tau/\tau$ envelope modulating a complex exponential carrier with temporal frequency ω_0 and wavenumber vector $\vec{k}(\omega_0)$. The peak in the envelope occurs at $\tau = 0$ or $t = \vec{x} \cdot d\vec{k}/d\omega$ evaluated at $\omega = \omega_0$. This peak, and indeed the whole envelope, propagates at the group velocity $\vec{v}_g$, which equals $\vec{x}/t$, implying that the group velocity vector must satisfy

$$\boxed{\vec{v}_g \cdot \left. \frac{d\vec{k}}{d\omega} \right|_{\omega=\omega_0} = 1} \tag{2.9}$$

For this and similar examples, we can easily show that the group velocity equals the derivative of the dispersion relation.

$$|\vec{v}_g| = \frac{d\omega}{dk}$$

The magnitude of the group velocity thus equals the slope of the dispersion relation (Eq. 2.8) at the point $\omega = \omega_0$, as illustrated in Fig. 2.5.

Example

Refer back to the dispersion relation Eq. 2.8.

$$\vec{k}(\omega) \cdot \vec{k}(\omega) = \frac{\omega^2 - \omega_c^2}{c^2}$$

Taking the derivative of both sides with respect to ω yields

$$2\vec{k}(\omega) \cdot \frac{d\vec{k}}{d\omega} = \frac{2\omega}{c^2}$$

A solution for $\vec{v}_g$ consistent with Eq. 2.9 is

$$\vec{v}_g = \frac{c^2}{\omega_0}\vec{k}(\omega_0)$$

The magnitude of the group velocity can be written in terms of the temporal frequency ω or the wavelength λ by using the dispersion relation (Eq. 2.8).

$$|\vec{v}_g| = c\sqrt{1 - (\omega_c/\omega_0)^2} \quad \text{and} \quad |\vec{v}_g| = \frac{c^2}{\sqrt{c^2 + (\lambda_{\omega_0}\omega_c/2\pi)^2}}, \; \lambda_{\omega_0} = \frac{2\pi}{|\vec{k}(\omega_0)|}$$

Note that $|\vec{v}_g||\vec{v}_p| = c^2$ and that $|\vec{v}_g| < c < |\vec{v}_p|$ for $\omega_0 > \omega_c$. The question becomes which velocity do we consider representing the speed at which information propagates through the medium? More provocatively, can information be propagated in the medium faster than the propagation velocity c? Because the information in a modulated carrier wave is normally associated with its envelope, the information propagates at the group velocity, the physically satisfying answer because it is less than c. Thus, for this dispersion relation, information propagates at a speed less than that characteristic of the medium, and this speed is frequency dependent.

In summary, the phase velocity and the group velocity are both needed to characterize the propagation of a field in a dispersive medium. They are given by

$$\boxed{\vec{v}_p = \frac{\omega\vec{k}}{|\vec{k}|^2} \qquad \vec{v}_g \cdot \frac{d\vec{k}}{d\omega} = 1}$$

The magnitude of the phase and group propagation speeds are expressed by

$$\boxed{|\vec{v}_p| = \frac{\omega}{k} \qquad |\vec{v}_g| = \frac{d\omega}{dk}}$$

Thus, as illustrated in Fig. 2.8, the phase propagation speed is the slope of a line extending from the origin to a point on the dispersion relation, whereas the group propagation speed is the dispersion relation's slope at the same point. Although the definition of phase propagation speed corresponds to one of the classic formulas in physics, the group propagation speed is far more important: It is that speed that determines how fast energy propagates from a source to the array.

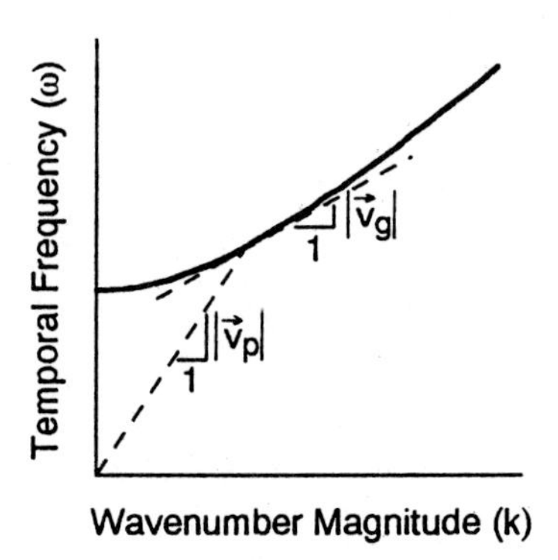

Figure 2.8 The group propagation speed $|\vec{v}_g|$ equals the slope of the dispersion relation at a particular wavelength. The phase propagation speed $|\vec{v}_p|$ equals the slope of the line joining a point on the dispersion relation to the origin.

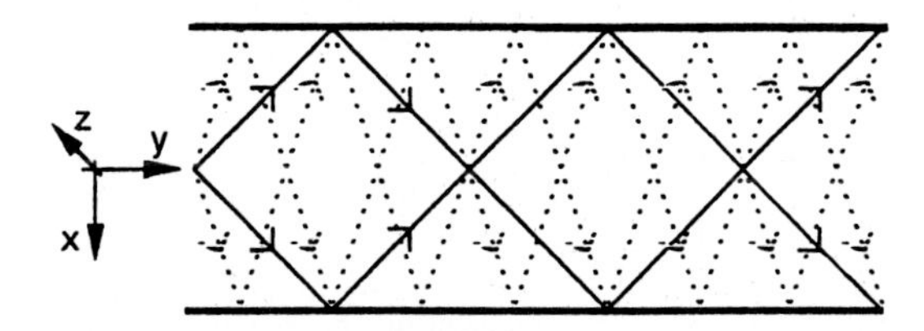

Figure 2.9 A simple metal waveguide can be used to control the direction of propagation of electromagnetic waves. The $TE_{m,0}$ solution to the wave equation may be interpreted as a plane wave bouncing from side to side down the waveguide.

Example

Waveguides are an interesting example of a dispersive medium. A rectangular metal tube such as the one sketched in Fig. 2.9 can be used to guide electromagnetic radiation from point to point. Because the walls of the waveguide are conducting, the electric field inside the waveguide is constrained to be normal to the walls. Similarly, the magnetic field is constrained to be parallel to the walls at the walls. A waveguide such as this can sustain a variety of propagating electromagnetic fields. These can be found by solving Maxwell's equations with the appropriate boundary conditions. One family of solutions, called the $TE_{m,0}$ family,† has its electric field perpendicular to the long axis of the waveguide. For this family, the electric field vector is parallel to the z axis and is given by

$$E_z(x, y, z, t) = A \sin \frac{m\pi x}{a} \exp\{j(\omega t - k_y y)\}, \quad m = 1, 2, \ldots$$

The term $m\pi/a$ corresponds to the x component k_x of the wavenumber vector. When $m = 1$, the electric field amplitude varies as half a period of a sine wave as x increases from 0 to a. Larger values of m correspond to more half-periods across the waveguide. It is straightforward to verify that $E_z(x, y, z, t)$ satisfies the wave equation provided

$$\frac{m^2\pi^2}{a^2} + k_y^2 = \frac{\omega^2}{c^2} \tag{2.10}$$

† *TE* stands for "transverse electric."

where $c = 1/\sqrt{\epsilon\mu}$. Solving for the magnitude of k_y, we obtain

$$|k_y| = \sqrt{\frac{\omega^2}{c^2} - \frac{m^2\pi^2}{a^2}}$$

Eq. 2.10 is similar to the usual dispersion relation for a propagating plane wave:

$$k_x^2 + k_y^2 + k_z^2 = \frac{\omega^2}{c^2}$$

For the waveguide, $k_z = 0$ and k_x is constrained by the boundary conditions to be $\pm m\pi/a$ independent of the temporal frequency ω. We may interpret the $TE_{m,0}$ solution as a plane wave traveling down the waveguide by bouncing from side to side, as indicated in Fig. 2.9. The angle of incidence, θ, is a function of the temporal frequency ω.

$$\tan\theta = \left|\frac{k_y}{k_x}\right| = \sqrt{\frac{\omega^2}{c^2}\frac{a^2}{m^2\pi^2} - 1} \tag{2.11}$$

For very high frequencies, $\tan\theta \approx \omega a/cm\pi$. Thus, as the frequency increases, the propagating wave grazes the sides of the waveguide at increasingly shallower angles ($\theta \to \pi/2$). As ω is reduced to the critical frequency $\omega = m\pi c/a$, θ becomes zero. In this case, the wave simply bounces back and forth between the walls without making any progress down the waveguide ($k_y = 0$).

Fig. 2.9 and Eq. 2.11 demonstrate that higher-frequency wave components travel down the waveguide faster than lower-frequency components (for a fixed value of m). Lower-frequency components bounce off the walls more times and hence must travel further to make the same headway.† As before, we can derive a phase velocity $\vec{v}_p$ and a group velocity $\vec{v}_g$. The magnitude of the phase velocity is given by

$$|\vec{v}_p| = \frac{\omega}{|\vec{k}_y|} = \frac{c}{\sqrt{1 - (\lambda/\lambda_x)^2}}$$

where $\lambda_x = 2\pi/k_x$. The phase velocity measures the rate of progress of constant-phase planes down the waveguide. The magnitude of the group velocity is

$$|\vec{v}_g| = \frac{1}{\left|\frac{dk_y}{d\omega}\right|} = c\sqrt{1 - (\lambda/\lambda_x)^2}$$

and, as before, can be interpreted as the rate at which information propagates down the waveguide. Note that $|\vec{v}_p||\vec{v}_g| = c^2$.

Evanescent waves. In some circumstances, no real-valued wavenumber vector $\vec{k}$ can be found that satisfies the dispersion relation for some values of temporal frequency. For example, consider the dispersion relation expressed by Eq. 2.10 for the waveguide example rewritten here as

$$k_y^2 = \frac{\omega^2}{c^2} - \frac{m^2\pi^2}{a^2}$$

†The picture of a sailboat tacking into the wind comes to mind.

Suppose that $\omega = m\pi c/2a$. The solutions for k_y are purely imaginary.

$$k_y = \pm\frac{m\pi}{a} j\sqrt{3}$$

Substituting these values for k_y back into the $TE_{m,0}$ solution for E_z gives

$$E_z(x, y, z, t) = A \sin\frac{m\pi x}{a} \exp\left\{\frac{\mp m\pi y\sqrt{3}}{a}\right\} e^{j\omega t}$$

Only the solution with the negative exponential factor makes physical sense for this situation; otherwise the field grows exponentially as it propagates down the waveguide, gaining energy as it propagates. The decaying solution is a wave field that oscillates in the x direction and in time, but decays exponentially along the waveguide and does not propagate. This solution is called an *evanescent wave*. It results when the waveguide is driven at a frequency below the so-called cutoff frequency $\omega = m\pi c/a$.

2.3.2 Attenuation

The medium imposes a solution to the wave equation that decays spatially when the wave does work as it passes through the medium. For example, if we have a conducting medium so that one of Maxwell's equations becomes $\nabla\times\vec{H} = \sigma\vec{E} + \partial(\epsilon\vec{E})/\partial t$, the wave equation for the magnetic field becomes

$$\nabla^2\vec{H} = \frac{1}{c^2}\frac{\partial^2\vec{H}}{\partial t^2} + \sigma\mu\frac{\partial\vec{H}}{\partial t}$$

From fundamental physics and the properties of the solutions of differential equations, a first-order temporal derivative usually means damping; this equation is no exception. Assume for simplicity that the equation under consideration is the scalar wave equation with a damping term.

$$\nabla^2 s = \frac{1}{c^2}\frac{\partial^2 s}{\partial t^2} + \gamma\frac{\partial s}{\partial t}, \quad \gamma > 0$$

Assuming a solution of the form $s(\vec{x}, t) = A\exp\{j(\omega t - \vec{k}\cdot\vec{x})\}$ yields the dispersion relation

$$\vec{k}\cdot\vec{k} = \frac{\omega^2}{c^2} - j\omega\gamma$$

The wavenumber vector must be complex ($\vec{k} = \vec{k}_{\text{Re}} + j\vec{k}_{\text{Im}}$) to satisfy this equation. Equating real and imaginary parts of the dispersion relation, we derive two simultaneous equations for $\vec{k}_{\text{Re}}$ and $\vec{k}_{\text{Im}}$.

$$\vec{k}_{\text{Re}}\cdot\vec{k}_{\text{Re}} - \vec{k}_{\text{Im}}\cdot\vec{k}_{\text{Im}} = \frac{\omega^2}{c^2} \qquad 2\vec{k}_{\text{Re}}\cdot\vec{k}_{\text{Im}} = -\omega\gamma$$

Assuming that the real and imaginary parts of the wavenumber vector are parallel, $\vec{k}_{\text{Im}} = \beta\vec{k}_{\text{Re}}$, we find that

$$|\vec{k}_{\text{Re}}|^2 = \frac{\omega^2}{(1-\beta^2)c^2} \qquad \frac{2\beta\omega}{(1-\beta^2)c^2} = -\gamma \tag{2.12}$$

Thus, $\vec{k}_{\text{Re}}$ and $\vec{k}_{\text{Im}}$ are antiparallel—β must be negative—and must obey the nondispersive dispersion relation $|\vec{k}_{\text{Re}}| = \omega/c\sqrt{1-\beta^2}$. Because of the nonzero imaginary part of the wavenumber vector, the monochromatic solution has an exponentially weighted amplitude.

$$s(\vec{x},t) = Ae^{\vec{k}_{\text{Im}}\cdot\vec{x}}\exp\{j(\omega t - \vec{k}_{\text{Re}}\cdot\vec{x})\}$$

The negative value for β means that the wave amplitude decays in the direction of propagation. Note that because of the dispersion relation, the rate of decay depends on frequency, most strongly so at low frequencies (see Prob. 2.6).

The physics of most realistic lossy media also demand dispersion. Equations of the form

$$\nabla^2 s = \frac{1}{c^2}\frac{\partial^2 s}{\partial t^2} + \gamma_1\frac{\partial s}{\partial t} + \gamma_2 s$$

are common, describing signals that not only attenuate as they propagate, but also spread temporally and do not retain their original waveforms.

2.3.3 Array Processing Implications

Because virtually all real-world media are lossy, we should expect some dispersion and attenuation to occur in every measurement we make. To exploit the properties of dispersion and attenuation, array processing algorithms need to face harsh realities.

- *Lossy media cause signals to decay more rapidly than predicted by the ideal wave equation.* For example, the amplitude of a wave radiating into a homogeneous medium from a point signal source decays faster than $1/r$ as predicted by the spherical wave equation, exponentially, in fact.

$$s(r) \propto \frac{\exp\{-|\vec{k}_{\text{Im}}|r\}}{r}$$

  Consequently, an array's range is limited: The further the source is located from the array, the more difficult it becomes to extract its radiation from ever-present noise. An array's effective range, as determined by the spatial decay constant $|\vec{k}_{\text{Im}}|$, is frequency dependent, usually decreasing with increasing frequency ($|\vec{k}_{\text{Im}}|$ increases with ω).

- *In dispersive media, narrowband sources propagate to the array at speeds different than the medium's characteristic speed.* The group velocity determines how quickly information propagates through space and time. Because this speed differs from c, we must incorporate this difference when inferring the slowness vector from the spatiotemporal linkage relation $t - \vec{\alpha}\cdot\vec{x}$. Otherwise, we introduce errors in locating sources from array data.

- *Dispersion causes the received waveform emanating from a broadband source to vary with range.* Each frequency component of a broadband source propagates at a different speed, meaning that the signal received at various locations differs even for a homogeneous medium. Consequently, if a source is located close to the array,† the distance from each sensor to the source can differ significantly, requiring some compensation for dispersion. Remote sources are essentially located the same distance from each sensor, meaning that intersensor waveforms differences owing to dispersion are minimal: Array processing algorithms need not consider different waveforms at each sensor. Be aware that the source's distance affects the received waveform: If we focus on one particular broadband source in our processing, the array processing algorithm must employ dispersion compensation. Meaningful compensation requires some form of source-location tracking; see Chap. 8.
- *Dispersion can remove some frequency components entirely.* Evanescent waves are one example; For some media, low-frequency waves may not propagate at all, meaning the array receives no energy at all at these frequencies.

Thus, we can model a medium's action on source waveforms like a linear filter: The filter's transfer function is range dependent, can be zero at low frequencies, produces significant attenuation at higher frequencies, and introduces phase shifts that tend to spread the waveform.

2.4 Refraction and Diffraction

Rays of light, as well as rays of sound, may be *bent* from straight-line propagation if the medium is *inhomogeneous.* Refraction and diffraction are names given to two physical phenomena that cause rays to bend. *Refraction* is typically caused by spatial changes in the medium's speed of propagation. For electromagnetic radiation, changes in the medium, say from vacuum to glass, cause changes in the dielectric permittivity ϵ or the magnetic permeability μ, which result in different speeds of light. Similarly, for acoustic radiation, spatial variations in the medium's ambient pressure, temperature, and density cause variations in the speed of sound that bend sound waves. *Diffraction* describes the phenomenon of waves bending around objects and represents a departure from the simple geometrical model of wave propagation.‡ This phenomenon can be traced to the nonzero wavelength of propagating waves.

A detailed discussion of refraction and diffraction would represent a digression too long for this text. Many books have been written discussing these two physical phenomena for the cases of acoustic waves, electromagnetic waves (optical waves in particular), and even the wave functions of quantum mechanics. In this section, we shall simply highlight the most important results to remind the reader that the simple equations describing physical laws can lead to very complex physical behavior.

---

†When located close to an array, a source is said to be in the "near field." What close means is defined in a later section.

‡If the geometrical model were accurate, it would be very difficult to hear around corners.

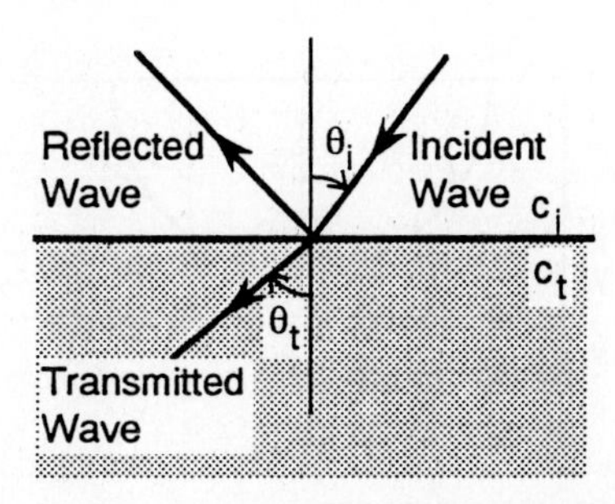

Figure 2.10 An incident wave striking a discontinuity in the medium results in a reflected wave and a transmitted wave. The angle of reflection equals the angle of incidence, and the angle of refraction of the transmitted wave obeys Snell's Law. In this example, the propagation speed in the lower medium is greater than in the upper.

2.4.1 Refraction

The most easily treated case of refraction occurs when a plane wave approaches a boundary between two media that have constant, but unequal, speeds of propagation. We are all familiar with the refraction of light as it passes from air to water. One need only stick a pole into a pond on a bright summer's day to observe refraction. As a wave encounters a discontinuity in the medium, it generally splits into a *reflected* wave and a *transmitted* wave. The physics of the situation generally require that $\vec{k}_i \cdot \vec{x} = \vec{k}_r \cdot \vec{x} = \vec{k}_t \cdot \vec{x}$, where $\vec{k}_i$, $\vec{k}_r$, and $\vec{k}_t$ represent the wavenumber vectors for the incident, reflected, and transmitted waves, respectively, and $\vec{x}$ is a point on the discontinuity. Considering the diagram shown in Fig. 2.10, this relationship reduces to

$$|\vec{k}_i| \sin\theta_i = |\vec{k}_r| \sin\theta_r = |\vec{k}_t| \sin\theta_t$$

Because the speed of propagation is the same for both the incident and reflected waves, we know that $|\vec{k}_i| = |\vec{k}_r|$, which implies $\theta_i = \theta_r$: The angle of incidence equals the angle of reflection.

The magnitude of the wavenumber vector for the transmitted wave depends on the speed of propagation on the other side of the boundary. Let's assume that the speed of propagation on the incident side to be c_i and on the transmitted side c_t. Because we have been tacitly assuming that the wave in question is monochromatic with frequency ω, the previous equation becomes

$$\frac{\sin\theta_i}{c_i} = \frac{\sin\theta_t}{c_t}$$

This equation represents one version of Snell's Law.†

In several situations, especially underwater acoustics, the medium undergoes gradual changes in the speed of propagation. In the ocean, for example, the speed of sound

†In optics, Snell's Law is usually written in terms of the *index of refraction* n. This index equals the ratio of the speed of light in a vacuum to the speed of light in the medium in question: $n = c_0/c$. For example, light travels slower in glass, which has an approximate refraction index of $n = 1.5$. The exact value depends on the frequency of light and the type of glass.

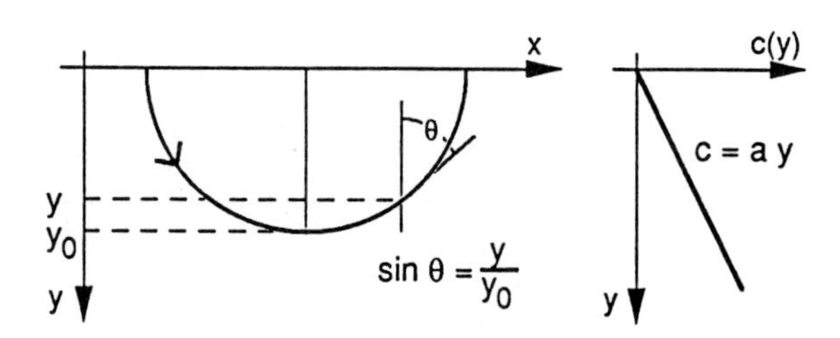

Figure 2.11 A linear change in the speed of sound with depth results in circular rays. The linear sound profile is shown in the right panel.

changes with depth, yielding what is known as a *sound speed profile* [150: §5.5]. This variation is caused by gradually changing temperature, salinity, and pressure in the ocean. These gradual changes cause the propagation path to bend as sound propagates through the water.

Example

We can demonstrate that a linear speed of sound profile results in rays that form circular arcs. Suppose that the speed of sound, as a function of the depth y, is given by $c(y) = ay$, where a is the slope of the velocity change with depth. Let us further suppose that a wave is propagating horizontally at a depth y_0. The ray representing the wave is bent upward, as shown in Fig. 2.11, according to Snell's Law because of the changing speed of sound. At any particular point, the ray makes an angle θ with the vertical axis. Because of Snell's Law, we know that

$$\frac{\sin\theta(y)}{c(y)} = \frac{\sin\theta(y-\delta y)}{c(y-\delta y)}$$

where δy is a small decrement in depth. To solve for the angle θ, or equivalently $\sin\theta$, as a function of the depth y, we derive the differential equation

$$\frac{1}{\sin\theta}\frac{d\sin\theta}{dy} = \frac{1}{c}\frac{dc}{dy}$$

that governs the relation we seek from Snell's Law. Integrating both sides of this equation with respect to y shows us that $\ln(\sin\theta) = \ln(c) + C_0$, where C_0 is a constant of integration. Equivalently, $\sin\theta(y) = C_1 c(y)$ where $C_1 = \exp\{C_0\}$. By assumption, we know that the wave is propagating horizontally at the depth $y = y_0$, leading us to conclude that $\sin\theta(y_0) = 1$ that in turn implies that $C_1 = 1/ay_0$. The relation for $\sin\theta$ satisfying Snell's Law and the boundary conditions is

$$\sin\theta = \frac{y}{y_0}$$

This relation describes the y component of a circle having radius y_0.

If in this example the medium were discontinuous and the velocity were not zero at the surface ($y = 0$), at least a portion of the wave would reflect back into the inhomogeneous medium and would thereby create another circular ray, which would bend back to the surface, reflect, and so forth. This infinite set of refractions and reflections captures the essential nature of *channeling*: An inhomogeneous medium creates guided waves. A source located at some depth would emit acoustic radiation in all directions, thereby

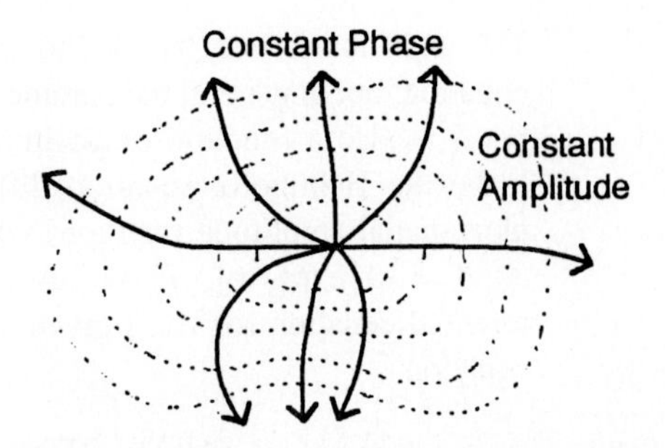

Figure 2.12 Rays of energy leaving a point source are bent by propagation through an inhomogeneous medium.

creating a continuum of initial angles that diffract and follow semicircular paths (see Prob. 2.8). This channeling phenomenon means sound in the ocean rarely propagates in a straight line, obviating spherical radiation models, for example, and leads to some great sea stories.†

2.4.2 Ray Theory

In many practical wave propagation problems, it may be difficult or impossible to solve the wave equation exactly, and sometimes when an exact solution is available it may be difficult to interpret that solution in terms of one's intuition about propagating waves. To visualize propagation phenomena under these conditions, researchers use ray-tracing computer programs to follow the path of a wavefront through a nonhomogeneous medium. These computer programs are based on the *ray theory of wave propagation*. As we shall soon see, ray theory is a *high-frequency approximation* to the wave equation; the results produced by the ray-tracing programs can only predict wave propagation when this approximation is valid.

The basic idea behind ray theory is that energy is visualized as propagating along curves in space, *rays*, that are perpendicular to the wavefront. In the case of a plane wave, the curve is a straight line orthogonal to the planes of constant phase. This notion of a ray corresponds closely to the geometrical definition of a ray for plane waves. The surfaces of constant phase of a propagating wave, often called the *wavefronts* or phase fronts, become distorted as a wave propagates through an inhomogeneous medium, but we may still define a ray as a continuous path that is orthogonal to all the wavefronts through which it passes. Fig. 2.12 shows a simple picture of rays of energy leaving a point source.

The basic equations of ray theory are most easily derived if we assume a monochromatic solution: $s(\vec{x}, t) = A(\vec{x}) \exp\{j\omega t\}$. If we substitute $s(\vec{x}, t)$ into the ideal wave equation (Eq. 2.1 {11}), we get a related partial differential equation, the *Helmholtz equation*, for the spatial amplitude function $A(\vec{x})$.

$$\nabla^2 A + k^2 A = 0\,, \;\; k = \omega / c$$

†Is it possible for a ship to "hear" a submarine during the morning, keep a fixed distance away from it as the day progresses, and have it become inaudible in the afternoon?

This equation is very similar to those arising from our earlier discussions of the wave equation, but in general the parameter c may be a function of position, meaning wavenumber k is also a function of position. Spatial inhomogeneities leading to variations in c make the Helmholtz equation difficult to solve in the general case. $A(\vec{x})$ is a complex spatial amplitude function; we may write it in terms of its magnitude and phase as $A(\vec{x}) = |A| \exp\{j\psi\}$. If we substitute this representation for $A(\vec{x})$ into the partial differential equation for $A(\vec{x})$ given above and expand, we obtain the cumbersome-looking equation

$$\nabla^2 |A| + |A|\{k^2 - (\nabla \psi \cdot \nabla \psi)\} + j\{|A|\, \nabla^2 \psi + 2(\nabla |A| \cdot \nabla \psi)\} = 0$$

Ray theory follows by making the approximation that ω is very large compared with the spatial variations of c [19: §2.6], which means that k becomes very large.† With this assumption, the second and third terms of the preceding equation dominate. Note that the second term is purely real, and the third term is purely imaginary. For their sum to be zero, each must be zero. Under this approximation, we are left with the *eikonal equation*‡

$$\nabla \psi \cdot \nabla \psi = k^2$$

and the *transport equation*

$$|A|\, \nabla^2 \psi + 2(\nabla |A| \cdot \nabla \psi) = 0$$

A curve in space, a ray, may be defined parametrically as a set of points $\vec{x}(\ell)$ with ℓ being distance along the curve. The tangent vector $\vec{e}$ for this curve is given simply as

$$\frac{d\vec{x}}{d\ell} \propto \vec{e}$$

where the tangent vector is understood to have unit length: $|\vec{e}| = 1$. Because rays are curves that are orthogonal to surfaces of constant phase, a ray is defined to be the vector orthogonal to the wavefront ψ. Because the gradient of the phase function is also orthogonal to wavefronts, we can use the eikonal equation to relate the phase gradient and the ray tangent vector. Thus, the ray is defined by equating its tangent vector to the gradient of the phase.

$$\nabla \psi = \frac{\omega}{c}\vec{e} = k\vec{e}$$

In general, both k and $\vec{e}$ vary with position $\vec{x}$.

How can we determine what paths the rays of energy follow? If we knew how the tangent vector $\vec{e}$ varied along the path as a function of the distance ℓ, we could integrate it with respect to ℓ to obtain

$$\vec{x}(\ell) = \int_0^{\ell} \vec{e}(\nu)\, d\nu \; + \; \vec{x}_0$$

---

†Increasing wavenumber corresponds to assuming that the wavelength $\lambda = 2\pi/k$ goes to zero.

‡*Eikonal* is derived from the Greek word for "image." Note that the eikonal equation can also be expressed as $\nabla \psi \cdot \nabla \psi = n^2$, where n is the index of refraction.

Because propagation speed and, consequently, index of refraction vary as functions of position $\vec{x}$, rays follow a curved path. Therefore, we need a relationship between the gradient ∇n and the tangent vector $\vec{e}(\ell)$. After some manipulation, the eikonal equation gives us [19: §2.6]

$$\frac{d}{d\ell}(n\vec{e}) = \nabla n \tag{2.13}$$

Thus, the gradient of the index of refraction is related to the ray's tangent vector by

$$\boxed{\vec{e}(\ell) = \frac{1}{n}\left[\int_0^\ell \nabla n(\nu)\,d\nu + n_0\vec{e}_0\right]} \tag{2.14}$$

Here, n_0 is the index of refraction and $\vec{e}_0$ is the tangent vector at the starting point of the ray ($\ell = 0$).

Rays can be determined by integrating the gradient of the index of refraction to obtain the ray tangent and then integrating the ray tangent to obtain the ray itself.† This procedure becomes a computer algorithm by implementing some variation of the following iteration:

1. Choose the ray starting point $\vec{x}(0) = \vec{x}_0$ and the initial ray direction $\vec{e}(0) = \vec{e}_0$; set $\ell = 0$.
2. Let $\vec{f}(\ell) = n(\ell)\vec{e}(\ell)$ represent the gradient of the index of refraction
3. Compute the new ray position: $\vec{x}(\ell + \delta\ell) = \vec{x}(\ell) + \vec{e}(\ell)\delta\ell$.
4. Compute the gradient at the new point: $\vec{f}(\ell + \delta\ell) = \vec{f}(\ell) + \nabla n(\ell)\delta\ell$.
5. Compute the ray tangent vector at the new point: $\vec{e}(\ell+\delta\ell) = \vec{f}(\ell + \delta\ell)/n(\ell + \delta\ell)$.
6. Increment ℓ by $\delta\ell$ and go to step 3.

It is also possible to use time, rather than distance along the ray, as the independent variable. We accomplish this change of variable by using the relation $d\ell/dt = c$.

A gradient in the speed of propagation, or equivalently, a gradient in the index of refraction causes rays to curve. We can relate the amount of curvature to the gradient by starting with the relation

$$\frac{d(n\vec{e})}{d\ell} = \nabla n$$

Expanding the derivative on the left side, we find

$$n\frac{d\vec{e}}{d\ell} + \vec{e}\frac{dn}{d\ell} = \nabla n$$

†A ray may also be derived using the calculus of variations: Given two points in the medium, a ray is that curve for which the propagation time between those two points is minimized. The problem becomes one of selecting a curve such that the integral

$$T = \int_{\vec{x}_0}^{\vec{x}_1} \frac{d\ell}{c}$$

is minimized over all curves connecting $\vec{x}_0$ and $\vec{x}_1$.

The rate of change of the tangent vector with respect to distance along the ray, $d\vec{e}/d\ell$, is equal to $\kappa\vec{N}$, where $\vec{N}$ is a unit vector orthogonal to the tangent vector $\vec{e}$ and κ is the *curvature* of the ray at that point.† Using this relationship in the above equation and dotting the result with the tangent vector $\vec{e}$, we obtain

$$\kappa = \frac{1}{n}\vec{N}\cdot\nabla n \quad \text{or equivalently} \quad \kappa = -\frac{1}{c}\vec{N}\cdot\nabla c$$

The curvature of the ray at any point is determined by the speed of propagation and its gradient at that point. This relation can be interpreted as a continuous version of Snell's Law. For a nondispersive inhomogeneous medium, ray paths do not depend on the frequency ω, or equivalently, the wavelength λ. For a dispersive inhomogeneous medium, index of refraction and, equivalently, wavenumber depend on both position and frequency. Consequently, *wave components with different wavelengths follow different paths in a dispersive medium.* In optics, this effect is called chromatic aberration.

We shall not discuss in detail the use of the transport equation mentioned earlier. In ray theory, the transport equation can be used to calculate the strength of the wave amplitude function $|A|$ [19: §2.6]. In simplistic terms, as energy propagates along rays, the amplitude is related to the increase or decrease in the cross-sectional area of a ray *tube*, a collection of closely spaced rays forming a tube, as it evolves in space.

$$|A| \propto (nJ)^{-1/2}$$

Here, J represents the Jacobian of the transformation from Cartesian coordinates to ray coordinates.

2.4.3 Diffraction

When ray theory is applied to the propagation of light, we obtain the *geometrical model of optics*: Light rays travel in straight lines through a homogeneous medium. When a ray of light strikes a surface, it scatters but the reflected ray also travels in a straight line. When light meets a discontinuity in the medium, the rays are bent by refraction at the interface, but they continue to travel in straight lines on either side of the discontinuity. An inhomogeneous medium causes light rays to bend, but locally the ray is traveling in a straight line.

Unfortunately, the geometrical model of propagation is too simple: It is based on ray theory, which applies only in the high-frequency–small-wavelength limit. For light and sound waves that have wavelengths comparable with the structures they meet, another propagation phenomena, *diffraction*, can cause the wave to deviate from straight-line propagation. Diffraction theory explains (or attempts to explain) why the edges of shadows are not perfectly sharp and how we can hear "around corners." A detailed discussion of diffraction theory is well beyond the scope of this text (see Goodman, Chaps. 3 and 4, for example). In this section, we shall simply try to indicate some of the important consequences of diffraction.

†The curvature is equal to the inverse of the radius of curvature. A circle of radius R has a curvature of $\kappa = 1/R$ at all points on the circle.

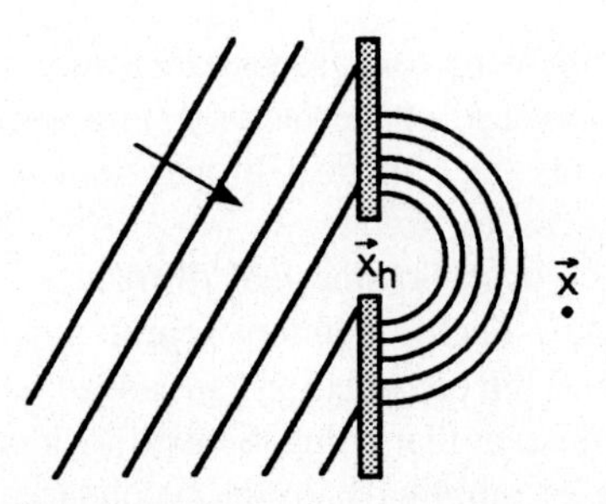

Figure 2.13 A wave is shown impinging on a hole in a planar screen. The Rayleigh-Sommerfeld diffraction formula tells us what the wavefield at the point $\vec{x}$ is in terms of the wavefield at the aperture.

In the late seventeenth century, Christian Huygens developed a simple theory to try to explain the observation that shadow edges were not sharp. He postulated that each point on a traveling wavefront could be considered as a secondary source of spherical radiation. The position of the wavefront at some later time would correspond to the envelope of the radiation from the secondary sources. This simple idea known as *Huygens's Principle*, was refined and later given a mathematical foundation by a series of renowned physicists including Fresnel, Kirchoff, Lord Rayleigh, and Sommerfeld. One result of this evolution in thought about diffraction is the *Rayleigh-Sommerfeld diffraction formula.* Consider a wave impinging on a hole cut from a planar screen as shown in Fig. 2.13. Let $s(\vec{x})$ represent a monochromatic wave with a single, well-defined wavelength λ[†] located at some point $\vec{x}$ remote from the aperture. The Rayleigh-Sommerfeld diffraction formula states that this wave can be represented as a superposition of fields originating within the hole, with each component field weighted by a spherical spreading function.

$$\boxed{s(\vec{x}) = \frac{1}{j\lambda}\iint_A s(\vec{x}_h)\frac{\exp\{jkr\}}{r}\cos\theta\, dA}$$

The double integral is taken over the aperture A, and dA represents an infinitesimal patch of area located at the position $\vec{x}_h$ within the hole. The variable θ represents the angle between the vector normal to the plane and the vector joining the aperture patch dA and the position $\vec{x}$. This angle is the angle of the difference vector $\vec{x} - \vec{x}_h$. The constant k represents the magnitude of the wavenumber and equals $2\pi/\lambda$. Finally, r is the distance from the patch dA to the position $\vec{x}$: $r = |\vec{x} - \vec{x}_h|$.

The Rayleigh-Sommerfeld diffraction formula can be interpreted in terms of the Huygens's Principle of secondary sources. The wavefield at $\vec{x}$ is indeed a summation of secondary sources in the aperture, but the summation includes several factors. The amplitude of the secondary source at $\vec{x}_h$ is proportional to the wavefield amplitude $s(\vec{x}_h)$ divided by the wavelength λ. It is also multiplied by the obliquity factor $\cos\theta$: This factor equals one when $\vec{x}$ is located along the vector normal to the aperture that passes

[†]We have temporarily suppressed denoting the time variation of the wave. It has the form $\exp\{j\omega t\}$ where $\omega = 2\pi c/\lambda$.

through $\vec{x}_h$. The phase of each secondary source is shifted by $\pi/2$ from that of the incident wavefield because of the factor $1/j$ in front of the integral. Finally, the factor $\exp\{jkr\}/r$ represents the spherical propagation of each infinitesimal secondary wave from $\vec{x}_h$ out to the position $\vec{x}$.

Note that the Rayleigh-Sommerfeld diffraction formula can also be interpreted as a superposition integral. Each secondary source makes a contribution to the wavefield at the position $\vec{x}$, and all the contributions are added together to determine the complex amplitude of the wavefield at that point. Several incident waves would be added together to determine the complex amplitudes of the secondary sources within the aperture. Because of the linearity of the wave equation, it is possible to analyze a complicated situation as the sum, the superposition, of several simpler situations.

In many cases of practical interest, the Rayleigh-Sommerfeld diffraction formula is difficult to evaluate. Two classical approximations to the diffraction formula have been developed. The first, the *Fresnel approximation*, results from the assumptions that the obliquity factor $\cos\theta$ is approximately one and that r is approximately equal to the distance d between the aperture plane and a parallel observation plane containing the point $\vec{x}$. The second assumes that spherical surfaces that represent the wavefronts of the secondary waves can be approximated by *quadratic surfaces*. It is easiest to express these approximations if we change notation slightly. Let the point $\vec{x}_h$ in the hole be represented by the Cartesian coordinates $(\tilde{x}, \tilde{y})$ and the point $\vec{x}$ in the observation plane by (x, y). Using these assumptions, we can obtain the Fresnel approximation to the Rayleigh-Sommerfeld diffraction formula.

$$\boxed{s(x, y) = \frac{\exp\{jkd\}}{j\lambda d} \iint_A s(\tilde{x}, \tilde{y}) \exp\left\{\frac{jk[(x-\tilde{x})^2 + (y-\tilde{y})^2]}{2d}\right\} d\tilde{x}\, d\tilde{y}} \tag{2.15}$$

This equation can be interpreted as a two-dimensional convolution between the wavefield in the aperture $s(\tilde{x}, \tilde{y})$ and the function

$$h(x, y) = \frac{\exp\{jkd\}}{j\lambda d} \exp\left\{\frac{jk(x^2 + y^2)}{2d}\right\}$$

Note that $h(x, y)$ is essentially a two-dimensional quadratic phase function. It represents the phase shift that a secondary wave encounters in propagating from the point $(\tilde{x}, \tilde{y})$ in the plane of the hole to the point (x, y) in the observation plane.

A second approximation to the Rayleigh-Sommerfeld diffraction formula, the *Fraunhofer approximation*, applies when the observation plane is far away from the aperture plane. It can be derived from the Fresnel approximation by expanding the quadratic phase term in Eq. 2.15 and using the approximation $d \gg D^2/\lambda$, where D represents the maximum linear dimension of the hole. This approximation allows us to neglect the quadratic phase factor across the hole $\exp\{jk(\tilde{x}^2 + \tilde{y}^2)/2d\} \approx 1$. For this distant imaging plane case, the Fraunhofer approximation results.

$$\boxed{s(x, y) = \frac{\exp\{jkd\}}{j\lambda d} \exp\left\{\frac{jk(x^2 + y^2)}{2d}\right\} \iint_A s(\tilde{x}, \tilde{y}) \exp\left\{\frac{jk(x\tilde{x} + y\tilde{y})}{d}\right\} d\tilde{x}\, d\tilde{y}}$$

The double integral, with a little imagination, can be interpreted as a two-dimensional Fourier Transform of $s(\tilde{x}, \tilde{y})$: Substitute the frequency-like variables $\omega_{\tilde{x}} = k\tilde{x}/d$ and $\omega_{\tilde{y}} = k\tilde{y}/d$ into the integral to obtain a more familiar form. The Fraunhofer approximation formula leads us to some interesting conclusions about wave propagation through holes. Because of the Fourier Transform relation between the wavefield at the hole $s(\tilde{x}, \tilde{y})$ and the wavefield at the observation plane $s(x, y)$, we can conclude that a small hole leads to a broadly spread wavefield at the observation plane. To reduce the spread of the wavefield, one must *enlarge* the hole.

2.4.4 Array Processing Implications

Inhomogeneous media lead to refraction, which changes the direction of propagation throughout space. Furthermore, the diffraction phenomenon means that sources "hidden" from view can be seen under appropriate conditions.

- *Spatial inhomogeneities must be taken into account by array processing algorithms.* Because the direction of propagation varies spatially, each sensor notes different propagation directions. These variations depend not only on homogeneities near the array, but also on the medium's variations in the intervening space between the source and the array. Thus, attempting to glean one source direction from the sensors must take into account the medium's spatial characteristics. This realization lies at the heart of *matched field processing* [9, 21]. Failure to take spatial variations into account results in a loss of array processing gain.

- *Waves propagating in an inhomogeneous medium rarely travel in a straight line.* By inhomogeneities, we mean spatial variations in those properties of the medium that affect propagation speed. These variations induce refractive effects, which bend the rays along which the wave propagates. To determine a source's location in an inhomogeneous medium taxes most array processing algorithms unless the characteristics of the medium are known quite well.

- *Refraction can lead to multipath.* In the sound channel case, a variety of rays can be traced between a source and the array. Since each path has a different length, propagation delay differs and several replicas of the radiated signal reach the array. The effects of multipath can be modeled as a lowpass filter (see Prob. 2.11); thus refraction can have the effect of reducing a signal's high frequency components.

- *Diffraction means that opaque objects located between the source and the array can induce complicated wavefields.* Not only do large barriers induce diffractive fields, but so do small, localized, opaque objects. Consideration of the effects of schools of fish on acoustic propagation and of rain on electromagnetic radiation leads to the theory of *scattering*. From the brief presentation of diffraction presented here, the reader can safely assume that scattering theory is extremely complicated but important to understand.

2.5 Wavenumber-Frequency Space

The Fourier Transform has proved to be an enormously useful tool for analyzing time series. For example, we use the Fourier Transform to estimate the frequency content of a signal or to represent a signal as a weighted sum of sinusoidal functions. The Fourier Transform also relates the frequency response of a linear, time-invariant system to its impulse response. Because the inverse Fourier Transform has a form very similar to that of the forward Fourier Transform, a duality can be established between the time-domain and frequency-domain representations of signals. The concept of Fourier analysis is straightforwardly extended to multidimensional signals [49: §1.3] and its utility does not diminish. In particular, we can write the four-dimensional† Fourier Transform of a spatiotemporal signal $s(\vec{x}, t)$ as

$$\boxed{S(\vec{k}, \omega) = \int_{-\infty}^{\infty} \int_{-\infty}^{\infty} s(\vec{x}, t) \exp\{-j(\omega t - \vec{k} \cdot \vec{x})\} \, d\vec{x} \, dt} \tag{2.16}$$

The Fourier Transform $S(\vec{k}, \omega)$ of a spatiotemporal signal can be interpreted as an alternative representation of the signal in terms of the familiar temporal frequency variable ω and the wavenumber vector variable $\vec{k}$. As discussed in §2.2.1 {13}, the wavenumber vector $\vec{k}$ can be thought of as a spatial frequency variable. Thus, $\vec{x}$ and $\vec{k}$ are dual variables in the same sense that t and ω are in one dimension. Note, however, that the sign of the spatial and temporal frequency variables in the exponent are different. This convention follows naturally from our concern with propagating waves.

2.5.1 Fourier Transform of a Uniform Plane Wave

Consider the uniform plane wave

$$s(\vec{x}, t) = \exp\{j(\omega_0 t - \vec{k}^o \cdot \vec{x})\}$$

where $\vec{k}^o$ and ω_0 represent a particular wavenumber vector and temporal frequency. This spatiotemporal signal is a complex sinusoid with temporal frequency ω_0 and corresponding temporal period $T = 2\pi/\omega_0$, with spatial frequency $|\vec{k}^o| = k^o$ and corresponding wavelength $\lambda = 2\pi/k^o$, with speed of propagation $c_0 = \omega_0/k^o$, and with direction of propagation parallel to the wavenumber vector $\vec{k}_0$. The signal's Fourier Transform is given by

$$S(\vec{k}, \omega) = \int_{-\infty}^{\infty} \int_{-\infty}^{\infty} \exp\{-j(\omega - \omega_0)t + j(\vec{k} - k^o) \cdot \vec{x}\} \, d\vec{x} \, dt$$

In this particular case, the four-dimensional integral is separable: It can be written as the product of four one-dimensional integrals, each of which has the form

$$I(\nu) = \int_{-\infty}^{\infty} \exp\{j\nu y\} \, dy$$

† We have represented the three-dimensional integral over the vector variable $\vec{x}$ as a single vector integral.

It can be shown that $I(\nu)$ acts operationally as an impulse function:[†] $I(\nu) \equiv \delta(\nu)$, where $\delta(\nu)$ is the idealized impulse function. Using this equivalence, we see that the wavenumber-frequency spectrum of the uniform plane wave signal can be written as

$$S(\vec{k}, \omega) = \delta(\vec{k} - k^o)\delta(\omega - \omega_0)$$

where $\delta(\vec{k})$ is shorthand notation for $\delta(k_x)\delta(k_y)\delta(k_z)$ and (k_x, k_y, k_z) are the Cartesian components of the vector $\vec{k}$. A uniform plane wave thus corresponds to a single point in wavenumber-frequency space in the same manner that a complex sinusoid $\exp\{j\omega_0 t\}$ corresponds to a single point along the frequency axis of a conventional one-dimensional spectrum. The values of the point's coordinates (k^o, ω_0) dictate the temporal frequency, the spatial wavelength, and the velocity of propagation. The plane wave's amplitude and phase are represented by the complex area of the impulse in wavenumber-frequency space.

When our space-time signal is a propagating wave, $s(\vec{x}, t) = s(t - \vec{\alpha}^o \cdot \vec{x})$, its Fourier Transform is found to be

$$S(\vec{k}, \omega) = S(\omega)\delta(\vec{k} - \omega\vec{\alpha}^o)$$

Thus, this propagating wave contains energy only along the line $\vec{k} = \omega\vec{\alpha}^o$ in wavenumber-frequency space.

Any plane-wave signal $s(\vec{x}, t)$ whose Fourier Transform $S(\vec{k}, \omega)$ exists and is well behaved[‡] may be written in terms of the inverse Fourier Transform.

$$\boxed{s(\vec{x}, t) = \frac{1}{(2\pi)^4} \int_{-\infty}^{\infty}\int_{-\infty}^{\infty} S(\vec{k}, \omega) \exp\{j(\omega t - \vec{k} \cdot \vec{x})\}\, d\vec{k}\, d\omega}$$

We can interpret this equation as equating any reasonable spatiotemporal signal to a superposition of infinitely many plane waves, each with an infinitesimally small complex amplitude. The complex amplitude for the plane wave with parameters $\vec{k}^o$ and ω_0 is $S(\vec{k}^o, \omega_0)\, d\vec{k}\, d\omega/(2\pi)^4$. For *any* reasonable spatiotemporal signal, this representation holds, not just for propagating ones. Its interpretation is therefore quite profound: An arbitrary, nonperiodic, nonpropagating distribution of energy in space and time can be represented as a superposition of periodic, propagating plane waves. On top of this, the physics of the situation, as embodied in the appropriate wave equation, dictates which plane waves can coexist in a given situation. The wave equation can be envisioned as a linear filter that restricts the set of plane waves that may exist in a given situation in the same way that a conventional bandpass filter for time-series analysis restricts the frequency content of its output signal.

[†] This equivalence can be shown rigorously by using limiting arguments to generate a series of functions whose integrals converge in the limit to the integral of an idealized impulse function [Siebert, Chap.11].

[‡] This caveat is included because not all signals have well-defined Fourier Transforms. For example, the Fourier Transform must be generalized to the Laplace Transform to analyze growing exponential signals. For a discussion of the subtleties of the existence of Fourier Transforms in the one-dimensional case, see Siebert (Chap. 13).

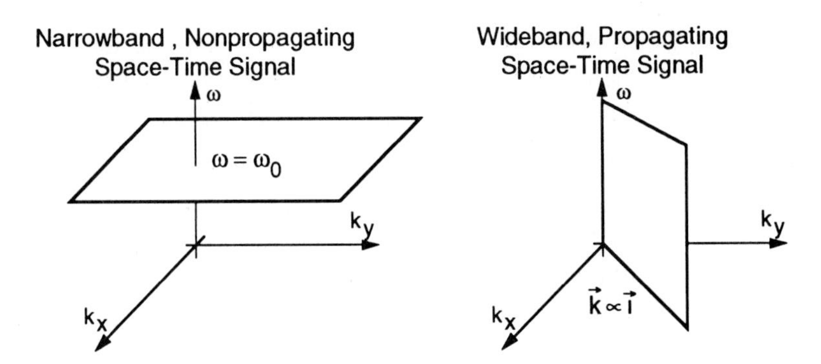

Figure 2.14 When $s(\vec{x}, t)$ contains temporal frequencies near ω_0 only (left portion), the wavenumber-frequency spectrum $S(\vec{k}, \omega)$ has significant energy only near the plane $\omega = \omega_0$ in $(\vec{k}, \omega)$ space. Here we have displayed the three-dimensional space (k_x, k_y, ω) rather than the full four-dimensional space (k_x, k_y, k_z, ω) for the purposes of illustration. If a signal consists of components propagating in a particular direction $\vec{\imath}$ (right portion), then its wavenumber-frequency spectrum is zero except for the half-plane where $\vec{k}$ is proportional to $\vec{\imath}$.

2.5.2 Subsets of Wavenumber-Frequency Space

We can visualize how various constraints on spatiotemporal signals are represented in wavenumber-frequency space. Suppose that a particular signal $s(\vec{x}, t)$ contains temporal frequencies only in a narrow band around $\omega = \omega_0$. Its wavenumber-frequency spectrum $S(\vec{k}, \omega)$ has its energy concentrated in a particular region of wavenumber-frequency space (Fig. 2.14). Similarly, if a signal contains only components propagating in a certain direction, $\vec{\imath}$, for example, then its wavenumber-frequency spectrum is zero except where $\vec{k}$ is proportional to $\vec{\imath}$ (Fig. 2.14). Finally, let's consider the case in which the wave equation describing the physics of propagation for a particular medium leads to a dispersion relation of the general form $\omega = f(k)$. In the simplest case, the dispersion is linear: $f(k) = ck$, where c is the speed of propagation for the medium. For a signal consisting of components propagating at the same speed (but in various arbitrary directions), the wavenumber-frequency spectrum has its energy distributed along the surface formed by revolving the curve $\omega = ck$ around the ω axis. Fig. 2.15 shows the cone-shaped spectral support that arises in this case.

These three basic subsets result from three constraints that may apply to particular spatiotemporal signals in certain situations. Other subsets may be generated by applying more than one constraint simultaneously. For example, if a particular signal has components that travel at the same speed (but arbitrary directions) over a narrow band of temporal frequencies, then its spectral support is a circular ring created by the intersection of a right circular cone with a plane perpendicular to the conic axis. Similarly a signal with many temporal frequency components propagating in a single direction $\vec{\imath}$ at a constant speed c has its spectrum lying along the line at the intersection of the cone $\omega = c|\vec{k}|$ and the half-plane in which $\vec{k}$ is proportional to $\vec{\imath}$ (Fig. 2.15).

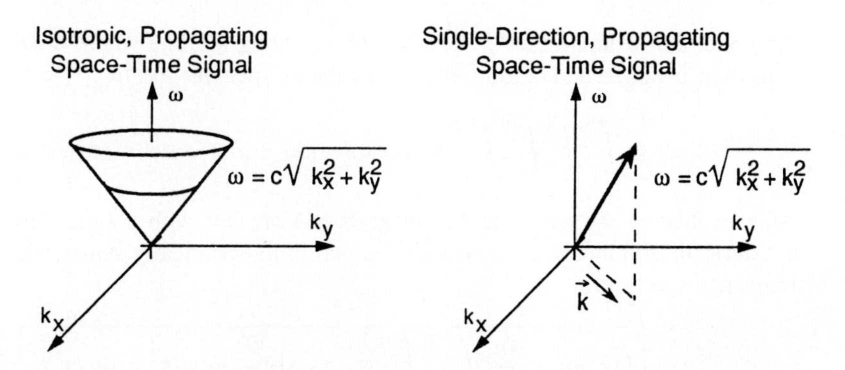

Figure 2.15 An isotropic signal, propagating in all directions with equal amplitudes and equal speeds, has a cone-shaped wavenumber-frequency spectrum. In contrast, a signal having a specific direction of propagation has spectrum defined on a line in the $(\vec{k}, \omega)$ plane.

2.5.3 Spectrum of a Spherical Wave

We saw in §2.2.2 {16} that spherical waves of the form

$$s(r, t) = \frac{A \exp\{j(\omega_0 t - k^o r)\}}{r}$$

are solutions to the three-dimensional wave equation.† To compute the wavenumber-frequency spectrum of such a spherically propagating wave, we must exploit the spherical symmetry of the problem. First, write the Fourier Transform in spherical coordinates in both spatiotemporal and wavenumber-frequency space using the relations

$$\begin{aligned} x &= r \sin\phi \cos\theta & \quad k_x &= k \sin\Phi \cos\Theta \\ y &= r \sin\phi \sin\theta & \quad k_y &= k \sin\Phi \sin\Theta \\ z &= r \cos\phi & \quad k_z &= k \cos\Phi \end{aligned}$$

This coordinate transformation yields the rather cumbersome formula for the Fourier Transform of a signal expressed in spherical coordinates.

$$\begin{aligned} S(k, \Phi, \Theta, \omega) = & \int_0^\infty \int_0^\pi \int_0^{2\pi} \int_{-\infty}^\infty s(r, \phi, \theta, t) \exp\{-j\omega t\} \\ & \cdot \exp\{jkr(\sin\phi \cos\theta \sin\Phi \cos\Theta + \sin\phi \sin\theta \sin\Phi \sin\Theta + \cos\phi \cos\Phi)\} \\ & r^2 \sin\phi \, dt \, d\theta \, d\phi \, dr \end{aligned}$$

A spherically symmetric signal $s(r, t)$ has a spherically symmetric spectrum $S(k, \omega)$. This fact allows us to evaluate the spectrum along one particular line emanating from the origin of $(\vec{k}, \omega)$ space and reason that the values along any other radial line must be

†As before, we use r to represent the radial distance $\sqrt{x^2 + y^2 + z^2}$ in space. Similarly k^o represents the radial distance in wavenumber space.

the same. For example, let's evaluate $S(k, \omega)$ along the line where $\Phi = 0$. Using this constraint in spherical coordinates reduces the Fourier integral to

$$S(k, \omega) = \int_0^{\infty} \int_0^{\pi} \int_0^{2\pi} \int_{-\infty}^{\infty} s(r, t) \exp\{-j\omega t + jkr \cos\phi\} r^2 \sin\phi \, dt \, d\theta \, d\phi \, dr$$

After evaluating the easy angular integrals, we are left with a simple formula relating a spherically symmetric spatiotemporal signal to its spherically symmetric wavenumber-frequency spectrum.

$$\boxed{S(k, \omega) = \frac{4\pi}{k} \int_0^{\infty} \left[\int_{-\infty}^{\infty} s(r, t) \exp\{-j\omega t\} \, dt \right] r \sin kr \, dr} \tag{2.17}$$

Although we shall not derive it here, the inverse formula has the form

$$s(r, t) = \frac{1}{2\pi^2 r} \int_0^{\infty} \left\{ \frac{1}{2\pi} \int_{-\infty}^{\infty} S(k, \omega) \exp\{j\omega t\} \, d\omega \right\} k \sin kr \, dk$$

For the monochromatic spherical wave given previously, the wavenumber-frequency spectrum takes the form

$$S(k, \omega) = \frac{4\pi A}{k} \int_0^{\infty} \exp\{-jk^o r\} \sin kr \, dr \cdot \int_{-\infty}^{\infty} \exp\{-j(\omega - \omega_0)t\} \, dt$$

The second integral is operationally equal to an impulse in temporal frequency located at $\omega = \omega_0$. The remaining integral over r has the form of a one-dimensional Fourier Transform of a sine wave multiplied by the unit step function. Care must be taken when evaluating it, because it does not converge in the usual sense. It does converge in an operational sense to

$$\frac{2\pi^2}{jk} [\delta(k - k^o) - \delta(k + k^o)] + \frac{4\pi}{k^2 - (k^o)^2}$$

Because k takes on nonnegative values only, we may ignore the impulse at $-k^o$. The wavenumber-frequency spectrum of the spherical wave is given by

$$S(k, \omega) = \left[\frac{2\pi^2}{jk} \delta(k - k^o) + \frac{4\pi}{k^2 - (k^o)^2} \right] \delta(\omega - \omega_0)$$

As one might expect, this spectrum is impulsive at the temporal frequency ω_0 and wavenumbers whose magnitudes equal k^o. There is, however, an additional term, $4\pi/\big(k^2 - (k^o)^2\big)$, singular at k^o and dropping off as $1/k^2$ for large values of k.

2.5.4 Filtering in Wavenumber-Frequency Space

In many array processing problems, we are interested in separating the components of a spatiotemporal waveform. We may want to retain signal components at certain temporal frequencies or those propagating in certain directions while filtering out noise as well as other interfering signals. Furthermore, the physics of wave propagation may exclude the existence of signal components in some regions of wavenumber-frequency space, in effect filtering it by constraining the form of the propagating waves. It is straightforward to extend the concept of a linear, time-invariant filter to spatiotemporal signals. For example, we can think of an arbitrary wavenumber-frequency spectrum $S(\vec{k}, \omega)$ being multiplied by a filter frequency response $H(\vec{k}, \omega)$ to obtain a modified wavenumber-frequency spectrum.

$$Y(\vec{k}, \omega) = H(\vec{k}, \omega) S(\vec{k}, \omega)$$

The modified spectrum $Y(\vec{k}, \omega)$ corresponds to a modified spatiotemporal signal $y(\vec{x}, t)$ given by the inverse Fourier Transform relation. This signal is related to the original spatiotemporal signal by the four-dimensional convolution integral

$$y(\vec{x}, t) = \int_{-\infty}^{\infty} \int_{-\infty}^{\infty} h(\vec{x} - \vec{\xi}, t - \tau)\, s(\vec{\xi}, \tau)\, d\vec{\xi}\, d\tau \tag{2.18}$$

where $h(\vec{x}, t)$ is the filter's *impulse response*: The modified signal $y(\vec{x}, t)$ equals $h(\vec{x}, t)$ when the original signal $s(\vec{x}, t)$ equals the spatiotemporal impulse $\delta(\vec{x})\delta(t)$. The impulse response $h(\vec{x}, t)$ also equals the inverse Fourier Transform of the filter's wavenumber-frequency response $H(\vec{k}, \omega)$.

The concept of a four-dimensional linear space- and time-invariant filter can express the direction-finding aim of array signal processing. For example, suppose we wanted to build a system to pass a narrow temporal-frequency band of signals propagating within a particular range of directions at a certain speed. By mapping the areas of $(\vec{k}, \omega)$ space corresponding to these components, we would design, conceptually at least, a filter whose wavenumber-frequency response is nearly one in those areas and nearly zero elsewhere. The resulting four-dimensional filter suffers from some practical problems, however. First of all, the input waveform $s(\vec{x}, t)$ is not known for all space and time; while to gather the temporal information the filter need only "wait" for the signal to pass by, spatially our filter would need signal values throughout all space, an unlikely circumstance.† In addition, an arbitrary filter response $H(\vec{k}, \omega)$ may be impossible to realize with a finite amount of computation. Succeeding chapters of this book describe array processing algorithms that correspond to realistic spatiotemporal filters.

2.6 Random Fields

In previous sections, we have discussed spatiotemporal signals as if all of them were deterministic: They could be described precisely as a function of space and time. *Noise* does play an important (but usually counterproductive) role in determining the characteristics of propagating signals. The theoretical foundation of modeling noise rests on the theory of probability and stochastic processes; appendix A {471} provides a summary of

†In a homogeneous medium, propagating signals would eventually be carried to a filter awaiting them at one spatial location; however, for the filter to remove nonpropagating signals, the signal's waveform throughout space would be needed.

basic probability theory and one-dimensional stochastic process theory important for array processing problems. Spatiotemporal stochastic processes are known as *random fields*. We need to rephrase notions of one-dimensional random process theory for random fields.

Let $f(\vec{x}, t)$ denote a *random field*: a mapping from sample space onto the space of spatiotemporal signals. For particular values of $\vec{x} = \vec{x}_0$ and $t = t_0$, $f(\vec{x}_0, t_0)$ represents a random variable that is characterized by a probability density function $p_{f(\vec{x}_0,t_0)}(\cdot)$. The expected value, correlation function, and covariance function of a random field are defined in the obvious way. Letting $f_0 = f(\vec{x}_0, t_0)$ and $f_1 = f(\vec{x}_1, t_1)$,

$$\mathcal{E}[f(\vec{x}_0, t_0)] = \int \alpha p_{f_0}(\alpha)\, d\alpha$$

$$R_f(\vec{x}_0, \vec{x}_1, t_0, t_1) = \mathcal{E}[f_0 f_1] = \iint \alpha\beta\, p_{f_0,f_1}(\alpha, \beta)\, d\alpha\, d\beta$$

$$K_f(\vec{x}_0, \vec{x}_1, t_0, t_1) = R_f(\vec{x}_0, \vec{x}_1, t_0, t_1) - \mathcal{E}[f_0]\,\mathcal{E}[f_1]$$

The mean m_f and variance σ_f^2 at a particular spatiotemporal location are defined according to the usual relations.

$$m_f(\vec{x}, t) = \mathcal{E}[f(\vec{x}, t)] \qquad \sigma_f^2(\vec{x}, t) = R_f(\vec{x}, \vec{x}, t, t) - m_f^2(\vec{x}, t)$$

Stationary random fields represent an important and very special case for which these and other statistical averages do *not* vary with absolute position or time.† Thus, all higher-order joint probability density functions of random variables generated by the random field depend only on differences in positions and times, never on absolute position or time. In particular, for a random field to be stationary implies that the correlation function depends only on the differences $\vec{x}_1 - \vec{x}_0$ and $t_1 - t_0$. If we denote these differences, lags, by $\vec{\chi}$ and τ, respectively, we can represent the correlation function of a stationary random field as

$$R_f(\vec{\chi}, \tau) = \mathcal{E}\big[f(\vec{x}_0, t_0) f(\vec{x}_0 + \vec{\chi}, t_0 + \tau)\big]$$

2.6.1 Power Spectra of Random Fields

We saw previously that a spatiotemporal signal could be characterized by its wavenumber-frequency spectrum. Similarly, a random field may be characterized by its *power spectral density function*. The power spectral density function $\mathcal{S}_f(\vec{k}, \omega)$ of a random field $f(\cdot, \cdot)$, which is defined only for stationary random processes, equals the Fourier Transform of the field's correlation function.

$$\mathcal{S}_f(\vec{k}, \omega) = \int_{-\infty}^{\infty}\int_{-\infty}^{\infty} R_f(\vec{\chi}, \tau) \exp\{-j(\omega\tau - \vec{k}\cdot\vec{\chi})\}\, d\vec{\chi}\, d\tau$$

$$R_f(\vec{\chi}, \tau) = \frac{1}{(2\pi)^4}\int_{-\infty}^{\infty} \mathcal{S}_f(\vec{k}, \omega) \exp\{+j(\omega\tau - \vec{k}\cdot\vec{\chi})\}\, d\vec{k}\, d\omega$$

† A more rigorous definition of stationarity for random fields requires special theoretical considerations [164].

Example

Consider a random field that has a separable correlation function of the form

$$R_f(\vec{\chi}, \tau) = \frac{\exp\{-\alpha|\tau| - \beta|\vec{\chi}|\}}{|\vec{\chi}|} = \frac{\exp\{-\beta|\vec{\chi}|\}}{|\vec{\chi}|} \cdot \exp\{-\alpha|\tau|\}$$

A random signal that results from this random process becomes less correlated with itself as either temporal lag τ or the vector spatial lag $\vec{\chi}$ increases: The field's amplitude at one position and time becomes less correlated with the amplitudes recorded at increasingly distant times or positions. The power spectral density of this random process can be computed by evaluating the Fourier Transform of the correlation function $R_f(\vec{\chi}, \tau)$. Because the correlation function depends only on the magnitude of the position lag $\vec{\chi}$, we can perform the spatial part of the Fourier Transform in spherical coordinates using Eq. 2.17 {44}. Substituting the scalar variable χ for $|\vec{\chi}|$ and using the spherically symmetric forms $R_f(\chi, \tau)$ and $\mathcal{S}_f(k, \omega)$ for the correlation function and the power spectral density function, respectively, gives us

$$\mathcal{S}_f(k, \omega) = \int_{-\infty}^{\infty} \exp\{-\alpha|\tau| - j\omega\tau\}\, d\tau\, \frac{4\pi}{k} \int_0^{\infty} \frac{\exp\{-\beta\chi\}}{\chi} \chi \sin k\chi\, d\chi$$

These integrals can be evaluated straightforwardly to find that

$$\mathcal{S}_f(k, \omega) = \frac{2\alpha}{\alpha^2 + \omega^2} \cdot \frac{4\pi}{\beta^2 + k^2}$$

The power density of this random process decreases as either the temporal frequency ω or the spatial frequency k increases.

Example

Now let's consider a power spectral density $\mathcal{S}_f(\vec{k}, \omega)$ that is flat for temporal frequencies and wavenumbers within some band. To be more precise, let

$$\mathcal{S}_f(\vec{k}, \omega) = \begin{cases} 1 & \text{for } |\omega| \le \Omega,\ |k_x| \le K_x,\ |k_y| \le K_y,\ |k_z| \le K_z \\ 0 & \text{otherwise} \end{cases}$$

where Ω, K_x, K_y, and K_z are all positive constants. The correlation function for this random process may be derived from the inverse Fourier Transform of the power spectral density.

$$R_f(\vec{\chi}, \tau) = \frac{\sin K_x\chi_x}{\pi\chi_x} \cdot \frac{\sin K_y\chi_y}{\pi\chi_y} \cdot \frac{\sin K_z\chi_z}{\pi\chi_z} \cdot \frac{\sin \Omega\tau}{\pi\tau}$$

Note that for certain special values of the position and time lags, the correlation function is zero:

$$\chi_x = \frac{l}{K_x}, \quad \chi_y = \frac{m}{K_y}, \quad \chi_z = \frac{n}{K_z}, \quad \tau = \frac{i}{\Omega}$$

where l, m, n, and i are nonzero integers. For positions and times separated by these values, the field's amplitudes are uncorrelated.

As the constants K_x, K_y,K_z, and Ω in the last example approach infinity, the power spectral density becomes a constant for all frequencies and wavenumbers. Such an idealized power spectrum is said to correspond to a *white* random field. White noise random fields have a correlation function proportional to

$$R_f(\vec{\chi}, \tau) \propto \delta(\vec{\chi}, \tau)$$

which equals zero for any nonzero lags in position or time and consists of an impulse when both the position and time lags are zero. The white-noise random process has *infinite* power; this result can be seen by integrating its power spectral density function over all frequencies and wavenumbers.† Despite the physical impossibility of white noise, we can use it to great advantage.

2.6.2 Filtered Stationary Random Fields

The signals that comprise the ensemble of a stationary random field may be filtered by linear, space-, and time-invariant filters. Suppose we have a stationary random field $f(\vec{x}, t)$, which when filtered by a linear, space-, and time-invariant filter, gives us another stationary random field $y(\vec{x}, t)$. The filter $\mathcal{H}$ is characterized by either its impulse response $h(\vec{x}, t)$ or, equivalently, its wavenumber-frequency response $H(\vec{k}, \omega)$. Given statistical averages of $f(\vec{x}, t)$, such as its mean, variance, and correlation function, we can derive the corresponding statistical averages for $y(\vec{x}, t)$. In particular, the correlation function of $y(\vec{x}, t)$ is given by

$$R_y(\vec{\chi}, \tau) = \int_{-\infty}^{\infty}\int_{-\infty}^{\infty} R_{hh}(\vec{a}, b) R_f(\vec{\chi} - \vec{a}, \tau - b)\, d\vec{a}\, db$$

where $R_{hh}(\vec{a}, b)$ represents the deterministic autocorrelation function of the impulse response $h(\vec{x}, t)$:

$$R_{hh}(\vec{a}, b) = \int_{-\infty}^{\infty}\int_{-\infty}^{\infty} h(\vec{x}, t) h(\vec{x} + \vec{a}, t + b)\, d\vec{x}\, dt$$

The correlation function of the output random field thus equals the convolution between the deterministic autocorrelation function of the filter's impulse response and the input correlation function. The power spectral density function of a random field filtered in space and time equals

$$\boxed{\mathcal{S}_y(\vec{k}, \omega) = \left|H(\vec{k}, \omega)\right|^2 \mathcal{S}_f(\vec{k}, \omega)}$$

The input power spectral density is multiplied by the squared magnitude of the filter's frequency response to obtain the power spectral density function of the output stationary random process.

2.6.3 Isotropic Noise

For many array processing problems, we can reasonably assume that the observed noise field consists of many random waves propagating in all possible directions with equal probability: an *isotropic* noise field. The power spectral density function of a *spherically isotropic* field would take the form

$$\mathcal{S}(\vec{k}, \omega) = G(\omega)\delta(|\vec{k}| - \omega/c)$$

†The value obtained by this integration, when it converges, equals $R_f(\vec{0}, 0)$.

The function $G(\omega)$ represents the distribution of power with respect to temporal frequency. The impulse embodies the physics of propagation by relating the magnitude of the wavenumber vector (or equivalently wavelength), temporal frequency, and speed of propagation c. In this case, we have assumed the linear dispersion relation $|\vec{k}| = \omega/c$.

To compute the correlation function $R(\vec{\chi}, \tau)$ of spherically isotropic noise, we need only evaluate the inverse Fourier Transform. Because the power spectral density function depends only on the magnitude of the wavenumber vector, we can address this problem, like the one previously, with spherical coordinates for position and wavenumber. Making the familiar substitutions $\chi = |\vec{\chi}|$, $k = |\vec{k}|$, the correlation function takes the form

$$R(\chi, \tau) = \frac{1}{2\pi} \int_{-\infty}^{\infty} G(\omega) e^{j\omega\tau} \frac{1}{2\pi^2 \chi} \int_{0}^{\infty} Ak \sin k\chi \, \delta\left(k - \frac{\omega}{c}\right) dk \, d\omega$$

Performing the integral over k and simplifying the resulting expression gives us

$$R(\chi, \tau) = \frac{A}{4\pi^3} \int_{-\infty}^{\infty} \frac{\omega}{c\chi} \sin \frac{\omega\chi}{c} G(\omega) e^{j\omega\tau} \, d\omega$$

The integrand is zero when $\omega\chi/c$ equals some nonzero integer multiple of π. Furthermore, because $\omega/c = k = 2\pi/\lambda$, the integrand equals zero whenever the position lag χ equals a nonzero integer number of half-wavelengths.

$$R(\chi, \tau) = 0 \text{ for } \chi = n\frac{\lambda}{2}, n \neq 0$$

If the isotropic noise were constrained by the physics of the situation or by temporal filtering to be monochromatic with a temporal frequency ω_0, the correlation function would have the form of a spatial sinc function.

$$R(\chi, \tau) = \frac{A\omega_0^2}{2\pi^2 c^2} \cos \omega_0 \tau \cdot \frac{\sin \frac{\omega_0 \chi}{c}}{\frac{\omega_0 \chi}{c}}$$

Thus, sampling at spatial locations separated by $\lambda/2 = \pi c/\omega_0$ would yield uncorrelated noise components. When the isotropic noise is Gaussian, such spatial sampling grids would yield statistically independent waveforms, which, as we shall see, can result in improved array processing performance.

2.6.4 Correlation Matrices of Sensor Outputs

Many array processing algorithms depend on the measured field via the correlation function computed from the sensor outputs. The resulting *correlation matrices* play a key role in this book, making it worthwhile considering their structure, how this structure expresses propagating signals and noise fields, and what structure is "naturally" present (induced by stationarity assumptions, for example). Correlation matrices fall into two classes: *spatiotemporal* correlation matrices and *spatiospectral* correlation matrices. They differ only by whether we defer the Fourier Transform of each sensor's output until after the correlation calculation or evaluate it before. Surprisingly, we shall find striking similarities between the structural aspects of the two kinds of spatial correlation matrices.

Spatiotemporal correlation matrix. Despite the correlation function $R_f(\vec{x}_0, \vec{x}_1, t_0, t_1)$ of the random field $f(\vec{x}, t)$ depending on four variables, its values can be arranged in a matrix in a way convenient for signal processing. First of all, sensors are usually statically positioned, meaning that the spatial variables $\vec{x}_0$, $\vec{x}_1$ are evaluated at the sensor locations $\{\vec{x}_m\}$, $m = 0, \ldots, M-1$. Similarly, we take sequences of outputs from the sensors over time, with all sensors sampled simultaneously. Thus, we need to evaluate the correlation function at the times $\{t_n\}$, $n = 0, \ldots, N-1$. The correlation matrix $\mathbf{R}(t_0)$ is formed from these $(MN)^2$ values by blocking the matrix into M^2 submatrices, each of which equals a cross-correlation matrix of the temporal values between pairs of sensor outputs. We define the $MN \times MN$ *spatiotemporal correlation matrix* $\mathbf{R}(t_0)$ as

$$\mathbf{R}(t_0) = \begin{bmatrix} \mathbf{R}_{0,0} & \mathbf{R}_{0,1} & \cdots & \mathbf{R}_{0,M-1} \\ \mathbf{R}_{1,0} & \mathbf{R}_{1,1} & \mathbf{R}_{1,2} & \vdots \\ \vdots & & \ddots & \vdots \\ \mathbf{R}_{M-1,0} & \cdots & \cdots & \mathbf{R}_{M-1,M-1} \end{bmatrix}$$

where $\mathbf{R}_{m_1,m_2}$ denotes the $N \times N$ cross-covariance matrix of the field recorded by sensors m_1 and m_2.

$$\left[\mathbf{R}_{m_1,m_2}\right]_{n_1,n_2} = R_f(\vec{x}_{m_1}, \vec{x}_{m_2}, t_{n_1}, t_{n_2}), \quad n_1, n_2 = 0, \ldots, N-1$$

The matrices $\mathbf{R}_{m,m}$ on the main diagonal thus all equal the temporal correlation matrix of the observations measured at the mth sensor. The off-diagonal matrices $\mathbf{R}_{m_1,m_2}$ represent both spatial and temporal correlation. Because of symmetry, $\mathbf{R}_{m_1,m_2}$ equals $\mathbf{R}'_{m_2,m_1}$, making $\mathbf{R}$ a block Hermitian matrix. When the field is stationary, each submatrix is Toeplitz. Unless the array itself has symmetries so that a constant index "lag" of $m_1 - m_2$ implies a constant physical separation (as in a linear array), the remainder of the terms in the spatiotemporal matrix for stationary fields lacks further symmetries. When the field is stationary, this matrix does not depend on the observations' time origin t_0.

An important special case occurs when the spatial and temporal correlations are separable: The spatiotemporal correlation function can be written as $R_f(\vec{\chi}, \tau) = R_f^{(s)}(\vec{\chi}) \cdot R_f^{(t)}(\tau)$. Here, $R_f^{(s)}(\cdot)$ is the spatial correlation function and $R_f^{(t)}(\cdot)$ the temporal correlation function. In terms of the spatiotemporal covariance matrix, this situation implies that $\mathbf{R}_{m_1,m_2} = \rho_{m_1,m_2}\mathbf{R}_{0,0}$, where ρ_{m_1,m_2} denotes the spatial correlation coefficient between sensors m_1 and m_2.† This decomposition means that the spatiotemporal covariance matrix can be expressed as a Kronecker product between the spatial correlation coefficient matrix $\boldsymbol{\rho}$ and the temporal correlation matrix $\mathbf{R}_{0,0}$.

$$\boxed{\mathbf{R} = \boldsymbol{\rho} \otimes \mathbf{R}_{0,0}} \tag{2.19}$$

†We have arbitrarily placed the power expressed by spatiotemporal correlation with the temporal correlation matrix. Consequently, ρ_{m_1,m_2} is dimensionless and assumes values in the interval $[-1, 1]$. Note that $\rho_{m,m} = 1$.

When the sampled field is either spatially or temporally white, the correlation matrix's structure simplifies greatly. If spatially white, the correlation function becomes $\delta_{m_2-m_1} R_f(\vec{0}, t_{n_2} - t_{n_1})$. Thus, $\boldsymbol{\rho} = \mathbf{I}$, yielding the spatiotemporal covariance matrix

$$\begin{bmatrix} \mathbf{R}_{0,0} & \mathbf{0} & \cdots & \mathbf{0} \\ \mathbf{0} & \mathbf{R}_{0,0} & \mathbf{0} & \vdots \\ \vdots & \ddots & \ddots & \vdots \\ \mathbf{0} & \cdots & \mathbf{0} & \mathbf{R}_{0,0} \end{bmatrix}$$

If the field is temporally white, the spatiotemporal correlation function simplifies to $\delta_{n_2-n_1} R_f(\vec{x}_{m_2} - \vec{x}_{m_1}, 0)$. The matrices $\mathbf{R}_{m_1,m_2}$ are diagonal, equaling $\rho_{m_1,m_2}\sigma^2\mathbf{I}$, which results in a spatiotemporal correlation matrix organized as

$$\mathbf{R} = \begin{bmatrix} \sigma^2\mathbf{I} & \rho_{0,1}\sigma^2\mathbf{I} & \cdots & \rho_{0,M-1}\sigma^2\mathbf{I} \\ \rho_{1,0}\sigma^2\mathbf{I} & \sigma^2\mathbf{I} & \rho_{1,2}\sigma^2\mathbf{I} & \vdots \\ \vdots & \ddots & \ddots & \vdots \\ \rho_{M-1,0}\sigma^2\mathbf{I} & \cdots & \cdots & \sigma^2\mathbf{I} \end{bmatrix}$$

Because of spatial stationarity, $\rho_{m_1,m_2} = \rho^*_{m_2,m_1}$.

Spatiospectral correlation matrix. Frequency-domain techniques find obvious application when signals are narrowband (a sinusoid, for example), but also produce viable algorithms when signals are sustained, narrowband or not. The Fourier Transform converts propagation delay at each sensor to a sensor-dependent phase shift. Algorithms described in subsequent chapters take advantage of this fact. Computing the transform over a finite portion of the field at a specific location yields the "spectrum"† we wish to correlate.

$$\mathcal{F}[f(\vec{x}_m, t_0)] = \int_{t_0}^{t_0+D} f(\vec{x}_m, t)e^{-j\omega t}\, dt$$

Because of the statistical properties of spectral estimates, we need not be concerned with correlations between different frequencies. $\mathcal{E}\{\mathcal{F}[f(\vec{x}_{m_1}, t_0)] \cdot \mathcal{F}^*[f(\vec{x}_{m_2}, t_0)]\}$ expresses the correlation between the Fourier Transforms of the outputs of sensors m_1 and m_2 at the *same* temporal frequency. We term the $M \times M$ matrix formed from these quantities the *spatiospectral correlation matrix* and denote it by the $M \times M$ matrix $\mathbf{R}(t_0, \omega)$.

$$\begin{aligned} [\mathbf{R}(t_0, \omega)]_{m_1,m_2} &= \int_{t_0}^{t_0+D}\int_{t_0}^{t_0+D} \mathcal{E}[f m_1(t) f^*_{m_2}(u)]e^{-j\omega(t-u)}\, dt\, du \\ &= \int_{t_0}^{t_0+D}\int_{t_0}^{t_0+D} R_f(\vec{x}_{m_1}, \vec{x}_{m_2}, t, u)e^{-j\omega(t-u)}\, dt\, du \end{aligned}$$

†Defining this spectrum is actually quite complicated. The authors are being somewhat sloppy here, but the final result is correct. A simple explanation of the difficulties encountered in defining the spectrum of a random process can be found in [152: 212–20].

When the field is stationary, the spatiospectral correlation matrix simplifies greatly, with the matrix no longer depending on the segment's time origin.

$$[\mathbf{R}(\omega)]_{m_1,m_2} = \int_{t_0}^{t_0+D}\int_{t_0}^{t_0+D} R_f(\vec{x}_{m_2} - \vec{x}_{m_1}, u - t)e^{-j\omega(t-u)}\,dt\,du$$
$$= \int_{-D}^{D} (D - |\tau|)R_f(\vec{x}_{m_2} - \vec{x}_{m_1}, \tau)e^{-j\omega\tau}\,d\tau$$

When we have a separable correlation function, the spatiospectral correlation matrix equals the spatial correlation coefficient matrix $\boldsymbol{\rho}$ times the temporal power density spectrum of the field:

$$\boxed{\mathbf{R}(\omega) = \boldsymbol{\rho} \cdot \widetilde{\mathcal{S}}_y(\omega)} \tag{2.20}$$

where $\widetilde{\mathcal{S}}_y(\omega)$ represents the Fourier Transform of the linearly weighted (in lag) temporal correlation function.

$$\widetilde{\mathcal{S}}_y(\omega) = \int_{-D}^{D} (D - |\tau|)R_f^{(t)}(\tau)e^{-j\omega\tau}\,d\tau$$

The spatial correlation coefficient matrix determines entirely the structure of the spatiospectral correlation matrix.

A spatiospectral correlation matrix's dimension is usually much smaller than a spatiotemporal one. We do need, however, a family of spatiospectral matrices that spans temporal frequency to represent the general field. If the signal is narrowband, we need only one, which illustrates the economies frequency-domain techniques offer in the right circumstances. If not, a collection of spatiospectral correlation matrices is needed to cover the frequency band of interest.

2.7 Signal and Noise Assumptions

We shall assume throughout that observations $\{f(\vec{x}_m, t)\}_{m=0}^{M-1}$ made at sensor outputs consist of an additive combination of signals and noise. Each sensor applies some linear transformation to its measured field to yield its output signal $y_m(t)$. We describe the portion labeled "noise" as a stochastic sequence, usually stationary, zero-mean Gaussian, and statistically independent of the field's signal(s). The stationarity assumption extends to both its temporal and spatial properties. For purposes of the current discussion, "stationarity" means that noise portion of the observations recorded at each sensor have identical statistical characteristics and any cross-correlations between sensor outputs depends only on the physical distance separating the sensors. These assumptions implicitly rely on the *isotropic* noise assumption: The noise component of the observations consists of waveforms propagating toward the array from all directions in the far field. Thus, from the array's viewpoint, noise is coming from everywhere and has a well-defined, but possibly unknown, power spectrum (spatially and temporally).

Each signal has a definite or a restricted range of directions from which it propagates toward the array. Using the word "signal" does not always mean that it represents the focus of our signal processing. Some signals are *jammers*, unwanted signals propagating toward the array that (usually) interfere with our ability to focus algorithms on the signals we do want. The source locations for jammers or wanted signals may be known before processing. For example, we know where the quarterback is on the field; we want to remove crowd noise and shouts from the sideline, enabling us to eavesdrop on the next play. Signals can be realizations of a stochastic process (in which case their amplitude distribution and temporal power spectrum is known) or they can be deterministic. In either case, signals are more than likely parameterized by numbers unknown to us. For example, one parameter is the location of the signal's source, whose value we probably want to know. Others, such as signal amplitude, may be unknown when we have little idea how "loud" the source is. In short, signals in this book convey information or have some structure making them appear to convey information. Effective array processing algorithms exploit signal structure.

We express the sampled signal obtained from the mth sensor as a linear combination of signals and stationary noise.

$$y_m(t_n) = \sum_{i=0}^{N_s-1} s_i(m, t_n) + n_m(t_n), \quad n = 0, \ldots, N-1$$

The signals s_i *usually* represent a source of propagating energy. Because of the possibility of multipath, signals can represent a source's reflections from various objects or a relatively direct path. Thus, some signal components in the observations may be related to others; the signal processing computations should reveal these relationships if possible. Each signal depends on sensor index m via a propagation-induced delay-per-sensor $\Delta_i(m)$: $s_i(m, t_n) = s_i(t_n - \Delta_i(m))$. This quantity equals the scaled dot product between the source's apparent direction vector and the sensor's location.

$$\Delta_i(m) = \frac{\vec{\zeta}_i^{\,o} \cdot \vec{x}_m}{c}$$

In some cases, the propagation speed c varies over the spatial extent of the array because of inhomogeneities in the medium. This variation complicates our ability to determine direction of propagation; given the sensor locations, signal processing algorithms can *only* estimate the ratio of the signal's direction vector and the speed of propagation. *To find direction unambiguously, the speed of propagation must be known; to find the speed of propagation, the source's direction must be known.*

The signal term in the spatial correlation matrices depends on whether we assume the signals to be stochastic or not. These matrices consist of all possible cross-correlations between the various signals and their observed values at each sensor. To evaluate the spatiotemporal correlation matrix, define a $MN \times N_s$ signal matrix $\mathbf{S}(t_0)$ with each column containing a sequence of observations for a specific signal.

$$\mathbf{S}(t_0) = \begin{bmatrix} \mathbf{s}_0(0) & \mathbf{s}_1(0) & \cdots & \mathbf{s}_{N_s-1}(0) \\ \mathbf{s}_0(1) & \mathbf{s}_1(1) & \cdots & \mathbf{s}_{N_s-1}(1) \\ \vdots & \vdots & \ddots & \vdots \\ \mathbf{s}_0(M-1) & \mathbf{s}_1(M-1) & & \mathbf{s}_{N_s-1}(M-1) \end{bmatrix}$$

In this notation, $\mathbf{s}_i(m) = \text{col}[s_i(m, t_0), \ldots, s_i(m, t_{N-1})]$. Using this matrix, the observation vector is written $\mathbf{y}(n_0T) = \mathbf{S}(n_0T)\mathbf{1}$, where $\mathbf{1} = \text{col}[1, \ldots, 1]$. The spatiotemporal correlation matrix $\mathbf{R}(t_0)$ for the deterministic signal case equals

$$\boxed{\mathbf{R}(t_0) = \mathbf{S}(t_0)\mathbf{1}\mathbf{1}'\mathbf{S}'(t_0) + \mathbf{R}_n}$$

The quantity $\mathbf{R}_n$ denotes the correlation matrix of the noise. The $M \times M$ block components of the signal-related matrix term are given by $\sum_{i,j} \mathbf{s}_i(m_1)\mathbf{s}_j^*(m_2)$. If stochastic, the spatiotemporal correlation matrix of the signals emerges, containing components equaling the expected value of terms such as this. Thus, this matrix's structure expresses intersignal correlations; these correlations are highly problem dependent. Those signals arising from distinct sources are presumably statistically independent, making their cross-correlation zero. Multipath induces correlation, with propagation effects determining the degree of correlation.

Evaluation of the spatiospectral correlation matrix reveals a strikingly similar mathematical form. Using the definitions, we find each element of $\mathbf{R}(\omega)$ to be

$$[\mathbf{R}(\omega)]_{m_1m_2} = \sum_{i_1,i_2} \mathcal{E}[S_{i_1}(\omega)S_{i_2}^*(\omega)]e^{-j\omega[\Delta_{i_1}(m_1)-\Delta_{i_2}(m_2)]} + [\mathbf{R}_N(\omega)]_{m_1m_2}$$

The matrix $\mathbf{R}_N$ denotes the spatiospectral correlation matrix of the noise and has Hermitian symmetry. To exploit matrix notation fully, we define the $M \times N_s$ signal matrix $\mathbf{S}$ as

$$\mathbf{S} = \begin{bmatrix} e^{-j\omega\Delta_0(0)} & \cdots & e^{-j\omega\Delta_{N_s-1}(0)} \\ \vdots & & \vdots \\ e^{-j\omega\Delta_0(M-1)} & \cdots & e^{-j\omega\Delta_{N_s-1}(M-1)} \end{bmatrix}$$

Each column of this matrix has the form $\mathbf{e}(\Delta_i) = \text{col}\left[e^{-j\omega\Delta_i(0)}, \ldots, e^{-j\omega\Delta_i(M-1)}\right]$, which represents the propagation-induced phase shift experienced by a unit-amplitude complex exponential signal and measured by the sensors. By defining the matrix $\mathbf{C}$ as the $N_s \times N_s$ *intersignal coherence matrix* with $c_{i_1,i_2} = \mathcal{E}[S_{i_1}(\omega)S_{i_2}^*(\omega)]$, we express the signal term as $\mathbf{SCS}'$. The correlation between two signals is the average of their complex spectral values at the frequency ω. When the signals' Fourier Transforms have nonrandom amplitudes (although they may be unknown), $\mathcal{E}[S_{i_1}(\omega)S_{i_2}^*(\omega)]$ expresses the degree of their phase coherence: c_{i_1,i_2} is a complex number having a magnitude ranging between 0 (no coherence) and $\mathcal{E}\left[|S_i(\omega)|^2\right]$, the expected value of the squared magnitude of the ith signal's Fourier Transform at frequency ω. The extreme case of the largest magnitude intersignal coherence implies a direct relationship between the signals concerned. This situation suggests that one member of the pair represents a strong multipath version of the other. Smaller magnitudes express a weaker degree of phase coherence, suggesting that weak multipath is present. The spatiospectral correlation matrix thus becomes

$$\boxed{\mathbf{R}(\omega) = \mathbf{SCS}' + \mathbf{R}_N} \tag{2.21}$$

The similarity between the spatiotemporal and spatiospectral correlation matrices allows the following discussions to disregard whether a time-domain or frequency-domain algorithm is in question. This unifying formulation, a typical byproduct of possibly overly succinct matrix notation, allows us to obtain general results in a straightforward manner. Applying these results is *very* domain dependent, however. The size of the matrix is quite large in the temporal case; there, no assumptions on signal bandwidth are made and using the spatiotemporal correlation matrix typically falls under the guise of wideband algorithms. The spatiospectral correlation matrix applies at each of possibly several temporal frequencies. If the signal is narrowband, permitting us to consider only one matrix, the matrix's vastly smaller size reduces the amount of computation required enormously; if wideband, several matrices are required to integrate fully the signal components in the observations and the number of computations increases proportionally. Thus, the signal's characteristics dramatically affect the algorithm *type*—wideband or narrowband, time domain or frequency domain—and the resulting complexity.

Summary

In this chapter, we discussed spatiotemporal signals (and noise, for that matter) and how nature can change them as they propagate from some source to a receiver. Propagation, as described by the wave equation, can give rise to many effects. Depending on the medium, the signal we wish to observe may be refracted, reflected, diffracted, and dispersed in frequency. It may be embedded in noise with various spatial and temporal correlation values. The challenge for the signal processing expert is to design receiving arrays and processing algorithms to extract the signal components from the noise, and put them back together so that meaningful information can be gleaned from them.

Problems

2.1 Derive the wave equation from Maxwell's equations. You might need the identity $\nabla\times\nabla\times\vec{E} = \nabla(\nabla\cdot\vec{E}) - \nabla^2\vec{E}$.

2.2 Assume that a pulse is propagating in an ideal medium. The pulse $p(t)$, given by the expression $\mathrm{u}(t) - \mathrm{u}(t-1)$, propagates with a slowness vector given by $\vec{\alpha} = (3/10, 2/5, 0)$. Assume that the formula for our propagating pulse is $p(t - \vec{\alpha}\cdot\vec{x})$.

- **(a)** What is the speed of propagation c?
- **(b)** What is the direction of propagation?
- **(c)** Where is the advancing edge of the pulse at time $t = 2$?
- **(d)** Sketch the pulse's locations at times $t = 0, 1, 2$. Show that indeed the formula given previously does correspond to a propagating signal.
- **(e)** Now assume the pulse is propagating as a spherical wave originating from the spatial origin at $t = 0$ with the same speed of propagation. What formula expresses the propagating pulse? Sketch its spatial location in the (x, y) plane for $t = 1, 2, 3$.

2.3 Assume a unit-amplitude sinusoid $s(\vec{x}, t)$ propagates in the (x, y) plane. We sample it spatially at two locations $\vec{x}_1 = (-d/2, 0)$ and $\vec{x}_2 = (d/2, 0)$, where $d = 2$.

(a) Assume that the two spatial samples are taken simultaneously: $s(\vec{x}_1, 0) = s_1$ and $s(\vec{x}_2, 0) = s_2$. Show that we cannot find the direction of propagation unless more information is provided.

The remainder of this problem investigates just what is needed to find $\vec{\zeta}$ from these samples.

(b) If the speed of propagation equals 1, show that the direction of propagation cannot be found uniquely.

(c) Assume the speed of propagation equals 1 and that the sinusoid's frequency equals 0.25 Hz. Can we find the direction of propagation now?

(d) Now assume the source is located somewhere in the upper half-plane ($y \geq 0$). Can the direction of propagation be found uniquely? For concreteness, let $s_1 = 0$ and $s_2 = 1/\sqrt{2}$. Explicitly find all possible directions of propagation.

(e) Summarize what is needed to find the direction of propagation uniquely.

2.4 Assume a cosinusoid is propagating outwardly from a point (spherical symmetry). We sample the spatiotemporal waveform at time t_0: $s(t_0, r) = \cos(\omega t_0 - kr)/r$.

(a) Find the distance between successive zeros.

(b) Find the distance between successive extrema of the propagating wave. Specifically show that the interzero distance does not depend on t_0 while the intermaxima distance does.

(c) Under what conditions are the separations between extrema equal?

2.5 In radar, the return from a moving source experiences a Doppler shift: A sinusoid emitted from an antenna strikes the target, reflects from it, and returns to the antenna at a different frequency.

(a) How is the propagation delay between emission and return related to the target's range?

(b) Assuming the stationary antenna transmits at frequency ω, find the received frequency in terms of the target's velocity vector $\vec{v}_{\text{target}}$.

(c) Although the field generated by an antenna may be directive, focused on a particular solid angle, the amplitude of the propagating wave usually decreases with range: It is proportional to $1/r$. How is the power of a radar return related to the target's range?

2.6 For the attenuation example described in §2.3.2 {28}, interesting details are not discussed.

(a) To find the proportionality constant β relating the real and imaginary parts of the wavenumber vector, show that Eq. 2.12 yields a unique solution.

(b) Show that, for small values of the ratio $\gamma c^2/\omega$, the spatial decay rate $|\vec{k}_{\text{Im}}|$ is small (in magnitude) relative to a wavelength and does *not* depend on frequency.

(c) Assuming $\gamma = 1$ and $c = 300$, plot the decay rate as a function of frequency, describing the important regions of the curve.

2.7 The propagation of sound through a viscous fluid is governed by the augmented wave equation [109: §6.4];[128: Vol. II, Chap. 19]

$$\nabla^2 s = \frac{1}{c^2}\frac{\partial^2 s}{\partial t^2} - \frac{4\mu}{3\rho_0 c^2}\frac{\partial}{\partial t}\nabla^2 s$$

where μ is the coefficient of viscosity and ρ_0 is the ambient density.

(a) Assuming a monochromatic solution, find the dispersion relation.

(b) Is the solution dispersive, attenuative, or both?

(c) Find the spatial decay constant. How does it depend on frequency and viscosity coefficient? Interpret your answer.

2.8 Let's make the example {32} describing sound channeling more realistic. Consider speed variations of the form $c(y) = c_0(1 + ay)$ (the speed at the surface is no longer zero) and the source (located at depth y_0 and horizontal coordinate of zero) radiates in all directions, producing spherical waves infinitesimally nearby.

(a) Show that a ray having *any* initial angle θ_0 follows a circular path of radius $(y_0 + 1/a)/\sin\theta_0$ and origin having y coordinate equaling the (virtual) depth where the speed equals zero $(-1/a)$.

(b) Show that the sine of the angle at which a ray strikes the surface is bounded by $1/(1 + ay_0)$.

(c) Rays that strike the surface reflect from it and continue to reflect from the surface forever as the propagation repeats. What is the distance between strikes of the surface?

(d) Suppose your sensor is located just under the surface some distance D horizontally from the source. Characterize the waveform you would measure: Is it dispersed because of refraction? Do you always receive a signal?

2.9 Eq. 2.13 {35} describes the relation between a ray's unit tangent vector and the medium's index of refraction. Show that this relation holds for the example {32} that yielded semicircular rays for linear variations with depth of sound speed.

2.10 Consider the situation of a ship approaching the open ocean from the Panama Canal. As the locks open, the ship radiates acoustic waves through the narrow slit between the lock's doors into the ocean. For this problem, assume the ocean to be homogeneous and that the distance between the canal and the measurement point to be large compared with the slit width W and the sound's wavelength.

(a) Assuming the ship's engines produce the sound and are turning the propellers at 30 rpm, what is the wavelength?

(b) What is the complex amplitude of an acoustic wave some distance away from the canal?

(c) How does this amplitude change as the doors open (as W increases)?

2.11 Consider a signal propagating toward an array that experiences a continuum of multipath delays. Using τ to represent propagation delay, assume that each sensor in the array receives a superposition of delayed signals, each component of which we describe as $s(t - \tau)$.

(a) Assume the multipath delays are uniformly distributed over $[\Delta_{\min}, \Delta_{\max}]$. Find an expression for the received signal.

(b) Assume each ray yields a signal having a different amplitude. Letting $A(\tau)$ denote this ray-dependent amplitude, show that multipath is equivalent to linearly filtering the signal. Find the transfer function of this filter.

(c) Assume that the biggest amplitude occurs at the average multipath delay $(\Delta_{\min} + \Delta_{\max})/2$ and decreases smoothly and symmetrically, becoming zero for $\tau = \Delta_{\min}$ and $\Delta_{\max}$. Show that multipath acts like a *lowpass* filter in this case.

2.12 The Fourier Transform pairs for spatiotemporal signals should become as familiar as they are in one dimension. Find the spectra for the following interesting spatiotemporal signals. Assume only two spatial dimensions so that you can sketch the results.

(a) A signal having constant wavelength across a narrow band of frequencies.

(b) A sinusoidal plane wave propagating parallel to and within the boundaries defined by $-W < y - x < W$.

(c) A signal having components propagating in all directions and obeying the dispersion relation $\omega = \sqrt{c^2k^2 + \omega_c^2}$.

2.13 When refraction creates channeling, we can no longer assume that isotropic noise has spherical symmetry. Instead, *cylindrically* symmetric noise would be much more plausible.

(a) Derive the spatiotemporal Fourier Transform for cylindrical coordinates (r, z, θ).

(b) The power spectrum of cylindrically isotropic noise—spatial variations occur only radially—should have what form?

(c) Derive the correlation function of monochromatic cylindrically isotropic noise. What spatial sampling separations yield uncorrelated signals?

(d) In the spherically isotropic noise case, the spatial correlations are bandlimited. Are they similarly bandlimited in the case of cylindrically isotropic noise?

2.14 A common form of spatiotemporal filter separates spatial from temporal filtering. Consider a system that spatially samples a spatiotemporal signal $y(\vec{x}, t)$ at locations $\vec{x}_m$, $m = 0, \ldots, M-1$, forms the sum

$$z(t) = \sum_m w_m y(\vec{x}_m, t)$$

then passes $z(t)$ through a linear, time-invariant filter having frequency response $H(\omega)$.

(a) Is the composite filter space invariant? linear?

(b) Find the spatiotemporal transfer function equivalent to this operation.

(c) Assume the signal to be a propagating monochromatic signal and the spatial sampling locations to be regularly spaced along the x axis: $(-(M-1)/2 + m)\, d\vec{i}_x$. What is the spatiotemporal transfer function for this important special case?

2.15 The spatiospectral matrix given by Eq. 2.20 describes, when we assume separability, the correlation structure of the noise at one temporal frequency. Subsequently, we need to consider the spectral characteristics of the observations at all frequencies. Show that the spatiospectral matrix has the form $\boldsymbol{\rho} \otimes \mathbf{D}$, where the elements of the cross-spectral matrix $\mathbf{D}$ have the form

$$D_{\omega_1,\omega_2} = \int_{t_0}^{t_0+D} \int_{t_0}^{t_0+D} R_y^{(t)}(t-u) e^{-j(\omega_1 t - \omega_2 u)}\, dt\, du$$

2.16 The spatiospectral correlation matrix has the form $\mathbf{R}(\omega) = \mathbf{SCS}' + \mathbf{K}_N$, where $\mathbf{K}_N$ is the noise covariance matrix. Assume that only one signal is present in the measured field; consequently, $\mathbf{C} = [A^2]$ and $\mathbf{S} = \mathrm{col}\left[e^{-j\omega\Delta(0)}, \ldots, e^{-j\omega\Delta(M-1)}\right]$.

(a) Assume the noise is spherically isotropic. What is the form of the noise covariance matrix? What symmetries does it have?

(b) What matrix symmetries does the signal term $\mathbf{SS}'$ have?

(c) Calculate Im[$\mathbf{R}$], the imaginary part of the correlation matrix. Show that the noise term is completely eliminated, which means we have a method of removing noise without specific knowledge of the signal. Can the original signal-related term $A^2\mathbf{SS}'$ be reconstructed from Im[$\mathbf{R}$]?

Chapter 3

Apertures and Arrays

To provide waveforms that represent physical signals, a *sensor* must be designed that interacts with propagating energy to produce an electrical signal. Thus, a sensor's primary function is to serve as a *transducer*, converting one energy form into another. Because of this transformation, we distinguish between a field's value $f(\vec{x}_m, t)$ at the mth sensor's location $\vec{x}_m$ and the sensor's output $y_m(t)$. Assuming the sensor is perfect—its transformation is linear and has infinite bandwidth—these two signals differ only by a constant that expresses the transduction ratio. Whether the transformation is this simple or not, each sensor must be calibrated so that we can determine physical parameters from electrical ones.

Because propagating waves vary in space and time, sensors are often designed to have significant spatial extent, gathering energy propagating from specific directions. In such cases, a sensor does more than sample the field at a particular location; it spatially integrates energy that, as described later, focuses the sensor on particular propagation directions. Such sensors are said to be *directional*; sensors that simply sample the field are *omnidirectional*. Examples of directional sensors are parabolic radar dishes and acoustic listening devices used along the sidelines in football games. In this chapter, we shall examine the effects of sensors that gather signal energy over finite areas: *apertures*. An array consists of a group of sensors, whether they are directional or not, combined to produce a single output. We begin by discussing sensors with finite apertures, concentrating on linear and circular apertures. After reviewing sampling theory, we then analyze sensor arrays as sampled apertures.

3.1 Finite Continuous Apertures

What are the effects of a finite aperture on a space-time signal $f(\vec{x}, t)$? The *aperture function* $w(\vec{x})$ embodies two kinds of information about the aperture. First of all, the spatial extent of $w(\vec{x})$ reflects the size and shape of the aperture. In many important cases, the aperture function is equal to one inside some closed region where the sensor integrates the field and zero outside of that region. In signal processing terms, the aperture acts like a window through which we observe the wavefield. In addition, aperture functions can take on any real value between 0 and 1 inside the aperture. This second aspect of aperture functions allows us to represent the relative weighting of the field within the aperture. Aperture weighting is sometimes also referred to as shading, tapering, or apodization. For example, suppose we have an aperture that covers the spherical region $|\vec{x}| \leq R$. Thus, the aperture function for this sensor would be

$$w(\vec{x}) = \begin{cases} 1, & |\vec{x}| \leq R \\ 0, & \text{otherwise} \end{cases}$$

In the absence of other constraints, nothing can be said about the behavior of $f(\vec{x}, t)$ outside this region based on the sensor's output. One type of shading that might be applied to the aperture would be $w(\vec{x}) = 1 - |\vec{x}|/R$.

3.1.1 Aperture Smoothing Function

When we observe a field through our finite aperture, the sensor's output is

$$z(\vec{x}, t) = w(\vec{x}) f(\vec{x}, t)$$

Calculating the space-time Fourier Transform of this relationship, we obtain

$$\boxed{Z(\vec{k}, \omega) = \frac{1}{(2\pi)^3} \int_{-\infty}^{\infty} W(\vec{k} - \vec{l}) F(\vec{l}, \omega)\, d\vec{l}} \tag{3.1}$$

$Z(\vec{k}, \omega)$ is a convolution over wavenumber between the Fourier Transform of the field

$$F(\vec{k}, \omega) = \frac{1}{(2\pi)^4} \int_{-\infty}^{\infty} \int_{-\infty}^{\infty} f(\vec{x}, t) \exp\{j(\vec{k} \cdot \vec{x} - \omega t)\}\, d\vec{x}\, dt$$

and the *aperture smoothing function*

$$\boxed{W(\vec{k}) = \frac{1}{(2\pi)^3} \int_{-\infty}^{\infty} w(\vec{x}) \exp\{j\vec{k} \cdot \vec{x}\}\, d\vec{x}} \tag{3.2}$$

This convolution means that the wavefield's spectrum becomes smoothed by the kernel $W(\vec{k})$ once we observe it through an aperture. Employing a spatial aperture to observe a space-time signal is analogous to the effect of measuring a one-dimensional temporal signal $s(t)$ only during a short observation period (a "time" window).

For our spherical aperture,

$$W(\vec{k}) = \frac{4\pi}{k^2} \left(\frac{\sin kR}{k} - R \cos kR \right)$$

where, as in Chap. 2 {16}, $k = |\vec{k}| = \sqrt{k_x^2 + k_y^2 + k_z^2}$. This smoothing kernel, shown in Fig. 3.1, exhibits spherical symmetry in wavenumber space because of the aperture's spherical symmetry.

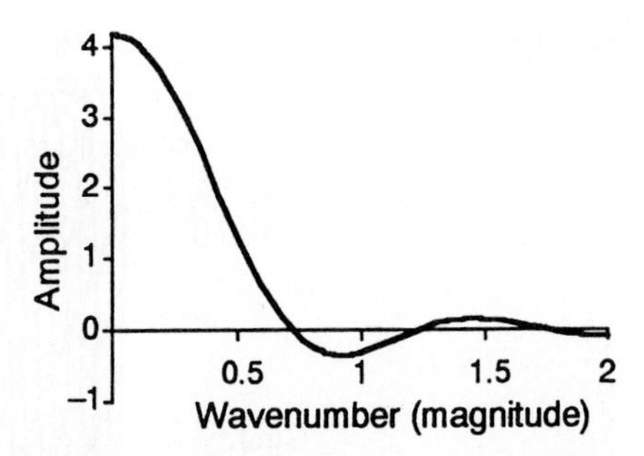

Figure 3.1 The spherical aperture smoothing function $W(k)$ resulting from a continuous, spherical spatial aperture is plotted as a function of k for $R = 1$.

Let's consider the case of a single plane wave propagating in a particular direction $\vec{\zeta}^o$. As in Chap. 2, we can write such signals in the form $f(\vec{x}, t) = s(t - \vec{\alpha}^o \cdot \vec{x})$, where $\vec{\alpha}^o$ is the slowness vector pointing in the direction $\vec{\zeta}^o$ with magnitude $1/c$: $\vec{\alpha}^o = \vec{\zeta}^o/c$.[†] The wavenumber-frequency spectrum of a propagating field is given by

$$\boxed{F(\vec{k}, \omega) = S(\omega)\delta(\vec{k} - \omega\vec{\alpha}^o)}$$

where $S(\omega)$ denotes the Fourier Transform of the source waveform $s(t)$ and $\delta(\vec{k})$ represents a three-dimensional impulse function.

By substituting this relation into Eq. 3.1, we can find the effect of observing a propagating signal through a finite aperture. The resulting spectrum $Z(\vec{k}, \omega)$ of the output is

$$Z(\vec{k}, \omega) = S(\omega)W(\vec{k} - \omega\vec{\alpha}^o) \tag{3.3}$$

This complicated relation expresses a simple but important result. When $\vec{k}$ equals $\omega\vec{\alpha}^o$, we see that $Z(\omega\vec{\alpha}^o, \omega) = S(\omega)W(\vec{0})$. Along this particular path in wavenumber-frequency space, the output spectrum equals the signal spectrum multiplied by a constant: the value of $W(\cdot)$ at the origin. *All of the information concerning the propagating signal is present in the aperture's output.* For other values of $\vec{k}$, the signal spectrum S is multiplied by a frequency-dependent gain $W(\vec{k} - \omega\vec{\alpha}^o)$ that distorts the relative strengths and phases of the frequency components in the signal spectrum, effectively filtering it. Fig. 3.2 illustrates these distortions in a simplified situation.

This result is easily generalized to the case in which $f(\vec{x}, t)$ is a superposition of plane waves.

$$f(\vec{x}, t) = \sum_i s_i(t - \vec{\alpha}_i^o \cdot \vec{x})$$

This space-time signal's spectrum is $\sum_i S_i(\omega)\delta(\vec{k} - \omega\vec{\alpha}_i^o)$ and the result of observing this field through the array aperture is given by the spectrum $\sum_i S_i(\omega)W(\vec{k} - \omega\vec{\alpha}_i^o)$. Let's set $\vec{k} = \omega\vec{\alpha}_j^o$, where j equals the index corresponding to one of the propagating signals. The output spectrum can then be written as

$$Z(\omega\vec{\alpha}_j^o, \omega) = S_j(\omega)W(0) + \sum_{i \neq j} S_i(\omega)W(\omega[\vec{\alpha}_j^o - \vec{\alpha}_i^o])$$

[†]As before, c is the propagation speed.

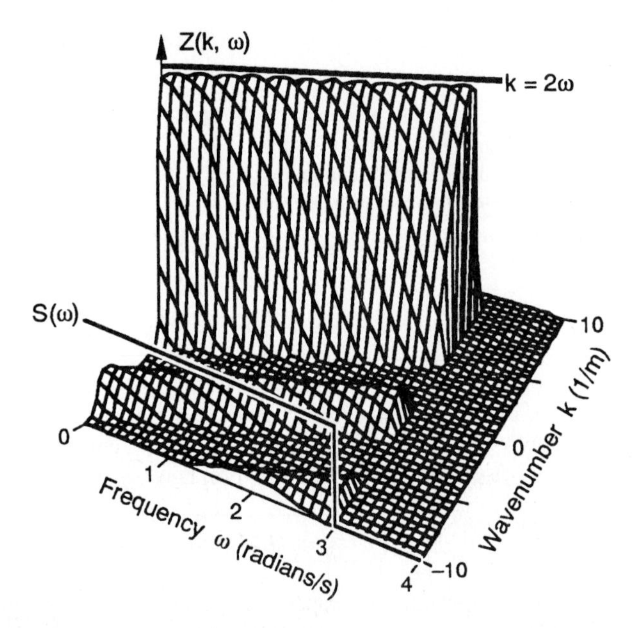

Figure 3.2 This complex figure illustrates the effect of a linear aperture's ($D = 2$) smoothing function on a baseband signal propagating with slowness $\alpha^o = 2$. The perspective plot depicts $Z(k, \omega)$ with the signal spectrum shown along the temporal frequency axis. The array output spectrum resembles the signal spectrum only along the line in wavenumber-frequency space corresponding to the signal's propagation characteristics. For all other wavenumber-frequency contours, the array output is a filtered version of the signal spectrum.

If the aperture smoothing function $W(\cdot)$ can be designed so that $W(\omega[\vec{\alpha}_j^o - \vec{\alpha}_i^o])$ is small compared to $W(0)$, then the sensor acts as a *spatial filter* that passes signals propagating from the direction represented by $\vec{\alpha}_j^o$ while rejecting others.

To exemplify subsequently defined attributes that summarize an aperture's spatial filtering characteristics, we shall examine two important apertures: the linear aperture, a one-dimensional example, and the circular aperture, a two-dimensional example. The linear aperture is often used in sonar and seismic signal processing. In these applications, the linear array is used rather than a continuous aperture. The discussion of sampled apertures is deferred until §3.3 {84}. The second important aperture is the two-dimensional circular aperture, which is commonly used in optics and in imaging systems. The characteristics of these arrays are summarized in Table 3.1 {67}.

Example

A *linear aperture* has an aperture function that is nonzero only along a finite-length line segment in three-dimensional space. For example, if we let

$$b(x) = \begin{cases} 1, & |x| \leq D/2 \\ 0, & \text{otherwise} \end{cases} \tag{3.4}$$

the three-dimensional aperture function $w(x, y, z)$ can be written as $w(x, y, z) = b(x)\delta(y)\delta(z)$. Fig. 3.3 illustrates the geometry associated with a typical linear aperture.

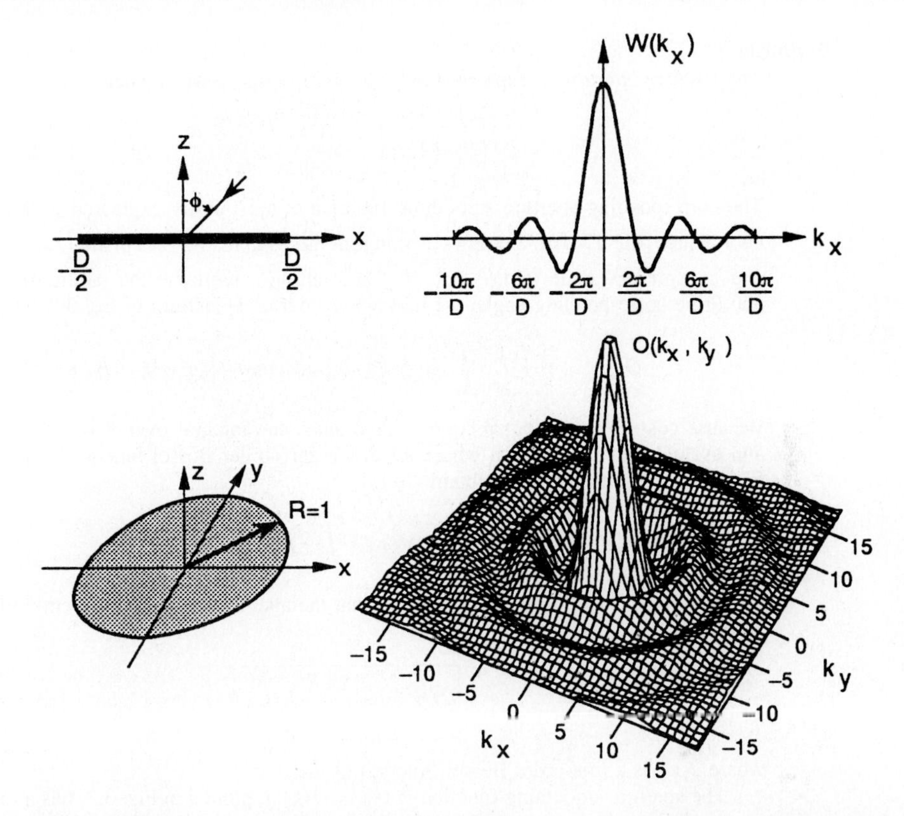

Figure 3.3 The commonest apertures, the linear and circular, are illustrated on the left, and their aperture smoothing functions on the right. A linear aperture lies along the x axis and spans D spatial units; the z axis intersects the midpoint of the aperture. The angle ϕ is formed by the z axis and the direction of propagation of a plane wave. Positive values of ϕ are taken clockwise from the z axis. Because the linear aperture lies along the x axis, its aperture smoothing function depends only on k_x. The circular aperture's aperture smoothing function $O(\cdot)$ is plotted in two-dimensional wavenumber space (k_x, k_y). Because of the aperture's circular symmetry, the aperture smoothing function depends only on the radius k_{xy}. First derived by G. B. Airy, $O(k_{xy})$ is often termed the "Airy disk" in the optical literature [61: 64].

For notational simplicity, define the aperture signal as $z(x,t) = b(x)f((x,0,0),t)$ for x within the aperture: The sensor is sensitive only to x components of the field. The aperture smoothing function for the linear aperture is given by

$$W(\vec{k}) = \frac{\sin k_x D/2}{k_x/2}$$

Because the aperture function is nonzero only along a small segment of the x axis, the aperture smoothing function W depends only on the x component of the wavenumber vector $\vec{k}$. The central portion of this function is plotted in Fig. 3.3. The aperture smoothing function has a central mainlobe of height D, $W(\vec{0}) = D$, and an infinite number of sidelobes of decreasing amplitude.

Example

The *circular aperture* is expressed by $w(x, y, z) = o(x, y)\delta(z)$, where

$$o(x, y) = \begin{cases} 1, & \sqrt{x^2 + y^2} \leq R \\ 0, & \text{otherwise} \end{cases} \tag{3.5}$$

The corresponding aperture smoothing function is most easily expressed in terms of polar coordinates [61: 12–16]. Letting k_{xy} represent radius in the (k_x, k_y) plane $\left(k_{xy} = \sqrt{k_x^2 + k_y^2}\right)$ and ψ represent angle ($\tan\psi = k_y/k_x$) and letting r represent the radius in (x, y) plane and θ the corresponding angle, we may write Fourier Transform of Eq. 3.2 {60} as

$$O(k_{xy}, \psi) = \int_0^{2\pi}\int_0^R \exp\{jk_{xy}r(\cos\psi\cos\theta + \sin\psi\sin\theta)\}\, r\, dr\, d\theta$$

Because $\cos(\theta - \psi) = \cos\psi\cos\theta + \sin\psi\sin\theta$, the integral over θ is independent of ψ and evaluates to $2\pi J_0(k_{xy}r)$, where $J_0(\cdot)$ is a zero-order Bessel function of the first kind. From this simplification, we obtain

$$O(k_{xy}) = 2\pi \int_0^R r J_0(k_{xy}r)\, dr$$

Using a Bessel function identity, we find that the aperture smoothing function of a circular array can be concisely written as

$$\boxed{O(k_{xy}) = \frac{2\pi R}{k_{xy}} J_1(k_{xy}R)}$$

where $J_1(\cdot)$ is a first-order Bessel function of the first kind.

The aperture smoothing function $W(\vec{k}) = O(k_{xy})$, plotted in Fig. 3.3, has a form similar to the $\sin x/x$ dependence of the linear aperture with the numerator being a first-order Bessel function instead of the sine function. Consequently, the zeros of $O(k_{xy})$, which are concentric circles centered on the origin in the (k_x, k_y) plane, are not evenly spaced. The squared magnitude of the circular aperture's aperture smoothing function is closely related to the intensity pattern of light produced when a plane wave strikes a circular hole.

Classical resolution. In general, an aperture $w(\vec{x})$ with a large spatial extent has a narrow aperture smoothing function $W(\vec{k})$, which leads to minimal spectral smoothing. Thus, the spatial extent of an aperture determines the *resolution* with which two plane waves can be separated: The larger the extent, the more focused the aperture can be on any specific direction. From this perspective, the perfect aperture smoothing function is the three-dimensional impulse function $W(\vec{k}) = \delta(\vec{k})$. Since the corresponding aperture function is constant over *all* space, we must compromise between perfection and the practicality of finite apertures.

Recalling the spatial filtering interpretation of the aperture smoothing function, the mainlobe can be considered the filter's passband and the sidelobes as the stopband. Thus, the mainlobe's width defines the aperture's ability to separate propagating waves. To define resolution precisely, consider two plane waves propagating past an aperture. The resulting spectrum is a superposition of the form

$$Z(\vec{k}, \omega) = S_1(\omega)W(\vec{k} - \omega\vec{\alpha}_1^o) + S_2(\omega)W(\vec{k} - \omega\vec{\alpha}_2^o)$$

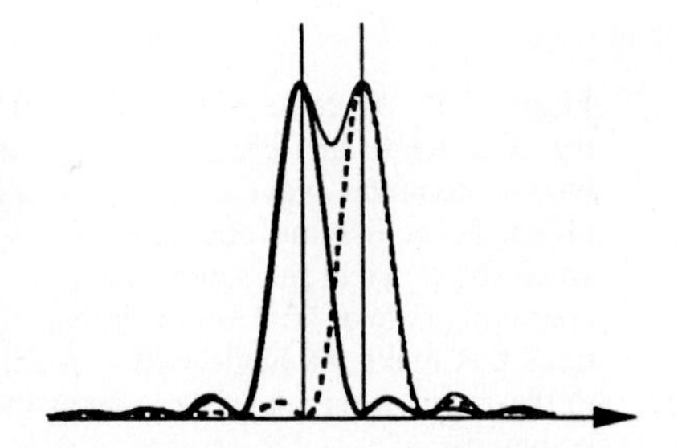

Figure 3.4 The Rayleigh criterion for resolution states that two waves are resolved if the peak of the aperture smoothing function due to one (solid) falls on the first zero of the aperture smoothing function due to the other (dashed). The combined output wavenumber spectrum due to both waves is shown as a light solid curve.

Each plane wave causes a replica of the aperture smoothing function to appear in the wavenumber-frequency spectrum $Z(\vec{k}, \omega)$. One classical definition of resolution, the Rayleigh† criterion, states that two incoherent‡ plane waves, propagating in two slightly different directions, are resolved if the mainlobe peak of one aperture smoothing function replica falls on the first zero of the other aperture smoothing function replica (Fig. 3.4). Thus, the resolution equals the smallest wavenumber that produces a zero aperture smoothing function or, said another way, half the mainlobe width. For a linear aperture, this zero occurs at $k_x = 2\pi/D$, giving the mainlobe a width of $4\pi/D$.¶ The first-zero contour for the circular aperture's smoothing function occurs at a radius of $k_{xy} \approx 1.22\pi/R$. Resolution can also be measured in terms of the just noticeable difference in direction of propagation. As $\vec{k} = 2\pi\vec{\zeta}/\lambda$, directional resolution $\delta\zeta = \delta k \cdot \lambda/2\pi$.

To understand the basis of this criterion, consider the monochromatic plane wave $\exp\{j(\omega^o t - \vec{k}^o \cdot \vec{x})\}$. At time $t = 0$ along a linear aperture, this plane wave yields the signal $z(x, 0) = \exp\{-jk_x^o x\}$, where k_x^o is the x component of the wavenumber vector $\vec{k}^o$. We can define a wavelength λ_x^o along the x axis as $2\pi/k_x^o$. Note that λ_x^o is actually greater than or equal to the true wavelength $\lambda^o = 2\pi/k^o$. The number of periods P of the output $z(x, 0)$ that occur over the extent of the aperture equals D/λ_x^o. According to the Rayleigh criterion, the x component of a second monochromatic plane wave's wavenumber vector just resolvable from the first equals $k_x^1 = k_x^o + 2\pi/D$. The number of spatial periods of this second plane wave that occur over the extent of the aperture is $P+1$. Thus, the Rayleigh resolution criterion in this case may be interpreted as requiring the two plane waves differ by at least one period over the sensor aperture (according to the x components of their wavelengths).

Sidelobe height. Again using the spatial filtering metaphor for the aperture smoothing function, the height of the highest sidelobe relative to the mainlobe measures an aperture's

†Lord Rayleigh (1842–1919) made major contributions to the study of diffraction theory, acoustics, and other topics.

‡The adjective "incoherent" means that the two plane waves have no common phase reference. If they did, the Rayleigh resolution criterion may not suffice to separate the propagating signals. More on this effect is provided in §4.6 {142}.

¶For more general aperture smoothing function shapes, the width is sometimes defined as the distance between the two points where the mainlobe equals half the height of its maximum. This width is referred to as full width, half maximum (*FWHM*). This measure can also be applied to the squared magnitude of the aperture smoothing function. For the simple linear aperture, *FWHM* approximately equals $2.41\pi/D$.

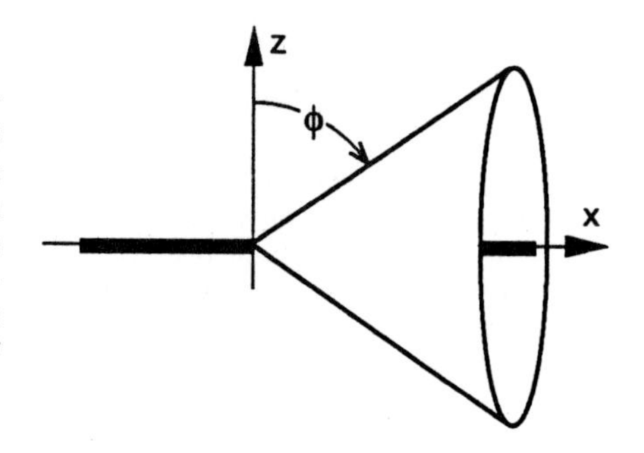

Figure 3.5 Because of the cylindrical symmetry of a linear aperture, monochromatic waves having common wavelengths and speeds that propagate across the aperture with a common value for α_x^o yield the same signal $z(x, t)$ at the aperture's output. The set of propagation directions that make the angle $\pi/2 - \phi$ with respect to the x axis form what is known as the cone of ambiguity.

ability to reject unwanted noise and signals, and focus on particular propagating signals. By differentiating the aperture smoothing function, we can find the location and height of the highest sidelobe. For the linear aperture, the first sidelobe maximum occurs at $k_x \approx 8.9868/D$ or $2.86\pi/D$. The value of $|W(k_x)|$ at this point is approximately $0.2172D$. Thus,

$$\frac{\text{Mainlobe height}}{\text{Sidelobe height}} \approx \frac{D}{0.2172D} \approx 4.603$$

Note that this ratio is independent of the aperture length D. A longer aperture gives better resolution—a narrower mainlobe/spatial passband—but does not lead to lower sidelobes. For the circular aperture, the highest sidelobe occurs at $k_{xy} = 5.14/R$, yielding a mainlobe-to-sidelobe height ratio of 7.56.

Aperture ambiguities. Because of the symmetries of many apertures, waves propagating at the same speed but from different directions can yield exactly the same output. The set of directions yielding identical responses is termed the *ambiguity set*. This set is defined entirely by the aperture smoothing function according to those slowness vectors $\vec{\alpha}^o$ that yield identical values for $W(\vec{k} - \omega\vec{\alpha}^o)$ under the constraint that $|\vec{\alpha}^o|$ is fixed. For example, because of the cylindrical symmetry about a linear aperture's axis, the aperture smoothing function depends only on the x component of the slowness vector. Thus, all monochromatic plane waves having the same slowness magnitude (which equals the reciprocal of the propagation speed c) that propagate across the aperture with identical values for α_x^o define the linear aperture's ambiguity set. Geometrically, this ambiguity set is a cone (Fig. 3.5), thereby defining the *cone of ambiguity* for a linear aperture: Signals propagating from above, below, or to the side of the linear aperture cannot be distinguished. The aperture smoothing function of a circular aperture $O(k_{xy})$ depends only on k_x and k_y. Because a common propagation speed defines the ambiguity set, choosing values for α_x^o and α_y^o means that only the sign of α_z^o is ambiguous. Thus, the circular aperture's surface of ambiguity is the z axis: Signals propagating directly above or below the aperture are ambiguous.

Characteristic	Linear Aperture	Circular Aperture
Aperture smoothing function	$\frac{\sin k_x D/2}{k_x/2}$	$\frac{2\pi R}{k_{xy}} J_1(k_{xy}R)$
Mainlobe width	$k_x = 4\pi/D$	$k_{xy} = 2.44\pi/R$
Resolution	$\delta\zeta_x = \lambda/D$	$\delta\zeta_{xy} = 0.61\lambda/R$
Relative sidelobe height	0.22	0.13
Ambiguity surface	Cone; axis parallel to array	Axis perpendicular to array plane

Many characteristics of an aperture depend on its spatial extent. For the linear array, this quantity is D and for the circular aperture, R. The variable k_x represents the x component of the wavenumber vector and k_{xy} the magnitude of the wavenumber vector in the (x, y) plane. Rayleigh resolution depends on the ratio of wavelength λ and aperture. Sidelobe height is the height of the highest sidelobe relative to the mainlobe.

TABLE 3.1 Fundamental Characteristics of Uniform Linear and Circular Apertures.

3.1.2 Apparent Velocity

If we observe signals propagating in three-dimensional space with a planar aperture, the apparent velocity (speed and direction) of a monochromatic plane wave across the aperture is related to its true speed of propagation in an interesting way. Let's assume, for example, that the aperture lies in the (x, y) plane and that a monochromatic plane wave of the form $f(\vec{x}, t) = \exp\{j(\omega^o t - \vec{k}^o \cdot \vec{x})\}$ is traveling past it. As usual, the propagation speed c is equal to ω^o/k and the slowness vector $\vec{\alpha}^o$ is equal to $\vec{k}^o/\omega^o$. If the wave is traveling in the $-z$ direction so that $\vec{k}^o = (0, 0, -k_z)$, then the (x, y) plane is a plane of constant phase and the wave's complex amplitude at each point in the aperture is the same: The amplitude changes value with time simultaneously at all points in the aperture, and the wave appears to be everywhere in the aperture simultaneously. The x and y components of the slowness vector $\vec{\alpha}^o$ are 0, meaning the speed of the wave in the x and y directions is infinite. Taking another extreme, consider a wave propagating in the $+x$ direction, $\vec{k}^o = (k_x, 0, 0)$. The velocity across the aperture in the x direction equals ω^o/k_x because the x component of the slowness vector is k_x/ω^o. The speed across the aperture in the y direction is infinite because $\alpha_y^o = 0$.

In general, the components of the slowness vector determine the slowness across the aperture. Because slowness is inversely related to speed, the components of velocity across the aperture are greater than or equal to the true speed c of wave propagation. Within the plane of the aperture, the *apparent propagation speed* c_a is given by

$$c_a = \frac{1}{\sqrt{(\alpha_x^o)^2 + (\alpha_y^o)^2}} = \frac{\omega^o}{\sqrt{(k_x^o)^2 + (k_y^o)^2}}$$

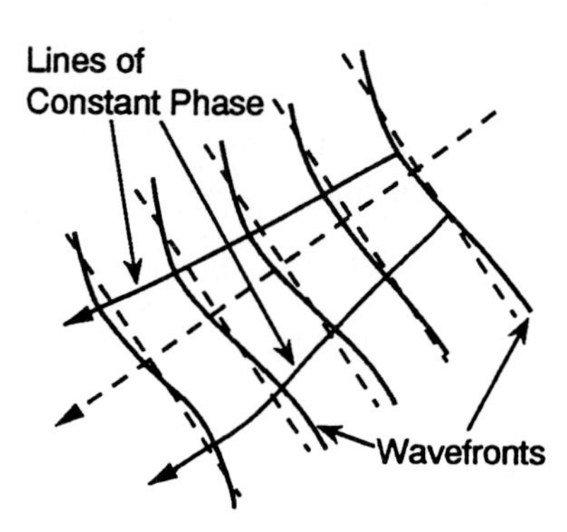

Figure 3.6 A turbulent medium or an error in location of some part of the aperture can cause a wavefront to be distorted. The dashed lines represent the ideal wavefront and its direction of propagation, and the solid lines represent the distorted wavefront and lines of constant phase. Lines of constant phase, indicators of the local direction of propagation, clearly show the medium's inhomogeneities that, in this case, cause the aberrations.

Because $(k^o)^2 = (k_x^o)^2 + (k_y^o)^2 + (k_z^o)^2$, we can also write

$$c_a = \frac{\omega^o}{\sqrt{(k^o)^2 - (k_z^o)^2}} \geq c = \frac{\omega^o}{k^o}$$

We can define an angle of incidence ϕ as the angle between the wave's direction of propagation vector $\vec{\zeta}^o$ (or equivalently its slowness vector $\vec{\alpha}^o$ or its wavenumber vector $\vec{k}^o$) and the normal to the plane of the aperture, in this case the z axis. As seen in Fig. 3.3 {63}, the cosine of this angle is given by $\cos\phi = -\zeta_x^o$. A little trigonometry reveals that $c_a = c/\sin\phi$.

3.1.3 Aberrations

In optics, the deviation of a lens from its ideal shape leads to an *aberration*: deviation in the wavefront from its intended form. More generally, turbulence in the medium or position errors in the aperture can cause a wavefront to be delayed (or advanced) relative to the ideal case, as in Fig. 3.6. Consider a propagating signal $f(\vec{x}, t) = s(t - \vec{\alpha}^o \cdot \vec{x})$ that suffers delays $\Delta(\vec{x})$ as a function of position in the aperture.† When position-dependent delays occur, the aperture output signal becomes

$$z(\vec{x}, t) = w(\vec{x})s(t - \Delta(\vec{x}) - \vec{\alpha}^o \cdot \vec{x})$$

By following the derivation that leads to Eq. 3.3 {61}, we find the output signal's spectrum to be

$$Z(\vec{k}, \omega) = S(\omega)W_\omega(\vec{k} - \omega\vec{\alpha}^o)$$

where the aperture smoothing function $W_\omega(\vec{k})$ is given by the spatial Fourier Transform

$$\boxed{W_\omega(\vec{k}) = \int_{-\infty}^{\infty} w(\vec{x}) \exp\{-j\omega\Delta(\vec{x})\} \exp\{j\vec{k} \cdot \vec{x}\}\, d\vec{x}} \qquad (3.6)$$

†Modeling aberration delays as depending solely on position is a simplification. In general, such delays depend on the direction of propagation as well. Thus, two waves propagating across a misplaced aperture from different directions suffer different delays.

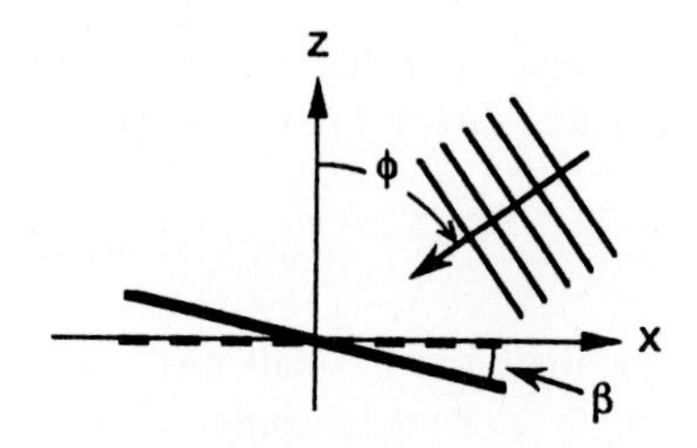

Figure 3.7 A linear aperture ideally lying along the x axis (dashed line) is inclined at an angle β instead (solid line). A plane wave propagates across the aperture at an angle ϕ to the negative z direction.

The aperture function $w(\vec{x})$ can be combined with the phase function $\exp\{-j\omega\Delta(\vec{x})\}$ to define a *generalized aperture function* $w_\omega(\vec{x})$ as the inverse spatial Fourier Transform of $W_\omega(\vec{k})$.

The generalized aperture function models wavefront distortion owing to aberration delays. The dependence on the temporal frequency ω can be interpreted in the following way. A delay of Δ corresponds to a distance $\delta x = c\Delta$. This distance, measured in fractions of a wavelength λ, equals the phase difference $\omega\Delta$ that appears in the complex exponential in the generalized aperture smoothing function $W_\omega(\cdot)$. At different frequencies, the delay Δ corresponds to different fractions of a wavelength.

Example

Consider a linear aperture that should, ideally, lie along the x axis but which is negatively inclined with respect to the x axis at an angle β (Fig. 3.7). When a plane wave propagates across the misplaced aperture in the negative z direction, it suffers a delay given by

$$\Delta(x) = \frac{x \tan\beta}{c}$$

In this simple case, the delay function $\Delta(x)$ causes the generalized aperture smoothing function $W_\omega(\cdot)$ to be a shifted version of the ideal aperture smoothing function: $W_\omega(k_x) = W(k_x - \omega\tan\beta/c)$.

More generally, a plane wave may propagate across the aperture at an angle ϕ from vertical. The amount of delay caused by aperture misplacement can be derived by the straightforward application of trigonometry, tedious though it is.

$$\Delta(x) = \frac{x\tan\beta}{c}\left[\frac{\cos(\beta+\phi)\,\cos\beta}{\cos^2\beta - \sin^2\phi}\right]$$

This result demonstrates that, in general, the delay depends not only on aperture misplacement, but also on the approaching wave's direction of propagation. The generalized aperture smoothing function again equals a shifted version of the ideal, but now it depends on the wave's characteristics as well as the aperture's.

This example shows that uniform aberration does not affect the mainlobe-sidelobe structure of the aperture smoothing function. More generally, the effects of aberrations well modeled as delays do distort the aperture smoothing function. Ideally, $W(\cdot)$ is a very narrow pulse with relatively low sidelobe levels. The distorted aperture smoothing function $W_\omega(\cdot)$ generally exhibits a wider mainlobe than the ideal mainlobe, a smaller peak value, and higher sidelobe levels. To illustrate the degradation in the aperture smoothing function induced by aperture errors, consider an ideal linear aperture of length D located along the x axis having a z axis sinusoidal distortion $h\sin\nu x$. For a plane wave propagating in the negative z direction, the time delay $\Delta(x)$ induced by this distortion equals $-(h/c)\sin \nu x$. The generalized aperture smoothing function for this case equals

$$W_\omega(k_x) = \int_{-D/2}^{D/2} e^{-j\omega\Delta(x)} e^{jk_x x}\, dx$$

For the case of small distortions—the deviation amplitude h is a fraction of a wavelength λ or equivalently $\omega\Delta(x) \ll 1$—we can use the approximation $\exp\{-j\omega\Delta(x)\} \approx 1 - j\omega\Delta(x)$. The generalized aperture smoothing function caused by sinusoidal aperture distortion then approximately equals

$$\boxed{W_\omega(k_x) \approx W(k_x) + \frac{\omega h}{2c}\big[W(k_x + \nu) - W(k_x - \nu)\big]}$$

where $W(k_x)$ denotes the undistorted linear array's aperture smoothing function {63}. Fig. 3.8 shows the effects of various sinusoidal aberrations on the aperture smoothing function. Clearly, the advantages of a large aperture, high gain and good directionality, can be lost if aberrations become large enough. Consequently, care must be taken to build radio telescopes and radar antennas within tight tolerances.

In the foregoing discussion of aberration, we found it necessary to consider the propagating signal's spatial frequency (or equivalently its wavelength). If the propagating signal consists of several spectral components, the overall effect of aberration can be analyzed by considering the individual effects on the various frequency components of the propagating signal. However, in many practical cases, these components are limited to a narrow band of frequencies distributed about some nominal center frequency ω^o corresponding to a nominal wavelength λ^o. Because of these narrow distributions, we can reexpress the generalized aperture smoothing function in terms of center frequencies, center wavelengths, and propagation directions.

In those cases where the propagating signals of interest have a narrow band of components centered on ω^o, expressing distances, such as sensor aperture, in units of λ^o simplifies calculations. Variables with units of inverse distance, such as wavenumber, can be normalized by multiplying them by λ^o. For example, let's reformulate the aberrated aperture smoothing function derived previously by defining the new dimensionless variables

$$D' = D/\lambda^o, \quad k'_x = k_x\lambda^o, \quad h' = h/\lambda^o, \quad \nu' = \nu\lambda^o$$

The variable k'_x acts as a normalized wavenumber variable; it equals 2π times the number of waves per nominal wavelength λ^o just as k_x indicates 2π times the number of waves per meter.

Defining $W'(k'_x) = W(k'_x\lambda^o)$ to be the unaberrated aperture smoothing function expressed in normalized variables, we have for a linear aperture

$$W'(k'_x) = \lambda^o \frac{\sin k'_x D'/2}{k'_x/2}$$

Similarly, we find the aberrated aperture smoothing function to be

$$W'_{\omega^o}(k'_x) \approx W'(k'_x) + \pi h'\big[W'(k'_x + \nu') - W(k'_x - \nu')\big]$$

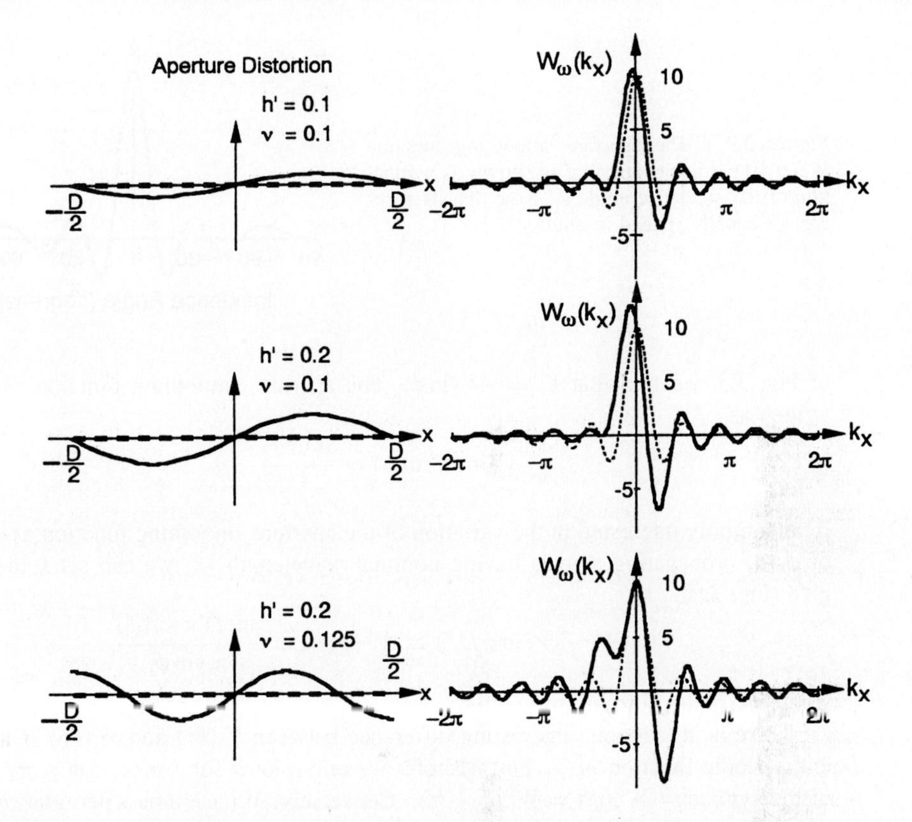

Figure 3.8 Physical distortions of an aperture, if uncorrected, lead to distorted aperture smoothing functions. On the left, sinusoidal aperture distortions, parameterized by normalized distortion amplitude $h' = h/\lambda$ and spatial frequency ν, are shown. The corresponding aperture smoothing functions are shown on the right by the solid lines. The aperture smoothing function of the ideal aperture ($D = 10$) is shown by the dashed line for comparison.

Thus, the normalized aperture smoothing function becomes independent of the nominal wavelength (and nominal frequency ω^o) save for an overall amplitude factor of λ^o: *An aperture smoothing function designed for a given value of D′ has the same shape independent of the center frequency* ω^o. The normalization allows expressing the width of the mainlobe, which determines the resolution, and the ratio of the mainlobe height to the sidelobe height independently of the nominal frequency.

The wavenumber vector $\vec{k}$ represents two kinds of information: Its magnitude represents the number of waves per meter, and its normalized vector represents the wave's direction of propagation. For situations in which the propagating signals have only a narrow band of spectral components, we may care more about how the aperture smoothing function varies with the direction of the wavenumber vector than with wavenumber magnitude. For example, recall the simple situation shown in Fig. 3.3 {63}. The aperture smoothing function for the linear array depends only on k_x. Using the geometry

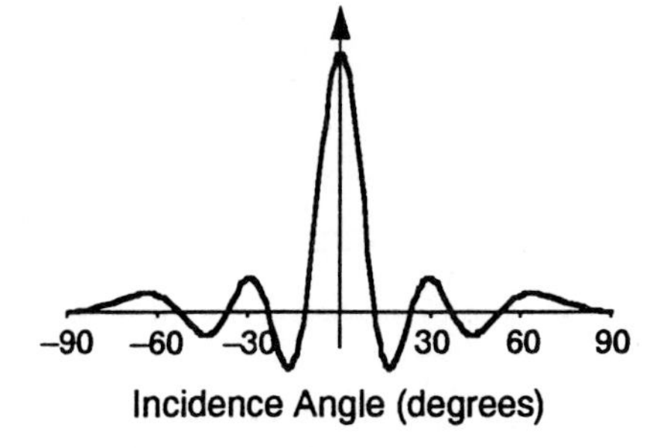

Figure 3.9 The aperture smoothing function $W''(\phi)$ for a linear aperture is shown as a function of the incidence angle ϕ. Note that its zeros are not equally spaced in angle.

of Fig. 3.3, we find that $k_x = -k \sin\phi$. The aperture smoothing function can then be written as

$$W(-k\sin\phi) = \frac{\sin\frac{kD\sin\phi}{2}}{\frac{k\sin\phi}{2}}$$

If we are only interested in the variation of the aperture smoothing function as a function of ϕ for propagating signals having nominal wavelength λ^o, we can set k to $2\pi/\lambda^o$ to give (Fig. 3.9)

$$W(-2\pi\sin\phi/\lambda^o) \equiv W''(\phi) = \lambda^o \frac{\sin(D'\pi\sin\phi)}{\pi\sin\phi}$$

where D' is the normalized aperture size.

There is at least one interesting difference between $W(k_x)$ and $W''(\phi)$. Clearly, W'' is a periodic function of ϕ. Furthermore, we only plot it for $-\pi/2 < \phi \le \pi/2$ because of the symmetry $W''(\phi) = W''(\phi + \pi)$. Conversely, $W(\cdot)$ is not a periodic function of its argument: The variable k_x can take on any real value. Thus, expressing the aperture smoothing function in terms of propagation direction reveals more about the aperture's ability to filter spatially narrowband signals. In general, the aperture smoothing function is a function of k_x, k_y, and k_z. To write the aperture smoothing function in terms of angles of azimuth θ and incidence ϕ, we can use the transformations given on page 10.†

$$k_x = -k\sin\phi\,\cos\theta, \quad k_y = -k\sin\phi\,\sin\theta, \quad k_z = -k\cos\phi$$

3.1.4 Co-Array for Continuous Apertures

The co-array is defined as the autocorrelation of the aperture function $w(\vec{x})$ [73].

$$\boxed{c(\vec{\chi}) \equiv \int w(\vec{x})w(\vec{x}+\vec{\chi})d\vec{x}}$$

The variable $\vec{\chi}$ is called a *lag*, and we term its domain *lag space*. The co-array becomes important when array processing algorithms employ the wave's spatiotemporal correlation function to characterize the wave's energy content.

†The minus signs are needed because a wave propagating into the origin from the first octant (where x, y, and z are all positive) would have negative wavenumber vector components.

To illustrate the importance of the co-array, consider a stationary random wavefield $f(\vec{x}, t)$ whose correlation function $R_f(\vec{\chi}, \tau)$ and power spectrum $\mathcal{S}_f(\vec{k}, \omega)$ we wish to estimate. We observe the wavefield through a finite aperture function $w(\vec{x})$, which gives us the signal

$$z(\vec{x}, t) = w(\vec{x}) f(\vec{x}, t)$$

The correlation function of $z(\vec{x}, t)$ equals

$$\begin{aligned} R_z(\vec{x}_1, \vec{x}_2; \tau) &= \mathcal{E}\{z(\vec{x}_1, t) z^*(\vec{x}_2, t + \tau)\} \\ &= w(\vec{x}_1) w^*(\vec{x}_1 + \vec{\chi}) R_f(\vec{\chi}, \tau) \end{aligned}$$

where we have defined $\vec{\chi} = \vec{x}_2 - \vec{x}_1$. Note that $z(\vec{x}, t)$ is stationary in time but not in space: $R_f(\vec{x}_1, \vec{x}_2; \tau)$ is not a function of $\vec{x}_2 - \vec{x}_1$ because of the finite aperture we used to observe the field. Because the field is stationary in both domains, we could average $R_z(\cdot; \tau)$ over all baselines† in the aperture with a length and direction given by $\vec{\chi}$ to produce an estimate of the field's correlation function.

$$\begin{aligned} \int_a R_z(\vec{x}_1, \vec{x}_1 + \vec{\chi}; \tau) \, d\vec{x}_1 &= \int_a w(\vec{x}_1) w^*(\vec{x}_1 + \vec{\chi}) R_f(\vec{\chi}, \tau) \, d\vec{x}_1 \\ &= R_f(\vec{\chi}, \tau) \int_a w(\vec{x}_1) w^*(\vec{x}_1 + \vec{\chi}) \, d\vec{x}_1 \end{aligned}$$

Here the integration region is the aperture in question. The integral on the right side is, of course, the co-array function $c(\vec{\chi})$, giving us

$$\boxed{\int_a R_z(\vec{x}_1, \vec{x}_1 + \vec{\chi}; \tau) \, d\vec{x}_1 = c(\vec{\chi}) R_f(\vec{\chi}, \tau)}$$

We see that our attempt to estimate $R_f(\vec{\chi}, \tau)$ by spatially averaging the aperture's correlation function over the aperture yields an estimate corrupted by multiplication with the co-array. In the frequency domain, the corresponding estimate of the power spectrum $\mathcal{S}_f(\vec{k}, \omega)$ would be smoothed by the Fourier Transform of $c(\vec{\chi})$, which equals $|W(\vec{k})|^2$, the squared magnitude of the aperture smoothing function.

Example

Consider the linear aperture along the x axis, as shown in Fig. 3.3. If the aperture function is uniformly shaded as described by Eq. 3.4 {62}, the co-array is the triangular function

$$c(\chi) = \begin{cases} D - |\chi|, & -D \le \chi \le D \\ 0, & \text{otherwise} \end{cases}$$

Because the aperture function is essentially one-dimensional, the co-array function is also.

In the case of the uniform circular aperture (Eq. 3.5 {64}), the co-array is a function of two lag variables. Because of the circular symmetry of the aperture, the co-array is also

†A baseline is simply a line segment connecting two points of the aperture. Consequently, it has a length and an orientation.

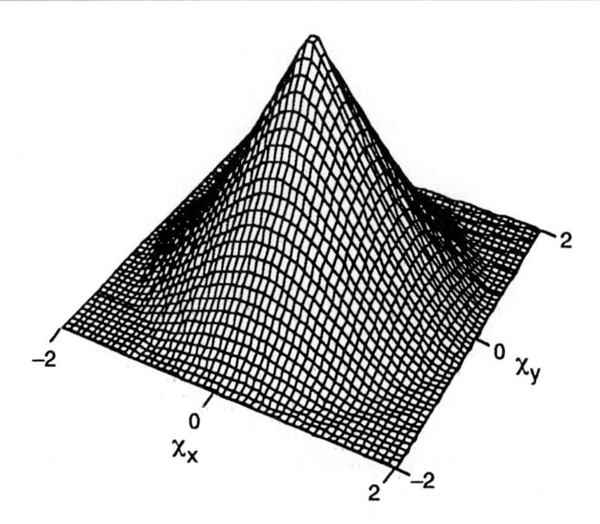

Figure 3.10 The cross-section of the co-array of a circular aperture of radius $R = 1$ is plotted as a function of radial distance χ in lag space.

circularly symmetric, allowing us to express the co-array as a function of radial distance $\chi = |\vec{\chi}|$ in lag space. The co-array is given by[†]

$$c(\chi) = \begin{cases} 2R^2 \left[\cos^{-1}\left(\frac{\chi}{2R}\right) - \frac{\chi}{2R}\sqrt{1 - \left(\frac{\chi}{2R}\right)^2} \right], & \chi \leq 2R \\ 0, & \text{otherwise} \end{cases}$$

This co-array function is almost cone shaped; its cross-section is plotted in Fig. 3.10.

The co-array can be interpreted as a density function that measures the relative number of baselines of various lengths that can be fitted into the sensor aperture. Because a propagating wave is characterized by its behavior at different points in space and the aperture is a collection of such points, the density of available baselines can be very useful. For example, in the case of the linear aperture, the co-array demonstrates the relatively large number of short baselines and small number of long baselines.

3.1.5 Focusing

Large apertures are used to collect radiation over a significant area to increase the signal-to-noise ratio and to measure the direction of propagation accurately. In general, the radiation collected over an area must be brought together to measure its intensity. For example, a parabolic antenna or a lens is used to collect propagating signals and focus them (ideally) at the point where a transducer is located. In addition to focusing the rays of a propagating wave, the aperture should bring them together without delaying portions of the wavefront with respect to others: The focusing operation should be *coherent*. If not, information conveyed by the propagating signal is distorted.

[†]This cumbersome-looking formula is easier to derive than you may think.

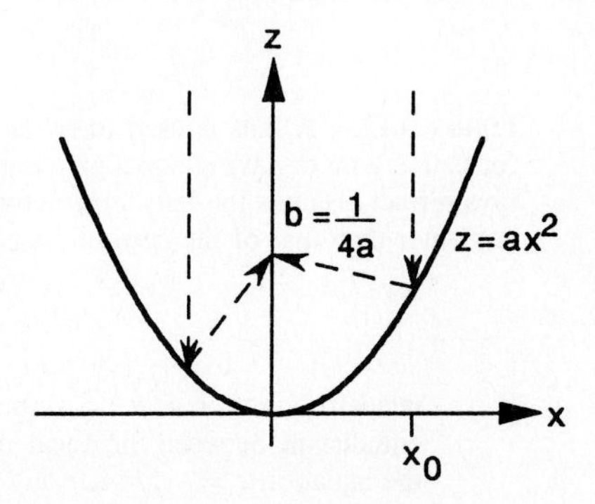

Figure 3.11 A parabolic dish is used to collect and focus waves propagating in the $-z$ direction. Parallel rays strike the parabolic surface and converge on the parabola's focal point located at $z = b$. Because of parabola's special characteristics, these rays travel the same distance to the focal point once they enter the aperture.

Pathlength differences are often measured in terms of wavelength. The ability to extract information degrades if pathlength errors become a significant fraction of a propagating signal's wavelength. Consequently, pathlength tolerances are generally much tighter for optical telescopes ($\lambda \approx 4 \times 10^{-7}$ m) than for radio telescopes ($\lambda \approx 10^{-2}$ m), microphones ($\lambda \approx 33$ cm at 1 kHz), or hydrophone arrays ($\lambda \approx 15$ m at 100 Hz). Exceeding these tolerances can result in aberrations as discussed in §3.1.3 {68}.

Example

Consider a parabolic dish for focusing radiation. Because the shape of the dish is rotationally symmetric, we can simplify the analysis by examining a two-dimensional cross-section. Suppose the dish has the form $z = ax^2$ and radiation is propagating in the $-z$ direction as shown in Fig. 3.11. At a point x_0, the slope of the dish is $m_0 \equiv 2ax_0$ and the slope of the normal at that point is $-1/m_0$. Note the line that represents the reflection of the incoming ray off the dish's surface: It passes through the point (x_0, ax_0^2) and has the general form $z = mx + b$. Because the angle of reflection equals the angle of incidence, basic geometry and some simple trigonometric formulas reveal that the slope of the reflected ray is given by

$$m = \frac{1}{2}\left(m_0 - \frac{1}{m_0}\right) = ax_0 - \frac{1}{4ax_0}$$

Consequently, the equation of the reflected ray is

$$z = \left(ax_0 - \frac{1}{4ax_0}\right)x + \frac{1}{4a}$$

Because the intercept b is equal to $1/4a$ independent of the value of x_0, all rays reflecting off the parabolic dish intersect at the point $(0, 1/4a)$.

Somewhat surprisingly, each ray's pathlength equals the others. We can calculate the pathlengths as a function of reflecting point (indexed by x_0) from an arbitrary height z_0 on the incoming ray to the focal point on the reflected ray. The distance from the point (x_0, z_0) to the parabola along the incident ray is simply $\ell_i = z_0 - ax_0^2$. The distance ℓ_r along the reflected ray from (x_0, ax_0^2) to the focus at $(0, 1/4a)$ can be written as

$$\ell_r = \sqrt{x_0^2 + \left(\frac{1}{4a} - ax_0^2\right)^2}$$

Expanding the expression under the square root and combining terms, we see that the expression simplifies to $\ell_r = \frac{1}{4a} + ax_0^2$. The total distance $\ell_i + \ell_r$ from an arbitrary height z on an incident ray to parabola's focal point is simply $z + 1/4a$, independent of x. This

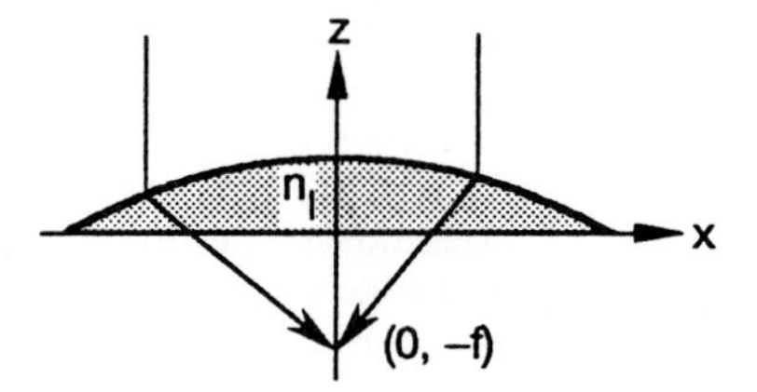

Figure 3.12 A lens is used to refract light and focus it at a point having equal path delays. Light rays refract because the lens's refractive index n_l is greater than that of the surrounding medium.

result should come as no surprise: A parabola has the property that each of its points is equidistant between the focal point and a line called the *directrix*, which in this case has the equation $z = -1/4a$.

Lenses are used to focus light. We can think of a lens's focusing property in several ways. One explanation is that the lens bends the light so that a planar wavefront impinging on the front surface of the lens results in a spherical wave emerging from the back surface. A signal processor would view the lens as a variable delay; because the speed of light in glass is slower than in vacuum (or air), the rays passing through the thickest part of the lens are delayed with respect to those passing through thinner parts. These delays equalize the time it takes each ray to travel from the lens's front surface to the focal point.

Example

Consider a lens as shown in Fig. 3.12. As in the previous example, waves are propagating in the $-z$ direction. In this case, however, the intervening lens delays rays by different amounts, bringing them together at a focal point below the x axis. Assume that the upper surface of the lens is parabolic, and the lower surface is flat. The upper surface is given by the equation $z = -ax^2 + b$. Therefore the lens's thickness is simply $T(x) \equiv b - ax^2$.

The speed of light c_l in the lens material is usually written as the speed of light in vacuum c divided by the material's index of refraction {31} n_l: $c_l \equiv c/n_l$. Therefore, in propagating from the plane $z = b$ through the lens to the plane $z = 0$, the wavefront undergoes a delay Δ_l that depends on x.

$$\Delta_l(x) = \frac{T(x)}{c_l} + \frac{b - T(x)}{c}$$

The first term corresponds to the delay through the lens material and the second term to the delay through the remaining vacuum. Using the definitions of T and c_l, we find that†

$$\Delta_l(x) = \frac{bn_l - a(n_l - 1)x^2}{c}$$

If a lens is constructed properly, the rays emerging from the lens are refracted so that they converge at a single focal point. For this example, assume that the focal point is located at $(0, -f)$. To obtain the total delay, we must add the delay encountered by the wavefront in propagating from the point $(x, 0)$ to the focal point. Using simple geometry, we find that this delay $\Delta_f(x)$ equals $\left(\sqrt{x^2 + f^2}\right)/c$. The total delay from the front plane

†Typically in geometrical optics, the assumption is made that the lens is "thin," meaning that the x coordinate of the point where a ray emerges approximately equals the point where it enters. The slight refraction that results in a displacement of the ray within the lens is ignored [61: 77].

of the lens to the focal point is

$$\Delta \equiv \Delta_l(x) + \Delta_f(x) = \frac{1}{c}\left[bn_l + a(n_l - 1)x^2 + \sqrt{x^2 + f^2}\right]$$

If we assume that x is small compared with f, we find that

$$\Delta \approx \frac{1}{c}\left[bn_l + a(n_l - 1)x^2 + f + \frac{x^2}{2f}\right]$$

If we choose the lens focus f to be $1/[2a(n_l - 1)]$, the dependence of the delay Δ on x disappears and parallel rays entering the lens are focused with equal delays on the focal point $(0, -f)$. Goodman [61: 77–96] describes a somewhat more general derivation of lens properties.

3.2 Spatial Sampling

In contrast to a continuous aperture, an *array* consists of individual sensors that sample the environment spatially. Each sensor could be an aperture or an omnidirectional transducer. As one might expect from digital signal processing theory, sampling introduces some complications that can trip the unwary. This section introduces the notion of spatial sampling, first in one dimension and then more generally.

3.2.1 Periodic Spatial Sampling in One Dimension

When we sample a signal $f(x, t)$ that is a function of one spatial dimension and time, we obtain a sequence of temporal signals $\{y_m(t)\}$ given by $y_m(t) = f(md, t)$, where d is the spatial sampling interval. Under what circumstances can the signal $f(x, t_0)$, thought of as a continuous function of space at a particular time t_0, be reconstructed from the set of samples $\{y_m(t_0)\}$? Let k be the spatial frequency (or equivalently, wavenumber) variable corresponding to the continuous spatial variable x. When $f(x, t_0)$ has no spatial frequency components in the range $|k| \geq k_0$ and the sampling interval d is less than or equal to π/k_0, the Sampling Theorem yields the interpolation formula

$$\boxed{f(x, t_0) = \sum_{m=-\infty}^{\infty} y_m(t_0) \frac{\sin \pi\left(\frac{x}{d} - m\right)}{\pi\left(\frac{x}{d} - m\right)}} \tag{3.7}$$

that gives the value of $f(x, t_0)$ given the values of $y_m(t_0)$.†

This formula may be derived by using the definitions of the Fourier Transforms for continuous-variable signals, such as $s_c(x)$, and for discrete-variable signals, such as $s(m)$.

†At this point, we are not exploiting values of $y_m(t)$ for $t \neq t_0$ that could in principle help determine $f(x, t_0)$.

For continuous-variable signals, the Fourier Transform and inverse transform are given by[†]

$$S_c(k) = \int_{-\infty}^{\infty} s_c(x)\exp\{jkx\}\,dx \qquad s_c(x) = \frac{1}{2\pi}\int_{-\infty}^{\infty} S_c(k)\exp\{-jkx\}\,dk$$

Using $\breve{k}$ to denote the spatial frequency (or wavenumber) variable for discrete-variable signals, we can write the Fourier Transform of $s(m)$ and its inverse as

$$S(\breve{k}) = \sum_{m=-\infty}^{\infty} s(m)\exp\{j\breve{k}m\} \qquad s(m) = \frac{1}{2\pi}\int_{-\pi}^{\pi} S(\breve{k})\exp\{-j\breve{k}m\}\,d\breve{k}$$

Note that the Fourier Transform for discrete-variable signals is periodic with period 2π: $S(\breve{k}) = S(\breve{k}+2\pi l)$ for any integer value of l.

The spatial frequency variable k is measured in radians/meter and equals 2π times spatial frequency measured in cycles/meter. The digital frequency variable $\breve{k}$ usually ranges from 0 to 2π or from $-\pi$ to π and is measured in radians per sampling period. Alternatively, $\breve{k}$ equals 2π times a normalized frequency variable measured in cycles per sampling period. Thus, the Fourier Transform of a discrete-variable signal can be expressed as a function of the frequency variable k: $S(kd) = S(\breve{k})$. The spectrum's period in terms of k equals $2\pi/d$.

Because $s(m) = s_c(md)$, we can write

$$s(m) = \frac{1}{2\pi}\int_{-\infty}^{\infty} S_c(k)\exp\{-jkmd\}\,dk$$

With a change of variables $\breve{k} = kd$, this expression becomes

$$s(m) = \frac{1}{2\pi d}\int_{-\infty}^{\infty} S_c\left(\frac{\breve{k}}{d}\right)\exp\{-j\breve{k}m\}\,d\breve{k}$$

The integral over the infinite domain of the $\breve{k}$ variable may be decomposed into a sequence of integrals over intervals of length 2π. If we let $\mathcal{I}_p$ denote the pth interval $-\pi+2\pi p \le \breve{k} \le \pi+2\pi p$, then with some manipulation our inverse transform expression becomes

$$s(m) = \frac{1}{2\pi d}\sum_{p=-\infty}^{\infty}\int_{\mathcal{I}_p} S_c\left(\frac{\breve{k}-2\pi p}{d}\right)\exp\{-j\breve{k}m\}\,\exp\{j2\pi pm\}\,d\breve{k} \tag{3.8}$$

Because the second exponential factor always equals 1, this equation now has the form of an inverse Fourier Transform for discrete-variable signals. By applying the appropriate

[†]To be consistent with the notation introduced in Chap. 2, the signs in the exponents of the spatial Fourier Transforms are reversed from the conventional signs for temporal Fourier Transforms. Because the forward and inverse transforms are so much alike, the convention of defining the spatial transform "backward" does not affect the resulting interpolation formulas.

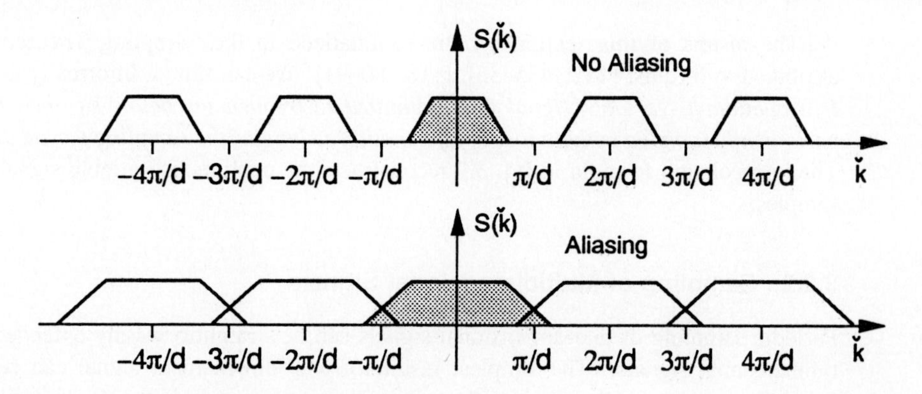

Figure 3.13 The periodic spectrum $S(\check{k})$ is equal to the sum of periodic replications of the spectrum $S_c(k)$. In this case the periodic replications do not overlap because $S_c(k)$ is bandlimited to a frequency $k_0 \leq \pi/d$. When the spectrum $S_c(k)$ is not bandlimited to frequencies below π/d, one period of the periodic spectrum $S(\check{k})$ does not equal $S_c(k)$. This phenomenon is called *aliasing*.

Fourier Transform to both sides, we conclude that

$$\boxed{S(\check{k}) = \frac{1}{d} \sum_{p=-\infty}^{\infty} S_c\left(\frac{\check{k} - 2\pi p}{d}\right)} \tag{3.9a}$$

or alternatively

$$\boxed{S(kd) = \frac{1}{d} \sum_{p=-\infty}^{\infty} S_c\left(k - \frac{2\pi p}{d}\right)} \tag{3.9b}$$

The sampled signal's spectrum $S(\check{k})$ thus equals the sum of periodic replications of the continuous-variable signal's spectrum.

If the continuous-variable signal $s_c(x)$ has no frequency components outside the domain $|k| \leq \pi/d$, then none of the periodic replications in the sum overlaps another—no aliasing occurs—and, as shown in Fig. 3.13

$$S(kd) = \frac{1}{d} S_c(k) \quad \text{for } |k| \leq \pi/d$$

In this case, the continuous-variable signal's spectrum can be easily obtained from the spectrum of the samples. The inverse Fourier Transform for continuous-variable signals can be applied to this latter spectrum to derive the interpolation formula (Eq. 3.7) given earlier.

When $s_c(x)$ is not bandlimited to frequencies below π/d, the periodic replications in the sum overlap; *aliasing* occurs. In this case, one period of the discrete-variable signal's spectrum does not equal the continuous-variable signal's spectrum: Spectral components of $S_c(k)$ outside the domain $|k| \leq \pi/d$ become spectral components of $S(\check{k})$ inside the domain $|\check{k}| \leq \pi$. Fig. 3.13 demonstrates this effect.

The results of this section can be summarized in the Sampling Theorem usually attributed to Nyquist [142: 435–36], [118: 80–91]. We can state it informally as follows: *If a continuous-variable signal is bandlimited to frequencies below k_0, then it can be periodically sampled without loss of information so long as the sampling period $d \leq \pi/k_0$.* The interpolation formula in Eq. 3.7 reconstructs the continuous-variable signal from its samples.

3.2.2 Sampling of Multidimensional Signals

Periodic sampling of one-dimensional signals can be straightforwardly extended to multidimensional signals. The simplest result for a D-dimensional signal can be derived by applying the one-dimensional Sampling Theorem to each of the D dimensions. For example, if a two-dimensional continuous-variable signal $s_c(x, y)$ has no frequency components outside the rectangle defined by $|k_x| \leq k_{x0}$, $|k_y| \leq k_{y0}$, $s_c(x, y)$ may be sampled in x with a period $d_x \leq \pi/k_{x0}$ and in y with a period $d_y \leq \pi/k_{y0}$ without loss of information. Letting $s(m, n) = s_c(md_x, nd_y)$, the two-dimensional continuous-variable signal may be reconstructed from its samples using the interpolation formula

$$s_c(x, y) = \sum_{n=-\infty}^{\infty} \sum_{m=-\infty}^{\infty} s(m, n) \frac{\sin \pi \left(\frac{x}{d_x} - m\right)}{\pi \left(\frac{x}{d_x} - m\right)} \cdot \frac{\sin \pi \left(\frac{y}{d_y} - n\right)}{\pi \left(\frac{y}{d_y} - n\right)} \tag{3.10}$$

This result extends in the obvious way to D-dimensional signals.

The two-dimensional Fourier Transform $S(\breve{k}_x, \breve{k}_y)$ of the discrete-variable signal $s(m, n)$ is a complex-valued function periodic in each of its independent variables with period 2π. When $s(m, n)$ represents the periodic sampling of a continuous-variable signal $s_c(x, y)$, their Fourier Transforms are related by

$$S(\breve{k}_x, \breve{k}_y) = \frac{1}{d_x d_y} \sum_{p=-\infty}^{\infty} \sum_{q=-\infty}^{\infty} S_c\left(\frac{\breve{k}_x - 2\pi p}{d_x}, \frac{\breve{k}_y - 2\pi q}{d_y}\right) \tag{3.11}$$

We can thus interpret the transform $S(\breve{k}_x, \breve{k}_y)$ as the periodic extension (in both the $\breve{k}_x$ and $\breve{k}_y$ directions) of the transform $S_c(k_x, k_y)$. Fig. 3.14 shows this relationship when $s_c(x, y)$ is bandlimited and the sampling intervals d_x and d_y are chosen small enough to avoid aliasing.

3.2.3 Alternative Sampling Grids

For continuous-variable signals defined over higher than one-dimensional spaces, sampling need not be restricted to the usual Cartesian grid. Instead, consider sampling points represented as an integer-weighted linear combination of D linearly independent sampling

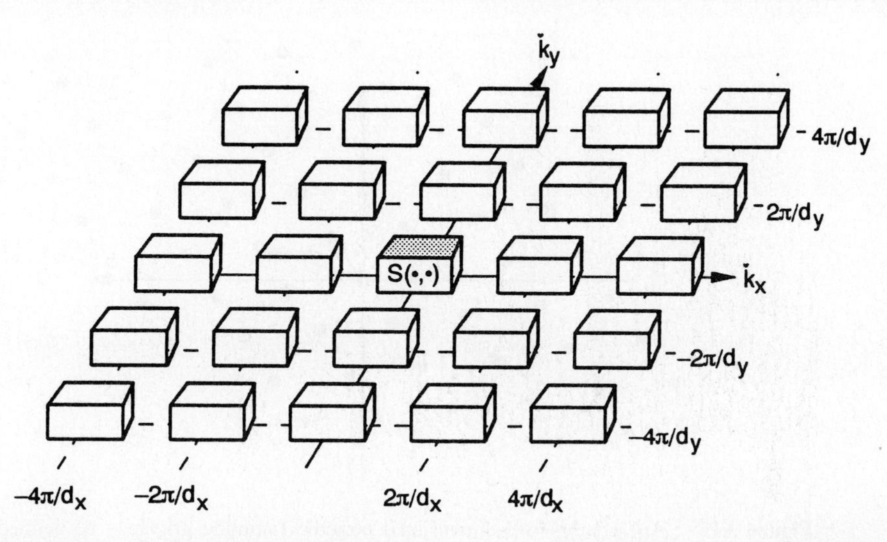

Figure 3.14 $S(\check{k}_x, \check{k}_y)$ can be interpreted as the periodic extension of $S_c(k_x, k_y)$. When $s_c(x, y)$ is bandlimited and the sampling intervals d_x and d_y are chosen small enough (as shown here), $s_c(x, y)$ may be reconstructed from the samples $s(m, n)$.

vectors $\{\mathbf{v}_1, \ldots, \mathbf{v}_D\}$. Each sample point $\mathbf{x}$ in D-dimensional space is thus represented as [19; 39, 41]

$$\mathbf{x} = \sum_{i=1}^{D} m_i \mathbf{v}_i$$

where the choice of the integer coefficients m_i determines the sample's location. The sampling vectors, when taken together, form the columns of the *sampling matrix* $\mathbf{V}$: $\mathbf{V} = [\mathbf{v}_1, \cdots, \mathbf{v}_D]$.

A D-dimensional continuous-variable signal $s_c(\mathbf{x})$ can be sampled using a D-dimensional grid, which is defined by the sampling matrix, to obtain the signal $s(\mathbf{m})$, with $\mathbf{m}$ representing a D-dimensional vector of the integer coefficients (Fig. 3.15).

$$s(\mathbf{m}) = s_c(\mathbf{V}\mathbf{m})$$

The D-dimensional Fourier Transforms of the continuous-variable signal $s_c(\mathbf{x})$ and its samples $s(\mathbf{m})$, $S_c(\mathbf{k})$ and $S(\check{\mathbf{k}})$ respectively, are related by the aliasing formula

$$\boxed{S(\check{\mathbf{k}}) = \frac{1}{|\det \mathbf{V}|} \sum_{\mathbf{p}} S_c\left(\mathbf{U}\left[\check{\mathbf{k}} - 2\pi \mathbf{p}\right]\right)}$$

where $\mathbf{p}$ is an integer vector. The *periodicity matrix* $\mathbf{U}$ is related to the sampling matrix $\mathbf{V}$ by

$$\mathbf{V}^t \mathbf{U} = \mathbf{I} \tag{3.12}$$

The matrix $\mathbf{U}$ is so named because its columns, taken as vectors, indicate the directions in multidimensional frequency space in which $S(\check{\mathbf{k}})$ exhibits its fundamental periodicities.

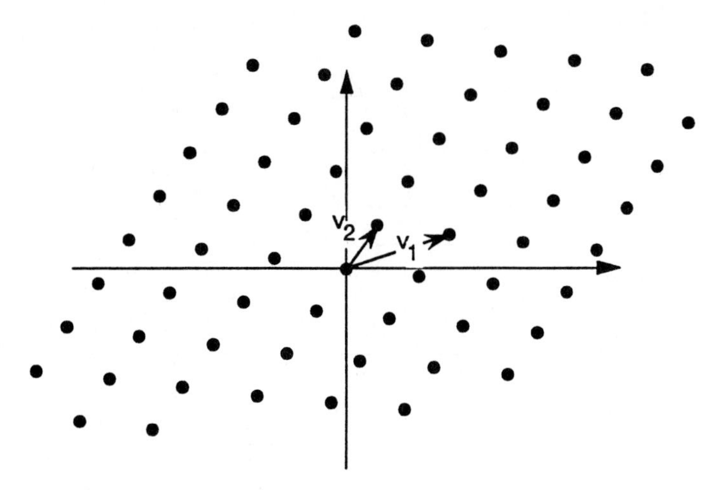

Figure 3.15 An arbitrary two-dimensional periodic sampling grid may be defined in terms of two linearly independent vectors $\mathbf{v}_1$ and $\mathbf{v}_2$.

Example

Consider the case of rectangular periodic sampling in two dimensions. Let the sample spacing in the x direction be d_x and in the y direction be d_y. The sampling vectors are given by

$$\mathbf{v}_1 = \begin{bmatrix} d_x \\ 0 \end{bmatrix}, \quad \mathbf{v}_2 = \begin{bmatrix} 0 \\ d_y \end{bmatrix}$$

yielding the sampling and periodicity matrices

$$\mathbf{V} = \begin{bmatrix} d_x & 0 \\ 0 & d_y \end{bmatrix} \quad \mathbf{U} = \begin{bmatrix} d_x^{-1} & 0 \\ 0 & d_y^{-1} \end{bmatrix}$$

The columns of $\mathbf{U}$, taken as vectors, indicate that spatial frequency aliasing can occur in the k_x direction with a period of d_x^{-1} cycles per meter (or equivalently $2\pi/d_x$ radians per meter) and in the k_y direction with a period of d_y^{-1} cycles per meter. The interpolation formula (Eq. 3.12) becomes

$$S(\breve{k}_x, \breve{k}_y) = \frac{1}{d_x d_y} \sum_{p,q} S_c\left(\frac{\breve{k}_x - 2\pi p}{d_x}, \frac{\breve{k}_y - 2\pi q}{d_y}\right)$$

which is the same as Eq. 3.11 {80}.

The interpolation formula for the D-dimensional case takes the general form

$$\boxed{s_c(\mathbf{x}) = \sum_{\mathbf{m}} s(\mathbf{m}) \Im(\mathbf{x} - \mathbf{Vm})} \tag{3.13}$$

The summation represents a D-dimensional sum over all components of the integer vector $\mathbf{m}$. The interpolation function $\Im(\mathbf{x})$ depends on the shape of the baseband region $\Re$.

$$\Im(\mathbf{x}) = \frac{|\det[\mathbf{V}]|}{(2\pi)^D} \int_{\Re} \exp\{j\mathbf{k}^t\mathbf{x}\}\, d\mathbf{k}$$

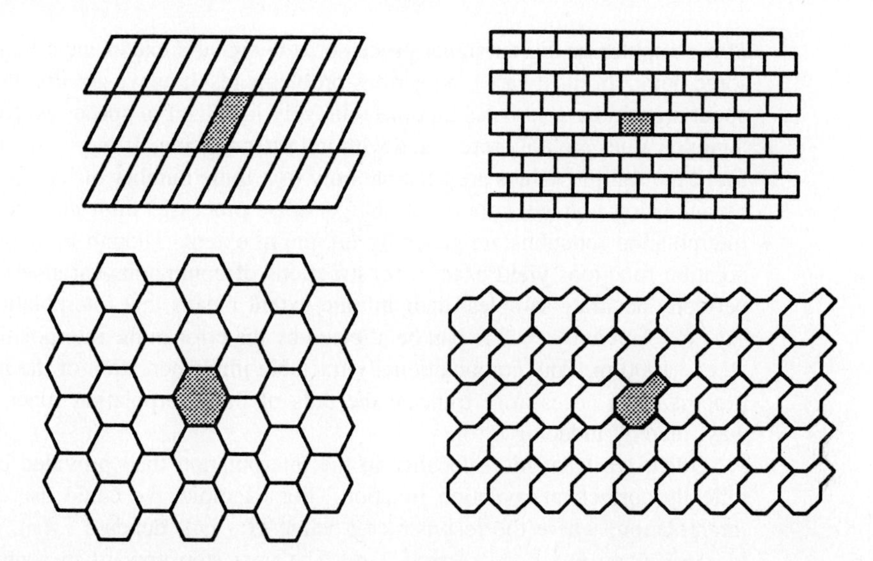

Figure 3.16 Several differently shaped baseband regions $\Re$, when periodically extended in a manner consistent with Cartesian sampling, completely cover the two-dimensional frequency plane.

This D-dimensional integral spans the region $\Re$ in frequency space. This region is defined as that part of the D-dimensional frequency space $\mathbf{k}$ within which the bandlimited signal $s_c(\mathbf{x})$ contains its only nonzero frequency components. For the two-dimensional case with conventional Cartesian sampling, the region $\Re$ is usually taken to be a rectangle, with the interpolation function equaling the product of $\sin x/x$ functions whose arguments correspond to the axes of the rectangle (Eq. 3.10 {80}). Other interesting possibilities can be defined as Fig. 3.16 illustrates. Note that the region $\Re$ when periodically extended need *not* cover all points in the frequency plane. As long as there is no overlap among the replicated versions of $\Re$, no aliasing occurs. This result means that we can use polygonal shapes that more closely delineate a multidimensional signal's spectral support to define carefully tuned sampling strategies.

3.2.4 Realities of Sampling

As the preceding expressions show, we can reconstruct a signal exactly from its samples under the appropriate conditions. Several realities, however, can cause reconstructions to fall short of perfection.

- Rarely is a signal perfectly bandlimited. For example, no finite-extent, continuous-variable signal can be perfectly bandlimited.
- An infinite number of samples is usually required to represent an infinitely long signal.† In practice, only a finite number of samples can be stored in the memory

†Can you think of a bandlimited, multidimensional signal that can be represented by a single (carefully chosen) sample? Such signals do exist.

of a computer or digital signal processor; the signal's extent must be truncated at some point. In most cases, no serious problems result as we are usually concerned about signal characteristics that are relatively localized in space and time.

- Sample values are not represented with infinite precision. If represented in a digital form, the sample values are accurate only to a finite number of bits; if represented by an analog voltage, current, or charge, noise processes limit their accuracy.
- Interpolation functions are generally infinite in extent. Though in theory the interpolation functions yield exact reconstructions of continuous-variable signals from perfectly accurate samples, their infinite extent means that interpolation between, say, $x = 1$ m and $x = 2$ m can be affected by the error in the interpolation function at $x = 1000$ m. Any computationally tractable implementation of the interpolation process must necessarily truncate the tails of the interpolation function in some way, thereby inducing error.

Often, we want more locality in the interpolation than provided by the theoretically correct interpolation function. For example, we could use *zeroth-order* interpolation, where the reconstructed value of $s_c(x)$, denoted $\widehat{s}_c(x)$, is set equal to $s(m)$ when $md \leq x < (m+1)d$. The next step upward in sophistication is *linear* interpolation, where, for $md \leq x < (m+1)d$, straight-line connection of the samples yields interpolated signal values.

$$\widehat{s}_c(x) = s(m)\left(1 - \frac{x - md}{d}\right) + s(m+1)\left(\frac{x - md}{d}\right)$$

Higher-order interpolation schemes, such as cubic spline functions [123: 94–98], can be used to reduce computation, obtain a smooth interpolation, and maintain some degree of locality in the interpolation.

Some thought has been given to other discrete-variable representations for continuous-variable signals [117]. Another approach is *parametric descriptions*: If a signal consists of a sum of decaying exponential functions, for example, the signal can be represented by the amplitudes and decay constants of its exponential components. Estimating the values of these parameters from the signal itself can require much more computation than simply sampling the signal at regular intervals. Periodic sampling also allows continuous-domain operations such as addition, amplification, convolution, and squaring to be straightforwardly mapped into discrete-domain operations. Viable alternatives to sampling have not yet emerged.

3.3 Arrays of Discrete Sensors

In many practical applications, arrays composed of individual sensors sample the wavefield at discrete spatial locations. For example, in exploration seismology, oil companies use geophone arrays to record the low-frequency acoustic waves to probe the Earth's subsurface structure. Similarly, sonar applications employ hydrophone arrays, and acoustic applications use microphone arrays. Phased-array radar antennas consist of many radiating elements that both transmit and receive electromagnetic waves.

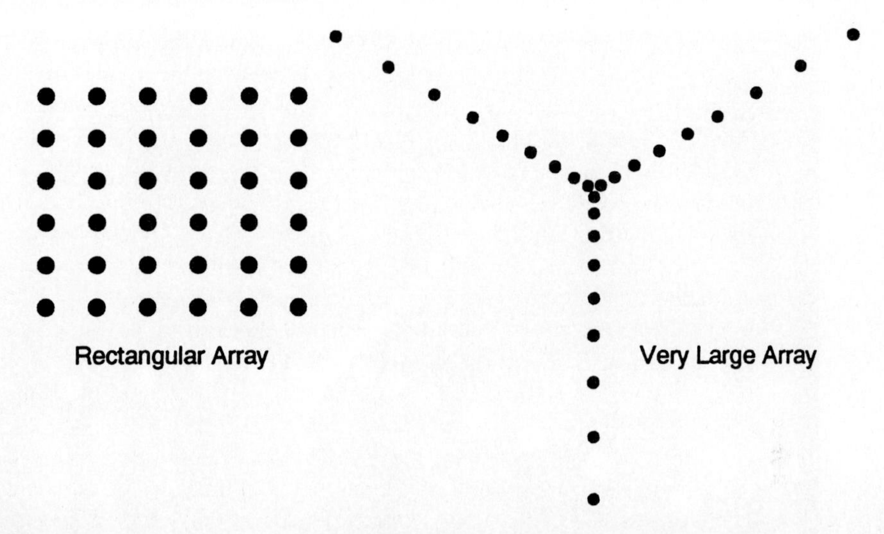

Figure 3.17 Two arrays of discrete sensors are shown. On the left, rectangular array has periodically spaced sensors. The *VLA* shown on the right has power-law spaced sensors along each arm (see Prob. 3.16).

Sensors can be positioned in *regular* or *irregular* patterns (Fig. 3.17). Advantages accrue in both approaches. For example, the Very Large Array (*VLA*) radio telescope, shown in Fig. 3.18, consists of 27 large parabolic reflectors positioned along three arms several miles in length. Fig. 3.19 shows an observation made with the *VLA* of extraterrestrial radio sources.† To begin our analysis of sensor patterns, we assume that the sensors are infinitesimally small, each occupies a single point in space, and they are equally sensitive to signals propagating in all directions (omnidirectional). Directional sensors, such as employed in the *VLA*, are discussed later.

3.3.1 Regular Arrays

One of the most straightforward ways of producing a sampled array aperture—an array—is to place the individual sensors on a regular grid. This approach produces aperture smoothing functions that are relatively simple to analyze and permits the use of fast algorithms, such as the fast Fourier Transform, to compute spatial spectra and array output signals. As with any periodic sampling scheme, care must be taken to avoid aliasing problems.

For simplicity, consider a propagating wavefield $f(x, t)$ with a single spatial dimension x and time t. We want to analyze this wavefield from measurements made by M equally spaced ideal sensors separated by d units along the x dimension. There are two effects with which we must be concerned: The sampling implicit in using arrays of discrete sensors can introduce aliasing, and, because we do not have measurements for all regions of space, the spatial windowing of the wavefield needs attention.

†Clearly, the *VLA*'s sampling grid is irregular. Why was this particular array geometry chosen?

Figure 3.18 The *VLA* radio telescope in New Mexico consists of 27 identical 25-m antennas arranged in a Y-shaped array. Two arms are 21 km long, and the third arm is 19 km long. The individual antennas can be moved along the three arms to obtain a variety of array configurations. *Photo courtesy of NRAO/AUI.*

To illustrate the results of spatial sampling and windowing, begin by defining a set of signals $\{y_m(t)\}$ that corresponds to the wavefield's values sampled every d meters: $y_m(t) = f(md, t)$. The wavefield's wavenumber-frequency representation is given by

$$F(k, \omega) = \int_{-\infty}^{\infty} \int_{-\infty}^{\infty} f(x, t) \exp\{-j(\omega t - kx)\}\, dx\, dt$$

and the sampled wavefield's wavenumber-frequency representation is

$$Y(k, \omega) = \int_{-\infty}^{\infty} \sum_{m=-\infty}^{\infty} y_m(t) \exp\{-j(\omega t - kmd)\}\, dt \tag{3.14}$$

The wavenumber-frequency representation $Y(k, \omega)$ is a little unusual because space has been sampled, but time has not. For our current purposes, we shall continue to think of time as a continuous variable to focus on spatial sampling effects; temporal sampling effects are discussed in §4.7 {152}. The wavenumber variable k is expressed in units of radians per meter,† requiring that the exponent on the right side include the sampling

†In classical digital signal processing texts, the frequency used in the Fourier Transform of a sampled signal (or sequence) is generally normalized to the sampling frequency, thus having units of radians per sampling interval. The authors could have followed that convention here, as in §3.2, but problems would arise when nonperiodic spatial sampling is analyzed.

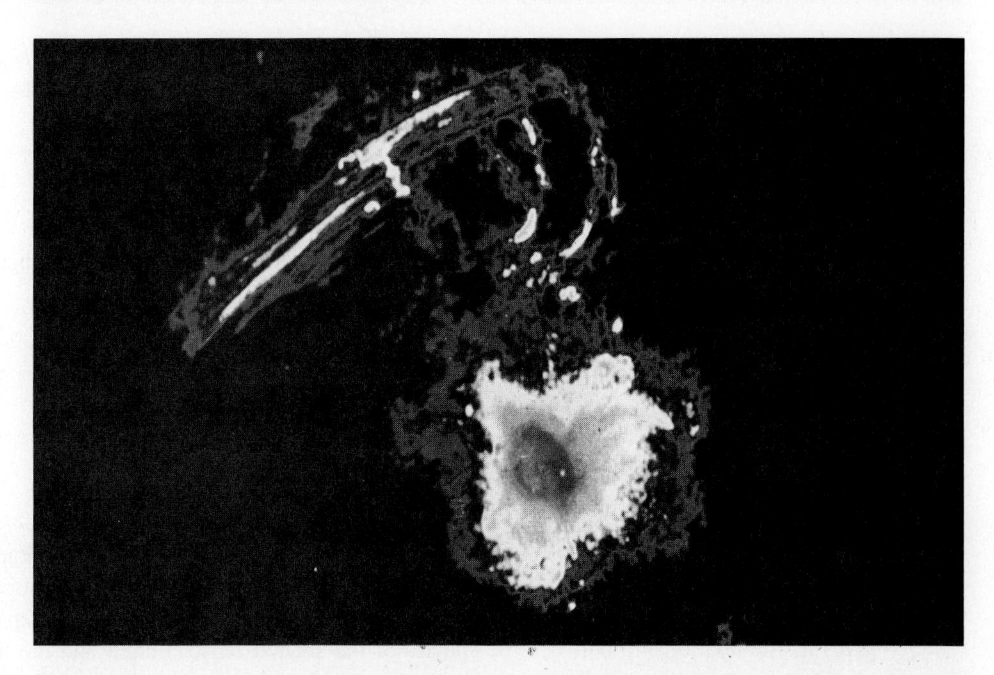

Figure 3.19 Shown are the radio continuum observations of Sagittarius A plus the continuum arc in the direction of the Galactic Center at a wavelength of 20 cm, corresponding to a frequency of 1.44 GHz. The resolution obtained by the *VLA* was 5×8 arcseconds in a field of view of 25×25 arcminutes ($1,500 \times 1,500$ arcseconds). *Photo courtesy of NRAO/AUI; observations made by F. Yusef-Zadeh, M. R. Morris, and D. R. Chance.*

interval d explicitly. We can now apply the results of §3.2 to relate the two wavenumber-frequency representations. By applying the relationship between discrete- and continuous-variable signal spectra Eq. 3.9b {79}, we find that

$$Y(k,\omega) = \frac{1}{d} \sum_{p=-\infty}^{\infty} F\left(k - \frac{2\pi p}{d}, \omega\right) \tag{3.15}$$

If the wavefield is spatially bandlimited, $F(k, t)$ is zero for $|k| \geq \pi/d$, only one term in the sum contributes to the spectrum. Under this condition, no spatial aliasing occurs. If this condition does not hold, energy at higher spatial frequencies appears in $Y(k,\omega)$ at spatial frequencies in the range $-\pi/d \leq k \leq \pi/d$.

Turning to the effects resulting from a finite number of sensors, assume that the number of sensors M is odd and that the middle sensor is located at the origin where $x = 0$. Number the sensors from $m = -M_{1/2}$ to $m = M_{1/2}$ inclusive, where $M_{1/2} = (M-1)/2$; thus, the mth sensor is located at $x = md$. Define another set of signals $\{z_m(t)\}$ that include sensor weights $\{w_m\}$: $z_m(t) = w_m y_m(t)$. This notation conveniently allows us to represent the absence of a sensor at a given location $x = md$ by setting w_m equal to zero for that sample. It also allows us to weight the sensor outputs before further processing; advantages of this weighting are described in Chap. 4.

By defining a wavenumber-frequency representation for $\{z_m(t)\}$ as in Eq. 3.14,

$$Z(k,\omega) = \int_{-\infty}^{\infty} \sum_{m=-M_{1/2}}^{M_{1/2}} z_m(t) \exp\{-j(\omega t - kmd)\}\, dt$$

we find this wavenumber-frequency spectrum to be a circular convolution between $Y(k,\omega)$ and the Fourier Transform of the weighting sequence w_m:

$$Z(k,\omega) = \frac{d}{2\pi} \int_{-\pi/d}^{\pi/d} Y(l,\omega) W(k-l)\, dl$$

where

$$W(k) \equiv \sum_m w_m e^{jkmd} \tag{3.16}$$

Interpreting the set of values $\{w_m\}$ as a *discrete aperture function*, $W(k)$ operates as an aperture smoothing function for the sampled wavefield. Exploiting the relationship between the spatiotemporal spectrum of the array output and the incident field (Eq. 3.15) gives us the relationship between the spectra of the shaded sensor output and the field.

$$\boxed{Z(k,\omega) = \frac{1}{2\pi} \int_{-\pi/d}^{\pi/d} \left[\sum_{p=-\infty}^{\infty} F\left(l - \frac{2\pi p}{d}\right) \right] W(k-l)\, dl}$$

This result is similar to that described in Eq. 3.1 {60}: The wavenumber-frequency spectrum $Z(k,\omega)$ is a smoothed version of the wavefield's spectrum $F(k,\omega)$ with $W(k)$ acting as the smoothing kernel. The sum over shifted replicas of $F(k,\omega)$ describes aliasing effects and becomes important only if the spatial sampling interval d is too large for the field's spatial bandwidth.

Example

Let's compute the aperture smoothing function for the case of uniform sensor weights. Formally,

$$w_m = \begin{cases} 1, & |m| \le M_{1/2} \\ 0, & |m| > M_{1/2} \end{cases}$$

The aperture smoothing function computed according to Eq. 3.16 is

$$W(k) = \frac{\sin \frac{kMd}{2}}{\sin \frac{kd}{2}}$$

This result shows that, unlike the case of a continuous linear aperture, the aperture smoothing function $W(k)$ for the linear array is a periodic function of k, the period equaling $2\pi/d$. Some basic characteristics of this aperture smoothing function can be derived. As shown in Fig. 3.20, each period of $W(k)$ consists of a *mainlobe*, a centrally located peak having the largest amplitude, and a number of *sidelobes*, smaller amplitude peaks not located at zero wavenumber. The height of the mainlobe is given by $W(0)$, which, in this uniform-weight

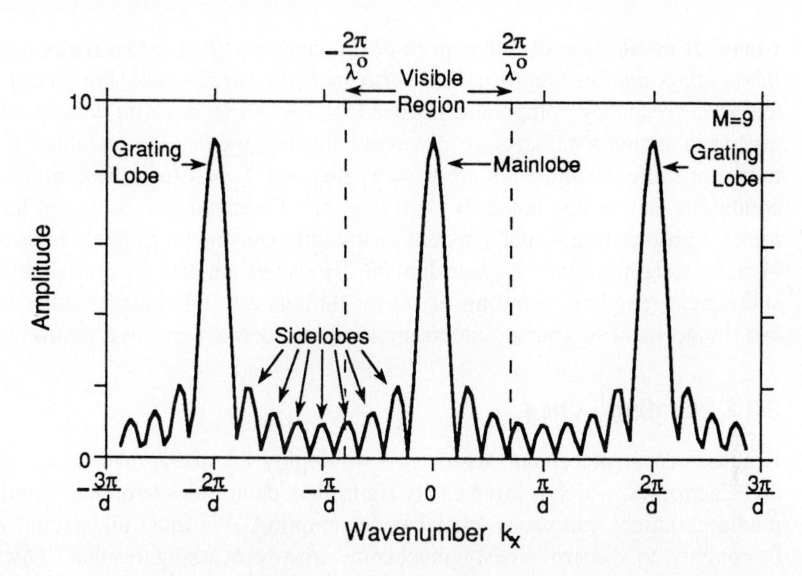

Figure 3.20 The aperture smoothing function magnitude $|W(k)|$ for uniform shading is plotted for a nine-sensor regular linear array. This spatial spectrum has period $k = 2\pi/d$. The visible region of the aperture smoothing function is that part for which $-2\pi/\lambda^o \leq k_x^o \leq 2\pi/\lambda^o$. What might be called secondary mainlobes—those not located at the origin—are termed grating lobes.

case, equals the number of sensors in the array M.[†] The first zero of $W(k)$ occurs when the argument of the numerator's sine function equals π: $k = 2\pi/Md$. Thus, the mainlobe width is $4\pi/Md$,[‡] meaning that the mainlobe width decreases as the number of sensors increases (holding constant intersensor spacing) or as the intersensor spacing increases (for the same number of sensors). Either action serves to increase the array's spatial extent—its *aperture*[¶]—and to improve its resolution.

In the case of two- and three-dimensional arrays of periodically spaced sensors, we can usually derive an expression that shows that $Z(\vec{k}, \omega)$ is a smoothed version of $Y(\vec{k}, \omega)$ when no spatial aliasing occurs. If the corresponding discrete aperture function w_{m_x,m_y,m_z} can be written as a separable function of m_x, m_y, and m_z, $w_{m_x,m_y,m_z} = w_{m_x}^{(x)}\, w_{m_y}^{(y)}\, w_{m_z}^{(z)}$, the aperture smoothing function is also separable: $W(\vec{k}) = W^{(x)}(k_x)W^{(y)}(k_y)W^{(z)}(k_z)$. Multidimensional aperture smoothing functions of this type are particularly straightforward to analyze.

For propagating signals, wavenumber and frequency are coupled by the dispersion relation (see §2.3.1 {21}). In the nondispersive case,

$$|\vec{k}|^2 = k_x^2 + k_y^2 + k_z^2 = \left(\frac{\omega}{c}\right)^2$$

[†] What is the mainlobe's amplitude when nonuniform weights are applied?

[‡] The *FWHM* approximately equals $7.56/Md$ or $2.4\pi/Md$.

[¶] The use of the term aperture to describe an array's spatial extent is confusing but widespread. This usage is meant to convey an analogy between finite continuous apertures and arrays.

where, as usual, c denotes the speed of propagation. *If $f(\vec{x}, t)$ is temporally bandlimited, this relation implies that the field is also spatially bandlimited.* For example, if f has no temporal frequency component greater than ω_0, it can have no wavenumber component greater in magnitude than ω_0/c. To avoid aliasing, the maximum values of the spatial and temporal sampling intervals are $d = \pi c/\omega_0$ and $T = \pi/\omega_0$; these minimal antialiasing conditions can be combined to yield $d = cT$. Consequently, temporal bandwidth determines a propagating signal's spatial bandwidth, causing an intimate relationship between Nyquist sampling rates in each domain. However, an array's spatial sampling rate and subsequent temporal sampling of sensor outputs are individually under our control; we can, by accident of course, undersample in one domain and oversample in the other.

3.3.2 Grating Lobes

Because sensor placement, fixed once we deploy the array, determines spatial sampling and electronics, which is more easily controlled, determines temporal sampling, signal and media variations can cause spatial undersampling. We must understand spatial aliasing thoroughly to discern grossly inaccurate array processing results. One manifestation of inadequate spatial sampling is the appearance of spurious mainlobes in the aperture smoothing function (Fig. 3.20). These lobes are often called *grating lobes* by their analogy to diffraction orders produced by an optical diffraction grating. Signals propagating from directions corresponding to spatial frequencies at which grating lobes occur would be indistinguishable from signals propagating from the mainlobe direction. We would become horribly confused as the following example shows.

Example

Consider the case of a linear array in three-dimensional space on which a monochromatic plane wave is impinging at incidence angle ϕ^o (Fig. 3.3 {63}). The array's response is $W(k_x^o)$, where $k_x^o = -(2\pi/\lambda^o)\sin\phi^o$. Because $W(k_x^o)$ is periodic with period $2\pi/d$, spurious mainlobes—grating lobes—occur at (nonzero) integer multiples of this period. The first grating lobe appears at $k_x^o = \pm 2\pi/d$. The curious would ask the question: For a given wavelength λ^o, what incidence angle ϕ_g^o corresponds to the first grating lobe? Simple calculations show that $\sin\phi_g^o = \lambda^o/d$. Since the sine of a real angle must lie between -1 and $+1$, this equation has a physically realizable solution only if $\lambda^o \le d$. Borderline sampling means $\lambda^o = d$ or $\phi_g^o = \pi/2$: A wave propagating in the $-x$ direction (from the right) yields an array response identical to that of a wave propagating in the $-z$ direction (from the top). Wouldn't this be confusing? In the latter case, each sensor measures the same value at the same time because the plane wave arrives at each sensor simultaneously. In the former case, each sensor again measures the same value, but now from the same point in a different wave cycle (Fig. 3.21).

As another simple example, take the case where $d = 4\lambda^o$. Because four wavelengths occur between sensor locations, we should expect a grossly undersampled wave, leading to considerable spatial aliasing. Grating lobes occur for $\phi^o = 0.08\pi$, $\pi/6$, 0.27π, and $\pi/2$, or equivalently for incidence angles equal to 14.5°, 30°, 48.6°, and 90°, thereby confirming our predicted confusion.

For values of λ^o between d and $2d$, spatial aliasing remains but without grating lobes (Fig. 3.20). Here, sidelobes alias into the mainlobe, creating multiple peaks and valleys that confuse the poor signal processor trying to determine the direction of propagation.

Because $k_x^o = -2\pi\sin\phi^o/\lambda^o$ and $|\sin\phi^o| \le 1$, k_x^o takes on real values only between

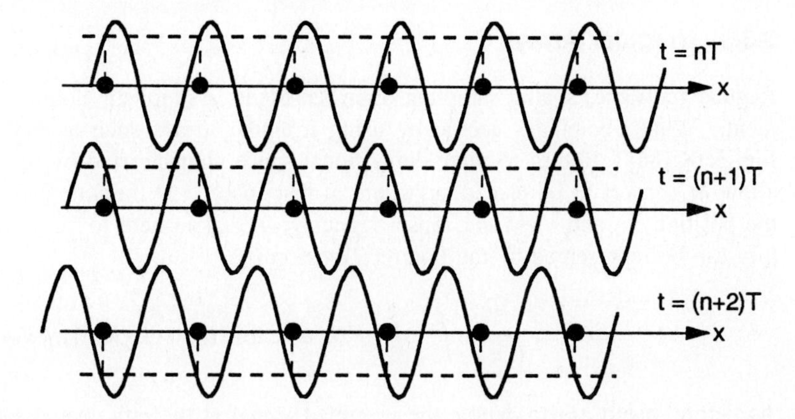

Figure 3.21 When the wavelength λ^o equals the intersensor spacing d, the response of the array to a single-frequency plane wave propagating in the $-x$ direction (endfire) equals the response to a wave propagating in the $-z$ direction (broadside). Here, the sinusoidal waveform depicts the propagating wave at three successive time instants. Note that the dashed lines also correspond to values of a sinusoid, but one propagating from broadside. This endfire-broadside ambiguity demonstrates aliasing in the time domain.

$\pm 2\pi/\lambda^o$. This region is sometimes called the *visible region*: the range of real angles of incidence (between $\pm 90°$) for a given wavelength λ^o (Fig. 3.20). If $\lambda^o \leq 2d$, the visible region occupies more than one period of the aperture smoothing function $W(k_x^o)$, indicating that spatial aliasing occurs. A wave with a spatial frequency k_x^o above π/d, say $\pi/d + \kappa$, is indistinguishable from one with a spatial frequency of $\pi/d - \kappa$. In analogy to temporal sampling, the spatial frequency $k_x^o = \pi/d$ is often called the *folding frequency*. The notions of grating lobes and their relationship to spatial aliasing carry over in a straightforward manner to multidimensional arrays.

Just as interesting as aliasing produced by spatial undersampling are the consequences of oversampling. For our example uniform linear array, oversampling occurs when $d \leq \lambda^o/2$. The full range of incidence angles are covered, but spatial frequencies (in magnitude) higher than $2\pi/\lambda^o$ do not correspond to a propagating signal; these spatial frequencies define the *invisible region* where no propagating signal energy can be found but where a spectrum can be calculated. The combination of signal characteristics (wavelength) and array properties (intersensor spacing) define the invisible region. Spectral energy can occur in the invisible region because of nonpropagating noise, amplifier noise in sensor electronics, for example. As always in the case of temporal oversampling, the wise signal processor disregards that portion of the spectrum to concentrate on propagating signals. Be aware, however, that some algorithms are sensitive to energy in the invisible region, particularly those that explicitly assume no energy *can* be there (Prob. 4.28 {198}).

3.3.3 Irregular Arrays

Regular (periodic) spatial sampling often represents a simplistic notion of sensor placement. What advantages accrue by using irregular arrays such as the *VLA* shown in Fig. 3.18 {86}? Return to three-dimensional space characterized by the variable $\vec{x}$ and allow M sensors to be placed anywhere in that space. Let the mth sensor be located at the position $\vec{x}_m$ and yield the sensor signal $y_m(t)$, taken here to be $f(\vec{x}_m, t)$, which in turn can be represented by the Fourier Transform

$$y_m(t) = \frac{1}{(2\pi)^4} \int_{-\infty}^{\infty} \int_{-\infty}^{\infty} F(\vec{k}, \omega) \exp\{j(\omega t - \vec{k} \cdot \vec{x}_m)\} \, d\vec{k} \, d\omega$$

As before, use $z_m(t)$ to denote the weighted signal at the mth spatial sample: $z_m(t) = w_m y_m(t)$. The wavenumber-frequency spectrum $Z(\vec{k}, \omega)$ of the array's output is given by

$$\boxed{Z(\vec{k}, \omega) = \frac{1}{(2\pi)^3} \int_{-\infty}^{\infty} F\left(\vec{l}, \omega\right) W(\vec{k} - \vec{l}) \, d\vec{l}} \tag{3.17}$$

As before, $W(\vec{k})$ is the aperture smoothing function.

$$\boxed{W(\vec{k}) = \sum_{m=0}^{M-1} w_m e^{j\vec{k} \cdot \vec{x}_m}} \tag{3.18}$$

We can interpret the aperture smoothing function $W(\vec{k})$ as the spatial Fourier Transform of an aperture function $w(\vec{x})$ consisting of a set of M impulse functions located at the sensor positions.

$$w(\vec{x}) = \sum_{m=0}^{M-1} w_m \delta(\vec{x} - \vec{x}_m)$$

Nonperiodic sensor placement, the locations of which are the spatial sampling points, makes analysis of spatial aliasing difficult. Depending on the choice of sensor locations, the aperture smoothing function $W(\vec{k})$ may or may not exhibit periodicity. From its definition, the aperture smoothing function is periodic, $W(\vec{k}) = W(\vec{k} + \vec{k}_o)$, if $\vec{k}_o \cdot \vec{x}_m = 2\pi n_m$, n_m a positive integer, for *each* value of m. Let's assume that sensor 0 is located at the origin, $\vec{x}_0 = (0, 0, 0)$, and that we place three other sensors so that the array does not form a plane. We can represent the preceding periodicity condition by the matrix equation $\mathbf{V}^t \vec{k}_o = 2\pi \mathbf{n}$, where $\mathbf{V} = [\vec{x}_1, \vec{x}_2, \vec{x}_3]$ and $\mathbf{n} = \text{col}[n_1, n_2, n_3]$.† Because $\vec{x}_1$, $\vec{x}_2$, and $\vec{x}_3$ are linearly independent, the 3×3 matrix $\mathbf{V}^t$ is invertible, and we can solve this equation for $\vec{k}_o$.

$$\vec{k}_o = 2\pi (\mathbf{V}^t)^{-1} \mathbf{n}$$

†The authors have chosen to use the symbol **V** because of its analogy to the sampling matrix {81}.

Example

We can examine the periodicities of a four-sensor array's $W(\vec{k})$ using this relationship. The sensors are located at $\vec{x}_0 = (0, 0, 0)$, $\vec{x}_1 = (1, 0, 0)$, $\vec{x}_2 = (1, 2, 0)$, and $\vec{x}_3 = (0, 1, 1/2)$. Consequently,

$$\mathbf{V}^t = \begin{bmatrix} 1 & 0 & 0 \\ 1 & 2 & 0 \\ 0 & 1 & \frac{1}{2} \end{bmatrix} \implies (\mathbf{V}^t)^{-1} = \begin{bmatrix} 1 & 0 & 0 \\ -\frac{1}{2} & \frac{1}{2} & 0 \\ 1 & -1 & 2 \end{bmatrix}$$

Thus, the aperture smoothing function is periodic in the directions

$$\vec{k}_o = 2\pi(\mathbf{V}^t)^{-1}\mathbf{n} = \begin{bmatrix} 2\pi n_1 \\ \pi(n_2 - n_1) \\ \pi(n_1 - n_2 + 2n_3) \end{bmatrix}$$

The three columns of $(\mathbf{V}^t)^{-1}$, or equivalently the three rows of $\mathbf{V}^{-1}$, can be interpreted as *periodicity vectors* that indicate the three directions (and periods) in which $W(\vec{k})$ is periodic. In this way, $(\mathbf{V}^t)^{-1}$ is analogous to the periodicity matrix $\mathbf{U}$ defined on page 81.

When we have more than four sensors, the position vectors can no longer be linearly independent. Assuming $\vec{x}_1$, $\vec{x}_2$, and $\vec{x}_3$ are linearly independent, we can write the remaining sensors' locations as linear combinations of these three positions:

$$\vec{x}_m = a_m^{(1)}\vec{x}_1 + a_m^{(2)}\vec{x}_2 + a_m^{(1)}\vec{x}_3$$

The dot product between sensor location and the presumed wavenumber period becomes

$$\vec{k}_o \cdot \vec{x}_m = 2\pi a_m^{(1)} n_1 + 2\pi a_m^{(2)} n_2 + 2\pi a_m^{(3)} n_3$$

If all the coefficients $\{a_m^{(i)}\}$ are integers, $W(\vec{k})$ exhibits the same periodicities as those induced by the positions of the first three sensors. This very special case may not occur often. More generally, if the coefficients $\{a_m^{(i)}\}$ are rational numbers, we can find their least common denominator D. Because we only require n_m to be an integer, we can reformulate the problem with $n'_m = n_m D$. Employing these integers results in a set of integer-valued coefficients denoted $a_m^{\prime(i)}$, and we have satisfied the requirement that $\vec{k}_o \cdot \vec{x}_m$ be an integer multiple of 2π for all m. The aperture smoothing function periodicities may be examined from the relationship

$$\vec{k}_o = 2\pi(\mathbf{V}^t)^{-1}\mathbf{N}' = 2\pi(\mathbf{V}^t)^{-1}\mathbf{N}D$$

This result means that we can construct a periodic spatial sampling grid in three dimensions from the columns of $\mathbf{V}$, but not every grid point need be occupied by a sensor. An array of this type is often called a *sparse* array. In contrast, a *filled* array has a sensor located at every grid point within its aperture. Conversely, if the coefficients $a_m^{(i)}$ are not rational, *no* periodicities may be evident in the aperture smoothing function. In such cases, grating lobe aliasing *in the strict sense* does not occur. As developed in the next section, aliasing effectively occurs when the aperture smoothing function's peaks have amplitudes nearly equal to that of the mainlobe.

3.3.4 Correlation Sampling and the Co-Array

As we discussed in §3.1.4 {72}, we often need to characterize a random wavefield. Suppose, as before, that we have a random wavefield $f(\vec{x}, t)$, but that it is propagating across an array of M sensors. We would like to use the signals captured by the sensors to estimate the wavefield's correlation function $R_f(\vec{\chi}, \tau)$. Allow shading on the signal obtained from the mth sensor: $y_m(t) = w_m f(\vec{x}_m, t)$. The collection of sensor weights $\{w_m\}$ coupled with the sensor locations $\{\vec{x}_m\}$ can be interpreted as an impulsive aperture function {92}. Define a spatiotemporal correlation matrix $\mathbf{R}_y(\tau)$ whose (m_1, m_2)th entry is $\mathcal{E}\{y_{m_1}(t) y^*_{m_2}(t + \tau)\}$. Assuming a spatially and temporally stationary wavefield, this entry equals

$$R^y_{m_1 m_2}(\tau) = w_{m_1} w^*_{m_2} R_f(\vec{x}_{m_2} - \vec{x}_{m_1}, \tau)$$

Note that this entry depends on the separation $\vec{x}_{m_2} - \vec{x}_{m_1}$ between the two sensor locations. These separations define where, in essence, the array spatially samples the wavefield's correlation function.

In the case of a continuous aperture, we averaged the correlation function over all baselines of the same length and direction present in the aperture. We can attempt to do the same for a discrete array, though we may not find many baselines for any particular value of the baseline $\vec{x}_{m_2} - \vec{x}_{m_1}$. The sum of $w(\vec{x}_{m_1}) w^*(\vec{x}_{m_2})$ over indices of equal baselines yields the *discrete co-array function* [73].

$$c(\vec{\chi}) = \sum_{(m_1, m_1) \in \vartheta(\vec{\chi})} w_{m_1} w^*_{m_2}$$

$\vartheta(\vec{\chi})$ denotes the set of indices (m_1, m_2) for which $\vec{x}_{m_2} - \vec{x}_{m_1} = \vec{\chi}$. Co-array values can vary between zero (no baselines of a particular length) and M, where M is the number of sensors in the array.† The co-array value for zero lag is always M: $c(\vec{0}) = M$. The discrete co-array function also equals the inverse Fourier Transform of $|W(\vec{k})|^2$, the squared magnitude of the aperture smoothing function. The expected value of the spatiotemporal correlation function estimate formed by summing $y_m(t) y^*_n(t + \tau)$ over baselines can be expressed in terms of the co-array and the wavefield's correlation function.

$$\boxed{\mathcal{E}\left[\sum_{(m_1, m_2) \in \vartheta(\vec{\chi})} y_{m_1}(t) y^*_{m_2}(t + \tau)\right] = c(\vec{\chi}) R_f(\vec{\chi}, \tau)}$$

This rather cumbersome notation becomes simpler when arrays are laid out on regular grids; see Prob. 3.18.

The co-array assumes values *only* at the discrete set of locations given by pairwise differences of the sensor locations. The correlation function spatial sampling pattern it represents may be regular or irregular. When regular, estimating the correlation function by baseline averaging corresponds to sampling the correlation function $R_f(\vec{\chi}, \tau)$ regularly and multiplying the result by a window corresponding to the co-array's values.

†We are tacitly assuming that there is at most one sensor located at any given point in the array aperture.

Because the Fourier Transform of the correlation function is the spatial power spectrum, sample spacing in the co-array domain must be small enough to avoid aliasing in the spatial power spectrum domain. For regular arrays, the co-array sample spacing, which equals the minimum separation between two sensors in the array, must be at most half the wavelength corresponding to the wavefield's highest spatial frequency (shortest wavelength) component. The subsequent windowing owing to the co-array's finite extent implies smoothing of the spatial power spectrum analogous to the smoothing of the wavenumber-frequency spectrum $F(\vec{k}, \omega)$ by the aperture smoothing function $W(\vec{k})$ for continuous apertures.

Example

When we have a regular linear array lying along the x-axis with uniform shading, the co-array is the discrete triangular function

$$c(ld\vec{i}_x) = \begin{cases} M - |l|, & 0 \leq |l| < M \\ 0, & \text{otherwise} \end{cases}$$

As in the continuous case, the co-array is indicative of the number of distinct baselines of a given length present in the aperture. For example, if there are four sensors spaced d meters apart, then $c(\vec{0}) = 4$ and $c(d\vec{i}_x) = 3$, corresponding to baselines of length d (sensor 0 to sensor 1, sensor 1 to sensor 2, and sensor 2 to sensor 3). We have two distinct baselines of length $2d$ (sensor 0 to sensor 2 and sensor 1 to sensor 3) and one distinct baseline of length $3d$ (sensor 0 to sensor 3). The smoothing function applied to the spatial power spectrum equals the squared magnitude of the aperture smoothing function, which for the uniform linear array equals

$$|W(k)|^2 = \left(\frac{\sin \frac{kMd}{2}}{\sin \frac{kd}{2}} \right)^2$$

If the number of distinct baselines of a given length, or spatial lag, is greater than one, that lag is said to be *redundant* and the array is said to be a *redundant array*. For any M-sensor array, the maximum number of distinct lags is $M(M-1)/2$. Thus, arrays that produce redundancies in their co-arrays sacrifice having values at other lags. Redundancy yields nonconstant co-array values that multiply the wavefield's correlation function we are trying to estimate. Although it might seem simpler for all co-array values (save at the origin) to equal 1—to have a *nonredundant array*—a lag's redundancy also indicates how many baselines had that lag and contributed to the spatial averaging. We would like constant-valued co-arrays having large values, but these are rare.

The co-arrays of arrays having similar geometries can vary widely. Consider the case of the planar array shown in Fig. 3.22. Known as a Haubrich array [67], this array has no redundant baselines, which means that all co-array values equal 1 except at the origin. Co-arrays for rectangular and circular array geometries are shown in Fig. 3.23. Despite the regularity of all these arrays, some co-arrays are irregular, indicating irregular sampling of the spatial correlation function. Most striking are the differences between the co-arrays shown in Fig. 3.23 for the two circular arrays. The eight-sensor array can be considered as two four-element square arrays, one rotated with respect to the other by 45°; the nine-element array has no such underlying regular grid, and its rather complex

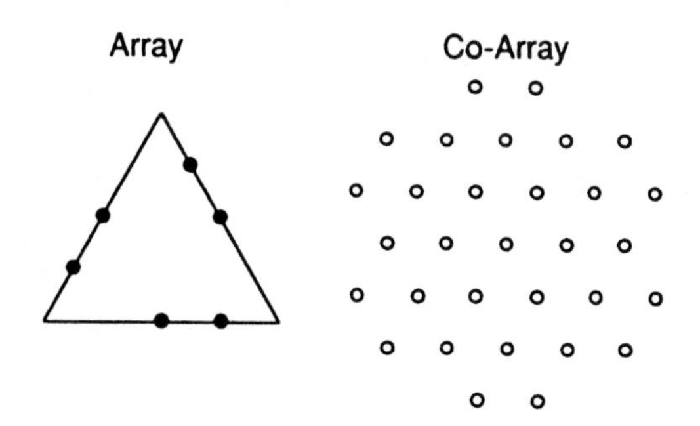

Figure 3.22 The Haubrich array shown on the left has the co-array on the right. Because there are no redundant baselines in the array, the co-array values are all equal to one except at the origin (zero lag), where the co-array value is M.

co-array mirrors that irregularity: No redundant lags exist save for the origin. Three-dimensional arrays have even greater complexity. Fig. 3.24 shows the co-arrays of a cubic array and a spherical array.

These multidimensional arrays have rather beautiful co-arrays, but predicting their spatial aliasing characteristics yields some surprises. The power-density spectrum of the array's output exhibits periodicities if the squared magnitude of the co-array's Fourier Transform, given by

$$|W(\vec{k})|^2 = \sum_{\vec{\chi}} c(\vec{\chi}) e^{j\vec{k}\cdot\vec{\chi}}$$

has a periodic structure. This periodic structure is induced by regularities, an equally spaced sampling grid, in the co-array. Linear arrays that have an underlying regular grid (sensors need not be present at all grid points) yield regular co-arrays. Irregular linear arrays always generate irregular co-arrays. Somewhat surprisingly, we have seen that regular† two- and three-dimensional arrays can have irregular co-arrays (an example is the nine-sensor circular array of Fig. 3.23), which means that the co-array's Fourier Transform does not have periodicities. This situation differs greatly from the more familiar one-dimensional case. In the case of regular linear arrays and *some* regular multidimensional arrays, spectral peaks emerge that equal mainlobe values; these peaks correspond to grating lobes. The separation between these peaks, which determines the array's spatial aliasing characteristics, is governed by the smallest separation between *co-array samples*, not the separation between array samples (Fig. 3.25). For array geometries yielding irregular co-arrays, peaks in $|W(\vec{k})|^2$ still emerge at the expected wavelengths but do not have amplitudes equal to mainlobe values. Thus, despite such transforms not being periodic in a rigid sense, grating-like lobes do emerge and restrict the spatial frequencies an array can uniquely determine. Although we may not encounter spatial aliasing in absolute terms, the wise signal processor should be wary of overextending the array's capabilities. By comparing co-array Fourier Transforms of eight- and nine-sensor circular arrays, grating like lobes occur at much greater wavenumbers in the latter case. This effect

†Note that "regular arrays" have regular sensor locations, not regular co-arrays.

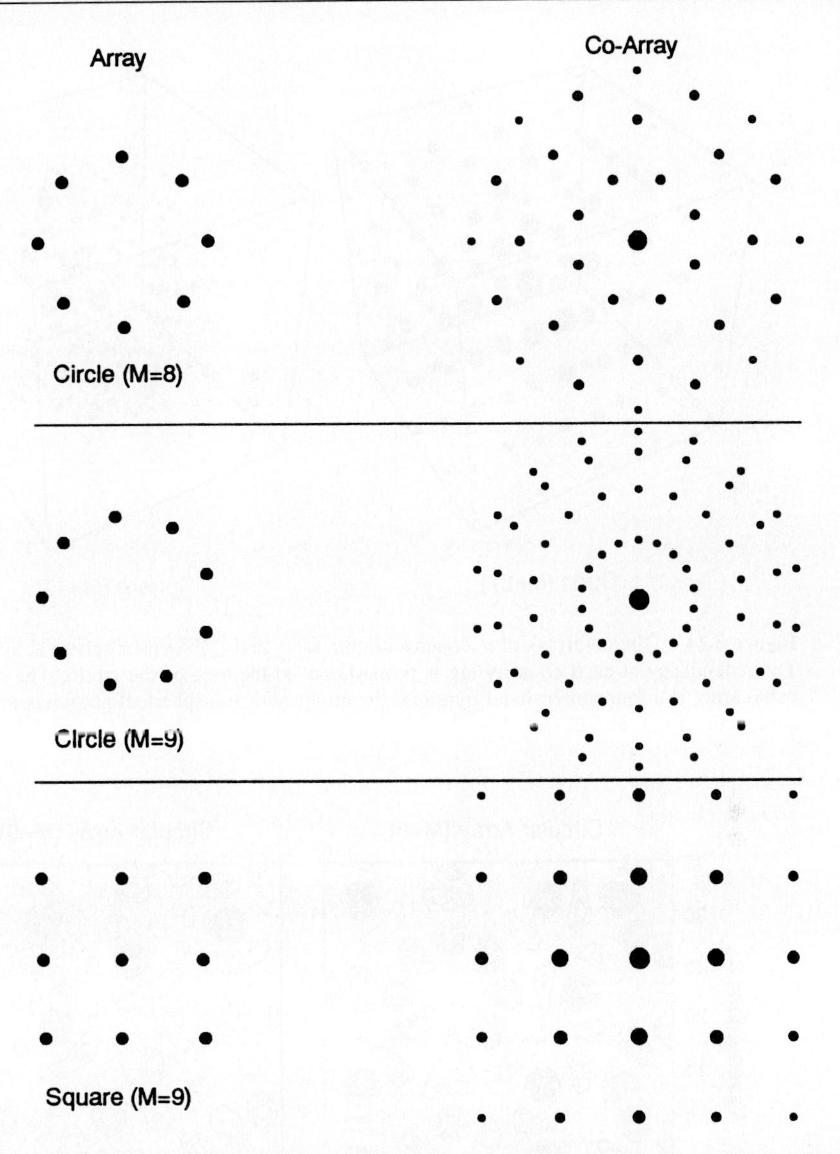

Figure 3.23 The sensor locations for two circular arrays and a square array are shown in the left column. The first of the circular arrays contains eight sensors; the square array and the remaining circular array each contain nine sensors. Their corresponding co-arrays are shown in the right column. The area of the circles denoting co-array locations is proportional to the redundancy at that lag. The redundancy at the origin of a co-array always equals M. Note how these regular array geometries lead to co-arrays spanning complicated spatial regions.

can easily be explained by the co-array structure (see Fig. 3.23): The minimum spacing between co-array samples is much smaller in the nine-sensor array. This result means that

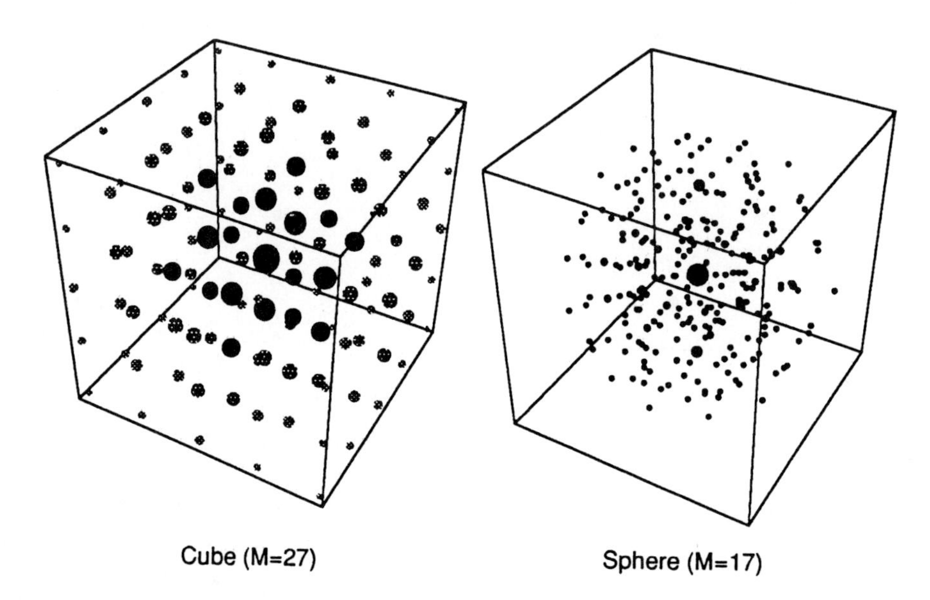

Figure 3.24 The co-arrays of a 27-sensor cubic array and a 17-sensor spherical array are shown. The redundancy at each co-array lag is proportional to the area of the circle. The co-array of the cubic array is a four-dimensional pyramid; the co-array of the spherical array has no redundancies save at the origin and at four other locations.

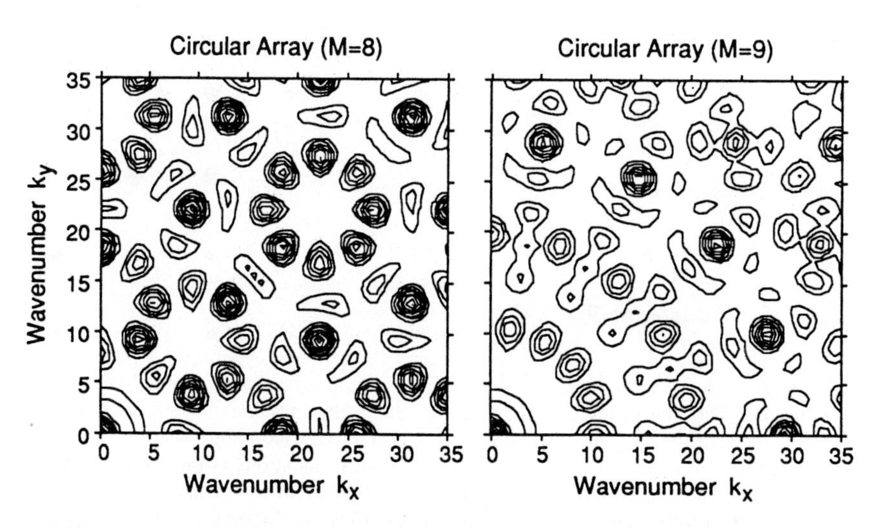

Figure 3.25 The panels depict the Fourier Transforms of the co-arrays for the circular arrays depicted in Fig. 3.23. The computations for the eight-sensor array are shown on the left, the nine-sensor on the right. The spectra are plotted only over the first quadrant of wavenumber space. Peak height can be judged by the number of contours encircling the peak.

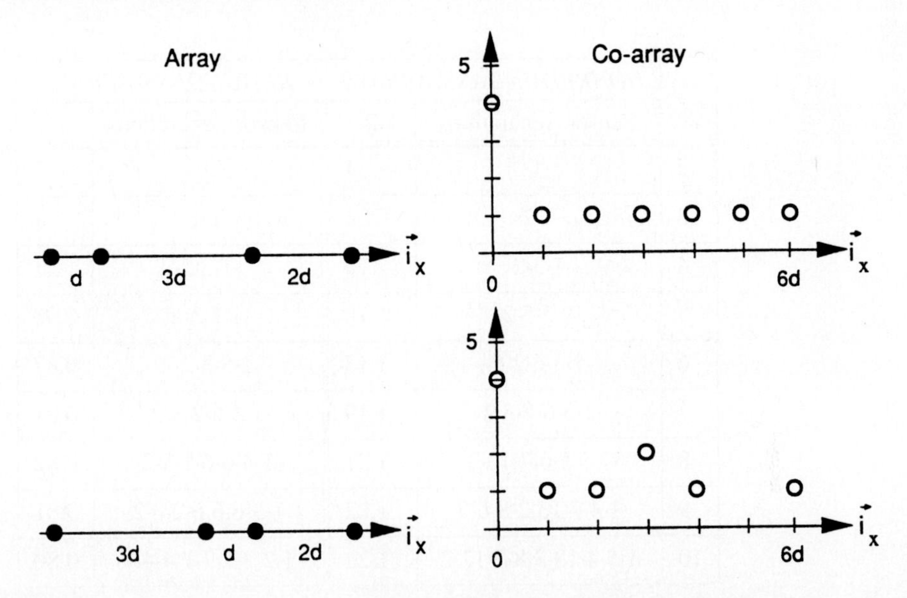

Figure 3.26 Two four-sensor arrays having a fundamental spacing d are depicted on the left and their co-arrays on the right. Note the array in the top row has a filled co-array just like that of a regular linear array, but with unit redundancy at each lag and the total extent of the co-array is greater. The second array has a gap (a lag with a zero value) in its co-array that cannot be removed. Only positive lags for the co-arrays are shown; the negative lags mirror the positive ones.

increasing the number of sensors by one while maintaining the same aperture provides additional processing capabilities—a wider range of wavenumbers can be supported—that may offset the complexity of having more sensors. For example, if the propagating signal has a wide bandwidth, which results in a wide range of spatial frequencies, an array such as the nine-sensor circular one might be more appropriate. Irregular co-arrays, however, are not redundant; a nonredundant co-array does not provide *any* spatial averaging for estimating the spatiotemporal correlation function. Thus, irregular co-arrays exhibit a tradeoff between spatial aliasing and spatial averaging.

3.3.5 Sparse Linear Arrays

Sparse arrays are laid out on an underlying regular grid, with strategically placed sensors located so as to produce the desired co-array. For linear arrays, the underlying grid simply becomes a set of points separated by a fundamental spacing d. Sparse linear arrays can have co-arrays for which every lag value within the co-array aperture has a redundancy of at least one: A filled co-array can be produced by a sparse array. In Fig. 3.26, the depicted four-sensor sparse array has this characteristic. Many arrays have some fundamental grid underlying their sensor locations. If the sensor locations are judiciously chosen with respect to this grid, well-structured co-arrays can result.

For a given number of sensors and a fundamental spacing d, the aperture of a sparse array is greater than that of a filled one. This increased aperture suggests that greater

	NONREDUNDANT ARRAYS		*REDUNDANT ARRAYS*	
M	Sensor Separations	D'	Sensor Separations	D'
2	·1·	1	·1·	1
3	·1·2·	1	·1·2·	1
4	·1·3·2·	1	·1·3·2·	1
5	·1·3·5·2·	1.10	·1·3·3·2·	0.90
6	·1·3·6·2·5·	1.13	·1·5·3·2·2·	0.87
7	·1·3·6·8·5·2·	1.19	·1·3·6·2·3·2·	0.81
8	·1·3·5·6·7·10·2·	1.21	·1·3·6·6·2·3·2·	0.82
9	·1·4·7·13·2·8·6·3·	1.22	·1·3·6·6·6·2·3·2·	0.81
10	·1·5·4·13·3·8·7·12·2·	1.22	·1·2·3·7·7·7·4·4·1·	0.80

Sparse arrays whose co-arrays either have minimum numbers of gaps with all nonzero co-array values equaling one (except for the origin)—nonredundant arrays [154]—or have no gaps with the largest possible apertures—redundant arrays [107]. For each array, the number of sensors M, the separation between sensors in grid units d, and the ratio D' of the array's spatial extent to that of the perfect array, $M(M-1)d/2$, is given. Asymptotically, this ratio approaches 3/4 for redundant arrays [12]. The geometries given here are not unique; other intersensor spacings yield the same co-array.

TABLE 3.2 Sparse Linear Array Geometries.

resolution can be obtained with a sparse array. Because of the common underlying grid, the aliasing properties of the sparse linear array and the filled linear array are the same: Grating lobes occur at the same incidence angles.

We can construct sparse linear arrays having co-arrays equaling 1 or 0 at all nonzero lags. Such arrays are called *nonredundant*; Table 3.2 lists minimum aperture nonredundant arrays. In some rare cases, we can construct a nonredundant array with no gaps (lags where the co-array values are zero) among the nonzero co-array values. Unfortunately, linear arrays having this property do not exist for $M > 4$. Such arrays are called *perfect arrays*;[†] examples of perfect and imperfect arrays are shown in Fig. 3.26. One can thus conceive of co-arrays that do not exist; synthesizing an array that corresponds to a given co-array cannot only be difficult but sometimes impossible.

Sparse arrays can also be designed so that their co-arrays have some redundant lags but no gaps. Such an array has a smaller aperture than the unattainable ideal of a perfect array. The filled linear array falls into this class, but other arrays with the same number of

[†]The only thing "perfect" about perfect arrays is that their co-arrays have no gaps and are filled with unit values, excluding, as usual, the zero-lag value. In all other respects, they have no special characteristics.

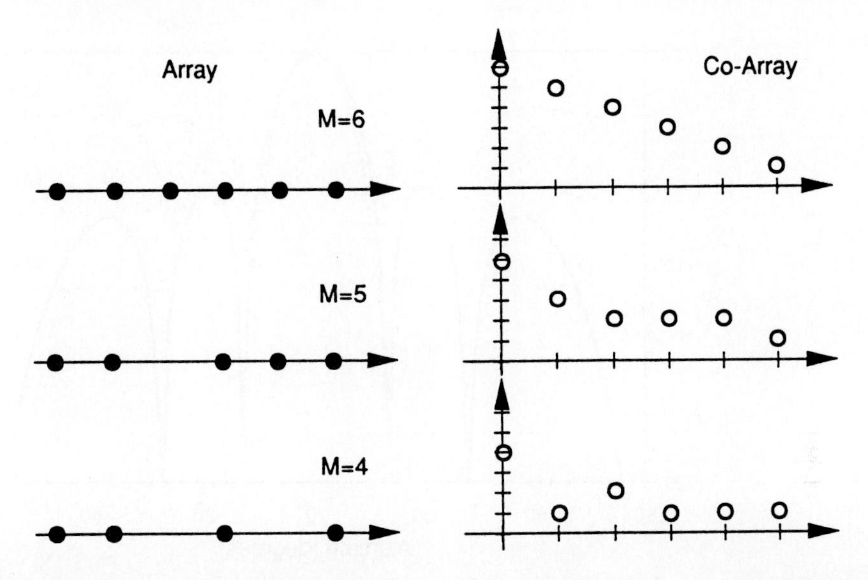

Figure 3.27 A six-sensor, filled array and its co-array are shown. Two arrays having the same aperture are derived by successively removing sensors from the array. This thinning procedure results in the depicted co-arrays. To derive the four-sensor perfect array, one must start with a seven-sensor filled array.

sensors and larger apertures also exist. Starting with the filled linear array, arrays having the same aperture but with fewer sensors can be obtained by judiciously *thinning* or *pruning* the array (Fig. 3.27). Finding arrays that have the maximum aperture (minimum redundancies) without gaps forms a discrete optimization problem. Although finding these array geometries remains computationally difficult, Bedrosian showed that the spatial extent spanned by an M-sensor minimum redundancy array equals $\lfloor M(3M-2)/8 \rfloor + 1$. Table 3.2 displays some minimum redundancy arrays.

Filled linear arrays and sparse arrays can be compared in two ways. The first is to keep the aperture fixed (constant resolution) while reducing the number of sensors to obtain a sparse array. The second keeps the number of sensors fixed and extends the aperture (thereby increasing resolution) to create a sparse array. If the aperture is fixed, the arrays have aperture smoothing functions with roughly the same mainlobe shape. In contrast, the structure of the aperture smoothing function's sidelobes varies as the number of sensors in the array is changed. As shown in Fig. 3.28, decreasing the number of sensors increases the sidelobes' amplitudes. Here, the sidelobe problem is a severe one; sidelobe amplitude places fundamental limits on the range of signal amplitudes that can be easily distinguished by the array.

The second variation should lead to a narrower mainlobe, but the sidelobes also change. The narrower mainlobe can be directly related to the increased aperture; sidelobe changes are due to the changes in the array's geometry. Fig. 3.29 depicts the squared magnitude aperture smoothing functions of four-sensor arrays having different co-arrays.

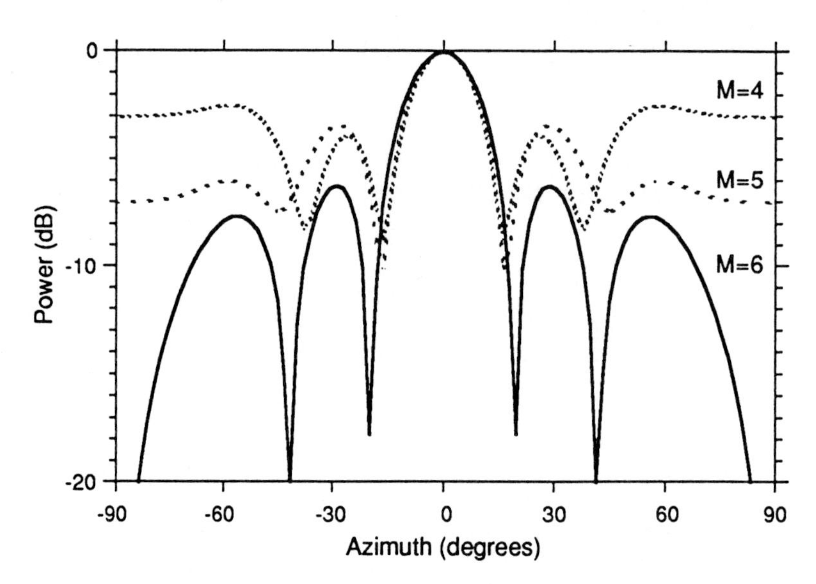

Figure 3.28 The squared magnitude aperture smoothing functions of the three linear arrays used in Fig. 3.27 are shown. Because the aperture is constant, the width of the mainlobe does not change. Note how the structure of the beampattern's sidelobes becomes more complicated as the number of sensors is reduced. More disturbing, however, is the increased amplitude of the sidelobes. The number of sensors rather than the aperture determines sidelobe levels.

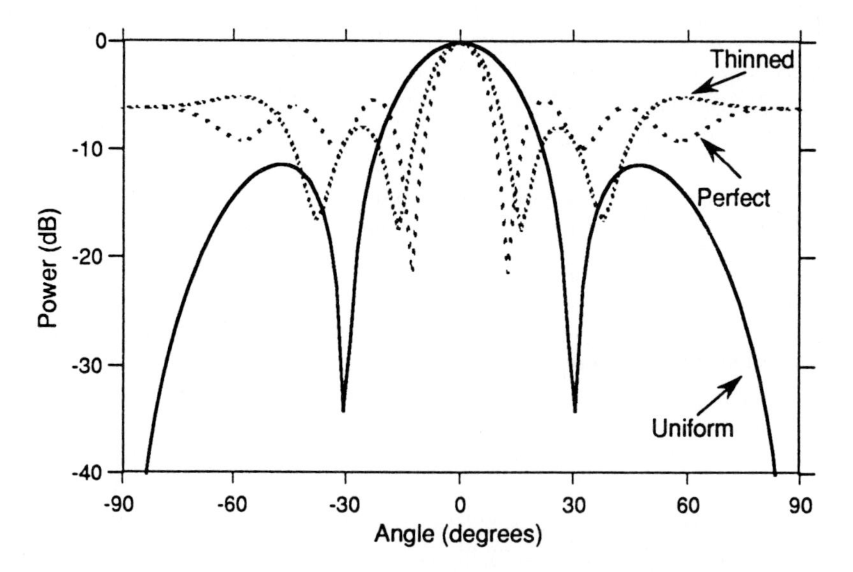

Figure 3.29 The squared magnitude aperture smoothing functions of a filled linear array, a perfect array, and a thinned array (the same as in Fig. 3.28) are shown. These arrays have the same number of sensors ($M = 4$) but the sensor locations are varied. The underlying grids of the arrays are identical.

These two figures illustrate fundamental principles that guide the selection of a particular array geometry.

- Aperture fundamentally determines the width of the aperture smoothing function's mainlobe, which in turn determines the spatial resolution of the array when conventional array processing algorithms are used.
- The number of sensors in the array determines the level of the aperture smoothing function's sidelobes.
- The sensor weight w_m, which may be different from unity, determines the detailed shape of the aperture smoothing function.

3.3.6 Random Arrays

The sensor positions of a *random array* are literally chosen at random according to some probability density defined over a specified aperture [146: 139–67]. Intelligent design of sensor placement (presumably) outperforms random sensor placement. In many circumstances, however, precise control of sensor locations is difficult. For example, sonobuoys are often dropped into the water from an airplane with little control over where they land. The aperture smoothing functions of such arrays can only be predicted in a statistical sense. As we have seen, aperture determines the width of the mainlobe and thereby the array's resolution. The sidelobe levels are strongly dependent on sensor location as evidenced by the array patterns that result from sparse arrays. Thus, a characterization of sidelobe amplitude relative to the mainlobe is the main consideration in random arrays.

Let the placements of M sensors each be governed by the probability density function $p_{\vec{x}}(\vec{x})$ that has finite support over some predefined aperture. The sensors are placed independently of each other and are assumed to be omnidirectional. In this model, the sensors are plopped into position randomly with no attention paid to the positions of previous sensors. With unit shading applied to the sensor outputs, the aperture smoothing function is given by

$$W(\vec{k}) = \sum_{m=0}^{M-1} \exp\{j\vec{k} \cdot \vec{x}_m\}$$

The expected value of the aperture smoothing function equals the number of sensors times the expected value of the phase factor for a random sensor placement.

$$\boxed{\mathcal{E}[W(\vec{k})] = M \cdot \mathcal{E}\left[\exp\{j\vec{k} \cdot \vec{x}_m\}\right]}$$

This expected value is the characteristic function {473} of the random variable governing sensor placement.

$$\mathcal{E}[W(\vec{k})] = M\Phi_{\vec{x}}(\vec{k})$$

In this average sense, the expected aperture smoothing function of a random array is the same as a *continuous* regular array where the array shading emulates the probability density of the sensor placement.

The expected squared magnitude of the aperture smoothing function for a random array is also related to the characteristic function of the sensor placement. From the expression

$$\mathcal{E}[|W(\vec{k})|^2] = \sum_{m_1,m_2} \mathcal{E}\left[\exp\{j\vec{k}\cdot(\vec{x}_{m_1} - \vec{x}_{m_2})\}\right]$$

we see that the co-array always has M values at the origin; M terms in this sum therefore equal 1. Because the sensor placements are statistically independent, the remaining $M(M-1)$ terms each equal the squared magnitude of the characteristic function $\Phi_{\vec{x}}(\vec{k})$. The expected value of a random array's squared magnitude aperture smoothing function is thus expressed by

$$\boxed{\mathcal{E}\left[|W(\vec{k})|^2\right] = M + M(M-1)\left|\Phi_{\vec{x}}(\vec{k})\right|^2}$$

This expression reveals the underlying probabilistic structure of the random array's aperture smoothing function. The first term, the constant M, determines the lower limit of the sidelobe level. The second term is proportional to the squared magnitude of the expected aperture smoothing function. The largest value of the characteristic function term corresponds to the mainlobe and equals $M(M-1)$. We find that value of the minimum sidelobe level is $(M-1)$ times smaller than the mainlobe's peak value. Clearly, the number of sensors rather than their placement determines how small the sidelobes can be in a random array.

The squared magnitude aperture smoothing function of a random linear array is shown in Fig. 3.30. Although the shape of the mainlobe is very well predicted, the levels of the sidelobes are not. To understand this inconsistent prediction, the variance of the squared magnitude aperture smoothing function must be evaluated. Tedious, but straightforward, evaluations similar to those used in deriving the expected squared magnitude aperture smoothing function lead to the following expression for the variance.

$$\boxed{\mathcal{V}\left[|W(\vec{k})|^2\right] = M(M-1)\begin{bmatrix} 1 + \left|\Phi_{\vec{x}}(2\vec{k})\right|^2 + 2(M-2)\left|\Phi_{\vec{x}}(\vec{k})\right|^2 \\ + 2(M-2)\,\mathrm{Re}\left[\Phi_{\vec{x}}(2\vec{k})\Phi_{\vec{x}}^2(-\vec{k})\right] \\ - 2(2M-3)\left|\Phi_{\vec{x}}(2\vec{k})\right|^4 \end{bmatrix}}$$

As shown in Fig. 3.30, the expected variability of a particular squared magnitude aperture smoothing function about its expected value is quite large in the sidelobe region. At the peak of the mainlobe, the variance is 0; in the sidelobes, the variance can be as large as $M(M-1)$. Thus, the squared magnitude aperture smoothing function of a particular random array differs little from its expected value near the mainlobe but may well deviate greatly in the sidelobes. For a given probabilistic description of how sensors are placed, the resolution of the resulting random array, as determined by the mainlobe's width, can be predicted quite well. However, the ability of the array to distinguish spurious sources of energy, as defined by sidelobe rejection levels, cannot be predicted nearly as accurately.

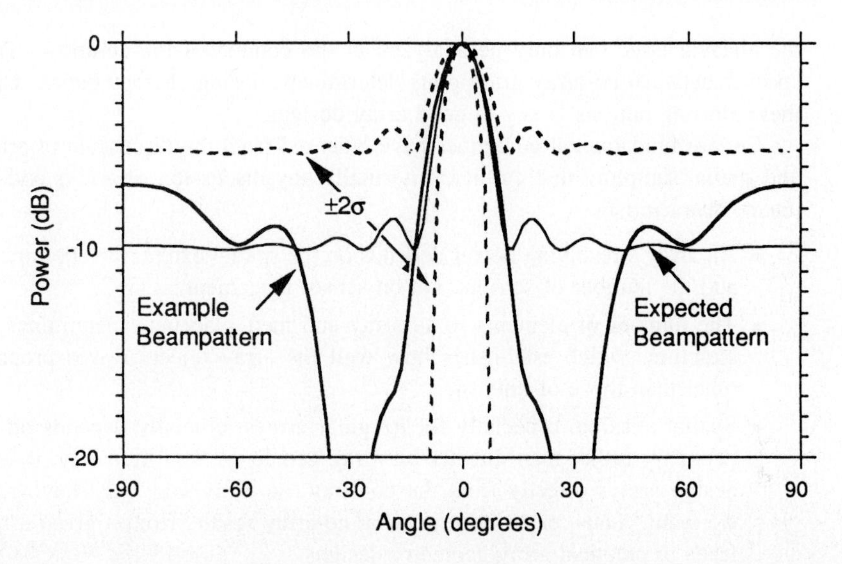

Figure 3.30 The squared magnitude aperture smoothing function of a random array having a uniform sensor placement within the aperture is shown. The positions of ten sensors within an aperture of $9\lambda/2$ (the aperture of a ten-sensor regular linear array having intersensor spacing $\lambda/2$) where chosen accordingly to a uniform probability density. A squared magnitude aperture smoothing function from such a random array is shown as the thick solid line. The thin solid line denotes the expected squared magnitude aperture smoothing function and the dashed lines the most likely variations ($\pm 2\sigma$) of a random array's squared magnitude aperture smoothing function about this mean. Clearly, actual squared magnitude aperture smoothing functions can deviate beyond these ranges; the denoted variation limits merely serve as guidelines.

Summary

In this chapter, both continuous and discrete apertures—arrays—have been discussed. Apertures are characterized spatially by their aperture function $w(\vec{x})$ and arrays by their discrete counterpart w_m. In both cases, the aperture can be thought of as a spatial window through which we view the universe. Propagating wavefields bring information about distant events to our localized window. Each sensor in an array is an aperture itself. In many cases, the aperture is quite simple: It corresponds to point evaluation, a spatial sample, of the field and has omnidirectional characteristics. In other cases, each "sensor" can be quite complicated; the *VLA* shown in Fig. 3.18, where each sensor is a parabolic antenna, is but one example. As yet, how such arrays of directional sensors behaves has not been presented; that topic is discussed in the next chapter.

Finite apertures limit our ability to distinguish—resolve—propagating waves. Furthermore, arrays can introduce spatial aliasing that would confuse naïve interpretation of an array's output. Spatial aliasing can be easily understood from familiar sampling notions if we use Cartesian grids. Sparse linear arrays, for example, have a wide spatial extent with their aliasing properties governed by the underlying grid structure. However, when we use non-Cartesian sampling, especially in planar and three-dimensional arrays, the aliasing issue can only be analyzed in the context of the co-array. The minimum spacing between co-array grid points determines aliasing characteristics. Understanding these aliasing notions is key to good array designs.

Certain fundamental guidelines have emerged from the discussion of array geometry and spatial sampling that cut across virtually any discussion of arrays and their performance characteristics.

- An array's resolving power depends on the spatial extent, the aperture, of the array and the number of sensors, not on sensor placement.
- The number of elements in an array and their placement determines the sidelobe structure, which establishes how well the array rejects waves propagating from other than those of interest.
- Spatial aliasing, especially for irregular arrays, crucially depends on the co-array geometry rather than directly on array geometry. Unfortunately, if we attempt to design arrays directly from the co-array, no array may exist having the co-array we want. Thus, experience in what co-array results from a given array geometry leads to practical array geometry designs.

Problems

3.1 Consider the continuous aperture function

$$w(x, y, z) = \begin{cases} \cos \dfrac{\pi x}{2D_x} \left(1 - \dfrac{|y|}{D_y}\right), & |x| \le D_x,\ |y| \le D_y,\ |z| \le D_z \\ 0, & \text{otherwise} \end{cases}$$

(a) Sketch $w(x, 0, 0)$, $w(0, y, 0)$, and $w(0, 0, z)$. Characterize this aperture function.

(b) Derive the aperture smoothing function $W(k_x, k_y, k_z)$.

3.2 Derive the aperture smoothing function for an equilateral triangular aperture. Assume that the aperture function $w(x, y)$ is 1 within the equilateral triangle and 0 outside it. Also find the co-array.

3.3 Suppose two monochromatic waves propagate across a linear aperture. The first wave arrives with incidence angle ϕ_1, at speed c_1, and with frequency ω_1. Similarly, the second wave is characterized by ϕ_2, c_2, and ω_2.

(a) If both waves have the same x component of wavelength, $\lambda_x = \lambda / \sin\phi$, how are their apparent speeds across the aperture related?

(b) If their apparent speeds are the same, how are their λ_x's related?

(c) Show that another interpretation of Snell's Law is that the apparent speed along an interface must be the same for waves on both sides of the interface.

3.4 Derive the co-array of a uniform circular aperture of radius R. Begin by using elementary planar geometry to compute the area of overlap of two circles whose centers are separated by χ.

3.5 Suppose the wavefield propagating toward a linear array of length D through a *random medium*. As a consequence of the medium's properties, the wavefield at the aperture is corrupted by a random aberration: When the incidence angle equals zero, the delay across the aperture is a Gaussian random process having zero mean, variance σ^2, and correlation function $R_\Delta(\chi)$.

(a) Find the expected value of generalized aperture smoothing function $W_\omega(k_x)$.

(b) Assuming the variance is small so that $\omega\Delta(x) \ll 1$ with high probability, find the expected value and variance of $W_\omega(k_x)$.

(c) Interpret your result: What is the effect of the spatial bandwidth of these random aberrations?

3.6 Consider a compound planar aperture consisting of two square regions, one centered at $(x_0, 0)$ and the other centered at $(-x_0, 0)$. Each side measures D meters.

(a) Sketch the compound aperture and derive its aperture smoothing function.

(b) Determine the co-array assuming that $x_0 > D$.

(c) How does the co-array change when $D/2 < x_0 < D$?

3.7 The ahead-of-its-time acoustic array depicted in Fig. 1.1 {2} consists of hexagonal apertures arranged on pairs of hexagonal grids.

(a) Deferring until Prob. 4.5 {191} consideration of the entire array's capabilities, what is the aperture smoothing function of each hexagonal aperture?

(b) What is the aperture smoothing function of each hexagonally arranged hexagonal aperture?

3.8 Consider a two-dimensional sampling grid wherein the sampling vectors are

$$\mathbf{v}_1 = \begin{bmatrix} 1 \\ 0 \end{bmatrix} \qquad \mathbf{v}_2 = \begin{bmatrix} a \\ b \end{bmatrix}$$

where a and b take on values between 0 and 1.

(a) What is the sampling matrix $\mathbf{V}$?

(b) Determine a and b so that a regular hexagonal sampling grid results.

(c) For this case, what is the corresponding periodicity matrix $\mathbf{U}$? Sketch the resulting periodic structure of (k_x, k_y) space.

3.9 One example of a regular planar array is the rectangular array. Assume that the array measures M_x sensors in the x direction (separation d_x meters) by M_y sensors in the y direction (separation d_y meters).

(a) Find this array's aperture smoothing function.

(b) Because the array is planar, defining the Rayleigh resolution becomes tricky. Characterize the region in wavenumber space delineated by the smallest wavenumbers at which the aperture smoothing function equals 0. From this result, determine the rectangular array's resolution.

(c) Assuming that $M_x > M_y$, what is the height of the highest sidelobe and where in wavenumber space is it located?

3.10 For nondispersive media, temporal bandwidth determines spatial bandwidth. What happens when the medium is dispersive?

(a) Consider the dispersion relation given in Eq. 2.8 {22} for a purely dispersive medium. Assuming the propagating signal is a sinusoid, what is the signal received at a linear aperture?

(b) Assuming the signal contains energy in a small band about center frequency ω_0, what is the spatial bandwidth?

(c) How do these results affect sensor placement for a regular linear array?

3.11 Some arrays may be thought of as *compound arrays*: They consist of a set of sites at which is located, not a sensor, but a subarray. For simplicity, let's assume that each site's subarray geometry is the same and that all sensors are weighted uniformly.

(a) Do any common array geometries have this symmetry? Consider common one- and two-dimensional geometries.

(b) Calculate the array's aperture smoothing function $W(\vec{k})$ in terms of the aperture smoothing function for the array of sites $W_1(\vec{k})$ and the aperture smoothing function for the subarray located at each site $W_2(\vec{k})$.

(c) Use this result to find the aperture smoothing function of the double-Y array shown in Fig. 7.26 {419}. Assume that the separation of the two subarrays exceeds their aperture and that the subarrays are centered on the x axis.

3.12 The *circular* array consists of M sensors equally spaced about the perimeter of a circle of radius R.

(a) Sketch the set of points in three-space where a source may be located such that the sensor signals cannot be used to disambiguate source location.

(b) Determine and sketch the aperture smoothing function of a circular array. Assume that sensor 0 is located at $\vec{x}_0 = (0, R)$ and that the sensors are numbered in a clockwise manner around the circle. You may find the Euler-MacLaurin summation formula useful.

$$\sum_{m=1}^{M-1} f(m) \approx \int_0^M f(x)dx - \frac{1}{2}\big(f(0) + f(M)\big)$$

(c) What is the mainlobe width and the amplitude of the first sidelobe?

(d) Given that the side of a square edge array equals the diameter of a circular array, which has the narrower mainlobe? Does this situation change when the circular array circumscribes the square edge array?

3.13 A *right-angle* array consists of two linear arrays sharing a sensor located at an end and separated by an angle of 90° (Fig. 3.31).

(a) Find the co-array and beampattern of a right-angle array having nine sensors where each leg of five sensors has regular intersensor spacing d.

(b) Describe the directional sensitivity of this array, comparing it with a square array having the same number of sensors (3×3) and intersensor spacing d.

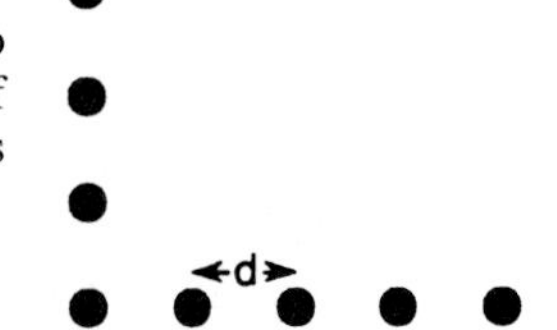

Figure 3.31 A right-angle array consists of two linear arrays sharing an end sensor. The axes of the two arrays form a right angle. Where is this array's spatial origin?

3.14 A commonly used three-dimensional array is the *cylindrical* array. It can be envisioned as a regular circular arrangement of linear arrays or as a stack of circular arrays. Let the number of linear arrays be M_c and the number of sensors in each linear array be M_l.

(a) Find the co-array for the cylindrical array parameterized by $M_c = 4$ and $M_l = 4$ where each linear array has regular intersensor spacing d_l and the radius of the cylinder is R.

(b) Find the co-array for a similar cylindrical array that differs only in that $M_c = 5$.

(c) Find the co-array for the array $M_c = 4$ and $M_l = 4$ where each linear array is perfect.

3.15 *Imaging* systems, such as the *VLA*, frequently employ arrays. How can we think of an array's output as an image?

(a) Consider a uniformly weighted square array illuminated (literally) by a monochromatic point source. What is the resulting "image" produced by the array? Assume that the point source (star) is known to be located near the axis perpendicular to the plane of the array.

(b) Interpret this result. What array parameters determine your point-source resolving capabilities? What kind of artifact could be traced to sidelobes?

3.16 The *VLA*'s geometry is shown in Fig. 3.18 {86}. The characteristics of this array and how it is used can be determined from its physical parameters [112]. Each sensor's output is passed through a narrowband filter to focus the array on sources radiating at some particular frequency.

(a) Each leg of the *VLA*'s symmetric Y configuration can be 0.59, 1.95, 6.4, or 21 km in length. Assuming fixed filters, how does changing the length affect the *VLA*'s output? Assuming fixed resolution, how does changing length affect the output?

(b) The sensors along each leg are arranged in a power law fashion: Assuming a leg coincides with the x axis, $\vec{x}_m = m^{1.716}d$, $m = 1, \ldots, 9$. Considering each leg as an irregular linear array, what is the corresponding aperture smoothing function?

(c) What is the aperture smoothing function of the entire *VLA*?

3.17 Consider a sparse array that has sensors located only at $x = \pm 3, \pm 4, \pm 5$.

(a) Sketch the co-array. Is the array redundant or nonredundant?

(b) What are the array's spatial aliasing characteristics?

(c) Derive the aperture smoothing function $W(\vec{k})$. Can you think of three different ways to derive the aperture smoothing function?

3.18 Derive a representation for the co-array of a discrete array by pretending that the array is a continuous aperture with an aperture function given by

$$w(\vec{x}) = \sum_{m=0}^{M-1} w_m \delta(\vec{x} - \vec{x}_m)$$

That is, calculate

$$c(\vec{\chi}) = \int w(\vec{x}) w^*(\vec{x} + \vec{\chi})\, d\vec{x}$$

Interpret your result.

3.19 Consider an equilateral triangular array with sensors at each vertex and edge midpoint (six sensors total).

(a) Sketch the co-array for this array.

(b) Now add a seventh sensor in the center of the triangle where the perpendicular bisectors of the three edges meet. Sketch the new co-array showing the points caused by the addition of the seventh sensor.

3.20 Consider two regular, linear arrays of eight sensors each. The sensors in both arrays are equally spaced and separated by d meters. The shading functions for the two arrays are different, however. In the first array, the middle four sensors are given weights of one while the flanking sensor pairs are given weights of $1/2$. In the second array, the middle four sensors are given weights of $1/2$, whereas the outer four sensors are given weights of 1. Derive the aperture smoothing function for each array.

Chapter 4

Beamforming

Propagating signals contain much information about the sources that produce them. Not only does each signal's waveform express the nature of the source, its temporal and spatial characteristics combined with the laws of physics allow us to determine the source's location. In the real world, modeling the field as a single propagating wave is usually naïve to say the least; several sources in addition to the one of interest usually bellow their presence, and noise always appears to contaminate measured signals. Thus, signal processing methods must *focus* on selected signals, allowing us to tease apart the cacophony into its components. One example of focusing is linear filtering: By applying bandpass filters, signals occupying different frequency bands can be separated. For propagating signals, more is needed; *spatiotemporal filtering* must be employed to separate signals according to their directions of propagation and their frequency content. This chapter describes the fundamental operations that produce this separation.

Can spatiotemporal filtering be accomplished with a single sensor? Clearly, omnidirectional sensors provide little spatial filtering. Continuous apertures having directional characteristics, however, can be used to advantage. For example, radar systems often use a single parabolic dish: The dish serves as the sensor and is designed to collect well fields propagating from the direction it is pointed and less well from others. The antenna's directional sensitivity, its *directivity pattern*, is governed by the antenna's physical construction [144: §6.3–6.4]; little can be changed once the antenna is built. To *steer* the antenna, it must be moved to point in the desired direction. Thus, the directivity pattern defines the bandwidth and stopband of our spatial filter while the antenna's physical axis determines its "center frequency."

The single-sensor approach has two drawbacks. First of all, spatial filtering characteristics are inflexible, requiring physical changes in the sensor. For example, tracking two radiating sources simultaneously is virtually impossible with one aperture. Second, to achieve fine enough resolution, apertures must not only be large but be designed to close tolerances lest aberrations inhibit the aperture's effectiveness. Achieving such large, high-precision apertures requires great initial expense and, when finally delivered, you have only one—it better not break!

In the search for cheaper, more flexible alternatives, arrays deserve consideration. As described in the previous chapter, an array processing system spatially samples the propagating field, and, if well-designed, no information is lost in the sampling process. The sensors can be directional, as in the *VLA*, or *omnidirectional*, having equal sensitivity to all directions of propagation. In either case, the array acts, as does the continuous aperture, as a spatial filter, attenuating all signals save those propagating from certain directions. The previous chapter details how an array's directivity pattern is determined by the aperture the sensors circumscribe and by the number of sensors. Sensors can be placed over a much wider area than the face of a parabolic dish without suffering the precision penalty. Consequently, the aperture of the array can be much greater than that of the single antenna, and, as we saw in the last chapter, increased aperture means greater angular resolution.

Augmenting an array with signal processing techniques not only enhances directivity but also allows us to aim the directivity pattern without physical changes. Furthermore, we can aim it in different directions simultaneously using signal processing notions. Thus, in addition to temporal filtering, array processing techniques can enhance an array's spatial filtering properties beyond those of a simple aperture filled with sensors. *Beamforming* is the name given to a wide variety of array processing algorithms that, by some means, focus the array's signal-capturing abilities in a particular direction. Based on the analogy of a flashlight, the mainlobe of an array's directivity pattern is called a *beam*.† Just as one might point a telescope or a radar dish, a beamforming algorithm "points" the array's spatial filter toward desired directions but algorithmically rather than physically. Beamforming algorithms generally perform the same operations on the sensors' outputs regardless of the number of sources or the character of the noise present in the wavefield. The signal processing algorithms developed for these *conventional* beamforming algorithms date from the early days of array processing and are well understood. Conventional beamforming algorithms use modern digital signal processing algorithms (such as the *FFT*) to great advantage, resulting in some of the most efficient array processing algorithms known. This chapter describes these algorithms and how they are implemented; later chapters describe more modern, adaptive algorithms. Array processing algorithms in conjunction with the array provide the best known spatiotemporal filters for decomposing the wavefield harmoniously into its components.

4.1 Delay-and-Sum Beamforming

Delay-and-sum beamforming, the oldest and simplest array signal processing algorithm, remains a powerful approach today. The underlying idea is quite simple: If a propagating signal is present in an array's aperture, the sensor outputs, delayed by appropriate amounts and added together, reinforce the signal with respect to noise or waves propagating in different directions. The delays that reinforce the signal are directly related to the length of time it takes for the signal to propagate between sensors (see Prob. 4.2). To be concrete,

†This nomenclature has its origins in arrays transmitting rather than receiving energy. Be that as it may, a beam now refers to the mainlobe of the directivity pattern whether the array transmits or receives.

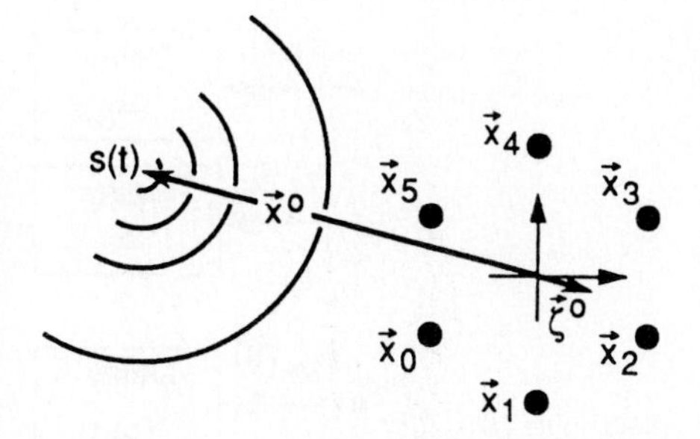

Figure 4.1 The array's phase center, defined by $\sum_m \vec{x}_m$, is taken as the origin of a coordinate system. A source located at $\vec{x}^o$ emits a signal $s(t)$ that propagates to the array. The direction of propagation $\vec{\zeta}^o$ is the unit vector that points from $\vec{x}^o$ to the origin. The distance between the source and the origin is $r^o \equiv |\vec{x}^o|$.

let $s(t)$ denote the signal emanating from a source located at the point $\vec{x}^o$. Several sources may be present, and their radiations sum to comprise the wavefield $f(\vec{x}, t)$ measured by the sensors. Let's consider an array of M sensors located at $\{\vec{x}_m\}$, $m = 0, \ldots, M-1$. The *phase center* of the array is defined as the vector quantity $\sum \vec{x}_m$. For convenience, we choose the origin of our coordinate system to coincide with the phase center (Fig. 4.1).

$$\sum_{m=0}^{M-1} \vec{x}_m = \vec{0}$$

The possible absence of a sensor at the array's phase center does not preclude using it as the origin and defining all the sensor positions $\{\vec{x}_m\}$ relative to it. Once the origin is thus established, the axes for the coordinate system can be chosen arbitrarily.

The waveform measured by the mth sensor is $y_m(t) = f(\vec{x}_m, t)$; the sensor samples the wavefield spatially at the sensor's location. The delay-and-sum beamformer consists of applying a delay Δ_m and an amplitude weight w_m to the output of each sensor, then summing the resulting signals (Fig. 4.2). We define the delay-and-sum beamformer's output signal to be

$$\boxed{z(t) \equiv \sum_{m=0}^{M-1} w_m y_m(t - \Delta_m)} \tag{4.1}$$

The amplitude weighting is sometimes called the array's *shading* or *taper*, and, as developed later, enhances the beam's shape and reduces sidelobe levels. The delays are adjusted to focus the array's beam on signals propagating in a particular direction $\vec{\zeta}^o$ or from a particular point $\vec{x}^o$ in space.

4.1.1 Near-Field and Far-Field Sources

Beamforming algorithms vary according to whether the sources are located in the *near field* or in the *far field* [61: 57–62]. If the source is located close to an array—in the near field—the wavefront of the propagating wave is perceptively curved with respect to the dimensions of the array and the wave propagation direction depends on sensor location.† If the direction of propagation is approximately equal at each sensor, then the source is

†In practice the relation between the local directions of propagation and the true direction of propagation may depend not only on the range of the source from the array's spatial origin, but also the properties of the

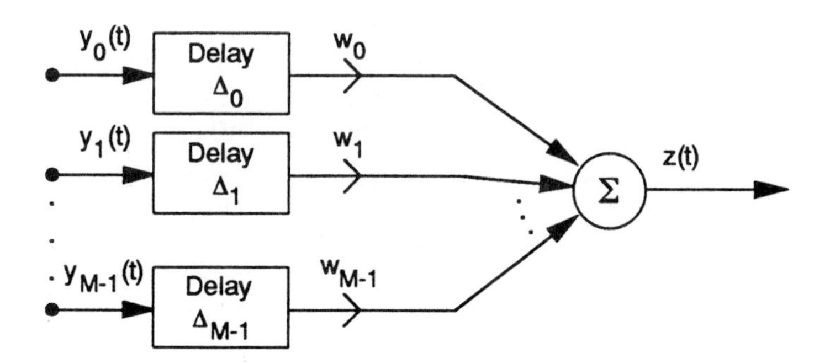

Figure 4.2 The beamformer's output $z(t)$ consists of adding together the sensors' outputs after a delay (and an amplitude weight) has been applied to each. The delays can be chosen to maximize the array's sensitivity to waves propagating from a particular direction or location in space.

located in the array's far field and the propagating field within the array aperture consists of plane waves. Thus, determining the location of far-field sources is quite difficult: We can determine direction of propagation but not range. With near-field sources, the direction vectors at each sensor emanate from a common location; we should be able to extract range and direction information from waves propagating from near-field sources.

A particular signal's direction of propagation relative to our coordinate system is denoted by $\vec{\zeta}^o$. For plane waves, this vector does not vary with sensor location. For near-field sources, however, the apparent direction of propagation varies across the array. The direction of propagation vector $\vec{\zeta}^o$ is thus defined relative to the array's phase center, but at the mth sensor it equals $\vec{\zeta}^o_m$. If the medium is nonrefractive, thereby yielding straight-line propagation, we see that $\vec{\zeta}^o = -\vec{x}^o/|\vec{x}^o|$. As $|\vec{x}^o|$ denotes the source's *range* from the array, denote this quantity by r^o. Similarly, $\vec{\zeta}^o_m = (\vec{x}_m - \vec{x}^o)/r^o_m$, where r^o_m is the distance from $\vec{x}^o$ to the mth sensor location $\vec{x}_m$. This geometry is illustrated in Fig. 4.3.

In many applications, we do not know *a priori* whether the source is located in the near or far field. To understand the errors induced by assuming far-field (plane wave) propagation instead of near-field, let ϵ_m be the angle between the rays emanating from the source to the array origin and to the mth sensor (Fig. 4.3). This angle represents the error we want to estimate. Simple application of the Law of Sines† yields $\sin\epsilon_m = \sin\psi_m \cdot (|\vec{x}_m|/r^o_m)$, with ψ_m denoting the angle between the vectors $\vec{x}_m$ and $\vec{x}^o$. When $r^o_m \gg |\vec{x}_m|$, the source is located well outside the array's aperture; we can make the approximation $r^o_m \approx r^o$ and assume the angle ϵ_m is small. Therefore,

$$\epsilon_m \approx \frac{|\vec{x}_m|}{r^o} \sin\psi_m$$

The largest value of this near-field–far-field error occurs for the most distant sensor

medium. For example, if the medium is dispersive, the local direction of propagation depends not only on sensor location but also on frequency. We consider here the case in which the medium is not dispersive or refractive.

†The Law of Sines for triangles states that $\sin a/A = \sin b/B = \sin c/C$, where A represents the length of the side of a triangle opposite the vertex having angle a, B is opposite b, and so forth.

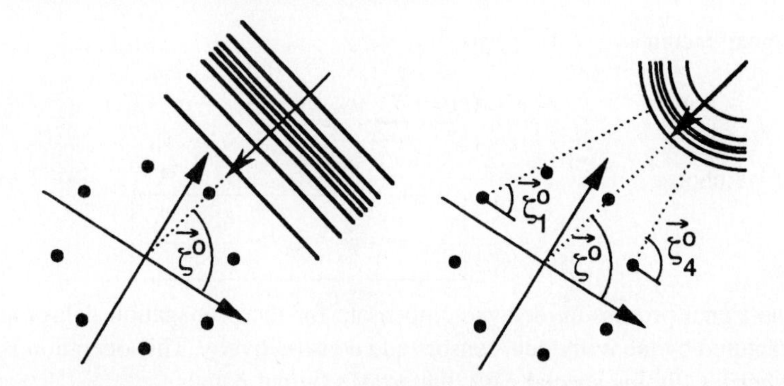

Figure 4.3 In near-field geometry, the source's angle with respect to the array's phase differs from those measured with respect to each sensor. The near-field–far-field discrepancy equals the angular difference between $\vec{\zeta}^o$ and $\vec{\zeta}^o_m$.

located at a right angle relative to the source direction vector ($\psi_m = \pi/2$). Defining the borderline between near-field and far-field sources becomes one of deciding what "negligible error" is, which is determined by the ratio of the array's spatial extent to source range. To achieve a maximum error of 1°, for example, the source would need to be located some 57 times the "radius" ($\max_m |\vec{x}_m|$) of the array's aperture.

4.1.2 Beamforming for Plane Waves

A judicious choice of sensor delays $\{\Delta_m\}$ in the delay-and-sum beamformer (Eq. 4.1) allows the array to "look" for signals propagating in a particular direction. Ideally, the *direction of look* is the opposite of a signal's direction of propagation: When we look north, we are trying to detect waves going south. By adjusting the delays, the array's direction of look can be steered toward the source of radiation.† Thus, we define an assumed propagation direction, denoted by the unit vector $\vec{\zeta}$, that is, the negative of the direction of look. Rather than saying we are looking north, it is more convenient to say we are looking for waves going south.

Assume a far-field source radiates a plane wave having waveform $s(t)$ that propagates across our array of M sensors in the direction $\vec{\zeta}^o$. The wavefield within the array's aperture is expressed by

$$f(\vec{x}, t) = s(t - \vec{\alpha}^o \cdot \vec{x})$$

where the slowness vector is defined by $\vec{\alpha}^o = \vec{\zeta}^o/c$. The mth sensor spatially samples the wavefield, yielding $y_m(t) = s(t - \vec{\alpha}^o \cdot \vec{x}_m)$;‡ the delay-and-sum beamformer's output

†Obviously, the beam is not physically steered; it is steered electronically. Common jargon for this electronic steering is *phase steering*. How phase is related to choosing sensor delays is described in §4.4 {132}.

‡This equation expresses the plane-wave or far-field assumption: The slowness vector does not depend on sensor location.

signal becomes

$$z(t) = \sum_{m=0}^{M-1} w_m s(t - \Delta_m - \vec{\alpha}^o \cdot \vec{x}_m)$$

If we choose

$$\boxed{\Delta_m = -\vec{\alpha}^o \cdot \vec{x}_m = \frac{-\vec{\zeta}^o \cdot \vec{x}_m}{c}}$$

the signal processing delays compensate for the propagation delays and the waveforms captured by the individual sensors add constructively. This operation is sometimes called *stacking*. In this special case, the array's output equals

$$z(t) = s(t) \cdot \left[\sum_{m=0}^{M-1} w_m\right]$$

and the beamformer's signal equals a constant times the waveform radiated by the source. We can thus steer the array's beam to an assumed propagation direction $\vec{\zeta}$ by using the set of delays given by

$$\boxed{\Delta_m = \frac{-\vec{\zeta} \cdot \vec{x}_m}{c} = -\vec{\alpha} \cdot \vec{x}_m}$$

where as usual we have defined the slowness vector $\vec{\alpha}$ corresponding to the assumed propagation direction as $\vec{\zeta}/c$. The beamformer signal $z(t)$ that results from a plane wave propagating in the direction $\vec{\zeta}^o$ is given by

$$\boxed{z(t) = \sum_{m=0}^{M-1} w_m s\big(t + (\vec{\alpha} - \vec{\alpha}^o) \cdot \vec{x}_m\big)} \tag{4.2}$$

If we look in the wrong direction, $\vec{\alpha} \neq \vec{\alpha}^o$, we obtain a degraded version of the propagating signal. In such cases, we say that the beamformer is *mismatched* to the propagating wave. This mismatch can occur in one of two ways.

- If the speed of propagation is known, mismatch means that the assumed propagation direction does not equal the true direction of propagation. Knowing the speed of propagation implies that the medium is relatively stable and that its characteristics can be predicted or measured. Such is the situation in many sonar and most radar applications.
- If the direction of propagation is known, we assumed the wrong speed of propagation. Precise knowledge of source locations occurs when we place them and calibrate their positions as in seismic applications. Assuming a propagation direction thus becomes equivalent to assuming a speed of propagation. Seismologists use arrays for just this purpose because medium properties can be inferred from sound speed measurements.

Thus, assuming a slowness vector for the delay-and-sum beamformer means that we are presuming a direction of propagation and a propagation speed. If one of these is known, we can find the other by scanning across wavenumber with a beamformer, searching for a maximum energy output. Without knowledge of either, the beamformer can only be used to infer the slowness vector of the propagating wave *unless* we employ a three-dimensional array having no ambiguities.

Example

Let's consider a linear array of $M = 2M_{1/2} + 1$ sensors spaced equally by d m. The array's phase center is located at the center sensor. The origin of our coordinate system coincides with phase center and the x axis lies along the line of sensors. The mth sensor's location is $\vec{x}_m = (m - M_{1/2})d\vec{i}_x$. If we choose the sensor weights to be uniform, $w_m = 1$, the delay-and-sum beamformer's response to a plane wave having slowness vector $\vec{\alpha}^o$ is

$$z(t) = \sum_{m=0}^{M-1} s(t - (\alpha_x - \alpha_x^o)(m - M_{1/2})d)$$

$$= \sum_{m=0}^{M-1} s\left(t - \frac{\zeta_x - \zeta_x^o}{c}(m - M_{1/2})d\right)$$

where ζ_x represents the x component of the array's assumed propagation direction vector and ζ_x^o represents the x component of the plane wave's direction of propagation.

Now let's assume a monochromatic source signal: $s(t) = \exp\{j\omega^o t\}$. The beamformer's output becomes

$$z(t) = e^{j\omega^o t} \exp\left\{-j\frac{M}{2}\beta\right\} \sum_{m=0}^{M-1} \exp\{jm\beta\}$$

where the angle $\beta \equiv k^o(\zeta_x - \zeta_x^o)d$ and the assumed wavenumber $k^o \equiv \omega^o/c$ streamline the notation. As the sum is a truncated geometric series, which has a closed form expression,[†] we can determine how mismatch affects the array's output.

$$z(t) = e^{j\omega^o t} \frac{\sin \frac{M}{2} k^o(\zeta_x - \zeta_x^o)d}{\sin \frac{1}{2} k^o(\zeta_x - \zeta_x^o)d}$$

If we have chosen the array's assumed propagation direction correctly, $\vec{\zeta} = \vec{\zeta}^o$, the beamformer output equals

$$z(t) = Me^{j\omega^o t}$$

If the assumed propagation direction and the wave's direction of propagation are mismatched, then $z(t)$ is a smaller constant times $\exp\{j\omega^o t\}$. Thus, the delay-and-sum beamformer yields the source waveform, but with an amplitude that depends on how much the assumed propagation direction differs from the direction of propagation. Maximum response occurs when they coincide; as described subsequently, this fact allows us to search for the source.

4.1.3 Beamforming for Spherical Waves

Now consider the case of a source located in the array's near field. Assume that the source is emitting a signal $s(t)$ that spreads spherically into space. From §2.2.2 {16}, we

† $\sum_{n=0}^{N-1} a^n = (1 - a^N)/(1 - a)$.

know that a spherically symmetric solution to the wave equation has the form

$$f(\vec{x}, t) = \frac{s(t - |\vec{x} - \vec{x}^o|/c)}{|\vec{x} - \vec{x}^o|}$$

The mth sensor thus measures the signal $y_m(t) = s(t - r_m^o/c)/r_m^o$, where r_m^o is the distance between the source and the sensor, and, as usual, c represents wave propagation speed. By choosing

$$\boxed{\Delta_m = \frac{r^o - r_m^o}{c}}$$

we can "stack" the signal replicas captured by all M sensors so that they reinforce each other. The beamformer's response to a spherically propagating wave becomes

$$\boxed{z(t) = \frac{1}{r^o} s(t - r^o/c) \left[\sum_{m=0}^{M-1} w_m \frac{r^o}{r_m^o} \right]}$$

The source signal emerges, delayed and attenuated[†] as if it had been received at phase center, times a weighting factor that, for large values of r^o, approaches the sum of the sensor weights. Let r and r_m denote range parameters presumed by the beamformer that do not equal actual values: Mismatch has occurred. The beamformer's output in this case becomes

$$\boxed{z(t) = \sum_{m=0}^{M-1} \frac{w_m}{r_m^o} s\left(t - \frac{r - (r_m^o - r_m)}{c}\right)} \tag{4.3}$$

The delay of r/c is common to all terms and represents a fixed time-delay that does not affect the received waveform. Just as in the plane-wave, far-field case, the mismatched terms yield a beamformer output that is a degraded version of the propagating signal, but here with amplitude as well as delay variations in the summation.

The beamformer's output $z(t)$ thus depends on the choice of the sensor delays $\{\Delta_m\}$. The delays maximally reinforce the signal only when they "cancel"[‡] the intersensor propagation delays. By matching assumed and actual propagation delays, the beamformer's output essentially equals the waveform of the propagating signal (Fig. 4.4). This choice of beamforming parameters maximizes the integrated energy in the output (Prob. 4.2).

[†]The attenuation is of course due to the spherical spreading that we've assumed. Note that for far-field sources, the authors did not assume any such attenuation or delay. The reason for this seeming sloppiness is that source range cannot be inferred by the array when the source is located in the far field. Assigning an attenuation would amount to clairvoyance on our part; we cannot be so presumptuous, at least most of the time!

[‡]The quotation marks are needed because a constant, nonzero delay could result from the "cancellation." The beamforming operation, in this case, introduces an overall delay that does not affect the measurement of the source's position. If, however, several arrays' outputs were being combined for some reason (as in Chap. 8), each array's delay would have to be compensated.

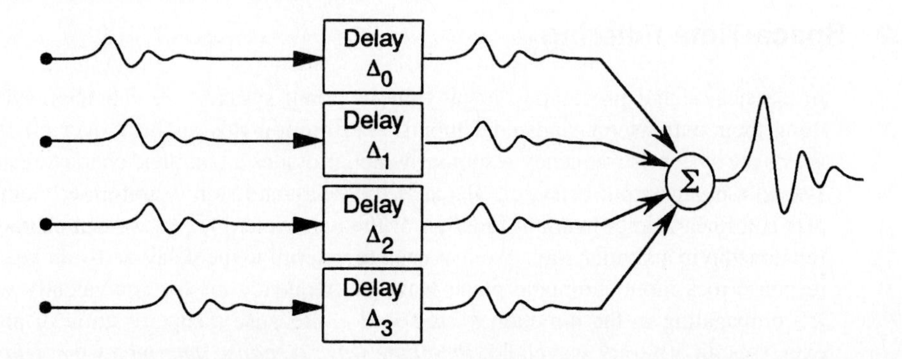

Figure 4.4 Adjusting the signal processing delays Δ_m to reinforce the signal is sometimes called *stacking*. As shown, the maximum-energy choice for Δ_m aligns the sensor outputs so that the sensors' waveforms resemble the signal waveform stacked on itself.

The ideal beamforming delays for spherical propagation depend on the difference $r^o - r^o_m$. Straightforward calculations show that

$$\boxed{\frac{r^o - r^o_m}{c} = \frac{r^o}{c}\left[1 - \sqrt{1 + \frac{|\vec{x}_m|^2}{(r^o)^2} - 2\frac{\vec{x}^o \cdot \vec{x}_m}{(r^o)^2}}\right]}$$

This expression determines the near-field beamformer's delays in terms of the assumed source location $\vec{x}^o$. Using these delays, we can *focus* the array on a particular spatial location, producing maximal output energy when the assumed location corresponds to the correct one.

Whether located in the near or far field, the expressions in Eq. 4.2 {116} and Eq. 4.3 {118} determine the beamformer's response to an ideally propagating wave as a function of an expected point-source's location. More generally, the delays could be set for other types of sources, such as extended sources, or for nonideal propagation. Often, we are interested in the inverse problem—determining from which directions or locations waves are propagating across our array. Beamformers may be used to look for plane waves propagating in a particular direction or for spherical waves emanating from a particular point, just as by tuning a bandpass filter we can search for energy in a particular frequency band. We accomplish this search by systematically adjusting the beamformer delays so as to steer the beam to different directions of look or to focus the array on points in space. By maximizing the integrated energy in the beamformer's output signal, we are focusing the array, just as if we had a camera or telescope, on the source. How we efficiently accomplish this focusing and characterize it are subsequent topics in this chapter.

4.2 Space-Time Filtering

In classical signal processing, linear time-invariant systems are characterized by examining their outputs for sinusoidal inputs. Performing this analysis over all frequencies yields the system's frequency response, which provides a complete characterization of the system's input-output behavior. Because the delay-and-sum beamformer's output signal $z(t)$ is a linear, time-invariant function of the wavefield $f(\vec{x}, t)$, we can characterize this relationship in a similar way. We just need to determine the delay-and-sum beamformer's response to a monochromatic plane wave of frequency ω^o (or equivalently wavelength λ^o) propagating in the direction $\vec{\zeta}^o$ at speed c. Because a superposition of plane waves expresses an arbitrary wavefield, *the plane wave response determines the delay-and-sum beamformer's output for the general case.* The delay-and-sum beamformer's response to a monochromatic wave is often called the *array pattern*: It corresponds to the wavenumber-frequency response of a spatiotemporal filter. As derived subsequently, the array pattern determines the array's directivity pattern.

4.2.1 Array Pattern

Assume we have a monochromatic plane wave with a temporal frequency ω^o propagating with a slowness vector $\vec{\alpha}^o$. The resulting wavefield is given by

$$f(\vec{x}, t) = s(t - \vec{\alpha}^o \cdot \vec{x}) = \exp\{j\omega^o(t - \vec{\alpha}^o \cdot \vec{x})\}$$

This wavefield propagates past an array of M sensors located at positions $\{\vec{x}_m\}$; a delay-and-sum beamformer combines the signals from these sensors to produce the output given by Eq. 4.2 {116}.

$$z(t) = \sum_{m=0}^{M-1} w_m s\big(t + (\vec{\alpha} - \vec{\alpha}^o) \cdot \vec{x}_m\big)$$

When our monochromatic test signal serves as the propagating signal, the output becomes

$$z(t) = W(\omega^o \vec{\alpha} - \vec{k}^o) e^{j\omega^o t}$$

where $\vec{k}^o = \omega^o \vec{\alpha}^o = (\omega^o/c)\vec{\zeta}^o$ and $W(\cdot)$ denotes the Fourier Transform of the sensor weights (shading sequence).

$$\boxed{W(\vec{k}) = \sum_{m=0}^{M-1} w_m \exp\{j\vec{k} \cdot \vec{x}_m\}} \tag{4.4}$$

In Chap. 3 {92}, we encountered $W(\vec{k})$, calling it the aperture smoothing function because it determined how the wavefield's space-time Fourier Transform $F(\vec{k}, \omega)$ is smoothed by observation through a finite aperture. This same function is also the *array pattern* because, through the quantity $W(\omega^o \vec{\alpha} - \vec{k}^o)$, it determines the amplitude and phase of the beamformed signal when the wavefield consists of a single plane wave. Using the inverse space-time Fourier Transform to express an arbitrary wavefield $f(\vec{x}, t)$,

$$f(\vec{x}, t) = \frac{1}{(2\pi)^4} \int_{-\infty}^{\infty} \int_{-\infty}^{\infty} F(\vec{k}, \omega) \exp\{j(\omega t - \vec{k} \cdot \vec{x})\} \, d\vec{k} \, d\omega$$

we find that the delay-and-sum beamformer's output to this general field can be expressed in terms of the array pattern as

$$\boxed{z(t) = \frac{1}{(2\pi)^4} \int_{-\infty}^{\infty} \int_{-\infty}^{\infty} F(\vec{k}, \omega) W(\omega\vec{\alpha} - \vec{k}) \exp\{j\omega t\} \, d\vec{k} \, d\omega} \tag{4.5}$$

The integral over $\vec{k}$ can be interpreted as a smoothing in wavenumber similar to that imposed by a finite aperture on a propagating wavefield. The integral over ω accumulates the array responses at all the relevant temporal frequencies to form the wideband signal $z(t)$.

Plane wave (far-field) array pattern. When the wavefield consists of a plane wave, $f(\vec{x}, t) = s(t - \vec{\alpha}^o \cdot \vec{x})$, its spatiotemporal Fourier Transform equals $F(\vec{k}, \omega) = S(\omega)\delta(\vec{k} - \omega\vec{\alpha}^o)$. The spectrum of the resulting beamformer output equals

$$Z(\omega) = S(\omega) W\big(\omega(\vec{\alpha} - \vec{\alpha}^o)\big)$$

When the beamformer selects the slowness vector accurately so that $\vec{\alpha} = \vec{\alpha}^o$, the beamformer signal becomes an undistorted replica of the propagating signal: $z(t) = W(0)s(t)$. When the propagation speed c is known, equality of the slowness vectors implies that the array's assumed propagation direction and the wave's direction of propagation are also equal. Thus, when we look in the right direction and assume the correct speed, the beamformer yields the signal we want. Conversely, if $\vec{\alpha} \neq \vec{\alpha}^o$ for any reason, the output signal's spectrum equals the temporal spectrum $S(\omega)$ of the source signal multiplied by a factor that depends on the array pattern $W(\cdot)$. The beamforming "distortion" owing to this mismatch is a linear filtering operation and typically results in attenuation of the source signal's high-frequency components.

Example

As in the previous example {117}, consider a linear array of $M = 2M_{1/2} + 1$ equally spaced sensors separated by d m. We found there that when the sensor weights all equal 1, the array pattern for this uniform linear array equals

$$W(\vec{k}) = \frac{\sin \frac{M}{2} k_x d}{\sin \frac{1}{2} k_x d}$$

This particular function's characteristics are examined in some detail in §3.3.1 {85}. Recall that the width of the mainlobe is inversely proportional to M, but that the relative height of the first sidelobe does not decrease as M becomes large.

As previously discussed, the important quantity in assessing beamformer performance is $W(\omega^o\vec{\alpha} - \vec{k}^o)$. For the case at hand, its expression depends only on the x coordinates.

$$W(\omega^o\vec{\alpha} - \vec{k}^o) = \frac{\sin \frac{M}{2}(\omega^o\alpha_x - k_x^o)d}{\sin \frac{1}{2}(\omega^o\alpha_x - k_x^o)d}$$

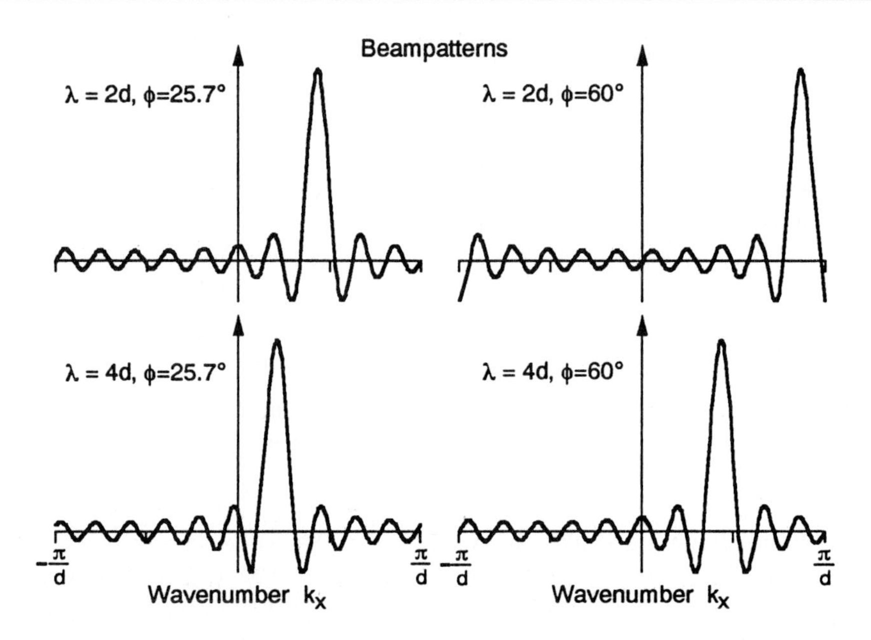

Figure 4.5 A shifted array pattern $W(k_x - k_x^o)$ plotted as a function of k_x^o, the x component of the propagating field's wavenumber vector, for two nominal wavelengths: $\lambda = 2d$ ($k = \pi/d$) on the top row, $\lambda = 4d$ ($k = \pi/2d$) on the bottom. This variation in wavelength could occur because of differing source frequencies, a factor of two lower in the bottom row than in the top, or by differing propagation speeds, a factor of two slower in the bottom than the top. Despite this signal/medium difference, the wavenumber periodicity of the beampattern depends on the intersensor separation d only. The left column displays the beampattern for a beam looking for waves propagating from an angle of 25.66°, an angle of 60° on the right. The authors chose this odd combination of directions of propagation because $\sin 25.66° = 1/2 \sin 60°$. Note that the top-left and bottom-right beampatterns are identical: This combination of source directions and frequencies results in the same beampattern. This correspondence again illustrates the point that source direction or frequency must be known to determine the other. In all cases, an $M = 21$ regular linear array having uniform weights produced the beampatterns.

By using $k_x = \omega^o \alpha_x = k\zeta_x$, we find that $W(k_x - k_x^o)$ is just a reversed and shifted version of the aperture smoothing function discussed in §3.3.1. The reversal doesn't matter in this case because W is symmetric. Fig. 4.5 shows the shifted array patterns for two wavelengths. These patterns are plotted as a function of k_x^o, the x component of a monochromatic plane wave's wavenumber vector. They show the response of a delay-and-sum beamformer steered to a particular assumed propagation direction.†

In applications, we may want to plot the array pattern as a function of incidence angle rather than wavenumber. For our linear array, the argument of the array pattern can be written as

$$\begin{aligned} \omega^o \vec{\alpha} - \vec{k}^o &= k(\zeta_x - \zeta_x^o) \\ &= k(\sin\phi^o - \sin\phi) \end{aligned}$$

†More accurately, a linear array's beam is steered toward a particular x component of the assumed propagation direction.

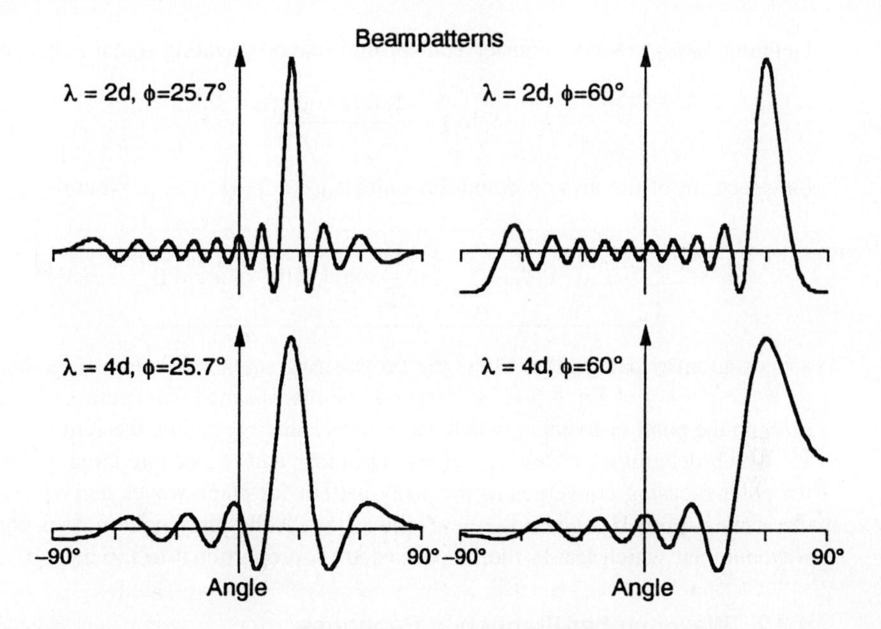

Figure 4.6 The array patterns displayed in the previous figure are now plotted as a function of incidence angle ϕ^o. Note the different number of sidelobes in the top and bottom rows. This effect occurs because the visible region in the bottom row corresponds to $|k_x^o| \leq \pi/2d$. Also note the larger mainlobe asymmetry for the larger off-broadside incidence angle. This effect occurs because, for a linear array, wavenumber is related to the sine of incidence angle.

where the incidence angles ϕ and ϕ^o are measured clockwise from the perpendicular to the array axis (Fig. 3.3 {63}). This clockwise definition leads to the sign change in the previous expression. Because the array pattern value depends on the difference of the *sines* of the angles rather than on the difference of the angles themselves, the array pattern plots have different shapes depending on the value of the steered angle ϕ (Fig. 4.6).

Note there are two periodicities that come into play in this example. The first is the inherent periodicity of $W(k_x)$ as a function of k_x because of the regular spatial sampling by the sensors. This type of periodicity was discussed in detail in §3.3 {84}. The second periodicity emerges when we plot the array pattern as a function of ϕ or ϕ^o; the angular argument forces these plots to be periodic.

Point focusing (near-field) array pattern. We can derive an explicit expression for the array pattern produced by focused arrays. From Eq. 4.3 {118} for the focused delay-and-sum beamformer, we can derive the beamformer's output in the frequency domain.

$$
\begin{aligned}
z(t) = \frac{1}{r^o} \int_{-\infty}^{\infty} & S(\omega) \exp\{j\omega(t - r^o/c)\} \\
& \times \sum_{m=0}^{M-1} w_m \frac{r^o}{r_m^o} \exp\left\{jk[(r^o - r) - (r_m^o - r_m)]\right\} d\omega
\end{aligned}
\tag{4.6}
$$

Defining $S_0(\omega)$ to be the Fourier Transform of the propagating signal at the spatial origin,

$$S_0(\omega) = \frac{S(\omega)\exp\{-j\omega r^o/c\}}{r^o}$$

the spectrum of the array's output becomes $S_0(\omega) \cdot \mathcal{W}(k, \vec{x}, \vec{x}^o)$, where

$$\boxed{\mathcal{W}(k, \vec{x}, \vec{x}^o) \equiv \sum_{m=0}^{M-1} w_m \frac{r^o}{r_m^o} \exp\left\{jk[(r^o - r) - (r_m^o - r_m)]\right\}} \tag{4.7}$$

acts as an array pattern that filters the propagating signal. This filter function, analogous to $W(\omega^o\vec{\alpha} - \vec{k}^o)$ of Eq. 4.5 {121}, depends on the assumed wavenumber k, which equals ω/c, on the point of focus $\vec{x}$, which determines r and r_m, and on the actual source location $\vec{x}^o$, which determines r^o and r_m^o. If we let both r and r^o become large, the array pattern for point focusing converges to the array pattern for plane waves derived earlier. When the array is properly focused, $r = r^o$ and $r_m = r_m^o$, the array pattern does not depend on wavenumber, which leaves the output spectrum proportional to the input's.

4.2.2 Wavenumber-Frequency Response

We would like to consider the beamformer's signal $z(t)$ as the output of a space-time filter operating on the wavefield $f(\vec{x}, t)$. Unfortunately, a minor problem emerges. The output signal of a spatiotemporal filter is generally a function of both space and time, but $z(t)$ is only a function of time. Subsequently we shall find that some underlying space-time output signal $z(\vec{x}, t)$ can be associated with the beamformer's output. If we assume that our space-time filter is linear and time invariant, its output signal equals the input signal $f(\vec{x}, t)$ convolved with the system impulse response $h(\vec{x}, t)$.

$$z(\vec{x}, t) = \iint h(\vec{\chi}, \tau) f(\vec{x} - \vec{\chi}, t - \tau)\, d\vec{\chi}\, d\tau$$

In the frequency domain, the output equals the inverse Fourier Transform of product between the wavenumber-frequency spectra of $h(\vec{x}, t)$ and $f(\vec{x}, t)$.

$$z(\vec{x}, t) = \frac{1}{(2\pi)^4} \iint H(\vec{k}, \omega) F(\vec{k}, \omega) \exp\{j(\omega t - \vec{k} \cdot \vec{x})\}\, d\vec{k}\, d\omega$$

Evaluating this equation at $\vec{x} = \vec{0}$ gives us

$$z(t) = \frac{1}{(2\pi)^4} \iint H(\vec{k}, \omega) F(\vec{k}, \omega) \exp\{j\omega t\}\, d\vec{k}\, d\omega \tag{4.8}$$

The delay-and-sum beamformer computes

$$z(t) = \sum_{m=0}^{M-1} w_m f(\vec{x}_m, t - \Delta_m)$$

As described earlier, to find the frequency response of the beamformer we let our test wavefield be the monochromatic plane wave $\exp\{j(\omega^o t - \vec{k}^o \cdot \vec{x})\}$. For this important special case, the output equals

$$z(t) = \left[\sum_{m=0}^{M-1} w_m \exp\{-j(\omega^o \Delta_m + \vec{k}^o \cdot \vec{x}_m)\}\right] \exp\{j\omega^o t\}$$

Comparing this expression with the one just derived for the frequency-domain expression of a spatiotemporal filter whose output is sampled at the spatial origin (Eq. 4.8), we can identify the term in brackets as $H(\vec{k}^o, \omega^o)$, the complex coefficient that characterizes the delay-and-sum beamformer at the wavenumber-frequency value $(\vec{k}^o, \omega^o)$. Thus, the wavenumber-frequency response for the delay-and-sum beamformer may be generally written as

$$\boxed{H(\vec{k}, \omega) = \sum_{m=0}^{M-1} \left[w_m \exp\{-j\omega\Delta_m\}\right] \exp\{-j\vec{k} \cdot \vec{x}_m\}} \tag{4.9}$$

The quantity in brackets is a *complex* weight for the mth sensor that describes the temporal filtering component. When the sensor delays $\{\Delta_m\}$ are chosen to look for plane waves propagating with a slowness vector $\vec{\alpha} = \vec{\zeta}/c$,

$$\boxed{H(\vec{k}, \omega) = W(\omega\vec{\alpha} - \vec{k})} \tag{4.10}$$

which is consistent with Eq. 4.4 {120}.

This expression is reminiscent of the generalized aperture function defined in Chap. 3 (see Eq. 3.6 {68}). Take careful note of the signs in the exponent, however. The aperture smoothing function (or equivalently the array pattern) is defined using $\exp\{j\vec{k}\cdot\vec{x}\}$, conforming to the way we define the four-dimensional spatiotemporal Fourier Transform using the kernel $\exp\{-j(\omega t - \vec{k} \cdot \vec{x})\}$. The wavenumber-frequency response $H(\vec{k}, \omega)$ is defined in terms of the beamformer's response to the plane wave $\exp\{j(\omega t - \vec{k} \cdot \vec{x})\}$. Consequently, if we write the generalized aperture smoothing function corresponding to Eq. 3.6 {68} for arrays as

$$W_\omega(\vec{k}) \equiv \sum_{m=0}^{M-1} w_m \exp\{-j\omega\Delta_m\} \exp\{j\vec{k} \cdot \vec{x}_m\}$$

we see that $H(\vec{k}, \omega) = W_\omega(-\vec{k})$.

Example

Continuing with the linear array discussed in the earlier example, we can infer that its wavenumber-frequency response is

$$H(k_x, \omega) = \frac{\sin \frac{M}{2}(\omega\alpha_x - k_x)d}{\sin \frac{1}{2}(\omega\alpha_x - k_x)d}$$

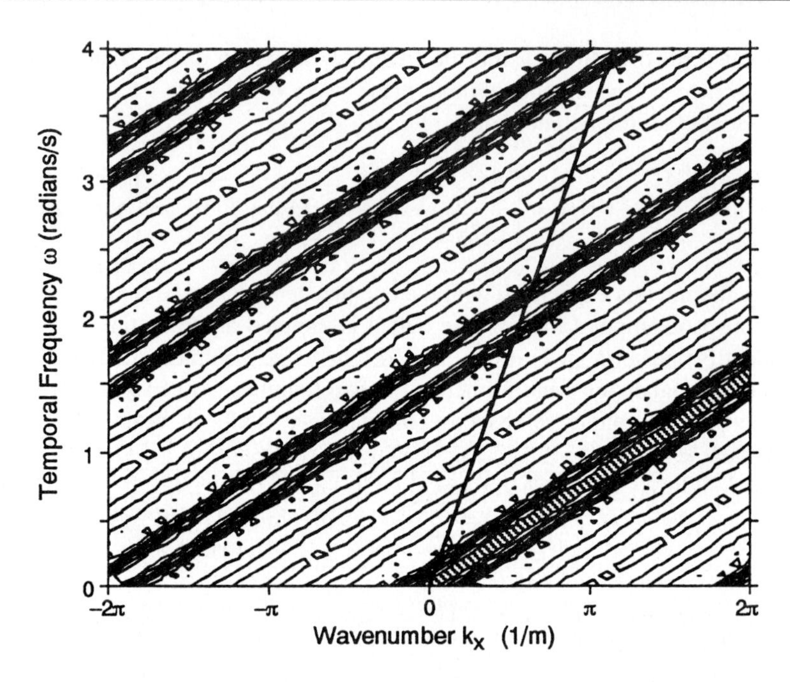

Figure 4.7 The wavenumber-frequency response of a linear array of seven sensors is shown in a crude contour plot. The ω and k_x axes are shown. The response's mainlobe is shaded; it is angled with respect to the axes because it is being steered in the direction $\alpha_x = 4$. Because of spatial aliasing, the mainlobe is repeated as a series of grating lobes, indicated here as a series of parallel ridges having the same amplitude as the mainlobe. The solid line represents the components of a polychromatic plane wave propagating with slowness $\alpha_x = 1$.

A contour plot of $H(k_x, \omega)$ (Fig. 4.7) reveals its mainlobes† and its passband/stopband structure. The mainlobes are centered along lines for which $(\omega\alpha_x - k_x)d = 2\pi N$ or, equivalently,

$$\omega = \frac{c}{\zeta_x}\left(k_x + \frac{2\pi N}{d}\right)$$

where N is an integer. These lines have k_x intercepts spaced $2\pi/d$ apart. The zeros are lines with the same slope but spaced only $2\pi/Md$ apart.

The solid line represents a polychromatic plane wave signal propagating in a direction such that the x component of its slowness vector equals $\alpha_x = \zeta_x^o/c$. This propagating wave's components fall along the line $\omega = k_x c/\zeta_x^o$. If the array is matched to the propagation direction of the wave, $\zeta_x = \zeta_x^o$, this line falls inside the first mainlobe for all frequencies ω. If not (as shown in the figure), the line leaves the mainlobe at some point. Frequencies higher than this are attenuated, but at an even higher frequency, the line crosses into the array's first grating lobe. The grating lobes thus pass spatially aliased components of the propagating wave, and these appear in the beamformer's output.

†More than one occurs because of spatial aliasing.

4.2.3 Beampattern and Steered Response

The wavenumber-frequency response allows us to derive the beamformer's response to an *arbitrary* wavefield just as the frequency response of a bandpass filter can be used to derive the response to an arbitrary signal. The wavenumber-frequency response summarizes the effect of sensor delays and shading on the array's spatiotemporal filtering properties. Because it determines the wavenumber-frequency response, the array pattern becomes the *primary* quantity used to evaluate array and algorithm designs, but in two conceptually different ways.

- The *beampattern* has been described: Assume the array looks for signals propagating with a specific slowness vector $\vec{\alpha}$. How does the array's output vary with the wavefield's parameters? By selecting the delays $\{\Delta_m\}$ for a particular assumed propagation direction, we want to know how the array responds to a variety of propagating fields. Previous sections have shown that when a beamformer is matched to the signal's propagation characteristics, maximal output results. Through the beampattern, we can analyze how the output is disturbed by signals different from the one of our focused attention.
- The second, the *steered response*, differs from the beampattern in important conceptual ways: Assuming that the wavefield's propagation parameters are fixed, how does the array's output vary with assumed propagation direction? With the steered response, we assume a fixed field, scan it by systematically varying the beamformer's delays and possibly its shading, and measure the resulting response.

For the delay-and-sum beamformer, these quantities are equivalent as the array pattern completely defines the spatiotemporal frequency response (Eq. 4.10). In that expression, $W(\omega^o\vec{\alpha} - \vec{k}^o)$, the assumed frequency and propagation direction are entirely represented by $\vec{\alpha}$, whereas the wavefield's propagation parameters are represented by ω^o and $\vec{k}^o$.

Array pattern:	$W(\vec{k}) = \sum_{m=0}^{M-1} w_m \exp\{j\vec{k} \cdot \vec{x}_m\}$
Wavenumber-frequency response:	$H(\vec{k}, \omega) = W(\omega\vec{\alpha} - \vec{k})$
Beampattern:	$W(\omega^o\vec{\alpha} - \vec{k}^o)$ for fixed $\vec{\alpha}$
Steered response:	$W(\omega^o\vec{\alpha} - \vec{k}^o)$ for fixed $\omega^o, \vec{k}^o$

We can also write $W(\omega^o\vec{\alpha} - \vec{k}^o) = W\big(k^o(\vec{\zeta} - \vec{\zeta}^o)\big)$. Thus, the beampattern and the steered response differ only by a sign change in the array pattern's argument when wavenumber is held fixed:† These two conceptually different descriptions are in fact equivalent for the delay-and-sum beamformer. For other, more complicated, algorithms, this equivalence may not hold.

†How these two quantities are related for the known direction of propagation but unknown wavelength is addressed in Prob. 4.12.

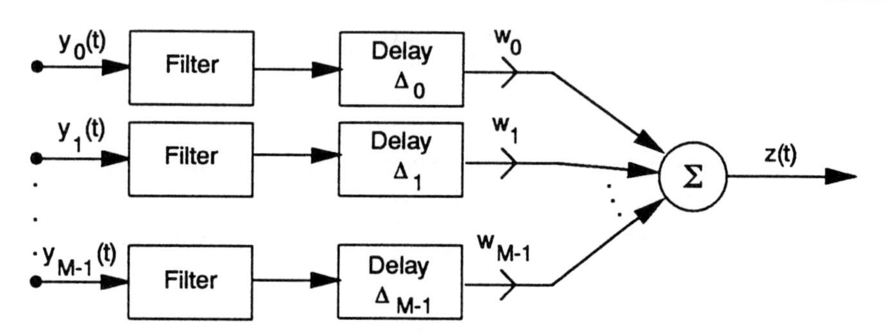

Figure 4.8 The configuration of the filter-and-sum beamformer consists of linear filters preceding the delay-and-sum beamforming stage. The transfer functions of these filters are usually identical and select the range of frequencies corresponding to the signal of interest.

4.3 Filter-and-Sum Beamforming

In describing the delay-and-sum beamformer, we assumed that the sensors sampled the wavefield without distortion or filtering: The mth sensor signal was given by $y_m(t) = f(\vec{x}_m, t)$. As described previously {59}, sensors in actuality act as transducers, frequently filtering the sensor signals both temporally and spatially in the process. We should also realize, somewhat painfully, that more than one signal may be present in the wavefield measured by the sensors and that noise can disturb the observations. To help remove these unwanted disturbances, we may want to insert additional linear filtering to focus the array more than the array pattern suggests. For example, we may know that the signals of interest possess energy in only a narrow frequency band. In addition, digitizing sensor signals means they must first be filtered with antialiasing filters before sampling and analog-to-digital conversion. For all these reasons, we must consider each sensor's output to be a filtered version of the local wavefield. Combining these filtered outputs to form a beam is known as *filter-and-sum beamforming*. How does this combined filtering impact the beamformer's spatiotemporal filtering characteristics?

4.3.1 Temporal Filtering

In the filter-and-sum beamformer depicted in Fig. 4.8, linear filters are placed on the sensor outputs to concentrate the later delay-and-sum beamforming operations on a range of temporal frequencies. Clearly, each filter's passband should correspond to the desired signal's spectral support. These filters' transfer functions can be optimally related to the signal's spectrum and to noise characteristics. The so-called Wiener filter discussed in §6.4.2 {297} is one such optimal filter. For now, we can consider the filter to be ideal, passing with unity gain the desired frequencies and rejecting the rest.

Passing the wavefield through a linear, time-invariant filter means that each sensor's signal equals the convolution of the spatially sampled wavefield with the filter's impulse response.

$$y_m(t) = \int_0^\infty h_m(\tau) f(\vec{x}_m, t - \tau)\, d\tau$$

From this set of sensor signals, filter-and-sum beamformers form beams in the usual way.

$$z(t) = \sum_{m=0}^{M-1} w_m y_m(t - \Delta_m)$$

The linear filters inserted on each sensor do not alter the fact that beamforming is a linear process. We can therefore characterize the filter-and-sum beamformer by determining its response to a monochromatic plane wave. Each sensor's output equals

$$y_m(t) = \exp\{j(\omega^o t - \vec{k}^o \cdot \vec{x}_m)\} \int_0^\infty h_m(\tau) e^{-j\omega^o \tau}\, d\tau$$

The integral equals the filter's frequency response, denoted by $H_m(\omega^o)$, which represents the complex coefficient that multiplies the monochromatic waveform to give the mth sensor signal. The filter-and-sum beamformer's response to the monochromatic plane wave thus becomes

$$z(t) = e^{j\omega^o t} \sum_{m=0}^{M-1} H_m(\omega^o) w_m \exp\{-j(\omega^o \Delta_m + \vec{k}^o \cdot \vec{x}_m)\}$$

Because the sum represents the complex coefficient that multiplies $\exp\{j\omega^o t\}$ to give the beamformer output signal, it equals the wavenumber-frequency response for the filter-and-sum beamformer.

$$\boxed{H(\vec{k}^o, \omega^o) = \sum_{m=0}^{M-1} w_m H_m(\omega^o) \exp\{-j(\omega^o \Delta_m + \vec{k}^o \cdot \vec{x}_m)\}}$$

When the sensor filters are identical with $H_m(\omega) = H_o(\omega)$, the wavenumber-frequency response equals $H_o(\omega)$ times what it would have been in the absence of any sensor filtering. If we set the sensor delays $\{\Delta_m\}$ to look for a plane wave propagating with a slowness vector $\vec{\alpha}$, the wavenumber-frequency response of the filter-and-sum beamformer becomes

$$H(\vec{k}^o, \omega^o) = H_o(\omega^o) W(\omega^o \vec{\alpha} - \vec{k}^o)$$

The result conveyed by this formula—the filtering properties of the beamformer and the sensors accentuate each other—is *very* important; it becomes a theme for all filter-and-sum beamformers.

Fig. 4.9 portrays an example filter-and-sum beamformer reminiscent of Fig. 4.7 {126}, differing only in that a bandpass temporal filter limits the range of frequencies that are passed to the beamforming computation. Such filtering can reduce problems caused by out-of-band noise and other propagating waves in the wavefield. Notice that energy from a mismatched wave, which might corrupt the beamformer's output by leaking in through

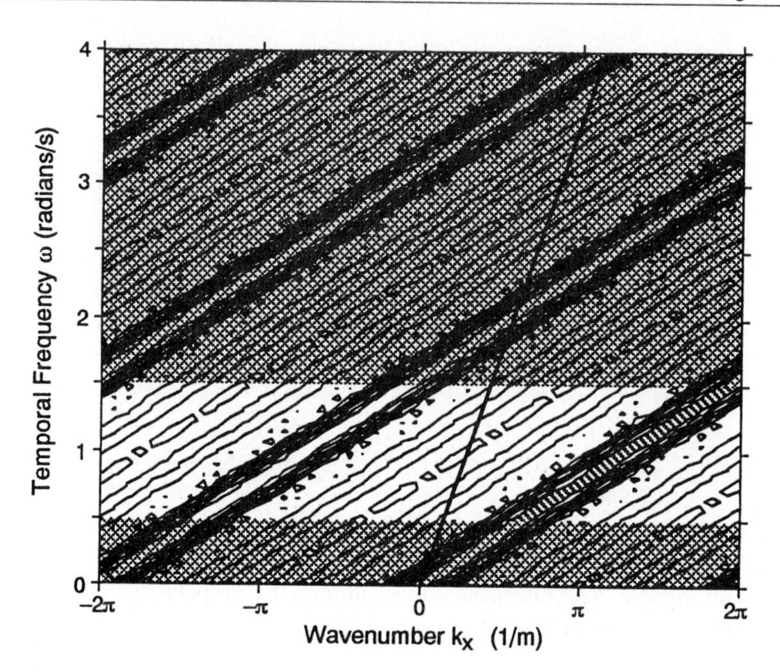

Figure 4.9 This figure shows a plot similar to that of Fig. 4.7 except that the temporal frequency response is limited by filtering at the sensor level to the frequency band [0.5, 1.5]. The filtering restricts the mainlobe of the beamformer to the unshaded limited range of temporal frequencies. Now the wave propagating with a different slowness than expected does not lie in any of the filter-and-sum beamformer's "passbands" (grating lobes).

a grating lobe, is attenuated by the temporal filter. In other words, the wavenumber-frequency values that would have been passed by the grating lobe are not in the range of frequencies passed by the temporal filter.

Because delay-and-sum beamforming is a linear, time-invariant operation, filtering sensor signals before beamforming is equivalent to filtering the beamformer's output signal. In some applications, it may make sense to implement such postbeamforming filtering because only one filter is needed. If array processing components were always linear, this structure and the associated economy should be realized. In practice, however, the prebeamforming structure described here is usually employed because of subtle nonlinearities in physical systems. An A/D converter to digitize sensor outputs is not linear; once inserted into a cascade of linear systems, ordering of the linear systems within the cascade now matters. By filtering before digitization, we eliminate uninteresting signals that would otherwise limit the available amplitude range for undistorted digitization of the interesting signal. By devoting more of the A/D converter's dynamic range to the signal, we reduce amplitude quantization noise.

4.3.2 Spatial Filtering

The simple concept of the filter-and-sum beamformer can be generalized to include spatial filtering of sensor signals. This filtering most frequently occurs when the sensors are not omnidirectional, having greater sensitivity to waves propagating in certain directions. We find this situation when the sensors are continuous apertures, as in the case of telescopes (Fig. 3.18 {86}) or World War I listening devices (Fig. 1.1 {2}), or consist of small subarrays, as in the case of some sonar and seismic arrays. When the sensors provide such local spatial filtering, the signal from the mth sensor can be written as a spatial convolution.

$$y_m(t) = \int_{-\infty}^{\infty} h_m(\vec{\chi}) f(\vec{x}_m - \vec{\chi}, t)\, d\vec{\chi}$$

Again using a monochromatic plane wave to characterize the resulting beamformer, the sensor signals become

$$y_m(t) = H_m(\vec{k}^o) \exp\{j(\omega^o t - \vec{k}^o \cdot \vec{x}_m)\}$$

where $H_m(\vec{k}^o)$ denotes the spatial frequency response of the mth sensor's filter.† Thus, the beamformer's output equals

$$z(t) = \exp\{j\omega^o t\} \sum_{m=0}^{M-1} H_m(\vec{k}^o) w_m \exp\{-j(\omega^o \Delta_m + \vec{k}^o \cdot \vec{x}_m)\}$$

meaning that the resulting wavenumber-frequency response for the spatial filter-and-sum beamformer equals

$$H(\vec{k}^o, \omega^o) = \sum_{m=0}^{M-1} H_m(\vec{k}^o) w_m \exp\{-j\omega^o \Delta_m - j\vec{k}^o \cdot \vec{x}_m\}$$

If identical spatial filtering occurs at each sensor and sensor delays are adjusted to steer the beam to look for plane waves propagating with a slowness vector $\vec{\alpha}$, then the wavenumber-frequency response of the filter-and-sum beamformer becomes

$$\boxed{H(\vec{k}^o, \omega^o) = H_o(\vec{k}^o)\, W(\omega^o \vec{\alpha} - \vec{k}^o)}$$

4.3.3 Spatiotemporal Filtering

Temporal and spatial filtering of sensor signals need not be be applied separately; we can combine them to enhance the beamformer's filtering properties. Filtering a sensor signal in space and time is represented as a space-time convolution.

$$y_m(t) = \iint h_m(\vec{\chi}, \tau)\, f(\vec{x} - \vec{\chi}, t - \tau)\, d\vec{\chi}\, d\tau$$

†The authors hope the reader is not confused by the notational similarities between $H(\omega)$ and $H(\vec{k})$. The first denotes the frequency response of a temporal filter and equals a complex-valued function of the scalar frequency variable ω. The second is a complex-valued function of the vector spatial frequency $\vec{k}$. Mathematically, the two functions represent very different quantities.

As just described, by assuming that the wavefield is a monochromatic plane wave, we find that the wavenumber-frequency response for the filter-and-sum beamformer equals

$$\boxed{H(\vec{k}^o, \omega^o) = \sum_{m=0}^{M-1} H_m(\vec{k}^o, \omega^o)\, w_m \exp\{-j(\omega^o \Delta_m + \vec{k}^o \cdot \vec{x}_m)\}}$$

Common sensor filters $H_o(\vec{k}, \omega)$ and beam steerage for plane waves propagating with a slowness vector $\vec{\alpha}$ yields the wavenumber-frequency response

$$H(\vec{k}^o, \omega^o) = H_o(\vec{k}^o, \omega^o)\, W(\omega^o \vec{\alpha} - \vec{k}^o)$$

This result means that the beamformer's spatiotemporal filter is in cascade with that of each sensor. Thus, arrays of directional sensors, such as the *VLA*, the directionality of the sensors and the directionality of the array *complement* each other, creating a directivity pattern vastly superior to that provided by either alone.

Suppose that each sensor is in fact a subarray: a small array of omnidirectional sensors. What we have now is an array of subarrays with each subarray acting as a spatiotemporal filter. When the mth subarray's sensor delays are adjusted to look for plane waves propagating with slowness $\vec{\alpha}$, their wavenumber-frequency responses are given by

$$H_m(\vec{k}^o, \omega^o) = W_m(\omega^o \vec{\alpha} - \vec{k}^o)$$

where the array pattern for the mth subarray equals

$$W_m(\vec{k}) = \sum_{m_l=0}^{M_l-1} w_{m_l} \exp\{j\vec{k} \cdot \vec{x}_{m_l}\}$$

The position $\vec{x}_{m_l}$ denotes the location of the lth sensor of the mth subarray *relative to the phase center of the* mth *subarray*. Again, if each subarray has the same relative configuration of sensors and corresponding sensor weights, each array pattern equals $W_o(\vec{k})$, and the overall wavenumber-frequency response of the resulting filter-and-sum beamformer equals

$$H(\vec{k}^o, \omega^o) = W_o(\omega^o \vec{\alpha} - \vec{k})\, W(\omega^o \vec{\alpha} - \vec{k}) \tag{4.11}$$

We see that the overall array pattern of an array of subarrays equals the product of the subarray pattern $W_o(\omega^o \vec{\alpha} - \vec{k})$ with the array pattern $W(\omega^o \vec{\alpha} - \vec{k})$ that would result if the subarrays were replaced by omnidirectional point sensors having no spatiotemporal filtering. This property can be exploited to analyze a complex array structure that may be considered as an array of subarrays, of sub-subarrays, and so on.

4.4 Frequency-Domain Beamforming

Previous sections describe how computing the beamformer output signal by delaying and summing (or filtering and summing) the sensor signals corresponds to a spatiotemporal filter that selectively passes waves propagating from a narrow range of directions or locations. This spatiotemporal filtering is accomplished by delaying, then adding, the sensor outputs. This time domain view of beamforming has obvious frequency-domain counterparts: Delay corresponds to linear phase shift. *Frequency-domain beamforming* implements the calculations entirely in the frequency domain by Fourier transforming the inputs, applying the spatiotemporal filter, and inverse transforming the result. This seemingly more complicated strategy does have advantages in some situations. We only introduce frequency-domain ideas here, saving the major results until §4.8.5 {172} when we are prepared to employ fast digital computational methods such as the *FFT*.

We can conceptually decompose a general plane wave of the form $f(\vec{x}, t) = s(t - \vec{\alpha}^o \cdot \vec{x})$ into a superposition of monochromatic plane waves. Each monochromatic component can be treated independently and formed into a beam signal at each frequency by applying a phase shift appropriate to the desired delay. More precisely, letting $Y_m(\omega)$ denote the Fourier Transform of the mth sensor's output, the spectrum of the delay-and-sum beamformer's output equals

$$Z(\omega) = \sum_{m=0}^{M-1} Y_m(\omega) \exp\{-j\omega\Delta_m\}$$

The inverse Fourier Transform of $Z(\omega)$ can then be calculated to produce the time-domain output; in some applications, though, $Z(\omega)$ itself may be of primary importance. In practice, however, we can't compute $Y_m(\omega)$ because that would require integrating over all time. Even if we had the patience, the source would move, violating the implicit assumption that the wavefield's propagation characteristics do not change with time. We must consider for any practical frequency-domain system the ideas expressed by short-time Fourier analysis.

4.4.1 Short-Time Fourier Analysis

The key theoretical framework that describes computing spectra over finite time intervals is the *short-time Fourier Transform*.[†] Calculating this quantity means that we apply to a signal at time t a window of duration D, then evaluate the Fourier Transform of the product.

$$\boxed{Y_m(t, \omega) \equiv \int_t^{t+D} \widetilde{w}(\tau - t) y_m(\tau) e^{-j\omega\tau}\, d\tau} \tag{4.12}$$

Here, $\widetilde{w}(t)$ denotes the finite-duration window defined over $[0, D]$. The window's duration defines the frequency resolution of the short-time Fourier analysis: Because $Y_m(t, \omega)$ equals the Fourier Transform of the product between the sensor signal and the window origined at t, the signal's spectrum is smoothed by the window's spectrum. For the short-time spectrum calculated at one time instant to approximate the signal's spectrum well, we should choose the window's duration so that the spectral smoothing function's

[†] We can "calculate" the short-time spectrum with a bandpass filter bank. Prob. 4.10 explores the correspondence between filter banks and Fourier Transforms.

width is narrow compared with the signal's spectral variations. Reasonable windows have Fourier Transforms that resemble a single narrow spike; examples are shown in Fig. 6.6 {324ff}. Simple manipulations yield an equivalent representation of the short-time Fourier Transform.

$$\boxed{Y_m(t,\omega)e^{j\omega t} = \int_0^D \widetilde{w}(\tau)y_m(t+\tau)\exp\{-j\omega\tau\}\,d\tau}$$

This representation means that $Y_m(t,\omega)$ is a complex-valued lowpass signal that approximates the "local" spectrum of the sensor's output at time t and frequency ω.

The frequency-domain beamformer's output results after summing the short-time spectra once each has been linearly phase-shifted by an amount appropriate to the assumed propagation delay.

$$\boxed{Z(t,\omega) \equiv \sum_{m=0}^{M-1} w_m Y_m(t,\omega)e^{j\omega t}e^{-j\omega\Delta_m}} \tag{4.13}$$

The quantity $Z(t,\omega)$ can be regarded as an approximation to $Z(\omega)e^{j\omega t}$. Integrating $Z(t,\omega)$ over all frequencies (and normalizing by 2π) gives us an approximation to $z(t)$. A filter-and-sum beamformer having temporal filtering properties may be realized simply by allowing the sensor weights $\{w_m\}$ to be functions of frequency. In this way, filter-and-sum beamformers find their frequency-domain counterparts. In many practical applications, such as radar, the waves of interest are known to occupy a relatively narrow band (width B) of frequencies centered on ω^o. In this case, $Z(t,\omega)$ approximately equals zero except for $\omega = \omega^o$. The sensor delays needed to steer the beam can be implemented accurately with phase shifts of $\{\omega^o\Delta_m\}$.

$$z(t) \approx \frac{B}{2\pi}Z(\omega^o)\exp\{j\omega^o t\} \approx \frac{B}{2\pi}Z(t,\omega^o)$$

4.4.2 Spatial Correlation Matrix

An alternate formulation of frequency-domain beamforming introduces a fundamental quantity, the *spatial correlation matrix*. This quantity not only underlies many conventional beamforming algorithms but also the more modern ones discussed in Chap. 7.[†] Let the vector $\mathbf{Y}(t,\omega)$ denote the collection of short-time Fourier transforms across the array: $\mathbf{Y}(t,\omega) = \text{col}[Y_0(t,\omega),\ldots,Y_{M-1}(t,\omega)]$. Shading can be introduced as multiplication of this vector by a diagonal matrix $\mathbf{W}$ having the sensor weights $\{w_m\}$ running down the

[†] At this point, the reader may well want to review appendix B, which describes fundamental properties of matrices and operations on them. The expression of beamforming operations in matrix form is *the* key to developing a deeper understanding of conventional beamforming and of more modern algorithms.

main diagonal: $\mathbf{W} = \text{diag}[w_0(\omega), \ldots, w_{M-1}(\omega)]$.

$$\mathbf{WY} = \begin{bmatrix} w_0(\omega)Y_0(t,\omega) \\ w_1(\omega)Y_1(t,\omega) \\ \vdots \\ w_{M-1}(\omega)Y_{M-1}(t,\omega) \end{bmatrix}$$

Dependence of the weights on frequency describes any temporal filtering that may be applied to each sensor. Let $\mathbf{e}$ denote the *steering vector*: A sequence of phasors whose exponents are chosen to cancel the plane-wave signal's propagation-related phase shift. These phase shifts steer the beam's assumed propagation direction to the wave's direction of propagation (or focus the beam in the case of a near-field source). Because the wave's direction of propagation may be slightly different at each sensor, denote the local wavenumber at the mth sensor by $\vec{k}^o_m$.

$$\mathbf{e} = \begin{bmatrix} \exp\{-j\vec{k}^o_0 \cdot \vec{x}_0\} \\ \vdots \\ \exp\{-j\vec{k}^o_{M-1} \cdot \vec{x}_{M-1}\} \end{bmatrix}$$

In the case of a plane wave propagating across the array, all the local wavenumbers equal $\vec{k}^o$ and the phase shifts satisfy $\vec{k}^o \cdot \vec{x}_m = -\omega\Delta_m$.† The steering vector thus corresponds to the collection of the *signal's* phase shifts across the array under the supposition that the signal is propagating in a given direction or from a given point. Consequently, the vector $\mathbf{e}$ models the signal's propagation characteristics in the frequency domain.

The Fourier Transform of the beamformer output can be concisely written as the inner product of $\mathbf{WY}$ and $\mathbf{e}$. As described in appendix B, the inner product of complex-valued vectors equals the Hermitian transpose of one times the other {489}.

$$\boxed{Z(t,\omega)e^{-j\omega t} = \sum_{m=0}^{M-1} w_m Y_m(t,\omega) e^{j\vec{k}^o_m \cdot \vec{x}_m} = \mathbf{e}'\mathbf{WY}(t,\omega)}$$

The conjugate of the steering vector in this formula introduces the phase shifts necessary to compensate for those caused by propagation, thereby allowing the sensor signals to be stacked properly.

The *steered response power* is defined as the power in the beamformer's output spectrum considered as a function of the steering vector $\mathbf{e}$, which defines the assumed propagation direction.

$$\boxed{\mathcal{P}(\mathbf{e}) \equiv \int_{-\infty}^{\infty} \mathbf{e}'\mathbf{WYY}'\mathbf{W}'\mathbf{e}\, d\omega}$$

For conciseness, the notational dependence on t and ω has been suppressed. The integrand is a quadratic form and plays a prominent role in sequel. The heart of this quadratic form

†Applying linear phase shifts to produce a beam focused on certain directions of propagation is known as *phase steering*.

is the outer product of the sensor Fourier Transform vector with itself. This outer product is called the *spatial correlation matrix* **R**.[†]

$$\boxed{\mathbf{R} \equiv \mathbf{Y}\mathbf{Y}'}$$

This $M \times M$ matrix depends on temporal frequency, but this dependence is suppressed notationally. When a single plane wave $f(\vec{x}, t) = s(t - \vec{\alpha}^o \cdot \vec{x})$ propagates toward the array, each element of this matrix equals

$$[\mathbf{R}]_{m_1 m_2} = |S(\omega)|^2 \exp\{-j\omega\vec{\alpha}^o \cdot (\vec{x}_{m_1} - \vec{x}_{m_2})\}$$

Example

Let's consider a regular linear array laid out along the x axis with sensors separated by d m. For this array, $\vec{\alpha}^o \cdot \vec{x}_m = \alpha_x^o d(m - M_{1/2})$, which means that each element of the spatial correlation matrix depends only on the difference of its indices $m_1 - m_2$ rather than on them individually. The values along any diagonal of the matrix equal one another, a structure technically known as a Toeplitz matrix {487}. Furthermore, the element $[\mathbf{R}]_{m_1 m_2}$ equals the complex conjugate of $[\mathbf{R}]_{m_2 m_1}$; this conjugate symmetric structure defines a Hermitian matrix {487}. The spatial correlation matrix thus has the form of a Toeplitz, Hermitian matrix.

$$\mathbf{R} = \begin{bmatrix} R_0 & R_1 & R_2 & \cdots & R_{M-1} \\ R_1^* & R_0 & R_1 & \cdots & R_{M-2} \\ R_2^* & R_1^* & R_0 & \cdots & R_{M-3} \\ \vdots & \vdots & \vdots & \ddots & \vdots \\ R_{M-1}^* & R_{M-2}^* & R_{M-3}^* & \cdots & R_0 \end{bmatrix}$$

For the regular linear array,

$$R_{(m_1 - m_2)} = |S(\omega)|^2 \exp\{-j\omega\alpha_x^o (m_1 - m_2) d\}$$

Note that the number of times that R_l appears in the matrix equals the redundancy of the lth lag in the discrete co-array. See the discussion of co-arrays in §3.3.4 {94}. In general, the number of times a lag appears in the spatial correlation matrix equals the discrete co-array value at that lag. Each corresponding negative lag has the conjugate value of its positive counterpart. The main diagonal itself reflects the M-fold redundancy of any M-sensor array at zero lag.

The computation of the steered response power can be interpreted in the context of our new vector-based notation. The spatial correlation matrix displays all possible conjugate products between the temporal Fourier Transform of one sensor's output with another's transform. This display contains all of the information needed to compute a signal's direction of propagation. The fact that only the relative phase between each sensor pair is retained means that an overall propagation delay does not affect the beamforming operation. The quadratic form $\mathbf{e}'\mathbf{R}\mathbf{e}$ equals the squared magnitude of the spatiotemporal Fourier Transform at a specific spatiotemporal frequency. As the spatial correlation

[†]This definition for the spatial correlation matrix represents a more carefully defined expression of the spatiospectral correlation matrix defined in §2.6.4 {51}. Here, we use the full theory of short-time Fourier analysis rather than computing the transform of a frame selected from the sensors' observations.

matrix depends only on temporal frequency, assailing it with a steering vector on both sides to forge a quadratic form represents a spatial Fourier Transform operation. With this interpretation, the Fourier Transform of a vector is the inner product of the vector with the phase-shift vector **e** and the squared-magnitude Fourier Transform (the power spectrum) of a matrix is the quadratic form composed of the matrix and the phase-shift vector. The former is a complex-valued quantity containing absolute phase information; the latter produces the power in the array output at a particular spatiotemporal frequency, which when integrated across frequency yields the key quantity in some beamforming algorithms.

4.5 Array Gain

The issue of contaminating noise in the wavefield can be ignored no longer. Whether noise or unwanted field variations propagate or not, how they contaminate sensor and array outputs becomes an important design criterion. Because an array is composed of many sensors, each making an independent measurement of the wavefield, intelligent combination of the sensor signals should lead to an overall increase in the signal-to-noise ratio. By combining many measurements, we hope to reinforce our ideal propagating signal while reducing noise amplitude. In this way, we try to achieve one of the primary aims of array signal processing: enhancing the measured signal with an array beyond the capabilities of a single sensor. Array gain measures an array's signal-to-noise ratio enhancement and concisely summarizes how well it and subsequent signal processing reject noise [37];[144: §6.3–6.4].

Signal-to-noise ratio. If a single sensor were located at the spatial origin, its response to a noise-corrupted signal would be

$$y(t) = s(\vec{0}, t) + n(\vec{0}, t)$$

where $n(\vec{x}, t)$ represents the noise field. This noise may be attributed to the sensor (thermal noise in its electronics, for example) or to background radiation. For simplicity, assume that the desired signal is a wideband plane wave of the form $s(\vec{x}, t) = s(t - \vec{\alpha}^o \cdot \vec{x})$, which means that our lonely sensor's output signal is $s(t) + n(\vec{0}, t)$. In addition, assume that s and n are stationary random fields. The signal-to-noise ratio SNR is defined to be the ratio of mean-squared values of the signal and noise components, which conceptually expresses the ratio of signal and noise powers. For our single sensor,

$$\boxed{SNR_{\text{sensor}} = \frac{\mathcal{E}[s^2(t)]}{\mathcal{E}[n^2(\vec{0}, t)]} = \frac{R_s(0)}{R_n(\vec{0}, 0)}} \tag{4.14}$$

This simple calculation makes an important assumption: The signal and noise components are uncorrelated. If they were correlated, SNR should reflect the ratio of signal power to *uncorrelated* noise power (Prob. 4.13). For simplicity, signal and noise are assumed uncorrelated from this point on.

When we employ an array of M sensors, the signal measured by the mth sensor is $y_m(t) = s(t - \vec{\alpha}^o \cdot \vec{x}_m) + n(\vec{x}_m, t)$. The delay-and-sum beamformer's output signal thus equals

$$\begin{aligned} z(t) &= \sum_m w_m y_m(t - \Delta_m) \\ &= \sum_{m=0}^{M-1} w_m s(t - \Delta_m - \vec{\alpha}^o \cdot \vec{x}) + \sum_{m=0}^{M-1} w_m n(\vec{x}_m, t - \Delta_m) \end{aligned}$$

The *array signal-to-noise ratio* is the ratio of mean-squared values of these signal and noise terms. Again, regarding signal and noise as stationary random processes, these mean-squared values are

$$\text{Signal:} \quad \sum_{m_1=0}^{M-1} \sum_{m_2=0}^{M-1} w_{m_1} w_{m_2}^* R_s\big(\Delta_{m_2} - \Delta_{m_1} - \vec{\alpha}^o \cdot (\vec{x}_{m_1} - \vec{x}_{m_2})\big) \tag{4.15a}$$

$$\text{Noise:} \quad \sum_{m_1=0}^{M-1} \sum_{m_2=0}^{M-1} w_{m_1} w_{m_2}^* R_n(\vec{x}_{m_1} - \vec{x}_{m_2}, \Delta_{m_2} - \Delta_{m_1}) \tag{4.15b}$$

where $R_s(\cdot)$ denotes the signal's correlation function and $R_n(\cdot, \cdot)$ the spatiotemporal correlation function of the noise.

The array gain G is defined as the ratio of the array signal-to-noise ratio and the sensor signal-to-noise ratio.

$$\boxed{G \equiv \frac{SNR_{\text{array}}}{SNR_{\text{sensor}}}}$$

For the general case, the array gain equals the ratio of the normalized signal and noise powers.

$$G = \frac{\sum_{m_1=0}^{M-1} \sum_{m_2=0}^{M-1} w_{m_1} w_{m_2}^* R_s\big(\Delta_{m_2} - \Delta_{m_1} - \vec{\alpha}^o \cdot (\vec{x}_{m_1} - \vec{x}_{m_2})\big) / R_s(0)}{\sum_{m_1=0}^{M-1} \sum_{m_2=0}^{M-1} w_{m_1} w_{m_2}^* R_n(\vec{x}_{m_1} - \vec{x}_{m_2}, \Delta_{m_2} - \Delta_{m_1}) / R_n(\vec{0}, 0)}$$

Matched delays. If the beamformer's delays are properly matched to the wave's direction of propagation, that is, if $\Delta_m = -\vec{\alpha}^o \cdot \vec{x}_m$, the argument of R_s goes to 0 and the unnormalized mean-squared value of the signal term equals $R_s(0) \left|\sum_m w_m\right|^2$. If we assume that the noise is uncorrelated spatially from sensor to sensor, $R_n(\vec{x}_{m_1} - \vec{x}_{m_2}, \tau) = 0$, for $\vec{x}_{m_1} \neq \vec{x}_{m_2}$, we find that the expression for the unnormalized mean-squared noise power in the array output becomes $R_n(\vec{0}, 0) \sum_m |w_m|^2$. Thus, the signal power is multiplied by a factor $\left|\sum_m w_m\right|^2$, whereas the noise power is multiplied by $\sum_m |w_m|^2$. For this simplified situation—no array mismatch and spatially uncorrelated noise at the sensor locations—the array signal-to-noise ratio thus equals

$$SNR_{\text{array}} = \frac{R_s(0)\left|\sum_{m=0}^{M-1} w_m\right|^2}{R_n(\vec{0},0)\sum_{m=0}^{M-1}|w_m|^2} \tag{4.16}$$

In our simplified case, the array gain depends only on the shading applied to the sensor outputs.

$$\boxed{G = \frac{\left|\sum_{m=0}^{M-1} w_m\right|^2}{\sum_{m=0}^{M-1}|w_m|^2}}$$

Example

Consider the simple case of an M-sensor *uniform* array: Constant sensor weights are applied to each sensor so that $w_m = w$, $m = 0, \ldots, M-1$. The array gain is simply $G = \left(M^2|w|^2\right)/\left(M|w|^2\right)$, which equals M *independent* of the array's intersensor spacing and the sensor weights. Thus, by stacking M replicas of the signal corrupted by independent noise, we increase the signal-to-noise ratio by a factor of M.

Now suppose that one-half the sensors are weighted by 1.0, the other half by 0.5. Calculations show that the array gain now equals

$$G = \frac{\left|\sum_{m=0}^{M-1} w_m\right|^2}{\sum_{m=0}^{M-1}|w_m|^2} = \frac{\frac{9}{16}M^2}{\frac{5}{8}M} = 0.9M$$

The nonunity weights cause the array gain to be 10% less than before. Note that the array gain depends on how many sensors are half-weighted, not which ones. Of course, the array pattern, and in particular the height of the sidelobes, are critically dependent on which sensors are half-weighted (Prob. 3.20 {110}). Because of the Schwarz Inequality, maximum array gain occurs when we apply a uniform shading.

Mismatched delays. In the preceding derivation, we steered perfectly to capture the propagating wave. What happens if the steering delays are mismatched to the wave's direction of propagation? The numerator of the general expression given for array gain does not simplify in this case, whereas the denominator remains as before. As $R(0) \geq |R(\tau)|$ for any correlation function, mismatch leads to a smaller numerator than produced by perfectly matched sensor delays. In relation to the matched case, *mismatched delays reduce the signal-to-noise ratio at the beamformer output, thereby reducing the array gain.* Not only does mismatch yield distorted array outputs, they're also noisier.

Example

We can derive a simpler expression for the array gain in the case of mismatched delays if we assume that the desired signal is a monochromatic plane wave. In this case, the signal power spectrum is given by $2\pi\delta(\omega - \omega^o)$. The corresponding correlation function equals the inverse Fourier Transform of this quantity, giving $R_s(\tau) = \exp\{j\omega^o\tau\}$. Under the assumption that the noise field is stationary and uncorrelated from sensor to sensor, the array gain becomes

$$G = \frac{\left|\sum_{m=0}^{M-1} w_m \exp\{j\omega^o(\Delta_m + \vec{\alpha}^o \cdot \vec{x}_m)\}\right|^2}{\sum_{m=0}^{M-1} |w_m|^2}$$

The exponent in the numerator thus depends on the difference between the selected delays $\{\Delta_m\}$ and the ideal ones $\{-\vec{\alpha}^o \cdot \vec{x}_m\}$. Using the Schwarz Inequality, the array gain is maximized when the shading cancels the mismatch: Mismatch causes array gain to suffer by reducing the signal power in the array's output.

Wavenumber-frequency expressions. The signal-to-noise ratio at the beamformer output as well as the array gain can be expressed in terms of wavenumber-frequency power spectra. Because delay-and-sum beamforming acts as a spatiotemporal filter, the power spectrum of its output signal and noise components can be expressed in terms of the field's component power spectra and the wavenumber-frequency transfer function §2.6.2 {48}.

$$\mathcal{S}_z(\vec{k}, \omega) = |H(\vec{k}, \omega)|^2 \mathcal{S}_f(\vec{k}, \omega)$$

where $H(\vec{k}, \omega)$ is the wavenumber-frequency response for the delay-and-sum beamformer defined on page 125. As power simply equals the integral of the power spectrum, the signal-to-noise ratio in the array output equals

$$SNR_{\text{array}} = \frac{\int_{-\infty}^{\infty}\int_{-\infty}^{\infty} |H(\vec{k}, \omega)|^2 \mathcal{S}_s(\vec{k}, \omega)\, d\vec{k}\, d\omega}{\int_{-\infty}^{\infty}\int_{-\infty}^{\infty} |H(\vec{k}, \omega)|^2 \mathcal{S}_n(\vec{k}, \omega)\, d\vec{k}\, d\omega}$$

This expression reveals the principles of designing a spatiotemporal filter to increase the output signal-to-noise ratio. If, for some particular value of $(\vec{k}, \omega)$, the signal's power spectrum is significantly greater than the noise power spectrum, then $|H(\vec{k}, \omega)|$ should be large. Conversely, if the signal's power spectrum is much smaller than that of the noise, the frequency response should be smaller at that point. These design ideas form the essence of Eckhart filtering (Prob. 6.15 {344}). The frequency-domain expression for array gain becomes

$$G = \frac{\int_{-\infty}^{\infty}\int_{-\infty}^{\infty} |H(\vec{k}, \omega)|^2 \mathcal{S}_s(\vec{k}, \omega)\, d\vec{k}\, d\omega \Big/ \int_{-\infty}^{\infty}\int_{-\infty}^{\infty} \mathcal{S}_s(\vec{k}, \omega)\, d\vec{k}\, d\omega}{\int_{-\infty}^{\infty}\int_{-\infty}^{\infty} |H(\vec{k}, \omega)|^2 \mathcal{S}_n(\vec{k}, \omega)\, d\vec{k}\, d\omega \Big/ \int_{-\infty}^{\infty}\int_{-\infty}^{\infty} \mathcal{S}_n(\vec{k}, \omega)\, d\vec{k}\, d\omega}$$

Optimization of array gain. Given a set of sensors located at particular positions $\{\vec{x}_m\}$, what values of sensor weights $\{w_m\}$ would maximize the array gain for a given propagating wave and noise field combination? Interestingly, this problem can be solved using matrix algebra. To translate into this convenient framework, note that the expression for array gain can be written as the ratio of normalized quadratic forms.

$$G = \frac{\mathbf{w}'\mathbf{R}_s\mathbf{w}/\,\mathrm{tr}[\mathbf{R}_s]}{\mathbf{w}'\mathbf{R}_n\mathbf{w}/\,\mathrm{tr}[\mathbf{R}_n]} \tag{4.17}$$

Here, $\mathrm{tr}[\cdot]$ denotes the trace of a matrix, $\mathbf{w} = \mathrm{col}[w_0, \ldots, w_{M-1}]$, and the correlation matrices contain elements of the form[†]

$$[\mathbf{R}]_{m_1,m_2} = R(\vec{x}_{m_1} - \vec{x}_{m_2}, \Delta_{m_2} - \Delta_{m_1})$$

By using matrix expressions, we can optimize array gain with eigenanalysis. As shown in §B.5 {495}, the optimal array gain occurs when the shading vector satisfies

$$\mathbf{w}_{\mathrm{opt}} \propto \mathbf{R}_n^{-1/2}\tilde{\mathbf{v}}_{\max}$$

where $\tilde{\mathbf{v}}_{\max}$ denotes the eigenvector of $\left(\mathbf{R}_n^{-1/2}\right)' \mathbf{R}_s\mathbf{R}_n^{-1/2}$ having the largest eigenvalue. Clearly, this complicated solution could stand some simplification. When one source appears in the wavefield and the beamformer delays match the propagation delays, all elements of the signal correlation matrix $\mathbf{R}_s$ are equal. This matrix can thus be expressed as an outer product of the vector $\mathbf{1}$, each component of which equals 1, with itself.

$$\mathbf{R}_s = A^2\mathbf{1}\mathbf{1}^t$$

The matrix kernel, whose eigenstructure we need to determine, itself becomes an outer product.

$$\left(\mathbf{R}_n^{-1/2}\right)' \mathbf{R}_s\mathbf{R}_n^{-1/2} = A\left(\mathbf{R}_n^{-1/2}\right)'\mathbf{1}\left[A\left(\mathbf{R}_n^{-1/2}\right)'\mathbf{1}\right]'$$

This kernel has only one nonzero eigenvalue and the corresponding eigenvector simply equals the vector used to form the outer product. Therefore $\tilde{\mathbf{v}}_{\max} \propto A(\mathbf{R}_n^{-1/2})'\mathbf{1}$ and $\mathbf{w}_{\mathrm{opt}} \propto \mathbf{R}_n^{-1}\mathbf{1}$. In this case, the array gain's maximum value equals $\mathrm{tr}[\mathbf{R}_n]\mathbf{1}^t\mathbf{R}_n^{-1}\mathbf{1}/M$.

$$\boxed{G_{\mathrm{opt}} = \frac{\mathrm{tr}[\mathbf{R}_n]}{M}\mathbf{1}^t\mathbf{R}_n^{-1}\mathbf{1};\quad \mathbf{w}_{\mathrm{opt}} \propto \mathbf{R}_n^{-1}\mathbf{1}} \tag{4.18}$$

Knowledge of the noise correlation function can be invaluable in selecting the array shading coefficients *if* maximum noise rejection is the primary aim of the array design. This seemingly optimal shading may not the best possible when other aspects of the array are considered. For example, a particular shading may not yield an array pattern

[†]This matrix is a simplified version of the full spatiotemporal correlation matrix introduced in §2.6.4 {50}. Here, we only care about correlations at a particular set of times, namely, the relative delays between sensors rather than at all possible times.

having the best resolution (Prob. 4.19). In this case, the designer of the delay-and-sum beamformer must decide what is more important: array gain (minimizing sensitivity to noise) or resolution (ability to localize direction of propagation).

The form of the solution to the optimal array gain problem suggests an intriguing idea. We have chosen array shading coefficients without regard to the nature of the field into which the array is placed. The array shading yielding the optimal array gain depends solely on the characteristics of the noise. This result means that "good" array shading coefficients (if array gain is the primary design criterion) should depend on the array's environment and *cannot* be chosen before the array is placed in it. Thus, the idea emerges of *adapting* the array's signal processing algorithm to the environment. Such *adaptive* beamforming algorithms are described in Chap. 7.

4.6 Resolution

A beamformer's resolution represents its capability to measure the source's location or its direction accurately. The choice of array geometry, aperture, and sensor weights all affect resolution. In Chap. 3, resolution is expressed in terms of the smoothing effect that results from observing a wavefield through an aperture (or an array) of finite spatial extent. Because of the close relationship between the aperture smoothing function, the array pattern, and the delay-and-sum beamformer's wavenumber-frequency response, similar results obtain when we derive expressions for a beamformer's resolution.

Limited resolution means limited ability to determine a plane wave's direction of propagation and to separate two plane waves propagating in slightly different directions. Either capability can be used to define resolution precisely. The first definition assesses how well an array can localize a given source, the second how well sources can be distinguished. These definitions are not equivalent for conventional beamforming algorithms because of difficulties arising in multiple-source problems. For example, distinguishing two sources with slightly different wavenumbers not only depends on the difference in the directions of propagation but also depends on the relative phases of the two propagating waves. Using this criterion thus brings in signal characteristics in addition to the array characteristics we are trying to assess. Thus, resolution is defined according to the first standard and becomes equivalent to measuring the width of the array pattern's mainlobe in terms of either wavenumber, angle of propagation relative to the array, or range.

4.6.1 Wavenumber Resolution

The wavenumber vector represents two kinds of information: direction of propagation and wavelength. This dual aspect becomes evident when we express the wavenumber vector as $\vec{k}^o = (2\pi/\lambda^o)\vec{\zeta}^o$ where, as usual λ^o is wavelength and $\vec{\zeta}^o$ is a unit-length vector pointing in the direction of propagation. *Wavenumber resolution* therefore summarizes a beamformer's ability to localize a propagating wave in wavenumber. This localization could be in direction of propagation, in wavelength, or both.

Wavenumber resolution would seem to be a complicated measure of resolution; why is wavenumber resolution important? Recall that when a delay-and-sum beamformer's delays are adjusted to look for plane waves, the resulting wavenumber-frequency response is

$$H(\vec{k}^o, \omega^o) = W(\omega^o \vec{\alpha} - \vec{k}^o)$$

where the array pattern is given by

$$W(\vec{k}) = \sum_{m=0}^{M-1} w_m \exp\{j\vec{k} \cdot \vec{x}_m\}$$

Using the spatiotemporal filtering interpretation of the delay-and-sum beamformer, the wavenumber-frequency response's mainlobe is equivalent to a passband. The narrower this passband, the more selective the filter. Thus, mainlobe width determines wavenumber resolution. Because wavenumber-frequency response and array pattern are equivalent for the delay-and-sum beamformer, width of the array pattern's mainlobe defines resolution. Because the array pattern depends on wavenumber, wavenumber is the natural variable on which to base resolution performance.

An array pattern typically has a narrow mainlobe and a number of sidelobes. Mainlobe width measures how accurately the beamformer is able to determine a wave's direction of propagation. Previous sections have demonstrated that this width is inversely related to array aperture (Table 3.1 {67} for continuous apertures; the example on page 88 for a regular linear array). *Wavenumber resolution* is taken to be the inverse of the array pattern's mainlobe width. Resolution is thus proportional to the array's aperture. Generally, the array pattern depends on the wavenumber vector's components: k_x, k_y, and k_z. The mainlobe "width" may consist of several numbers, namely, the widths along lines parallel to the axes defined by these components. For simplicity, let's examine the case of the regular linear array lying along the x axis. This array's array pattern depends only on k_x.

Mainlobe width can be measured in several ways. If the array pattern has zeros located symmetrically about the mainlobe, the peak-to-zero distance can serve as one measure of width. This definition of width is consistent with the Rayleigh resolution criterion {65}. The full width of the mainlobe at one-half the peak value is another useful measure of mainlobe width. This *FWHM* measure is sometimes applied to the squared magnitude of the array pattern. In cases in which the array pattern is not symmetric or doesn't have conveniently located zeros, other measures of the mainlobe's width are used. Here, we normalize the array pattern to have unity gain at the peak, fit a parabola to the peak, and calculate the *FWHM* of the approximating parabola. This calculation yields the *parabolic width.* To illustrate how parabolic width is calculated, suppose the array pattern truly is a parabola: $W(k) = a - bk^2$. The second derivative evaluated at the peak ($k = 0$) equals $-2b$, and the *FWHM* of the parabola, denoted by *PW*, is easily computed to be

$$PW = 2\sqrt{\frac{a}{2b}}$$

For the general array pattern, we would use the mainlobe peak value in place of a and the negative of the array pattern's second derivative evaluated at the peak in place of

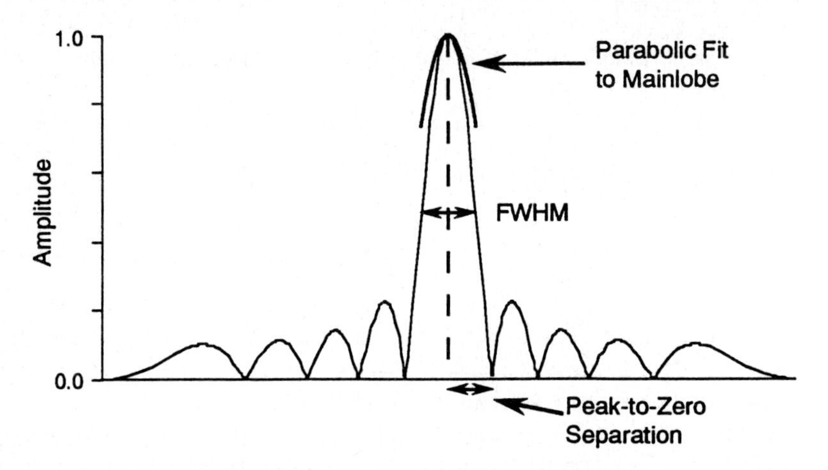

Figure 4.10 A typical array pattern is shown. The wavenumber resolution is defined as the reciprocal of the width of the mainlobe, but "width" may be defined in several ways. For symmetric array patterns, the simplest is the peak to first zero. A second measure is *FWHM* (full-width, half-maximum). Asymmetric array patterns are best characterized by the parabolic width.

2*b*. Thus, a mainlobe's parabolic width can be expressed directly in terms of the array pattern.

$$PW = 2\sqrt{\frac{W(\omega\alpha - k)}{-\dfrac{d^2 W(\omega\alpha - k)}{dk^2}}}\Bigg|_{\text{peak}}$$

No matter which measure of mainlobe width is used, beamformer resolution is defined to be the reciprocal of the mainlobe width. Figure 4.10 demonstrates a typical array pattern and how its wavenumber resolution is defined.

Example

Let's examine the wavenumber resolution of the linear array discussed in the examples provided in §4.2 {120}. With $M = 2M_{1/2} + 1$ sensors and unity shading, the wavenumber-frequency response equals

$$H(\vec{k}^o, \omega^o) = W(\omega^o \alpha_x - k_x^o) = \frac{\sin \frac{M}{2}\left(\omega^o \alpha_x - k_x^o\right) d}{\sin \frac{1}{2}\left(\omega^o \alpha_x - k_x^o\right) d}$$

This function exhibits a mainlobe peak when $k_x^o = \omega^o \alpha_x$, which occurs when the sensor delays are adjusted so that the array's assumed propagation direction and assumed speed of propagation c match the wave's direction and speed of propagation. Because we are dealing with a linear array in this example, only the x components of the vectors have to match. The characteristics of this array pattern are detailed in §3.3.1 {85}. There, the peak-to-zero width of the mainlobe is $2\pi/Md$ and the *FWHM* equals $2.4\pi/Md$.† In general, the parabolic width of a function is given in terms of its second derivative evaluated at a peak.

†As an exercise, show that the *FWHM* of the squared magnitude of the array pattern approximately equals $1.77\pi/Md$.

To simplify calculation of the parabolic width, we start with the expression for the array pattern before the summation has been reduced to a closed form, namely,

$$W(\omega^o\alpha_x - k_x^o) = \sum_{l=-M_{1/2}}^{M_{1/2}} \exp\{-j(\omega^o\alpha_x - k_x^o)ld\}$$

The required second derivative equals

$$\frac{d^2}{dk_x^{o2}} W(\omega^o\alpha_x - k_x^o) = -2\sum_{l=1}^{M_{1/2}} l^2 d^2 \cos\{(\omega^o\alpha_x - k_x^o)ld\}$$

Evaluation of the array pattern and its second derivative at the peak of the mainlobe, which occurs at $k_x^o = \omega^o\alpha_x$, results in

$$W(\omega^o\alpha_x - k_x^o)\Big|_{\text{peak}} = M \qquad \frac{d^2}{dk_x^{o2}} W(\omega^o\alpha_x - k_x^o)\Bigg|_{\text{peak}} = -\frac{M(M^2-1)}{12}d^2$$

The parabolic width thus equals

$$PW = 2\sqrt{\frac{M}{\frac{M(M^2-1)}{12}d^2}} = \frac{4\sqrt{3}}{d\sqrt{M^2-1}}$$

For large values of M,

$$PW \approx \frac{4\sqrt{3}}{Md} = \frac{6.93}{Md} = \frac{2.2\pi}{Md}$$

The three measures of mainlobe width differ only by a constant of proportionality: Peak to zero equals $2\pi/Md$, *FWHM* $2.4\pi/Md$, and *PW* $2.2\pi/Md$. The important result is that resolution is proportional to the product Md, which essentially equals the array's aperture.[†]

We may be interested in the resolution, not in terms of the wavenumber component k_x^o but rather in terms of the component ζ_x^o of the direction of propagation. When plotted as a function of this component, the mainlobe width is proportional to λ^o/Md. Thus, this direction of propagation resolution measure depends on wavelength: The shorter the wavelength, the narrower the mainlobe peak, and the better the resolution; for long wavelengths, the mainlobe width is large and poor resolution results. This discussion relates back to the notion of measuring array apertures in terms of wavelength {65}.

4.6.2 Angular Resolution

The previous section defined resolution as the reciprocal width of the array pattern's mainlobe when plotted against the components of either wavenumber or propagation vector. In many applications, *angular resolution* represents a practical definition: What is reciprocal mainlobe width when the array pattern is expressed as a function of angle of incidence ϕ^o? Because incidence angle is related nonlinearly to the propagation vector, $\zeta_x^o = -\sin\phi^o$ for a linear array, the width changes with incidence angle. This effect does not occur when we define resolution in terms of wavenumber or propagation direction.

[†]What is the aperture of our example regular linear array? Note that resolution depends on a slightly larger quantity than the array's actual spatial extent.

For our regular linear array, the wavenumber-frequency response expressed in terms of incidence angle becomes the *angular array pattern*

$$W_a(\phi, \phi^o) \equiv W\big(k^o(\sin\phi^o - \sin\phi)\big) = \frac{\sin\frac{M}{2}k^o(\sin\phi^o - \sin\phi)d}{\sin\frac{1}{2}k^o(\sin\phi^o - \sin\phi)d}$$

where ϕ is the angular direction to which the beamformer has been steered by adjusting its delays. The mainlobe's peak clearly occurs when $\phi = \phi^o$, but its width is no longer independent of the assumed propagation direction. For example, for a given value of ϕ, the array pattern's first zero occurs when ϕ^o satisfies†

$$\frac{Mdk}{2}(\sin\phi^o - \sin\phi) = \pi \quad \text{or} \quad \sin\phi = \sin\phi^o - \frac{2\pi}{Mdk}$$

If we define the angular peak-to-zero width to be $\delta\phi = \phi - \phi^o$, a surprisingly complicated formula emerges.

$$\delta\phi = \phi - \sin^{-1}\left(\sin\phi + \frac{2\pi}{Mdk}\right)$$

The mainlobe width varies with the assumed propagation direction of the beam.

Note that the array pattern W_a depends on the difference $(\sin\phi^o - \sin\phi)$. When $\delta\phi$ is small, $\sin\phi^o - \sin\phi \approx -\delta\phi\cos\phi$; in the vicinity of the mainlobe peak, then, the angular array pattern becomes

$$W_a(\phi, \phi - \delta\phi) \approx \frac{\sin\frac{M}{2}k^o\delta\phi\, d\cos\phi}{\sin\frac{1}{2}k^o\delta\phi\, d\cos\phi}$$

This approximation suggests that near the mainlobe, the array pattern considered as a function of $\delta\phi$ has the form of an array pattern for a uniform linear array of length $Md\cos\phi$. Angular resolution varying with assumed incidence angle translates into a fixed resolution for an array whose aperture varies with angle. This effective aperture is maximum at broadside ($\phi = 0°$) and smoothly decreases until the minimum is reached at endfire ($\phi = 90°$). The array's effective aperture equals the array baseline projected onto a line perpendicular to the assumed propagation direction.

Example

To derive the angular parabolic width of the mainlobe for a uniform linear array (M sensors), we must calculate

$$PW = 2\left.\sqrt{\frac{W_a(\phi, \phi^o)}{-\dfrac{\partial^2 W_a(\phi, \phi^o)}{\partial\phi^{o2}}}}\right|_{\phi=\phi^o}$$

†Because of the symmetry of this expression, resolution for the steered response equals that for the beampattern.

$W_a(\phi^o, \phi^o)$ equals M. The second derivative can be (tediously) evaluated directly, but we can use the results of the previous example {144} to advantage by exploiting the definition of the angular array pattern: $W_a(\phi, \phi^o) = W\big(k^o(\sin\phi^o - \sin\phi)\big)$. Calculations detailed in the example show that the second derivative of $W(\cdot)$ evaluated at the mainlobe peak equals

$$\left.\frac{d^2W(k)}{dk^2}\right|_{\text{peak}} = -\frac{M(M^2-1)d^2}{12}$$

Using the chain rule, the second derivative operator we need equals

$$\frac{\partial^2}{\partial\phi^{o2}} = \left(\frac{dk_x^o}{d\phi^o}\right)^2 \frac{d^2}{dk_x^{o2}} + \left(\frac{d^2k_x^o}{d\phi^{o2}}\right)\frac{d}{dk_x^o}$$

Because the first derivative of the array pattern equals 0 at a peak, the second term equals 0 and we only need the second derivative term from this expression, which the previous example conveniently provides. Using $k_x^o = -k^o \sin\phi^o$, we obtain

$$\left.\frac{\partial^2 W_a(\phi, \phi^o)}{\partial\phi^{o2}}\right|_{\phi=\phi^o} = -\frac{M(M^2-1)d^2}{12} \cdot (k^o)^2 \cos^2\phi$$

Finally, using the formula for parabolic width, we find that

$$PW = \frac{4\sqrt{3}}{k^o d \cos\phi \sqrt{M^2-1}}$$

The width varies inversely as the cosine of the steering angle ϕ. Fig. 4.11 shows two array patterns for a uniform linear array plotted in terms of incidence angle. Broadening of the beampattern is clearly evident.

The results obtained here apply to any linear array, regular or not, and to multidimensional geometries. Generally speaking, effective aperture determines angular resolution. For two- and three-dimensional arrays, azimuth and elevation must be considered. For planar arrays, such as circular arrays, the effective aperture analogy predicts constant azimuthal resolution with elevation resolution varying with incidence angle. A square-shaped perimeter array has similar elevation resolution, but azimuthal resolution should vary as the array appears to be wider from some perspectives than others. Such simple analogies allow ready calculation of how resolution varies with incidence angle, but nongeometrical proportionality constants must be calculated directly.

4.6.3 Range Resolution

When a source lies in the array's near field, range as well as direction of propagation can be measured (Fig. 4.12). Range measurements depend on curved wavefronts striking the array so that each sensor "sees" a different direction of propagation. We should expect that as range increases, the wavefront becomes less curved, and range resolution, the accuracy to which range can be measured, decreases.

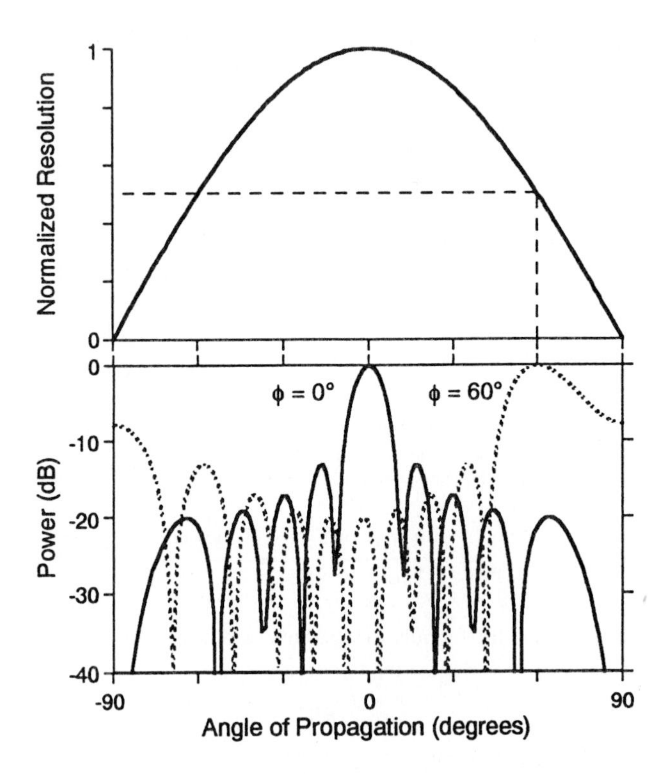

Figure 4.11 Two array patterns are shown for a linear array of ten equally spaced sensors. One corresponds to steering the beam to broadside, $\phi = 0°$, and the other to steering the beam to $\phi = 60°$. Because $\cos 60° = 0.5$ as indicated by the normalized resolution curve, the resolution at this assumed propagation direction is twice that at broadside. The broadside pattern's mainlobe is clearly sharper than that of the first when expressed as a function of angle.

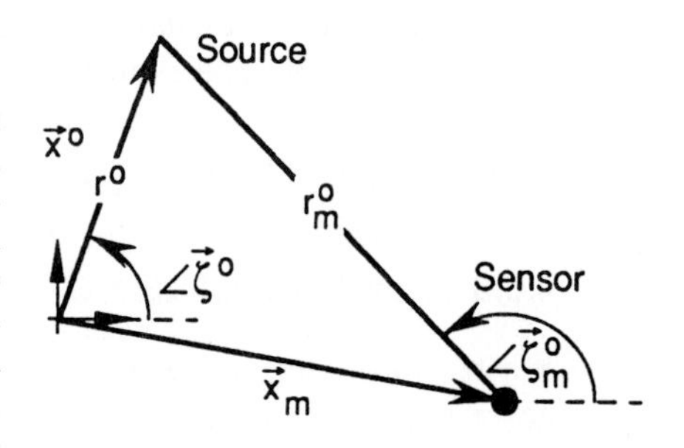

Figure 4.12 The points, distances, and angles that help define the range resolution problem are shown. The origin of the coordinate system is the array center, and the wave's source is located at the point $\vec{x}^o$. The range is the distance from the source to the origin, $r^o = |\vec{x}^o|$. Because the source is in the near field, the wave's direction of propagation varies across the array. At the array center it is given by $\vec{\zeta}^o = -\vec{x}^o/r^o$. The mth sensor is located at $\vec{x}_m$, and the range from the mth sensor to the source is $r_m^o = |\vec{x}_m - \vec{x}^o|$. The direction of propagation measured at the mth sensor is $\vec{\zeta}_m^o = (\vec{x}_m - \vec{x}^o)/r_m^o$.

On page 124, we defined an array pattern for point focusing.

$$\mathcal{W}(k, \vec{x}, \vec{x}^o) = \sum_{m=0}^{M-1} w_m \frac{r^o}{r_m^o} \exp\left\{jk[(r^o - r_m^o) - (r - r_m)]\right\}$$

The variables $\vec{x}$ and $\vec{x}^o$ denote the assumed and actual locations of the point source, r (r^o) the distance from the array's center to the assumed (actual) source location, and r_m (r_m^o) the distance from the mth sensor to the assumed (actual) source location. The crucial quantities $r^o - r_m^o$ and $r - r_m$ have the same, complicated, prototypic form.

$$r - r_m = r\left[1 - \sqrt{1 + \frac{|\vec{x}_m|^2}{r^2} - 2\frac{\vec{x} \cdot \vec{x}_m}{r^2}}\right]$$

Assumed and actual source location vectors can be expressed in terms of range and a direction vector: $\vec{x} = -r\vec{\zeta}$ and $\vec{x}^o = -r^o\vec{\zeta}^o$. For the moment, assume we know the direction of the point source so that $\vec{\zeta} = \vec{\zeta}^o$. We can then ask how the array pattern varies with r^o for fixed r. To the authors' knowledge, few interesting arrays yield tractable range array patterns. The issue of how to measure range resolution is best illustrated by a familiar example.

Example

Let's return to our usual example of a regular linear array of M sensors lying along the x axis. Assume the source is located on the array's broadside, $\vec{x}^o = r^o\vec{i}_z$, and that the source emits a monochromatic spherical wave with $\lambda = 1$ ($k = 2\pi$). The beamformer computes its output signal by applying unit sensor weights and setting the delays so that the assumed propagation direction corresponds to $\phi = 0°$ and range r. In this situation, the distance from the source to the mth sensor is

$$r_m^o = \sqrt{(r^o)^2 + (m - M_{1/2})^2 d^2}$$

where d is the intersensor spacing and $M_{1/2} = (M - 1)/2$. Fig. 4.13 illustrates steered responses—fixed source location and varying look ranges—for several broadside, near-field source locations. As this steered response clearly illustrates, the steered response (as well as the array pattern) as a function of range has a complicated structure that changes significantly with source location. As the source moves away from the array, new sidelobes appear, and, as expected, the mainlobe changes shape, significantly broadening.

In this case and virtually all others, defining mainlobe width (and hence resolution) as the distance between the zeros of the array pattern doesn't make sense. The parabolic width method can be used to measure resolution. The range resolution of a linear array computed this way is shown in Fig. 4.14. Several surprising features of the range resolution are evident. The range at which maximum resolution is achieved occurs near to the array but not at zero range. This peak-resolution range varies with the array aperture, approximately equaling one-fifth of the array aperture. Over most ranges, the resolution is greater for the larger aperture array. Note that the value of the maximum resolution does *not* change with aperture: Extending the spatial extent of an array does not improve its maximum range resolution.

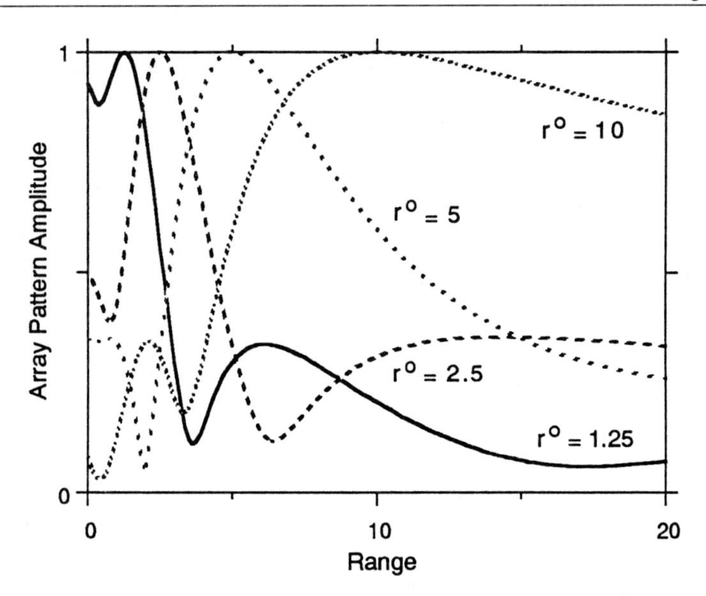

Figure 4.13 The magnitudes of range steered responses for a delay-and-sum beamformer with a ten-element uniform linear array having an intersensor spacing of $d = \lambda/2$ are shown for a narrowband source whose angular location is broadside to the array, $\phi = 0°$. For simplicity, the wavelength λ is taken to be 1. The array patterns for sources are located at ranges of 1.25λ, 2.5λ, 5λ, and 10λ. Note how the widths of the array patterns' mainlobes increase as distance to the source increases, indicating decreasing resolution.

Fig. 4.15 illustrates the full steered response for the linear array as the point of focus $\vec{x}$ varies in range and assumed propagation direction for a fixed source location $\vec{x}^o$. In this case, we have no *a priori* knowledge of the source's location, and adjust the beamformer delays to scan the point of focus over angle and range. In contrast to the angle-only array pattern (illustrated in Fig. 4.10) or the range-only array pattern (Fig. 4.13) where a peak can be easily distinguished, ridges appear in the steered response. Along these ridges, which form curves in the range-angle plane, the steered response's value changes little. This result means that we have a more difficult task in distinguishing sidelobes from the mainlobe.

Because the range resolution of the delay-and-sum beamformer is usually poor, especially when the source is not close and its incidence angle ϕ^o is unknown, we encounter difficulties in localizing a source in both range and angle with a single array. If the source is known to be in the far field, the beamformer can be focused at infinity (the delays are adjusted to look for plane waves) and the angular resolution thus afforded can be used to localize the direction of propagation of the source relative to the array. Thus, a single array is seldom used to localize sources in two or three spatial dimensions. The outputs of several arrays can be coalesced to locate such sources much more accurately; we can use triangulation. Trying to employ triangulation reveals a plethora of problems, however, requiring detailed attention to array processing and source properties; see Chap. 8.

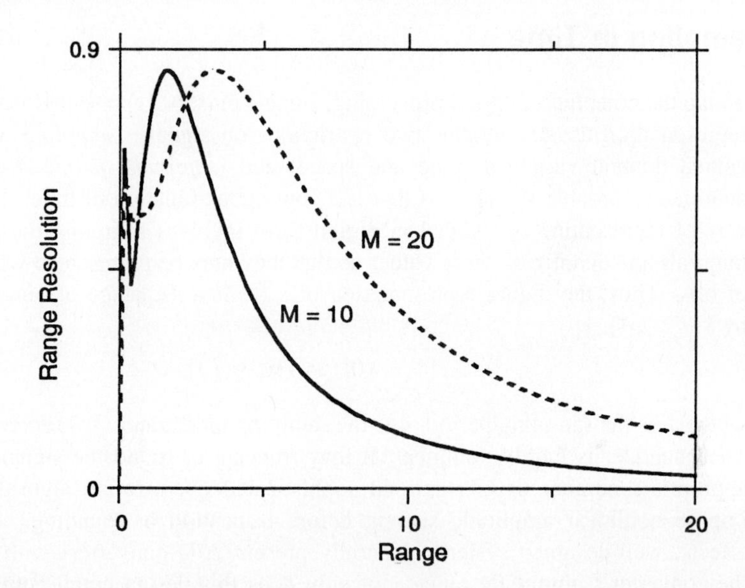

Figure 4.14 The range resolution, defined as the reciprocal of the mainlobe's parabolic width, is shown as a function of the broadside source's range. The linear array consisted of equally spaced sensors as described in the example and Fig. 4.13. The resolutions of a 10-element array and a 20-element array are shown. Maximum resolution occurs at an intermediate range that increases as the aperture increases.

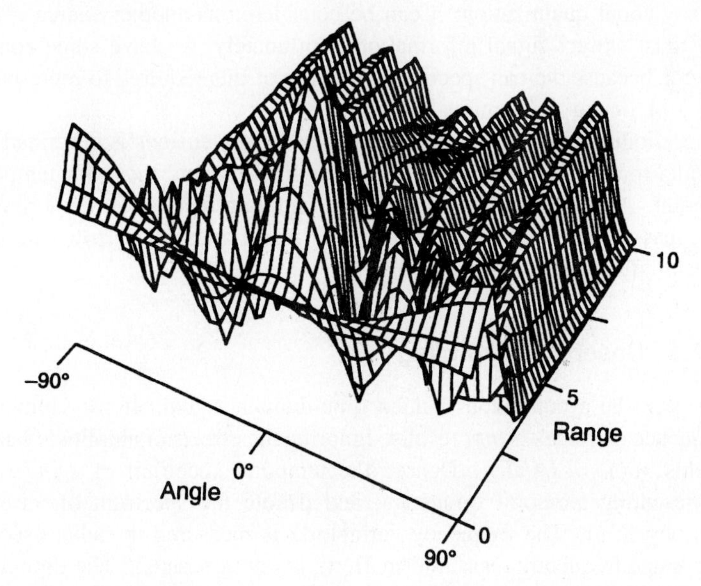

Figure 4.15 The array pattern for a 10-element linear array of equally spaced sensors is shown plotted as a function of range r and angle ϕ for a fixed source location. The source is located at an angle of 0° (broadside) at a normalized range of 2.5 (corresponding to five times the basic sensor spacing $d = 0.5$). The plot is shown with a linear vertical scale.

4.7 Sampling in Time

To tap the potential of digital processing, signals must be represented in digital form. This representation usually implies two restrictions on signals: *sampling* of the continuous-valued domain variables (time and space) and *amplitude quantization*. Consider the continuous-variable signal $s_c(t)$ that is a continuous function of time. The most common way of representing this signal in digital form involves sampling the signal at periodic intervals and quantizing these values so that they may be represented with a finite number of bits. Thus, the digital representation of $s_c(t)$ is a sequence of binary numbers given by

$$s(n) = \mathcal{Q}[s_c(nT)]$$

where T is the sampling period, n is the sampling index, and $\mathcal{Q}[\cdot]$ represents quantization. The quantizer is highly nonlinear; it may truncate or round the signal amplitude to the appropriate number of bits and, in sophisticated quantization algorithms, it may also impose nonlinear amplitude scaling before truncation or rounding. Because of these effects, well-designed systems generally present A/D converters with signals spanning the converter's amplitude range; not only does this design entail amplitude scaling but also noise removal by filtering to the greatest extent possible. As described on page 130, filter-and-sum beamformers generally apply filtering at each sensor before sampling for just this reason. Once quantized, the quantization error $s_c(nT)-s(n)$ is usually considered to be a random signal (noise) added to the sampled signal. We shall have little further to say about quantization; it can be considered as another source of noise out of which we must extract signal information. Fortunately, we have some control over this noise source because we can specify the number of bits required to represent signals accurately for a particular application.

Periodic sampling in one or more spatial dimensions is described in §3.2 {77ff}. The results for one-dimensional spatial sampling emulate those for temporal sampling up to a point. Although equivalent periodic sampling results can be derived in either domain, bandpass sampling, which has practical importance for narrowband signals, can only be meaningfully applied temporally.

4.7.1 Baseband Sampling

Let $s_c(t)$ be a continuous-valued time-domain signal. If we sample $s_c(t)$ every T, the sequence of values $\{s(n)\}$ results. Ignoring the effects of amplitude quantization, sampling yields $s(n) = s_c(nT)$. Denote the temporal spectrum of $s_c(nT)$ by $S_c(\Omega)$, with Ω representing temporal frequency, and denote the spectrum of the discrete-time signal $s(n)$ by $S(\omega)$. The frequency variable Ω is measured in radians/seconds, which equals 2π times frequency measured in Hertz (cycles/seconds). The digital frequency variable ω usually ranges from either 0 to 2π or from $-\pi$ to π and has units of radians per sampling period.

In the case in which $s_c(t)$ has no frequency components for $|\Omega| \geq \Omega_0$ and the sampling period T is less than or equal to π/Ω_0, the Sampling Theorem yields the same interpolation formula as derived for spatial sampling (Eq. 3.7 {77}).

$$s_c(t) = \sum_{n=-\infty}^{\infty} s(n) \frac{\sin \pi \left(\frac{t}{T} - n\right)}{\pi \left(\frac{t}{T} - n\right)} \tag{4.19}$$

Thus, the samples provide a complete representation of the continuous-time signal for *any* value of t. Linear transformations applied to the samples are equivalently applied to the original signal. Spectrally, the continuous-time and discrete-time signals are related by Eq. 3.9 {79}.

$$S(\omega) = \frac{1}{T} \sum_{l=-\infty}^{\infty} S_c\left(\frac{\omega - 2\pi l}{T}\right) \tag{4.20}$$

Alternatively, this equation may be written as

$$S(\Omega T) = \frac{1}{T} \sum_{l=-\infty}^{\infty} S_c\left(\Omega - \frac{2\pi l}{T}\right)$$

The Sampling Theorem states that as long as none of the periodic replications of $S_c(\cdot)$ in the sum overlaps another, the sampled signal's spectrum is proportional to the original.

$$S(\Omega T) = \frac{1}{T} S_c(\Omega) \quad |\Omega| \leq \pi/T$$

When $s_c(t)$ is not bandlimited to frequencies below π/T, the periodic replications in the sum overlap and *aliasing* occurs. The distortions to $S(\omega)$ introduced by aliasing prevent recovery of $s_c(t)$ exactly from its temporal samples $\{s(n)\}$.

4.7.2 Bandpass Sampling

In many practical cases, the signal of interest has a bandpass spectrum, containing energy only in a narrow band of frequencies from Ω_L, the lower band edge, to Ω_U, the upper band edge. The signal's bandwidth is denoted by $B = \Omega_U - \Omega_L$. Blind application of the Sampling Theorem to such signals implies a required sampling frequency of $2\Omega_U$ or greater. Because the signal's spectral support is much narrower, more efficient discrete-time representations of such signals would seem to exist. Such is indeed the case.

Analytic signal sampling. To illustrate the basic concept of *bandpass sampling*, first consider the complex bandpass signal $s_c^+(t)$ known as the *analytic signal*. Analytic signals have Fourier transforms identically equal to zero for negative frequencies: $S_c^+(\Omega) = 0$, $\Omega < 0$. Real-valued bandpass signals $s_c(t)$ can be represented as the real part of their corresponding analytic signals: $s_c(t) = \text{Re}[s_c^+(t)]$. The analytic signal is expressed in the time domain in terms of the bandpass signal and its Hilbert Transform or, better yet, in

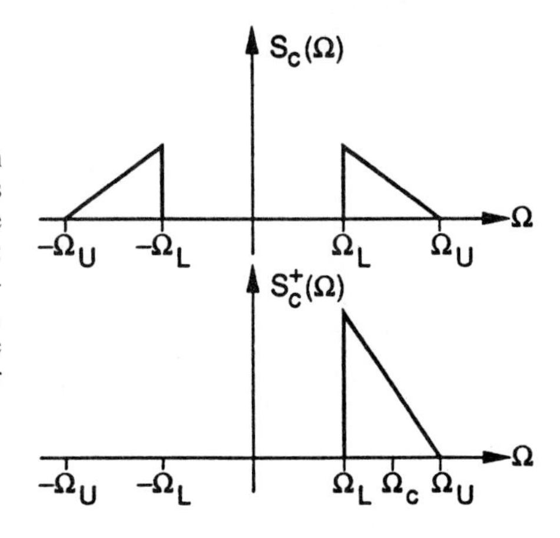

Figure 4.16 A bandpass signal's spectrum as well as that of its analytic counterpart is shown. The analytic signal contains only the positive frequency components of the original: $\Omega_L \leq \Omega \leq \Omega_U$. The bandwidth B of this signal equals $\Omega_U - \Omega_L$ and its center frequency Ω_c equals $(\Omega_U + \Omega_L)/2$. Note that convention has the positive-frequency spectrum amplified by a factor of 2.

the frequency domain as the positive frequency portion of the original signal's spectrum scaled by a factor of 2.[†]

$$s_c^+(t) = s_c(t) + j\hat{s}_c(t) \qquad \text{Analytic signal (time domain)} \tag{4.21a}$$

$$S_c^+(\Omega) = 2S_c(\Omega)\mathrm{u}(\Omega) \qquad \text{Analytic signal (frequency domain)} \tag{4.21b}$$

$$s_c(t) = \mathrm{Re}[s_c^+(t)] \qquad \text{Signal recovery} \tag{4.21c}$$

$$s_c^b(t) = s_c^+(t)e^{-j\Omega_c t} \qquad \text{Complex baseband signal} \tag{4.21d}$$

Here, $\mathrm{u}(\cdot)$ denotes the unit step function. Fig. 4.16 depicts a hypothetical spectrum for an analytic signal.

As indicated by the last equation, the analytic signal's narrow range of frequency support allows us to shift the spectrum of $s_c^+(t)$ down to baseband by multiplying it by the phasor $\exp\{-j\Omega_c t\}$, where Ω_c is defined as the band's center frequency.[‡] The resulting complex baseband signal $s_c^b(t) = s_c^+(t)e^{-j\Omega_c t}$ is bandlimited to the interval $\pm B/2$. It may thus be sampled without aliasing at a sampling frequency Ω_s greater than or equal to B or, equivalently, a sampling period T less than or equal to $2\pi/B$. Thus, to sample $s_c(t)$ efficiently, we can create and sample its complex baseband counterpart. Note, somewhat surprisingly, that if we choose the sampling frequency to be Ω_c/N, N an integer, we can sample the analytic signal directly without requiring modulation by the complex exponential. To derive this result, write the sampling period as $T = 2\pi N/\Omega_c$; at such sampling intervals, $s_c^b(nT) = s_c^+(nT)$. This sampling scheme causes in-band frequency components to alias down to the range $\pm B/2$ as well as into other frequency ranges; see Fig. 4.17. The relation between the lower band edge and the bandwidth determines the maximum allowed value of N.

[†]The notation $\hat{s}_c(t)$ denotes the Hilbert Transform of $s_c(t)$. The Hilbert Transform's definition is most easily expressed spectrally; the spectrum of a signal's Hilbert Transform equals the signal spectrum multiplied by $-j\,\mathrm{sign}\,\omega$.

[‡]Note that the "center" frequency can be chosen arbitrarily. In many cases, including throughout this book, the true center frequency is indeed chosen. However, in nonarray processing situations, such as single sideband modulation, the upper or lower band edge is chosen.

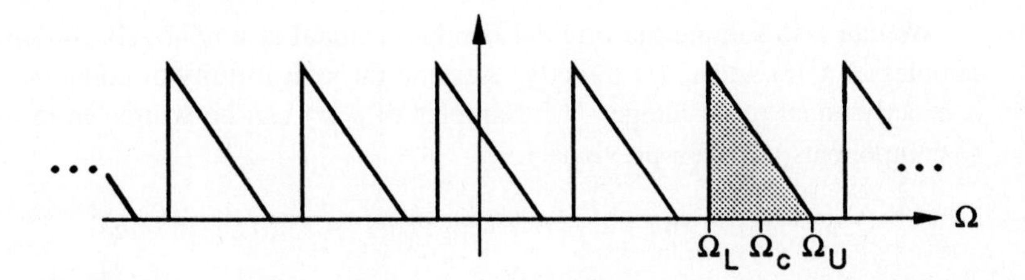

Figure 4.17 The spectrum of $s_c^+(t)$, indicated by the shaded area, is replicated and shifted by aliasing owing to sampling with a sampling frequency of $\Omega_s = \Omega_c/N$, where N is a positive integer. Here, $N = 2$. As shown, the replicas do not overlap; Ω_s satisfies the additional constraint $\Omega_s \geq B$, $B = \Omega_U - \Omega_L$.

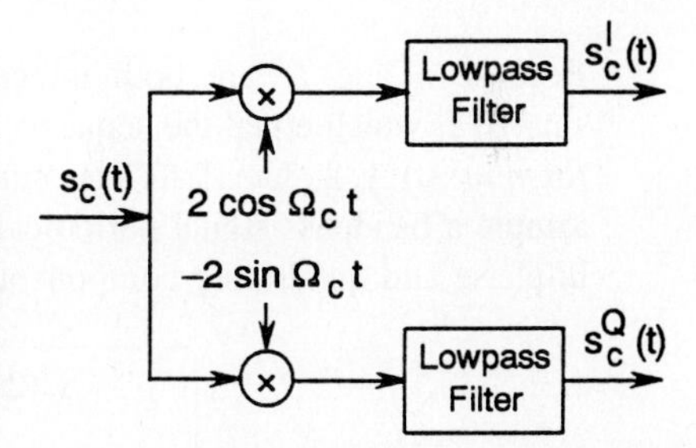

Figure 4.18 The in-phase and quadrature components, $s_c^I(t)$ and $s_c^Q(t)$, respectively, of a real-valued bandpass signal $s_c(t)$ can be obtained from the system shown. The bandpass signal $s_c(t)$ flows into two channels. In the upper channel, it is multiplied by $2\cos\Omega_c t$, then filtered by an ideal lowpass filter having cutoff frequency $B/2$; in the lower channel it is multiplied by $-2\sin\Omega_c t$ and then filtered similarly.

In-phase and quadrature signals. A common approach to analyzing demodulation and sampling of a real-valued signal involves isolating its *in-phase* and *quadrature* components. These components, often called "I" and "Q," respectively, equal the real and imaginary parts of the complex baseband counterpart signal $s_c^b(t)$:

$$s_c^I(t) = \text{Re}[s_c^b(t)] \quad s_c^Q(t) = \text{Im}[s_c^b(t)]$$

The baseband counterpart is related to the bandpass signal by Eqs. 4.21c and 4.21d. Because $s_c(t) = \text{Re}[s_c^b(t)e^{j\Omega_c t}]$, the bandpass signal $s_c(t)$ can be written in terms of the in-phase and quadrature components as

$$s_c(t) = s_c^I(t)\cos\Omega_c t - s_c^Q(t)\sin\Omega_c t$$

Thus, if we can sample these components efficiently, knowledge of the center frequency Ω_c and these samples allows us to recover the original bandpass signal.

Although this formalism does describe the relations between the various ways of representing bandpass signals, they give little indication of how a signal's in-phase and quadrature components can be extracted. The last equation provides a clue; the system shown in Fig. 4.18 demonstrates how modulation and lowpass filtering can extract these components. When extracted this way, these lowpass signals can each be sampled with period $T = 2\pi/B$ without suffering aliasing. Once we have successfully sampled these signals, the original bandpass signal can be recovered from them according to the preceding equation.

We can also sample the original bandpass signal at a relatively low rate and recover samples of $s_c^I(t)$ and $s_c^Q(t)$ directly. Assume through fortuity or connivance that $\Omega_L/2B$ is exactly equal to an integer N.† Samples of $s_c(t)$ can be written in terms of its I and Q components as given previously.

$$s_c(nT) = s_c^I(nT)\cos\Omega_c nT - s_c^Q(nT)\sin\Omega_c nT$$

Using the relation $\Omega_c = \Omega_L + B/2$ and sampling at the minimally acceptable rate $\Omega_L/N = 2B$ (so that $T = \pi/B$), we find that

$$\Omega_c nT = \left(\Omega_L + \frac{B}{2}\right)\frac{2\pi n}{2B} = 2\pi nN + \frac{\pi}{2}n$$

Because n and N are both integers, $\cos\Omega_c nT$ and $\sin\Omega_c nT$ become $\cos\pi n/2$ and $\sin\pi n/2$, which equal the sequences $\{1, 0, -1, 0, \ldots\}$ and $\{0, 1, 0, -1, \ldots\}$, respectively, for $n = \{0, 1, 2, 3, \ldots\}$. Thus, under the constraint that $\Omega_L/2B$ is an integer, we can sample a bandpass signal periodically with sampling interval $T = \pi/B$ so as to yield its in-phase and quadrature components as

$$\boxed{\begin{aligned} s_c(2nT) &= (-1)^n s_c^I(2nT) \\ s_c\big((2n+1)T\big) &= (-1)^n s_c^Q\big((2n+1)T\big) \end{aligned}}$$

Because $s_c(t)$ is sampled at a rate of $2B$ rad/s, the alternating sampling of $s_c^I(t)$ and $s_c^Q(t)$ means they are each sampled at half that rate. This rate corresponds to the minimum acceptable sampling rate because the full bandwidth of the complex baseband signal $s_c^b(t) = s_c^I(t) + js_c^Q(t)$, and therefore of the I and Q components, equals B.

Direct bandpass sampling. Although the analytic sampling scheme works, forming the analytic signal from a real-valued bandpass signal demands much preprocessing. Fig. 4.19 illustrates another bandpass sampling idea. We can choose to sample the original bandpass signal at a rate that shifts frequency Ω_L down to 0.‡ For example, if $\Omega_s = \Omega_L/N$, where N is a some positive integer, and Ω_s also satisfies the constraint $\Omega_s \geq 2B$, then the spectral components in the range $\Omega_L \leq \Omega \leq \Omega_U$ alias into the interval $0 \leq \Omega \leq B$. Simultaneously, components in the range $-\Omega_U \leq \Omega \leq -\Omega_L$ alias into the interval $-B \leq \Omega \leq 0$.

The minimum acceptable sampling frequency $\Omega_s^{\min}$ aliases the frequency band of interest down to baseband without overlapping any of the aliased replicas. If we take N to be the largest integer less than or equal to $\Omega_L/2B$, then generally $\Omega_s^{\min}$ equals Ω_L/N. If $\Omega_L < 2B$, however, N is 0. In this special case, we must sample at least

†We can always pretend that the lower band edge is a little lower than Ω_L or that the bandwidth is a little greater than B to achieve this result.

‡In practice, choosing a positive frequency near 0 rather than exactly equal to 0 might be more prudent. We have this flexibility because we can define a lower band edge frequency that is somewhat less than the lowest frequency containing energy.

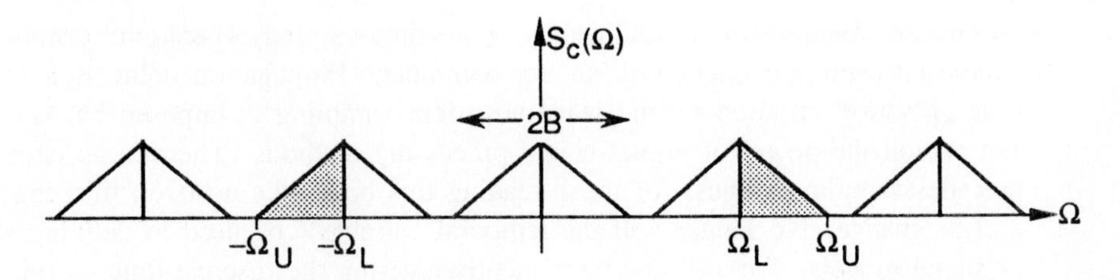

Figure 4.19 The spectrum of a real-valued, bandlimited signal $s_c(t)$ is shown by the shaded regions. It has nonzero frequency components for negative as well as positive frequencies. The spectrum that results by employing the sampling frequency $\Omega_s = \Omega_L/N$, where $\Omega_s \geq 2B$, has no aliasing and thus represents a viable bandpass sampling strategy.

at the full Nyquist sampling frequency (or employ analytic signal sampling as described previously). These rules for choosing the minimum acceptable sampling frequency for a bandpass signal are summarized by†

$$\boxed{\Omega_s^{\min} = \min\left[\frac{\Omega_L}{N},\, 2\Omega_U\right] \qquad N = \left\lfloor \frac{\Omega_L}{2B} \right\rfloor}$$

Consequently, judicious choice of sampling rate allows sampling of bandpass signals at a rate proportional to their bandwidths, which, after all, forms the information-bearing part of their spectra rather than at a rate proportional to their highest frequencies. For a more detailed treatment of bandpass sampling, see Vaughan et al.

All of this said, we should not seek efficient sampling methods as we would the Holy Grail. In array processing, what we seek is not only signal waveform but also the location of the source producing the signal. Location information is represented by the relative propagation delay between spatially sampled signals, which means we search for delays between sensor outputs. To base a beamforming system on bandpass sampled signals means excluding propagation delays present at the signal's center frequency. Take a cosinusoid as an example; its quadrature component equals zero, and its in-phase component equals a constant. Said sarcastically, it is somewhat difficult to measure the delay between two constant-valued signals. The next section begins the discussion of discrete-time beamforming methods; there, difficulties with this and other issues related to sampling are explored. Crystallizing these thoughts, it pays to optimize the *entire* system's performance rather than each component individually.

4.8 Discrete-Time Beamforming

Modern beamforming systems exploit the advantages of digital implementations. Once they are analog filtered to prevent temporal aliasing or to restrict the bandwidth of the wavefield to frequencies of interest, sensor signals are time sampled. This section contains the ramifications of sampling and how to overcome some of the problems sampling

†The notation $\lfloor x \rfloor$ denotes the *floor* of x: the largest integer less than or equal to x.

introduces. Sampling is a nonlinear-in-time (time-varying) operation: Continuous-time delay and sampling operations do not commute. Propagation delay is a continuous-time operation imposed by the wave equation; sampling is imposed by us so that we can exploit the power of digital signal processing methods. There is no choice: Delay precedes sampling. Thus, for those reading this book like a novel, this page demarks a clear change: No longer will the temporal variable t be used in defining algorithms or signal models. We are now permanently entering the discrete-time world, where all algorithms must be programmable, and computational complexity becomes an important design issue.

4.8.1 Sensor Sampling

The key problem in developing a delay-and-sum beamformer for sampled signals can be traced to the need for noninteger delays. The discrete-time delay-and-sum beamformer produces

$$\boxed{z(n) \equiv \sum_{m=0}^{M-1} w_m y_m(n - n_m)}$$

where n_m is the *integer* delay associated with the mth sensor. For example, suppose the wavefield consists of a single, monochromatic plane wave: $f(\vec{x}, t) = \exp\{j(\omega^o t - \vec{k}^o \cdot \vec{x})\}$. If the mth sensor is located, as usual, at $\vec{x}_m$ and its output sampled with period T, the discrete-time sensor signal equals

$$y_m(n) \equiv f(\vec{x}_m, nT) = \exp\{j(\omega^o nT - \vec{k}^o \cdot \vec{x}_m)\}$$

The beamformer's output becomes

$$\begin{aligned} z(n) &= e^{j\omega^o nT} \sum_{m=0}^{M-1} w_m \exp\{-j(\omega^o n_m T + \vec{k}^o \cdot \vec{x}_m)\} \\ &= e^{j\omega^o nT} H(\vec{k}^o, \omega^o) \end{aligned}$$

To steer a beam in a particular assumed propagation direction, we would ideally adjust the sensor delays $\{\Delta_m\}$ to equal $\{-\vec{\alpha}^o \cdot \vec{x}_m\}$. However, these ideal delays are generally not integer multiples of the sampling period T. Because we have sensor values only every T s, we cannot form sums that involve sensor signals delayed by noninteger multiples of T: The delay and sampling operations do not commute. *We can no longer steer beams in arbitrary directions.* For some particularly perverse choices of T and array geometry, it may be that no beam can be formed exactly.

Example

Consider our usual regular linear array of M sensors along the x axis, each of which is separated from its neighbors by d m. The position of the mth sensor is $x_m = (m - M_{1/2})d$, $m = 0, \ldots, M-1$, where $M_{1/2} = (M-1)/2$. If the sensor signals are sampled with a sampling interval T, then the allowed values of the sensor delays are given by

$$\Delta_m = n_m T = -(m - M_{1/2})qT$$

where q is an integer.† When $q = 0$, no relative delay occurs from one sensor to the next, resulting in a broadside beam: $\phi = 0°$. When $q = 1$, each sensor signal is delayed one sample with respect to the sensor on its right. To steer in arbitrary directions, the sensor delays would equal

$$\Delta_m = -\alpha_x(m - M_{1/2})d = \frac{\sin\phi}{c}(m - M_{1/2})d$$

Constraining the delays to be $n_m = -(m - M_{1/2})q$ samples corresponds to constraining the assumed propagation directions to have the form

$$\phi = \sin^{-1}\left(\frac{-qcT}{d}\right)$$

Assume we are only interested in signals whose highest frequency is B Hz. The largest temporal sampling interval we can tolerate without inducing aliasing equals $T = \pi/B$ s and the largest sensor separation we can have without spatial aliasing is $d = \lambda/2 = \pi c/B$ m. Thus, the spatial and temporal sampling intervals are linked by the relation $d = cT$. This relationship coupled with the relation of incidence angle to delay given earlier yields $\sin\phi = -q$: The only possible beams are the broadside beam ($q = 0$) and the two endfire beams ($q = \pm 1$). Using larger values of q requires the beamformer to look for waves traveling more slowly than the slowest propagating wave supported by the medium.

This analysis shows that sampling at the minimum acceptable rate in space and time leads to very coarse sampling of propagation delay. Clearly, this situation is unacceptable for any practical problem. To form more beams using integer delays, we must adjust d and T so that $d \gg cT$: Either T must be reduced, corresponding to oversampling the sensor signals temporally, or d must be increased, chancing the risks of spatial aliasing.

We must solve this problem to make digital processing of array outputs practical. If we don't insist that propagation delays equal integer multiples of T, what errors are introduced by employing only integer delays? As previously shown, the wavenumber-frequency response of the beamformer having integer-valued delays equals

$$H(\vec{k}^o, \omega^o) = \sum_{m=0}^{M-1} w_m \exp\{-j(\omega^o n_m T + \vec{k}^o \cdot \vec{x}_m)\}$$

Defining the delay error as $\epsilon_{\Delta_m} = n_m T - \Delta_m$, the wavenumber-frequency response can be written as

$$\boxed{H(\vec{k}^o, \omega^o) = \sum_{m=0}^{M-1} \left(w_m e^{-j\omega^o \epsilon_{\Delta_m}}\right) \exp\{-j(\omega^o \Delta_m + \vec{k}^o \cdot \vec{x}_m)\}}$$

Employing delays quantized to the sampling interval results in a complex-valued, position-dependent scaling of the sensor weights. Aberrations have exactly the same effect (Eq. 3.6 {68}); thus, the inability of discrete-time beamformers to delay sensor outputs exactly can be regarded as a wavefront aberration.

†The minus sign is introduced so that positive values of q correspond to the delays necessary to compensate for waves propagating in the x direction, that is, with $\alpha_x > 0$.

Example

Returning to the regular linear array of M sensors having intersensor spacing d m, make $M = 9$ for this example. Suppose that we sample the sensor signals with sampling interval T: $y_m(n) = f(\vec{x}_m, nT)$. As usual, $M_{1/2} = (M-1)/2$, which equals four in this example. The sampled delay-and-sum beamformer's output equals

$$z(n) = \sum_{m=0}^{8} y_m(n - n_m)$$

where the sensor delays $\{n_m\}$ are constrained to the integers. To steer the beam broadside, $\vec{\alpha} = (0, 0, -c^{-1})$ or $\sin\phi = 0$, we set all the sensor delays equal to 0. If we set the delays to effect a one-sample difference from sensor to sensor, $n_m = m - 4$, the resulting steering direction is the solution of the equation

$$n_m T = -\alpha_x (m-4)d = \frac{\sin\phi}{c}(m-4)d$$

which yields $\sin\phi = cT/d$.

To steer the beam in an intermediate direction, $\sin\phi = cT/(3d)$, for example, what happens to the wavenumber-frequency response when we use quantized ideal delays? The ideal delay Δ_m for the mth sensor equals $(m-4)d\sin\phi/c = (m-4)T/3$. The corresponding ideal wavenumber-frequency response equals

$$H(k_x^o, \omega^o) = \frac{\sin\frac{3}{2}\left(\omega^o T + 3k_x^o d\right)}{\sin\frac{1}{6}\left(\omega^o T + 3k_x^o d\right)}$$

and is portrayed in Fig. 4.20. When the delays are quantized to integer multiples of T, we can replace the ideal delays by quantized ones equaling rounded approximations.

$$n_m = \left[\frac{m-4}{3}\right]_r$$

The notation $[\cdot]_r$ means "rounded to the nearest integer." The resulting delay error equals

$$\epsilon_{\Delta_m} = \begin{cases} T/3 & m = 0, 3, 6 \\ 0 & m = 1, 4, 7 \\ -T/3 & m = 2, 5, 8 \end{cases}$$

The wavenumber-frequency response for the quantized-delay beamformer is most easily calculated by splitting the sum into three sums corresponding to the delay error values.

$$\sum_{m=0,3,6} = \exp\{jk_x^o d\}\frac{\sin\frac{3}{2}(\omega^o T + 3k_x^o d)}{\sin\frac{1}{2}(\omega^o T + 3k_x^o d)}$$

$$\sum_{m=1,4,7} = \frac{\sin\frac{3}{2}(\omega^o T + 3k_x^o d)}{\sin\frac{1}{2}(\omega^o T + 3k_x^o d)}$$

$$\sum_{m=2,5,8} = \exp\{-jk_x^o d\}\frac{\sin\frac{3}{2}(\omega^o T + 3k_x^o d)}{\sin\frac{1}{2}(\omega^o T + 3k_x^o d)}$$

Adding these three yields the quantized-delay wavenumber-frequency response (Fig. 4.20).

$$H_Q(k_x^o, \omega^o) = \frac{\sin\frac{3}{2}k_x^o d}{\sin\frac{1}{2}k_x^o d} \cdot \frac{\sin\frac{3}{2}(\omega^o T + 3k_x^o d)}{\sin\frac{1}{2}(\omega^o T + 3k_x^o d)}$$

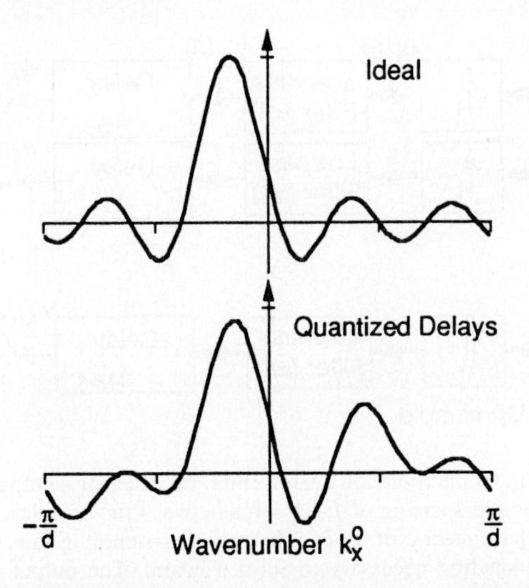

Figure 4.20 The ideal (upper panel) and quantized-delay (lower panel) wavenumber-frequency responses are plotted as a function of k_x^o. Here, $\omega^o T = \pi/2$, which corresponds to twice the Nyquist sampling rate, and the ideal steering delays equal $\Delta_m = (m - 4)T/3$. In the quantized case, the delays equal rounded approximations to these values.

Clearly, this wavenumber frequency response deviates significantly from the ideal; some improvement is required.

From the form of $H_Q(k_x^o, \omega^o)$, an idea emerges that can be used to advantage once the sampling problems are overcome. The wavenumber-frequency response equals the product of two similar terms, which resembles the subarray formula Eq. 4.11 {132}. For the particular case at hand, we can consider the array of nine sensors as three subarrays having three sensors each. The leftmost three sensors form one subarray, the middle three the next, and so on. To steer the beam, each subarray is steered broadside with the three subarray signals delayed one sample with respect to the adjacent subarray(s) to produce the array output. Relative to the center subarray, the right-most subarray is delayed one sample and the left-most subarray is advanced one sample. This partitioning is dependent on the desired propagation directions. Conceivably, to implement a set of fixed beams, this subarray decomposition could be effected as a parallel structure, with common delays shared in the partitioning to minimize the computational burden.

4.8.2 Time-Domain Interpolation Beamforming

To reduce the aberrations introduced by delay quantization, we can *interpolate* between the samples of the sensor signals. Two such approaches can be followed. In the first approach, all the sensor signals are interpolated by a factor I to obtain oversampled signals. As shown subsequently, some short-cuts can be taken to calculate only the needed interpolated values [124, 125]. In the second approach, we endeavor to resample the sensor outputs so that we effectively sample after imposing the sensor delay Δ_m rather than before.

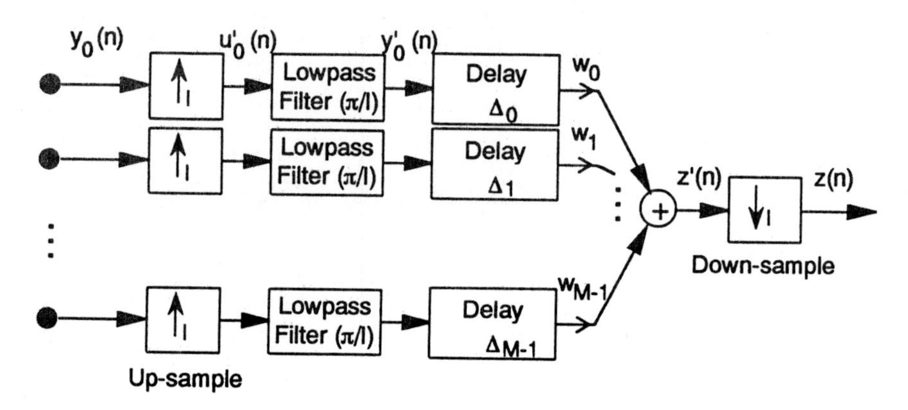

Figure 4.21 In an interpolation beamformer, each sensor's output is passed through an upsampler, consisting of an interspersing of $I-1$ zeros between the samples, and a lowpass interpolation filter having a cutoff frequency of π/I. These higher-sampling-rate signals can then be subjected to the delays and shading necessary to form a beam. The output of the beamformer may then be downsampled by a factor of I to obtain the original sampling rate.

Sensor signal interpolation. The conceptually simplest interpolation beamformer is shown in Fig. 4.21. Each sensor's output signal $y_m(n)$ must first be "upsampled": We intersperse $I-1$ zero-valued samples between the sensor's samples. We can represent this by writing

$$u'_m(n) = \uparrow_I [y_m(n)]$$

Because writing equations involving two sampling rates can get a little confusing, we put a prime (′) on signals having the smaller sampling interval $T' = T/I$. We then pass the upsampled signal through a lowpass interpolation filter having unit-sample response $h'(n)$ to yield the interpolated sensor signal $y'_m(n) = h'(n) \star u'_m(n)$. The interpolation filter has the effect of smoothing the upsampled signal so that the interspersed samples are no longer zero-valued; they take on values interpolated between the original sample values [133]. Interspersing $I-1$ zeros between samples in the time domain causes a periodic replication of the signal spectrum in the frequency domain, as indicated in Fig. 4.22. Lowpass filtering ideally eliminates all of the replications except the one at baseband ($-\pi/I < \omega \leq \pi/I$), resulting in a spectrum that is simply a compressed version of the original [49: 307–9];[116, 133].

The signals entering the beamformer have an effective sampling rate that is I times higher than the original. Conventional beamforming calculations—shading, sensor-dependent time delays, and summing—can then be performed as usual. Because the sample spacing has been reduced from T to T/I, a greater number of ideal beams can be steered for regular arrays and the wavenumber-frequency response errors for all beams are reduced. The beamformer's output signal equals

$$z'(n) = \sum_{m=0}^{M-1} w_m y'_m(n - n_m)$$

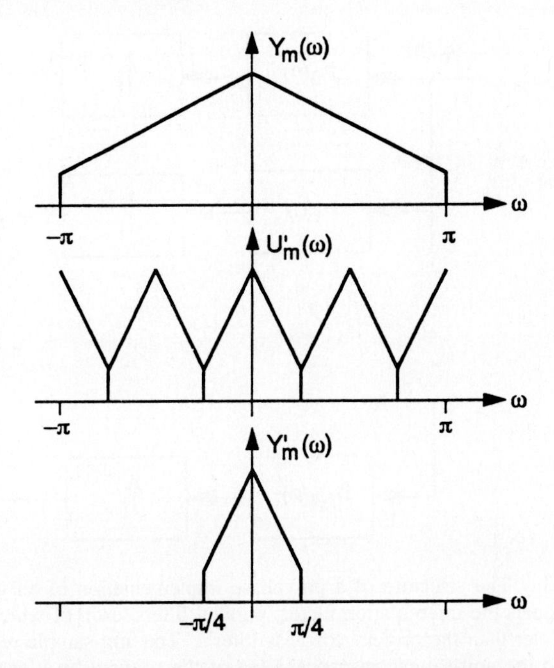

Figure 4.22 The upper panel shows a hypothetical frequency spectrum for a sensor signal $y_m(n)$ with an intersample interval of T s. The middle panel shows the resulting spectrum when $y_m(n)$ is upsampled by a factor $I = 4$ to obtain $u'_m(n)$. Finally, the lower panel shows the spectrum that results after lowpass filtering the upsampled signal to obtain the interpolated signal $y'_m(n)$. Ideally, the interpolated signal's spectrum is a compressed (by a factor I) version of its original.

Because the original sensor sampling rate did not induce temporal aliasing, the beamformer's output signal can be subsampled by a factor of I without error. Once subsampled, the beamformer output has the same sampling interval T as the sensor signals. These signal processing steps—interspersing zeros, lowpass filtering, conventional beamforming, and subsampling—constitute *time-domain interpolation beamforming*.

Each of the lowpass filters used in interpolation can be efficiently implemented using the so-called *polyphase* filter structure [43: Chap. 3];[125, 151]. This structure's efficiency is based on the large number of zero-valued samples in its input. The filter's output at the time $nI + r$ equals

$$y'_m(nI + r) = \sum_i u'_m(i)h'(nI + r - i)$$

The crucial observation is that $u'_m(i)$ equals zero unless $i = kI$, at which time $u'_m(kI) = y_m(k)$. Therefore, summing only over the potentially nonzero terms yields an efficient computation of the output.

$$\boxed{y'_m(nI + r) = \sum_k y_m(k)h'\big((n - k)I + r\big)} \tag{4.22}$$

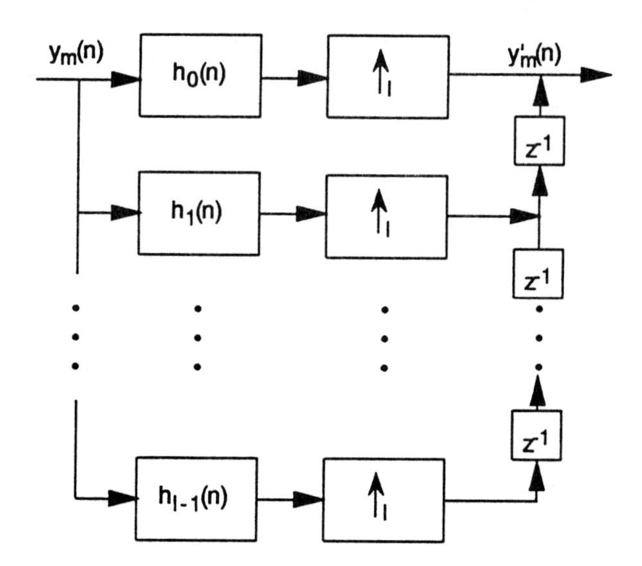

Figure 4.23 The structure of a polyphase implementation of an interpolation filter consists of I (which equals the interpolation factor) parallel filters, each of which has a unit-sample response I times shorter than the original lowpass filter's. The unit-sample response $h_r(n)$ of the rth filter is related to the unit-sample response $h'(n)$ of the original by $h_r(n) = h'(nI + r)$. The outputs of these filters are then upsampled and combined to produce the interpolated signal $y'_m(n)$. The systems labeled by z^{-1} denote one-sample delays, the label corresponding to their transfer function.

This expression is the convolution of $y_m(n)$ and $h'(nI+r)$. The samples produced by the interpolation filter between the actual sampling times are the convolution of the sensor output signals with a succession of filters, each of which has a unit-sample response equal to a downsampled version of the original filter's unit-sample response. If the interpolation filter's unit-sample response $h'(n)$ has N nonzero values (assuming a finite-duration impulse response or *FIR* implementation), each component filter's unit sample response has only N/I nonzero values, and these filters can be computed in parallel (Fig. 4.23). Each component filter computes a set of output samples spaced T s apart; the high-rate output (T/I-s intervals) is formed by interleaving these output samples. This last point is important because, for a regular linear array, the output of only one component filter in the polyphase implementation is needed when we steer the beam in particular directions. As shown subsequently, the samples needed to form each beam correspond to specific combinations of m and r. Consequently, for each beam, only one of the component polyphase filters need be implemented at each sensor's output. Though complicated organizationally, the polyphase scheme can reduce the calculations by a factor of M.

The beamformer output signal is given by

$$z'(nI + r) = \sum_{m=0}^{M-1} w_m y'_m(nI + r - n_m)$$

$$= \sum_{m=0}^{M-1} w_m \left[\sum_k y_m(k) h'\big((n-k)I + r - n_m\big) \right]$$

As $h_r(n) = h'(nI + r)$ and n_m is an integer, we can write

$$h'\big((n-k)I + r - n_m\big) = h_{(r-n_m)_I}(n-k)$$

where computing the difference $r - n_m$ modulo I means bringing the difference back into the range $[0, I-1]$ by adding or subtracting multiples of I from n_m and delaying or advancing $h_{r-n_m}(n)$ by the appropriate integer number of samples.[†] These manipulations give us an efficient polyphase structure for the interpolation beamformer.

$$z'(nI + r) = \sum_{m=0}^{M-1} w_m \left[\sum_k y_m(k) h_{(r-n_m)_I}(n-k) \right] \tag{4.23}$$

For each sensor and each output sample, only one component of the polyphase filter need operate for any particular steering direction. Note also the term in brackets can be interpreted as a filter; this implementation can be interpreted as a filter-and-sum beamformer. Further computational savings result if we downsample the beamformer output signal by a factor I. This downsampling corresponds simply to fixing the value of r in the preceding expression.

Because they are linear, we can interchange the lowpass filtering and beamforming operations, resulting in an approach called *postbeamforming interpolation*. Beamforming is performed on the upsampled sensor signals. Because of the upsampling, each of these signals has mostly zero-valued samples, and much of the computation to form the beam is trivial. The resulting beamformer output must be lowpass filtered, and, if desired, downsampled by any factor less than or equal to I. Fig. 4.24 provides a simple block diagram for the postbeamforming interpolation structure. Postbeamforming interpolation results in computational savings when the number of sensors M is significantly larger than the interpolation factor I.

Implementation of exact delays. For irregular arrays or for steering a regular array in particular directions,[‡] more precise beamforming results when we construct discrete-time versions of the delayed sensor signals $\{y_m(n)\}$. Conceptually, we must interpolate a discrete-time sensor signal to obtain its continuous-time version, delay it, and then resample it. For example, suppose we have a discrete-time signal $s(n)$ obtained by sampling the continuous-time signal $s_c(t)$ at a sampling interval of T s: $s(n) = s_c(nT)$. The continuous-time $s_c(t)$ is appropriately bandlimited so that we can recover it by using the interpolation formula (Eq. 4.19).

$$s_c(t) = \sum_{n=-\infty}^{\infty} s(n) h(t - nT) \quad h(t) = \frac{\sin \pi t/T}{\pi t/T}$$

[†] This procedure can be rationalized as breaking up the delay into coarse (multiples of T s) and fine (multiples of T/I s) adjustments.

[‡] Try steering an interpolation beamformer in a direction where the ideal delays are irrational numbers!

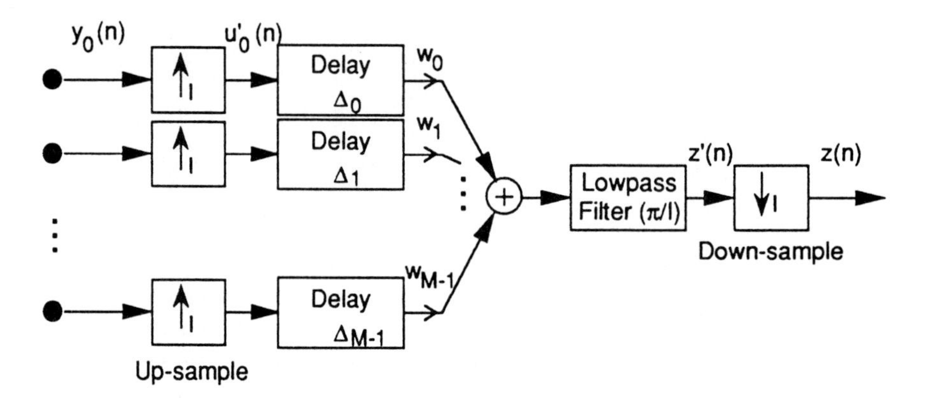

Figure 4.24 The linear, time-invariant portions of the interpolation beamformer (see Fig. 4.21 {162}), those occurring between the upsampling and downsampled operations, can be calculated in any order. In postbeamforming interpolation, the upsampled sensor signals are directly beamformed before lowpass filtering. The resulting beamformer output signal is then filtered and, if desired, downsampled.

Once we have recovered the continuous-time signal, we can delay it by Δ and resample it to obtain $s_\Delta(n) = s_c(nT - \Delta)$. Using the interpolation formula, we see that

$$s_\Delta(n) = \sum_k s(k) h\big((n-k)T - \Delta\big)$$

This technique can be applied individually to all M sensor signals to obtain the signals

$$y_{\Delta_m}(n) = \sum_k y_m(k) h\big((n-k)T - \Delta_m\big)$$

The beamformer's output is then generated by computing

$$\begin{aligned} z(n) &= \sum_{m=0}^{M-1} w_m y_{\Delta_m}(n) \\ &= \sum_{m=0}^{M-1} w_m \sum_k y_m(k) h\big((n-k)T - \Delta_m\big) \end{aligned}$$

This equation is similar in spirit to that describing the interpolation beamformer (Eq. 4.23). In fact, that beamformer simply creates discrete-time delayed signals for the case in which the delays are integer multiples of T/I seconds. Thus, the two approaches are equivalent save for the properties of the interpolation function $h(t)$, which has infinite extent. Any interpolation beamformer capable of being implemented must use finite interpolation functions. Designing such filters is best framed using the interpolation beamforming ideas of the previous section.

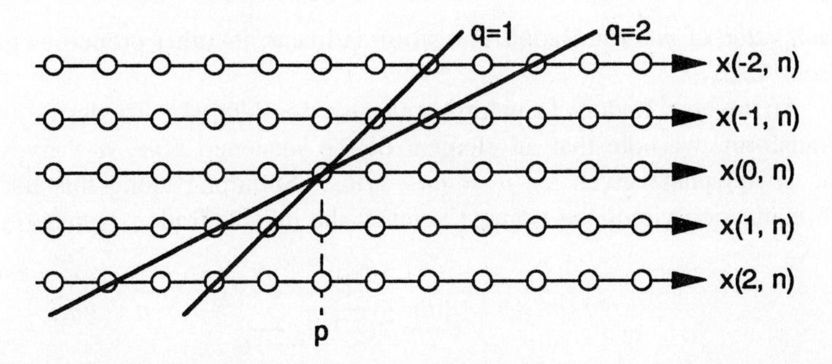

Figure 4.25 The rays along which a two-dimensional sequence is summed to form the Radon Transform are shown. The quantity q denotes the reciprocal slope (slowness) and p the time intercept about which the sum is taken.

4.8.3 Radon Transform

The properties of a sampled delay-and-sum beamformer for a linear array of equally spaced sensors can be analyzed in terms of the *Radon Transform*. Not named for the noble gas, this transform was first defined by the Bohemian mathematician Johann Radon in 1917 for continuous-valued functions [134]. The discrete Radon Transform described here maps a two-dimensional sequence into another two-dimensional sequence.

$$X_R(p,q) \equiv \mathcal{R}[x(m,n)] = \sum_{m=-\infty}^{\infty} x(m, p+qm) \quad p,q = \ldots,-1,0,1,\ldots$$

Consider for a moment the formula for the delay-and-sum beamformer with unity weighting:

$$z(p) = \sum_m y_m(p - n_m)$$

For a linear array, the sensor delay n_m is proportional to m, and since it is constrained to be an integer, we can write it as $n_m = -qm$. As discussed on page 158, the minus sign means that positive q corresponds to positive slowness (waves propagating in the x direction). Thus, the delay-and-sum beamformer corresponds to a Radon Transform

$$z(p,q) = \sum_m y_m(p+qm) = \mathcal{R}[y_m(n)]$$

where the output's dependence on the chosen value of q is made explicit. The Radon Transform variables p and q can be interpreted as discrete-time intercept and discrete slowness, respectively.

As shown in Fig. 4.25, the Radon Transform corresponds to the steered response of conventional beamforming, but is restricted to slownesses that can be defined with no error. Because q is an integer, each line along which we sum the two-dimensional signal to form the Radon Transform must pass through an element of the sequence for each value of m. The Radon Transform is linear; its other properties form the subject of Prob. 4.25.

An inverse Radon Transform can also be defined. To derive the inverse Radon Transform, we note that an element of the sequence $x(m_0, n_0)$ gives rise to a line in the (p, q) plane given by $p = n_0 - qm_0$. Summing along this line with the proper normalization yields the element's value; the inverse Radon Transform is found to be

$$x(m,n) = \lim_{Q\to\infty} \frac{1}{2Q+1} \sum_{q=-Q}^{Q} X_R(q, n - qm)$$

Example

The most important example of the Radon Transform is a unit sample that appears to be propagating across an M-sensor (M odd) regular linear array with slowness q_0 (an integer). This wave results in a set of sensor signals given by $y_m(n) = \delta(n - q_0 m)$, where $m = -(M-1)/2, \ldots, (M-1)/2$. The Radon Transform of this sequence is therefore given by

$$z(p,q) = \mathcal{R}[y_m(n)] = \sum_{m=-(M-1)/2}^{(M-1)/2} \delta(qm + p - q_0 m)$$

As the unit sample $\delta(\cdot)$ is nonzero only when its argument is zero, this sum is easily computed. Consider $q = q_0$; the Radon Transform becomes $\sum_m \delta(p)$, which equals M when $p = 0$ and is 0 otherwise. When $q \neq q_0$, the sum is nonzero *only* when $p = (q_0 - q)m$, which occurs at most once in the sum. This last equation defines the previously mentioned line in the (p, q) plane evoked by a single sample. As shown in Fig. 4.26, the dominant peak in the Radon Transform occurs at the signal's slowness and temporal centroid.

Example

The next interesting Radon Transform example occurs when the previous one is extended to a general signal $s(n)$ propagating across a regular linear array. This signal can be modeled as the temporal convolution of the propagating unit sample described in the previous example with the signal's waveform. Using the convolution property described in Prob. 4.25, the Radon Transform of the propagating signal equals the convolution with respect to the time-intercept variable p between the Radon Transform of the propagating unit sample and the Radon Transform of the signal $s(n)\delta(m)$. The latter Radon Transform is easily found to be $\mathcal{R}[s(n)\delta(m)] = s(p)$. Thus, the Radon Transform of the propagating signal equals the Radon Transform of $\delta(n - q_0 m)$ convolved with the signal for each value of q. As illustrated in Fig. 4.26, the signal waveform appears, centered at the slowness and time-intercept governing its propagation. Smaller, more erratic waveforms result at other slowness values. In relation to conventional beamforming, these waveforms correspond to sidelobes, with some slownesses yielding larger values than some closer to the correct one.

The calculations performed in conventional beamforming thus correspond to those summarized by the Radon Transform. In beamforming, we find the signal's direction of propagation by searching for the maximum energy assumed propagation direction; with the Radon Transform, the energy is calculated with respect to p for each slowness, with the maximum-energy slowness corresponding to the estimated direction of propagation. When applied to transient signals, this energy calculation is performed over a limited range of p. This range is slid along the time-intercept axis so that both the signal's slowness and its time of occurrence can be estimated.

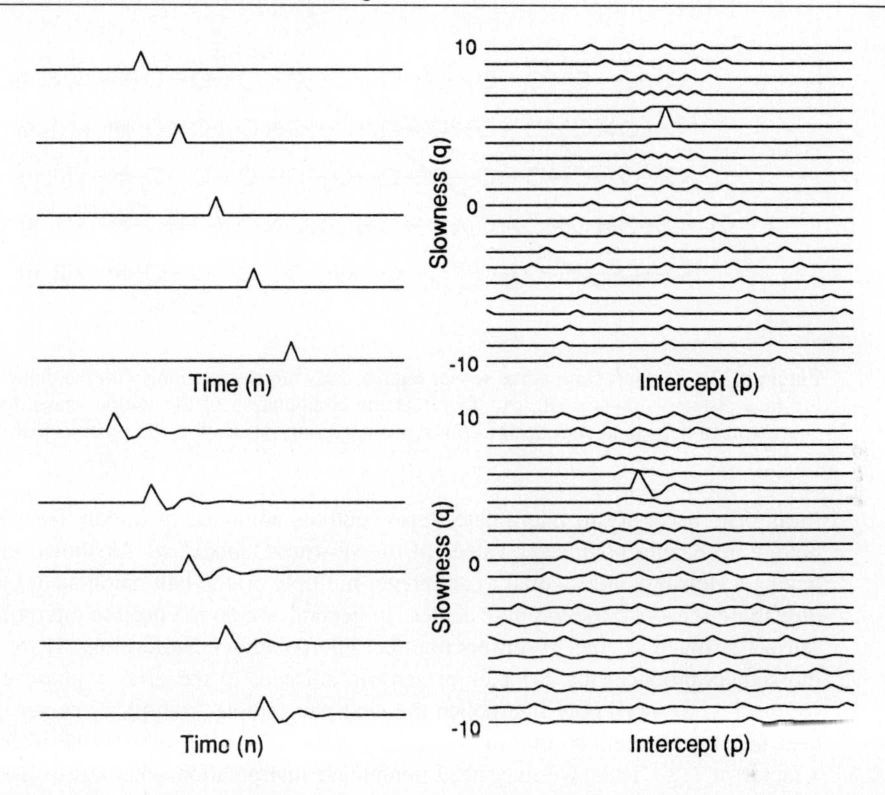

Figure 4.26 A single-sample signal is shown in the upper-left panel propagating across a five-sensor array with a slowness of five samples/sensor. A portion of the Radon Transform is shown in the upper right panel. This transform has a dominant peak at the slowness and central time origin of the signal. The other nonzero values extend forever, taking on unity values at all the suggested values of p and q. A transient signal is shown in the lower-left panel propagating with the same slowness as the unit sample. Its Radon Transform is the convolution of the transform shown in the upper-right panel with the signal. Again, only a portion of this transform is shown in the lower-right panel.

When several propagating signals are present, the Radon Transform consists of the sum of the Radon Transforms of each. Assuming first that the signals occur at nearly the same time but with different slownesses, the maximum energy output for *each* transform occurs at the correct slowness; behavior of the sum is less clear-cut. Because of the sidelobe-like structure of the Radon Transform, the maximum-energy peaks in the composite Radon Transform may not occur at the component's values. Furthermore, the waveforms corresponding to these peaks may not equal the signal propagating across the array at that slowness. These problems in conventional beamforming algorithms are fundamental and are very difficult, if not impossible, to surmount. If several propagating but nonoverlapping signals are present, the maximum energy can fluctuate as do the various signal energies.

Useful insights into interpolation beamforming can be gleaned from the Radon Trans-

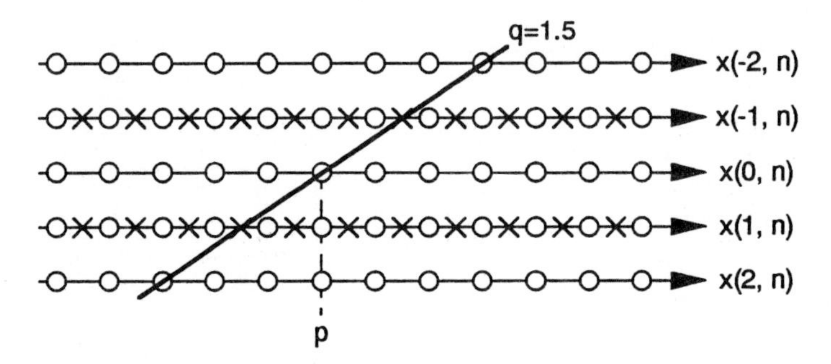

Figure 4.27 Outputs from a five-sensor regular array are shown along with the delay corresponding to a discrete slowness of 1.5. To effect the computation of the Radon Transform, which is equivalent to delay-and-sum beamforming, we need only interpolate the inner sensor outputs.

form. The necessity to interpolate sensor outputs amounts, in Radon Transform terminology, to having noninteger values of the slowness "index" q. As shown in Fig. 4.27, having a slowness index equal to an integer multiple of one-half amounts to interpolating *only* those sensors close to phase center. In general, we do not need to interpolate "outer" sensors as much as "inner" sensors to effect interpolation beamforming. If we interpolate the signals produced by the pair of sensors adjacent to the array's phase center by a factor of I, we need only interpolate the next pair remote from phase center by $I/2$, the next pair by $I/3$, and so on. In general, the kth such pair need only be interpolated by a factor of I/k. Thus, we may need noninteger interpolation schemes; to interpolate by the rational factor of I/J, $I > J$, we upsample by I, lowpass filter accordingly (cutoff frequency of π/I), then upsample by J [43: 39–42]. Precise interpolation schemes such as these lead to the most efficient time-domain algorithms known.

4.8.4 Slowness Aliasing

As we have seen, sampling a signal too slowly results in temporal frequency aliasing; sampling a wave too infrequently in space causes spatial frequency aliasing. *Slowness aliasing* is another problem that can occur in some cases, even when temporal and spatial frequency aliasing are absent. As its name suggests, slowness aliasing can cause a wave with one value of slowness to be interpreted as a wave with a different value of slowness. In other words, it is possible for the direction of propagation to be inaccurately computed by a delay-and-sum beamformer even when we obey spatiotemporal sampling rules.

Grating lobes, discussed in §3.3.2 {90; see also 126}, are a form of slowness aliasing in the sense that waves with high slowness numbers pass through a delay-and-sum beamformer as a wave with a low slowness number. We've already discussed this problem in terms of spatial frequency aliasing, so we won't say anything further here. In addition, a wave with a low slowness number can be mistaken for a wave with a high slowness number, even if it is adequately sampled in time and space. This phenomena can occur if we look for waves with impossibly large slowness values. Assume as usual that the medium supports waves that propagate with a constant speed c. The maximum slowness any wave can have simply equals $1/c$. Nothing, however, prevents us from adjusting the delays in a delay-and-sum beamformer to look for waves with larger slownesses. We would not expect to find any such waves, but nature is not so kind to the naïve.

Suppose we have our usual linear array of M sensors spread out along the x axis and that a periodic wave of the form $s(t + z/c)^{\dagger}$ is propagating down the z axis, broadside into the array. Assume that the waveform $s(t)$ has a period T_0. Setting the sensor delays $\Delta_m = (m - M_{1/2})jT_0$ for any integer j, the beamformer's output is identical to that produced when the array is steered to broadside ($\Delta_m = 0$).

$$z(t) = \sum_m w_m s(t - mjT_0) = \sum_m w_m s(t)$$

Slowness aliasing can thus occur when beamformer delays are adjusted to look for propagating waves having impossibly large slownesses (speeds that are impossibly small). Such "propagating" waves would fall into the invisible region defined in §3.3.2 {91}.

When we consider the discrete-time version of this problem, let the wave period be commensurate with the sampling period, yielding sensor signals having the form $y_m(n) = y_m(n + N)$, where N is the integer period. Because of this periodicity, we find that $y_m(n + mN) = y_m(n)$. Steering the beam to correspond to an integer slowness q and assuming unity shading, we have

$$z(p, q + N) = \sum_m y_m\big(p + m(q + N)\big) = z(p, q)$$

Because of the wavefield's periodicity, we can also show that $z(p+N, q) = z(p, q)$. The beamformer output, when considered as a Radon Transform of the sampled wavefield, is therefore doubly periodic in intercept and slowness.

This effect can also be shown schematically from the first Radon Transform example {168}. Instead of a single impulse propagating across the array, assume we have a periodic train of unit samples separated by N samples. Each impulse results in five‡ lines in the (p, q) plane, as shown schematically in Fig. 4.28. For example, a unit sample at time $p = jN$ causes five lines, one vertical and the others with slopes of ± 1 and $\pm 1/2$, crossing at $(p, q) = (jN, q_0)$. Where these five lines cross, the Radon Transform attains its maximal value, thereby indicating the presence of a signal at that slowness and time intercept. Note, however, that five lines cross in other places for other values of q. Indeed, crossings occur at $(p, q) = (jN, q_0 + iN)$ for all integer values of i and j.

Another type of slowness aliasing can occur when we consider determining the slowness that yields maximum beamformer output. In many cases, this maximum occurs at some noninteger value of q. If we find the maximum's location by interpolating $z(p, q)$ in the slowness variable, we may produce erroneous estimates: The effective sampling rate of some sensor signals, usually those furthermost from the array origin, in the Radon Transform may be below the Nyquist rate [134].

---

†This lapse back into continuous time shows that this point not a result of temporal sampling.

‡The plot shown in Fig. 4.26 corresponds to $M = 5$ sensors in the linear array.

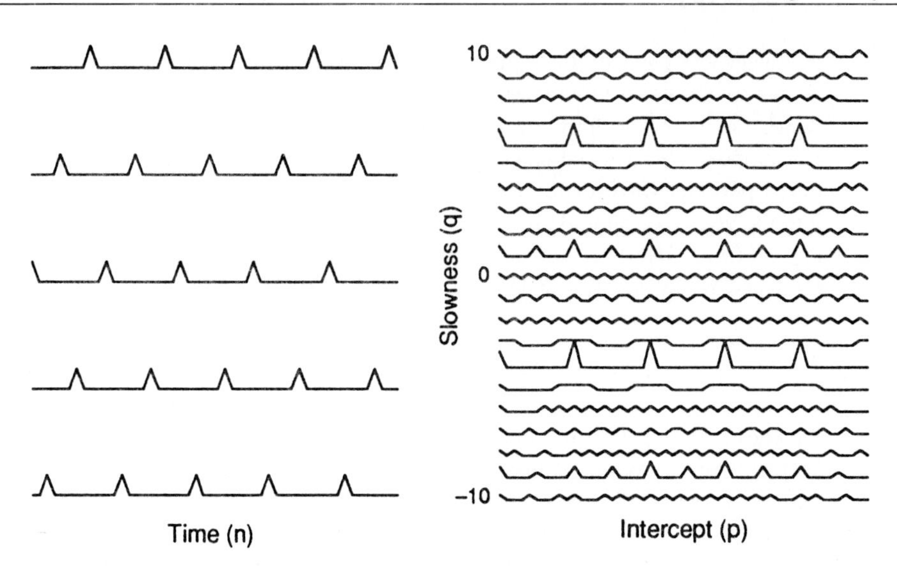

Figure 4.28 The Radon Transform, which corresponds to the output of a delay-and-sum beamformer, is shown schematically for a periodic impulse train with slowness $q_0 = 5$ that results in sensor signals of the form $y_m(n) = \sum_j \delta(n - jN + q_0 m)$. Each impulse in the train causes $M = 5$ lines in the (p, q) plane. Where these lines intersect, the Radon Transform yields its maximal response. Note that two slowness values, $q = 5$ and $q = -4$, yield maximal values. Can you determine why significant peaks occur at unit slowness with more frequent intercepts?

4.8.5 Frequency-Domain Beamforming

In §4.4 {132}, we discussed how, in theory, the signal processing operations required to form beams can be computed in the frequency domain. Save for the filter-bank implementations, none of those algorithms can actually be implemented: How would you compute a continuous-time signal's Fourier Transform? With discrete-time beamforming, the flexibility of digital processing supports these processing ideas well, and, in many cases, discrete-time beamformers implemented in the frequency domain have superior characteristics to their time-domain counterparts [48, 49].

Short-time Fourier analysis. Similar to the definition given in Eq. 4.12 {133} for the short-time Fourier Transform of a continuous-time signal, the short-time Fourier Transform of a sampled sensor signal $y_m(n)$ and the sampled array output $z(n)$ are given by

$$\boxed{\begin{aligned} Y_m(p,\omega) &= \sum_{l=p}^{p+D-1} \widetilde{w}(l-p)\, y_m(l) \exp\{-jl\omega T\} \\ Z(p,\omega) &= \sum_{m=0}^{M-1} w_m Y_m(p,\omega) e^{j\omega p} e^{-j\omega \Delta_m} \end{aligned}} \tag{4.24}$$

The D values of y_m following the time-sample p are used to calculate $Y_m(p, \omega)$. The function $\widetilde{w}(n)$ is a time-domain window equaling zero outside the interval $[0, D-1]$; thus, D represents the window's duration. Note that the product ωT denotes the usual digital frequency variable that has units of radians per sample period T. A simple change in the summation variable shows that the short-time Fourier Transform has the related expression

$$\boxed{Y_m(p, \omega) e^{jp\omega T} = \sum_{l=0}^{D-1} \widetilde{w}(l) y_m(p+l) \exp\{-jl\omega T\}} \tag{4.25}$$

The shape of the window function $\widetilde{w}(n)$ influences the resolution and detailed structure of the spectrum $Y_m(p, \omega)$. Its Fourier Transform $\widetilde{W}(\omega)$ effectively smoothes the desired but unobtainable spectrum $Y_m(\omega)$ to give us the short-time spectrum. A discussion of the various tradeoffs and properties of window functions occurs in §6.5 {318}.

If we restrict ourselves to discrete frequency values, namely $\omega T = 2\pi v/D$ for $v = 0, \ldots, D-1$, then the short-time Fourier Transform can be computed with a discrete Fourier Transform (*DFT*). Furthermore, if D is highly composite, an *FFT* algorithm can be used. In the interest of clarity, we shall simplify the notation somewhat by writing $Y_m(p, v)$ for $Y_m\big(p, 2\pi v/(DT)\big)$. Thus, the first integer variable, p, represents the time sample index and the second integer variable, v, represents the frequency sample index in units of $2\pi/DT$ rad/s. Using this simplified notation, we have

$$Y_m(p, v) \exp\left\{j\frac{2\pi vp}{D}\right\} = \sum_{l=0}^{D-1} \widetilde{w}(l) y_m(p+l) \exp\left\{-j\frac{2\pi vl}{D}\right\} \tag{4.26}$$

Thus, $Y_m(p, v)$ is a complex-valued lowpass signal approximating the spectrum of $y_m(l)$ at sample p and frequency index v.

The right side of this equation has several interesting interpretations. The simplest is that it describes the *DFT* of some signal. Because the formal definition of a signal's *DFT* is

$$X(v) = \sum_{l=0}^{D-1} x(l) \exp\left\{-j\frac{2\pi vl}{D}\right\}$$

then the right side of Eq. 4.26 represents the *DFT* of the signal $x(l) = \widetilde{w}(l) y_m(p+l)$ and equals $X(v)$.

A subtler interpretation leads to the filter-bank idea. The window function $\widetilde{w}(l)$ can be considered the unit-sample response of a narrowband lowpass filter; thus, the term $\widetilde{w}(l) \exp\{-j2\pi vl/D\}$ represents the unit-sample response of a bandpass filter centered at the frequency represented by the index v. Thus, the equation represents the response of this bandpass filter to the signal $y_m(n)$. By taking the frequency index v to also be a filter index, the equation actually represents a bank of D bandpass filters, each of which is driven by the mth sensor signal.†

†Note that these D filter outputs are not necessarily unrelated. For example, when $y_m(n)$ and $\widetilde{w}(n)$ are both real valued, the output signal from the $(D-v)$th bandpass filter equals the complex conjugate of the output of

Frequency-domain array output. The discrete-time, discrete-frequency version of Eq. 4.13 {134} that describes the frequency-domain beamformer's output equals

$$Z(p, v) = \sum_{m=0}^{M-1} w_m Y_m(p, v) \exp\left\{j\frac{2\pi v}{D}\left(p - \frac{\Delta_m}{T}\right)\right\} \tag{4.27}$$

As usual, the sensor delays $\{\Delta_m\}$ steer the beam in a desired direction or focus it on a particular point. The delay per sample now equals Δ_m/T, which could be fractional. For example, to steer the beam to a particular assumed propagation direction, we set $\Delta_m = -\vec{\zeta}\cdot\vec{x}_m/c = -\vec{\alpha}\cdot\vec{x}_m$. Substituting for $Y_m(p, v)$, the short-time discrete Fourier Transform of each sensor output, we find the frequency-domain discrete-time beamformer's output to be

$$\boxed{Z(p, v) = \sum_{m=0}^{M-1}\sum_{l=0}^{D-1} \widetilde{w}(l) w_m y_m(p+l) \exp\left\{-j\frac{2\pi vl}{D}\right\} \exp\left\{-j\frac{2\pi v}{D}\frac{\Delta_m}{T}\right\}}$$

At this point, let's specialize the analysis to a regular linear array of M sensors having intersensor spacing d. We shall find that regular linear arrays permit exploiting the computational advantages of the *FFT*. Geometries having sensors located on a three-dimensional Cartesian grid share this advantage; examples are thinned linear arrays, rectangular, and cubic arrays. Others, such as circular arrays, do not yield efficient computational frequency-domain methods. This deficiency does not mean they should not be used; rather, this property should be a factor in array design. Computational issues aside, the frequency-domain properties that follow apply to all arrays.

To yield a *DFT* formula for the sum over m, the last exponent in the boxed equation should have the form

$$\frac{2\pi v}{D}\frac{\Delta_m}{T} = \frac{2\pi u}{M}(m - M_{1/2}) \quad m = 0, \ldots, M-1$$

where the index u, $u = 0, \ldots, M-1$, denotes the spatial frequency index, which is related to the assumed propagation direction. This formula applies for each choice of temporal frequency index v and results in sensor delays having the somewhat mysterious form

$$\Delta_m = \frac{uDT}{vM}(m - M_{1/2}) \quad m = 0, \ldots, M-1 \tag{4.28}$$

The *only* one-dimensional array geometry yielding propagation delays such as these is the regular linear array. The exponential term in the expression for the array output's

---

the vth filter.

$$Y_m(p, D-v)\exp\left\{j\frac{2\pi(D-v)p}{D}\right\} = \left[Y_m(p, v)\exp\left\{j\frac{2\pi vp}{D}\right\}\right]^*$$

short-time spectrum becomes[†]

$$\exp\left\{-j\frac{2\pi v}{D}\frac{\Delta_m}{T}\right\} = \exp\left\{j\frac{2\pi u}{M}M_{1/2}\right\}\exp\left\{-j\frac{2\pi u}{M}m\right\}$$

This substitution gives the equation for $Z(p, v)$ a form similar to that of a two-dimensional *DFT*. Using the two-dimensional *DFT*'s definition,

$$X(u, v) \equiv \sum_{m=0}^{M-1}\sum_{l=0}^{D-1} x(m, l)\exp\left\{-j\frac{2\pi vl}{D}\right\}\exp\left\{-j\frac{2\pi um}{M}\right\}$$

and the association

$$x(m, l) = \widetilde{w}(l)w_m y_m(p+l)\exp\left\{j\frac{2\pi uM_{1/2}}{M}\right\}$$

the frequency-domain discrete-time beamformer for a regular linear array's output equals the two-dimensional *DFT* of this sequence.

$$\boxed{Z(p, v) = X(u, v)}$$

This equation demonstrates the fundamental relationship between a frequency-domain beamformer's output and the wavefield's wavenumber frequency spectrum. The index p, implicit in the *DFT* interpretation, denotes the position of the temporal window. The signal $x(m, l)$ represents a windowed-in-time, windowed-in-space version of the sampled original wavefield $f(\vec{x}, t)$. This two-dimensional *DFT* represents an attempt to estimate the wavefield's wavenumber-frequency spectrum, here indexed by the discrete variables v and u. They can be related back to frequency ω (in radians/seconds) and wavenumber component k_x (in radians/meter) as

$$\omega = \frac{2\pi v}{DT}, \qquad k_x = \frac{2\pi u}{Md}$$

The ratio v/D equals the frequency ω divided by the sampling frequency $2\pi/T$. Similarly, the ratio u/M equals the spatial frequency component k_x divided by the spatial sampling frequency $2\pi/d$. The fact that the discrete short-time Fourier spectrum of the array's output depends *only* on the frequency index v means that the beamformer is searching for propagating waves, for which temporal frequency and wavenumber are linked by a dispersion relation. To derive the two-dimensional *DFT* result, we have assumed the linear dispersion relation described by Eq. 4.28.

Fig. 4.29 depicts the sequence of calculations for a frequency-domain beamformer. Fig. 4.30 shows the result of computing the spatial *DFT* at one temporal frequency. The regular sampling of wavenumber provided by the *DFT* means that angle is irregularly

---

[†]The phase factor $\exp\{j2\pi uM_{1/2}/M\}$ originates with our convention that the array's spatial origin occurs at phase center. Thus, sensor delays for a regular linear array have the form $\Delta_m = \alpha_x(m - M_{1/2})d$. As far as computing the *DFT* is concerned, this phase shift amounts to a reindexing of the sensor outputs.

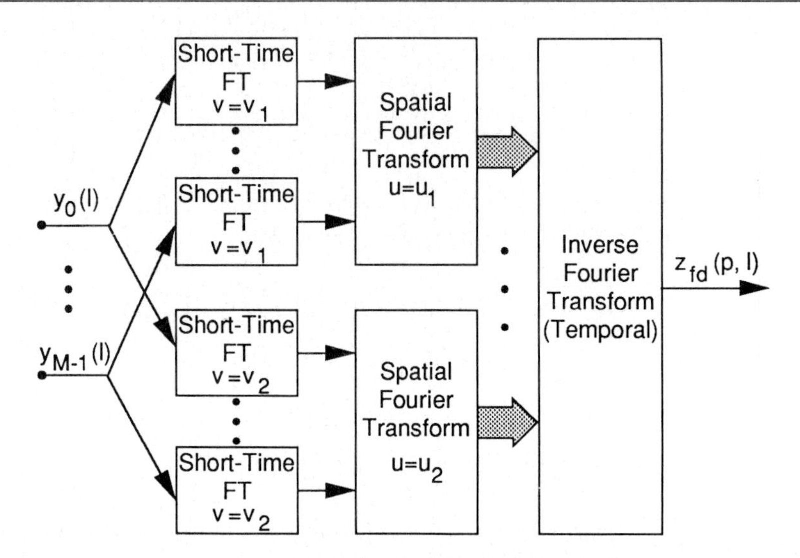

Figure 4.29 The *DFT* calculations of a frequency-domain beamformer consist of a series of one-dimensional Fourier Transform computations. The short-time Fourier Transform of each sensor's signal is computed and evaluated at a specific frequency index v. When we have a linear array, the spatial transform amounts to a *DFT* computation also, with frequency index u corresponding to the wavenumber $2\pi u/(Md)$. As expected, for a fixed direction of propagation, wavenumber varies with temporal frequency, requiring a different spatial frequency index for each choice of temporal frequency index v. Once the spatial transforms are evaluated over the signal's bandwidth, an inverse temporal transform can be evaluated to yield a steered response.

sampled, which leads to the nonlinear angular scale. Furthermore, spatial spectra can be calculated for wavenumbers that do not correspond to propagating signals: in other words, calculated in invisible regions. The beamformer's spatial frequency response is nonzero in such regions because the array's limited aperture permits only finite spatial resolution. The spatial spectrum need not be calculated in invisible regions, but any nonpropagating noise, such as amplifier noise, present in the invisible region can affect the spectrum in the visible region. Shading functions designed to exploit a presumed lack of energy in the invisible region so as to produce narrower mainlobes can be (cautiously) used; see Prob. 4.28.

Direction of propagation sampling. Using a two-dimensional *DFT* elegantly yields the frequency-domain beamformer's output for several steering directions and frequencies simultaneously. To take advantage of this computational opportunity, however, the sensor delays $\{\Delta_m\}$ *must* take on the particular form given by Eq. 4.28 rather than the more general $\Delta_m = \alpha_x md$. These convenient values for Δ_m imply that the assumed propagation directions in which beams can be steered are given by

$$\boxed{\phi = \sin^{-1}\left[\frac{-uDcT}{vMd}\right]} \tag{4.29}$$

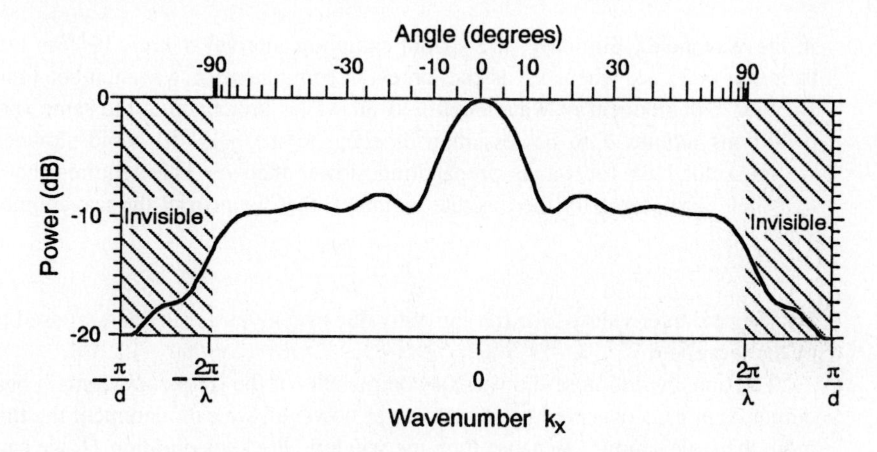

Figure 4.30 The panel depicts the steered response of a uniform regular linear array ($M = 10$) to a broadside plane wave and spherically isotropic noise. This steered response is computed by calculating the spatial Fourier Transform and the squared magnitude of that calculation is shown. Because wavenumber is equivalent to spatial frequency, the resulting spectrum is uniformly sampled in wavenumber, with angle a nonlinear function of wavenumber. Because the sensor spacing $d = 3\lambda/4$, invisible regions occur (see Fig. 3.20 {89}) where no propagating energy exists. The steered response is nonzero there because of the finite spatial resolution provided by the array pattern.

Thus, incidence angle is sampled irregularly in this curious way.

The sampling density can be varied by changing the window's duration D. Increasing D, which increases the sampling density, can be accomplished in two ways. We could simply use a longer window to give us longer data frames from each sensor's signal, or we could increase the value of D by appending zero-valued samples to the data frames. In the first case, we obtain higher temporal frequency resolution because more data are used. In the second, D is increased somewhat artificially; in effect, the window function $\widetilde{w}(n)$ has a number of zero-valued samples tacked onto the end. Circumstances under which zero padding becomes necessary to avoid circular-shifting problems are described subsequently. Although increasing the number of sensors is usually out of the question, we can also increase M without physically installing more sensors in the array. We can pretend more sensors exist, but their sensor weights equal 0. This idea is akin to the zero padding mentioned earlier. It does not increase the achievable spatial resolution—only a larger aperture would do that—but it does reduce the interval between spatial frequency samples, which determines the assumed propagation directions in which the beam can be steered. *These ways of increasing the sampling density of incidence angle by computing longer temporal and spatial transforms than suggested by window lengths are the frequency-domain counterparts to time-domain interpolation beamforming described earlier* {161ff}.

The frequency-domain beamformer for discrete-time signals is sampled in frequency and wavenumber. To avoid temporal frequency aliasing the sampling interval T must be less than or equal to π/ω_U, where ω_U is the highest frequency to be encountered in the wavefield. Similarly, the spatial sampling interval d must be less than or equal to $\pi/k_x^U = \lambda_x^U/2$, where k_x^U is the largest x component of wavenumber (and λ_x^U is the smallest x component of wavelength). If all waves propagate at the same speed c, these conditions require d to be less than or equal to $\pi c/\omega_U$. To avoid slowness aliasing, we must not look for waves propagating slower than c. This requires that we restrict $k_x \leq \omega/c$, or in terms of the discrete variables, that we restrict the wavenumber index as

$$u \leq \frac{Md}{DcT} v$$

Looking at larger values of u for a given value of v gives us slowness-aliased components of the wavefield.

The time evolution of the various components of the frequency-domain beamformer's output $Z(p, v)$ is observed by increasing p; however, we can increment the time index by more than one sample. Because the time window $\widetilde{w}(n)$ has duration D, we can increment p by as much as D samples at a time. However, the increment, or window *stride* as it is sometimes called, may be less than D for a particular application because of the window's shape. This issue is explored in Chap. 6 {327}.

Formation of time-domain output. For many applications, all that may be required is producing the frequency-domain values of the array's output at a given set of discrete wavenumbers. This situation might occur for sinusoidal propagating signals; however, when wider bandwidth signals pique our interest, we need to coalesce the frequency-domain beamformer's output across temporal frequency. To obtain a time-sampled beamformer output signal, we need to "sum" the various frequency components of $Z(p, v)$. For simplicity, let's assume that the sensor delays $\{\Delta_m\}$ equal integer multiples of the sampling period: $\Delta_m = n_m T$. The frequency-domain beamformer's output (Eq. 4.27 {174}) equals

$$Z(p, v) = \sum_{m=0}^{M-1} w_m Y_m(p, v) \exp\left\{j\frac{2\pi vp}{D}\right\} \exp\left\{-j\frac{2\pi vn_m}{D}\right\}$$

This reexpression of $Z(p, v)$, the short-time Fourier Transform of the array output at time index p and temporal frequency index v, explicitly demonstrates its relation to a *DFT*. Note that the phasor $\exp\{j2\pi vp/D\}$ represents the linear phase shift owing to the time origin p of the Fourier analysis. By computing the inverse *DFT*,

$$z_{\text{fd}}(p, l) = \frac{1}{D}\sum_{u=0}^{D-1} Z(p, v) \exp\left\{j\frac{2\pi vl}{D}\right\} = \text{DFT}^{-1}[Z(p, v)] \quad l = 0, \ldots, D-1$$

we obtain a time-domain version of the frequency-domain beamformer's output. Note the two "time" indices in $z_{\text{fd}}(p, l)$: The first denotes the time origin of the short-time analysis and the second time since that origin.

In more detail, this inverse Fourier Transform equals

$$z_{\text{fd}}(p, l) = \sum_{m=0}^{M-1} w_m \text{DFT}^{-1}\left[Y_m(p, v) \exp\left\{j\frac{2\pi vp}{D}\right\} \exp\left\{-j\frac{2\pi vn_m}{D}\right\}\right] \tag{4.30}$$

Multiplication of a *DFT* by a phase factor results in a *circular* shift of the corresponding time-domain signal [118: 536–38]. This effect amounts to "circular aliasing" wherein a frame's trailing values, once circular shifted to the right, appear as leading samples. Thus, $X(u)\exp\{-j2\pi n_0 u/N\} \leftrightarrow x(n-n_0)_N$, where $(n)_N$ denotes that the index n is evaluated modulo N. Here $Y_m(p,v)\exp\{j2\pi vp/D\}$ is the *DFT* of the sequence $\widetilde{w}(l)y_m(p+l)$ (see Eq. 4.26 {173}) and the phase factor is $\exp\{-j2\pi vn_m/D\}$. Thus,

$$\text{DFT}^{-1}\left[Y_m(p,v)\exp\left\{j\frac{2\pi vp}{D}\right\}\exp\left\{-j\frac{2\pi vn_m}{D}\right\}\right] = \widetilde{w}(l-n_m)_D\, y_m\big(p+(l-n_m)_D\big)$$

Consequently, the time-domain output of the frequency-domain beamformer contains circular shifts of the window and the sensor outputs.

$$z_{\text{fd}}(p,l) = \sum_{m=0}^{M-1} w_m \widetilde{w}(l-n_m)_D\, y_m\big(p+(l-n_m)_D\big)$$

Fig. 4.31 illustrates the effect of the circular shift. Time-domain beamforming on the windowed signals aligns the waveforms as desired but misaligns the windows, causing distortions in the beamformer output near the edges of the window. We can minimize this effect if the window is longer than the signal length (for transient signals) or if the window's Fourier Transform is narrower than the signal's spectrum (for sustained signals). Furthermore, the window should also be significantly longer than the longest sensor delay. The frequency-domain approach discussed here yields sensor signals that are circularly shifted *within* the temporal window. Again, if the window is long compared with the signal's duration and to the maximum sensor delay, distortions are confined to the window's edges. Because of the circular shifting property $z_{\text{fd}}(p,l)$ is not the same as its (uncomputable) continuous-frequency counterpart.

The problems of both time- and frequency-domain beamformers foisted on them by discrete-time realizations can be mitigated somewhat by padding frames with zeros and adjusting the stride of the window. Zero padding can be interpreted as using a window that has zero-valued samples. In this case, the effective window length, the region of nonzero values, is shorter than D. Attention must be paid to details, however; ideal placement of the window's zero-valued portion depends on the sign of n_m because it determines the direction of the circular shift. If the window size is such that the distortions occur only near the window edges, then the window stride should be adjusted so that the distorted samples in the current window correspond to undistorted samples in the previous or subsequent window. This approach is similar in spirit to the overlap-save method of high-speed convolution [118: 553–60].

The calculation of the *DFT* to form the short-time time-domain output can be interpreted in the two-dimensional *DFT* domain. The frequency-domain beamformer's expression as a two-dimensional *DFT* is $Z(p,v) = X(u,v)$, where $X(u,v)$ is the two-dimensional *DFT* of the signal $\widetilde{w}(l)w_m y_m(p+l)$ with respect to m and l. The points in the wavenumber-frequency plane (k_x,ω) corresponding to its discrete representation, the (u,v) plane, are shown schematically in Fig. 4.32. Also shown for reference are the lines indicating the maximum (positive) and minimum (negative) possible values of the

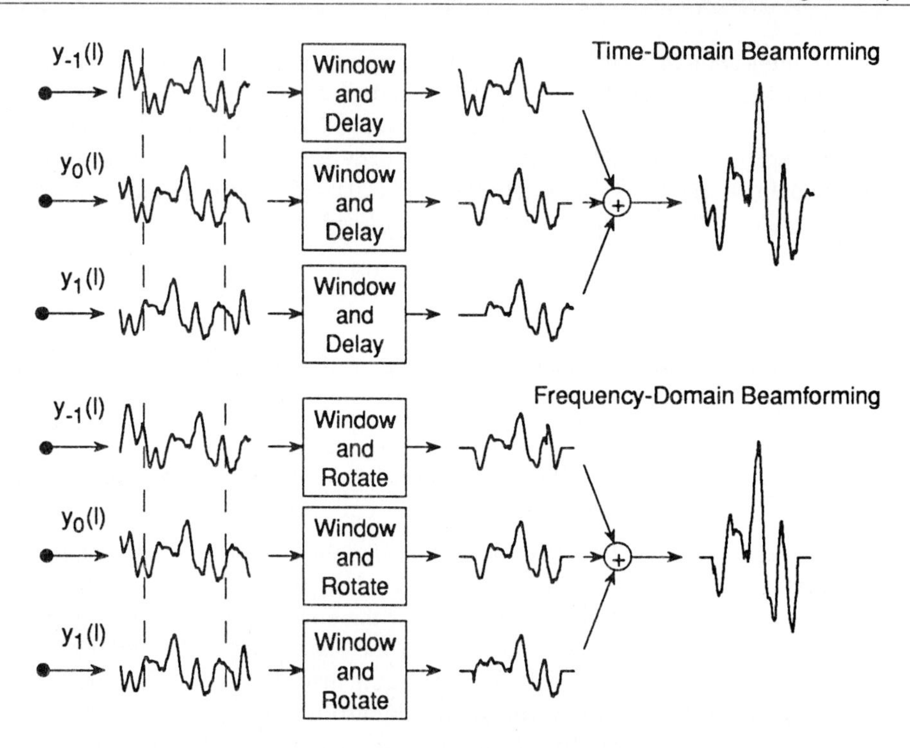

Figure 4.31 The upper panel shows delay-and-sum beamforming operations on three sensor signals. Signal samples falling outside the frames are not considered further in the beamforming operation for this window position. Delaying the windowed sensor signals appropriately to align the waveforms for the time-domain beamforming operation unfortunately misaligns the windows. As a consequence of adding windowed, then delayed, sensor signals, values near the beginning and end of the frame do not correspond to a beamforming operation. The lower panel shows, in the time domain, the result of frequency-domain beamforming followed by summing over the frequency components. Without zero padding the sensor signals, applying a linear phase shift in the frequency domain to effect beamforming delays results in a *circular* shift (rotation) of the windowed signals. Summing these signals yields errors at the ends of the frames.

slowness's x component for waves propagating with the speed c: $u = \pm(Mdv)/(DcT)$. To obtain the time-domain output, we can sum selected components from the (u, v) plane. Clearly, holding the wavenumber index u constant and summing only over the temporal frequency v is incorrect: To steer a coherent beam in a given direction, the wavenumber component k_x and the frequency ω must be proportional: $k_x = -\omega/c \sin\phi$.[†] Thus, u must be proportional to v: A sum over v involves changing the values for u as well.

$$u = -\frac{vMd}{DcT}\sin\phi$$

The components that when summed yield $z_{\text{fd}}(p, l)$ for a particular assumed propagation direction are those that lie along lines through the origin in the (u, v) plane.

[†]Recall that the minus sign occurs because we have defined ϕ as being positive clockwise from the z axis.

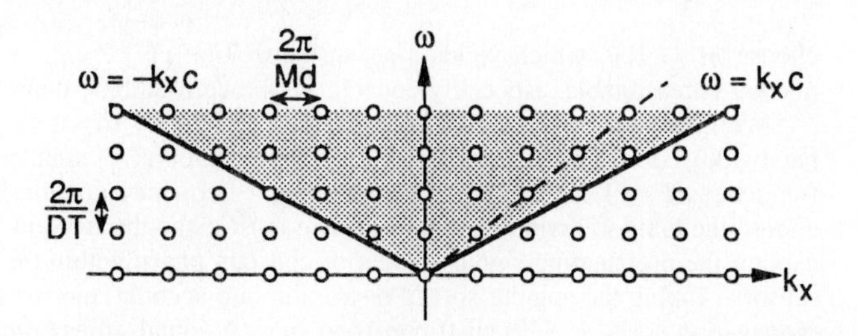

Figure 4.32 The discrete wavenumber-frequency plane is shown. The discrete variable v corresponds to temporal frequency ω and the discrete variable u corresponds to the x component k_x of the wavenumber vector. These indices increment across their respective variables with the indicated separations. Note that finer grids can be obtained by computing the underlying Fourier Transform with longer, zero-padded observations, be they in space or in time. Finer grids essentially interpolate wavenumber-frequency space. The shaded area shows where components of waves propagating with a velocity c may exist. The dashed line shows a path of summation that could be used to calculate the beamformer's output signal.

Example

To illustrate how we would design a conventional beamforming system, let's go through the initial stages of parameter selection for a "practical" problem. The coach of a football team sitting high in the stadium's press box wants to eavesdrop on the opposing coach standing on the sideline. His idea is to use an acoustic linear array to focus on the coach's voice. Here are his (naïve) specifications.

Signal spectral support	1–3 kHz
Array gain	20 dB
Angular resolution	10 m spanning 60 m at a 150 m range

The last requirement comes from wanting to listen to conversations occurring within 10 m of the opposing coach while rejecting more distant conversations. The coach wanders over a distance of 60 m along the sideline centered at midfield.

To begin, we need the propagation speed c; from Table 2.1 {12}, we take the speed of sound in air to be 330 m/s. The array gain specification suggests that at least $M = 100$ sensors are needed. To determine d, the sensor separation, and to refine the number of sensors, we turn to the resolution specification. Assuming the spying coach is located at midfield, he wants to scan over ± 30 m of the sideline, which corresponds to a ± 0.2 rad ($11.3°$) angular range. Achieving 10 m resolution at a range of 150 m translates into an angular resolution requirement of 0.07 rad. From an earlier example {146} in which we derive the parabolic width resolution of a linear array, the resolution constraint demands

$$\frac{4\sqrt{3}}{d\sqrt{M^2-1}} \cdot \frac{c}{\omega} \leq 0.07 \text{ rad}$$

Because resolution provided by the linear array varies inversely with temporal frequency, the lower-frequency spectral constraint must be used. We find that the array's aperture must satisfy $Md \geq 5.2$ m. This length means that the spying coach won't have a compact array, but at least it's not enormous. To avoid spatial aliasing, the sensor separation must satisfy $d \leq \lambda^o/2$, where λ^o denotes the wavelength of the propagating signal. As the signal has a wide bandwidth, we must use the upper-frequency limit to determine sensor separation: $d \leq 11$ cm. This separation implies that at least $M = 47$ sensors are needed to satisfy the spatial aliasing constraint. Thus, the array gain requirement dominates, and we must choose $M = 100$, which yields a sensor separation of 5.2 cm. The number of sensors is not too unreasonable, especially considering modern athletic department budgets.

We now turn to the signal processing parameters. To satisfy the temporal-frequency bandwidth constraint, we must use a sampling interval T smaller than twice the highest frequency: $T \leq 1.7 \times 10^{-4}$ s. To implement a frequency-domain beamformer, we need to choose the temporal window length D. We must make the window duration long enough to capture the propagating signal's plane of constant phase within the windowed output of all sensors. Taking the angular spread restriction into account, the worst-case (fastest) apparent propagation speed is $330/\sin 0.2 = 1684$ m/s. A signal propagating at this apparent speed takes 3.1 ms to propagate across the array. We should include several such propagation times within our window, but not so many that source motion becomes a significant factor; taking the number of propagation times to be 10, we have window and *FFT* lengths of at least 184 samples. To delay a signal sampled at a 6 kHz rate by the maximum delay (3.1 ms) means an 18-sample delay. To avoid circular convolution effects, a longer transform length of at least $184 + 18 = 202$ points is needed. We can thus conveniently choose the *FFT* length to be 256. One-third of the number of frequency samples (here 85) occur within the signal's bandwidth. Given no other signal properties other than its bandwidth, we must use *all* these samples in our frequency-domain beamforming calculations. In contrast, if the propagating signal's temporal characteristics were such that its spectrum did not vary rapidly with frequency, fewer samples would be needed in succeeding processing. Speech signals, however, generally have rapidly varying spectra [127: Chap. 3] from both temporal and frequency-domain viewpoints; no such savings can be realized. Again, in athletics money is no object: The coach calculates 100 256-point time-domain Fourier Transforms and 85 128-point[†] space-domain Fourier Transforms at least every 184 ms. These requirements demand that length-256 Fourier Transforms be computed within about 1.25 ms. More rapid transform calculation may be needed if the time-domain windows need overlapping (see Fig. 6.9 {328}).

To confirm that the angular resolution afforded by these parameters satisfies the original specifications, frequency-domain beamformers sample the direction of propagation according to Eq. 4.29 {176}. The coefficient DcT/Md in this relation has the value 1.99. With the temporal-frequency sample index v ranging from 43 to 128, the worst-case sampling of the direction-of-propagation angles afforded by the beamformer obey $\phi = \sin^{-1}(0.05u)$. Thus, uniform spatial transform samples correspond to angle changes of about 0.05 rad near broadside, satisfying the original specification of 0.07 rad resolution.

4.9 Averaging in Time and Space

One of the practical aspects of array signal processing is averaging received signals over time and space to reduce the variability of signal estimates. Previous sections (§4.5 {137ff}) have shown that the delay-and-sum beamforming operation of stacking multiple examples of a signal waveform leads to a signal-to-noise ratio enhancement beyond that provided by a single sensor. In very noisy situations, this enhancement is insufficient; we often require averaging of some sort to further remove variability from the observations. An array and its subsequent signal processing provides two additional averaging alternatives; we can average over multiple signal occurrences in time, and we can average over subarray responses in space. Such averaging comes with a tradeoff; we must usually sacrifice spatial or temporal frequency resolution to obtain less variable estimates.

[†]The length of 128 corresponds to choosing the power of two just larger than 100, the number of sensors.

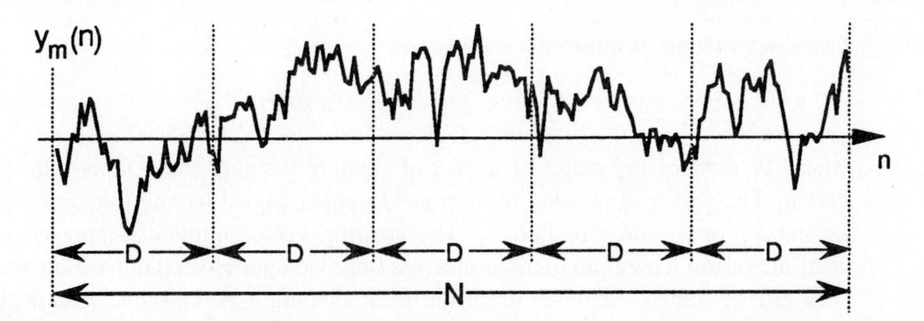

Figure 4.33 The observation interval is segmented into frames of duration D. The frames may be chosen to overlap each other or can be disjoint. The number L of disjoint frames (illustrated here to be five) in the observation interval equals the time-bandwidth product.

4.9.1 Time Averaging

When the propagating signal persists for long periods (it's not a transient occurring only once), we can consider breaking the observation interval into *frames* and averaging quantities of interest over these frames. Random variations in the sensor signals and in estimated quantities, such as the temporal frequency spectrum or the direction of propagation, have smaller variances when we average. As shown in Fig. 4.33, we can break an interval of N data samples into L frames, each having D samples. These quantities are related by $L = N/D$. In frequency-domain algorithms, we window each frame, compute the temporal Fourier Transform, and then *average* the spectra to form the spatiotemporal spectrum from which we compute the steered response power in a manner similar to that discussed in §4.4.2 {134}.

Let $\mathbf{Y}_l$ denote the vector of short-time Fourier transforms across the array for the lth frame.

$$\mathbf{Y}_l = \begin{bmatrix} Y_0(lD,\omega) \\ \vdots \\ Y_{M-1}(lD,\omega) \end{bmatrix} \qquad l = 0, \ldots, L-1$$

Each component of this vector is the short-time Fourier Transform $Y_m(p,\omega)$ of each sensor's output (Eq. 4.24 {172}), with observations assumed to begin at the zeroth sample. Each vector $\mathbf{Y}_l$ is colloquially known as a *snapshot*. The observations spanned by successive snapshots do not overlap in this formulation. In practice, overlap is common, with the window's stride being less than the window's duration. For statistical purposes, the number of independent frames determines the amount of averaging rather than the total number of frames. Thus, the number of nonoverlapping frames more closely approximates the number of independent frames; for now, having no overlap simplifies the succeeding discussion.

To compute the frequency-domain beamformer's output at the lth frame, we need to apply the shading sequence and the phasor representing beamforming delays (Eq. 4.27 {174}). Using the matrix notation developed earlier {134ff}, we can write the frequency-domain beamformer's output as

$$Z(lD, \vec{\zeta}) = \mathbf{e}'\mathbf{W}\mathbf{Y}$$

where $\mathbf{W}$ denotes the diagonal matrix of shading weights and $\mathbf{e}$ represents the steering vector: The vector of phasors for a monochromatic signal having frequency ω and some assumed propagation direction $\vec{\zeta}$. The steering vector implicitly represents where we steer the beam. To employ temporal averaging, we could average these quantities over a sequence of frames. "Could" does not mean "should": As repeatedly emphasized, each frequency-domain snapshot contains a linear phase shift owing to the varying temporal origin of the analysis. Although we could just cancel this phase shift, averaging complex spectral quantities is always fraught with complications. Instead, we turn to the steered response power.

The shaded steered response power for lth frame is $\mathbf{e}'\mathbf{W}\mathbf{Y}_l\mathbf{Y}_l'\mathbf{W}'\mathbf{e}$. The steered response power for the entire set of observations is the average of these components.†

$$\mathcal{P}(\mathbf{e}) = \frac{1}{L}\sum_{l=0}^{L-1} \mathbf{e}'\mathbf{W}\mathbf{Y}_l\mathbf{Y}_l'\mathbf{W}'\mathbf{e}$$

This expression suggests that rather than computing a beampattern for each frame and averaging, an *average* spatial correlation matrix can be computed from the snapshots followed by a single evaluation of the spatial Fourier Transform.

$$\boxed{\begin{aligned} Y_m(lD, \omega) &= \sum_{n=lD}^{(l+1)D-1} \widetilde{w}(n - lD)\, y_m(n) \exp\{-jn\omega T\} \\ \mathbf{Y}_l &= \mathrm{col}\big[Y_0(lD, \omega), \ldots, Y_{M-1}(lD, \omega)\big] \\ \widehat{\mathbf{R}} &= \frac{1}{L}\sum_{l=0}^{L-1} \mathbf{Y}_l\mathbf{Y}_l' \\ \mathcal{P}(\mathbf{e}) &= \mathbf{e}'\mathbf{W}\widehat{\mathbf{R}}\mathbf{W}'\mathbf{e} \end{aligned}} \tag{4.31}$$

When temporal averaging is used, these equations express the fundamental computations of discrete-time frequency-domain beamforming.

The number L of disjoint frames has a signal processing interpretation. Because the number of terms in the average equals L, it determines the variability of the spatial correlation matrix. Why not make it as large as possible? In many cases the length N of the observation interval cannot be increased arbitrarily; to increase L therefore means decreasing the duration D of each frame. We would then need to compute short-time spectra over a shorter span of data with a concomitant reduction in the temporal frequency resolution. A tradeoff immediately appears: As the bandwidth of the Fourier

†This technique falls into the general class of Bartlett spectral estimation procedures, which are discussed fully in §6.5.1 {327}.

transform is proportional to $1/D$, the product of the bandwidth and the extent of the data N defines a *time-bandwidth product*, which equals the number of frames ($L = N/D$). This quantity parameterizes the amount of temporal averaging available to us. If we want a less variable spectral estimate, we must either increase the observation interval, incurring performance penalties because the source's position changes significantly, or we must increase the bandwidth of the temporal Fourier Transforms, paying the price of reduced spectral resolution. Selecting the bandwidth is a signal processing decision based on how much spectral resolution we need to separate sources from one another and on how much noise is permissible: The greater the bandwidth, the greater the amount of broadband noise that leaks into the array's output.

The formula for the estimated spatial correlation matrix $\widehat{\mathbf{R}}$ has the form of an averaging operation. Averaging is a necessary evil in situations in which a significant amount of noise is present in the observations. As shown in Prob. 6.27 {347}, this average is the maximum likelihood estimate of the covariance matrix of a Gaussian random vector when the vectors are statistically independent. Because of the averaging interpretation, the time-bandwidth product affects the statistical characteristics of beamforming calculations in the obvious way: Assuming the array output vectors $\mathbf{Y}_l$ are statistically independent, the standard deviation of the spectral estimate is inversely proportional to $\sqrt{L}$. Consequently, the time-bandwidth product assesses the noise reduction in averaging operations beyond that implicit in beamforming.

Just as important as its effect on noise, if not more so, averaging's effects on the signals must be considered. The outer product of each snapshot expresses the signals' directions of propagation an interesting way. If a single signal is present in the field of a regular linear array, each snapshot expresses the direction of propagation as a phase shift that increases down the vector.

$$\mathbf{Y}_l = A_l e^{-jM_{1/2}\gamma} \begin{bmatrix} 1 \\ e^{+j\gamma} \\ \vdots \\ e^{+j(M-1)\gamma} \end{bmatrix}$$

Here, the phase shift γ equals $2k_x^o d$ for an array situated along the x axis with sensors separated by d m. The quantity A_l represents the complex amplitude of the signal's short-time spectrum during the lth frame at the frequency ω. The outer product of $\mathbf{Y}_l$ with itself has the form†

$$\mathbf{Y}_l \mathbf{Y}_l' = |A_l|^2 \begin{bmatrix} 1 & e^{-j\gamma} & \dots & e^{-j(M-1)\gamma} \\ e^{+j\gamma} & 1 & \ddots & \vdots \\ \vdots & \ddots & \ddots & e^{-j\gamma} \\ e^{+j(M-1)\gamma} & \dots & e^{+j\gamma} & 1 \end{bmatrix}$$

†The sharp reader notices that forming the outer product cancels the phase shift owing to the choice of the array's spatial origin. The really sharp reader notices that A_l also contains a linear phase shift related to the increasing time origin of the short-time Fourier analysis. The outer product also cancels this phase shift.

The spatial correlation matrix thus has a Toeplitz, Hermitian structure that expresses the signal's direction of propagation as a phase shift that increases in magnitude away from the main diagonal of the matrix but remains constant along the diagonals. When these outer products are averaged, this structure is not altered and, when no noise is present, $\widehat{\mathbf{R}}$ has the same form with an overall amplitude equaling the signal's mean-squared amplitude.

When two (or more) signals are present, the story becomes more complicated. Each signal has a complex amplitude and a direction of propagation; each snapshot is the sum of the component snapshots.

$$\mathbf{Y}_l = A_l^{(1)} e^{-jM_{1/2}\gamma^{(1)}} \begin{bmatrix} 1 \\ e^{+j\gamma^{(1)}} \\ \vdots \\ e^{+j(M-1)\gamma^{(1)}} \end{bmatrix} + A_l^{(2)} e^{-jM_{1/2}\gamma^{(2)}} \begin{bmatrix} 1 \\ e^{+j\gamma^{(2)}} \\ \vdots \\ e^{+j(M-1)\gamma^{(2)}} \end{bmatrix}$$

The outer product $\mathbf{Y}_l\mathbf{Y}_l'$ contains the sum of the individual outer products *and* a cross-term between the two signals. Denoting this cross-term by the matrix $\mathbf{C}$, its main diagonal is given by

$$[\mathbf{C}]_{i,i} = 2\,\mathrm{Re}\left[A_l^{(1)}A_l^{(2)*}\right]\cos\left[\left(\gamma^{(1)} - \gamma^{(2)}\right)i - \varphi\right]$$

where $\varphi = M_{1/2}\left(\gamma^{(1)} - \gamma^{(2)}\right)$. This matrix's main diagonal is not constant as it is for each individual signal's spatial correlation matrix. It varies sinusoidally with the frequency related to the spatial separation between the signals. The remainder of the cross-term matrix has a similar structure; each has a sinusoidal term multiplied by a complex phase that varies along the diagonal. Whether this complicated cross-term remains after averaging depends on the quantity $1/L\sum_l \mathrm{Re}\left[A_l^{(1)}A_l^{(2)*}\right]$. This quantity represents an empirical approximation to the real-part of the expected value $\mathcal{E}\left\{A_l^{(1)}A_l^{(2)*}\right\}$, known as the *intersignal coherence* $c_{1,2}$. Source signals are said to be *incoherent* if this expected value is 0. Incoherent signals have nothing in common, with the phase between them varying randomly. Increasingly coherent signals, corresponding to increasing $|c_{1,2}|$, have an increasingly stable phase relationship between them. This situation is prevalent when multipath reflections impinge on the array: A single source may have more than one propagation path to the array, and these paths have different directions of propagation. Intersignal coherence greatly affects spatial resolution, as illustrated in Fig. 4.34.

4.9.2 Spatial (Subarray) Averaging

Regular array geometries have advantages over irregular or random arrays when the time-bandwidth product is small. This lack of data can occur when the sources move significantly during the averaging time. Having few snapshots decreases the statistical accuracy of the averaging operation. Regular arrays based on rectangular geometries have many redundancies in the co-array; the redundancies can be exploited to achieve *spatial averaging* that can mitigate the lack of data [73]. The key idea is to decompose

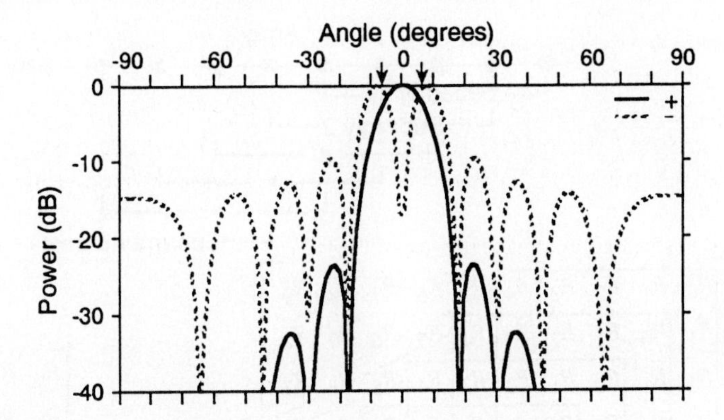

Figure 4.34 Two equal-amplitude monochromatic waves, separated in incidence angle by 11.53° impinge on a ten-sensor linear array. This angular separation equals the Rayleigh resolution limit in this case. The solid line denotes the steered response that results from the signals having the same phase; they are not resolved. In the dashed steered response, the signals are out of phase and twin peaks emerge: The signals are resolved.

an array into *subarrays*. As illustrated in Fig. 4.35, a regular linear array of M sensors can be decomposed into component regular linear subarrays having M_s sensors. If the subarrays are spatially disjoint, then we can average over nonoverlapping subapertures. Generally, however, the number of independent subarrays within a given array is small or else the resulting subapertures are small. The trick is to *overlap* the subarrays, compute the spatial correlation matrix for each, and average these to form a spatial correlation matrix that uses the data provided by the entire array. It is this final averaging step that requires that the subarray geometries be congruent. Said another way, the subarrays' correlation matrices can be averaged because the subarrays have identical co-arrays.

In frequency-domain algorithms, the *subarray averaging* computation can be performed on the array's spatial correlation matrix. The spatial correlation matrix of each subarray is a $M_s \times M_s$ submatrix sharing its main diagonal elements with those of the entire matrix. Thus, subarray averaging is the averaging of submatrices of the array's spatial correlation matrix (Fig. 4.35). Once this new correlation matrix is obtained, computation of the steered response power proceeds as if the signals were propagating across an M_s-sensor array.

Clearly, we compromise aperture when we use subarray averaging. To compensate, each element of the $M_s \times M_s$ spatial correlation matrix now has $L(M - M_s + 1)$ terms contributing to its computation instead of L. In this way, we sacrifice aperture to obtain accuracy in the correlation matrix elements. The steered response power that results from a subarray averaging example is shown in Fig. 4.36.

Subarray averaging reduces the intersignal coherence in the resulting spatial correlation matrix from that in the original. As noted on page 186, coherence induces a modulation along the diagonals of the correlation matrix. Subarray averaging coalesces a sequence of matrices taken from this matrix along its diagonals. Consequently, we average the values along the diagonals to create the smaller matrix that results from sub-

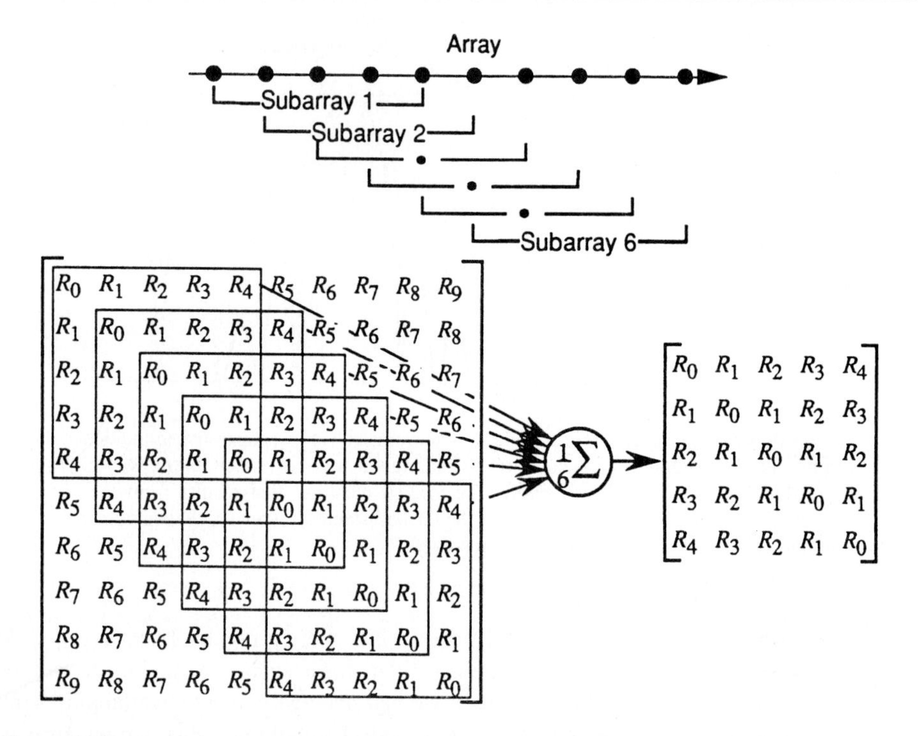

Figure 4.35 A regular linear array of M sensors can be conceptually decomposed into $M-M_s+1$ subarrays each having M_s sensors. Here, spatial correlation matrices are computed for each of the six subarrays formed from a ten-sensor array. These matrices, which are submatrices of the array's spatial correlation matrix, are then averaged to form a smaller spatial correlation matrix. In this way, time-bandwidth product is gained at the expense of aperture.

array averaging. This averaging operation can be represented as passing each diagonal through an *FIR* filter having a rectangular unit-sample response. Thus, to the degree that this filtering operation attenuates coherence-induced sinusoidal variations along the diagonals, the effective intersignal coherence is reduced.

Summary

This chapter describes beamforming, principally delay-and-sum beamforming and its variants, at length. The notion that underlies beamforming is fundamental to array signal processing: Adjust the relative delays between sensor signals so that waves propagating in a particular direction at a particular speed add constructively. Noise and other waves, be they propagating from other directions or nonpropagating, add destructively. Thus, the delay-and-sum beamformer acts as a spatiotemporal bandpass filter. This filter's "center frequency" corresponds to *propagating* waves having an assumed propagation direction defined by beamformer delays. The array's aperture and shading defines the filter's bandwidth with the number of sensors, their locations within the aperture, and the

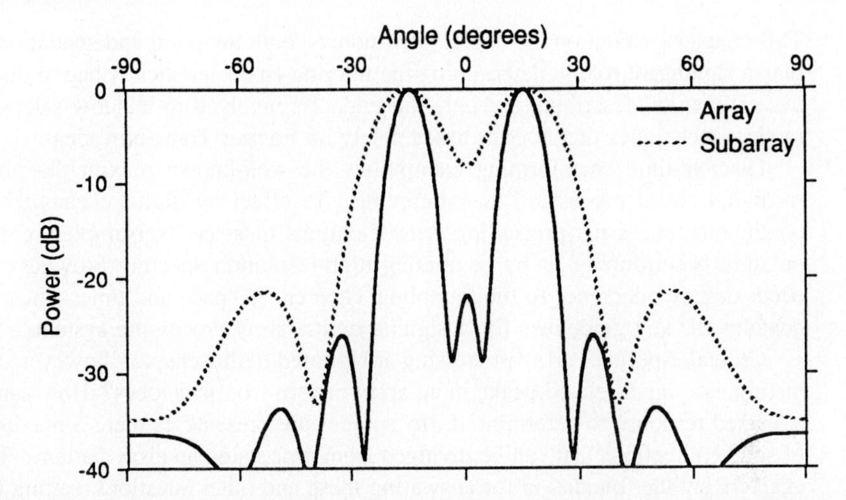

Figure 4.36 As shown in Fig. 4.35, a ten-sensor array is decomposed into six subarrays having five sensors each. Here, two equiamplitude, incoherent signals are present in the array's field. The steered response power of the full array as well as the that of the subarray-averaged array (dashed line) are shown. Note how much wider the mainlobes are for the subarray-averaged steered response than for the original array's. This broadening illustrates the fundamental tradeoff resulting from subarray averaging.

shading determining the "stopband" (sidelobe) structure. The filter's transfer function can be narrowed through judicious filtering at each sensor (filter-and-sum beamforming). This frontend filtering can be temporal, spatial, or both, and complements the delay-and-sum beamformer's inherent filtering properties.

Arrays can be used to two different modes. A beamformer, looking for propagating energy, can be scanned over several assumed propagation directions by systematically adjusting the sensor delays in the same manner that we tune a radio receiver. Usually, the result of this scan is the *steered response power*, a plot of beamformer output power as a function of assumed propagation direction. Alternatively, we can fix the beamformer's delays, defining a *beam* that filters all signals save that propagating from a specific direction. This mode is akin to your car radio, where the buttons define a fixed set of stations (beams) that correspond to your musical interests. The effect of other monochromatic plane waves on the beamformer's output as a function of the plane wave's wavenumber and frequency defines the array's *beampattern*. This chapter demonstrates that these two quantities, the steered response and the beampattern, are equivalent. This seemingly picky distinction becomes paramount in Chap. 7 where adaptive methods are discussed. No matter which beamforming mode we employ, its output is a time-domain signal that can be subjected to further signal analysis to extract information.

The discussion of frequency-domain beamforming illustrates another important interpretation. Beamforming can be regarded as an attempt to estimate the wavenumber-frequency spectrum of the wavefield, just as a scanned bandpass filter or a bank of fixed bandpass filters can be used to assess the distribution of energy as a function of frequency.

This chapter is silent on the choice of windows, both temporal and spatial (shading), that play a significant role in determining the filtering characteristics. Chap. 6 discusses *spectral estimation*, describing not only the tradeoffs involved in window selection, but also modern techniques that are not based solely on Fourier Transform ideas.

Discrete-time beamforming exemplifies the well-known maxim that there is more to digital signal processing than sampling. To effect by digital means the promise of continuous-time array processing, sensor outputs must be oversampled, either by interpolation beamforming or by computing high-resolution spectra. Slowness aliasing can occur despite adherence to the Sampling Theorem in space and time. Thus, this chapter contains the key guidelines for designing digital array processing systems.

Critical aspects of array processing are ignored in this chapter, however. How can we distinguish signal-related peaks in an array pattern from sidelobes? How can we process a steered response to determine if *any* sources are present? Is there a maximum number of source directions that can be divulged from processing an array pattern? The next two chapters lay the foundation for answering these and other questions, setting the stage for the discussion in Chap. 7 of adaptive array processing methods.

Problems

4.1 In §4.1, the signal measured at sensor m was defined to be $s(t - \vec{\alpha}^o_m \cdot \vec{x}_m)$. As the slowness vector $\vec{\alpha}^o_m$ is most conveniently defined as pointing *away* from the location of the source, has the sign of $\vec{\alpha}^o_m \cdot \vec{x}_m$ been chosen properly?

(a) Define the coordinate system for a five-element linear, regular array of sensors. Let the sensors lie along the x axis and have intersensor spacing d.

(b) If the source is in the far field of this array, the propagation vectors at each sensor all equal $\vec{\zeta}^o$. Assuming $\vec{\zeta}^o = -\vec{i}_z$, indicate the direction from which the wave is propagating and compute the propagation delays (relative to the origin that you defined) to each sensor location.

(c) Let $\vec{\zeta}^o = \frac{1}{\sqrt{2}}\vec{i}_x - \frac{1}{\sqrt{2}}\vec{i}_z$. Again compute the propagation delays. Does your answer make sense? Are sensors located further from the source receiving the signal later?

4.2 Consider a simple beamformer for a two-sensor array having an output given by $z(t) = y(t + \Delta) + y(t - \Delta)$.

(a) Show that the energy E_z in $z(t)$ as measured over the interval $0 \leq t \leq T$ is maximized when $\Delta = 0$.

$$E_z \equiv \int_0^T z^2(t)dt$$

(b) Under what circumstances do other values of Δ maximize the output energy?

(c) What implicit assumptions did you make in solving part a?

4.3 Consider an array of seven sensors located at $(x, y) = (-1, -1), (1, -1), (-2, 0), (0, 0), (2, 0), (-1, 1)$, and $(1, 1)$.

(a) Compute the array pattern as a function of k_x, k_y, and k_z.

(b) Compute the array pattern as a function of the angles of incidence ϕ and azimuth θ.

4.4 Suppose we no longer assume that the propagation speed is known. We can associate an assumed speed c with the beamformer delays and an actual speed c^o with the propagating wavefront. Rework the equations for the plane-wave case to describe an array pattern that depends on both c and c^o.

4.5 In Prob. 3.7 {107}, we calculated the aperture smoothing function of each hexagonal subaperture of the World War I listening device depicted in Fig. 1.1 {2}. Perrin arranged these apertures into a hexagonal pattern of hexagons, one for each ear.

(a) Determine the array pattern for Perrin's array.

(b) From the picture, each subaperture measures about 6 cm on a side with the subarrays separated by roughly 3 m; use these values to determine the array's resolution.

4.6 Consider a regular linear array of M sensors spaced d m apart. As usual, let the x axis coincide with the line of the array, take the spatial origin to coincide with the middle sensor's location, and index sensors by the integer m, $0 \leq m \leq M-1$. For convenience, consider only values of M such that $M = 2M_{1/2} + 1$, where $M_{1/2}$ is a positive integer.

(a) Derive the array patterns for the following shading sequences.

i. Uniform: $w_m = 1$

ii. Triangular: $w_m = 1 - \frac{|m-M_{1/2}|}{M_{1/2}+1}$

iii. Cosine: $w_m = \cos \frac{\pi(m-M_{1/2})}{M}$

iv. Cosine squared: $w_m = \cos^2 \frac{\pi(m-M_{1/2})}{M}$

These formulas may be of some help.

$$\sum_{m=0}^{M} m = \frac{M(M+1)}{2} \qquad \sum_{m=0}^{M} m^2 = \frac{M(M+1)(2M+1)}{6}$$

(b) Determine the peak gain, wavenumber resolution, and the ratio of the first sidelobe's peak value to that of the mainlobe.

4.7 We have seen that the steered response power is the result of two shading operations: explicit shading owing to the signal processing and implicit shading owing to array geometry.

(a) Find the general expression for the steered response power of an array with nontrivial sensor weights given by w_m. What is the implicit weighting related to array geometry?

(b) Perhaps the simplest shading for linear arrays (other than the trivial uniform shading) is triangular shading. What is the explicit shading component of the steered response power in this case? What is the total shading related to sensor weights and array geometry? Compare the resolution afforded by uniform and triangular shading.

(c) What shading for sparse linear arrays is analogous to triangular shading for filled linear arrays? Take a four-sensor perfect array as your example.

(d) What is the shading for square arrays analogous to triangular shading for linear arrays?

4.8 Situations do occur in which frequency-domain beamformers have distinct advantages over their time-domain counterparts and vice versa. For example, let's explore the relative merits of each implementation when the array lies in a dispersive medium.

(a) Assuming that the signal is monochromatic, show that either implementation suffices. How are the sensor delays related to propagation direction?

(b) Now assume the source emits a transient signal; which implementation can cope most easily with this wideband source? Assuming the dispersion relation given by Eq. 2.8 {22}, how are sensor delays/phase shifts related to the direction of propagation?

4.9 The steering vector defined on page 135 models an array's output to a propagating monochromatic plane wave. The steering vector can also be used to describe near-field signals as well.

(a) In terms of local directions of propagation and ranges, what steering vector describes the measured delays for a propagating spherical wave?

(b) Describe the calculations of a delay-and-sum beamformer using matrix notation.

(c) How do the calculations of the steered response power for the near-field case differ from those for the far-field (plane-wave) case?

4.10 Eq. 4.12 {133} defines the short-time Fourier Transform.

$$Y_m(t, \omega) \equiv \int_t^{t+D} \widetilde{w}(\tau - t) y_m(\tau) e^{-j\omega\tau} \, d\tau$$

Although this quantity can be considered the Fourier Transform of the windowed observations, it can also be cast as a filtering operation.

(a) Show that this expression can be rewritten as a convolution between $y_m(\cdot)$ and a quantity that can be interpreted as the impulse response of some filter.

(b) Assuming that all windows we could conceivably use have lowpass Fourier transforms, what is the frequency response of your equivalent filter?

(c) Because we want to consider $Y_m(t, \omega)$ as a function of ω, depict a filter structure that yields this short-time spectrum for four frequency values. This structure is known as a *filter bank*.

(d) Filter-bank structures do have an advantage over direct Fourier Transform calculation of the short-time spectrum: They support variable resolution frequency analysis. How would you modify the filter bank to accomplish this goal?

4.11 Consider the differences in the array patterns of a mechanically steered linear array and an electronically steered linear array.

(a) Compute the array pattern of a regular linear array (laid out along the x axis) of M sensors (assume M is odd for convenience) as a function of angle of incidence ϕ for narrowband signals whose wavelength λ is twice the intersensor separation d.

(b) In the mechanically scanned array, the sensor delays are all 0. The array is physically rotated in the (x, z) plane to steer its beam to an assumed propagation direction. Assuming the narrowband case with $d = \lambda/2$, derive a formula for $W(\phi^o, \phi)$, the array pattern as a function of steered direction ϕ when the plane wave angle of incidence is ϕ^o.

(c) Again assuming the narrowband case, derive the formula for the electronically steered array pattern $W(\phi^o, \phi)$. The electronically steered array pattern is just the array pattern for the delay-and-sum beamformer written in terms of angle of incidence. Compare the mainlobe widths and sidelobe amplitudes of the two array patterns for the case in which the assumed angle of incidence is $\phi = 60^\circ$.

4.12 Consider the usual regular linear array of M sensors lying along the x axis.

(a) If the array is steered broadside but a plane wave is arriving off-axis at $\zeta_x^o = -\sin\phi^o$, what is the beampattern as a function of wavelength λ? In other words, assume a monochromatic plane wave with $k_x^o = 2\pi\zeta_x^o/\lambda$ and compute the complex amplitude of the resulting array output signal.

(b) What is the beampattern as a function of frequency ω?

4.13 When what we consider as signal is statistically correlated to what we consider noise, defining signal-to-noise ratio requires careful thought. Assume a signal consists of the sum of signal and zero-mean noise: $y(t) = s(t) + n(t)$. One method defines the signal power P_s in the sum to be $\mathcal{E}[s(t)y(t)]$ and the noise power P_n as $\mathcal{E}[n(t)y(t)]$.

(a) Assuming the signal and noise are statistically uncorrelated, does this definition yield the conventional definition expressed by Eq. 4.14 {137}?

(b) Show that each power measure has the properties of an inner product {489}, which can be considered a projection of the observed signal onto either the signal or the noise. Can we interpret this signal-to-noise ratio definition as the ratio of signal power to uncorrelated noise power?

(c) Find the signal and noise powers when correlation exists. Does their sum equal the power in the observed signal $y(t)$?

(d) Rewrite your expression for SNR in terms of the signal-to-noise ratio SNR_i calculated under the independence assumption and the correlation coefficient ρ_{sn} between signal and noise. Explore interesting extremes ($SNR_i \to \infty$, $\rho_{sn} \to 0, \pm 1$); interpret your results.

4.14 Suppose a source radiating with an average power P_s is located at a range r^o from a receiving array of M sensors and that this range is large compared to the array's aperture. The sensor delays are adjusted perfectly to focus the array on the source's location $\vec{x}^o$ and the sensor weights equal unity. Assume that a noise field with average constant power P_n exists throughout space.

(a) What is the signal-to-noise ratio at the source?

(b) What is the signal-to-noise ratio at each sensor?

(c) What is the signal-to-noise ratio at the output of our properly focused delay-and-sum beamformer?

(d) Now suppose we transmit a wave of average power P_t from a location near the array. It propagates spherically and reflects from an object located at a range r^o far from the array (at least far compared to the aperture). What is the signal-to-noise ratio at the object?

(e) Assuming that the object has a "cross-section" ρ_c, the object reflects ρ_c times the power incident on it, what is the signal-to-noise ratio back at the receiving array? This relation is often called the radar equation (or the sonar equation, depending on the application).

4.15 Consider a noise field with a correlation function of the form

$$R_n(\vec{\chi}, \tau) = \exp\{-\gamma|\vec{\chi}|\}\exp\{-\beta|\tau|\}$$

where γ and β are positive constants. Assume a regular linear array of four sensors separated by d m steered in the broadside direction ($\Delta_m = 0$) senses this noise field.

(a) Show that the normalized noise correlation matrix has the form

$$\mathbf{R}_n = \begin{bmatrix} 1 & a & a^2 & a^3 \\ a & 1 & a & a^2 \\ a^2 & a & 1 & a \\ a^3 & a^2 & a & 1 \end{bmatrix}$$

(b) Demonstrate that the inverse of this matrix equals

$$\mathbf{R}_n^{-1} = \frac{1}{1-a^2} \begin{bmatrix} 1 & -a & 0 & 0 \\ -a & 1+a^2 & -a & 0 \\ 0 & -a & 1+a^2 & -a \\ 0 & 0 & -a & 1+a^2 \end{bmatrix}$$

(c) What is the optimal set of weights to maximize the array gain when the noise field is spatially uncorrelated ($a = 0$)? when the noise field is partially correlated ($a = 1/2$)? when the noise field is highly correlated ($a = 1 - \epsilon$, $\epsilon \ll 1$)?

(d) What is the optimal array gain as a function of a?

4.16 The optimal array gain can be calculated in the frequency domain with a bit more generality than in the time domain. Consider a frequency-domain beamformer where the shading sequence $\{w_m\}$ represents both the amplitude scalings and phase shifts applied by the beamformer. A single monochromatic signal propagates toward the array and is modeled by the vector $\mathbf{s} = A\mathbf{e}$, where A represents the signal's complex amplitude.

(a) Find the signal's spatial correlation matrix.

(b) Using the definition of array gain expressed by Eq. 4.17 {141}, find the shading vector that yields the optimal array gain.

(c) Based on this result, what shading sequence should be used to optimize array gain when the noise is spatially white? What does this result suggest about delay-and-sum beamforming?

(d) Can you bound the array gain? For what propagating signal and noise correlation matrix combinations is the array gain maximized?

4.17 The signal's correlation structure has an effect on array gain. Consider the matrix formula for array gain (Eq. 4.17 {141}).

$$G = \frac{\mathbf{w}'\mathbf{R}_s\mathbf{w}/\operatorname{tr}[\mathbf{R}_s]}{\mathbf{w}'\mathbf{R}_n\mathbf{w}/\operatorname{tr}[\mathbf{R}_n]}$$

To explore this effect, consider a three-sensor array wherein the signal's correlation matrix has the form

$$\mathbf{R}_s = \begin{bmatrix} 1 & \beta & \beta^2 \\ \beta & 1 & \beta \\ \beta^2 & \beta & 1 \end{bmatrix}$$

Recall that each element of this matrix represents the *spatial* correlation of the signal values recorded at sensor pairs. Thus, when $\beta = 1$, the signal at all three sensors is perfectly coherent, implying that identical signal replicas are present at each sensor. At the other extreme, $\beta = 0$ means that the signals are mutually incoherent: Signals measured by the sensors are uncorrelated.

(a) For what values of β does beamforming work best? Explain your reasoning.

(b) Show that the three eigenvalues of $\mathbf{R}_s$ are given by

$$1-\beta^2 \quad 1+\frac{1}{2}\beta^2 \pm \frac{1}{2}\beta\sqrt{\beta^2+8}$$

Which eigenvalue is the largest when $|\beta| < 1$?

(c) If the noise field surrounding the array is spatially white, $\mathbf{R}_n = \mathbf{I}$, what is the optimal array gain as a function of β, $0 < \beta < 1$? Summarize this result: How does signal coherence affect optimal array gain?

4.18 The array gain, as described on page 140, can be expressed in the frequency domain.

$$G = \left[\frac{\int_{-\infty}^{\infty}\int_{-\infty}^{\infty} |S(\vec{k},\omega)|^2\, |H(\vec{k},\omega)|^2\, d\vec{k}\, d\omega \Big/ \int_{-\infty}^{\infty}\int_{-\infty}^{\infty} |S(\vec{k},\omega)|^2\, d\vec{k}\, d\omega}{\int_{-\infty}^{\infty}\int_{-\infty}^{\infty} S_n(\vec{k},\omega)\, |H(\vec{k},\omega)|^2\, d\vec{k}\, d\omega \Big/ \int_{-\infty}^{\infty}\int_{-\infty}^{\infty} S_n(\vec{k},\omega)\, d\vec{k}\, d\omega} \right]$$

(a) Suppose that $S(\vec{k},\omega)$ is the wavenumber-frequency spectrum of a plane wave of the form $s(t - \vec{\alpha}^o \cdot \vec{x})$ where $S(\omega)$ equals

$$S(\omega) = \begin{cases} S_0, & \omega^o - B/2 \le |\omega| \le \omega^o + B/2 \\ 0, & \text{otherwise} \end{cases}$$

What does $|S(\vec{k},\omega)|^2$ equal?

(b) Suppose that an array with an array pattern given by

$$W(\vec{k}) = \begin{cases} M, & |\vec{k}| \le K_0 \\ 0, & \text{otherwise} \end{cases}$$

is used to receive the propagating plane wave. What does $|H(\vec{k},\omega)|^2$ equal when the sensor delays are set properly to steer the beam in the propagation direction $\vec{\zeta}^o$?

(c) Suppose the noise field's power spectrum $S_n(\omega)$ is given by

$$S_n(\omega) = \begin{cases} N_0, & |\omega| \le \Omega_0 \\ 0, & \text{otherwise} \end{cases}$$

Derive the array gain using the formula given previously, assuming that $\Omega_0 > \omega^o + B/2$.

(d) Suppose now that a filter-and-sum beamformer is used. Assume that each sensor signal is temporally filtered with a frequency response $H_s(\omega)$ given by

$$H_s(\omega) = \begin{cases} H_0, & \omega_L \le |\omega| \le \omega_U \\ 0, & \text{otherwise} \end{cases}$$

Derive the array gain for the case where $\omega_L < \omega^o - B/2$ and $\omega_U > \omega^o + B/2$. What happens if $\omega^o - B/2 < \omega_L < \omega_U < \omega^o + B/2$?

4.19 Shading coefficients that optimize array gain do not always yield array patterns that appeal from other aspects.

(a) What is the array pattern produced by an optimal-array-gain delay-and-sum beamformer when the noise is spatially white?

(b) Assume a regular linear array measures wavefields corrupted by noise having a spatial correlation function $R_n(\vec{\chi}) = \exp\{-\gamma|\vec{\chi}|\}$. Compute the array pattern for a ten-sensor array employing optimal-array-gain shading when the intersensor spacing d equals $1/\gamma$, $2/\gamma$.

4.20 A *perimeter* array has sensors lying only along the perimeter of the geometrical shape defining the array.

(a) Calculate the array pattern for a square perimeter array having M sensors equally spaced along each edge with a spacing d along the perimeter equal to one-half wavelength. Plot the array pattern as a function of azimuth and elevation angles.

(b) How do the sidelobes compare with that of a filled square array (that is, an $M \times M$ array) having the same intersensor spacing?

(c) How do the mainlobe widths of the perimeter array and the square array compare?

4.21 The asymptotic form of an array's range resolution defines the variability of range estimation errors as the source moves farther from the array. This information is crucial to modeling array processing estimation error for tracking applications described in Chap. 8. Taking the parabolic width measure of range resolution, how does an array's resolution vary with source range when the assumed propagation direction matches the source's direction of propagation? Array geometry should enter into your expression *only* as a proportionality constant.

4.22 A two-sensor array consists of directional sensors separated by d m. These sensors have identical directivity patterns, but they may be positioned to point in different fixed directions. Assume that the sensors and the mainlobes of their directivity patterns lie in the same plane, allowing direction to be characterized by a single angle ϕ measured with respect to the normal of the array axis. Let $g_0(\phi)$ and $g_1(\phi)$ represent the directivity patterns of the two sensors. The two sensor outputs are equally weighted.

(a) Find the formula for the array pattern as a function of ϕ for the case of narrowband signals with wavelengths equal to $2d$.

(b) Let the sensors' directivity patterns be *symmetric*: $g_0(\phi) = g_1(\phi)$. Calculate the resulting array pattern.

(c) Let the sensors' directivity patterns be *antisymmetric*: $g_0(\phi) = g_1(-\phi)$.† Calculate the resulting array pattern.

(d) Assume that $g_0(\phi) = \exp\{-\sin\phi\}$. Compare the antisymmetric and symmetric array patterns for $\phi^o = 0°$ and $\phi^o = 90°$. Which would you judge to be better? Why?

4.23 Suppose $s(t)$ is a real-valued bandpass signal with nonzero frequency components in the interval $\Omega_c \pm \delta\Omega$.

(a) Because $S(-\Omega) = S^*(\Omega)$, it follows that $S(\Omega)$ is also nonzero in the interval $-\Omega_c \pm \delta\Omega$. Show that $s(t)$ can thus be written in the form

$$s(t) = \frac{1}{2}\left(s_b(t)e^{j\Omega_c t} + s_b^*(t)e^{-j\Omega_c t}\right) = \mathrm{Re}\{s_b(t)e^{j\Omega_c t}\}$$

where $s_b(t)$ is the complex-valued baseband counterpart.

(b) Suppose that a sampling frequency Ω_s is chosen so that $\Omega_s = \Omega_c/N$, N is a positive integer, and $\Omega_s \geq 2\delta\Omega$. What signal results from sampling $s(t)$ at this rate? Is it a useful representation of $s(t)$? Explain.

4.24 Can beamforming techniques be applied directly to the baseband counterpart of a bandpass sampled signal? As an example, consider a complex-valued‡ propagating signal $s(t)$ that has center frequency Ω_c, bandwidth B, and constant spectrum within this band. For simplicity, consider a two-sensor array lying along the x axis and that the propagating wave has incidence angle ϕ with respect to the z axis.

†Can you think of systems having two antisymmetric directional sensors? (Hint: They are *very* common).

‡Because beamforming operations are linear, we can always evaluate the real part of our solution to determine how the array responds to the real part of our complex propagating signal.

(a) What sensor spacing d guarantees that no spatial aliasing occurs?

(b) Suppose that the sensors measure $s(t - \Delta)$ and $s(t + \Delta)$ and that we form their baseband counterparts. If a delay-and-sum beamformer compensates properly for the propagation delays, does the output with the baseband signals serving as inputs equal the output's baseband counterpart when the bandpass signals serve as inputs?

(c) Sketch the locus in wavenumber-frequency space containing energy of the original signal. How does this locus change when we consider the baseband counterpart? Does this change provide some clue as to what difficulties are encountered with baseband systems?

4.25 Show that the Radon Transform defined below has the following properties.

$$\mathcal{R}[f(m,n)] = F_R(p,q) = \sum_{m=-\infty}^{\infty} f(m, p+qm)$$

i. Linearity: $\mathcal{R}[a\, f(m,n) + b\, g(m,n)] = a\, F_R(p,q) + b\, G_R(p,q)$

ii. Translation: $\mathcal{R}[f(m+m_0, n+n_0)] = F_R(p+n_0-m_0, q)$

iii. Convolution: $\mathcal{R}[f(m,n) \star g(m,n)] = F_R(p,q) \star_p G_R(p,q)$

Here, f and g are convolved in both dimensions, whereas $\star_p$ denotes convolution *only* along the p coordinate.

iv. Modulation:

$$(a) \quad \mathcal{R}\left[g(m,n)e^{-jk_0 m}\right] = \textstyle\sum_m g(m, p+qm)e^{-jk_0 m}$$

$$(b) \quad \mathcal{R}\left[g(m,n)e^{-j\omega_0 n}\right] = e^{-j\omega_0 p} \textstyle\sum_m g(m, qm)e^{-jq\omega_0 m}$$

Interpret the right sides in terms of Fourier transforms.

v. Average amplitude and energy

$$(a) \quad \sum_{p=-\infty}^{\infty} F_R(p,q) = \sum_{m=-\infty}^{\infty}\sum_{n=-\infty}^{\infty} f(m,n) \quad \text{for each } q$$

$$(b) \quad \sum_{p=-\infty}^{\infty} \mathcal{R}[f^2(m,n)] = \sum_{m=-\infty}^{\infty}\sum_{n=-\infty}^{\infty} f^2(m,n) \quad \text{for each } q$$

4.26 An alternative to the interpolation beamformer shown on page 162 is to upsample and lowpass filter *all* sensor outputs. Assume that the regular array has M sensors.

(a) What is the sampling rate of the beamformer in this scheme?

(b) Show that the lowpass filters can be coalesced into a single filter and moved from their current locations in the signal processing structure to the beamformer's output without affecting the input-output characteristics.

(c) Compare the number of signal processing operations involved in these two structures. Assume that the lowpass filter(s) are implemented as *FIR* filters of length $4M$.

4.27 In Fig. 4.23 {164}, the polyphase structure for an interpolation filter is given. There, a lowpass filter is decomposed into a bank of parallel subinterpolation filters. Because only one sample from each interpolated sensor output is needed by a delay-and-sum beamformer for any particular assumed propagation direction, additional computational savings may accrue by only computing *one* of the subinterpolation filters for each propagation direction.

(a) For a five-sensor regular linear array, determine the minimal polyphase filter configuration needed for slownesses $q = 1$ and $q = 1.5$. This should be a single structure, with the appropriate signals selected to form the two output signals.

(b) Assuming that the filter length is $N = 4M$, what is the computational savings of this scheme over a full-blown calculation of all interpolated values?

4.28 To give insight into the processing power that shading can give, consider what are known as *superdirective arrays*. Misnamed to some degree, superdirective arrays are not defined by their geometry, which is usually regular and linear, but by their shading sequences and intersensor spacings. To begin, when the intersensor spacing for a uniform, regular linear array becomes less than the Nyquist spacing $\lambda/2$, the resulting spatial oversampling results in reduced angular resolution since the aperture is reduced (see §4.6.2 {145}). Can this reduced resolution be mitigated by shading sequence selection? Superficially not, but, spatial oversampling leads to invisible regions as far as the array pattern is concerned. Superdirective shading sequences exploit the presumed lack of propagating energy in these regions by sharpening the mainlobe while allowing large (huge is more accurate) sidelobes in the invisible region [126]. Let the example linear array have five sensors with two intersensor separations d and corresponding shading sequences $\{w_m\}$.

$$d = \lambda/2: \quad \{0.117683, 0.235279, 0.294077, 0.235279, 0.117683\}$$
$$d = \lambda/8: \quad \{5.48723, -17.128, 24.2815, -17.128, 5.48723\}$$

(a) Compute the array pattern for each array by using the *FFT*, padding each sequence with 251 zeros. For each case, identify where on the spectral axis the spatial Nyquist frequency occurs. Note that, despite the discrepancy in the values for the shading sequences, the mainlobe gain equals one in both cases.

(b) Determine what portions of these transforms correspond to the invisible regions.

(c) Estimate the angular resolution for each array pattern by taking the mainlobe width to be the distance between the zeros straddling the mainlobe. Does the resolution in the $d = \lambda/8$ case exceed that provided by uniform shading?

(d) Calculate the array gain for each array when they are matched to the signal's propagation characteristics under two noise conditions: spatially white noise and spherically isotropic noise.

Superdirective arrays are *very* sensitive to their design assumptions. Furthermore, note the negative shading elements that hallmark superdirective shading. For more information on modern approaches to designing superdirective arrays and an extensive bibliography, see Cox *et al.* [38].

4.29 Given spectral information about the propagating signal, we can determine direction of propagation *without* using delays in the array processing algorithm [113]. Assume the signal $s(t - \vec{\alpha}^o \cdot \vec{x})$, modeled as a stationary random process, propagates toward our array.

(a) How is the spatial correlation function of the propagating signal related to the correlation function of $s(\cdot)$?

(b) Show that the spatial power spectrum equals $\mathcal{S}_s(\vec{k}/|\vec{\alpha}^o|)$.

(c) Assuming we know the signal's power spectrum, how can the signal's direction of propagation be determined to some degree?

4.30 For regular linear arrays, subarray averaging is frequently employed to reduce intersignal coherency. As described on page 187, this averaging can be described as a kind of linear filtering operation on the diagonals of the spatial correlation matrix.

(a) What is the transfer function of this filter in terms of the array and subarray lengths? What is the frequency axis of this transfer function?

(b) Assuming two spatially coherent far-field sources, under what conditions is subarray averaging most effective in reducing coherence?

Chapter 5

Detection Theory

The previous chapters have concentrated on fundamental signal models and the consequent signal properties. Based on these, signal processing algorithms were derived to extract information from these signals (direction of propagation, for example) *by exploiting the special properties of signals expected to arrive at the array*. Although approaching algorithm design this way is natural, it behooves the wise signal processor to check the veracity of his or her notion of signal properties. In many array processing problems, the noise is inescapable,[†] and signal processing algorithms should take unwanted disturbances into account. At worst, the presence of noise may well cause algorithms derived ignorant of noise to yield severely erroneous results; at best, signal-model–based algorithms degrade gracefully as the amount of signal relative to the noise decreases. Rather than the "cross-your-fingers" approach to algorithm design, this and succeeding chapters describe array processing methods that *explicitly* include noise as a fundamental component of their input. We assume that the reader is familiar with both probabilistic concepts and the theory of stochastic processes; §2.6 {45} and appendix A {471} should acquaint the reader with the notation used herein.

The intent of *detection theory* is to provide rational (instead of arbitrary) techniques for determining which of several conceptions—models—of data generation and measurement is most "consistent" with a given set of data. In digital communication, the received signal must be processed to determine whether it represented a binary "0" or "1"; in radar or sonar, the presence or absence of a target must be determined from measurements of propagating fields; in seismic problems, the presence of oil deposits must be inferred from measurements of sound propagation in the earth. Using detection theory, we can derive signal processing algorithms that give good answers to questions such as these when the information-bearing signals are corrupted by superfluous signals (noise). More to the point, we have found that propagating signals result in peaks in the beampattern but that sidelobes as well as the mainlobe are "peaks." How can a mainlobe be objectively distinguished from a sidelobe? Another application of detection is the signal classification problem. When the array

[†]In passive sonar, the signal-to-noise ratio at each sensor may well be 0 dB in "normal" situations.

is steered toward a source, does the source "fit" into one of several predefined categories?

5.1 Elementary Hypothesis Testing

The detection theory's foundation rests on statistical hypothesis testing [41: Chap. 35]; [98]; [122: Chap. 2]; [152: 19–52]. Given a probabilistic model (an event space Ω and the associated probabilistic structures), a random vector $\mathbf{y}$ expressing the observed data, and a listing of the probabilistic models, the *hypotheses*, which may have generated $\mathbf{y}$, we want a systematic, optimal method of determining which hypothesis produced the data. In the simple case in which only two hypotheses, H_0 and H_1, are possible, we ask, for each set of observations, what is the "best" method of deciding whether H_0 or H_1 was true? We have many ways of mathematically stating what "best" means: We shall initially choose the average cost of each decision as our criterion for correctness. This seemingly arbitrary choice of criterion is shown later *not* to impose rigid constraints on the algorithms that solve the hypothesis testing problem. Over a variety of reasonable criteria, one central solution to the hypothesis testing problem persistently emerges; this result forms the basis of *all* detection algorithms.

5.1.1 Likelihood Ratio Test

In a binary hypothesis testing problem, four possible outcomes can result. Hypothesis H_0 did in fact represent the best model for the data, and the decision rule said it was (a correct decision) or said it wasn't (an erroneous decision). The other two outcomes arise when hypothesis H_1 was in fact true with either a correct or incorrect decision made. The decision process operates by segmenting the range of observation values into two disjoint *decision regions* $\Re_0$ and $\Re_1$. All values of $\mathbf{y}$ fall into either $\Re_0$ or $\Re_1$. If a given $\mathbf{y}$ lies in $\Re_0$, for example, we announce our decision "hypothesis H_0 was true"; if in $\Re_1$, hypothesis H_1 would be proclaimed. To derive a rational method of deciding which hypothesis best describes the observations, we need a criterion to assess the quality of the decision process. Optimizing this criterion specifies the decision regions.

Bayes's decision criterion seeks to minimize a cost function associated with making a decision. Let C_{ij} be the cost of mistaking hypothesis j for hypothesis i ($i \neq j$) and C_{ii} the presumably smaller cost of correctly choosing hypothesis i: $C_{ij} > C_{ii}, i \neq j$. Let P_j be the a priori probability of hypothesis j. The so-called *Bayes's cost* $\overline{C}$ is the average cost of making a decision.

$$\begin{aligned}\overline{C} &= \sum_{i,j} C_{ij} P_j \Pr[\text{say } H_i \text{ when } H_j \text{ true}] \\ &= \sum_{i,j} C_{ij} P_j \Pr[\text{say } H_i | H_j \text{ true}]\end{aligned}$$

The Bayes's cost can be expressed as

$$\overline{C} = \sum_{i,j} C_{ij} P_j \Pr[\mathbf{y} \in \Re_i | H_j \text{ true}]$$

$$= \sum_{i,j} C_{ij} P_j \int_{\Re_i} p_{\mathbf{y}|H_j}(\mathbf{y}|H_j)\, d\mathbf{y}$$

$$= \int_{\Re_0} \{C_{00} P_0 p_{\mathbf{y}|H_0}(\mathbf{y}|H_0) + C_{01} P_1 p_{\mathbf{y}|H_1}(\mathbf{y}|H_1)\}\, d\mathbf{y}$$

$$+ \int_{\Re_1} \{C_{10} P_0 p_{\mathbf{y}|H_0}(\mathbf{y}|H_0) + C_{11} P_1 p_{\mathbf{y}|H_1}(\mathbf{y}|H_1)\}\, d\mathbf{y}$$

$p_{\mathbf{y}|H_i}(\mathbf{y}|H_i)$ is the conditional probability density function of the observed data $\mathbf{y}$ given that hypothesis H_i was true. To minimize this expression with respect to the decision regions $\Re_0$ and $\Re_1$, ponder which integral would yield the smallest value if its integration domain included a specific observation vector. This selection process defines the decision regions; for example, we choose H_0 for those values of $\mathbf{y}$ that yield a smaller value for the first integral. The decision region $\Re_0$ is thus defined by

$$P_0 C_{00} p_{\mathbf{y}|H_0}(\mathbf{y}|H_0) + P_1 C_{01} p_{\mathbf{y}|H_1}(\mathbf{y}|H_1) < P_0 C_{10} p_{\mathbf{y}|H_0}(\mathbf{y}|H_0) + P_1 C_{11} p_{\mathbf{y}|H_1}(\mathbf{y}|H_1)$$

We choose H_1 when the inequality is reversed. This expression is easily manipulated to obtain the decision rule known as the *likelihood ratio test*.

$$\frac{p_{\mathbf{y}|H_1}(\mathbf{y}|H_1)}{p_{\mathbf{y}|H_0}(\mathbf{y}|H_0)} \underset{H_0}{\overset{H_1}{\gtrless}} \frac{P_0(C_{10} - C_{00})}{P_1(C_{01} - C_{11})} \tag{5.1}$$

The comparison relation means selecting hypothesis H_1 if the left-hand ratio exceeds the value on the right; otherwise, H_0 is selected. Thus, the *likelihood ratio* $p_{\mathbf{y}|H_1}(\mathbf{y}|H_1)/p_{\mathbf{y}|H_0}(\mathbf{y}|H_0)$, symbolically represented by $\Lambda(\mathbf{y})$, is computed from the observed value of $\mathbf{y}$ and then compared with a *threshold* η equaling $[P_0(C_{10} - C_{00})]/[P_1(C_{01} - C_{11})]$. Thus, for binary hypotheses, the likelihood ratio test can be succinctly expressed as the comparison of the likelihood ratio with a threshold.

$$\Lambda(\mathbf{y}) \underset{H_0}{\overset{H_1}{\gtrless}} \eta \tag{5.2}$$

The data processing operations are captured entirely by the likelihood ratio $p_{\mathbf{y}|H_1}(\mathbf{y}|H_1)/p_{\mathbf{y}|H_0}(\mathbf{y}|H_0)$. Furthermore, note that only the value of the likelihood ratio *relative* to the threshold matters; to simplify the computation of the likelihood ratio, we can perform *any* positively monotonic operations simultaneously on the likelihood ratio and the threshold without affecting the comparison. We can multiply the ratio by a positive constant, add any constant, or apply a monotonically increasing function that simplifies the expressions. We single out one such function, the logarithm, because it simplifies likelihood ratios that commonly occur in signal processing applications. Known as the log likelihood, we explicitly express the likelihood ratio test with it as

$$\ln \Lambda(\mathbf{y}) \underset{H_0}{\overset{H_1}{\gtrless}} \ln \eta \tag{5.3}$$

Useful simplifying transformations are problem dependent; by laying bare that aspect of the observations essential to the hypothesis testing problem, we reveal the *sufficient statistic* $\Upsilon(\mathbf{y})$: the scalar quantity that best summarizes the data [98: 18–22]. The likelihood ratio test is best expressed in terms of the sufficient statistic.

$$\boxed{\Upsilon(\mathbf{y}) \underset{H_0}{\overset{H_1}{\gtrless}} \gamma} \tag{5.4}$$

We denote the threshold value by γ when the sufficient statistic is used or by η when the likelihood ratio appears before its reduction to a sufficient statistic.

As we shall see, if we use a different criterion other than the Bayes's criterion, the decision rule often involves the likelihood ratio. The likelihood ratio is comprised of the quantities $p_{\mathbf{y}|H_i}(\mathbf{y}|H_i)$, termed the *likelihood function*, which is also important in estimation theory. It is this conditional density that portrays the probabilistic model describing data generation. The likelihood function completely characterizes the kind of "world" assumed under each hypothesis; for each model, we must specify the likelihood function to solve the hypothesis testing problem.

A complication, which arises in some cases, is that the sufficient statistic may not be monotonic. If monotonic, the decision regions $\Re_0$ and $\Re_1$ are simply connected (all portions of a region can be reached without crossing into the other region). If not, the regions are not simply connected and decision region islands are created (see Prob. 5.2). Such regions usually complicate calculations of decision performance. Monotonic or not, the decision rule proceeds as described: The sufficient statistic is computed for each observation vector and compared with a threshold.

Example

An instructor in a course in detection theory wants to determine if a particular student studied for his last test. The observed quantity is the student's grade, which we denote by y. Failure may not indicate studiousness: Conscientious students may fail the test. Define the hypotheses as

H_0: Did not study
H_1: Studied

The conditional densities of the grade are shown in Fig. 5.1. Based on knowledge of student behavior, the instructor assigns *a priori* probabilities of $P_0 = 1/4$ and $P_1 = 3/4$. The costs C_{ij} are chosen to reflect the instructor's sensitivity to student feelings: $C_{01} = 1 = C_{10}$ (an erroneous decision either way is given the same cost) and $C_{00} = 0 = C_{11}$. The likelihood ratio is plotted in Fig. 5.1 and the threshold value η, which is computed from the a priori probabilities and the costs to be $1/3$, is indicated. The calculations of this comparison can be simplified in an obvious way.

$$\frac{y}{50} \underset{H_0}{\overset{H_1}{\gtrless}} \frac{1}{3} \quad \text{or} \quad y \underset{H_0}{\overset{H_1}{\gtrless}} \frac{50}{3} = 16.7$$

The multiplication by the factor of 50 is a simple illustration of the reduction of the likelihood ratio to a sufficient statistic. Based on the assigned costs and a priori probabilities, the optimum decision rule says the instructor must assume that the student did not study if the student's grade is less than 16.7; if greater, the student is assumed to have studied

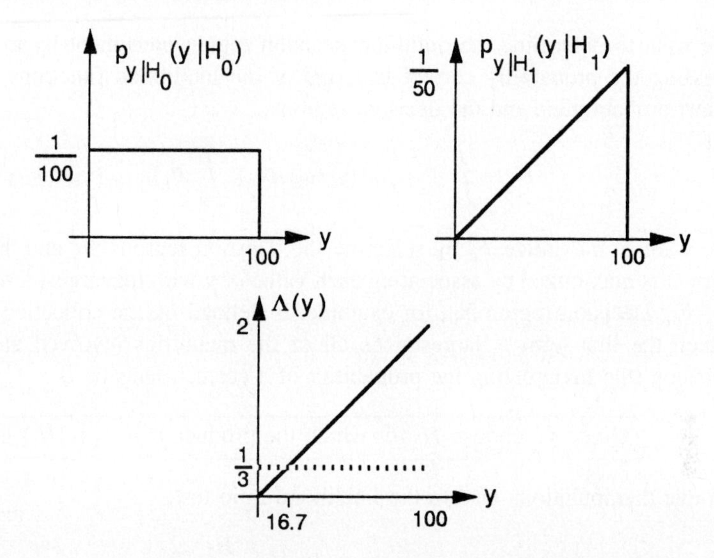

Figure 5.1 Conditional densities for the grade distributions assuming that a student did not study (H_0) or did (H_1) are shown in the top row. The lower portion depicts the likelihood ratio formed from these densities.

despite receiving an abysmally low grade such as 20. Note that as the densities under each hypothesis overlap entirely, the possibility of making the wrong interpretation *always* haunts the instructor. Be that as it may, no other procedure is better.

5.1.2 Criteria in Hypothesis Testing

The criterion used in the previous section—minimize the average cost of an incorrect decision—may seem to be a contrived way of quantifying decisions. Well, often it is. For example, the Bayesian decision rule depends explicitly on the a priori probabilities; a rational method of assigning values to these, either by experiment or through true knowledge of the relative likelihood of each hypothesis, may be unreasonable. In this section, we develop alternative decision rules that try to answer such objections. One essential point emerges from these considerations: *The fundamental nature of the decision rule does not change with choice of optimization criterion.* Even criteria remote from error measures can result in the likelihood ratio test (see Prob. 5.3). Such results do not occur often in signal processing and underline the likelihood ratio test's significance.

Maximum probability of correct decision. As only one hypothesis can be true for a given set of data (the hypotheses are mutually exclusive), the probability of being correct P_c in the binary hypothesis testing problem is given by

$$P_c = \Pr[\text{say } H_0 \text{ when } H_0 \text{ true}] + \Pr[\text{say } H_1 \text{ when } H_1 \text{ true}]$$

We wish to determine the optimum decision region placement by maximizing P_c. Expressing the probability correct in terms of the likelihood functions $p_{\mathbf{y}|H_i}(\mathbf{y}|H_i)$, the a priori probabilities, and the decision regions,

$$P_c = \int_{\Re_0} P_0 p_{\mathbf{y}|H_0}(\mathbf{y}|H_0)\,d\mathbf{y} + \int_{\Re_1} P_1 p_{\mathbf{y}|H_1}(\mathbf{y}|H_1)\,d\mathbf{y}$$

We want to maximize P_c by selecting the decision regions $\Re_0$ and $\Re_1$. The probability correct is maximized by associating each value of $\mathbf{y}$ with the largest term in the expression for P_c. Decision region $\Re_0$, for example, is defined by the collection of values of $\mathbf{y}$ for which the first term is largest. As all of the quantities involved are nonnegative, the decision rule maximizing the probability of a correct decision is

Given $\mathbf{y}$, choose H_i for which the product $P_i p_{\mathbf{y}|H_i}(\mathbf{y}|H_i)$ is largest.

Simple manipulations lead to the likelihood ratio test.

$$\frac{p_{\mathbf{y}|H_1}(\mathbf{y}|H_1)}{p_{\mathbf{y}|H_0}(\mathbf{y}|H_0)} \underset{H_0}{\overset{H_1}{\gtrless}} \frac{P_0}{P_1}$$

Note that if the Bayes's costs were chosen so that $C_{ii} = 0$ and $C_{ij} = C$, $(i \neq j)$, we would have the same threshold as in the previous section.

To evaluate the quality of the decision rule, we usually compute the *probability of error* P_e rather than the probability of being correct. This quantity can be expressed in terms of the observations, the likelihood ratio, and the sufficient statistic.

$$\begin{aligned} P_e &= P_0 \int_{\Re_1} p_{\mathbf{y}|H_0}(\mathbf{y}|H_0)\,d\mathbf{y} + P_1 \int_{\Re_0} p_{\mathbf{y}|H_1}(\mathbf{y}|H_1)\,d\mathbf{y} \\ &= P_0 \int_{\Lambda>\eta} p_{\Lambda|H_0}(\Lambda|H_0)\,d\Lambda + P_1 \int_{\Lambda<\eta} p_{\Lambda|H_1}(\Lambda|H_1)\,d\Lambda \\ &= P_0 \int_{\Upsilon>\gamma} p_{\Upsilon|H_0}(\Upsilon|H_0)\,d\Upsilon + P_1 \int_{\Upsilon<\gamma} p_{\Upsilon|H_1}(\Upsilon|H_1)\,d\Upsilon \end{aligned}$$

When the likelihood ratio is nonmonotonic, the first expression is most difficult to evaluate. When monotonic, the middle expression proves the most difficult. Furthermore, these expressions point out that the likelihood ratio and the sufficient statistic can be considered functions of the observations $\mathbf{y}$; hence, they are random variables and have probability densities under each hypothesis. Another aspect of the resulting probability of error is that *no other decision rule can yield a lower probability of error.* This statement is obvious as we minimized the probability of error in deriving the likelihood ratio test. The point is that these expressions represent a lower bound on performance (as assessed by the probability of error). This probability is nonzero if the conditional densities overlap over some range of values of $\mathbf{y}$, such as occurred in the previous example. In this region of overlap, the observed values are ambiguous: Either hypothesis is consistent with the observations. Our "optimum" decision rule operates in such regions by selecting that model that is most likely (has the highest probability) of generating any particular value.

Neyman-Pearson criterion. Situations occur frequently in which assigning or measuring the a priori probabilities P_i is unreasonable. For example, just what is the a priori probability of a supernova occurring in any particular region of the sky? We clearly need a hypothesis testing procedure that can function without a priori probabilities. This kind of test results when the so-called Neyman-Pearson criterion is used to derive the decision rule. The ideas behind and decision rules derived with the Neyman-Pearson criterion [114] serve us well in sequel; their result is important.

Using nomenclature from radar, where hypothesis H_1 represents the presence of a target and H_0 its absence, the various types of correct and incorrect decisions have the following names [165: 127–29].†

$P_D = \Pr[\text{say } H_1 \mid H_1 \text{ true}]$	*Detection*–we say it's there when it is
$P_F = \Pr[\text{say } H_1 \mid H_0 \text{ true}]$	*False alarm*–we say it's there when it's not
$P_M = \Pr[\text{say } H_0 \mid H_1 \text{ true}]$	*Miss*–we say it's not there when it is

The remaining probability $\Pr[\text{say } H_0 \mid H_0 \text{ true}]$ has historically been left nameless and equals $1 - P_F$. We should also note that the detection and miss probabilities are related by $P_M = 1 - P_D$. As these are conditional probabilities, they do not depend on the a priori probabilities and the two probabilities P_F and P_D characterize the errors when *any* decision rule is used.

These two probabilities are related to each other in an interesting way. Expressing these quantities in terms of the decision regions and the likelihood functions, we have

$$P_F = \int_{\Re_1} p_{\mathbf{y}|H_0}(\mathbf{y}|H_0)\, d\mathbf{y}\,, \qquad P_D = \int_{\Re_1} p_{\mathbf{y}|H_1}(\mathbf{y}|H_1)\, d\mathbf{y}$$

As the region $\Re_1$ shrinks, *both* of these probabilities tend toward zero; as $\Re_1$ expands to engulf the entire range of observation values, they both tend toward unity. This rather direct relationship between P_D and P_F does not mean that they equal each other; in most cases, as $\Re_1$ expands, P_D increases more rapidly than P_F (we had better be right more often than we are wrong!). The "ultimate" situation in which a rule is always right and never wrong ($P_D = 1$, $P_F = 0$) cannot occur when the conditional distributions overlap. Thus, to increase the detection probability we must also allow the false-alarm probability to increase. This behavior represents the fundamental tradeoff in hypothesis testing and detection theory.

One can attempt to impose a performance criterion that depends only on these probabilities with the consequent decision rule not depending on the a priori probabilities. The Neyman-Pearson criterion assumes that the false-alarm probability is constrained to

† In hypothesis testing, a false alarm is known as a type I error and a miss a type II error.

be less than or equal to a specified value α while we attempt to maximize the detection probability P_D.

$$\max_{\Re_1} P_D \text{ subject to } P_F \leq \alpha$$

A subtlety of the succeeding solution is that the underlying probability distribution functions may not be continuous, with the result that P_F can never equal the constraining value α. Furthermore, an (unlikely) possibility is that the optimum value for the false-alarm probability is somewhat less than the criterion value. Assume, therefore, that we rephrase the optimization problem by requiring that the false-alarm probability equal a value α' that is less than or equal to α.

This optimization problem can be solved using Lagrange multipliers [72: 93–94];[152: 33] (see appendix C {502}); we seek to find the decision rule that maximizes

$$F = P_D + \lambda(P_F - \alpha')$$

where λ is the Lagrange multiplier. This optimization technique amounts to finding the decision rule that maximizes F, then finding the value of the multiplier that allows the criterion to be satisfied. As is usual in the derivation of optimum decision rules, we maximize these quantities with respect to the decision regions. Expressing P_D and P_F in terms of them, we have

$$\begin{aligned} F &= \int_{\Re_1} p_{\mathbf{y}|H_1}(\mathbf{y}|H_1)\, d\mathbf{y} + \lambda\left(\int_{\Re_1} p_{\mathbf{y}|H_0}(\mathbf{y}|H_0) d\mathbf{y} - \alpha'\right) \\ &= -\lambda\alpha' + \int_{\Re_1} \left[p_{\mathbf{y}|H_1}(\mathbf{y}|H_1) + \lambda p_{\mathbf{y}|H_0}(\mathbf{y}|H_0)\right] d\mathbf{y} \end{aligned}$$

To maximize this quantity with respect to $\Re_1$, we need only to integrate over those regions of $\mathbf{y}$ where the integrand is positive. The region $\Re_1$ thus corresponds to those values of $\mathbf{y}$ where $p_{\mathbf{y}|H_1}(\mathbf{y}|H_1) > -\lambda p_{\mathbf{y}|H_0}(\mathbf{y}|H_0)$ and the resulting decision rule is

$$\frac{p_{\mathbf{y}|H_1}(\mathbf{y}|H_1)}{p_{\mathbf{y}|H_0}(\mathbf{y}|H_0)} \underset{H_0}{\overset{H_1}{\gtrless}} -\lambda$$

The ubiquitous likelihood ratio test again appears; it *is* indeed the fundamental quantity in hypothesis testing. Using the logarithm of the likelihood ratio or the sufficient statistic, this result can be expressed as either

$$\ln \Lambda(\mathbf{y}) \underset{H_0}{\overset{H_1}{\gtrless}} \ln(-\lambda) \quad \text{or} \quad \Upsilon(\mathbf{y}) \underset{H_0}{\overset{H_1}{\gtrless}} \gamma$$

We have not as yet found a value for the threshold. The false-alarm probability can be expressed in terms of the Neyman-Pearson threshold in two (useful) ways.

$$\boxed{\begin{aligned} P_F &= \int_{-\lambda}^{\infty} p_{\Lambda|H_0}(\Lambda|H_0)\, d\Lambda \\ &= \int_{\gamma}^{\infty} p_{\Upsilon|H_0}(\Upsilon|H_0)\, d\Upsilon \end{aligned}}$$

One of these implicit equations must be solved for the threshold by setting P_F equal to α'. The selection of which to use is usually based on pragmatic considerations: the easiest to compute. From the previous discussion of the relationship between the detection and false-alarm probabilities, we find that to maximize P_D we must allow α' to be as large as possible while remaining less than α. Thus, we want to find the *smallest* value of $-\lambda$ (note the minus sign) consistent with the constraint. Computation of the threshold is problem dependent, but a solution always exists.

Example

An important application of the likelihood ratio test occurs when $\mathbf{y}$ is a Gaussian random vector under each hypothesis. Suppose the hypotheses correspond to Gaussian random vectors having different mean values but sharing the same identity covariance.

$$\begin{aligned} H_0&: \mathbf{y} \sim \mathcal{N}(\mathbf{0},\, \sigma^2\mathbf{I}) \\ H_1&: \mathbf{y} \sim \mathcal{N}(\mathbf{m},\, \sigma^2\mathbf{I}) \end{aligned}$$

Thus, $\mathbf{y}$ is of dimension L and has statistically independent, equal variance components. The vector of means $\mathbf{m} = \text{col}[m_0, \ldots, m_{L-1}]$ distinguishes the two hypotheses. The likelihood functions associated this problem are

$$\begin{aligned} p_{\mathbf{y}|H_0}(\mathbf{y}|H_0) &= \prod_{l=0}^{L-1} \frac{1}{\sqrt{2\pi\sigma^2}} \exp\left\{-\frac{1}{2}\left(\frac{y_l}{\sigma}\right)^2\right\} \\ p_{\mathbf{y}|H_1}(\mathbf{y}|H_1) &= \prod_{l=0}^{L-1} \frac{1}{\sqrt{2\pi\sigma^2}} \exp\left\{-\frac{1}{2}\left(\frac{y_l - m_l}{\sigma}\right)^2\right\} \end{aligned}$$

The likelihood ratio $\Lambda(\mathbf{y})$ becomes

$$\Lambda(\mathbf{y}) = \frac{\displaystyle\prod_{l=0}^{L-1} \exp\left\{-\frac{1}{2}\left(\frac{y_l - m_l}{\sigma}\right)^2\right\}}{\displaystyle\prod_{l=0}^{L-1} \exp\left\{-\frac{1}{2}\left(\frac{y_l}{\sigma}\right)^2\right\}}$$

This expression for the likelihood ratio is complicated. In the Gaussian case (and many others), we use the logarithm the reduce the complexity of the likelihood ratio and form a sufficient statistic.

$$\begin{aligned} \ln \Lambda(\mathbf{y}) &= \sum_{l=0}^{L-1} \left\{-\frac{1}{2}\frac{(y_l - m_l)^2}{\sigma^2} + \frac{1}{2}\frac{y_l^2}{\sigma^2}\right\} \\ &= \frac{1}{\sigma^2}\sum_{l=0}^{L-1} m_l y_l - \frac{1}{2\sigma^2}\sum_{l=0}^{L-1} m_l^2 \end{aligned}$$

The likelihood ratio test then has the much simpler, but equivalent form

$$\sum_{l=0}^{L-1} m_l y_l \underset{H_0}{\overset{H_1}{\gtrless}} \sigma^2 \ln \eta + \frac{1}{2}\sum_{l=0}^{L-1} m_l^2$$

To focus on the hypothesis testing aspects of this problem, let's assume means be equal to a positive constant: $m_l = m\ (> 0)$.†

$$\sum_{l=0}^{L-1} y_l \underset{H_0}{\overset{H_1}{\gtrless}} \frac{\sigma^2}{m}\ln \eta + \frac{Lm}{2}$$

Note that all that need be known about the observations $\{y_l\}$ is their sum. This quantity is the sufficient statistic for the Gaussian problem: $\Upsilon(\mathbf{y}) = \sum y_l$ and $\gamma = \sigma^2 \ln \eta / m + Lm/2$.

When trying to compute the probability of error or the threshold in the Neyman-Pearson criterion, we must find the conditional probability density of one of the decision statistics: the likelihood ratio, the log likelihood, or the sufficient statistic. The log likelihood and the sufficient statistic are quite similar in this problem, but clearly we should use the latter. One practical property of the sufficient statistic is that it usually simplifies computations. For this Gaussian example, the sufficient statistic is a Gaussian random variable under each hypothesis.

$$H_0\colon \Upsilon(\mathbf{y}) \sim \mathcal{N}(0, L\sigma^2)$$
$$H_1\colon \Upsilon(\mathbf{y}) \sim \mathcal{N}(Lm, L\sigma^2)$$

To find the probability of error from the expressions found on page 204, we must evaluate the area under a Gaussian probability density function. These integrals are succinctly expressed in terms of $Q(x)$, which denotes the probability that a unit-variance, zero-mean Gaussian random variable exceeds x (see appendix A {476}). As $1 - Q(x) = Q(-x)$, the probability of error can be written as

$$P_e = P_1 Q\left(\frac{Lm - \gamma}{\sqrt{L}\sigma}\right) + P_0 Q\left(\frac{\gamma}{\sqrt{L}\sigma}\right)$$

An interesting special case occurs when $P_0 = 1/2 = P_1$. In this case, $\gamma = Lm/2$ and the probability of error becomes

$$P_e = Q\left(\frac{\sqrt{L}m}{2\sigma}\right)$$

As $Q(\cdot)$ is a monotonically decreasing function, the probability of error decreases with increasing values of the ratio $\sqrt{L}m/2\sigma$. As depicted in appendix Fig. A.1 {477}, note that $Q(\cdot)$ decreases in a nonlinear fashion. Thus, increasing m by a factor of two may decrease the probability of error by a larger *or* a smaller factor; the amount of change depends on the initial value of the ratio.

To find the threshold for the Neyman-Pearson test from the expressions given on page 207, we need the area under a Gaussian density.

$$P_F = Q\left(\frac{\gamma}{\sqrt{L\sigma^2}}\right) = \alpha' \tag{5.5}$$

As $Q(\cdot)$ is a monotonic and continuous function, we can now set α' equal to the criterion value α with the result

$$\gamma = \sqrt{L}\sigma Q^{-1}(\alpha)$$

†Why did the authors assume that the mean was positive? What would happen if it were negative?

where $Q^{-1}(\cdot)$ denotes the inverse function of $Q(\cdot)$. The solution of this equation cannot be performed analytically as no closed form expression exists for $Q(\cdot)$ (much less its inverse function); the criterion value must be found from tables or numerical routines. Because Gaussian problems arise frequently, Table A.1 {477} provides numeric values for this quantity at the decade points. The detection probability is given by

$$P_D = Q\left(Q^{-1}(\alpha) - \frac{\sqrt{L}m}{\sigma}\right)$$

We could well translate this Gaussian hypothesis testing example into an array processing problem: Consider using L sensors to determine if the received field contains either zero-mean (H_0) or nonzero-mean (H_1) random disturbances, the latter presumably related to a propagating signal of some sort. By assuming identical components in the vector of means, the signal must be identical at each sensor, which for a linear array corresponds to a signal source located at broadside. Our assumption that the observation vector's covariance matrix equals identity means that the sensors make statistically independent observations, with the standard deviation σ serving as an indication of measurement variability owing to additive Gaussian noise. By assuming the same variance under each hypothesis, the noise characteristics are assumed unchanged whether a signal is present or not. If equally likely a priori probabilities are assumed, we are stating that the presence of a signal is just as likely as one not being there. The importance of this kind of assumption in framing the hypothesis testing problem cannot be overemphasized. The so-called equally likely assumption does *not* represent a noncommittal statement about the presence of the signal. By fixing the a priori probabilities, we are not claiming ignorance; we are *clearly* stating that over many sets of observations, we expect a signal to be present half of the time. In contrast, the Neyman-Pearson criterion *is* noncommittal because it allows development of an optimal hypothesis testing procedure without specifying the a priori probabilities. The Neyman-Pearson criterion thus emerges as the natural hypothesis testing approach for array processing problems, and we shall concentrate on it in sequel. Its drawback is the need to evaluate the false-alarm probability, which can lead to difficulties in calculating the test's threshold.

The decision rule we derived in the example suggests that the proper use of the sensor outputs is computing their average and comparing that quantity with a threshold. The expression for the detection probability suggests that to achieve a given level of performance (achieve as large a detection probability P_D as possible), we must increase the product of the number of sensors and the ratio m^2/σ^2, which we may term a signal-to-noise ratio in this problem. The dependence of P_D on this product occurs frequently in statistics: If the signal-to-noise ratio is a factor of two smaller than some nominal value, twice as many sensors are needed to compensate. Although the array can be used to overcome a low signal-to-noise ratio at each sensor, a prohibitively large number may be required if performance requirements are set too stringently.

5.1.3 Model Consistency Testing

In many situations, we seek to check consistency of the observations with some preconceived model. Alternative hypotheses are usually difficult to describe parametrically because inconsistency may be beyond our modeling capabilities. We need a hypothesis test that accepts consistency of observations with a model or rejects the model without pronouncing a more favored alternative. Assuming we know (or presume to know) the probability distribution of the observations under H_0, the hypotheses are

$$\begin{aligned} H_0&: \mathbf{y} \sim p_{\mathbf{y}|H_0}(\mathbf{y}|H_0) \\ H_1&: \mathbf{y} \not\sim p_{\mathbf{y}|H_0}(\mathbf{y}|H_0) \end{aligned}$$

Null hypothesis testing seeks to determine if the observations are consistent with this description. The best procedure for consistency testing amounts to determining whether the observations lie in a highly probable region as defined by the null probability distribution. No one region, however, defines a probability that is less than unity. We must restrict the size of the region so that it best represents those observations maximally consistent with the model while satisfying a performance criterion. Letting P_F be a false-alarm probability established by us, we define the decision region $\Re_0$ to satisfy

$$\Pr[\mathbf{y} \in \Re_0 | H_0] = \int_{\Re_0} p_{\mathbf{y}|H_0}(\mathbf{y}|H_0)\, d\mathbf{y} = 1 - P_F \quad \text{and} \quad \min_{\Re_0} \int_{\Re_0} d\mathbf{y}$$

Usually, this region is located about the mean, but may not be symmetrically centered if the probability density is skewed. Our null hypothesis test for model consistency becomes

$$\boxed{\begin{aligned} \mathbf{y} \in \Re_0 &\Longrightarrow \text{Say observations are consistent} \\ \mathbf{y} \notin \Re_0 &\Longrightarrow \text{Say observations are not consistent} \end{aligned}}$$

Example

Consider the problem of determining whether the sequence y_l, $l = 1, \ldots, L$, is white and Gaussian with zero mean and unit variance. Stated this way, the alternative model is not provided: Is this model correct or not? We could estimate the probability density function of the observations and test the estimate for consistency. Here we take the null-hypothesis testing approach of converting this problem into a one-dimensional one by considering the statistic $y = \sum_l y_l^2$, which has a χ_L^2 distribution. Because this probability distribution is unimodal, the decision region can be safely assumed to be an interval $[y', y'']$.† In this case, we can find an analytic solution to the problem of determining the decision region. Letting $Y = y'' - y'$ denote the width of the interval, we seek the solution of the constrained optimization problem

$$\min_{y'} Y \quad \text{subject to} \quad P_y(y' + Y) - P_y(y') = 1 - P_F$$

As described in appendix C.2 {502}, we convert the constrained problem into an unconstrained one using Lagrange multipliers.

$$\min_{y'} \left\{ Y + \lambda \left[P_y(y' + Y) - P_y(y') - (1 - P_F) \right] \right\}$$

†This one-dimensional result for the consistency test extends to the multidimensional case in the obvious way.

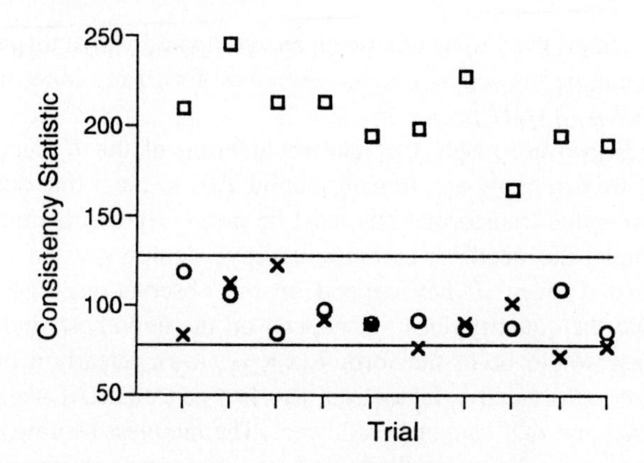

Figure 5.2 Ten trials of testing a 100-element sequence for consistency with a white, Gaussian model, $y_l \sim \mathcal{N}(0, 1)$, for three situations. In the first (shown by the circles), the observations do conform to the model. In the second (boxes), the observations are zero-mean Gaussian but with variance two. Finally, the third example (crosses) has white observations with a density closely resembling the Gaussian: a hyperbolic secant density having zero mean and unit variance. The sum of squared observations for each example are shown with the optimal χ^2_{100} interval displayed. Note how dramatically the test statistic departs from the decision interval when parameters disagree.

Evaluation of the derivative of this quantity with respect to y' yields the result $p_y(y' + Y) - p_y(y')$: To minimize the interval's width, the probability density function's values at the interval's endpoints must be equal. Finding these endpoints to satisfy the constraints amounts to searching the probability distribution at such points for increasing values of Y until the required probability is contained within. For $L = 100$ and $P_F = 0.05$, the optimal decision region for the χ^2_L distribution is [78.82, 128.5]. Fig. 5.2 demonstrates ten testing trials for observations that fit the model and for observations that don't.

5.1.4 Beyond Two Hypotheses

Frequently, more than two viable models for data generation can be defined for a given situation. The *classification* problem is to determine which of several models best "fits" a set of measurements. For example, determining the type of airplane from its radar returns forms a classification problem. The hypothesis testing framework has the right structure if we can allow more than two hypotheses. We happily note that in deriving the likelihood ratio test we did not need to assume that only two possible descriptions exist. Go back and examine the expression for the maximum probability correct decision rule {204}. If K hypotheses seem appropriate for a specific hypothesis testing problem, the decision rule maximizing the probability of making a correct choice is

$$\boxed{\text{Choose the largest of } P_i\, p_{\mathbf{y}|H_i}(\mathbf{y}|H_i), \quad i = 1, \ldots, K}$$

To determine the largest of K quantities, exactly $K - 1$ numeric comparisons need be made. For the binary hypothesis test ($K = 2$), this decision rule reduces to the computation of the likelihood ratio and its comparison to a threshold. In general, $K - 1$

likelihood ratios need to be computed and compared with a threshold. Thus the likelihood ratio test can be viewed as a specific method for determining the largest of the decision statistics $P_i p_{\mathbf{y}|H_i}(\mathbf{y}|H_i)$.

Because we need only the relative ordering of the K decision statistics to make a decision, we can apply any transformation $T(\cdot)$ to them that does not affect ordering. In general, possible transformations must be positively monotonic to satisfy this condition. For example, the needless common additive components in the decision statistics can be eliminated, even if they depend on the observations. Mathematically, "common" means that the quantity does not depend on the hypothesis index i. The transformation in this case would be of the form $T(x_i) = x_i - a$, clearly a monotonic transformation. A *positive* multiplicative factor can also be "canceled"; if negative, the ordering would be reversed and that cannot be allowed. The simplest resulting expression becomes the sufficient statistic $\Upsilon_i(\mathbf{y})$ for the hypothesis. Expressed in terms of the sufficient statistic, the maximum probability correct or the Bayesian decision rule becomes

$$\boxed{\text{Choose the largest of } C_i + \Upsilon_i(\mathbf{y}), \quad i = 1, \ldots, K}$$

where C_i summarizes all additive terms that do not depend on the observation vector $\mathbf{y}$. The quantity $\Upsilon_i(\mathbf{y})$ is termed the *sufficient statistic associated with hypothesis* i. In many cases, the functional form of the sufficient statistic varies little from one hypothesis to another and expresses the necessary operations that summarize the observations. The constants C_i are usually lumped together to yield the threshold against which we compare the sufficient statistic. For example, in the binary hypothesis situation the decision rule becomes

$$\Upsilon_1(\mathbf{y}) + C_1 \underset{H_0}{\overset{H_1}{\gtrless}} \Upsilon_0(\mathbf{y}) + C_0 \quad \text{or} \quad \Upsilon_1(\mathbf{y}) - \Upsilon_0(\mathbf{y}) \underset{H_0}{\overset{H_1}{\gtrless}} C_0 - C_1$$

Thus, the sufficient statistic for the decision rule is $\Upsilon_1(\mathbf{y}) - \Upsilon_0(\mathbf{y})$ and the threshold γ is $C_0 - C_1$.

Example

In the Gaussian problem just discussed, the logarithm of the likelihood function is

$$\ln p_{\mathbf{y}|H_i}(\mathbf{y}|H_i) = -\frac{L}{2}\ln 2\pi\sigma^2 - \frac{1}{2\sigma^2}\sum_{l=0}^{L-1}(y_l - m^{(i)})^2$$

where $m^{(i)}$ is the mean under hypothesis i. After appropriate simplification that retains the ordering, we have

$$\Upsilon_i(\mathbf{y}) = \frac{m^{(i)}}{\sigma^2}\sum_{l=0}^{L-1} y_l \qquad C_i = -\frac{1}{2}\frac{Lm^{(i)2}}{\sigma^2} + c_i$$

The term c_i is a constant defined by the error criterion; for the maximum probability correct criterion, this constant is $\ln P_i$.

When employing the Neyman-Pearson test, we need to specify the various error probabilities Pr[say $H_i|H_j$ true]. These specifications amount to determining the constants c_i

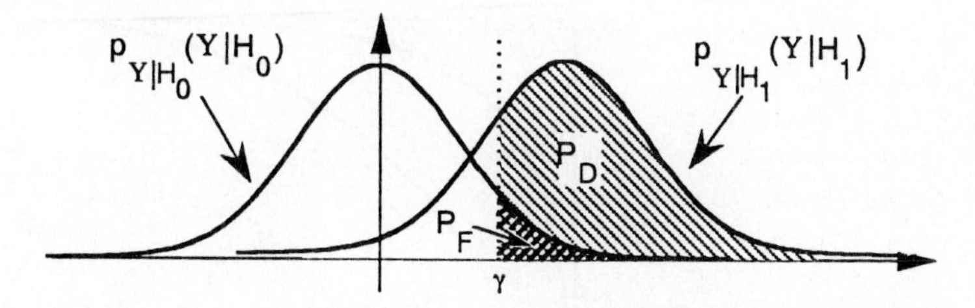

Figure 5.3 The densities of the sufficient statistic $\Upsilon(\mathbf{y})$ conditioned on two hypotheses are shown for the Gaussian example. The threshold γ that is used to distinguish between the two hypotheses is indicated. The false-alarm probability is the area under the density corresponding to H_0 to the right of the threshold; the detection probability is the area under the density corresponding to H_1.

when the sufficient statistic is used. Because $K-1$ comparisons are used to home in on the optimal decision, only $K-1$ error probabilities need be specified. Typically, the quantities Pr[say $H_i|H_0$ true], $i=1,\dots,K-1$, are used, particularly when the hypothesis H_0 represents the situation when no signal is present (see Prob. 5.4).

5.1.5 Performance Evaluation

We alluded earlier {205} to the relationship between the false-alarm probability P_F and the detection probability P_D as one varies the decision region. Because the Neyman-Pearson criterion depends on specifying the false-alarm probability to yield an acceptable detection probability, we need to examine carefully how the detection probability is affected by a specification of the false-alarm probability. The usual way these quantities are discussed is through a parametric plot of P_D versus P_F: the *receiver operating characteristic* (*ROC*).

As we discovered in the Gaussian example {207}, the sufficient statistic provides the simplest way of computing these probabilities; thus, they are usually considered to depend on the threshold parameter γ. In these terms, we have

$$P_D = \int_\gamma^\infty p_{\Upsilon|H_1}(\Upsilon|H_1)\,d\Upsilon \quad \text{and} \quad P_F = \int_\gamma^\infty p_{\Upsilon|H_0}(\Upsilon|H_0)\,d\Upsilon$$

These densities and their relationship to the threshold γ are shown in Fig. 5.3. We see that the detection probability is greater than or equal to the false-alarm probability. Because these probabilities must decrease monotonically as the threshold is increased, the *ROC* curve must be concave down and must *always* exceed the equality line (Fig. 5.4). The degree to which the *ROC* departs from the equality line $P_D = P_F$ measures the relative "distinctiveness" between the two hypothesized models for generating the observations. In the limit, the two hypotheses can be distinguished perfectly if the *ROC* is discontinuous and consists of the point $(1,0)$. The two are totally confused if the *ROC* lies on the equality line (this would mean, of course, that the two models are identical); distinguishing the two in this case would be "somewhat difficult."

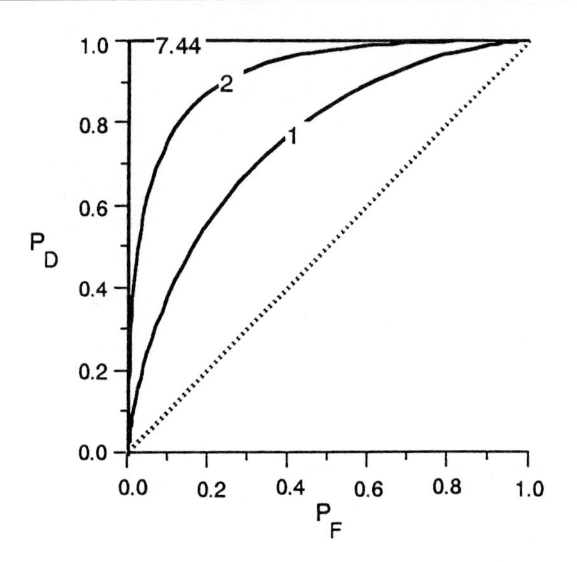

Figure 5.4 A plot of the receiver operating characteristic for the densities shown in the previous figure. Three *ROC* curves are shown corresponding to different values for the parameter $\sqrt{L}m/\sigma$.

Example

Consider the Gaussian example we have been discussing where the two hypotheses differ only in the means of the conditional distributions. In this case, the two hypothesis testing probabilities are given by

$$P_F = Q\left(\frac{\gamma}{\sqrt{L}\sigma}\right) \quad \text{and} \quad P_D = Q\left(\frac{\gamma - Lm}{\sqrt{L}\sigma}\right)$$

By re-expressing γ as $(\sigma^2/m)\gamma' + Lm/2$, we discover that these probabilities depend only on the ratio $\sqrt{L}m/\sigma$.

$$P_F = Q\left(\frac{\gamma'}{\sqrt{L}m/\sigma} + \frac{\sqrt{L}m}{2\sigma}\right), \qquad P_D = Q\left(\frac{\gamma'}{\sqrt{L}m/\sigma} - \frac{\sqrt{L}m}{2\sigma}\right)$$

As this signal-to-noise ratio increases, the *ROC* curve approaches its "ideal" form: The northwest corner of a square as illustrated in Fig. 5.4 by the value of 7.44 for $\sqrt{L}m/\sigma$, which corresponds to a signal-to-noise ratio of $7.44^2 \approx 17$ dB. If a small false-alarm probability (say, 10^{-4}) is specified, a large detection probability (0.9999) can result. Such values of signal-to-noise ratios can thus be considered "large" and the corresponding hypothesis testing problem relatively easy. If, however, the signal-to-noise ratio equals 4 (6 dB), the figure illustrates the worsened performance: A 10^{-4} specification on the false-alarm probability would result in a detection probability of essentially zero. Thus, in a fairly small signal-to-noise ratio range, the likelihood ratio test's performance capabilities can vary dramatically. No other decision rule, however, can yield better performance.

Specification of the false-alarm probability for a new problem requires experience. Choosing a "reasonable" value for the false-alarm probability in the Neyman-Pearson criterion depends strongly on the problem difficulty. Too small a number results in small detection probabilities; too large and the detection probability is close to unity, suggesting that fewer false alarms could have been tolerated. Problem difficulty is assessed by the degree to which the conditional densities $p_{\mathbf{y}|H_0}(\mathbf{y}|H_0)$ and $p_{\mathbf{y}|H_1}(\mathbf{y}|H_1)$ overlap, a problem dependent measurement. If we are testing whether a distribution has one of two possible mean values as in our Gaussian example, a quantity like a signal-to-noise ratio probably emerges as determining performance. The performance in this case can vary drastically depending on whether the signal-to-noise ratio is large or small. In other kinds of problems, the best possible performance provided by the likelihood ratio test can be poor. For example, consider the problem of determining which of two zero-mean probability densities describes a given set of data consisting of statistically independent observations (Prob. 5.2). Presumably, the variances of these two densities are equal as we are trying to determine which density is most appropriate. In this case, the performance probabilities can be quite low, especially when the general shapes of the densities are similar. Thus a single quantity, like the signal-to-noise ratio, does *not* emerge to characterize problem difficulty in all hypothesis testing problems. In sequel, we analyze each hypothesis testing and detection problem in a standard way. After the sufficient statistic has been found, we seek a value for the threshold that attains a specified false-alarm probability. The detection probability is then determined as a function of "problem difficulty," the measure of which is problem dependent. We can control the choice of false-alarm probability; we cannot control problem difficulty. Confusingly, the detection probability varies with *both* the specified false-alarm probability and the problem difficulty.

We are implicitly assuming that we have a rational method for choosing the false-alarm probability criterion value. In signal processing applications, we usually make a sequence of decisions and pass them to systems making more global determinations. For example, in digital communications problems the hypothesis testing formalism could be used to "receive" each bit. Each bit is received in sequence and then passed to the decoder which invokes error-correction algorithms. The important notions here are that the decision-making process occurs at a given *rate* and that the decisions are presented to other signal processing systems. The rate at which errors occur in system input(s) greatly influences system design. Thus, the selection of a false-alarm probability is usually governed by the *error rate* that can be tolerated by succeeding systems. If the decision rate is one per day, then a moderately large (say, 0.1) false-alarm probability might be appropriate. If the decision rate is a million per second as in a 1-megabit communication channel, the false-alarm probability should be much lower: 10^{-12} would suffice for the one-tenth per day error rate.

5.2 Hypothesis Testing in Presence of Unknowns

We assumed in the previous sections that we have a few well-specified models (hypotheses) for a set of observations. These models were probabilistic; to apply the techniques of statistical hypothesis testing, the models take the form of conditional probability densities. In many interesting circumstances, the exact nature of these densities may not be known. For example, we may know a priori that the mean is either zero or some constant (as in the Gaussian example), but the variance of the observations may not be known or the value of the nonzero mean may be in doubt. In an array processing context, these respective situations could occur when the background noise level is unknown (a likely possibility in applications) or when the signal amplitude is not known because of far-field range uncertainties (the further the source of propagating energy, the smaller its received energy at each sensor). In an extreme case, we can question the exact nature of the probability densities (everything is not necessarily Gaussian!). The hypothesis-testing problem can still be posed for these situations; we classify the "unknown" aspects of a hypothesis-testing problem as either *parametric* (the variance is not known, for example) or *nonparametric* (the formula for the density is in doubt). The former situation has a relatively long history compared with the latter; many techniques can be used to approach parametric problems while the latter is a subject of current research [59]. We concentrate on parametric problems here.

We describe the dependence of the conditional density on a set of parameters by incorporating the parameter vector $\boldsymbol{\xi}$ as part of the condition. We write the likelihood function as $p_{\mathbf{y}|H_i,\boldsymbol{\xi}}(\mathbf{y}|H_i,\boldsymbol{\xi})$ for the parametric problem. In statistics, this situation is said to be a *composite hypothesis* [41: 528]. Such situations can be further categorized according to whether the parameters are *random* or *nonrandom*. For a parameter to be random, we have an expression for its a priori density, which could depend on the particular hypothesis. As stated many times, a specification of a density usually expresses some knowledge about the range of values a parameter may assume *and* the relative probability of those values. Saying that a parameter has a uniform distribution implies that the values it assumes *are* equally likely, *not* that we have no idea what the values might be and express this ignorance by a uniform distribution. If we are ignorant of the underlying probability distribution that describes the values of a parameter, we characterize them simply as being *unknown* (not random). Once we have considered the random parameter case, nonrandom but unknown parameters are discussed.

5.2.1 Random Parameters

When we know the density of $\boldsymbol{\xi}$, the likelihood function can be expressed as

$$p_{\mathbf{y}|H_i}(\mathbf{y}|H_i) = \int p_{\mathbf{y}|H_i,\boldsymbol{\xi}}(\mathbf{y}|H_i,\boldsymbol{\xi})\,p_{\boldsymbol{\xi}}(\boldsymbol{\xi})\,d\boldsymbol{\xi}$$

and the likelihood ratio in the random parameter case becomes

$$\boxed{\Lambda(\mathbf{y}) = \frac{\int p_{\mathbf{y}|H_1,\boldsymbol{\xi}}(\mathbf{y}|H_1,\boldsymbol{\xi})\,p_{\boldsymbol{\xi}}(\boldsymbol{\xi})\,d\boldsymbol{\xi}}{\int p_{\mathbf{y}|H_0,\boldsymbol{\xi}}(\mathbf{y}|H_0,\boldsymbol{\xi})\,p_{\boldsymbol{\xi}}(\boldsymbol{\xi})\,d\boldsymbol{\xi}}}$$

Unfortunately, there are many examples where either the integrals involved are intractable or the sufficient statistic is virtually the same as the likelihood ratio, which can be difficult to compute.

Example

A simple, but interesting, example that results in a computable answer occurs when the mean of Gaussian random variables is either zero (hypothesis 0) or is $\pm m$ with equal probability (hypothesis 1). The second hypothesis means that a nonzero mean is present, but its sign is not known. We are therefore stating that if hypothesis one is in fact valid, the mean has fixed sign for each observation; what is random is its a priori value. As before, L statistically independent observations are made.

$$\begin{aligned} H_0&: \quad \mathbf{y} \sim \mathcal{N}(\mathbf{0}, \sigma^2\mathbf{I}) \\ H_1&: \quad \mathbf{y} \sim \mathcal{N}(\mathbf{m}, \sigma^2\mathbf{I}), \quad \mathbf{m} = \begin{cases} \text{col}[m, \ldots, m] & \text{Prob} = 1/2 \\ \text{col}[-m, \ldots, -m] & \text{Prob} = 1/2 \end{cases} \end{aligned}$$

The numerator of the likelihood ratio is the sum of two Gaussian densities weighted by $1/2$ (the a priori probability values), one having a positive mean, the other negative. The likelihood ratio, after simple cancellation of common terms, becomes

$$\Lambda(\mathbf{y}) = \frac{1}{2}\exp\left\{\frac{2m\left(\sum_{l=0}^{L-1} y_l\right) - Lm^2}{2\sigma^2}\right\} + \frac{1}{2}\exp\left\{\frac{-2m\left(\sum_{l=0}^{L-1} y_l\right) - Lm^2}{2\sigma^2}\right\}$$

and the decision rule takes the form

$$\cosh\left(\frac{m}{\sigma^2}\sum_{l=0}^{L-1} y_l\right) \underset{H_0}{\overset{H_1}{\gtrless}} \eta \exp\left\{\frac{Lm^2}{2\sigma^2}\right\}$$

where $\cosh(x)$ is the *hyperbolic cosine* given simply as $(e^x + e^{-x})/2$. As this quantity is an even function, the sign of its argument has no effect on the result. The decision rule can be written more simply as

$$\left|\sum_{l=0}^{L-1} y_l\right| \underset{H_0}{\overset{H_1}{\gtrless}} \frac{\sigma^2}{|m|}\cosh^{-1}\left[\eta \exp\left\{\frac{Lm^2}{2\sigma^2}\right\}\right]$$

The sufficient statistic equals the *magnitude* of the sum of the observations in this case. While the right side of this expression, which equals γ, is complicated, it need only be computed once. Calculation of the performance probabilities can be complicated; in this case, the false-alarm probability is easy to find and the others more difficult.

5.2.2 Nonrandom Parameters

In those cases where a probability density for the parameters cannot be assigned, the hypothesis testing problem can be solved in several ways; the methods used depend on the form of the likelihood ratio and the way in which the parameter(s) enter the problem. In the Gaussian problem we have discussed so often, the threshold used in the likelihood ratio test η may be unity. In this case, examination of the resulting computations reveals that implementing the test *does not require knowledge of the variance of the observations* (see Prob. 5.11). Thus, if the common variance of the underlying Gaussian distributions is not known, this lack of knowledge has *no effect* on the optimum decision rule. This happy situation, knowledge of the value of a parameter is not required by the optimum decision rule, occurs rarely but should be checked before using more complicated procedures.

A second fortuitous situation occurs when the sufficient statistic as well as its probability density under one of the hypotheses do *not* depend on the unknown parameter(s). Although the sufficient statistic's threshold γ expressed in terms of the likelihood ratio's threshold η depends on the unknown parameters, γ may be computed as a single value using the Neyman-Pearson criterion *if the computation of the false-alarm probability does not involve the unknown parameters.*

Example

Continuing the example of the previous section, let's consider the situation in which the value of the mean of each observation under hypothesis H_1 is not known. The sufficient statistic is the sum of the observations (that quantity doesn't depend on m) and the distribution of the observation vector under hypothesis H_0 does not depend on m (allowing computation of the false-alarm probability). A subtlety emerges, however; in the derivation of the sufficient statistic, we had to divide by the value of the mean. The critical step occurs once the logarithm of the likelihood ratio is manipulated to obtain

$$m \sum_{l=0}^{L-1} y_l \underset{H_0}{\overset{H_1}{\gtrless}} \sigma^2 \ln \eta + \frac{Lm^2}{2}$$

Recall that only *positively* monotonic transformations can be applied; if a negatively monotonic operation is applied to this inequality (such as multiplying both sides by -1), the *inequality reverses.* If the sign of m is known, it can be taken into account explicitly, and a sufficient statistic results. If, however, the sign is not known, the previous expression cannot be manipulated further, and the left side constitutes the sufficient statistic for this problem. The sufficient statistic then depends on the unknown parameter, and we cannot develop a decision rule in this case. If the sign is known, we can proceed. Assuming the sign of m is positive, the sufficient statistic is the sum of the observations and the threshold γ is found by

$$\gamma = \sqrt{L}\sigma Q^{-1}(P_F)$$

Note that if the variance σ^2 instead of the mean were unknown, we could not compute the threshold. The difficulty lies not with the sufficient statistic (it doesn't depend on the variance) but with the false-alarm probability as the expression indicates. Another approach is required to deal with the unknown-variance problem.

When the sufficient statistic *and* the false-alarm probability can be computed without needing the parameter in question, we have devised what is known as a *uniformly most powerful hypothesis test* (*UMP* test) [41: 529–31];[152: 89ff]. If a *UMP* test does not exist, which can only be demonstrated by explicitly finding the sufficient statistic and evaluating its probability distribution, then the composite hypothesis testing problem cannot be solved without inserting some value for the parameter.

This seemingly impossible situation—we need the value of parameter that is assumed unknown—can be approached by noting that some data are available for "guessing" the value of the parameter. If a reasonable guess could be obtained, it could then be used in our hypothesis testing procedures developed in this chapter. *The data available for estimating unknown parameters are precisely the data used in the decision rule.* Procedures intended to yield "good" guesses of the value of a parameter are said to be *parameter estimates.* Estimation procedures are the topic of the next chapter; there we explore a variety of estimation techniques and develop measures of estimate quality. For the moment, these issues are secondary; even if we knew the size of the estimation error, the more pertinent issue is how the imprecise parameter value affects the performance probabilities. We can compute these probabilities *without* explicitly determining the estimate's error characteristics.

One parameter estimation procedure that fits nicely into the composite hypothesis testing problem is the *maximum likelihood estimate*.† Letting $\mathbf{y}$ denote the vector of observables and $\boldsymbol{\xi}$ a vector of parameters, the maximum likelihood estimate of $\boldsymbol{\xi}$, $\widehat{\boldsymbol{\xi}}_{\text{ML}}$, is that value of $\boldsymbol{\xi}$ that maximizes the conditional density $p_{\mathbf{y}|\boldsymbol{\xi}}(\mathbf{y}|\boldsymbol{\xi})$ of the observations given the parameter values. To use $\widehat{\boldsymbol{\xi}}_{\text{ML}}$ in our decision rule, we estimate the parameter vector *separately* for each hypothesis, use the estimated value in the conditional density of the observations, and compute the likelihood ratio. This procedure is termed the *generalized likelihood ratio test* for the unknown parameter problem in hypothesis testing [98: 16];[152: 92ff].

$$\boxed{\Lambda(\mathbf{y}) = \frac{\max\limits_{\boldsymbol{\xi}} p_{\mathbf{y}|H_1,\boldsymbol{\xi}}(\mathbf{y}|H_1,\boldsymbol{\xi})}{\max\limits_{\boldsymbol{\xi}} p_{\mathbf{y}|H_0,\boldsymbol{\xi}}(\mathbf{y}|H_0,\boldsymbol{\xi})}} \tag{5.6}$$

Note that we do *not* find that value of the parameter that (necessarily) maximizes the likelihood ratio. Rather, we estimate the parameter value most consistent with the observed data in the context of each assumed model (hypothesis) of data generation. In this way, the estimate conforms with each potential model rather than being determined by some amalgam of supposedly mutually exclusive models.

Example

Returning to our Gaussian example, assume that the variance σ^2 is known but that the mean under H_1 is unknown.

$$\begin{aligned} H_0&: \quad \mathbf{y} \sim \mathcal{N}(\mathbf{0}, \sigma^2\mathbf{I}) \\ H_1&: \quad \mathbf{y} \sim \mathcal{N}(\mathbf{m}, \sigma^2\mathbf{I}) \quad \mathbf{m} = \text{col}[m, \dots, m],\ m = ? \end{aligned}$$

The unknown quantity occurs only in the exponent of the conditional density under H_1; to maximize this density, we need only to maximize the exponent. Thus, we consider the derivative of the exponent with respect to m.

$$\begin{aligned} \frac{\partial}{\partial m}\left[-\frac{1}{2\sigma^2}\sum_{l=0}^{L-1}(y_l - m)^2\right]\Bigg|_{m=\widehat{m}_{\text{ML}}} &= 0 \\ \Longrightarrow \quad \sum_{l=0}^{L-1}(y_l - \widehat{m}_{\text{ML}}) &= 0 \end{aligned}$$

The solution of this equation is the average value of the observations.

$$\widehat{m}_{\text{ML}} = \frac{1}{L}\sum_{l=0}^{L-1} y_l$$

To derive the decision rule, we substitute this estimate in the conditional density for H_1. The critical term, the exponent of this density, is manipulated to obtain

$$-\frac{1}{2\sigma^2}\sum_{l=0}^{L-1}\left(y_l - \frac{1}{L}\sum_{k=0}^{L-1} y_k\right)^2 = -\frac{1}{2\sigma^2}\left[\sum_{l=0}^{L-1} y_l^2 - \frac{1}{L}\left(\sum_{l=0}^{L-1} y_l\right)^2\right]$$

†The maximum likelihood estimation procedure and its characteristics are fully described in §6.2.4 {275}.

Noting that the first term in this exponent is identical to the exponent of the denominator in the likelihood ratio, the generalized likelihood ratio becomes

$$\Lambda(\mathbf{y}) = \exp\left\{+\frac{1}{2L\sigma^2}\left(\sum_{l=0}^{L-1} y_l\right)^2\right\}$$

The sufficient statistic thus becomes the square (or equivalently the magnitude) of the summed observations. Compare this result with that obtained in the example {218}. There, a *UMP* test existed *if* we knew the sign of m and the sufficient statistic was the sum of the observations. Here, where we employed the generalized likelihood ratio test, we made no such assumptions about m; this generality accounts for the difference in sufficient statistic. Which test do you think would lead to a greater detection probability for a given false-alarm probability?

Once the generalized likelihood ratio is determined, we need to determine the threshold. If the a priori probabilities P_0 and P_1 are known, the evaluation of the threshold proceeds in the usual way. If they are not known, all of the conditional densities must not depend on the unknown parameters lest the performance probabilities also depend on them. In most cases, the original hypothesis testing problem is posed in such a way that one of the hypotheses does not depend on the unknown parameter; a criterion on the performance probability related to that hypothesis can then be established via the Neyman-Pearson procedure. If not the case, the threshold cannot be computed and the threshold must be set experimentally: We force one of the hypotheses to be true and modify the threshold on the sufficient statistic until the desired level of performance is reached. Despite this nonmathematical approach, the overall performance of the hypothesis testing procedure is optimum because of the results surrounding the Neyman-Pearson criterion.

5.3 Detection of Signals in Gaussian Noise

Detection theory applies hypothesis testing to signals [72, 122, 152]. Usually, we measure a signal in the presence of additive noise over some finite number of samples. Each observed datum is of the form $s(l)+n(l)$, where $s(l)$ denotes the lth signal value and $n(l)$ the lth noise value. Because of the linearity of conventional beamforming algorithms, if the field at each sensor consists of signal and additive noise, so does the array's output. In this and in succeeding sections of this chapter, we focus the general methods of hypothesis testing on the kind of observations prevalent in array processing. We first apply hypothesis testing to the array's output signal z. We thereby assume that the propagation direction and the shading have been chosen and that we are only interested in determining a signal's presence or absence (the common meaning of detection) or whether one of several signals is propagating from that direction (the classification problem). Once we have addressed these problems and detector structures, we will question the *entire* structure of array signal processing, and derive algorithms *inherently* designed to detect or classify signals. In this way, we pass from ad hoc conventional array processing algorithms to those tailored to the signal-noise environment and to the task at hand.

For the moment, we assume we know the joint distribution of the noise values. In most cases, the various models for the form of the observations—the hypotheses—do not differ because of noise characteristics. Rather, the signal component determines model variations and the noise is statistically independent of the signal; such is the specificity of detection problems in contrast to the generality of hypothesis testing. For example, we may want to determine whether a signal characteristic of a particular ship is present in a sonar array's output (the signal is known) or whether no ship is present (zero-valued signal).

To apply optimal hypothesis testing procedures previously derived, we first obtain a finite number L of observations from the array's output, $z(l)$, $l = 0, \ldots, L-1$, and form them into a vector: $\mathbf{z} = \mathrm{col}[z(0), \ldots, z(L-1)]$. The binary detection problem is to distinguish between two possible signals present in the noisy output waveform.

$$\begin{aligned} H_0 &: \mathbf{z} = \mathbf{s}_0 + \mathbf{n} \\ H_1 &: \mathbf{z} = \mathbf{s}_1 + \mathbf{n} \end{aligned}$$

To apply the hypothesis testing results, we need the probability density of $\mathbf{z}$ under each hypothesis. As the only probabilistic component of the observations is the noise, the required density for the detection problem is given by

$$\boxed{p_{\mathbf{z}|H_i}(\mathbf{z}|H_i) = p_{\mathbf{n}}(\mathbf{z} - \mathbf{s}_i)}$$

and the corresponding likelihood ratio by

$$\boxed{\Lambda(\mathbf{z}) = \frac{p_{\mathbf{n}}(\mathbf{z} - \mathbf{s}_1)}{p_{\mathbf{n}}(\mathbf{z} - \mathbf{s}_0)}}$$

Much of detection theory revolves about interpreting this likelihood ratio and deriving the detection threshold (either η or γ).

5.3.1 White Gaussian Noise

By far the easiest detection problem to solve occurs when the noise vector consists of statistically independent, identically distributed, Gaussian random variables. In this book, a "white" sequence consists of statistically independent random variables. The white sequence's mean is usually taken to be zero,[†] and each component's variance is σ^2. The equal-variance assumption implies the noise characteristics are unchanging throughout the entire set of observations. The probability density of the zero-mean noise vector evaluated at $\mathbf{z} - \mathbf{s}_i$ equals that of Gaussian random vector having independent components ($\mathbf{K} = \sigma^2 \mathbf{I}$) with mean $\mathbf{s}_i$.

$$p_{\mathbf{n}}(\mathbf{z} - \mathbf{s}_i) = \left(\frac{1}{2\pi\sigma^2}\right)^{L/2} \exp\left\{-\frac{1}{2\sigma^2}(\mathbf{z} - \mathbf{s}_i)^t(\mathbf{z} - \mathbf{s}_i)\right\}$$

[†]The zero-mean assumption is realistic for the detection problem. If the mean were nonzero, simply subtracting it from the observed sequence results in a zero-mean noise component.

The resulting detection problem is similar to the Gaussian example examined so frequently in the hypothesis testing sections, with the distinction here being a nonzero mean under both hypotheses. The logarithm of the likelihood ratio becomes

$$(\mathbf{z}-\mathbf{s}_0)^t(\mathbf{z}-\mathbf{s}_0)-(\mathbf{z}-\mathbf{s}_1)^t(\mathbf{z}-\mathbf{s}_1) \underset{H_0}{\overset{H_1}{\gtrless}} 2\sigma^2 \ln \eta$$

and the usual simplifications yield

$$\left(\mathbf{z}^t\mathbf{s}_1-\frac{\mathbf{s}_1^t\mathbf{s}_1}{2}\right)-\left(\mathbf{z}^t\mathbf{s}_0-\frac{\mathbf{s}_0^t\mathbf{s}_0}{2}\right) \underset{H_0}{\overset{H_1}{\gtrless}} \sigma^2 \ln \eta$$

The quantities in parentheses express the signal processing operations for each hypothesis. If more than two signals were assumed possible, quantities such as these would need to be computed for each signal and the largest selected. This decision rule is optimum for the additive, white Gaussian noise problem.

Each term in the computations for the optimum detector has a signal processing interpretation. When expanded, the term $\mathbf{s}_i^t\mathbf{s}_i$ equals $\sum_{l=0}^{L-1} s_i^2(l)$, which is the *signal energy* E_i. The remaining term, $\mathbf{z}^t\mathbf{s}_i$, is the only one involving the observations and hence constitutes the sufficient statistic $\Upsilon_i(\mathbf{z})$ for the additive white Gaussian noise detection problem.

$$\boxed{\Upsilon_i(\mathbf{z})=\mathbf{z}^t\mathbf{s}_i}$$

An abstract, but physically relevant, interpretation of this important quantity comes from the theory of linear vector spaces. There, the quantity $\mathbf{z}^t\mathbf{s}_i$ would be termed the *dot product* between $\mathbf{z}$ and $\mathbf{s}_i$ or the *projection* of $\mathbf{z}$ onto $\mathbf{s}_i$. By employing the Schwarz Inequality, the largest value of this quantity occurs when these vectors are proportional to each other. Thus, a dot product computation measures how much alike two vectors are: They are completely alike when they are parallel (proportional) and completely dissimilar when orthogonal (the dot product is zero). More precisely, the dot product removes those components from the observations that are orthogonal to the signal. The dot product thereby generalizes the familiar notion of filtering a signal contaminated by broadband noise. In filtering, the signal-to-noise ratio of a bandlimited signal can be drastically improved by lowpass filtering; the output would consist only of the signal and "in-band" noise. The dot product serves a similar role, ideally removing those "out-of-band" components (the orthogonal ones) and retaining the "in-band" ones (those parallel to the signal).

Expanding the dot product, $\mathbf{z}^t\mathbf{s}_i=\sum_{l=0}^{L-1} z(l)s_i(l)$, another signal processing interpretation emerges. The dot product now describes a *FIR* filtering operation evaluated at a specific index. To demonstrate this interpretation, let $h(l)$ be the unit-sample response of a linear, shift-invariant filter where $h(l)=0$ for $l<0$ and $l\geq L$. Letting $z(l)$ be the filter's input sequence, the convolution sum expresses the output.

$$z(k)\star h(k)=\sum_{l=k-(L-1)}^{k} z(l)h(k-l)$$

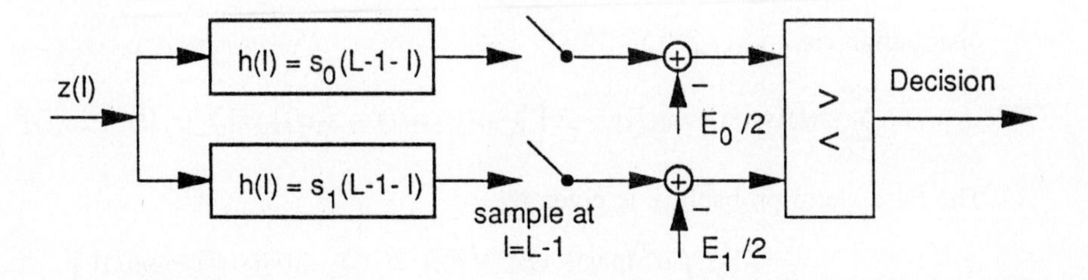

Figure 5.5 The detector for signals contained in additive, white Gaussian noise consists of a matched filter, whose output is sampled at the duration of the signal and half of the signal energy is subtracted from it. The optimum detector incorporates a matched filter for each signal and compares their outputs to determine the largest.

Letting $k = L - 1$, the index at which the unit-sample response's last value overlaps the input's value at the origin, we have

$$z(k) \star h(k)|_{k=L-1} = \sum_{l=0}^{L-1} z(l)h(L-1-l)$$

If we set the unit-sample response equal to the index reversed, then delayed signal $\big(h(l) = s_i(L-1-l)\big)$, we have

$$z(k) \star s_i(L-1-k)|_{k=L-1} = \sum_{l=0}^{L-1} z(l)s_i(l)$$

which equals the observation-dependent component of the optimal detector's sufficient statistic. Fig. 5.5 depicts these computations graphically. The sufficient statistic for the ith signal is thus expressed in signal processing notation as $z(k) \star s_i(L-1-k)|_{k=L-1} - E_i/2$. The filtering term is called a *matched filter* because the observations are passed through a filter whose unit-sample response "matches" that of the signal being sought. We sample the matched filter's output at the precise moment when all of the observations fall within the filter's memory and then adjust this value by half the signal energy. The adjusted values for the two assumed signals are subtracted and compared with a threshold.

To compute the performance probabilities, the expressions should be simplified in the ways discussed in the hypothesis-testing sections. As the energy terms are known *a priori*, they can be incorporated into the threshold with the result

$$\sum_{l=0}^{L-1} z(l)[s_1(l) - s_0(l)] \underset{H_0}{\overset{H_1}{\gtrless}} \sigma^2 \ln \eta + \frac{E_1 - E_0}{2}$$

The left term constitutes the sufficient statistic for the binary detection problem. Because the additive noise is presumed Gaussian, the sufficient statistic is a Gaussian random variable no matter which hypothesis is assumed. Under H_i, the specifics of this probability distribution are

$$\sum_{l=0}^{L-1} z(l)[s_1(l) - s_0(l)] \sim \mathcal{N}\left(\sum s_i(l)[s_1(l) - s_0(l)], \sigma^2 \sum [s_1(l) - s_0(l)]^2\right)$$

The false-alarm probability is given by

$$P_F = Q\left(\frac{\sigma^2 \ln \eta + (E_1 - E_0)/2 - \sum s_0(l)[s_1(l) - s_0(l)]}{\sigma \cdot \left\{\sum [s_1(l) - s_0(l)]^2\right\}^{1/2}}\right)$$

The signal-related terms in the numerator of this expression can be manipulated to yield succinct expressions for the false-alarm and detection probabilities provided by the optimal white Gaussian noise detector.

$$P_F = Q\left(\frac{\ln \eta + \frac{1}{2\sigma^2}\sum [s_1(l) - s_0(l)]^2}{\frac{1}{\sigma}\left\{\sum [s_1(l) - s_0(l)]^2\right\}^{1/2}}\right)$$

$$P_D = Q\left(\frac{\ln \eta - \frac{1}{2\sigma^2}\sum [s_1(l) - s_0(l)]^2}{\frac{1}{\sigma}\left\{\sum [s_1(l) - s_0(l)]^2\right\}^{1/2}}\right)$$

Note that the *only* signal-related quantity affecting this performance probability (and all of the others) is the *ratio of energy in the difference signal to the noise variance.* The larger this ratio, the better (smaller) the performance probabilities become. Note that the details of the signal waveforms do not greatly affect the energy of the difference signal. For example, consider the case in which the two signal energies are equal ($E_0 = E_1 = E$); the energy of the difference signal is given by $2E - 2\sum s_0(l)s_1(l)$. The largest value of this energy occurs when the signals are negatives of each other, with the difference-signal energy equaling $4E$. Thus, equal-energy but opposite-signed signals such as sine waves, square waves, Bessel functions, and so on *all* yield exactly the same performance levels. The essential signal properties that do yield good performance values are elucidated by an alternate interpretation. The term $\sum [s_1(l) - s_0(l)]^2$ equals $\|\mathbf{s}_1 - \mathbf{s}_0\|^2$, the L_2 norm of the difference signal. Geometrically, the difference-signal energy is the same quantity as the square of the Euclidean distance between the two signals. In these terms, a larger distance between the two signals means better performance.

Example

A common detection problem in array processing is to determine whether a signal is present (H_1) or not (H_0) in the array output. In this case, $s_0(l) = 0$. The optimal detector relies on filtering the array output with a matched filter having an impulse response based on the assumed signal. Letting the signal under H_1 be denoted simply by $s(l)$, the optimal detector consists of

$$z(l) \star s(L-1-l)|_{l=L-1} - E/2 \underset{H_0}{\overset{H_1}{\gtrless}} \sigma^2 \ln \eta$$

$$\text{or} \quad z(l) \star s(L-1-l)|_{l=L-1} \underset{H_0}{\overset{H_1}{\gtrless}} \gamma$$

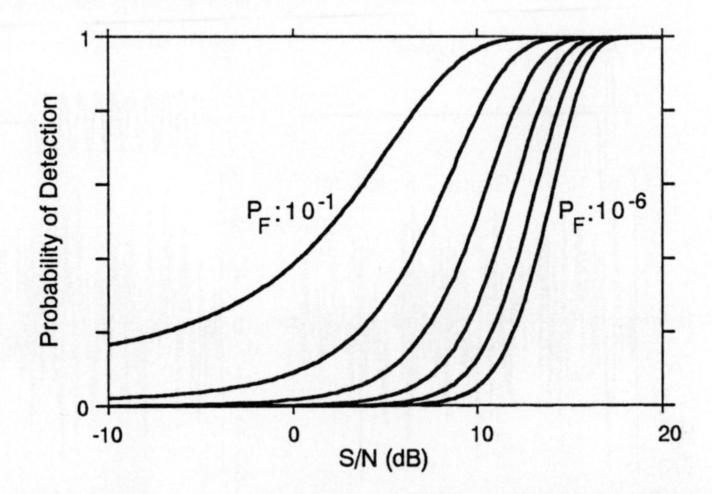

Figure 5.6 The probability of detection is plotted versus signal-to-noise ratio for various values of the false-alarm probability P_F. False-alarm probabilities range from 10^{-1} down to 10^{-6} by decades. The matched filter receiver was used because the noise is white and Gaussian. Note how the range of signal-to-noise ratios over which the detection probability changes shrinks as the false-alarm probability decreases. This effect is a consequence of the nonlinear nature of the function $Q(\cdot)$.

The false-alarm and detection probabilities are given by

$$P_F = Q\left(\frac{\gamma}{E^{1/2}/\sigma}\right) \quad P_D = Q\left(Q^{-1}(P_F) - \sqrt{\frac{E}{\sigma}}\right)$$

Fig. 5.6 displays the probability of detection as a function of the signal-to-noise ratio E/σ^2 for several values of false-alarm probability. Given an estimate of the expected signal-to-noise ratio, these curves can be used to assess the tradeoff between the false-alarm and detection probabilities.

The important parameter determining detector performance derived in this example is the *signal-to-noise ratio* E/σ^2: The larger it is, the smaller the false-alarm probability is (generally speaking). Signal-to-noise ratios can be measured in many different ways. For example, one measure might be the ratio of the root-mean-squared (RMS) signal amplitude to the RMS noise amplitude. Note that the important one for the detection problem is much different. The signal portion is the *sum* of the squared signal values over the *entire* set of observed values, the signal energy; the noise portion is the variance of *each* noise component, the noise power. Thus, energy can be increased in two ways that increase the signal-to-noise ratio: The signal can be made larger *or* the observations can be extended to encompass a larger number of values.

To illustrate this point, two signals having the same energy are shown in Fig. 5.7. When these signals are shown in the presence of additive noise, the signal is visible on the left because its amplitude is larger; the one on the right is much more difficult to discern. The instantaneous signal-to-noise ratio, the ratio of signal amplitude to average

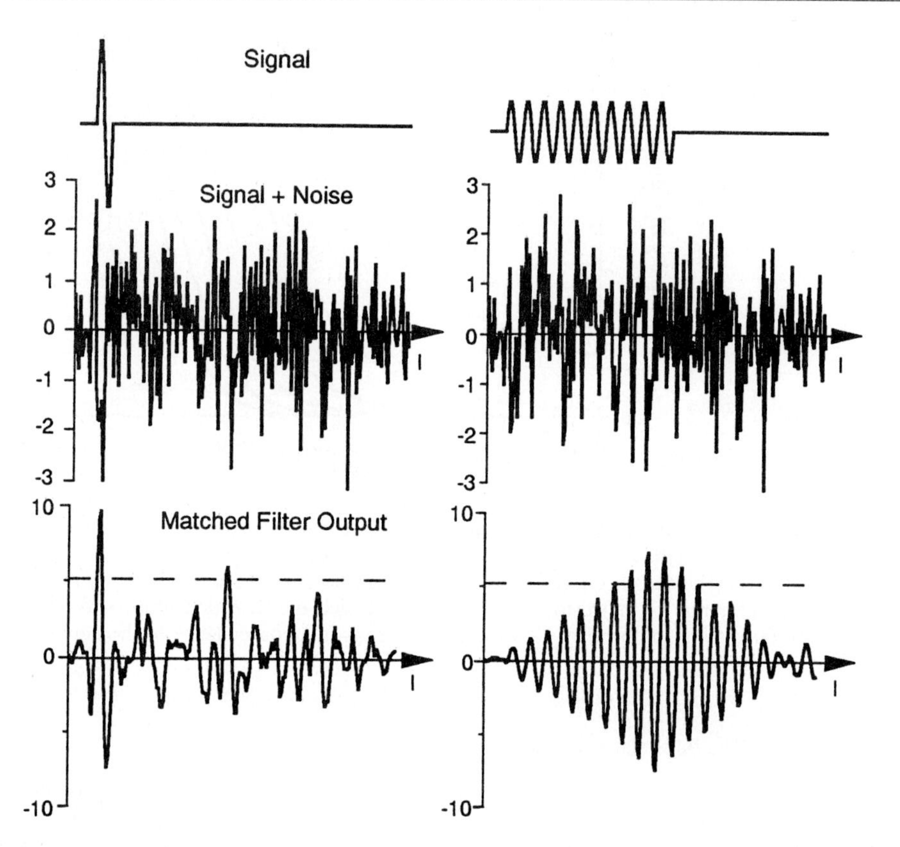

Figure 5.7 Two signals having the same energy are shown at the top of the figure. The one on the left equals one cycle of a sinusoid having ten samples/period ($\sin(\omega_0 l)$ with $\omega_0 = 2\pi \times 0.1$). On the right, ten cycles of similar signal is shown, with an amplitude a factor of $\sqrt{10}$ smaller. The middle portion of the figure shows these signals with the same noise signal added; the duration of this signal is 200 samples. The lower portion depicts the outputs of matched filters for each signal. The detection threshold was set by specifying a false-alarm probability of 10^{-2}.

noise amplitude, is the important visual cue, but the kind of signal-to-noise ratio that determines detection performance belies the eye. The matched filter outputs have similar maximal values, indicating that total signal energy rather than amplitude determines the performance of a matched filter detector.

5.3.2 Validity of White-Noise Assumptions

The optimal detection paradigm for the additive, white Gaussian noise problem has a relatively simple solution: Construct *FIR* filters whose unit-sample responses are related to the presumed signals and compare the filtered outputs with a threshold. We may well wonder which assumptions made in this problem are most questionable in "real-world" applications. Noise is additive in most cases. In many situations, the additive noise present in observed data is Gaussian. Because of the Central Limit Theorem, if numerous noise sources impinge on a measuring device, their superposition is Gaussian to a great extent. As we know from the discussion in §A.1.8 {480}, glibly appealing to the Central Limit Theorem is not without hazards; the non-Gaussian detection problem is discussed in some detail later. Interestingly, the weakest assumption is the "whiteness" of the noise. Note that the observation sequence is obtained as a result of *sampling* the sensor outputs. Assuming white-noise samples does *not* mean that the continuous-time noise was white. White noise in continuous time has infinite variance and cannot be sampled; discrete-time white noise has a finite variance with a constant power spectrum. The Sampling Theorem suggests that a signal is represented accurately by its samples only if we choose a sampling frequency commensurate with the signal's bandwidth. One should note that fidelity of representation does *not* mean that the sample values are independent. In most cases, satisfying the Sampling Theorem means that the samples are correlated. As shown in §A.2.3 {483}, the correlation function of sampled noise equals samples of the original correlation function. For the sampled noise to be white, $\mathcal{E}[n(l_1T)n(l_2T)] = 0$ for $l_1 \neq l_2$: The samples of the correlation function at locations other than the origin must all be zero. Although some correlation functions have this property, *many examples satisfy the sampling theorem but do not yield uncorrelated samples.* In many practical situations, *undersampling* the noise reduces intersample correlation. Thus, we obtain uncorrelated samples either by deliberately undersampling, which wastes signal energy, or by imposing antialiasing filters that have a bandwidth larger than the signal and sampling at the signal's Nyquist rate. Because the noise power spectrum usually extends to higher frequencies than the signal, this intentional undersampling can result in larger noise variance. In either case, by trying to make the problem at hand match the solution, we are actually reducing performance. We need a *direct* approach to attacking the correlated noise issue that arises in virtually *all* sampled-data detection problems rather than trying to work around it.

5.3.3 Colored Gaussian Noise

When the additive Gaussian noise in the sensors' outputs is colored, the noise values are correlated in some fashion, the linearity of beamforming algorithms means that the array processing output z also contains colored noise. The solution to the colored-noise, binary detection problem remains the likelihood ratio but differs in the form of the a priori densities. The noise is again assumed zero mean, but the noise vector has a nontrivial covariance matrix $\mathbf{K}$: $\mathbf{n} \sim \mathcal{N}(\mathbf{0}, \mathbf{K})$.

$$p_{\mathbf{n}}(\mathbf{n}) = \frac{1}{\sqrt{\det[2\pi \mathbf{K}]}} \exp\left\{-\frac{1}{2}\mathbf{n}^t\mathbf{K}^{-1}\mathbf{n}\right\}$$

In this case, the logarithm of the likelihood ratio is

$$(\mathbf{z}-\mathbf{s}_1)^t\mathbf{K}^{-1}(\mathbf{z}-\mathbf{s}_1) - (\mathbf{z}-\mathbf{s}_0)^t\mathbf{K}^{-1}(\mathbf{z}-\mathbf{s}_0) \underset{H_0}{\overset{H_1}{\gtrless}} 2\ln\eta$$

which, after the usual simplifications, is written

$$\left[\mathbf{z}^t\mathbf{K}^{-1}\mathbf{s}_1 - \frac{\mathbf{s}_1^t\mathbf{K}^{-1}\mathbf{s}_1}{2}\right] - \left[\mathbf{z}^t\mathbf{K}^{-1}\mathbf{s}_0 - \frac{\mathbf{s}_0^t\mathbf{K}^{-1}\mathbf{s}_0}{2}\right] \underset{H_0}{\overset{H_1}{\gtrless}} \ln\eta$$

The sufficient statistic for the colored Gaussian noise detection problem is

$$\boxed{\Upsilon_i(\mathbf{z}) = \mathbf{z}^t\mathbf{K}^{-1}\mathbf{s}_i} \tag{5.7}$$

The quantities computed for each signal have a similar, but more complicated interpretation than in the white-noise case. $\mathbf{z}^t\mathbf{K}^{-1}\mathbf{s}_i$ is a dot product but with respect to the so-called *kernel* $\mathbf{K}^{-1}$. The effect of the kernel is to weight certain components more heavily than others. In appendix B {495}, we find an expression for a positive-definite symmetric matrix (the covariance matrix is one such example) in terms of its eigenvectors and eigenvalues.

$$\mathbf{K}^{-1} = \sum_{k=1}^{L} \frac{1}{\lambda_k}\mathbf{v}_k\mathbf{v}_k^t$$

The sufficient statistic can thus be written as the complicated summation

$$\mathbf{z}^t\mathbf{K}^{-1}\mathbf{s}_i = \sum_{k=1}^{L} \frac{1}{\lambda_k}(\mathbf{z}^t\mathbf{v}_k)(\mathbf{v}_k^t\mathbf{s}_i)$$

where λ_k and $\mathbf{v}_k$ denote the kth eigenvalue and eigenvector of the covariance matrix $\mathbf{K}$. Each of the constituent dot products is largest when the signal and the observation vectors have strong components parallel to $\mathbf{v}_k$. The product of these dot products is weighted by the reciprocal of the associated eigenvalue, however. Thus, components in the observation vector parallel to the signal tend to be accentuated; those components parallel to the eigenvectors having the *smaller* eigenvalues receive greater accentuation than others. The usual notions of parallelism and orthogonality become "skewed" because of the presence of the kernel. A covariance matrix's eigenvalue has "units" of variance; these accentuated directions thus correspond to small noise variances. We can therefore view the weighted dot product as a computation that is simultaneously trying to select components in the observations similar to the signal, but concentrating on those where the noise variance is small.

The second term in the expressions constituting the optimal detector are of the form $\mathbf{s}_i^t\mathbf{K}^{-1}\mathbf{s}_i$. This quantity is a special case of the dot product just discussed. The two vectors involved in this dot product are identical; they are parallel by definition. The weighting of the signal components by the reciprocal eigenvalues remains. Recalling the units of the eigenvectors of $\mathbf{K}$, $\mathbf{s}_i^t\mathbf{K}^{-1}\mathbf{s}_i$ has the units of a signal-to-noise ratio, which is computed in a way that enhances the contribution of those signal components parallel to the "low-noise" directions.

To compute the performance probabilities, we express the detection rule in terms of the sufficient statistic.

$$\mathbf{z}^t\mathbf{K}^{-1}(\mathbf{s}_1 - \mathbf{s}_0) \underset{H_0}{\overset{H_1}{\gtrless}} \ln\eta + \frac{1}{2}(\mathbf{s}_1^t\mathbf{K}^{-1}\mathbf{s}_1 - \mathbf{s}_0^t\mathbf{K}^{-1}\mathbf{s}_0)$$

The distribution of the sufficient statistic on the left side of this equation is Gaussian because it consists as a linear transformation of the Gaussian random vector $\mathbf{z}$. Assuming the ith hypothesis to be true,

$$\mathbf{z}^t\mathbf{K}^{-1}(\mathbf{s}_1 - \mathbf{s}_0) \sim \mathcal{N}\left(\mathbf{s}_i^t\mathbf{K}^{-1}(\mathbf{s}_1 - \mathbf{s}_0), (\mathbf{s}_1 - \mathbf{s}_0)^t\mathbf{K}^{-1}(\mathbf{s}_1 - \mathbf{s}_0)\right)$$

The false-alarm probability for the optimal Gaussian colored-noise detector is given by

$$\boxed{P_F = Q\left(\frac{\ln\eta + \frac{1}{2}(\mathbf{s}_1 - \mathbf{s}_0)^t\mathbf{K}^{-1}(\mathbf{s}_1 - \mathbf{s}_0)}{[(\mathbf{s}_1 - \mathbf{s}_0)^t\mathbf{K}^{-1}(\mathbf{s}_1 - \mathbf{s}_0)]^{1/2}}\right)} \tag{5.8}$$

As in the white-noise case, the important signal-related quantity in this expression is the signal-to-noise ratio of the difference signal. The distance interpretation of this quantity remains, but the distance is now warped by the kernel's presence in the dot product.

The sufficient statistic computed for each signal can be given two signal processing interpretations in the colored-noise case. Both of these rest on considering the quantity $\mathbf{z}^t\mathbf{K}^{-1}\mathbf{s}_i$ as a simple dot product but with different ideas on grouping terms. The simplest is to group the kernel with the signal so that the sufficient statistic is the dot product between the observations and a *modified* version of the signal $\tilde{\mathbf{s}}_i = \mathbf{K}^{-1}\mathbf{s}_i$. This modified signal thus becomes the equivalent to the unit-sample response of the matched filter. In this form, the observed data are unaltered and passed through a matched filter whose unit-sample response depends on both the signal and the noise characteristics. The size of the noise covariance matrix, equal to the number of observations used by the detector, is usually large: Hundreds if not thousands of samples are possible. Thus, computation of the inverse of the noise covariance matrix becomes an issue. This problem needs to be solved only once if the noise characteristics are static; the inverse can be precomputed on a general-purpose computer using well-established numerical algorithms. The signal-to-noise ratio term of the sufficient statistic is the dot product of the observations with the modified signal $\tilde{\mathbf{s}}_i$. This view of the receiver structure is shown in Fig. 5.8.

A second and more theoretically powerful view of the computations involved in the colored-noise detector emerges when we *factor* the covariance matrix. The *Cholesky factorization* of a positive-definite, symmetric matrix (such as a covariance matrix or its inverse) has the form $\mathbf{K} = \mathbf{L}\mathbf{D}\mathbf{L}^t$ (see §B.4 {492}). With this factorization, the sufficient statistic can be written as

$$\mathbf{z}^t\mathbf{K}^{-1}\mathbf{s}_i = \left(\mathbf{D}^{-1/2}\mathbf{L}^{-1}\mathbf{z}\right)^t\left(\mathbf{D}^{-1/2}\mathbf{L}^{-1}\mathbf{s}_i\right)$$

The components of the dot product are multiplied by the same matrix ($\mathbf{D}^{-1/2}\mathbf{L}^{-1}$), which is lower triangular. *If* this matrix were also Toeplitz, the product of this kind between a Toeplitz matrix and a vector would be equivalent to the convolution of the components of the vector with the first column of the matrix. If the matrix is not Toeplitz (which, inconveniently, is the typical case), a convolution also results, but with a unit-sample response that varies with the index of the output, a time-varying, linear filtering operation. The variation of the unit-sample response corresponds to the different rows of the matrix $\mathbf{D}^{-1/2}\mathbf{L}^{-1}$ running *backward* from the main-diagonal entry. What is the

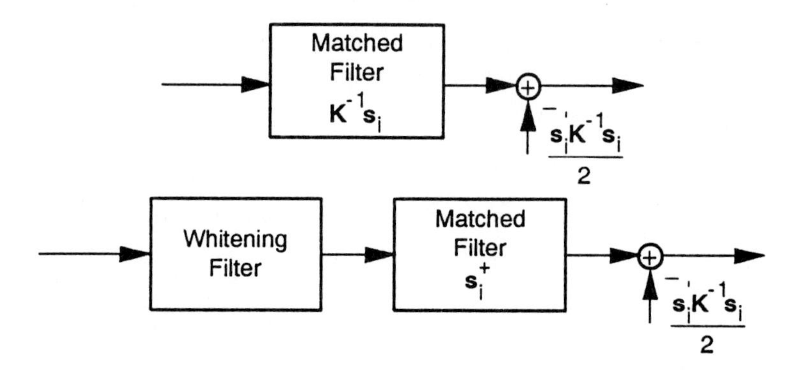

Figure 5.8 These diagrams depict the signal processing operations involved in the optimum detector when the additive noise is not white. The upper diagram shows a matched filter whose unit-sample response depends both on the signal and the noise characteristics. The lower diagram is often termed the whitening filter structure, where the noise components of the observed data are first whitened, then passed through a matched filter whose unit-sample response is related to the "whitened" signal.

physical interpretation of the action of this filter? The covariance of the random vector $\mathbf{x} = \mathbf{A}\mathbf{z}$ is given by $\mathbf{K}_x = \mathbf{A}\mathbf{K}_z\mathbf{A}^t$. Applying this result to the current situation, we set $\mathbf{A} = \mathbf{D}^{-1/2}\mathbf{L}^{-1}$ and $\mathbf{K}_z = \mathbf{K} = \mathbf{L}\mathbf{D}\mathbf{L}^t$ with the result that the covariance matrix $\mathbf{K}_x$ is the identity matrix! Thus, the matrix $\mathbf{D}^{-1/2}\mathbf{L}^{-1}$ corresponds to a (possibly time-varying) *whitening filter*: We have converted the colored-noise component of the observed data to white noise. As the filter is always linear, the Gaussian observation noise remains Gaussian at the output. Thus, the colored-noise problem is converted into a simpler one with the whitening filter: The whitened observations are first match filtered with the "whitened" signal $\mathbf{s}_i^+ = \mathbf{D}^{-1/2}\mathbf{L}^{-1}\mathbf{s}_i$ (whitened with respect to noise characteristics only) then half the energy of the whitened signal is subtracted (see Fig. 5.8).

Example

To demonstrate the interpretation of the Cholesky factorization of the covariance matrix as a time-varying whitening filter, consider the covariance matrix

$$\mathbf{K} = \frac{1}{1-a^2}\begin{bmatrix} 1 & a & a^2 & a^3 \\ a & 1 & a & a^2 \\ a^2 & a & 1 & a \\ a^3 & a^2 & a & 1 \end{bmatrix}$$

This covariance matrix indicates that the noise was produced by passing white Gaussian noise through a first-order filter having coefficient a: $n(l) = an(l-1) + w(l)$, where $w(l)$ is unit-variance white noise. Thus, we would expect that if a whitening filter emerged from the matrix manipulations (derived just below), it would be a first-order *FIR* filter having an unit-sample response proportional to

$$h(l) = \begin{cases} 1, & l = 0 \\ -a, & l = 1 \\ 0, & \text{otherwise} \end{cases}$$

Simple arithmetic calculations of the Cholesky decomposition suffice to show that the matrices $\mathbf{L}$ and $\mathbf{D}$ are given by

$$\mathbf{L} = \begin{bmatrix} 1 & 0 & 0 & 0 \\ a & 1 & 0 & 0 \\ a^2 & a & 1 & 0 \\ a^3 & a^2 & a & 1 \end{bmatrix} \qquad \mathbf{D} = \begin{bmatrix} \frac{1}{1-a^2} & 0 & 0 & 0 \\ 0 & 1 & 0 & 0 \\ 0 & 0 & 1 & 0 \\ 0 & 0 & 0 & 1 \end{bmatrix}$$

and that their inverses are

$$\mathbf{L}^{-1} = \begin{bmatrix} 1 & 0 & 0 & 0 \\ -a & 1 & 0 & 0 \\ 0 & -a & 1 & 0 \\ 0 & 0 & -a & 1 \end{bmatrix} \qquad \mathbf{D}^{-1} = \begin{bmatrix} 1-a^2 & 0 & 0 & 0 \\ 0 & 1 & 0 & 0 \\ 0 & 0 & 1 & 0 \\ 0 & 0 & 0 & 1 \end{bmatrix}$$

Because $\mathbf{D}$ is diagonal, the matrix $\mathbf{D}^{-1/2}$ equals the term-by-term square root of the inverse of $\mathbf{D}$. The product of interest here is therefore given by

$$\mathbf{D}^{-1/2}\mathbf{L}^{-1} = \begin{bmatrix} \sqrt{1-a^2} & 0 & 0 & 0 \\ -a & 1 & 0 & 0 \\ 0 & -a & 1 & 0 \\ 0 & 0 & -a & 1 \end{bmatrix}$$

Let $\tilde{\mathbf{z}}$ express the product $\mathbf{D}^{-1/2}\mathbf{L}^{-1}\mathbf{z}$. This vector's elements are given by

$$\tilde{z}_0 = \sqrt{1-a^2}z_0\,, \quad \tilde{z}_1 = z_1 - az_0\,, \quad \text{and so on}$$

Thus, the expected *FIR* whitening filter emerges after the first term. The expression could *not* be of this form as no observations were assumed to precede z_0. This edge effect is the source of the time-varying aspect of the whitening filter. If the system modeling the noise generation process has only poles, this whitening filter always stabilizes—the unit-sample response ceases to vary with time—once sufficient data are present within the memory of the *FIR* inverse filter. In contrast, the presence of zeros in the generation system would imply an *IIR* whitening filter. With finite data, the unit-sample response would then change on each output sample.

5.3.4 Spectral Detection

Based on the results presented in the previous sections, we found the colored-noise problem to be pervasive but requiring a computationally difficult detector. The simplest detector structure occurs when the additive noise is white; this notion leads to the idea of whitening the observations, thereby transforming the data into a simpler form (as far as detection theory is concerned). The required whitening filter is often time-varying and can have a long-duration unit-sample response. Other, more computationally expedient, approaches to whitening are worth considering. An only slightly more complicated detection problem occurs when we have a diagonal noise covariance matrix, as in the white-noise case, but unequal values on the diagonal. In terms of the observations, this situation means that they are contaminated by noise having statistically independent but unequal variance components: The noise would thus be nonstationary. Few problems fall directly into this category; however, the colored-noise problem can be recast into the white, unequal-variance problem by calculating the *DFT* of the observations and basing the detector on the resulting spectrum. The resulting *spectral detectors* greatly simplify detector structures for discrete-time problems *if* the qualifying assumptions described in sequel hold.

Let $\mathbf{W}$ be the so-called $L \times L$ "*DFT* matrix"

$$\mathbf{W} = \begin{bmatrix} 1 & 1 & 1 & \cdots & 1 \\ 1 & W & W^2 & \cdots & W^{L-1} \\ 1 & W^2 & W^4 & \cdots & W^{2(L-1)} \\ \vdots & \vdots & \vdots & \vdots & \vdots \\ 1 & W^{L-1} & W^{2(L-1)} & \cdots & W^{(L-1)(L-1)} \end{bmatrix}$$

where W is the elementary complex exponential $\exp\{-j\ 2\pi/L\}$. The *DFT* of the sequence $z(l)$, usually written as $Z(k) = \sum_{l=0}^{L-1} z(l)\exp\{-j\ (2\pi lk)/L\}$, can be written in matrix form as $\mathbf{Z} = \mathbf{Wz}$. To analyze the effect of evaluating the *DFT* of the observations, we describe the computations in matrix form for analytic simplicity. The first critical assumption has been made: Take special note that the length of the transform *equals* the duration of the observations. In many signal processing applications, the transform length can differ from the data length, being either longer or shorter. The statistical properties developed in the following discussion are critically sensitive to the equality of these lengths. The covariance matrix $\mathbf{K}_Z$ of $\mathbf{Z}$ is given by $\mathbf{W}\mathbf{K}_z\mathbf{W}'$. Symmetries of these matrices, the Vandermonde form of $\mathbf{W}$ and the Hermitian, Toeplitz form of $\mathbf{K}_z$, lead to many simplifications in evaluating this product. The entries on the main diagonal are given by†

$$K_{kk}^Z = \sum_{l=-(L-1)}^{(L-1)} (L - |l|) K_{1,|l|+1}^z e^{-j\frac{2\pi lk}{L}}$$

The *variance* of the kth term in the *DFT* of the noise thus equals the *DFT* of the *windowed* covariance function. This window has a triangular shape; colloquially termed the "rooftop" window, its technical name is the Bartlett window, and it occurs frequently in array processing and spectral estimation. We have found that the variance equals the smoothed noise power spectrum evaluated at a particular frequency. The off-diagonal terms of $\mathbf{K}_Z$ are not as easily written; the complicated result is

$$K_{k_1k_2}^Z = \sum_{l=0}^{L-1} K_{1,l+1}^z \frac{(-1)^{k_1-k_2+1} \sin \dfrac{\pi l(k_1-k_2)}{L}}{\sin \dfrac{\pi(k_1-k_2)}{L}} \left(e^{+j\frac{2\pi lk_1}{L}} + e^{-j\frac{2\pi lk_2}{L}}\right), \quad k_1 \neq k_2$$

The complex exponential terms indicate that each off-diagonal term consists of the sum of two Fourier Transforms: one at the frequency index k_2 and the other negative index $-k_1$. In addition, the transform is evaluated only over nonnegative lags. The transformed quantity again equals a windowed version of the noise covariance function, but with a

†The curious index $l+1$ on the matrix arises because rows and columns of matrices are traditionally indexed beginning with 1 instead of 0.

sinusoidal window whose frequency depends on the indices k_1 and k_2. This window can be negative valued. In contrast to the Bartlett window encountered in evaluating the on-diagonal terms, the maximum value achieved by the window is not large ($1/\sin[\pi(k_1 - k_2)/L]$ compared with L). Furthermore, this window is *always* zero at the origin, the location of the maximum value of any covariance function. The largest magnitudes of the off-diagonal terms tend to occur when the indices k_1 and k_2 are nearly equal. Let their difference be 1; if the covariance function of the noise tends toward 0 well within the number of observations L, then the Bartlett window has little effect on the covariance function while the sinusoidal window greatly reduces it. This condition on the covariance function can be interpreted physically: The noise in this case is wideband and any correlation between noise values does not extend over significant portion of the observation record. Conversely, if the width of the the covariance function is comparable with L, the off-diagonal terms are significant. This situation occurs when the noise bandwidth is smaller than or comparable with the reciprocal of the observation interval's duration. This condition on the duration of the observation interval relative to the width of the noise correlation function forms the second critical assumption of spectral detection. The off-diagonal terms are thus much smaller than corresponding terms on the main diagonal ($|K^Z_{k_1k_2}|^2 \ll K^Z_{k_1k_1}K^Z_{k_2k_2}$).

In the simplest case, the covariance matrix of the discrete Fourier Transform of the observations can be well approximated by a diagonal matrix.

$$\mathbf{K}_Z = \begin{bmatrix} \sigma_0^2 & 0 & \cdots & 0 \\ 0 & \sigma_1^2 & 0 & \vdots \\ \vdots & 0 & \ddots & 0 \\ 0 & \cdots & 0 & \sigma_{L-1}^2 \end{bmatrix}$$

The nonzero components σ_k^2 of this matrix constitute the noise power spectrum at the various frequencies. The signal component of the transformed observations $\mathbf{Z}$ is represented by $\mathbf{S}_i$, the *DFT* of the signal $\mathbf{s}_i$, whereas the noise component has this diagonal covariance matrix structure. *Thus, in the frequency domain, the colored-noise problem can be approximately converted to a white-noise problem where the components of the noise have unequal variances.* To recap, the critical assumptions of spectral detection are

- The transform length equals that of the observations. In particular, the observations cannot be "padded" to force the transform length to equal a "nice" number (like a power of 2): See Prob. 5.20.
- The noise's correlation structure should be much less than the duration of the observations. Equivalently, a narrow correlation function means the corresponding power spectrum varies slowly with frequency. If either condition fails to hold, calculating the Fourier Transform of the observations does not necessarily yield a simpler noise covariance matrix.

The optimum spectral detector computes, for each possible signal, the quantity

$\text{Re}[\mathbf{Z}'\mathbf{K}_Z^{-1}\mathbf{S}_i] - \mathbf{S}_i'\mathbf{K}_Z^{-1}\mathbf{S}_i/2$.† Because of the covariance matrix's simple form, this sufficient statistic for the spectral detection problem has the simple form

$$\boxed{\text{Re}[\mathbf{Z}'\mathbf{K}_Z^{-1}\mathbf{S}_i] - \frac{1}{2}\mathbf{S}_i'\mathbf{K}_Z^{-1}\mathbf{S}_i = \sum_{k=0}^{L-1}\left(\frac{\text{Re}[Z^*(k)S_i(k)]}{\sigma_k^2} - \frac{1}{2}\frac{|S_i(k)|^2}{\sigma_k^2}\right)} \tag{5.9}$$

Each term in the dot product between the *DFT* of the observations and the signal is weighted by the reciprocal of the noise power spectrum at that frequency. This computation is much simpler than the equivalent time domain version, and, because of algorithms such as the *FFT* the initial transformation (the multiplication by $\mathbf{W}$ or the *DFT*) can be evaluated expeditiously.

In frequency-domain beamforming, the Fourier Transform of the each sensor's output is computed before the beamformer's spatial filtering operation. Thus, for each propagation direction, the temporal spectrum is available, making the application of spectral detection techniques particularly natural. Sinusoidal signals are particularly well suited to the spectral detection approach. *If* the signal's frequency equals one of the analysis frequencies in the Fourier Transform ($\omega_0 = 2\pi k/L$ for some k), then the sequence $S_i(k)$ is nonzero only at this frequency index, only one term in the sufficient statistic's summation need be computed, and the noise power is no longer explicitly needed by the detector (it can be merged into the threshold).

$$\text{Re}[\mathbf{Z}'\mathbf{K}_Z^{-1}\mathbf{S}_i] - \frac{1}{2}\mathbf{S}_i'\mathbf{K}_Z^{-1}\mathbf{S}_i = \frac{\text{Re}[Z^*(k)S_i(k)]}{\sigma_k^2} - \frac{1}{2}\frac{|S_i(k)|^2}{\sigma_k^2}$$

If the signal's frequency does not correspond to one of the analysis frequencies, spectral energy is maximal at the nearest analysis frequency but extends to nearby frequencies also. This effect is termed "leakage" and has been well studied. Exact formulation of the signal's *DFT* is usually complicated in this case; approximations that use only the maximal-energy frequency component is suboptimal (yield a smaller detection probability). The performance reduction may be small, however, justifying the reduced amount of computation.

5.3.5 Particulars for Array Processing

We have phrased the detection problem for array processing applications in terms of determining whether a known signal may or may not be observed in additive noise. To refocus the general results we have derived on the array processing problem, recall that a conventional beamformer's output is given by

$$z(l) = \sum_m w_m s_m\big(l - \Delta_m\big) + \sum_m w_m n_m\big(l - \Delta_m\big)$$

where $\{w_m\}$ denote the shading weights and Δ_m the signal processing delay applied to the mth sensor's output. For the moment, assume that the assumed propagation direction and

†The real part in the statistic emerges because $\mathbf{Z}$ and $\mathbf{S}_i$ are complex quantities.

the actual direction of propagation agree;† in this case, the array processing has properly "stacked" the signals to yield

$$z(l) = s(l)\sum_m w_m + \sum_m w_m n_m\bigl(l - \Delta_m\bigr)$$

Our detection theoretical results for the generic problem of detecting a signal in additive noise demand that, in this case, the matched filter's unit-sample response be derived from the propagating signal $s(l)$. The covariance function of the noise term $\tilde{n}(l) = \sum_m w_m n_m(l - \Delta_m)$ depends in a complicated way on the spatiotemporal correlation function $R_n(\vec{\chi}, \tau)$ introduced in §2.6 {45}.

$$K_{\tilde{n}}(k) = \sum_{m_1,m_2} w_{m_1} w_{m_2} R_n\bigl(\vec{x}_{m_1} - \vec{x}_{m_2}, (k + \Delta_{m_1} - \Delta_{m_2})T\bigr)$$

The optimal time-domain detector depends on the covariance matrix derived from this quantity in the obvious way: $\mathbf{K}^{\tilde{n}}_{ij} = K_{\tilde{n}}(i - j)$. Note that this covariance matrix not only depends on the spatiotemporal correlation function of the noise (over which we have little control), but also on array geometry through the sensor positions $\vec{x}_m$, the shading weights $\{w_m\}$, and the delays $\{\Delta_m\}$ applied to sensor outputs. The geometry and shadings are (usually) fixed and do not depend on propagation direction. However, the delays are varied systematically to determine a propagation direction for signal processing purposes. In this way, the noise covariance matrix used by the detector can change with the assumed propagation direction.

Clearly, many simplifying assumptions would be needed to force this covariance matrix to be proportional to the identity matrix. Hence, the results for the colored-noise detector must be used to derive a detector and to assess its performance. As indicated by Eq. 5.8 {229}, the dependence of the false-alarm probability on signal and noise parameters are summarized by $\left(\sum w_m\right)^2 \mathbf{s}^t\mathbf{K}_{\tilde{n}}^{-1}\mathbf{s}$, which we have interpreted as a signal-to-noise ratio. Other than the rather obvious conclusions that bigger signals or smaller noise contributions increase the value of this quantity, the shading weights remain as the sole variables to enhance detector performance. We have considered the problem of choosing the shading weights in an "optimal" fashion in §4.5 {141}. There, we maximized the array gain G, defined to be the ratio of the signal-to-noise ratio in the array output to the signal-to-noise ratio at a single sensor. Using notation of the current discussion, the array gain formula found in Eq. 4.16 {139} fundamentally differs from that for signal-to-noise ratio of import for optimal detection performance, even when it is normalized by the single-sensor SNR $(\mathbf{s}^t\mathbf{s}/M)/R_n(\vec{0}, 0)$.

$$\left(\sum w_m\right)^2 R_n(\vec{0}, 0)\cdot\frac{1}{K^{\tilde{n}}_{1,1}} \quad \text{vs.} \quad \left(\sum w_m\right)^2 R_n(\vec{0}, 0)\cdot\frac{\mathbf{s}^t\mathbf{K}_{\tilde{n}}^{-1}\mathbf{s}}{\mathbf{s}^t\mathbf{s}/M}$$

$K^{\tilde{n}}_{1,1}$ is the first entry on the main diagonal of the noise covariance matrix.‡ The key difference is the assumed coupling between the signal and the noise: The array gain

†The more complicated situation of inaccuracies in the assumed propagation direction, and therefore in the applied delays, is discussed in §5.4.1 {242}.

‡All of the main diagonal terms are equal because the noise has been assumed stationary. We chose the first diagonal element arbitrarily; any of them could have represented the noise variance in the array output.

assumes none, whereas the optimal detector depends on the *combined* characteristics of the signal and the noise covariance function. Using matrix notation, particularly that of the Kronecker product (see §B.3 {489}), the dependence of these expressions on the shading can be made explicit. The covariance matrix $\mathbf{K}_{\tilde{n}}$ of the noise in the array output can be expressed as $(\mathbf{w} \otimes \mathbf{I})' \mathbf{K}_n (\mathbf{w} \otimes \mathbf{I})$, where $\mathbf{K}_n$ denotes the block Toeplitz $LM \times LM$ spatiotemporal covariance matrix of the noise defined in §2.6.4 {50}. Grouping terms that depend on the shading weights, our two optimality criteria become

$$R_n(\vec{0}, 0) \frac{\mathbf{w}' \mathbf{1}\mathbf{1}' \mathbf{w}}{[(\mathbf{w} \otimes \mathbf{I})' \mathbf{K}_n (\mathbf{w} \otimes \mathbf{I})]_{1,1}} \quad \text{vs.} \quad \frac{R_n(\vec{0}, 0)}{\mathbf{s}'\mathbf{s}/M} \mathbf{w}' \mathbf{1}\mathbf{1}' \mathbf{w} \cdot \mathbf{s}' \left[(\mathbf{w} \otimes \mathbf{I})' \mathbf{K}_n (\mathbf{w} \otimes \mathbf{I})\right]^{-1} \mathbf{s} \tag{5.10}$$

Here, the identity matrix has dimension L and the constant vector $\mathbf{1}$ has components all equaling 1. The array gain optimization can now be seen as a (convenient) approximation to the more pertinent second expression derived from optimizing detector performance. These expressions agree under the (boring) circumstance that the noise is spatially *and* temporally white (see Prob. 5.22). Optimizing the detection-based criterion is a difficult task analytically; numerical optimization can be attempted, but signal waveform uncertainties may make such computations a difficult academic exercise.

5.4 Detection in Presence of Uncertainties

5.4.1 Unknown Signal Parameters

Applying the techniques described in the previous section may be difficult to justify when the signal or noise models are uncertain. For example, we must "know" a signal down to the precise value of every sample. In other cases, we may know the signal's waveform, but not the waveform's *amplitude* as measured by a sensor. A ubiquitous example of this uncertainty is propagation loss: The range of a far-field signal can only be lower bounded, which leads to the known waveform, unknown amplitude detection problem. Another array processing-specific problem is the signal's *time origin*: The propagation direction might be properly chosen, but conventional beamformers do not consider the time origin, and we may have no control on the signal's arrival at the origin of the array's coordinate system. Without this information, we do not know when to start the matched filtering operation. In other circumstances, the noise may have a white power spectrum but its variance is unknown. Much worse situations (from the viewpoint of detection theory) can be imagined: The signal may not be known at all and one may want to detect the presence of *any* disturbance in the observations other than that of well-specified noise. These problems are very realistic, but the detection schemes as presented are inadequate to attack them. The detection results we have derived to date need to be extended to incorporate the presence of unknowns just as we did in hypothesis testing (§5.2 {215}).

Unknown signal amplitude. Assume that a signal's waveform is known exactly, but the amplitude is not. We need an algorithm to detect the presence or absence of this signal observed in additive noise at an array's output. The hypotheses can be formally stated as

$$\begin{array}{ll} H_0: & z(l) = n(l) \\ H_1: & z(l) = As(l) + n(l), \quad A = ? \end{array} \quad l = 0, \ldots, L-1$$

As usual, L observations are available and the noise is Gaussian. This problem is equivalent to an unknown parameter problem described in §5.2 {215}. We learned there that the first step is to ascertain the existence of a uniformly most powerful test. For each value of the unknown parameter A, the logarithm of the likelihood ratio is written

$$A\,\mathbf{z}^t\mathbf{K}^{-1}\mathbf{s} - A^2\,\mathbf{s}^t\mathbf{K}^{-1}\mathbf{s} \underset{H_0}{\overset{H_1}{\gtrless}} \ln \eta$$

Assuming that $A > 0$, a typical assumption in array processing problems, we write this comparison as

$$\mathbf{z}^t\mathbf{K}^{-1}\mathbf{s} \underset{H_0}{\overset{H_1}{\gtrless}} \frac{1}{A} \ln \eta + A\,\mathbf{s}^t\mathbf{K}^{-1}\mathbf{s} = \gamma$$

As the sufficient statistic does not depend on the unknown parameter and one of the hypotheses (H_0) does not depend on this parameter, a uniformly most powerful test exists: The threshold term, despite its explicit dependence on a variety of factors, can be determined by specifying a false-alarm probability. If the noise is not white, the whitening filter or a spectral transformation may be used to simplify the computation of the sufficient statistic.

Example

Assume that the waveform but not the amplitude of a signal is known. The Gaussian noise is white with a variance of σ^2. The decision rule expressed in terms of a sufficient statistic becomes

$$\mathbf{z}^t\mathbf{s} \underset{H_0}{\overset{H_1}{\gtrless}} \gamma$$

The false-alarm probability is given by

$$P_F = Q\left(\frac{\gamma}{\sqrt{E\sigma^2}}\right)$$

where E is the *assumed* signal energy that equals $\|\mathbf{s}\|^2$. The threshold γ is thus found to be

$$\gamma = \sqrt{E\sigma^2}Q^{-1}(P_F)$$

The probability of detection for the matched filter detector is given by

$$\begin{aligned} P_D &= Q\left(\frac{\gamma - AE}{\sqrt{E\sigma^2}}\right) \\ &= Q\left(Q^{-1}(P_F) - \sqrt{\frac{A^2E}{\sigma^2}}\right) \end{aligned}$$

where A is the signal's *actual* amplitude relative to the assumed signal having energy E. Thus, the observed signal, when it is present, has energy A^2E. The probability of detection is shown in Fig. 5.9 as a function of the observed signal-to-noise ratio. For any false-alarm probability, the signal must be sufficiently energetic for its presence to be reliably determined.

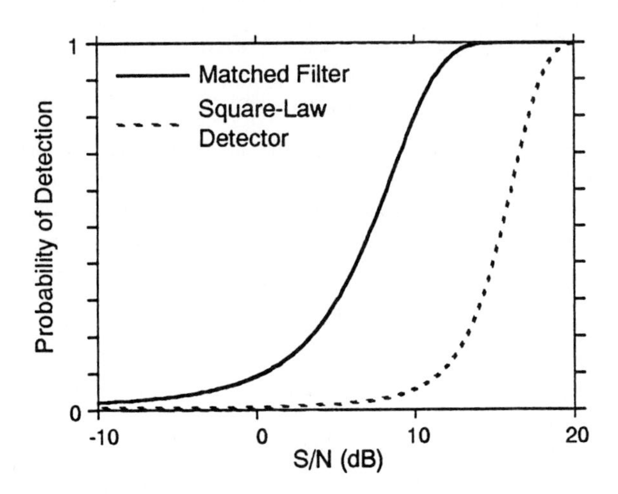

Figure 5.9 The false-alarm probability of the detector was fixed at 10^{-2}. The signal equaled $A\sin(\omega_0 l)$, $l = 0, \ldots, L-1$, where ω_0 was $2\pi \times 0.1$ and $L = 100$; the noise was white and Gaussian. The detection probabilities that result when a matched filter detector and a square-law detector are shown. The latter detector makes few assumptions about the signal, but yields smaller detection probabilities at each signal-to-noise ratio.

Sometimes the parameter must be known. All too many interesting problems exist where a uniformly most powerful decision rule cannot be found. Suppose in the problem just described that the amplitude is known ($A = 1$, for example), but the variance of the noise is not. Writing the covariance matrix as $\sigma^2\widetilde{\mathbf{K}}$, where we normalize the covariance matrix to have unit variance entries by requiring $\text{tr}[\widetilde{\mathbf{K}}] = L$, unknown values of σ^2 express the known correlation structure, unknown noise power problem. From the results just given, the decision rule can be written so that the sufficient statistic does not depend on the unknown variance.

$$\mathbf{z}^t\widetilde{\mathbf{K}}^{-1}\mathbf{s} \underset{H_0}{\overset{H_1}{\gtrless}} \sigma^2 \ln\eta + \mathbf{s}^t\widetilde{\mathbf{K}}^{-1}\mathbf{s} = \gamma$$

As *both* hypotheses depend on the unknown parameter, performance probabilities cannot be computed, and we cannot design a detection threshold.

Hypothesis testing ideas show the way out; estimate the unknown parameter(s) under each hypothesis separately and then use these estimates in the likelihood ratio (Eq. 5.6 {219}). Using the maximum likelihood estimates for the parameters results in the generalized likelihood ratio test for the detection problem [90, 91, 152]. Letting $\boldsymbol{\xi}$ denote the vector of unknown parameters, be they for the signal or the noise, the generalized likelihood ratio test for detection problems is expressed by

$$\Lambda(\mathbf{z}) = \frac{\max\limits_{\boldsymbol{\xi}} p_{\mathbf{n}|\boldsymbol{\xi}}(\mathbf{z} - \mathbf{s}_1(\boldsymbol{\xi}))}{\max\limits_{\boldsymbol{\xi}} p_{\mathbf{n}|\boldsymbol{\xi}}(\mathbf{z} - \mathbf{s}_0(\boldsymbol{\xi}))} \underset{H_0}{\overset{H_1}{\gtrless}} \eta$$

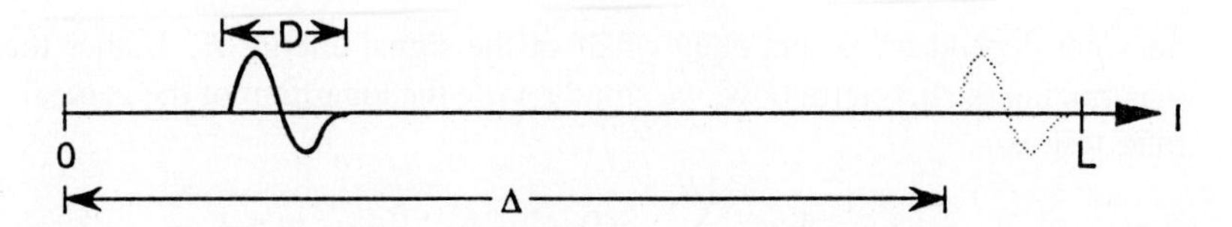

Figure 5.10 Despite uncertainties in the signal's delay Δ, the signal is assumed to lie entirely within the observation interval. Hence the signal's duration D, the duration L of the observation interval, and the maximum expected delay are assumed to be related by $\max \Delta + D - 1 < L$. The figure shows a signal falling properly within the allowed window and a gray one falling just outside.

Again, the use of *separate* estimates for each hypothesis (rather than for the likelihood ratio as a whole) must be stressed. Unknown signal-related parameter problems and unknown noise-parameter problems have different characteristics; the signal may not be present in one of the observation models. This simplification allows a threshold to be established objectively. In contrast, the noise is present in each hypothesized model; establishing a threshold value objectively forces new techniques to be developed. We first continue our adventure in unknown–signal-parameter problems, deferring the more challenging unknown–noise-parameter ones to §5.4.2 {246}.

Unknown signal delay. A uniformly most powerful decision rule may not exist when an unknown parameter appears in a nonlinear way in the signal model. Most pertinent to array processing is the unknown time origin case: The signal has been subjected to an unknown delay ($s(l - \Delta)$, $\Delta = ?$) and we must determine the signal's presence. The likelihood ratio cannot be manipulated so that the sufficient statistic can be computed without having a value for Δ. Thus, the search for a uniformly most powerful test ends in failure and other methods must be sought. As expected, we resort to the generalized likelihood ratio test.

More specifically, consider the binary test where a signal is either present (H_1) or not (H_0). The signal waveform is known, but its time origin is not. For all possible values of Δ, the delayed signal is assumed to lie *entirely* in the observations (see Fig. 5.10). This signal model is ubiquitous in active sonar and radar, where the reflected signal's exact time of arrival is not known and we want to determine whether a return is present or not *and* the value of the delay.† Additive white Gaussian noise is assumed present. The conditional density of the observations under H_1 is

$$p_{\mathbf{z}|H_1,\Delta}(\mathbf{z}|H_1, \Delta) = \frac{1}{\left(2\pi\sigma^2\right)^{L/2}} \exp\left\{-\frac{1}{2\sigma^2}\sum_{l=0}^{L-1}[z(l) - s(l-\Delta)]^2\right\}$$

The exponent contains the only portion of this conditional density that depends on the unknown quantity Δ. Maximizing the conditional density with respect to Δ is equivalent to maximizing $\sum_{l=0}^{L-1}[z(l)s(l-\Delta) - 1/2\, s^2(l-\Delta)]$. As the signal is assumed to be contained entirely in the observations for all possible values of Δ, the second term

†For a much more realistic (and more difficult) version of the active radar/sonar problem, see Prob. 5.24.

does not depend on Δ and equals half of the signal energy E. Rather than analytically maximizing the first term now, we simply write the logarithm of the generalized likelihood ratio test as

$$\max_{\Delta} \sum_{l=\Delta}^{\Delta+D-1} z(l)s(l-\Delta) \underset{H_0}{\overset{H_1}{\gtrless}} \sigma^2 \ln \eta + \frac{E}{2}$$

where the nonzero portion of the summation is expressed explicitly. Using the matched filter interpretation of the sufficient statistic, this decision rule is expressed by

$$\max_{\Delta} [z(l) \star s(D-1-l)] \Big|_{l=D-1+\Delta} \underset{H_0}{\overset{H_1}{\gtrless}} \gamma$$

This formulation suggests that the matched filter having a unit-sample response equal to the zero-origin signal be evaluated for each possible value of Δ and that we use the maximum value of the resulting output in the decision rule. In the known-delay case, the matched-filter output is sampled at the "end" of the signal; here, the filter, which has a duration D less than the observation interval L, is allowed to continue processing over the allowed values of signal delay with the maximum output value chosen. The result of this procedure is illustrated in Fig. 5.7 {226}. There two signals, each having the same energy, are passed through the appropriate matched filter. Note that the index at which the maximum output occurs is the maximum likelihood estimate of Δ. Thus, *the detection and the estimation problems are solved simultaneously*. Furthermore, *the amplitude of the signal need not be known* as it enters in the expression for the sufficient statistic in a linear fashion and a *UMP* test exists in that case. We can easily find the threshold γ by establishing a criterion on the false-alarm probability; the resulting simple computation of γ can be traced to the lack of a signal-related quantity or an unknown parameter appearing in H_0.

We have argued the doubtfulness of assuming that the noise is white in discrete-time detection problems. The approach for solving the colored-noise problem is to use spectral detection. Handling the unknown delay problem in this way is relatively straightforward. Because a sequence can be represented equivalently by its values or by its *DFT*, maximization can be calculated in either the time or the frequency domain without affecting the final answer. Thus, the spectral detector's decision rule for the unknown delay problem is (from Eq. 5.9 {234})

$$\boxed{\max_{\Delta} \sum_{k=0}^{L-1} \left(\frac{\operatorname{Re}[Z^*(k)S(k)e^{-j2\pi k\Delta/L}]}{\sigma_k^2} - \frac{1}{2}\frac{|S(k)|^2}{\sigma_k^2} \right) \underset{H_0}{\overset{H_1}{\gtrless}} \gamma}$$

where, as usual in unknown delay problems, the observation interval captures the entire signal waveform no matter what the delay might be. The energy term is a constant and can be incorporated into the threshold. The maximization amounts to finding the best linear phase fit to the observations' spectrum once the signal's phase has been removed. A more interesting interpretation arises by noting that the sufficient statistic is itself a

Fourier Transform; the maximization amounts to finding the location of the maximum of a sequence given by

$$\mathrm{Re}\left[\sum_{k=0}^{L-1} \frac{Z^*(k)S(k)}{\sigma_k^2} e^{-j2\pi k\Delta/L}\right]$$

The spectral detector thus becomes a succession of two Fourier Transforms with the final result determined by the maximum of a sequence.

Unfortunately, the solution to the unknown–signal-delay problem in either the time or frequency domains is confounded when two or more signals are present. Assume two signals are known to be present in the array output, each of which has an unknown delay: $z(l) = s_1(l - \Delta_1) + s_2(l - \Delta_2) + n(l)$. Using arguments similar those used in the one-signal case, the generalized likelihood ratio test becomes

$$\max_{\Delta_1, \Delta_2} \sum_{l=0}^{L-1} z(l)s_1(l - \Delta_1) + z(l)s_2(l - \Delta_2) - s_1(l - \Delta_1)s_2(l - \Delta_2) \underset{H_0}{\overset{H_1}{\gtrless}} \sigma^2 \ln \eta + \frac{E_1 + E_2}{2}$$

Not only do matched filter terms for each signal appear but also a cross-term between the two signals. It is this latter term that complicates the multiple signal problem: If this term is not zero for *all* possible delays, a nonseparable maximization process results and both delays must be varied in concert to locate the maximum. If, however, the two signals are orthogonal regardless of the delay values, the delays can be found separately, and the structure of the single signal detector (modified to include matched filters for each signal) suffices. This seemingly impossible situation can occur, at least approximately. Using Parseval's Theorem, the cross-term can be expressed in the frequency domain.

$$\sum_{l=0}^{L-1} s_1(l - \Delta_1)s_2(l - \Delta_2) = \frac{1}{2\pi} \int_{-\pi}^{\pi} S_1(\omega)S_2^*(\omega)e^{j\omega\cdot(\Delta_2 - \Delta_1)}\, d\omega$$

For this integral to be zero for all Δ_1, Δ_2, the product of the spectra must be zero. Consequently, if the two signals have disjoint spectral support, they are orthogonal no matter what the delays may be.[†] Under these conditions, the detector becomes

$$\max_{\Delta_1} \left[z(l) \star s_1(D - 1 - l)\right]\Big|_{l=D-1+\Delta_1} + \max_{\Delta_2} \left[z(l) \star s_2(D - 1 - l)\right]\Big|_{l=D-1+\Delta_2} \underset{H_0}{\overset{H_1}{\gtrless}} \gamma$$

with the threshold again computed independently of the received signal amplitudes.[‡]

$$P_F = Q\left(\frac{\gamma}{\sqrt{(E_1 + E_2)\sigma^2}}\right)$$

This detector has the structure of two parallel, independently operating, matched filters, each of which is tuned to the specific signal of interest.

[†] We stated earlier that this situation happens "at least approximately." Why the qualification?

[‡] We emphasize that E_1 and E_2 are the energies of the signals $s_1(l)$ and $s_2(l)$ used in the detector, *not* those of their received correlates $A_1 s_1(l)$ and $A_2 s_2(l)$.

Reality is insensitive to mathematically simple results. The orthogonality condition on the signals that yielded the relatively simple two-signal, unknown-delay detector is often elusive. The signals often share similar spectral supports, thereby violating the orthogonality condition. In fact, we may be interested in detecting the *same* signal repeated twice (or more) within the observation interval. Because of the complexity of incorporating intersignal correlations, which are dependent on the relative delay, the idealistic detector is often used in practice. In the repeated signal case, the matched filter is operated over the entire observation interval, and the number of *excursions* above the threshold noted. An excursion is defined to be a portion of the matched filter's output that exceeds the detection threshold over a contiguous interval. Because of the signal's nonzero duration, the matched filter's response to just the signal has a nonzero duration, implying that the threshold can be crossed at more than a single sample. When one signal is assumed, the maximization step automatically selects the peak value of an excursion. As shown in lower panels of Fig. 5.7 {226}, a low-amplitude excursion may have a peak value less than a nonmaximal value in a larger excursion. Thus, when considering multiple signals, the important quantities are the times at which excursion peaks occur, not all of the times the output exceeds the threshold.

Fig. 5.7 illustrates the two kinds of errors prevalent in multiple signal detectors. In the left panel, we find two excursions, the first of which is due to the signal, the second to noise. This kind of error cannot be avoided; we never said that detectors could be perfect! The right panel illustrates a more serious problem: The threshold is crossed by four excursions, all of which are owing to a single signal. Hence, excursions must be sorted through, taking into account the nature of the signal being sought. In the example, excursions surrounding a large one should be discarded if they occur in close proximity. This requirement means that closely spaced signals cannot be distinguished from a single one.

Unknown direction of propagation. Not knowing the direction of propagation exactly—*the* direction-finding problem of array processing—means that the matched filter detector of §5.3.5 {234} cannot be used. We can treat the direction of propagation as an unknown parameter, however, and, by applying the ideas of the generalized likelihood ratio test, not only detect the signal's presence but also determine the direction of propagation. Using the usual formula for the conventional beamformer's output and treating each sensor's delay as an unknown, we arrive at the following decision rule in the white-noise case.

$$\max_{\{\Delta_m\}} \sum_l \sum_m w_m s(l-\Delta_m) z(l) - \frac{1}{2} \sum_l \left[\sum_m w_m s(l-\Delta_m) \right]^2 \underset{H_0}{\overset{H_1}{\gtrless}} \gamma$$

This rule is an extension of the two-signal, unknown delay case just discussed. Here, the individual signals have the same waveform, meaning that they cannot be orthogonal. Consequently, the delays need to be considered jointly in the maximization process. Because no analytic formulas for this problem exists, the maximum likelihood estimates must be found numerically. An initial guess for numerical solution could be made by assuming orthogonality, which amounts to matched filtering the output and using the M largest output values.

Because the signal we seek to detect is a propagating wave, we know that the delays $\{\Delta_m\}$ are coupled, being a function of the source's position $\vec{x}^o$ in the array's coordinate system. The optimal detector for the white-noise case now becomes

$$\max_{\vec{x}^o} \sum_l \sum_m w_m s\big(l - \Delta_m(\vec{x}^o)\big) z(l) - \frac{1}{2} \sum_l \left[\sum_m w_m s\big(l - \Delta_m(\vec{x}^o)\big) \right]^2 \underset{H_0}{\overset{H_1}{\gtrless}} \gamma$$

Although little in the signal processing has changed, the maximization has become easier as only one parameter, the source location, needs to be varied.

For the unknown direction of propagation problem, spectral detection methods shine when compared with time-domain methods. The approximate orthogonality of *DFT* outputs *greatly* simplifies the optimal detector.

$$\max_{\vec{x}^o} \sum_{k=0}^{L-1} \left(\frac{\sum_m w_m \operatorname{Re}[Z^*(k) S(k) e^{-j2\pi k \Delta_m(\vec{x}^o)/L}]}{\sigma_k^2} - \frac{1}{2} \left(\textstyle\sum_m w_m^2\right) \frac{|S(k)|^2}{\sigma_k^2} \right) \underset{H_0}{\overset{H_1}{\gtrless}} \gamma$$

Now the energy term does not depend on the unknown delays and it can be absorbed into the threshold. The explicit formula for the sensor delays resembles that described for the Radon Transform in §4.8.3 [167]: $\Delta_m(\vec{x}^o) = q_m(\vec{x}^o) + p$ where $q_m(\cdot)$ and p are integers. $q_m(\vec{x}^o)$ is the propagation delay taken so that it equals 0 when evaluated at the array's spatial origin, and p is the overall delay.† Letting $W_q(k)$ denote the quantity $\sum_m w_m \exp\{-j2\pi k q_m/L\}$, the spectral decision rule for the unknown direction of propagation problem itself becomes a Fourier Transform.

$$\boxed{\max_{p,q} \operatorname{Re}\left[\sum_{k=0}^{L-1} \left(\frac{W_q(k) Z^*(k) S(k)}{\sigma_k^2} \right) e^{-j2\pi kp/L} \right] \underset{H_0}{\overset{H_1}{\gtrless}} \gamma} \tag{5.11}$$

Interestingly, $q(\vec{x}^o)$, the parameter needed to locate the source, enters into this expression through the *DFT* of the shading. Thus, the optimal detection rule amounts to assuming a value for the slope q, evaluating the shading sequence's *DFT* at the appropriate frequency index, computing the indicated transform, and finding the temporal index at which the transform is maximum.

As simple as the detector's structure might be, we have not completely specified the decision rule: What is the threshold γ? In the problems we have encountered to date, this calculation has been relatively straightforward; now, it's not. The usual

†Note that the delay parameter p has been assumed *not* to depend on the source's location. In many problems, such a delay cannot be related to propagation from a given direction. This assumption is clearly untrue in active radar/sonar problems because overall delay represents distance between the array and the source. Assuming a far-field source, the slope q would represent direction of propagation (the angle of $\vec{x}^o$) and p the distance (the magnitude of $\vec{x}^o$). Thus, even in this situation each component of the parameterized delay can be varied independently.

procedure of assuming no signal is present and finding the false-alarm probability remains correct. Finding P_F now becomes delicate. Under H_0, the *DFT* $Z(k)$ of the observations has a Gaussian distribution because linear transformations of Gaussian sequences yield Gaussian random variables. The assumed independence of distinct spectral values implies that the detector's Fourier Transform is a weighted linear combination of independent Gaussian random variables, also a Gaussian quantity. Now for the tricky part. We now need to compare the *maximum* of the transform's real part with a threshold; that maximum does *not* have a Gaussian distribution. However, as discussed in §A.1.5 {475}, all is not lost; the probability that the maximum of L statistically independent, zero-mean, Gaussian random variables $x_1, \ldots, x_L$ exceeds γ is given by

$$P_F = \prod_{l=1}^{L} Q\left(\frac{\gamma}{\sigma_l}\right)$$

where σ_l^2 is the variance of x_l. Each of these random variables represents the spectral detector's sufficient statistic for a particular choice of the pair p, q. *If* these quantities are identically distributed, the threshold is ultimately determined by a formula not too different from what we have found so frequently.

$$\gamma = \sigma_x Q^{-1}(P_F^{1/L})$$

Unfortunately, the variation of $W_q(k)$ with q results in unequal variances, requiring us to evaluate the much more complicated previous expression. Exploring how these results apply to the unknown direction of propagation spectral detector is the subject of Probs. 5.25 and 5.26.

Unknown signal waveform. The most general unknown–signal-parameter problem occurs when the signal itself is unknown. This phrasing of the detection problem can be applied to two different sorts of situations. The signal's waveform may not be known precisely because of propagation effects (rapid multipath, for example) or because of source uncertainty. Another situation is "Hello, is anyone out there?": You want to determine if any nonnoise-like quantity is present in the observations. These problems impose severe demands on the detector, which must function with little a priori knowledge of the signal's structure. Consequently, we cannot expect superlative performance.

$$\begin{aligned} H_0&: z(l) = n(l) \\ H_1&: z(l) = s(l) + n(l), \quad s(l) = ? \end{aligned}$$

The noise is assumed to be Gaussian with a covariance matrix $\mathbf{K}$. The conditional density under H_1 is given by

$$p_{\mathbf{z}|H_1,\mathbf{s}}(\mathbf{z}|H_1, \mathbf{s}) = \frac{1}{\sqrt{\det[2\pi\mathbf{K}]}} \exp\left\{-\frac{1}{2}(\mathbf{z}-\mathbf{s})^t\mathbf{K}^{-1}(\mathbf{z}-\mathbf{s})\right\}$$

Using the generalized likelihood ratio test, the maximum value of this density with respect to the unknown "parameters," the signal values, occurs when $\mathbf{z} = \mathbf{s}$.

$$\max_{\mathbf{s}} p_{\mathbf{z}|H_1,\mathbf{s}}(\mathbf{z}|H_1, \mathbf{s}) = \frac{1}{\sqrt{\det[2\pi\mathbf{K}]}}$$

The other hypothesis does not depend on the signal, and the generalized likelihood ratio test for the unknown signal problem, often termed the *square-law* detector, is

$$\boxed{\mathbf{z}^t\mathbf{K}^{-1}\mathbf{z} \underset{H_0}{\overset{H_1}{\gtrless}} \gamma}$$

For example, if the noise were white, the sufficient statistic is the sum of the squares of the observations.

$$\sum_{l=0}^{L-1} z^2(l) \underset{H_0}{\overset{H_1}{\gtrless}} \gamma$$

If the noise is not white, the detection problem can be formulated in the frequency domain, as in §5.3.4 {231}, where the decision rule becomes

$$\sum_{k=0}^{L-1} \frac{|Z(k)|^2}{\sigma_k^2} \underset{H_0}{\overset{H_1}{\gtrless}} \gamma$$

Computation of the false-alarm probability in, for example, the white noise case is relatively straightforward. The probability density of the sum of the squares of L statistically independent zero-mean, unit-variance, Gaussian random variables is termed chi-squared with L degrees of freedom {479}: χ_L^2. The percentiles of this density are tabulated in many standard statistical references (see Abramowitz and Stegun).

Example

Assume that the additive noise is white and Gaussian, having a variance σ^2. The sufficient statistic $\Upsilon = \sum z_l^2$ of the square-law detector has the probability density

$$\Upsilon/\sigma^2 \sim \chi_L^2$$

when no signal is present. The threshold γ for this statistic is established by $\Pr(\chi_L^2 > \gamma/\sigma^2) = P_F$. The probability of detection is found from the density of a noncentral chi-squared random variable having $\chi_L'^2(\lambda)$ having L degrees of freedom and "centrality" parameter $\lambda = \sum_{l=0}^{L-1} \mathcal{E}^2[z(l)]$. In this example, $\lambda = E$, the energy of the observed signal. In Fig. 5.9 {238}, the false-alarm probability was set to 10^{-2}, and the resulting probability of detection shown as a function of signal-to-noise ratio. Clearly, the inability to specify the signal waveform leads to a significant reduction in performance. In this example, roughly 10 dB more signal-to-noise ratio is required by the square-law detector than the matched-filter detector, which assumes knowledge of the waveform but not of the amplitude, to yield the same performance.

5.4.2 Unknown Noise Parameters

When aspects of the noise, such as the variance or power spectrum, are in doubt, the detection problem becomes more difficult to solve. Although a decision rule can be derived for such problems using the techniques we have been discussing, establishing a rational threshold value is impossible in many cases. The reason for this inability is simple: *All* hypotheses depend on the noise, thereby disallowing a computation of a threshold based on a performance probability. The solution is innovative: Derive decision rules and accompanying thresholds that do not depend on false-alarm probabilities.

Consider the case where the variance of the noise is not known and the noise covariance matrix is written as $\sigma^2\widetilde{\mathbf{K}}$ where the trace of $\widetilde{\mathbf{K}}$ is normalized to L. The conditional density of the observations under a signal-related hypothesis is

$$p_{\mathbf{z}|H_i,\sigma^2}(\mathbf{z}|H_i,\sigma^2) = \frac{1}{(\sigma^2)^{L/2}\sqrt{\det[2\pi\widetilde{\mathbf{K}}]}} \exp\left\{-\frac{1}{2\sigma^2}(\mathbf{z}-\mathbf{s}_i)^t\widetilde{\mathbf{K}}^{-1}(\mathbf{z}-\mathbf{s}_i)\right\}$$

Using the generalized likelihood ratio approach, the maximum value of this density with respect to σ^2 occurs when

$$\widehat{\sigma^2_{\text{ML}}} = \frac{(\mathbf{z}-\mathbf{s}_i)^t\widetilde{\mathbf{K}}^{-1}(\mathbf{z}-\mathbf{s}_i)}{L}$$

This seemingly complicated answer is easily interpreted. The presence of $\widetilde{\mathbf{K}}^{-1}$ in the dot product can be considered a whitening filter. Under the ith hypothesis, the expected value of the observation vector is the signal. This computation amounts to subtracting the expected value from the observations, whitening the result, then averaging the squared values, the usual form for a variance's estimate. Using this estimate for each hypothesis, the logarithm of the generalized likelihood ratio becomes

$$\frac{L}{2}\ln\left(\frac{(\mathbf{z}-\mathbf{s}_0)^t\widetilde{\mathbf{K}}^{-1}(\mathbf{z}-\mathbf{s}_0)}{(\mathbf{z}-\mathbf{s}_1)^t\widetilde{\mathbf{K}}^{-1}(\mathbf{z}-\mathbf{s}_1)}\right) \underset{H_0}{\overset{H_1}{\gtrless}} \ln\eta$$

Computation of the threshold remains. A threshold can be computed only if the probability density of the sufficient statistic does *not* depend on the variance of the noise. If this property held, we would have what is known as *constant false-alarm rate* (*CFAR*) detector [33];[72: 317ff]. If a detector has this property, the value of the statistic does not change if the observations are scaled about their presumed mean. Let there be no signal under H_0. The scaling property can be checked in this zero-mean case by replacing $\mathbf{z}$ by $c\mathbf{z}$. With this substitution, the statistic becomes $c^2\mathbf{z}^t\widetilde{\mathbf{K}}^{-1}\mathbf{z}/(c\mathbf{z}-\mathbf{s}_1)^t\widetilde{\mathbf{K}}^{-1}(c\mathbf{z}-\mathbf{s}_1)$. The constant c cannot be eliminated and the detector does not have the *CFAR* property. In fact, both hypotheses depend on the unknown variance, meaning that a threshold cannot be computed. If, however, we assume that the signal amplitude is also in doubt, a *CFAR* detector emerges. Express the signal component of hypothesis i as $A\mathbf{s}_i$, where A is an unknown constant. The maximum likelihood estimate of this amplitude under hypothesis i is

$$\widehat{A}_{\text{ML}} = \frac{\mathbf{z}^t\widetilde{\mathbf{K}}^{-1}\mathbf{s}_i}{\mathbf{s}_i^t\widetilde{\mathbf{K}}^{-1}\mathbf{s}_i}$$

Using this estimate in the likelihood ratio, we find the decision rule for the *CFAR* detector.

$$\frac{L}{2}\ln\left(\frac{\mathbf{z}^t\widetilde{\mathbf{K}}^{-1}\mathbf{z}-\dfrac{\left(\mathbf{z}^t\widetilde{\mathbf{K}}^{-1}\mathbf{s}_0\right)^2}{\mathbf{s}_0^t\widetilde{\mathbf{K}}^{-1}\mathbf{s}_0}}{\mathbf{z}^t\widetilde{\mathbf{K}}^{-1}\mathbf{z}-\dfrac{\left(\mathbf{z}^t\widetilde{\mathbf{K}}^{-1}\mathbf{s}_1\right)^2}{\mathbf{s}_1^t\widetilde{\mathbf{K}}^{-1}\mathbf{s}_1}}\right)\underset{H_0}{\overset{H_1}{\gtrless}}\ln\eta$$

Now we find that when $\mathbf{z}$ is replaced by $c\mathbf{z}$, the statistic is unchanged. Thus, the probability distribution of this statistic does *not* depend on the unknown variance σ^2. In most array processing applications, no signal is assumed present in hypothesis H_0; in this case, H_0 does not depend on the unknown amplitude A and a threshold can be found to ensure a specified false-alarm rate for *any* value of the unknown variance. For this specific problem, the likelihood ratio can be manipulated to yield the *CFAR* decision rule

$$\frac{\left(\mathbf{z}^t\widetilde{\mathbf{K}}^{-1}\mathbf{s}_1\right)^2}{\mathbf{z}^t\widetilde{\mathbf{K}}^{-1}\mathbf{z}\cdot\mathbf{s}_1^t\widetilde{\mathbf{K}}^{-1}\mathbf{s}_1}\underset{H_0}{\overset{H_1}{\gtrless}}\gamma$$

Example

Let's extend the previous example to the *CFAR* statistic just discussed to the white-noise case. The sufficient statistic is

$$\Upsilon(\mathbf{z})=\frac{\left[\sum z(l)s(l)\right]^2}{\sum z^2(l)\cdot\sum s^2(l)}$$

We first need to find the false-alarm probability as a function of the threshold γ. Using the techniques described by Wishner [163], the probability density of $\Upsilon(\mathbf{z})$ under H_0 is given by a β density (see §A.1.7 {480}), the parameters of which do *not* depend on either the noise variance (expectedly) or the signal values (unexpectedly).

$$p_{\Upsilon|H_0}(\Upsilon|H_0)=\beta\left(\Upsilon,\frac{1}{2},\frac{L-1}{2}\right)$$

We express the false-alarm probability derived from this density as an incomplete β function [1], resulting in the curves shown in Fig. 5.11. The statistic's density under hypothesis H_1 is related to the noncentral F distribution, expressible by the fairly simple, quickly converging, infinite sum of β densities

$$p_{\Upsilon|H_1}(\Upsilon|H_1)=\sum_{k=0}^{\infty}e^{-d^2}\frac{(d^2)^k}{k!}\beta\left(\Upsilon,k+\frac{1}{2},\frac{L-1}{2}\right)$$

where d^2 equals a signal-to-noise ratio: $d^2=\sum_l s^2(l)/2\sigma^2$. The results of using this *CFAR* detector are shown in Fig. 5.12.

5.5 Detection-Based Array Processing Algorithms

As shown in previous sections of this chapter, devising optimal detectors for an array processor's output is, generally speaking, a straightforward task. The resulting post-processing detectors are widely used, providing for example, algorithms for scanning

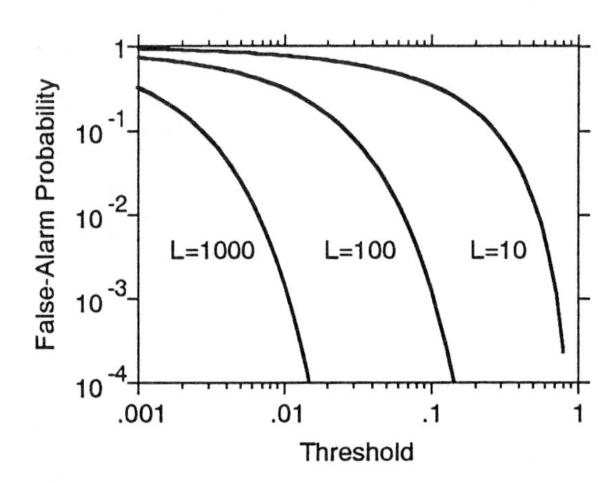

Figure 5.11 The false-alarm probability for the *CFAR* receiver is plotted against the threshold value γ for several values of L, the number of observations. Note that the test statistic, and thereby the threshold, does not exceed 1.

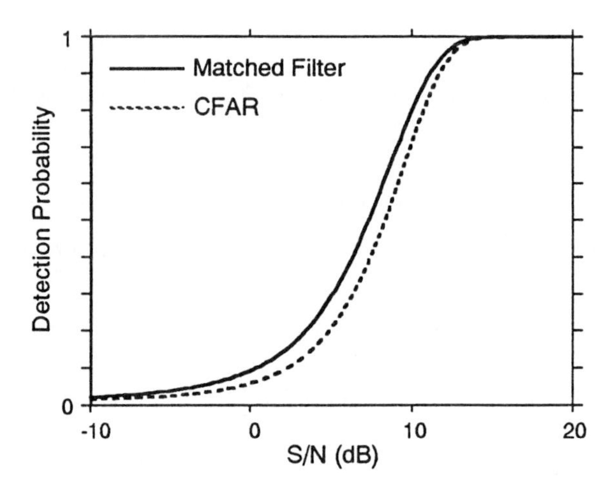

Figure 5.12 The probability of detection for the *CFAR* detector and the matched filter detector is shown as a function of signal-to-noise ratio. The signal and the false-alarm criterion are the same as in Fig. 5.9 {238}. Note how little performance has been lost in this case.

a beampattern to determine a signal's direction of propagation. The array processing algorithm and the detector were derived separately, however, leaving behind a nagging question: Can we do better with an encompassing algorithm that considers both direction of propagation processing and detection? The obvious answer is yes, but we shall see that the rather ad hoc conventional array processing algorithms of Chap. 4 do have some rational basis that, unfortunately, misses the mark.

The uniqueness of the array processing problem is the hypothesized presence of not only a signal, but of multiple versions of it in the observations measured by the array.

The signal is presumed present at every sensor; because of the directionality of the propagation, the signal's time origin in a sensor's output usually depends on the sensor's spatial location and the source's location. Thus, unknown parameters enter the problem, precisely the ones we are trying to extract from array measurements. We first need to recast the observations that are functions of both space (sensor location) and time (sample index) into functions of one variable so as to fit into our detection theoretic framework. We thus form the concatenated observation vector $\mathbf{y}^c$ defined as

$$\mathbf{y}^c = \text{col}[\mathbf{y}_0, \mathbf{y}_1, \ldots, \mathbf{y}_{M-1}]$$

Here $\mathbf{y}_m = \text{col}\big[y_m(0), y_m(T), \ldots, y_m\big((N-1)T\big)\big]$ denotes the observations obtained from the mth sensor.

5.5.1 Detection of Known Signals

The simplest situation occurs when the time at which a propagating signal arrives at each sensor is known. This seemingly artificial problem arises when the location of the energy source is known, but we are uncertain whether it is emitting energy or not. Assume the noise is Gaussian for the moment. The correlation structure of the noise, both temporally and spatially, wields great influence on the complexity of the optimum detector.

White noise. First we assume the noise spatially and temporally white. The detector maximizing the probability of detection under a constraint on the false-alarm probability is the matched filter.

$$\sum_{l=0}^{L-1} y^c(l) s^c(l) \underset{H_0}{\overset{H_1}{\gtrless}} \gamma$$

$s^c(l)$ is the concatenated version of the propagating signal that expresses the signal's known propagation delays, amplitudes, and waveforms at each sensor. Decomposing this summation on a per-sensor basis, we find that the detector can be expressed as the sum of the matched-filtered outputs at each sensor.

$$\boxed{\sum_{m=0}^{M-1} \sum_{n=0}^{N-1} y_m(n) s(n - \Delta_m) \underset{H_0}{\overset{H_1}{\gtrless}} \gamma}$$

This optimum detector does *not* make a decision at each sensor, then somehow merge the individual decisions. Rather, as shown in Fig. 5.13, each sensor's observations are match-filtered and the resulting values summed to produce the sufficient statistic.

The threshold γ is found by specifying a false-alarm probability, then solving the equation

$$P_F = Q\left(\frac{\gamma}{\sqrt{ME\sigma^2}}\right)$$

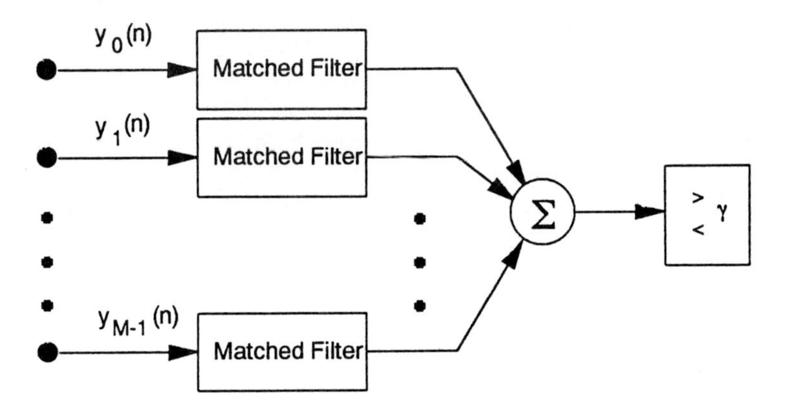

Figure 5.13 The fundamental form of the detector for array processing problems is to perform a computation—match filtering—on each sensor's output, then combine these filtered outputs to form the sufficient statistic. This basic structure applies over a wide range of noise characteristics including non-Gaussian problems.

where E is the energy of the signal measured at a single sensor. The resulting probability of detection is

$$P_D = Q\left(Q^{-1}(P_F) - \sqrt{\frac{ME}{\sigma^2}}\right)$$

The performance of the detector thus depends on the ratio of the signal energy *integrated over the entire array* to the variance of an individual noise value. An array can clearly enhance detection performance well beyond that of a single sensor. The probability of detection is shown in Fig. 5.14 for several values of M. This effect of boosting the signal-to-noise ratio through use of an array is termed *array gain*. As shown in §4.5 {137}, the maximum array gain occurs when the noise is spatially and temporally white, resulting in a signal-to-noise ratio increase of a factor of M over that measured at a single sensor. The presence of sensor-to-sensor noise correlation reduces the best obtainable array gain from this value.

We have been (subtly) assuming that the common time origin of the signals is known. For transient signals, this origin may not be known and must be incorporated into the detection problem. Letting n_0 be the common delay of the signal at each sensor, the signal component of each sensor's output is $s(n - \Delta_m - n_0)$. As discussed in §5.4.1 {236}, this uncertainty can be modeled as an unknown parameter, with the value of the parameter replaced by some estimate of it. The maximum likelihood estimate fits detection problems well. As H_0, the noise-only hypothesis, does not depend on the unknown parameter, the detector for the unknown signal-arrival problem becomes

$$\max_{n_0} \sum_{m=0}^{M-1} \sum_{n=0}^{N-1} y_m(n) s(n - \Delta_m - n_0) \underset{H_0}{\overset{H_1}{\gtrless}} \gamma$$

The threshold value does not change from the situation where the delay n_0 is known. The

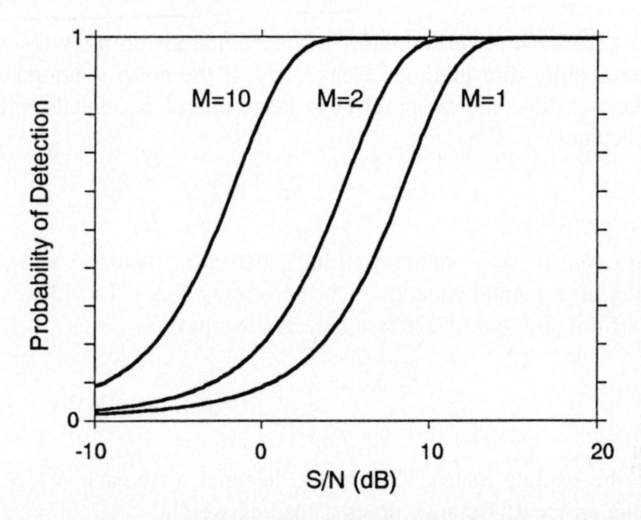

Figure 5.14 The probability of detection for an array of M sensors is shown as a function of the signal-to-noise ratio E/σ^2 found at a single sensor. The false-alarm probability criterion was 0.01.

detector employs a matched filter at each sensor, each using the same value of the delay. The maximization occurs *after* summing the outputs of the matched filters at each sensor, *not* before. In this way, the noise is "averaged" across the array to improve detection performance.

Colored noise. When nonwhite noise is present, the optimum detector compensates by first whitening the observation signal, then match-filtering using the "whitened" signal as the template. Because of the covariance matrix's structure in the array processing problem, this general detection rule simplifies to a specific form that explicitly demonstrates temporal and spatial whitening. Using vector notation, the general detector consists of

$$(\mathbf{y}^c)^t \mathbf{K}^{-1} \mathbf{s}^c \underset{H_0}{\overset{H_1}{\gtrless}} \gamma$$

As discussed in §2.6.4 {50}, separable spatiotemporal correlations result in a covariance matrix that can be expressed as a Kronecker product (Eq. 2.19 {50}): $\mathbf{K} = \boldsymbol{\rho} \otimes \mathbf{K}_0$, where $\boldsymbol{\rho}$ is the matrix of spatial correlation coefficients, and $\mathbf{K}_0$ is the temporal covariance matrix of each sensor's output. As shown in appendix B {489}, the inverse of a Kronecker product equals the Kronecker product of the inverses: $\mathbf{K}^{-1} = \boldsymbol{\rho}^{-1} \otimes \mathbf{K}_0^{-1}$. If the noise is spatially white, $\boldsymbol{\rho} = \mathbf{I}$, which results in the detection rule

$$\sum_{m=0}^{M-1} (\mathbf{y}_m)^t \mathbf{K}_0^{-1} \mathbf{s}_m \underset{H_0}{\overset{H_1}{\gtrless}} \gamma \tag{5.12}$$

In this case, each sensor output is whitened and match-filtered *separately* and the results summed before thresholding. Conversely, if the noise is temporally white but spatially correlated so that the temporal covariance matrix $\mathbf{K}_0$ equals $\sigma^2\mathbf{I}$, the optimal decision rule becomes†

$$(\mathbf{y}^c)^t\left(\boldsymbol{\rho}^{-1}\otimes\mathbf{I}\right)\mathbf{s}^c \underset{H_0}{\overset{H_1}{\gtrless}} \gamma \tag{5.13}$$

To compare these optimal (from a detection theoretic viewpoint) array processors with the fundamental equation of beamforming (Eq. 4.1 {113}), we need to recast it into matrix form. Interpreting it as a detector, we have

$$\sum_{n=0}^{N-1}\sum_{m=0}^{M-1} w_m y_m(n) s_m(n) = (\mathbf{y}^c)^t\mathbf{W}\mathbf{s}^c$$

where the shading matrix $\mathbf{W}$ is block diagonal, with each $N\times N$ block on the diagonal equaling an identity matrix times a shading weight.

$$\mathbf{W} = \begin{bmatrix} w_0\mathbf{I} & \mathbf{0} & \cdots & \mathbf{0} \\ \mathbf{0} & w_1\mathbf{I} & \mathbf{0} & \vdots \\ \vdots & \mathbf{0} & \ddots & \mathbf{0} \\ \mathbf{0} & \cdots & \mathbf{0} & w_{M-1}\mathbf{I} \end{bmatrix}$$

This array processor contrasts sharply with that of the optimal detector: $(\mathbf{y}^c)^t\mathbf{W}\mathbf{s}^c$ vs. $(\mathbf{y}^c)^t\mathbf{K}^{-1}\mathbf{s}^c$. This comparison reveals that *shading weights should be designed with noise characteristics in mind rather than the signal's.* The shading matrix $\mathbf{W}$ fulfills the role of the *inverse* of the spatiotemporal covariance matrix $\mathbf{K}^{-1}$. For this covariance matrix to be block diagonal, we must have the spatial correlation matrix $\boldsymbol{\rho}$ equaling identity: *The conventional beamformer is thus implicitly assuming the noise to be spatially white.* Under this assumption, the optimal detector employs the inverse $\mathbf{K}_0^{-1}$ of the temporal covariance matrix in the inner product between a sensor's output and the signal (Eq. 5.12). Drawing the two formulas together requires this covariance matrix to be proportional to an identity matrix, the proportionality constant equaling the *reciprocal* of the shading weight: $\mathbf{K}_0 = (1/w_m)\mathbf{I}$. Again using an optimal detector interpretation, this equation reveals that *the conventional beamformer is tacitly assuming to the noise to be temporally white with a variance that changes systematically with sensor position: The noise is presumed spatially nonstationary.* This detection-theoretic interpretation reveals conventional beamforming's ad hoc nature, but shows that the right ideas are present. The sensor outputs should be weighted, but both temporally and spatially. Only if the noise is spatially and temporally white and the shading sequence is unity (rectangular shading) do the two results agree.

Spectral techniques. Rather than using whitening filters, we have found spectral techniques computationally efficient for *some* colored-noise detection problems. The

†Where is the variance term σ^2?

critical factor in their utility is how the noise power spectrum varies with frequency: It must vary slowly for underlying assumptions to be valid. As linear wave propagation theory imposes no restriction on the kind of signals that can propagate in a medium, the temporal correlation function of the noise is similarly unrestricted. The variability of the noise's temporal spectrum must, therefore, be evaluated in each situation. In contrast, the models of ambient noise propagating from all directions—isotropic noise—result in spatial spectra that are relatively constant over the range of spatial frequencies corresponding to propagating signals. Outside that range, the spectrum changes rapidly and the slowly varying spectrum requirement is violated. Thus, from a detection theoretic viewpoint, *Fourier transform techniques can be fruitfully applied to each sensor's output, but spatial Fourier Transforms have little utility*: They do not create spatially uncorrelated sensor outputs.

Assuming that the power spectrum of the noise varies slowly with temporal frequency, evaluating the Fourier Transform of each sensor's output results in a diagonal covariance matrix $\mathbf{D}$ for the noise component. The entries σ_k^2 on its diagonal equal the discrete Fourier Transform of the triangularly windowed temporal correlation function, which approximately equals the power spectrum of the noise {232}. When spatial and temporal correlations are separable, the spatiospectral covariance matrix has the form $\rho \otimes \mathbf{D}$, where ρ represents spatial correlation and $\mathbf{D}$ temporal correlation (Eq. 2.20, {52}). The sufficient statistic of the spectral detector thus equals the real part of

$$\begin{bmatrix} \mathbf{Y}'_0 & \cdots & \mathbf{Y}'_{M-1} \end{bmatrix} [\rho^{-1} \otimes \mathbf{D}^{-1}] \begin{bmatrix} \mathbf{S}_0 \\ \vdots \\ \mathbf{S}_{M-1} \end{bmatrix}$$

where $\mathbf{S}_m$ and $\mathbf{Y}_m$ denote the *DFT*s of the signal and observations, respectively, at the mth sensor. Because of the partitioned matrix structures, the matrix $\mathbf{D}^{-1}$ can be merged with the $\mathbf{S}_m$ vectors to yield the sufficient statistic for the array-based spectral detector given by the real part of

$$\boxed{\begin{bmatrix} \mathbf{Y}'_0 & \cdots & \mathbf{Y}'_{M-1} \end{bmatrix} \rho^{-1} \begin{bmatrix} \mathbf{D}^{-1}\mathbf{S}_0 \\ \vdots \\ \mathbf{D}^{-1}\mathbf{S}_{M-1} \end{bmatrix}} \tag{5.14}$$

The matrix $\mathbf{D}^{-1}$ weights each spectral component of the presumed signal by the reciprocal of the noise power spectrum. This spectral weighting by $1/\sigma_k^2$ has occurred in every spectral detection problem we have discussed and should not be surprising. However, the ever-present dot product between the spectra of the observations with weighted signal spectra is complicated by the inverse ρ^{-1} of the spatial correlation matrix serving as the dot product's kernel. Only when the spatial correlation is trivial (white so that $\rho^{-1} = \mathbf{I}$) is this combining step computationally simple.

Example

The structure of this important decision rule can be elucidated by considering a narrowband example. When the *DFT* of the signal has only one nonzero component (at frequency index k_0), Eq. 5.14 simplifies greatly. The kernel ρ^{-1} affects how the signals from the sensors are combined and leaves unaffected the dot product between the *DFT*s of the signal and each sensor's output. Consequently, the sufficient statistic becomes

$$\mathrm{Re}\left\{ \begin{bmatrix} Y_0^*(k_0) & \cdots & Y_{M-1}^*(k_0) \end{bmatrix} \rho^{-1} \begin{bmatrix} e^{-j2\pi k_0 \Delta_0/N} \\ \vdots \\ e^{-j2\pi k_0 \Delta_{M-1}/N} \end{bmatrix} \frac{S(k_0)}{\sigma_{k_0}^2} \right\}$$

where $S(k)$ is the *DFT* of the zero-delay signal $s(n)$. The *magnitude* of the ratio $S(k_0)/\sigma_{k_0}^2$ need not be calculated as it can be incorporated into the detection threshold; however, the phase of the signal must be retained. By explicitly writing the matrix products, the remaining terms can be interpreted as a spatial Fourier Transform of the spatially combined sensor outputs.

$$\mathrm{Re}\left\{ e^{j\arg[S(k_0)]} \sum_m \left(\sum_l \left[\rho^{-1}\right]_{l,m} Y_l^*(k_0) \right) e^{-j2\pi k_0 \Delta_m/N} \right\} \underset{H_0}{\overset{H_1}{\gtrless}} \gamma$$

For a regular linear array, $\Delta_m = q \cdot (m - M_{1/2})$ where q is an integer expressing the assumed propagation delay. The Fourier Transform interpretation of the decision rule now becomes more apparent.

$$\mathrm{Re}\left\{ e^{j\left(\arg[S(k_0)] + 2\pi k_0 q M_{1/2}/N\right)} \sum_{m=0}^{M-1} \left(\sum_{l=0}^{M-1} \left[\rho^{-1}\right]_{l,m} Y_l^*(k_0) \right) e^{-j2\pi k_0 \Delta_m/N} \right\} \underset{H_0}{\overset{H_1}{\gtrless}} \gamma$$

5.5.2 Signals with Unknown Parameters

Determining the signal intersensor delays owing to wave propagation is perhaps the most important problem in array processing. Thus, the common time origin n_0 and the propagation delays Δ_m are always problem unknowns. In addition, the signal's amplitude, which depends on various propagation considerations, is ultimately unknown. The imposition of an a priori density on these parameters is, in most cases, unjustified. We have found that determining the presence or absence of a signal does not require knowledge of the signal's amplitude. The remaining quantities must somehow be determined and these estimates used in the optimal detector. For example, assuming white noise the optimal decision rule would be

$$\sum_{m=0}^{M-1} \sum_{n=0}^{N-1} y_m(n) s(n - \widehat{\Delta}_m) \underset{H_0}{\overset{H_1}{\gtrless}} \gamma$$

where $\widehat{\Delta}_m$ denotes an estimate of the propagating signal's timing parameters at the mth sensor. In the array processing problem, wave propagation yields intersensor delays that are related to each other, resulting in an *orderly* set of delays. Rather than M independent delay parameters, only two, the overall delay and the delay relative to the array's coordinate origin, are related to the signal source's direction of propagation (or equivalently its location). Letting $\vec{x}^o$ denote the location of the source relative to the array, we saw in §5.4.1 {243} that the intersensor delays depend on this single quantity and are expressed by $\Delta_m = q_m(\vec{x}^o) + p$, where q_m is the propagation delay at the mth sensor relative to array origin, and p is the overall propagation delay. The optimal white-noise detector for array processing applications thus computes

$$\boxed{\max_{\vec{x}^o,p} \sum_{m=0}^{M-1}\sum_{n=0}^{N-1} y_m(n)s(n - q_m(\vec{x}^o) - p) \underset{H_0}{\overset{H_1}{\gtrless}} \gamma}$$

This computation can be laborious: Many different values of $\vec{x}^o$ must be used in each sensor's matched filter, the outputs of these summed, and the maximum of the sum used in the threshold test. Alternatively, matched filters can be employed at each sensor *as if* the signal delay at that sensor was to be determined without regard to the other sensors. These match filtered outputs would be delayed according to the presumed value of $\vec{x}^o$, summed, and the maximum found. In any case, the sensor outputs must be combined so that the noise is suppressed and the signal enhanced; array processing algorithms improve their performance by *spatial averaging*: averaging carefully selected quantities across the array.

When the noise is colored, the sufficient statistic expressed by Eq. 5.14 {253} remains approximately valid. The diagonal matrix $\mathbf{D}$ rather than the signal's *DFT* is indexed by sensor number and its elements modified to be

$$\mathbf{D}_m = \text{diag}\left[\sigma_0^2, \ldots, \sigma_k^2 e^{j2\pi k(q_m+p)/N}, \ldots, \sigma_{N-1}^2 e^{j2\pi(N-1)(q_m+p)/N}\right]$$

and the decision rule for the colored-noise detector in the unknown direction of propagation array processing problem becomes

$$\boxed{\max_{p,q} \text{Re}\left\{\begin{bmatrix} \mathbf{Y}_0' & \cdots & \mathbf{Y}_{M-1}' \end{bmatrix} \rho^{-1} \begin{bmatrix} \mathbf{D}_0^{-1}\mathbf{S} \\ \vdots \\ \mathbf{D}_{M-1}^{-1}\mathbf{S} \end{bmatrix}\right\} \underset{H_0}{\overset{H_1}{\gtrless}} \gamma} \qquad (5.15)$$

When this result is compared with the colored-noise detector that operated only on the array's output (Eq. 5.11 {243}), we again see that the naïve use of shading only approximates what the optimal detector exploits from using the spatial correlation. In most cases, the spatial correlation coefficient matrix is not diagonal, the simpler case being implicitly assumed in conventional beamforming. To find the detection threshold, ideas similar to that expressed earlier {244} and in Prob. 5.26 must be followed.

When more than one signal is present, the complexity of array processing algorithms based on optimal detection can dramatically increase. The considerations described in §5.4.1 {239} remain valid: Only when the signals are "orthogonal" can the detector focus on each signal independently. Orthogonality occurs *only* when signal spectra have disjoint supports. The important case of common support, two signals at the same temporal frequency propagating from different directions, for example, remains to haunt us. The detector's structure can be broadly outlined: Each signal's location $\vec{x}^o$ and overall delay p must be searched with the fundamental result given previously modified to remove signal cross-terms. This structure is so complicated that it is rarely used; one must know the number of signals before it can be implemented. Determining the number of signals can be approached using detection theoretic ideas: After all, it's just another parameter to be estimated. Now the detector becomes *very* complicated, so much so that this situation is not even a homework problem![†]

The approach we have generally described in Chap. 4 can now be properly analyzed. There, we assumed a one-signal model and searched for prominent peaks in beampatterns. While this approach is valid when the signals are orthogonal, narrowband processing situations exclude this possibility within a single beampattern. A detection threshold can be set with a single signal model, but it is computed *regardless of the sidelobe height.* The detection-based strategy therefore relates all significantly large peaks to propagating signals. This algorithm might have some uses, but leaves us unsatisfied. What is needed is an algorithm for estimating the number of signals without the extra baggage of signal models, signal cross-correlations, and so on. Such algorithms exist (see §7.3.5 {390}); we describe them once other ideas are explored.

Summary

Hypothesis testing provides the foundation for solving the many detection problems we have encountered in this chapter. Neither the authors nor the array processing community have exhausted the possible applications of the ideas described here to real problems. It behooves the engineer seeking detection algorithms for problems more difficult than those presented here to follow the statistics literature. The most prominent and respected journal in this area is the *Annals of Statistics.*

The key result of array processing-prejudiced detection theory is the matched filter. When we have Gaussian noise spoiling our measurements, the matched filter expresses the signal processing operations, known signal parameters or not. Even when noise uncertainties arise, the matched filtering idea remains in the optimal detector, sometimes in a nonlinear form. Clearly, this idea must be mastered to develop insights into detecting signals in situations not described herein.

Detection algorithms can be successfully applied to the array's output so as not to disturb an array processing algorithm carefully designed for some particular problem. If, however, the array is employed to serve a detection function, we have found that allowing the detector to adjust the array processing algorithm's structure can enhance performance. The key idea elucidated by detection theory is the explicit inclusion of the noise covariance function in the array processing algorithm. If known, including it certainly boosts performance; if partially known, robust detection ideas may be useful; if unknown, the adaptive array processing algorithms described in Chap. 7 prove their merits.

[†]For the curious and courageous, formulate your own homework problem. Note how the complexity of the signal cross-terms increases with number of presumed signals. How would you find the detection threshold?

Problems

5.1 Consider the following binary hypothesis testing problem [152: Prob. 2.2.1].

$$\begin{aligned} H_0&: y = n \\ H_1&: y = s + n \end{aligned}$$

where s and n are statistically independent, positively valued, random variables having the densities

$$p_s(s) = ae^{-as}\mathrm{u}(s) \quad \text{and} \quad p_n(n) = be^{-bn}\mathrm{u}(n)$$

(a) Prove that the likelihood ratio test reduces to

$$y \underset{H_0}{\overset{H_1}{\gtrless}} \gamma$$

(b) Find γ for the minimum probability of error test as a function of the a priori probabilities.

(c) Now assume that we need a Neyman-Pearson test. Find γ as a function of P_F, the false-alarm probability.

5.2 The two hypotheses describe different equivariance statistical models for the observations [152: Prob. 2.2.11].

$$H_0: p_y(y) = \frac{1}{\sqrt{2}} e^{-\sqrt{2}|y|}$$

$$H_1: p_y(y) = \frac{1}{\sqrt{2\pi}} e^{-\frac{1}{2}y^2}$$

(a) Find the likelihood ratio test.

(b) Compute the decision regions for various values of the threshold in the likelihood ratio test.

(c) Assuming these two densities are equally likely, find the probability of making an error in distinguishing between them.

5.3 A hypothesis testing criterion radically different from those discussed in §5.1.1 and §5.1.2 is *minimum equivocation*. In this information theoretic approach, the binary hypothesis testing problem is modeled as a digital channel (Fig. 5.15). The channel's inputs, generically represented by the $\mathbf{x}$, are the hypotheses and the channel's outputs, denoted by $\mathbf{y}$, are the decisions.

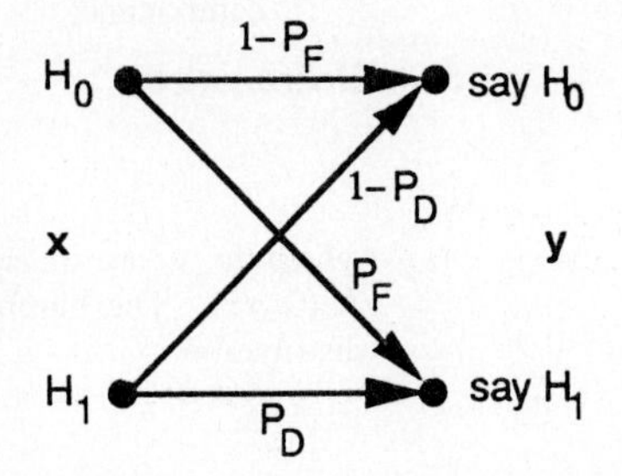

Figure 5.15 The binary hypothesis testing problem can be abstractly described as a communication channel where the inputs are the hypotheses and the outputs are the decisions. The transition probabilities are related to the false-alarm (P_F) and detection (P_D) probabilities.

The quality of such information theoretic channels is quantified by the *mutual information* $I(\mathbf{x}; \mathbf{y})$ defined to be difference between the entropy of the inputs and the *equivocation*.

$$I(\mathbf{x}; \mathbf{y}) = H(\mathbf{x}) - H(\mathbf{x}|\mathbf{y})$$
$$H(\mathbf{x}) = -\sum_i P(x_i) \log P(x_i)$$
$$H(\mathbf{x}|\mathbf{y}) = -\sum_{i,j} P(x_i, y_j) \log \frac{P(x_i, y_j)}{P(y_j)}$$

Here, $P(x_i)$ denotes the a priori probabilities, $P(y_j)$ the output probabilities, and $P(x_i, y_j)$ the joint probability of input x_i resulting in output y_j. For example, $P(x_0, y_0) = P(x_0)(1 - P_F)$ and $P(y_0) = P(x_0)(1 - P_F) + P(x_1)(1 - P_D)$. For a fixed set of a priori probabilities, show that the decision rule that maximizes the mutual information is the likelihood ratio test. What is the threshold when this criterion is employed?

Note: This problem is relatively difficult. The key to its solution is to exploit the concavity of the entropy function.

5.4 Developing a Neyman-Pearson decision rule for more than two hypotheses has not been detailed. Assume K distinct hypotheses are required to account for the observations. We seek to maximize the probability of correctly announcing H_i under the constraint that the probability of announcing H_i when hypothesis H_0 was indeed true does not exceed a specified value.

(a) Formulate the optimization problem that simultaneously maximizes $\Pr[\text{say } H_i | H_i]$ under the constraint $\Pr[\text{say } H_i | H_0] \leq \alpha_i$. Find the solution using Lagrange multipliers.

(b) Show that your solution can be expressed as choosing the largest of the sufficient statistics $\Upsilon_i(\mathbf{y}) + C_i$.

5.5 Pattern recognition relies heavily on ideas derived from the principles of statistical hypothesis testing. Measurements are made of a test object and these compared with those of "standard" objects to determine which the test object most closely resembles. Assume that the measurement vector $\mathbf{y}$ is jointly Gaussian with mean $\mathbf{m}_i$ $(i = 1, \ldots, K)$ and covariance matrix $\sigma^2\mathbf{I}$ (the components are statistically independent). Thus, there are K possible objects, each having an "ideal" measurement vector $\mathbf{m}_i$ and probability P_i of being present.

(a) How is the minimum probability of error choice of object determined from the observation of $\mathbf{y}$?

(b) Assuming that only two objects are possible $(K = 2)$, what is the probability of error of your decision rule?

(c) The expense of making measurements is always a practical consideration. Assuming each measurement costs the same to perform, we would want to select those measurements that are most effective: They influence the probability of error more than the others. How would you determine the effectiveness of a measurement vector's component?

5.6 Define y to be

$$y = \sum_{k=0}^{L} x_k$$

where the x_k are statistically independent random variables, each having a Gaussian density $\mathcal{N}(0, \sigma^2)$. The number L of variables in the sum is a random variable with a Poisson distribution.

$$\Pr[L = l] = \frac{\lambda^l}{l!} e^{-\lambda}, \quad l = 0, 1, \ldots$$

Based on the observation of y, we want to decide whether $L \leq 1$ or $L > 1$. Write an expression for the minimum P_e likelihood ratio test.

5.7 One observation of the random variable y is obtained. This random variable is either uniformly distributed between -1 and $+1$ or expressed as the sum of statistically independent random variables, each of which is also uniformly distributed between -1 and $+1$.

(a) Suppose there are two terms in the aforementioned sum. Assuming that the two hypotheses are equally likely, find the minimum probability of error decision rule.

(b) Compute the resulting probability of error of your decision rule.

(c) Show that the decision rule found in part a applies no matter how many terms are assumed present in the sum.

5.8 The observed random variable y has a Gaussian density on each of five hypotheses.

$$p_{y|H_k}(y|H_k) = \frac{1}{\sqrt{2\pi\sigma^2}} \exp\left\{-\frac{(y-m_k)^2}{2\sigma^2}\right\}, \quad k = 1, 2, \ldots, 5$$

where $m_1 = -2m$, $m_2 = -m$, $m_3 = 0$, $m_4 = +m$, and $m_5 = +2m$. The hypotheses are equally likely and the criterion of the hypothesis test is to minimize P_e.

(a) Draw the decision regions on the y axis.

(b) Compute the probability of error P_e. What is the maximum value of P_e for any reasonable hypothesis test?

5.9 We want to choose which of the following four hypotheses is true on reception of the three dimensional vector $\mathbf{y}$ [152. Prob. 2.6.6]. Under H_i, $\mathbf{y} = \mathbf{m}_i + \mathbf{n}$, where

$$\mathbf{m}_0 = \begin{bmatrix} a \\ 0 \\ b \end{bmatrix}, \quad \mathbf{m}_1 = \begin{bmatrix} 0 \\ a \\ b \end{bmatrix}, \quad \mathbf{m}_2 = \begin{bmatrix} -a \\ 0 \\ b \end{bmatrix}, \quad \mathbf{m}_3 = \begin{bmatrix} 0 \\ -a \\ b \end{bmatrix}$$

The noise vector $\mathbf{n}$ is a Gaussian random vector having statistically independent, identically distributed components, each of which has zero mean and variance σ^2. We have L independent observations of the received vector $\mathbf{y}$.

(a) Assuming equally likely hypotheses, find the minimum P_e decision rule.

(b) Calculate the resulting error probability.

(c) Show that neither the decision rule nor the probability of error do not depend on b. Intuitively, why is this fact true?

5.10 You decide to flip coins with Sleazy Sam. If heads is the result of a coin flip, you win one dollar; if tails, Sam wins a dollar. However, Sam's reputation has preceded him. You suspect that the probability of tails, p, may not be $1/2$. You want to determine whether a fair coin is being used or not after observing the results of three coin tosses.

(a) You suspect that $p = 3/4$. Assuming that the probability of a fair coin equals that of an unfair coin, how would you decide whether a fair coin is being used or not in a "good" fashion?

(b) Using your decision rule, what is the probability that your determination is incorrect?

(c) One potential flaw with your decision rule is that a specific value of p was assumed. Can a reasonable decision rule be developed without knowing p? If so, demonstrate the rule; if not, show why not.

5.11 A common situation in statistical signal processing problems is that the variance of the observations is unknown (there is no reason that noise should be nice to us!). Consider the binary Gaussian hypothesis testing problem where the hypotheses differ in their means and have a common, but unknown variance.

$$\begin{array}{ll} H_0: & \mathbf{y} \sim \mathcal{N}(\mathbf{0}, \sigma^2 \widetilde{\mathbf{K}}) \\ H_1: & \mathbf{y} \sim \mathcal{N}(\mathbf{m}, \sigma^2 \widetilde{\mathbf{K}}) \end{array} \qquad \sigma^2 = ?$$

where $\widetilde{\mathbf{K}}$ denotes the normalized covariance matrix: $\text{tr}[\widetilde{\mathbf{K}}] = \dim[\mathbf{y}]$.

(a) Show that the unknown variance enters into the optimum decision *only* in the threshold term.

(b) In the (happy) situation where the threshold η equals 1, show that the optimum test does not depend on σ^2 and that we did not need to know its value in the first place. When does η equal 1?

5.12 Consider the following composite hypothesis testing problem [152: Prob. 2.5.2].

$$\begin{array}{l} H_0: y \sim \mathcal{N}(0, \sigma_0^2) \\ H_1: y \sim \mathcal{N}(0, \sigma_1^2) \end{array}$$

where σ_0 is known but σ_1 is known only to be greater than σ_0. Assume that the false-alarm probability P_F must not exceed 10^{-2}.

(a) Does a *UMP* test exist for this problem? If it does, find it.

(b) Construct a generalized likelihood ratio test for this problem. Under what conditions can the requirement on the false-alarm probability be met?

5.13 A paper has been published recently by S. O. Balderdash about a new detection scheme. The results contained in the paper apply to problems where unknown parameter(s) are introduced into the characterization of the signal. His idea, which he claims is new, is to derive a detection rule that minimizes the maximum value that the average probability of error attains with respect to the possible parameter vector $\boldsymbol{\xi}$.

$$\min \left[\max_{\boldsymbol{\xi}} P_e(\boldsymbol{\xi}) \right]$$

Derive the decision rule for a binary hypothesis test. Is this rule novel?

5.14 A common problem in detecting signals in an array's output occurs when the signal processing delays cannot be adjusted to precisely equal the propagation delays. This situation occurs whenever sampling is used. Consequently, the signal that may appear in the array output may not equal its ideal: There is *signal mismatch.* Assume a two hypothesis problem in which either no signal is present or the signal $\tilde{s}(l)$ is found in the array output contaminated by additive white Gaussian noise. The matched filter employed in the detector uses $s(l)$, which (it is hoped) bears some resemblance to that produced by the beamformer. Assume that the observation interval is extended so that the mismatched signal can be completely used.

(a) Assuming that the detector and its threshold were determined under a no-mismatch assumption (what else could you do?), how is the detection probability affected?

(b) Assume the signal model is a sinusoid having P samples/period with L/P periods in the observation interval. The beamformed signal is improperly aligned by an amount equivalent to a phase error of ϕ/sensor. Assuming that $\phi < 2\pi/P$, how is detection performance affected?

5.15 The additive noise in a detection problem consists of a sequence of statistically independent Laplacian random variables. The probability density of $n(l)$ is therefore

$$p_{n(l)}(n) = \frac{1}{2}e^{-|n|}$$

The two possible signals are constant throughout the observation interval, equaling either $+1$ or -1.

(a) Find the optimum decision rule that could be used on a *single* value of the observation signal.

(b) Find an expression for the threshold in your rule as a function of the false-alarm probability.

(c) What is the threshold value for the "symmetric" error situation ($P_F = 1 - P_D$)?

(d) Now assume that two values of the observations are used in the detector (patience; to solve this problem, we need to approach it gradually). Find the optimum detector for $\eta = 1$. How does this result generalize to L observations?

5.16 Assume that observations of a sinusoidal signal $s(l) = A\sin(\omega l)$, $l = 0, \ldots, L-1$, are contaminated by first-order colored noise as described in the example {230}.

(a) Assuming that the alternative hypothesis is the sole presence of the colored Gaussian noise, what is the probability of detection for the optimum detector?

(b) How does this probability vary with signal frequency ω when the first-order coefficient is positive? Does your result make sense? Why?

5.17 Noise generated by a system having zeros complicates the calculations for the colored-noise detection problem. To illustrate these difficulties, assume the observation noise is produced as the output of a filter governed by the difference equation

$$n(l) = an(l-1) + w(l) + bw(l-1), \quad a \neq -b,$$

where $w(l)$ is white, Gaussian noise. Assume an observation duration sufficiently long to capture the colored-noise effects.

(a) Find the covariance matrix of this noise process. Compare it with the matrix that results when $b = 0$.

(b) Calculate the Cholesky factorization of the covariance matrix.†

(c) Find the unit-sample response of the optimal detector's whitening filter. If it weren't for the finite observation interval, would it indeed have an infinite-duration unit-sample response as claimed on page 231? Describe the edge-effects of your filter, contrasting them with the case when $b = 0$.

5.18 The calculation of the sufficient statistic in spectrally based detectors (Eq. 5.9 {234}) can be simplified using signal processing notions.

(a) When the observations are real-valued, their spectra are conjugate symmetric. Exploiting this symmetry, how can the calculations of Eq. 5.9 be simplified. Be careful to note "spectral edge-effects."

(b) Can your formula be manipulated to depend on the power spectra of the signals and the observations? Why or why not?

---

†This part is computationally complex if done symbolically. If a symbolic mathematics package such as Maple™ or Mathematica™ is not available, use numerically oriented Matlab™ for the case $a = 1/2$, $b = -1/3$.

5.19 On page 233, we claimed that the relation between noise bandwidth and reciprocal duration of the observation interval played a key role in determining whether *DFT* values were approximately uncorrelated. Although the statements sound plausible, their veracity should be checked. Let the covariance function of the observation noise be $K_n(l) = a^{|l|}$.

(a) How is the bandwidth (defined by the half-power point) of this noise's power spectrum related to the parameter a? How is the duration (defined to be two time constants) of the covariance function related to a? Is the product of these quantities approximately constant?

(b) Find the variance of the length-L *DFT* of this noise process as a function of the frequency index k. This result should be compared with the power spectrum calculated in part a; they *should* resemble each other when the "memory" of the noise, expressed by the duration of the covariance function, is much less than L while demonstrating differences as the memory becomes comparable to or exceeds L.

(c) Calculate the covariance between adjacent frequency indices. Under what conditions are they approximately uncorrelated? Compare your answer to the relation between a and L found in part b.

5.20 The results derived in Prob. 5.19 assumed that a length-L Fourier Transform was computed from a length-L segment of the noise process. What happens if the transform has a length longer than the observation interval? For simplicity, assume that the observations consist of white noise† and that a length-$2L$ transform is computed from length-L observations.

(a) Find the variance of *DFT* values at index k.

(b) Show that the spectral values at frequency indices separated by even integers are uncorrelated while those separated by odd integers are correlated. Compute the correlation coefficient between adjacent spectral values.

5.21 A common application of spectral detection is the detection of sinusoids in noise. Not only can one sinusoidal signal be present but several. The problem is to determine which combination of sinusoids is present. Let's first assume that no more than two sinusoids can be present, the frequencies of which are integer fractions of the number of observations: $\omega_i = 2\pi k_i/L, i = 1, 2$. Assume additive Gaussian noise is present.

(a) What is the detection procedure for determining if one of the sinusoids is present while ignoring the second? The amplitude of the sinusoid is known.

(b) How does the probability of detection vary with the ratio of the signal's squared amplitude to the noise variance at the frequency of the signal? Sketch this result for several false-alarm probabilities.

(c) Now assume that none, either, or both of the signals can be present. Find the detector that best determines the combination present in the observations assuming the amplitudes are known. Does your answer have a simple interpretation?

(d) Derive a detector that determines the number of sinusoids present in the observations.

5.22 From a detection theoretic viewpoint, the optimal choice of shading weights in general depends on the signal as well as the noise characteristics. In this problem, we explore the equation {236} for the normalized signal-to-noise ratio produced by a regular linear array in the somewhat unrealistic situation that the noise is spatially and temporally white.

(a) Rewrite the somewhat condensed expression $(\mathbf{w} \otimes \mathbf{I})'\mathbf{K}(\mathbf{w} \otimes \mathbf{I})$ for the case in which the noise is spatially homogeneous.

---

†Direct comparison with Prob. 5.19 is certainly instructive but analytically more tedious, obscuring what we need to understand. For "fun," assume $K_n(l) = a^{|l|}$.

(b) Simplify this expression further by assuming that the noise is also spatially white. Do you obtain the expression for the array gain G?

(c) In addition, (dubiously) assume the noise is temporally white; further simplify the expression and find the optimal shading vector.

5.23 Because sources rarely remain still for an array to determine their locations, a common problem in array processing applications is uncertainty of the signal frequency because of Doppler shifts. This effect is usually small, but conceivably could be large enough for a sinusoidal signal to "wander" from its presumed frequency index.

(a) Design a spectrally based detection strategy that examines a range of frequencies, testing for the presence of a sinusoid having known amplitude and phase in any *single* bin or no sinusoid in any bin.

(b) Predict the detection loss (or gain) relative to the known frequency case (no Doppler shift) by contrasting the signal-to-noise ratio terms in the detection probability expressions for each detector. Assume the noise has constant power across the frequency band of concern.

(c) Contrast the performance of this detector with a square-law detector that considers the entire frequency band.

(d) More realistically, the amplitude and phase of the sinusoid are not known. Redesign your spectrally based detector and derive an expression for the detection probability.

5.24 The authors confess that their expression of unknown-delay problem {239} is not terribly relevant to active sonar and radar ranging problems. In realistic cases, the signal's delay represents the round-trip propagation time from the source to the energy-reflecting object and back. What makes the example oversimplistic is the assumed independence of signal amplitude and delay.

(a) Because of attenuation owing to spherical propagation, show that the received signal energy is inversely related to the fourth power of the range. This result is known as the *radar equation*.

(b) Derive the detector that takes the dependence of delay and amplitude into account, thereby optimally determining the presence of a signal in active radar/sonar applications, and produces a delay estimate, thereby simultaneously providing the object's range. Not only determine the detector's structure, but also how the threshold and false-alarm probability are related.

(c) Does the object's reflectivity need to be known to implement your detector?

5.25 The unknown direction of propagation problem {242} can be extended to the case when the signal amplitude is unknown.

(a) Derive the optimal decision rule.

(b) Which case is simpler from the detector's viewpoint? Are the sensor delays Δ_m varied independently or are they jointly expressed in terms of the source's location $\vec{x}^o$? Which is more important for array processing?

5.26 The spectral detector given by Eq. 5.11 {243} for the unknown direction of propagation problem remains unsolved; we have not explicitly determined the threshold γ. This threshold depends on the variance of each component random variable considered in the maximization.

(a) Show that the variances of the component terms depend on the slope p.

(b) How is the false-alarm probability related to the threshold when the component variances are unequal?† Can your formula be inverted to express γ in terms of P_F?

(c) Develop a decision rule for the unknown direction of propagation detector for the white-noise case.

5.27 We did not assume any particular array geometry in deriving the spectral detector given by Eq. 5.11 {243} for the unknown direction of propagation problem. Furthermore, the Radon Transform interpretation given there suggests that interpolation beamforming might well be accommodated within our detection framework. Assume that the array geometry is linear with spacing d between adjacent sensors.

(a) Find and interpret the quantity $W_q(k) = \sum_m w_m \exp\{-j2\pi k q_m/L\}$ for the regular linear array geometry.

(b) Extend the spectral detector to the case $q_m = 1/2 \cdot (m - M_{1/2})$, $m = 0, \ldots, M-1$, where interpolation by a factor of two must be applied to some sensor outputs. Is the sufficient statistic changed in any fundamental way?

5.28 *CFAR* detectors are extremely important in applications because they automatically adapt to the value of noise variance during the observations, allowing them to be used in varying noise situations. As described in Prob. 5.24, unknown delays must also be considered in realistic problems.

(a) Derive a *CFAR* detector that also takes unknown signal delay into account.

(b) Show that your detector *automatically* incorporates amplitude uncertainties.

(c) Assuming the observations are corrupted by white Gaussian noise, what is the distribution of the sufficient statistic under the no-signal hypothesis?

5.29 Fresh insights into the *CFAR* detector are obtained by considering its frequency domain counterpart. Assume we are trying to detect the presence of a sinusoid, having a frequency we know, observed in Gaussian noise of unknown variance.

(a) Derive the frequency domain counterpart of the *CFAR* detector. Assume use of only one frequency bin in making the decision.

Note: Your answer should be degenerate. The sufficient statistic equals one regardless of the observations. Is this plausible? Why?

(b) Modify your *CFAR* receiver to consider surrounding frequency bins having no signal-related energy. Assume the noise power spectrum is constant over these bins.

(c) Contrast this detector to that derived in Prob. 5.23. How do they differ? How much penalty in performance is paid by uncertainty in the noise variance?

5.30 For the rather simplistic white-noise case {250}, detection performance is determined by the array's signal-to-noise ratio, which is a factor of M larger than that provided by a single-sensor detector. No wonder this enhancement is called the array gain. This result needs to be examined carefully.

(a) Assume that the noise is spatially white but not temporally white. Find the array gain.

(b) Assume that the array geometry is regular and linear and that the array aperture is fixed. As the sensor separation decreases, the number of sensors increases and, based on our "careful" analysis, the array gain also increases. A clever way of obtaining large array gain would be to locate a large number of sensors closely together. Would this approach work? Why or why not?

---

†This question has an interesting subtlety; be sure and consider *all* of the assumptions used to derive Eq. 5.11 {244} and what you need to assume to apply the "maximum of random variables" result.

5.31 Eq. 5.15 {255} expresses the decision rule for the optimal detector for array processing. Such an important result needs to be thoroughly understood.

(a) Translate this expression into a block diagram that depicts not only the signal processing transformations, but also the segmentation of the observations.

(b) Interpret the kernel ρ^{-1} of the dot product as a spatial whitening operation. Show that it is equivalent to a space-varying linear filtering operation.

(c) Assuming a 10-element regular linear array placed in spherically isotropic noise, find the unit-sample responses of these filters when the sensor spacing d equals first $\lambda/2$, then $3\lambda/8$.

(d) Finally, determine the threshold γ when the direction of propagation is known.

5.32 For a single signal, the optimal array processing detector's threshold ignored the sidelobe height of the array processing algorithm {255}. Consequently, using this algorithm when the number of signals is unknown can mean "detecting" sidelobes as signals. While this analysis seems intuitively correct, we do *not* yet have the details under control. One prominent flaw is that the array processing algorithm is *not* equivalent to conventional beamforming: The shading is much more complicated than scaling each sensor's output by a constant. Assume a 10-element regular linear array placed in a spherically isotropic noise field. For simplicity, assume a sinusoidal propagating signal having a wavelength of $3\lambda/8$ sampled to yield precisely P samples/period.

(a) Interpret the optimal detector as a beamformer, finding an explicit expression for the beamforming calculations.

(b) Find the beampattern for this algorithm. Discuss the algorithm's sensitivity to multiple sources.

(c) How noisy does it have to be for no sidelobe to exceed the detection threshold found in Prob. 5.31? Compute the beampattern at this signal-to-noise ratio.

5.33 In the discussion of the optimal detector for array processing {255}, we described the complexity of multiple signal problems. For *simplicity*, assume that we (magically) know that two signals having the same temporal spectral content but differing propagation directions are present in the field.

(a) Find the decision rule for the optimal spectral detector.

(b) Determine the complexity of the decision rule. In other words, how does the number of calculations vary with the number of sensors, the duration of the temporal observations, and the number of delay values to be searched?

5.34 Unfortunately, to realize the optimum array processing detector, we must know the noise variance to set the detector's threshold. Realistically, this parameter is at best unknown and, more than likely, varies slowly.

(a) Develop a *CFAR* receiver corresponding to the array processing problem for locating a sinusoidal signal using spectral techniques. Assume the detector considers only the signal's temporal frequency bin.

(b) Show that a degenerate detector results (as in Prob. 5.29) if the noise is spatially white.

(c) Assuming the conditions of Prob. 5.32, determine the beamforming operations and calculate the beampattern. Note whether the beamformer is linear or not.

Chapter 6

Estimation Theory

Continuing the search for methods of extracting information from noisy observations, this chapter describes *estimation theory*, which has the goal of *extracting from noise-corrupted observations the values of disturbance parameters (noise variance, for example), signal parameters (amplitude or propagation direction), or signal waveforms.* Estimation theory assumes that observations contain an information-bearing quantity, thereby tacitly assuming that detection-based preprocessing has been performed (in other words, do I have something in the observations worth estimating?). Conversely, detection theory often requires estimation of unknown parameters: Signal presence is assumed, parameter estimates are incorporated into the detection statistic, and consistency of observations and assumptions tested. Consequently, detection and estimation theory form a symbiotic relationship, each requiring the other to yield high-quality signal processing algorithms.

Despite a wide variety of error criteria and problem frameworks, the optimal detector is characterized by a single result: the likelihood ratio test. Surprisingly, optimal detectors thus derived are usually easy to implement, not often requiring simplification to obtain a feasible realization in hardware or software. In contrast to detection theory, no fundamental result in estimation theory exists to be summoned to attack the problem at hand. The choice of error criterion and its optimization heavily influences the form of the estimation procedure. Because of the variety of criterion-dependent estimators, arguments frequently rage about which of several optimal estimators is "better." Each procedure is optimum for its assumed error criterion; thus, the argument becomes which error criterion best describes some intuitive notion of quality. When more ad hoc, noncriterion-based procedures[†] are used, we cannot assess the quality of the resulting estimator relative to the best achievable. As shown later, bounds on the estimation error do exist, but their tightness and applicability to a given situation are always issues in assessing estimator quality. At best, estimation theory is less structured than detection theory. Detection is science, estimation art. Inventiveness coupled with an understanding of the problem (what types of errors are critically important, for example) are key elements to deciding which estimation procedure "fits" a given problem well.

[†]This governmentese phrase concisely means guessing.

6.1 Terminology in Estimation Theory

More so than detection theory, estimation theory relies on jargon to characterize the properties of estimators. Without knowing any estimation technique, let's use parameter estimation as our discussion prototype. The parameter estimation problem is to determine from a set of L observations, represented by the L-dimensional vector $\mathbf{y}$, the values of parameters denoted by the vector $\boldsymbol{\xi}$. We write the *estimate* of this parameter vector as $\widehat{\boldsymbol{\xi}}(\mathbf{y})$, where the "hat" denotes the estimate, and the functional dependence on $\mathbf{y}$ explicitly denotes the dependence of the estimate on the observations. This dependence is always present,[†] but we frequently denote the estimate compactly as $\widehat{\boldsymbol{\xi}}$. Because of the probabilistic nature of the problems considered in this chapter, a parameter estimate is itself a random vector, having its own statistical characteristics. The *estimation error* $\boldsymbol{\epsilon}(\mathbf{y})$ equals the estimate minus the actual parameter value: $\boldsymbol{\epsilon}(\mathbf{y}) = \widehat{\boldsymbol{\xi}}(\mathbf{y}) - \boldsymbol{\xi}$. It too is a random quantity and is often used in the criterion function. For example, the *mean-squared error* is given by $\mathcal{E}[\boldsymbol{\epsilon}^t\boldsymbol{\epsilon}]$; the minimum mean-squared error estimate would minimize this quantity. The mean-squared error matrix is $\mathcal{E}[\boldsymbol{\epsilon}\boldsymbol{\epsilon}^t]$; on the main diagonal, its entries are the mean-squared estimation errors for each component of the parameter vector, whereas the off-diagonal terms express the correlation between the errors. The mean-squared estimation error $\mathcal{E}[\boldsymbol{\epsilon}^t\boldsymbol{\epsilon}]$ equals the trace of the mean-squared error matrix $\text{tr}\{\mathcal{E}[\boldsymbol{\epsilon}\boldsymbol{\epsilon}^t]\}$ (see §B.3 {491}).

Bias. An estimate is said to be *unbiased* if the expected value of the estimate equals the true value of the parameter: $\mathcal{E}[\widehat{\boldsymbol{\xi}}\,|\boldsymbol{\xi}] = \boldsymbol{\xi}$. Otherwise, the estimate is said to be *biased*: $\mathcal{E}[\widehat{\boldsymbol{\xi}}\,|\boldsymbol{\xi}] \neq \boldsymbol{\xi}$. The *bias* $\mathbf{b}(\boldsymbol{\xi})$ is usually considered to be additive, so that $\mathbf{b}(\boldsymbol{\xi}) = \mathcal{E}[\widehat{\boldsymbol{\xi}}\,|\boldsymbol{\xi}] - \boldsymbol{\xi}$. When we have a biased estimate, the bias usually depends on the number of observations L. An estimate is said to be *asymptotically unbiased* if the bias tends to zero for large L: $\lim_{L\to\infty} \mathbf{b} = \mathbf{0}$. An estimate's variance equals the mean-squared estimation error *only* if the estimate is unbiased.

An unbiased estimate has a probability distribution where the mean equals the actual value of the parameter. Should the lack of bias be considered a desirable property? If many unbiased estimates are computed from statistically independent sets of observations having the same parameter value, the average of these estimates will be close to this value. This property does *not* mean that the estimate has less error than a biased one; there exist biased estimates whose mean-squared errors are smaller than unbiased ones. In such cases, the biased estimate is usually asymptotically unbiased. Lack of bias is good, but that is just one aspect of how we evaluate estimators.

Consistency. We term an estimate *consistent* if the mean-squared estimation error tends to zero as the number of observations becomes large: $\lim_{L\to\infty} \mathcal{E}[\boldsymbol{\epsilon}^t\boldsymbol{\epsilon}] = 0$. Thus, a consistent estimate must be at least asymptotically unbiased. Unbiased estimates do exist whose errors never diminish as more data are collected: Their variances remain

---

[†]Estimating the value of a parameter given no data may be an interesting problem in clairvoyance, but not in estimation theory.

nonzero no matter how much data are available. Inconsistent estimates may provide reasonable estimates when the amount of data are limited, but have the counterintuitive property that the quality of the estimate does not improve as the number of observations increases. Although appropriate in the proper circumstances (smaller mean-squared error than a consistent estimate over a pertinent range of values of L), consistent estimates are usually favored in practice.

Efficiency. As estimators can be derived in a variety of ways, their error characteristics must always be analyzed and compared. In practice, many problems and the estimators derived for them are sufficiently complicated to render analytic studies of the errors difficult, if not impossible. Instead, numerical simulation and comparison with lower bounds on the estimation error are frequently used instead to assess the estimator performance. An *efficient* estimate has a mean-squared error that equals a particular lower bound: the Cramér-Rao bound. If an efficient estimate exists (the Cramér-Rao bound is the greatest lower bound), it is optimum in the mean-squared sense: No other estimate has a smaller mean-squared error (see §6.2.4 {278} for details).

For many problems no efficient estimate exists. In such cases, the Cramér-Rao bound remains a lower bound, but its value is smaller than that achievable by any estimator. How much smaller is usually not known. However, practitioners frequently use the Cramér-Rao bound in comparisons with numerical error calculations. Another issue is the choice of mean-squared error as the estimation criterion; it may not suffice to pointedly assess estimator performance in a particular problem. Nevertheless, every problem is usually subjected to a Cramér-Rao bound computation and the existence of an efficient estimate considered.

6.2 Parameter Estimation

Determining signal parameter values or a probability distribution's parameters are the simplest estimation problems. Their fundamental utility in signal processing, much less array processing, is unquestioned. How do we estimate noise power? What is the best estimator of signal amplitude? How should array outputs be effectively combined to estimate propagation delay? Examination of useful estimators, and evaluation of their properties and performances constitute a case study of estimation problems. As expected, many of these issues are interrelated and serve to highlight the intricacies that arise in estimation theory.

All parameters of concern here have unknown values; we classify parameter estimation problems according to whether the parameter is stochastic or not. If so, then the parameter has a probability density and choosing the density, as we have said so often, narrows the problem considerably, suggesting that measurement of the parameter's density would yield something like what was assumed! If the density is not known, the parameter is termed "nonrandom," and its values range unrestricted over some interval. The resulting nonrandom-parameter estimation problem differs greatly from the random-parameter problem. We consider first the latter problem, letting ξ be a scalar parameter having the a priori (before any data are available) density $p_\xi(\xi)$. The impact of the a priori density becomes evident as various error criteria are established, and an "optimum" estimator is derived.

6.2.1 Minimum Mean-Squared Error Estimators

In terms of the densities involved in scalar random-parameter problems, the mean-squared error is given by

$$\mathcal{E}[\epsilon^2] = \iint (\xi - \widehat{\xi})^2 p_{\mathbf{y},\xi}(\mathbf{y}, \xi)\, d\mathbf{y}\, d\xi$$

where $p_{\mathbf{y},\xi}\, p(\mathbf{y}, \xi)$ is the joint density of the observations and the parameter. To minimize this integral with respect to $\widehat{\xi}$, we rewrite it using the laws of conditional probability as

$$\mathcal{E}[\epsilon^2] = \int p_{\mathbf{y}}(\mathbf{y})\, d\mathbf{y} \left(\int [\xi - \widehat{\xi}(\mathbf{y})]^2 p_{\xi|\mathbf{y}}(\xi|\mathbf{y})\, d\xi \right) d\mathbf{y}$$

The density $p_{\mathbf{y}}(\cdot)$ is nonnegative. To minimize the mean-squared error, we must minimize the inner integral for each value of $\mathbf{y}$ because the integral is weighted by a positive quantity. We focus attention on the inner integral, which is the conditional expected value of the squared estimation error. The condition, a fixed value of $\mathbf{y}$, implies that we seek that constant $[\widehat{\xi}(\mathbf{y})]$ derived from $\mathbf{y}$ that minimizes the second moment of the random parameter ξ. A well-known result from probability theory states that the minimum of $\mathcal{E}\left[(x-c)^2\right]$ occurs when the constant c equals the expected value of the random variable x (see §A.1.3 {473}). The inner integral and thereby the mean-squared error is minimized by choosing the estimator to be the conditional expected value of the parameter given the observations.

$$\boxed{\widehat{\xi}_{\text{MMSE}}(\mathbf{y}) = \mathcal{E}[\xi|\mathbf{y}]}$$

Thus, a parameter's minimum mean-squared error (*MMSE*) estimate is the parameter's a posteriori (after the observations have been obtained) expected value.

The associated conditional probability density $p_{\xi|\mathbf{y}}(\xi|\mathbf{y})$ is not often directly stated in a problem definition and must somehow be derived. In many applications, the likelihood function $p_{\mathbf{y}|\xi}(\mathbf{y}|\xi)$ and the a priori density of the parameter are a direct consequence of the problem statement. These densities can be used to find the joint density of the observations and the parameter, enabling us to use Bayes's Rule to find the a posteriori density *if* we knew the unconditional probability density of the observations.

$$p_{\xi|\mathbf{y}}(\xi|\mathbf{y}) = \frac{p_{\mathbf{y}|\xi}(\mathbf{y}|\xi)\, p_\xi(\xi)}{p_{\mathbf{y}}(\mathbf{y})}$$

This density $p_{\mathbf{y}}(\mathbf{y})$ is often difficult to determine. Be that as it may, to find the a posteriori conditional expected value, it need not be known. The numerator entirely expresses the a posteriori density's dependence on ξ; the denominator only serves as the scaling factor to yield a unit-area quantity. The expected value is the center-of-mass of the probability density and does *not* depend directly on the "weight" of the density, bypassing calculation of the scaling factor. If not, the *MMSE* estimate can be exceedingly difficult to compute.

Example

Let L statistically independent observations be obtained, each of which is expressed by $y(l) = \xi + n(l)$. Each $n(l)$ is a Gaussian random variable having zero mean and variance σ_n^2. Thus, the unknown parameter in this problem is the mean of the observations. Assume it to be a Gaussian random variable a priori (mean m_ξ and variance σ_ξ^2). The likelihood function is easily found to be

$$p_{\mathbf{y}|\xi}(\mathbf{y}|\xi) = \prod_{l=0}^{L-1} \frac{1}{\sqrt{2\pi\sigma_n^2}} \exp\left\{-\frac{1}{2}\left(\frac{y(l)-\xi}{\sigma_n}\right)^2\right\}$$

so that the a posteriori density is given by

$$p_{\xi|\mathbf{y}}(\xi|\mathbf{y}) = \frac{\dfrac{1}{\sqrt{2\pi\sigma_\xi^2}} \exp\left\{-\dfrac{1}{2}\left(\dfrac{\xi - m_\xi}{\sigma_\xi}\right)^2\right\} \displaystyle\prod_{l=0}^{L-1} \dfrac{1}{\sqrt{2\pi\sigma_n^2}} \exp\left\{-\dfrac{1}{2}\left(\dfrac{y(l)-\xi}{\sigma_n}\right)^2\right\}}{p_{\mathbf{y}}(\mathbf{y})}$$

In an attempt to find the expected value of this distribution, lump all terms that do not depend *explicitly* on the quantity ξ into a proportionality term.

$$p_{\xi|\mathbf{y}}(\xi|\mathbf{y}) \propto \exp\left\{-\frac{1}{2}\left[\frac{\sum(y(l)-\xi)^2}{\sigma_n^2} + \frac{(\xi - m_\xi)^2}{\sigma_\xi^2}\right]\right\}$$

After some manipulation, this expression can be written as

$$p_{\xi|\mathbf{y}}(\xi|\mathbf{y}) \propto \exp\left\{-\frac{1}{2\sigma^2}\left[\xi - \sigma^2\left(\frac{m_\xi}{\sigma_\xi^2} + \frac{\sum y(l)}{\sigma_n^2}\right)\right]^2\right\}$$

where σ^2 is a quantity that succinctly expresses the ratio $\sigma_n^2\sigma_\xi^2/(\sigma_n^2 + L\sigma_\xi^2)$. The form of the a posteriori density suggests that it too is Gaussian; its mean, and therefore the *MMSE* estimate of ξ, is given by

$$\widehat{\xi}_{\text{MMSE}}(\mathbf{y}) = \sigma^2\left(\frac{m_\xi}{\sigma_\xi^2} + \frac{\sum y(l)}{\sigma_n^2}\right)$$

More insight into the nature of this estimate is gained by rewriting it as

$$\widehat{\xi}_{\text{MMSE}}(\mathbf{y}) = \frac{\sigma_n^2/L}{\sigma_\xi^2 + \sigma_n^2/L} m_\xi + \frac{\sigma_\xi^2}{\sigma_\xi^2 + \sigma_n^2/L} \cdot \frac{1}{L}\sum_{l=0}^{L-1} y(l)$$

The term σ_n^2/L is the variance of the averaged observations for a given value of ξ; it expresses the squared error encountered in estimating the mean by simple averaging. If this error is much greater than the a priori variance of ξ ($\sigma_n^2/L \gg \sigma_\xi^2$), implying that the observations are noisier than the variation of the parameter, the *MMSE* estimate ignores the observations and tends to yield the a priori mean m_ξ as its value. If the averaged observations are less variable than the parameter, the second term dominates, and the average of the observations is the estimate's value. This estimate behavior between these extremes is very intuitive. The detailed form of the estimate indicates how the squared error can be minimized by a linear combination of these extreme estimates.

The conditional expected value of the estimate equals

$$\mathcal{E}[\widehat{\xi}_{\text{MMSE}}|\xi] = \frac{\sigma_n^2/L}{\sigma_\xi^2 + \sigma_n^2/L} m_\xi + \frac{\sigma_\xi^2}{\sigma_\xi^2 + \sigma_n^2/L}\xi$$

This estimate is biased because its expected value does not equal the value of the sought-after parameter. It is asymptotically unbiased as the squared measurement error σ_n^2/L tends to zero as L becomes large. The consistency of the estimator is determined by investigating the expected value of the squared error. Note that the variance of the a posteriori density is the quantity σ^2; as this quantity does not depend on $\mathbf{y}$, it also equals the unconditional variance. As the number of observations increases, this variance tends to zero. In concert with the estimate being asymptotically unbiased, the expected value of the estimation error thus tends to zero, implying that we have a consistent estimate.

6.2.2 Maximum a Posteriori Estimators

In those cases in which the expected value of the a posteriori density cannot be computed, a related but simpler estimate, the maximum a posteriori (*MAP*) estimate, can usually be evaluated. The estimate $\widehat{\xi}_{\text{MAP}}(\mathbf{y})$ equals the location of the maximum of the a posteriori density. Assuming that this maximum can be found by evaluating the derivative of the a posteriori density, the *MAP* estimate is the solution of the equation

$$\boxed{\left.\frac{\partial p_{\xi|\mathbf{y}}(\xi|\mathbf{y})}{\partial \xi}\right|_{\xi=\widehat{\xi}_{\text{MAP}}} = 0}$$

Any scaling of the density by a positive quantity that depends on $\mathbf{y}$ does not change the location of the maximum. Symbolically, $p_{\xi|\mathbf{y}} = p_{\mathbf{y}|\xi}\, p_\xi / p_{\mathbf{y}}$; the derivative does not involve the denominator, and this term can be ignored. Thus, the only quantities required to compute $\widehat{\xi}_{\text{MAP}}$ are the likelihood function and the parameter's a priori density.

Although not apparent in its definition, the *MAP* estimate does satisfy an error criterion. Define a criterion that is zero over a small range of values about $\epsilon = 0$ and a positive constant outside that range. Minimization of the expected value of this criterion with respect to $\widehat{\xi}$ is accomplished by centering the criterion function at the maximum of the density. The region having the largest area is thus "notched out," and the criterion is minimized. Whenever the a posteriori density is symmetric and unimodal, the *MAP* and *MMSE* estimates coincide. In Gaussian problems, such as the last example, this equivalence is always valid. In more general circumstances, they differ.

Example

Let the observations have the same form as the previous example, but with the modification that the parameter is now uniformly distributed over the interval $[\xi_1, \xi_2]$. The a posteriori mean cannot be computed in closed form. To obtain the *MAP* estimate, we need to find the location of the maximum of

$$p_{\mathbf{y}|\xi}(\mathbf{y}|\xi)p_\xi(\xi) = \frac{1}{\xi_2 - \xi_1}\prod_{l=0}^{L-1}\frac{1}{\sqrt{2\pi\sigma_n^2}}\exp\left\{-\frac{1}{2}\left(\frac{y(l)-\xi}{\sigma_n}\right)^2\right\}, \quad \xi_1 \le \xi \le \xi_2$$

Evaluating the logarithm of this quantity does not change the location of the maximum and simplifies the manipulations in many problems. Here, the logarithm is

$$\ln p_{\mathbf{y}|\xi}(\mathbf{y}|\xi)p_{\xi}(\xi) = -\ln(\xi_2 - \xi_1) - \sum_{l=0}^{L-1}\left(\frac{y(l)-\xi}{\sigma_n}\right)^2 + \ln C, \quad \xi_1 \le \xi \le \xi_2$$

where C is a constant with respect to ξ. Assuming that the maximum is interior to the domain of the parameter, the *MAP* estimate is found to be the sample average $\sum y(l)/L$. If the average lies outside this interval, the corresponding endpoint of the interval is the location of the maximum. To summarize,

$$\widehat{\xi}_{\mathrm{MAP}}(\mathbf{y}) = \begin{cases} \xi_1, & \sum_l y(l)/L < \xi_1 \\ \sum_l y(l)/L, & \xi_1 \le \sum_l y(l)/L \le \xi_2 \\ \xi_2, & \xi_2 < \sum_l y(l)/L \end{cases}$$

The a posteriori density is not symmetric because of the finite domain of ξ. Thus, the *MAP* estimate is not equivalent to the *MMSE* estimate, and the accompanying increase in the mean-squared error is difficult to compute. When the sample average is the estimate, the estimate is unbiased; otherwise it is biased. Asymptotically, the variance of the average tends to zero, with the consequences that the estimate is unbiased and consistent.

6.2.3 Linear Estimators

We derived the minimum mean-squared error estimator in the previous section with no constraint on the form of the estimator. Depending on the problem, the computations could be a linear function of the observations (which is always the case in Gaussian problems) or nonlinear. Deriving this estimator is often difficult, which limits its application. We consider here a variation of *MMSE* estimation by constraining the estimator to be linear while minimizing the mean-squared estimation error. Such *linear estimators* may not be optimum; the conditional expected value may be nonlinear and it *always* has the smallest mean-squared error. Despite this occasional performance deficit, linear estimators have well-understood properties, they interact well with other signal processing algorithms because of linearity, and they can always be derived, no matter what the problem.

Let the parameter estimate $\widehat{\boldsymbol{\xi}}(\mathbf{y})$ be expressed as $\mathcal{L}[\mathbf{y}]$, where $\mathcal{L}[\cdot]$ is a linear operator: $\mathcal{L}[a_1\mathbf{y}_1 + a_2\mathbf{y}_2] = a_1\,\mathcal{L}[\mathbf{y}_1] + a_2\,\mathcal{L}[\mathbf{y}_2]$, a_1, a_2 scalars. Although all estimators of this form are obviously linear, the term *linear estimator* denotes that member of this family that minimizes the mean-squared estimation error.

$$\arg\min_{\mathcal{L}[\mathbf{y}]} \mathcal{E}[\boldsymbol{\epsilon}^t\boldsymbol{\epsilon}] = \widehat{\boldsymbol{\xi}}_{\mathrm{LIN}}(\mathbf{y})$$

Because of the transformation's linearity, the theory of linear vector spaces can be fruitfully used to derive the estimator and to specify its properties. One result of that theoretical framework is the well-known *Orthogonality Principle* [120: 407–14]: The linear estimator is that particular linear transformation that yields an estimation error orthogonal to all linear transformations of the data. The orthogonality of the error to *all* linear transformations is termed the "universality constraint." This principle provides us not only with a formal definition of the linear estimator but also with the mechanism to derive it. To demonstrate this intriguing result, let $\langle\cdot,\cdot\rangle$ denote the abstract inner product between two vectors and $\|\cdot\|$ the associated norm (see §B.3 {489}).

$$\|\mathbf{x}\|^2 = \langle\mathbf{x},\mathbf{x}\rangle$$

For example, if $\mathbf{x}$ and $\mathbf{y}$ are each column matrices having only one column,[†] their inner product might be defined as $\langle\mathbf{x},\mathbf{y}\rangle = \mathbf{x}'\mathbf{y}$. Thus, the linear estimator as defined by the Orthogonality Principle must satisfy

$$\boxed{\mathcal{E}\big[\langle\widehat{\boldsymbol{\xi}}_{\mathrm{LIN}}(\mathbf{y})-\boldsymbol{\xi},\mathcal{L}[\mathbf{y}]\rangle\big]=0, \quad \text{for all linear transformations } \mathcal{L}[\cdot]} \tag{6.1}$$

To see that this principle produces the *MMSE* linear estimator, we express the mean-squared estimation error $\mathcal{E}[\boldsymbol{\epsilon}'\boldsymbol{\epsilon}] = \mathcal{E}[\|\boldsymbol{\epsilon}\|^2]$ for *any* choice of linear estimator $\widehat{\boldsymbol{\xi}}$ as

$$\begin{aligned}\mathcal{E}[\|\widehat{\boldsymbol{\xi}}-\boldsymbol{\xi}\|^2] &= \mathcal{E}[\|(\widehat{\boldsymbol{\xi}}_{\mathrm{LIN}}-\boldsymbol{\xi})-(\widehat{\boldsymbol{\xi}}_{\mathrm{LIN}}-\widehat{\boldsymbol{\xi}})\|^2]\\ &= \mathcal{E}[\|\widehat{\boldsymbol{\xi}}_{\mathrm{LIN}}-\boldsymbol{\xi}\|^2]+\mathcal{E}[\|\widehat{\boldsymbol{\xi}}_{\mathrm{LIN}}-\widehat{\boldsymbol{\xi}}\|^2]-2\,\mathcal{E}[\langle\widehat{\boldsymbol{\xi}}_{\mathrm{LIN}}-\boldsymbol{\xi},\widehat{\boldsymbol{\xi}}_{\mathrm{LIN}}-\widehat{\boldsymbol{\xi}}\rangle]\end{aligned}$$

As $\widehat{\boldsymbol{\xi}}_{\mathrm{LIN}}-\widehat{\boldsymbol{\xi}}$ is the difference of two linear transformations, it too is linear and is orthogonal to the estimation error resulting from $\widehat{\boldsymbol{\xi}}_{\mathrm{LIN}}$. As a result, the last term is zero and the mean squared estimation error is the sum of two squared norms, each of which is, of course, nonnegative. Only the second norm varies with estimator choice; we minimize the mean-squared estimation error by choosing the estimator $\widehat{\boldsymbol{\xi}}$ to be the estimator $\widehat{\boldsymbol{\xi}}_{\mathrm{LIN}}$, which sets the second term to zero.

The estimation error for the minimum mean-squared linear estimator can be calculated to some degree without knowledge of the form of the estimator. The mean-squared estimation error is given by

$$\begin{aligned}\mathcal{E}[\|\widehat{\boldsymbol{\xi}}_{\mathrm{LIN}}-\boldsymbol{\xi}\|^2] &= \mathcal{E}[\langle\widehat{\boldsymbol{\xi}}_{\mathrm{LIN}}-\boldsymbol{\xi},\widehat{\boldsymbol{\xi}}_{\mathrm{LIN}}-\boldsymbol{\xi}\rangle]\\ &= \mathcal{E}[\langle\widehat{\boldsymbol{\xi}}_{\mathrm{LIN}}-\boldsymbol{\xi},\widehat{\boldsymbol{\xi}}_{\mathrm{LIN}}\rangle]+\mathcal{E}[\langle\widehat{\boldsymbol{\xi}}_{\mathrm{LIN}}-\boldsymbol{\xi},-\boldsymbol{\xi}\rangle]\end{aligned}$$

The first term is zero because of the Orthogonality Principle. Rewriting the second term yields a general expression for the *MMSE* linear estimator's mean-squared error.

$$\boxed{\mathcal{E}[\|\boldsymbol{\epsilon}\|^2]=\mathcal{E}[\|\boldsymbol{\xi}\|^2]-\mathcal{E}[\langle\widehat{\boldsymbol{\xi}}_{\mathrm{LIN}},\boldsymbol{\xi}\rangle]}$$

This error is the difference of two terms. The first, the mean-squared value of the parameter, represents the largest value that the estimation error can be for any reasonable estimator. That error can be obtained by the estimator that ignores the data and has a

[†]There is a confusion as to what a vector is. "Matrices having one column" are colloquially termed vectors as are the field quantities such as electric and magnetic fields. "Vectors" and their associated inner products are taken to be much more general mathematical objects than these. Hence the prose in this section is rather contorted.

value of zero. The second term reduces this maximum error and represents the degree to which the estimate and the parameter agree on the average.

Note that the definition of the minimum mean-squared error *linear* estimator makes no explicit assumptions about the parameter estimation problem being solved. This property makes this kind of estimator attractive in many applications where neither the a priori density of the parameter vector nor the density of the observations is known precisely. Linear transformations, however, are homogeneous: A zero-valued input yields a zero output. Thus, the linear estimator is especially pertinent to those problems where the expected value of the parameter is zero. If the expected value is nonzero, the linear estimator would not necessarily yield the best result (see Prob. 6.9).

Example

Express the first example {270} in vector notation so that the observation vector is written as

$$\mathbf{y} = \mathbf{A}\xi + \mathbf{n}$$

where the matrix $\mathbf{A}$ has the form $\mathbf{A} = \text{col}[1, \ldots, 1]$. The expected value of the parameter is zero. The linear estimator has the form $\widehat{\xi}_{\text{LIN}} = \mathbf{L}\mathbf{y}$, where $\mathbf{L}$ is a $1 \times L$ matrix. The Orthogonality Principle states that the linear estimator satisfies

$$\mathcal{E}[(\mathbf{L}\mathbf{y} - \xi)'\mathbf{M}\mathbf{y}] = 0, \quad \text{for all } 1 \times L \text{ matrices } \mathbf{M}$$

To use the Orthogonality Principle to derive an equation implicitly specifying the linear estimator, the "for all linear transformations" phrase must be interpreted. Usually, the quantity specifying the linear transformation must be removed from the constraining inner product by imposing a very stringent but equivalent condition. In this example, this phrase becomes one about matrices. The elements of the matrix $\mathbf{M}$ can be such that each element of the observation vector multiplies each element of the estimation error. Thus, in this problem the Orthogonality Principle means that the expected value of the matrix consisting of all pairwise products of these elements must be zero.

$$\mathcal{E}[(\mathbf{L}\mathbf{y} - \xi)\mathbf{y}'] = \mathbf{0}$$

Thus, two terms must equal each other: $\mathcal{E}[\mathbf{L}\mathbf{y}\mathbf{y}'] = \mathcal{E}[\xi\mathbf{y}']$. The second term equals $\mathcal{E}[\xi^2]\mathbf{A}'$ as the additive noise and the parameter are assumed to be statistically independent quantities. The quantity $\mathcal{E}[\mathbf{y}\,\mathbf{y}']$ in the first term is the correlation matrix of the observations, which is given by $\mathbf{A}\mathbf{A}'\,\mathcal{E}[\xi^2] + \mathbf{K}_n$. Here, $\mathbf{K}_n$ is the noise covariance matrix, and $\mathcal{E}[\xi^2]$ is the parameter's variance. The quantity $\mathbf{A}\mathbf{A}'$ is a $L \times L$ matrix with each element equaling 1. The noise vector has independent components; the covariance matrix thus equals $\sigma_n^2\mathbf{I}$. The equation that $\mathbf{L}$ must satisfy is therefore given by

$$[\mathbf{L}_1 \cdots \mathbf{L}_L] \cdot \begin{bmatrix} \sigma_n^2 + \sigma_\xi^2 & \sigma_\xi^2 & \cdots & \sigma_\xi^2 \\ \sigma_\xi^2 & \sigma_n^2 + \sigma_\xi^2 & \ddots & \vdots \\ \vdots & \ddots & \ddots & \sigma_\xi^2 \\ \sigma_\xi^2 & \cdots & \sigma_\xi^2 & \sigma_n^2 + \sigma_\xi^2 \end{bmatrix} = [\,\sigma_\xi^2 \quad \cdots \quad \sigma_\xi^2\,]$$

The components of $\mathbf{L}$ are equal and are given by $\mathbf{L}_i = \sigma_\xi^2/(\sigma_n^2 + L\sigma_\xi^2)$. Thus, the minimum mean-squared error linear estimator has the form

$$\widehat{\xi}_{\text{LIN}}(\mathbf{y}) = \frac{\sigma_\xi^2}{\sigma_\xi^2 + \sigma_n^2/L} \frac{1}{L} \sum_l y(l)$$

Note that this result equals the minimum mean-squared error estimate derived earlier under the condition that $\mathcal{E}[\xi] = 0$. Mean-squared error, linear estimators, and Gaussian problems are intimately related to each other. The linear minimum mean-squared error solution to a problem is optimal if the underlying distributions are Gaussian.

6.2.4 Maximum Likelihood Estimators

When the a priori density of a parameter is not known or the parameter itself is inconveniently described as a random variable, techniques must be developed that make no presumption about the relative possibilities of parameter values. Lacking this knowledge, we can expect the error characteristics of the resulting estimates to be worse than those that can use it.

The maximum likelihood estimate $\widehat{\xi}_{\text{ML}}(\mathbf{y})$ of a nonrandom parameter is, simply, that value that maximizes the likelihood function (the a priori density of the observations). Assuming that the maximum can be found by evaluating a derivative, $\widehat{\xi}_{\text{ML}}(\mathbf{y})$ is defined by

$$\boxed{\left.\frac{\partial p_{\mathbf{y}|\xi}(\mathbf{y}|\xi)}{\partial \xi}\right|_{\xi=\widehat{\xi}_{\text{ML}}} = 0}$$

The logarithm of the likelihood function may also be used in this maximization.

Example

Let $y(l)$ be a sequence of independent, identically distributed Gaussian random variables having an unknown mean ξ but a known variance σ_n^2. Often, we cannot assign a probability density to a parameter of a random variable's density; we simply do not know what the parameter's value is. Maximum likelihood estimates are often used in such problems. In the specific case here, the derivative of the logarithm of the likelihood function equals

$$\frac{\partial \ln p_{\mathbf{y}|\xi}(\mathbf{y}|\xi)}{\partial \xi} = \frac{1}{\sigma_n^2}\sum_{l=0}^{L-1}[y(l) - \xi]$$

The solution of this equation is the maximum likelihood estimate, which equals the sample average.

$$\widehat{\xi}_{\text{ML}} = \frac{1}{L}\sum_{l=0}^{L-1} y(l)$$

The expected value of this estimate $\mathcal{E}[\widehat{\xi}_{\text{ML}}|\xi]$ equals the actual value ξ, showing that the maximum likelihood estimate is unbiased. The mean-squared error equals σ_n^2/L, and we infer that this estimate is consistent.

Parameter vectors. The maximum likelihood procedure (as well as the others being discussed) can be easily generalized to situations in which more than one parameter must be estimated. Letting $\boldsymbol{\xi}$ denote the parameter vector, the likelihood function is now expressed as $p_{\mathbf{y}|\boldsymbol{\xi}}(\mathbf{y}|\boldsymbol{\xi})$. The maximum likelihood estimate $\widehat{\boldsymbol{\xi}}_{\text{ML}}$ of the parameter vector is given by the location of the maximum of the likelihood function (or equivalently of its logarithm). Using derivatives, the calculation of the maximum likelihood estimate becomes

$$\boxed{\nabla_{\boldsymbol{\xi}} \ln p_{\mathbf{y}|\boldsymbol{\xi}}(\mathbf{y}|\boldsymbol{\xi})\big|_{\boldsymbol{\xi}=\widehat{\boldsymbol{\xi}}_{\text{ML}}} = \mathbf{0}}$$

where $\nabla_{\boldsymbol{\xi}}$ denotes the gradient with respect to the parameter vector (see §C.1 {500}). This equation means that we must estimate all of the parameters *simultaneously* by setting the partial of the likelihood function with respect to *each* parameter to zero. Given P parameters, we must solve in most cases a set of P nonlinear, simultaneous equations to find the maximum likelihood estimates.

Example

Let's extend the previous example to the situation in which neither the mean nor the variance of a sequence of independent Gaussian random variables is known. The likelihood function is, in this case,

$$p_{\mathbf{y}|\boldsymbol{\xi}}(\mathbf{y} \mid \boldsymbol{\xi}) = \prod_{l=0}^{L-1} \frac{1}{\sqrt{2\pi\xi_2}} \exp\left\{-\frac{1}{2\xi_2}[y(l) - \xi_1]^2\right\}$$

Evaluating the partial derivatives of the logarithm of this quantity, we find the following set of two equations to solve for ξ_1, representing the mean, and ξ_2, representing the variance.†

$$\frac{1}{\xi_2}\sum_{l=0}^{L-1}[y(l) - \xi_1] = 0$$

$$-\frac{L}{2\xi_2} + \frac{1}{2\xi_2^2}\sum_{l=0}^{L-1}[y(l) - \xi_1]^2 = 0$$

The solution of this set of equations is easily found to be

$$\widehat{\xi}_1^{\text{ML}} = \frac{1}{L}\sum_{l=0}^{L-1} y(l)$$

$$\widehat{\xi}_2^{\text{ML}} = \frac{1}{L}\sum_{l=0}^{L-1}\left(y(l) - \widehat{\xi}_1^{\text{ML}}\right)^2$$

The expected value of $\widehat{\xi}_1^{\text{ML}}$ equals the actual value of ξ_1; thus, this estimate is unbiased. In contrast, the expected value of the estimate of the variance equals $\xi_2 \cdot (L-1)/L$. The estimate of the variance is biased, but asymptotically unbiased. This bias can be removed by replacing the normalization of L in the averaging computation for $\widehat{\xi}_2^{\text{ML}}$ by $L-1$.

Cramér-Rao bound. The mean-squared estimation error for *any* estimate of a nonrandom parameter has a lower bound, the *Cramér-Rao bound* [41: 474–77], which defines the ultimate accuracy of *any* estimation procedure. This lower bound, as shown later, is intimately related to the maximum likelihood estimator.

We seek a "bound" on the mean-squared error matrix $\mathbf{M}$ defined to be

$$\mathbf{M} = \mathcal{E}[(\widehat{\boldsymbol{\xi}} - \boldsymbol{\xi})(\widehat{\boldsymbol{\xi}} - \boldsymbol{\xi})^t] = \mathcal{E}[\boldsymbol{\epsilon}\boldsymbol{\epsilon}^t]$$

†The variance rather than the standard deviation is represented by ξ_2. The mathematics is messier, and the estimator has less attractive properties in the latter case. Prob. 6.5 illustrates this point.

A matrix is "lower bounded" by a second matrix if the difference between the two is a nonnegative definite matrix. Define the column matrix $\mathbf{x}$ to be

$$\mathbf{x} = \begin{bmatrix} \widehat{\boldsymbol{\xi}} - \boldsymbol{\xi} - \mathbf{b}(\boldsymbol{\xi}) \\ \nabla_{\boldsymbol{\xi}} \ln p_{\mathbf{y}|\boldsymbol{\xi}}(\mathbf{y} \mid \boldsymbol{\xi}) \end{bmatrix}$$

where $\mathbf{b}(\boldsymbol{\xi})$ denotes the column matrix of estimator biases. To derive the Cramér-Rao bound, evaluate $\mathcal{E}[\mathbf{x}\mathbf{x}^t]$.

$$\mathcal{E}[\mathbf{x}\mathbf{x}^t] = \begin{bmatrix} \mathbf{M} - \mathbf{b}\mathbf{b}^t & \mathbf{I} + \nabla_{\boldsymbol{\xi}} \mathbf{b} \\ (\mathbf{I} + \nabla_{\boldsymbol{\xi}} \mathbf{b})^t & \mathbf{F} \end{bmatrix}$$

The matrix $\mathbf{F}$ is the *Fisher information matrix*

$$\boxed{\mathbf{F} = \mathcal{E}\left[\left(\nabla_{\boldsymbol{\xi}} \ln p_{\mathbf{y}|\boldsymbol{\xi}}(\mathbf{y} \mid \boldsymbol{\xi})\right)\left(\nabla_{\boldsymbol{\xi}} \ln p_{\mathbf{y}|\boldsymbol{\xi}}(\mathbf{y} \mid \boldsymbol{\xi})\right)^t\right]}$$

and $\nabla_{\boldsymbol{\xi}} \mathbf{b}$ represents the matrix of partial derivatives of the bias $[\partial b_i / \partial \xi_j]$. The matrix $\mathcal{E}[\mathbf{x}\mathbf{x}^t]$ is nonnegative definite because it is a correlation matrix. Thus, for any column matrix $\boldsymbol{\alpha}$, the quadratic form $\boldsymbol{\alpha}^t \, \mathcal{E}[\mathbf{x}\mathbf{x}^t]\boldsymbol{\alpha}$ is nonnegative. Choose a form for $\boldsymbol{\alpha}$ that simplifies the quadratic form. A convenient choice is

$$\boldsymbol{\alpha} = \begin{bmatrix} \boldsymbol{\beta} \\ -\mathbf{F}^{-1}\left(\mathbf{I} + \nabla_{\boldsymbol{\xi}} \mathbf{b}\right)^t \boldsymbol{\beta} \end{bmatrix}$$

where $\boldsymbol{\beta}$ is an arbitrary column matrix. The quadratic form becomes in this case

$$\boldsymbol{\alpha}^t \, \mathcal{E}[\mathbf{x}\mathbf{x}^t]\boldsymbol{\alpha} = \boldsymbol{\beta}^t \left[\mathbf{M} - \mathbf{b}\mathbf{b}^t - \left(\mathbf{I} + \nabla_{\boldsymbol{\xi}} \mathbf{b}\right)\mathbf{F}^{-1}\left(\mathbf{I} + \nabla_{\boldsymbol{\xi}} \mathbf{b}\right)^t\right]\boldsymbol{\beta}$$

As this quadratic form must be nonnegative, the matrix expression enclosed in brackets must be nonnegative definite. We thus obtain the well-known Cramér-Rao bound on the mean-square error matrix.

$$\boxed{\mathcal{E}[\boldsymbol{\epsilon}\boldsymbol{\epsilon}^t] \geq \mathbf{b}(\boldsymbol{\xi})\mathbf{b}^t(\boldsymbol{\xi}) + \left(\mathbf{I} + \nabla_{\boldsymbol{\xi}} \mathbf{b}\right)\mathbf{F}^{-1}\left(\mathbf{I} + \nabla_{\boldsymbol{\xi}} \mathbf{b}\right)^t}$$

This form for the Cramér-Rao bound does *not* mean that each term in the matrix of squared errors is greater than the corresponding term in the bounding matrix. As stated earlier, this expression means that the difference between these matrices is nonnegative definite. For a matrix to be nonnegative definite, each term on the main diagonal must be nonnegative. The elements of the main diagonal of $\mathcal{E}[\boldsymbol{\epsilon}\boldsymbol{\epsilon}^t]$ are the squared errors of the estimate of the individual parameters. Thus, for each parameter, the mean-squared estimation error can be no smaller than

$$\mathcal{E}[(\widehat{\xi}_i - \xi_i)^2] \geq b_i^2(\boldsymbol{\xi}) + \left[\left(\mathbf{I} + \nabla_{\boldsymbol{\xi}} \mathbf{b}\right)\mathbf{F}^{-1}\left(\mathbf{I} + \nabla_{\boldsymbol{\xi}} \mathbf{b}\right)^t\right]_{ii}$$

This bound simplifies greatly if the estimator is unbiased ($\mathbf{b} = \mathbf{0}$). In this case, the Cramér-Rao bound becomes

$$\mathcal{E}[(\widehat{\xi}_i - \xi_i)^2] \geq \left[\mathbf{F}^{-1}\right]_{ii}$$

Thus, the mean-squared error for each parameter in a multiple-parameter, unbiased-estimator problem can be no smaller than the corresponding diagonal term in the *inverse* of the Fisher information matrix. In such problems, the estimate's error characteristics of any parameter become intertwined with the other parameters in a complicated way. Any estimator satisfying the Cramér-Rao bound with equality is said to be *efficient.*

Example

Let's evaluate the Cramér-Rao bound for the example we have been discussing: the estimation of the mean and variance of a length L sequence of statistically independent Gaussian random variables. Let the estimate of the mean ξ_1 be the sample average $\widehat{\xi}_1 = \sum y(l)/L$; as shown in the last example, this estimate is unbiased. Let the estimate of the variance ξ_2 be the unbiased estimate $\widehat{\xi}_2 = \left[\sum\left(y(l) - \widehat{\xi}_1\right)^2\right]/(L-1)$. Each term in the Fisher information matrix $\mathbf{F}$ is given by the expected value of the paired products of derivatives of the logarithm of the likelihood function.

$$F_{ij} = \mathcal{E}\left[\frac{\partial \ln p_{\mathbf{y}|\boldsymbol{\xi}}(\mathbf{y} \mid \boldsymbol{\xi})}{\partial \xi_i} \frac{\partial \ln p_{\mathbf{y}|\boldsymbol{\xi}}(\mathbf{y} \mid \boldsymbol{\xi})}{\partial \xi_j}\right]$$

The logarithm of the likelihood function is

$$\ln p_{\mathbf{y}|\boldsymbol{\xi}}(\mathbf{y} \mid \boldsymbol{\xi}) = -\frac{L}{2} \ln 2\pi \xi_2 - \frac{1}{2\xi_2} \sum_{l=0}^{L-1} [y(l) - \xi_1]^2$$

and its partial derivatives are

$$\frac{\partial \ln p_{\mathbf{y}|\boldsymbol{\xi}}(\mathbf{y} \mid \boldsymbol{\xi})}{\partial \xi_1} = \frac{1}{\xi_2} \sum_{l=0}^{L-1} [y(l) - \xi_1] \tag{6.2a}$$

$$\frac{\partial \ln p_{\mathbf{y}|\boldsymbol{\xi}}(\mathbf{y} \mid \boldsymbol{\xi})}{\partial \xi_2} = -\frac{L}{2\xi_2} + \frac{1}{2\xi_2^2} \sum_{l=0}^{L-1} [y(l) - \xi_1]^2 \tag{6.2b}$$

The Fisher information matrix has the surprisingly simple form

$$\mathbf{F} = \begin{bmatrix} \frac{L}{\xi_2} & 0 \\ 0 & \frac{L}{2\xi_2^2} \end{bmatrix}$$

Its inverse is also a diagonal matrix with the elements on the main diagonal equaling the reciprocal of those in the original matrix. Because of the zero-valued off-diagonal entries in the Fisher information matrix, the errors between the corresponding estimates are not interdependent. In this problem, the mean-squared estimation errors can be no smaller than

$$\mathcal{E}[(\widehat{\xi}_1 - \xi_1)^2] \geq \frac{\xi_2}{L}$$

$$\mathcal{E}[(\widehat{\xi}_2 - \xi_2)^2] \geq \frac{2\xi_2^2}{L}$$

Note that *nowhere* in the preceding example did the form of the estimator enter into the computation of the bound. The only quantity used in the computation of the Cramér-Rao bound is the logarithm of the likelihood function, which is a consequence of the problem statement, not how it is solved. *Only in the case of unbiased estimators is the bound independent of the estimators used.*† Because of this property, the Cramér-Rao bound is frequently used to assess the performance limits that can be obtained with an unbiased estimator in a particular problem. When bias is present, the exact form of the estimator's bias explicitly enters the computation of the bound. All too frequently, the unbiased form is used in situations in which the *existence* of an unbiased estimator can be questioned. As we shall see, one such problem is time-delay estimation, presumably of some importance to the reader. This misapplication of the unbiased Cramér-Rao arises from desperation: The estimator is so complicated and nonlinear that computing the bias is nearly impossible. As shown in Prob. 6.6, biased estimators can yield mean-squared errors smaller as well as larger than the unbiased version of the Cramér-Rao bound. Consequently, desperation can yield misinterpretation when a general result is misapplied.

In the single parameter estimation problem, the Cramér-Rao bound incorporating bias has the well-known form‡

$$\boxed{\mathcal{E}[\epsilon^2] \geq b^2 + \frac{\left(1 + \frac{db}{d\xi}\right)^2}{\mathcal{E}\left[\left(\frac{\partial \ln p_{\mathbf{y}|\xi}(\mathbf{y} \mid \xi)}{\partial \xi}\right)^2\right]}}$$

Note that the sign of the bias's derivative determines whether this bound is larger or potentially smaller than the unbiased version, which is obtained by setting the bias term to 0. The denominator can be equivalently expressed as¶

$$\mathcal{E}\left[\left(\frac{\partial \ln p_{\mathbf{y}|\xi}(\mathbf{y} \mid \xi)}{\partial \xi}\right)^2\right] = -\mathcal{E}\left[\frac{\partial^2 \ln p_{\mathbf{y}|\xi}(\mathbf{y} \mid \xi)}{\partial \xi^2}\right]$$

In many circumstances, the right side is more easily evaluated than the left.

Efficiency. An interesting question arises: When, if ever, is the bound satisfied with equality? Recalling the details of the derivation of the bound, equality results when the quantity $\mathcal{E}[\boldsymbol{\alpha}^t\mathbf{x}\mathbf{x}^t\boldsymbol{\alpha}]$ equals 0. As this quantity is the expected value of the square of $\boldsymbol{\alpha}^t\mathbf{x}$, it can only equal zero if $\boldsymbol{\alpha}^t\mathbf{x} = 0$. Substituting in the form of the column matrices $\boldsymbol{\alpha}$ and $\mathbf{x}$, equality in the Cramér-Rao bound results whenever

$$\boxed{\nabla_{\boldsymbol{\xi}} \ln p_{\mathbf{y}|\boldsymbol{\xi}}(\mathbf{y} \mid \boldsymbol{\xi}) = \left[\mathbf{I} + (\nabla_{\boldsymbol{\xi}}\, \mathbf{b})^t\right]^{-1} \mathbf{F}[\widehat{\boldsymbol{\xi}}(\mathbf{y}) - \boldsymbol{\xi} - \mathbf{b}]} \tag{6.3}$$

†That's why we assumed in the example that we used an unbiased estimator for the variance.

‡Note that this bound differs somewhat from that originally given by Cramér [41: 480]; his derivation ignores the additive bias term $\mathbf{b}\mathbf{b}^t$.

¶This expression does not imply that the square of the first derivative equals the negative of the second derivative. Only when we evaluate the expected value of each does equality emerge.

This complicated expression means that only if estimation problems (as expressed by the a priori density) have the form of the right side of this equation can the mean-squared estimation error equal the Cramér-Rao bound. In particular, the gradient of the log likelihood function can *only* depend on the observations through the estimator. In all other problems, the Cramér-Rao bound is a lower bound but not a tight one; *no* estimator can have error characteristics that equal it. In such cases, we have limited insight into ultimate limitations on estimation error size with the Cramér-Rao bound. In contrast, consider the case in which the estimator is unbiased ($\mathbf{b} = \mathbf{0}$). In addition, note the maximum likelihood estimate occurs when the gradient of the logarithm of the likelihood function equals 0: $\nabla_{\boldsymbol{\xi}} \ln p_{\mathbf{y}|\boldsymbol{\xi}}(\mathbf{y} \mid \boldsymbol{\xi}) = 0$ when $\boldsymbol{\xi} = \widehat{\boldsymbol{\xi}}_{\mathrm{ML}}$. In this case, the condition for equality in the Cramér-Rao bound becomes

$$\mathbf{F}[\widehat{\boldsymbol{\xi}} - \widehat{\boldsymbol{\xi}}_{\mathrm{ML}}] = \mathbf{0}$$

As the Fisher information matrix is positive definite, we conclude that if the estimator equals the maximum likelihood estimator, equality in the Cramér-Rao bound can be achieved. To summarize, if the Cramér-Rao bound *can* be satisfied with equality, *only* the maximum likelihood estimate achieves it. To use estimation-theoretic terminology, *if an efficient estimate exists, it is the maximum likelihood estimate.* This result stresses the importance of maximum likelihood estimates, despite the seemingly ad hoc manner by which they are defined.

Example

Consider the Gaussian example being examined so frequently in this section. The components of the gradient of the logarithm of the likelihood function were given earlier by Eq. 6.2 {278}. These expressions can be rearranged to reveal

$$\begin{bmatrix} \dfrac{\partial \ln p_{\mathbf{y}|\boldsymbol{\xi}}(\mathbf{y} \mid \boldsymbol{\xi})}{\partial \xi_1} \\ \dfrac{\partial \ln p_{\mathbf{y}|\boldsymbol{\xi}}(\mathbf{y} \mid \boldsymbol{\xi})}{\partial \xi_2} \end{bmatrix} = \begin{bmatrix} \frac{L}{\xi_2}\left[\left(\frac{1}{L}\sum_l y(l)\right) - \xi_1\right] \\ -\frac{L}{2\xi_2} + \frac{1}{2\xi_2^2}\sum_l [y(l) - \xi_1]^2 \end{bmatrix}$$

The first component, which corresponds to the estimate of the mean, *is* expressed in the form required for the existence of an efficient estimate. The second component, the partial with respect to the variance ξ_2, *cannot* be rewritten in a similar fashion. No unbiased, efficient estimate of the variance exists in this problem. The mean-squared error of the variance's unbiased estimate equals $2\xi_2^2/(L-1)^2$, which is strictly greater than the Cramér-Rao bound of $2\xi_2^2/L^2$. As no unbiased estimate of the variance can have a mean-squared error equal to the Cramér-Rao bound (no efficient estimate exists for the variance in the Gaussian problem), we might be tempted to presume that the closeness of our unbiased estimator's mean-squared error to the bound implies that the estimator possesses the smallest squared error. This presumption is without foundation and may be incorrect.

Properties of maximum likelihood estimators. The maximum likelihood estimate is the most used estimation technique for nonrandom parameters, not only because of its close linkage to the Cramér-Rao bound but also because it has desirable asymptotic properties in the context of *any* problem [41: 500–6].

- *The maximum likelihood estimate is at least asymptotically unbiased.* It may be unbiased for any number of observations (as in the estimation of the mean of a sequence of independent random variables) for some problems.
- *The maximum likelihood estimate is consistent.*
- *The maximum likelihood estimate is asymptotically efficient.* As more and more data are incorporated into an estimate, the Cramér-Rao bound accurately projects the best attainable error and the maximum likelihood estimate has those optimal characteristics.
- *Asymptotically, the maximum likelihood estimate is distributed as a Gaussian random variable.*

Most would agree that a "good" estimator should have these properties. What these results do not provide is an assessment of how many observations are needed for the asymptotic results to apply to some specified degree of precision. Consequently, they should be used with caution; for instance, some other estimator may have a smaller mean-squared error than the maximum likelihood for a modest number of observations.

6.2.5 Detection Probability Estimators

Analytic calculation of the performance probabilities for complex detection algorithms can be virtually impossible. These difficulties increase when amplitude and time quantization may be a limiting factor. Rather than analytically computing the probability that a random variable exceeds a threshold, we often must resort to simulating the array processing system and estimating performance probabilities from simulation results.

Assume the array processing algorithm results in the output z, which is then compared with a threshold γ to determine, for instance, the presence of a propagating signal. Given a sequence of statistically independent outputs z_l, $l = 0, \ldots, L-1$, how can we estimate $p = \Pr[z > \gamma]$? Let the binary-valued random variable b_l denote when z_l exceeds the threshold: $b_l = I(z_l > \gamma)$, where $I(\cdot)$ is the indicator function.[†] The number of times the threshold is exceeded thus equals $\sum_l b_l$ and has a binomial probability distribution.

$$\Pr\left[\textstyle\sum_l b_l = n\right] = \binom{L}{n} p^n (1-p)^{L-n}$$

We can easily show that the maximum likelihood estimate of the probability p equals

$$\hat{p}_{\mathrm{ML}} = \frac{1}{L}\sum_{l=0}^{L-1} b_l$$

This estimate is unbiased and has variance $p(1-p)/L$.

We need to determine how many simulation trials L are required for this estimate (or any other) to achieve a specified level of precision. For positive-valued quantities, such

[†] The indicator function equals 1 over a set and 0 over the set's complement. In this application, the indicator function is defined as $I(x > x_0) = \begin{cases} 1, & x > x_0 \\ 0, & x \le x_0 \end{cases}$.

as probabilities, a measure of the percentage error known as the *coefficient of variation* is frequently employed. This quantity simply equals the standard deviation of the estimate divided by its expected value; for the problem at hand, the coefficient of variation equals $\sqrt{(1-p)/Lp}$. When p is very small, specifying a fixed level of precision means that the number of simulation trials must be inversely related to the probability p. For example, to achieve 10% precision, we must have (approximately) $100/p$ trials; when p is small, we therefore require an *enormous* number of trials, tasking computational resources, random number generators, and our patience!†

A class of alternative estimation procedures, which falls under the category of *importance sampling methods*, provides effective alternatives to classic estimation procedures. As described below, these procedures are ad hoc, having not been derived from minimizing an error criterion.‡ They have been developed over the years by people having a deep understanding of the problem and thereby creating estimates that have provably superior performance when compared with the maximum likelihood estimate. Importance sampling techniques illustrate the often-stated "art" of estimation, which is often contrasted with the "science" of detection (the likelihood ratio provides best performance in detection problems across a wide variety of criteria).

The key insight is relating the few times the threshold is crossed when p is small to the estimate's relatively large variance. Note that when p is near $1/2$, the coefficient of variation approximately equals $1/\sqrt{L}$, which does not vary with p. If we could somehow produce, in a controlled way, threshold crossings about half the time and then correct the resulting probability estimate, reduced variance for a given number of trials might be achieved. The importance sampling estimate $\hat{p}_{\text{IS}}$ of a detection probability is given by

$$\boxed{\hat{p}_{\text{IS}} = \frac{1}{L_{\text{IS}}} \sum_{l=0}^{L_{\text{IS}}-1} c_l \tilde{b}_l} \tag{6.4}$$

where the binary-valued variable $\tilde{b}_l$ equals $I(\tilde{z}_l > \gamma)$, with $\tilde{z}_l$ a to-be-determined random variable related to z_l, and c_l equals a weighting factor given by $p_z(\tilde{z}_l)/p_{\tilde{z}}(\tilde{z}_l)$. This estimate is unbiased, with the expected value of $\hat{p}_{\text{IS}}$ equaling p, while its variance equals

$$\mathcal{V}[\hat{p}_{\text{IS}}] = \frac{1}{L_{\text{IS}}} \left\{ \int_\gamma^\infty \left[\frac{p_z(\alpha)}{p_{\tilde{z}}(\alpha)} \right]^2 p_{\tilde{z}}(\alpha)\, d\alpha - p^2 \right\}$$

Equating the variance produced by the maximum likelihood estimate with that produced by importance sampling, we find that the so-called *importance sampling gain* Γ_{IS}, defined to be the ratio L/L_{IS}, equals

$$\Gamma_{\text{IS}} = \frac{p - p^2}{\bar{c} - p^2}$$

†At this point, we should note that the maximum likelihood estimate's guarantee of consistency did not state how hard we might have to work to achieve acceptably accurate answers. Estimating of detection probabilities is a prime example of the necessity to read the fine print.

‡We should note that maximum likelihood estimates were similarly derived; maximum likelihood estimates are not guaranteed to minimize *any* error criterion. Only when efficient estimates exist can we equate the maximum likelihood technique's performance with minimizing mean-squared error.

where $\bar{c}$ represents the integral in the expression for the importance sampling variance.[†] If we can find a random variable having a probability density so that $\bar{c} < p$, this gain exceeds unity, and fewer simulation trials are needed by importance sampling technique than by classic maximum likelihood.

The optimum choice for the random variable $\tilde{z}$ can be determined, but employing it means knowing p, which we are trying to find. Suppose $p_{\tilde{z}}(\tilde{z}) \propto p_z(\tilde{z})I(\tilde{z} > \gamma)$, which corresponds to the truncated density of the processed output z. The rub is that the constant of proportionality equals $1/p$; if we knew p, we would not need fancy estimation schemes. Pursuing the claim that this choice is optimum, note that $\bar{c}$ equals p^2 in this case, which means that the variance of the importance sampling estimate equals 0 and the gain is infinite. With this optimum, but unrealizable choice for $p_{\tilde{z}}(\cdot)$, crossings of the threshold *always* occur, and we can (theoretically) correct for $\tilde{b}_l$ always equaling 1 to yield a very accurate estimate. Because we must know p to produce this random variable and to calculate the weight c_l, this optimum answer is useless.[‡] The basic idea is correct, however: If we can somehow find a random variable $\tilde{z}$ related to z that causes $\tilde{b}$ to equal 1 more often, the importance sampling algorithm exemplified by Eq. 6.4 yields accurate estimates with fewer trials than maximum likelihood. The art of importance sampling is choosing the density for the random variable $\tilde{z}$ to yield significant importance sampling gain.

Example

Consider a hypothesis testing problem where one of the hypothesized densities for the observations is Gaussian: $z_l \sim \mathcal{N}(0, \sigma^2)$. We seek an importance sampling method to estimate $p = \Pr[z_l > \gamma]$. We artfully choose $\tilde{z}_l \sim \mathcal{N}(\gamma, \sigma^2)$: The mean of the Gaussian density now equals the threshold, thereby causing $\tilde{b}_l$ to equal 1 with probability 1/2. To estimate p using importance sampling, we simulate random variables having a Gaussian distribution having mean γ (rather than the hypothesized value of 0) and compare it to the threshold γ. If the threshold is exceeded, we incorporate c_l into the average rather than unity. The weight is given by $c_l = \exp\{-(2\gamma\tilde{z}_l - \gamma^2)/2\sigma^2\}$. Averaging the results yields $\hat{p}_{\text{IS}}$.

To analyze how many simulation trials are needed, we find that $\bar{c} = Q(2\gamma/\sigma)\exp\{\gamma^2/\sigma^2\}$. Employing the upper and lower bounds for $Q(\cdot)$ given in §A.1 {477} and assuming p^2 can be ignored with respect to p, we can bound the importance sampling gain.

$$\frac{1}{2\left[1+\left(\frac{\gamma}{\sigma}\right)^2\right]}\exp\{\gamma^2/\sigma^2\} \le \Gamma_{\text{IS}} \le \frac{1+\left(\frac{2\gamma}{\sigma}\right)^2}{2}\exp\{\gamma^2/\sigma^2\}$$

These bounds suggest we can obtain huge gains, and many fewer importance sampling trials are needed when the ratio γ/σ exceeds unity. For example, if this ratio equals 5, the importance sampling gain is at least 7×10^{10}. As $Q(5) = 2.9 \times 10^{-7}$, achieving 10% error with maximum likelihood techniques would require 3.5×10^8 simulation trials. The importance sampling gain translates into estimating the required probability to the same precision with only 200 trials. Such reductions in simulation computation cannot be ignored; the smaller the probability, the greater the gain.

[†] Simple manipulations show that this integral, which equals $\mathcal{E}_{\tilde{z}}\left[(c_l b_l)^2\right]$, becomes $\mathcal{E}_z[c_l b_l]$, the expected value of $c_l b_l$ with respect to the density for z. This fact inspires the notation $\bar{c}$ for this quantity.

[‡] This result demonstrates that seeking optimal answers can yield truly utopian results.

To apply the importance sampling idea to array processing problems, the decision portions of the algorithm remain fixed, whereas the signal amplitude is modified to accomplish the shift in mean. If the system is very complicated, applying importance sampling is limited by our ability to know the density of z, the array's output. This quantity must be known so that the weight c_l can be calculated.

6.3 Signal Parameter Estimation

One extension of parametric estimation theory necessary for its application to array processing is the estimation of signal parameters. We assume that we observe a signal $s(l, \boldsymbol{\xi})$, whose characteristics are known save a few parameters $\boldsymbol{\xi}$, in the presence of noise. Signal parameters, such as amplitude, time origin, and frequency if the signal is sinusoidal, must be determined in some way. In many cases of interest, we would find it difficult to justify a particular form for the unknown parameters' a priori density. For example, when a source's distance from an array determines the received signal's amplitude because of propagation effects, just what would the amplitude's a priori probability distribution be? Because of such uncertainties, the minimum mean-squared error and maximum a posteriori estimators *cannot* be used in many cases. The minimum mean-squared error *linear* estimator does not require this density, but it is most fruitfully used when the unknown parameter appears in the problem in a linear fashion (such as signal amplitude as we shall see).

6.3.1 Linear Minimum Mean-Squared Error Estimator

The only parameter that is linearly related to a signal is the amplitude. Consider, therefore, the problem where the observations at an array's output are modeled as

$$y(l) = \xi s(l) + n(l), \quad l = 0, \ldots, L-1$$

The signal waveform $s(l)$ is known and its energy normalized to be unity ($\sum s^2(l) = 1$). We want to estimate the signal's amplitude once the direction of propagation has been found. The linear estimate of the signal's amplitude is assumed to be of the form $\widehat{\xi} = \sum h(l)y(l)$, where $h(l)$ minimizes the mean-squared error. To use the Orthogonality Principle expressed by Eq. 6.1 {273}, an inner product must be defined for scalars. Little choice avails itself but multiplication as the inner product of two scalars. The Orthogonality Principle states that the estimation error must be orthogonal to all linear transformations defining the kind of estimator being sought.

$$\mathcal{E}\left[\left(\sum_{l=0}^{L-1} h_{\text{LIN}}(l)y(l) - \xi\right)\sum_{k=0}^{L-1} h(k)y(k)\right] = 0 \quad \text{for all } h(\cdot)$$

Manipulating this equation to make the universality constraint more transparent results in

$$\sum_{k=0}^{L-1} h(k)\cdot\mathcal{E}\left[\left(\sum_{l=0}^{L-1} h_{\text{LIN}}(l)y(l) - \xi\right)y(k)\right] = 0 \quad \text{for all } h(\cdot)$$

Written in this way, the expected value must be 0 for each value of k to satisfy the constraint. Thus, the quantity $h_{\text{LIN}}(\cdot)$ of the estimator of the signal's amplitude must satisfy

$$\sum_{l=0}^{L-1} h_{\text{LIN}}(l)\,\mathcal{E}[y(l)y(k)] = \mathcal{E}[\xi y(k)] \quad \text{for all } k$$

Assuming that the signal's amplitude has zero mean and is statistically independent of the zero-mean noise, the expected values in this equation are given by

$$\mathcal{E}[y(l)y(k)] = \sigma_\xi^2 s(l)s(k) + K_n(k,l)$$
$$\mathcal{E}[\xi y(k)] = \sigma_\xi^2 s(k)$$

where $K_n(k,l)$ is the covariance function of the noise. The equation that must be solved for the unit-sample response $h_{\text{LIN}}(\cdot)$ of the optimal linear *MMSE* estimator of signal amplitude becomes

$$\boxed{\sum_{l=0}^{L-1} h_{\text{LIN}}(l)K_n(k,l) = \sigma_\xi^2 s(k)\left[1 - \sum_{l=0}^{L-1} h_{\text{LIN}}(l)s(l)\right] \quad \text{for all } k}$$

This equation is easily solved once phrased in matrix notation. Letting $\mathbf{K}_n$ denote the covariance matrix of the noise, $\mathbf{s}$ the signal vector, and $\mathbf{h}_{\text{LIN}}$ the vector of coefficients, this equation becomes

$$\mathbf{K}_n\mathbf{h}_{\text{LIN}} = \sigma_\xi^2[1 - \mathbf{s}^t\mathbf{h}_{\text{LIN}}]\mathbf{s}$$

The matched filter for colored-noise problems consisted of the dot product between the vector of observations and $\mathbf{K}_n^{-1}\mathbf{s}$ (see the detector result {228}). Assume that the solution to the linear estimation problem is proportional to the detection theoretical one: $\mathbf{h}_{\text{LIN}} = c\mathbf{K}_n^{-1}\mathbf{s}$, where c is a scalar constant. This proposed solution satisfies the equation; the *MMSE* estimate of signal amplitude corresponds to applying a matched filter to the observations with

$$\boxed{\mathbf{h}_{\text{LIN}} = \frac{\sigma_\xi^2}{1 + \sigma_\xi^2\mathbf{s}^t\mathbf{K}_n^{-1}\mathbf{s}}\mathbf{K}_n^{-1}\mathbf{s}}$$

The mean-squared estimation error of signal amplitude is given by

$$\mathcal{E}[\epsilon^2] = \sigma_\xi^2 - \mathcal{E}\left[\xi\sum_{l=0}^{L-1} h_{\text{LIN}}(l)y(l)\right]$$

Substituting the vector expression for $\mathbf{h}_{\text{LIN}}$ yields the result that the mean-squared estimation error equals the proportionality constant c defined earlier.

$$\mathcal{E}[\epsilon^2] = \frac{\sigma_\xi^2}{1 + \sigma_\xi^2\mathbf{s}^t\mathbf{K}_n^{-1}\mathbf{s}}$$

Thus, the linear filter that produces the optimal estimate of signal amplitude is equivalent to the matched filter used to detect the signal's presence. We have found this situation to occur when estimates of unknown parameters are needed to solve the detection problem (see §5.4.1 {236}). If we had not assumed the noise to be Gaussian, however, this detection-theoretic result would be different, but the estimator would be unchanged. To repeat, this invariance occurs because the linear *MMSE* estimator requires *no* assumptions on the noise's amplitude characteristics.

Example

Let the noise be white so that its covariance matrix is proportional to the identity matrix ($\mathbf{K}_n = \sigma_n^2\mathbf{I}$). The weighting factor in the minimum mean-squared error linear estimator is proportional to the signal waveform.

$$h_{\text{LIN}}(l) = \frac{\sigma_\xi^2}{\sigma_n^2 + \sigma_\xi^2}s(l) \qquad \widehat{\xi}_{\text{LIN}} = \frac{\sigma_\xi^2}{\sigma_n^2 + \sigma_\xi^2}\sum_{l=0}^{L-1} s(l)y(l)$$

This proportionality constant depends only on the relative variances of the noise and the parameter. *If* the noise variance can be considered to be much smaller than the a priori variance of the amplitude, then this constant does not depend on these variances and equals unity. Otherwise, the variances must be known.

We find the mean-squared estimation error to be

$$\mathcal{E}[\epsilon^2] = \frac{\sigma_\xi^2}{1 + \sigma_\xi^2/\sigma_n^2}$$

This error is significantly reduced from its nominal value σ_ξ^2 only when the variance of the noise is small compared with the a priori variance of the amplitude. Otherwise, this admittedly optimum amplitude estimate performs poorly, and we might as well as have ignored the data and "guessed" that the amplitude was zero.[†]

6.3.2 Maximum Likelihood Estimators

Many situations are either not well suited to linear estimation procedures, or the parameter is not well described as a random variable. For example, signal delay is observed nonlinearly and usually no a priori density can be assigned. In such cases, maximum likelihood estimators are more frequently used. Because of the Cramér-Rao bound, fundamental limits on parameter estimation performance can be derived for *any* signal parameter estimation problem where the parameter is not random.

Assume that the data are expressed as a signal observed in the presence of additive Gaussian noise.

$$y(l) = s(l, \boldsymbol{\xi}) + n(l), \quad l = 0, \ldots, L-1$$

The vector of observations $\mathbf{y}$ is formed from the data in the obvious way. Evaluating the logarithm of the observation vector's joint density,

$$\ln p_{\mathbf{y}|\boldsymbol{\xi}}(\mathbf{y}|\boldsymbol{\xi}) = -\frac{1}{2}\ln\det[2\pi\mathbf{K}_n] - \frac{1}{2}[\mathbf{y} - \mathbf{s}(\boldsymbol{\xi})]'\mathbf{K}_n^{-1}[\mathbf{y} - \mathbf{s}(\boldsymbol{\xi})]$$

[†]In other words, the problem is difficult in this case.

where $\mathbf{s}(\boldsymbol{\xi})$ is the signal vector having P unknown parameters, and $\mathbf{K}_n$ is the covariance matrix of the noise. The partial derivative of this likelihood function with respect to the ith parameter ξ_i is, for real-valued signals,

$$\frac{\partial \ln p_{\mathbf{y}|\boldsymbol{\xi}}(\mathbf{y}|\boldsymbol{\xi})}{\partial \xi_i} = [\mathbf{y} - \mathbf{s}(\boldsymbol{\xi})]^t \mathbf{K}_n^{-1} \frac{\partial \mathbf{s}(\boldsymbol{\xi})}{\partial \xi_i}$$

and, for complex-valued ones,

$$\frac{\partial \ln p_{\mathbf{y}|\boldsymbol{\xi}}(\mathbf{y}|\boldsymbol{\xi})}{\partial \xi_i} = \mathrm{Re}\left[[\mathbf{y} - \mathbf{s}(\boldsymbol{\xi})]' \mathbf{K}_n^{-1} \frac{\partial \mathbf{s}(\boldsymbol{\xi})}{\partial \xi_i}\right]$$

If the maximum of the likelihood function can be found by setting its gradient to $\mathbf{0}$, the maximum likelihood estimate of the parameter vector is the solution of the set of equations

$$\boxed{[\mathbf{y} - \mathbf{s}(\boldsymbol{\xi})]^t \mathbf{K}_n^{-1} \frac{\partial \mathbf{s}(\boldsymbol{\xi})}{\partial \xi_i}\bigg|_{\boldsymbol{\xi}=\widehat{\boldsymbol{\xi}}_{\mathrm{ML}}} = 0, \quad i = 1, \ldots, P}$$

The Cramér-Rao bound depends on the evaluation of the Fisher information matrix $\mathbf{F}$. The elements of this matrix are found to be

$$F_{ij} = \frac{\partial \mathbf{s}^t(\boldsymbol{\xi})}{\partial \xi_i} \mathbf{K}_n^{-1} \frac{\partial \mathbf{s}(\boldsymbol{\xi})}{\partial \xi_j}, \quad i, j = 1, \ldots, P \tag{6.5}$$

Further computation of the Cramér-Rao bound's components is problem dependent if more than one parameter is involved, and the off-diagonal terms of $\mathbf{F}$ are nonzero. If only one parameter is unknown, the Cramér-Rao bound is given by

$$\boxed{\mathcal{E}[\epsilon^2] \geq b^2(\xi) + \frac{\left(1 + \dfrac{db(\xi)}{d\xi}\right)^2}{\dfrac{\partial \mathbf{s}^t(\xi)}{\partial \xi} \mathbf{K}_n^{-1} \dfrac{\partial \mathbf{s}(\xi)}{\partial \xi}}}$$

When the signal depends on the parameter nonlinearly (which constitute the interesting cases), the maximum likelihood estimate is usually biased. Thus, the numerator of the expression for the bound cannot be ignored. One interesting special case occurs when the noise is white. The Cramér-Rao bound becomes

$$\mathcal{E}[\epsilon^2] \geq b^2(\xi) + \frac{\sigma_n^2 \left(1 + \dfrac{db(\xi)}{d\xi}\right)^2}{\displaystyle\sum_{l=0}^{L-1} \left(\frac{\partial s(l, \xi)}{\partial \xi}\right)^2}$$

The derivative of the signal with respect to the parameter can be interpreted as the sensitivity of the signal to the parameter. The mean-squared estimation error depends on the "integrated" squared sensitivity: The greater this sensitivity, the smaller the bound.

For an efficient estimate of a signal parameter to exist, the estimate must satisfy the condition we derived earlier (Eq. 6.3 {279}).

$$[\nabla_\xi \mathbf{s}(\boldsymbol{\xi})]^t \mathbf{K}_n^{-1}[\mathbf{y}-\mathbf{s}(\boldsymbol{\xi})] \stackrel{?}{=} \left[\mathbf{I}+(\nabla_\xi \mathbf{b})^t\right]^{-1}[\nabla_\xi \mathbf{s}(\boldsymbol{\xi})]^t \mathbf{K}_n^{-1}[\nabla_\xi \mathbf{s}(\boldsymbol{\xi})][\widehat{\boldsymbol{\xi}}(\mathbf{y})-\boldsymbol{\xi}-\mathbf{b}]$$

Because of the complexity of this requirement, we quite rightly question the existence of any efficient estimator, especially when the signal depends nonlinearly on the parameter (see Prob. 6.10).

Example

Let the unknown parameter be the signal's amplitude; the signal is expressed as $\xi s(l)$ and is observed in an array's output in the presence of additive noise. The maximum likelihood estimate of the amplitude is the solution of the equation

$$[\mathbf{y}-\widehat{\xi}_{\mathrm{ML}}\mathbf{s}]^t \mathbf{K}_n^{-1}\mathbf{s}=0$$

The form of this equation suggests that the maximum likelihood estimate is efficient. The amplitude estimate is given by

$$\widehat{\xi}_{\mathrm{ML}}=\frac{\mathbf{y}^t\mathbf{K}_n^{-1}\mathbf{s}}{\mathbf{s}^t\mathbf{K}_n^{-1}\mathbf{s}}$$

The form of this estimator is precisely that of the matched filter derived in the colored-noise situation (see Eq. 5.7 {228}). The expected value of the estimate equals the actual amplitude. Thus the bias is zero and the Cramér-Rao bound is given by

$$\mathcal{E}[\epsilon^2] \geq \left(\mathbf{s}^t\mathbf{K}_n^{-1}\mathbf{s}\right)^{-1}$$

The condition for an efficient estimate becomes

$$\mathbf{s}^t\mathbf{K}_n^{-1}(\mathbf{y}-\xi\mathbf{s}) \stackrel{?}{=} \mathbf{s}^t\mathbf{K}_n^{-1}\mathbf{s}\cdot(\widehat{\xi}_{\mathrm{ML}}-\xi)$$

whose veracity we can easily verify.

In the special case where the noise is white, the estimator has the form $\widehat{\xi}_{\mathrm{ML}}=\mathbf{y}^t\mathbf{s}$, and the Cramér-Rao bound equals σ_n^2 (the nominal signal is assumed to have unit energy). The maximum likelihood estimate of the amplitude has *fixed* error characteristics that do not depend on the actual signal amplitude. A signal-to-noise ratio for the estimate, defined to be $\xi^2/\mathcal{E}[\epsilon^2]$, equals the signal-to-noise ratio of the observed signal.

When the amplitude is well described as a random variable, its linear minimum mean-squared error estimator has the form

$$\widehat{\xi}_{\mathrm{LIN}}=\frac{\sigma_\xi^2\mathbf{y}^t\mathbf{K}_n^{-1}\mathbf{s}}{1+\sigma_\xi^2\mathbf{s}\mathbf{K}_n^{-1}\mathbf{s}}$$

which we found in the white-noise case becomes a weighted version of the maximum likelihood estimate (see the example {286}).

$$\widehat{\xi}_{\mathrm{LIN}}=\frac{\sigma_\xi^2}{\sigma_\xi^2+\sigma_n^2}\mathbf{y}^t\mathbf{s}$$

Seemingly, these two estimators are being used to solve the same problem: Estimating the amplitude of a signal whose waveform is known. They make very different assumptions, however, about the nature of the unknown parameter; in one it is a random variable (and thus it has a variance), whereas in the other it is not (and variance makes no sense). Despite this fundamental difference, the computations for each estimator are equivalent. It is reassuring that different approaches to solving similar problems yield similar procedures.

6.3.3 Time-Delay Estimation

One of the most important signal parameter estimation problems in array signal processing is time-delay estimation. Here the unknown is the time origin of the signal: $s(l, \xi) = s(l - \xi)$. The duration of the signal (the domain over which the signal is defined) is assumed brief compared with the observation interval L. Although in continuous time the signal delay is a continuous-valued variable, in discrete time it is not. Consequently, the maximum likelihood estimate *cannot* be found by differentiation, and we must determine the maximum likelihood estimate of signal delay by the most fundamental expression of the maximization procedure. Assuming Gaussian noise, the maximum likelihood estimate of delay is the solution of

$$\min_{\xi} \, [\mathbf{y} - \mathbf{s}(\xi)]^t \mathbf{K}_n^{-1} [\mathbf{y} - \mathbf{s}(\xi)]$$

The term $\mathbf{s}^t \mathbf{K}_n^{-1} \mathbf{s}$ is usually assumed not to vary with the presumed time origin of the signal because of the signal's short duration. If the noise is white, this term is constant except near the "edges" of the observation interval. If not white, the kernel of this quadratic form is equivalent to a whitening filter. As discussed in the previous chapter (§5.3.3 {227}), this filter may be time varying. For noise spectra that are rational and have only poles, the whitening filter's unit-sample response varies only near the edges (see the example {231}). Thus, near the edges, this quadratic form varies with presumed delay and the maximization is analytically difficult. Taking the "easy way out" by ignoring edge effects, the estimate is the solution of

$$\max_{\xi} \, [\mathbf{y}^t \mathbf{K}_n^{-1} \mathbf{s}(\xi)]$$

Thus, the delay estimate is the signal time origin that maximizes the matched filter's output.

In addition to the complexity of finding the maximum likelihood estimate, the discrete-valued nature of the parameter also calls into question the use of the Cramér-Rao bound. One of the fundamental assumptions of the bound's derivation is the differentiability of the likelihood function with respect to the parameter. Mathematically, a sequence cannot be differentiated with respect to the integers. A sequence can be differentiated with respect to its argument if we consider the variable to be continuous valued. This approximation can be used only if the sampling interval, unity for the integers, is dense with respect to variations of the sequence. This condition means that the signal must be oversampled to apply the Cramér-Rao bound in a meaningful way. Under these conditions, the mean-squared estimation error for *unbiased estimators* can be no smaller than the Cramér-Rao bound, which is given by

$$\mathcal{E}[\epsilon^2] \geq \frac{1}{\sum_{k,l} \left[\mathbf{K}_n^{-1}\right]_{k,l} \dot{s}(k - \xi)\dot{s}(l - \xi)}$$

which, in the white-noise case, becomes

$$\mathcal{E}[\epsilon^2] \geq \frac{\sigma_n^2}{\sum_l [\dot{s}(l)]^2} \tag{6.6}$$

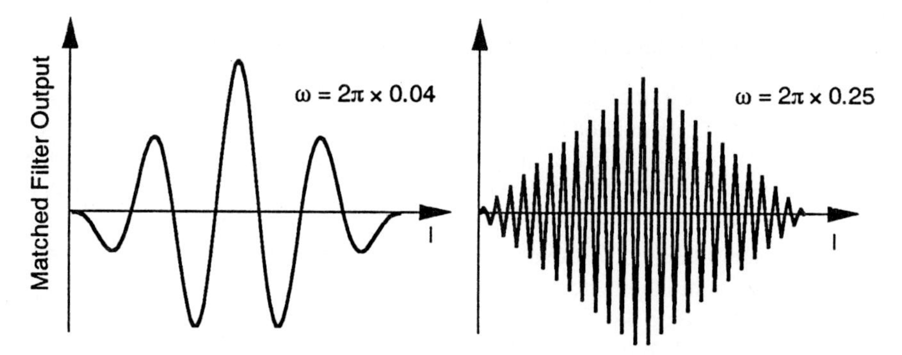

Figure 6.1 The matched filter outputs are shown for two separate signal situations. In each case, the observation interval (100 samples), the signal's duration (50 samples) and energy (unity) are the same. The difference lies in the signal waveform; both are sinusoids with the first having a frequency of $2\pi \times 0.04$ and the second $2\pi \times 0.25$. Each output is the signal's autocorrelation function. Few, broad peaks characterize the low-frequency example whereas many narrow peaks are found in the high frequency one.

Here, $\dot{s}(\cdot)$ denotes the "derivative" of the discrete-time signal. To justify using this Cramér-Rao bound, we must face the issue of whether an unbiased estimator for time delay *exists*. No general answer exists; each estimator, including the maximum likelihood one, must be examined individually.

Example

Assume that the noise is white. Because of this assumption, we determine the time delay by maximizing the match-filtered observations.

$$\arg\max_{\xi} \sum_{l} y(l)s(l-\xi) = \widehat{\xi}_{\text{ML}}$$

The number of terms in the sum equals the signal duration. Fig. 6.1 illustrates the match-filtered output in two separate situations; in one the signal has a relatively low-frequency spectrum as compared with the second. Because of the symmetry of the autocorrelation function, the estimate *should* be unbiased so long as the autocorrelation function is completely contained within the observation interval. Direct proof of this claim is left to the masochistic reader. For sinusoidal signals of energy E and frequency ω_0, the Cramér-Rao bound is given by $\mathcal{E}[\epsilon^2] = \sigma_n^2/\omega_0^2 E$. This bound on the error is accurate only if the measured maximum frequently occurs in the dominant peak of the signal's autocorrelation function. Otherwise, the maximum likelihood estimate "skips" a cycle and produces values concentrated near one of the smaller peaks. The interval between zero crossings of the dominant peak is $\pi/(2\omega_0)$; the signal-to-noise ratio E/σ_n^2 must exceed $4/\pi^2$ (about 0.5). Remember that this result implicitly assumed a low-frequency sinusoid. The second example demonstrates that cycle skipping occurs more frequently than this guideline suggests when a high-frequency sinusoid is used.

The size of the errors encountered in the time-delay estimation problem can be more accurately assessed by a bounding technique tailored to the problem: the Ziv-Zakai bound [159, 166]. The derivation of this bound relies on results from detection theory [34].† Consider the detection problem in which we must distinguish the signals

†This result is an example of detection and estimation theory complementing each other to advantage.

$s(l-\tau)$ and $s\big(l-(\tau+\Delta)\big)$ while observing them in the presence of white noise that is not necessarily Gaussian. Let hypothesis H_0 represent the case in which the delay, denoted by our parameter symbol ξ, is τ and H_1 the case in which $\xi = \tau + \Delta$. The *suboptimum* test statistic consists of estimating the delay, then determining the closest a priori delay to the estimate.

$$\widehat{\xi} \underset{H_0}{\overset{H_1}{\gtrless}} \tau + \frac{\Delta}{2}$$

By using this ad hoc hypothesis test as an essential part of the derivation, the bound can apply to many situations. Furthermore, by not restricting the type of parameter estimate, the bound applies to any estimator. The probability of error for the optimum hypothesis test (derived from the likelihood ratio) is denoted by $P_e(\tau, \Delta)$. Assuming equally likely hypotheses, the probability of error resulting from the ad hoc test must be greater than that of the optimum.

$$P_e(\tau,\Delta) \le \frac{1}{2}\Pr\left(\epsilon > \frac{\Delta}{2} \,\middle|\, H_0\right) + \frac{1}{2}\Pr\left(\epsilon < -\frac{\Delta}{2} \,\middle|\, H_1\right)$$

Here, ϵ denotes the estimation error appropriate to the hypothesis.

$$\epsilon = \begin{cases} \widehat{\xi} - \tau & \text{under } H_0 \\ \widehat{\xi} - \tau - \Delta & \text{under } H_1 \end{cases}$$

The delay is assumed to range uniformly between 0 and L. Combining this restriction to the hypothesized delays yields bounds on both τ and Δ: $0 \le \tau < L - \Delta$ and $0 \le \Delta < L$. Simple manipulations show that the integral of this inequality with respect to τ over the possible range of delays is given by†

$$\int_0^{L-\Delta} P_e(\tau,\Delta)\, d\tau \le \frac{1}{2}\int_0^L \Pr\left(|\epsilon| > \frac{\Delta}{2} \,\middle|\, H_0\right) d\tau$$

Note that if we define $(L/2)\widetilde{P}(\Delta/2)$ to be the right side of this equation so that

$$\widetilde{P}\left(\frac{\Delta}{2}\right) = \frac{1}{L}\int_0^L \Pr\left(|\epsilon| > \frac{\Delta}{2} \,\middle|\, H_0\right) d\tau$$

$\widetilde{P}(\cdot)$ is the complementary distribution function‡ of the magnitude of the average estimation error. Multiplying $\widetilde{P}(\Delta/2)$ by Δ and integrating, the result is

$$\int_0^L \Delta \widetilde{P}\left(\frac{\Delta}{2}\right) d\Delta = -2\int_0^{L/2} x^2 \frac{d\widetilde{P}}{dx}\, dx$$

---

†Here again, the issue of the discrete nature of the delay becomes a consideration; this step in the derivation implicitly assumes that the delay is continuous valued. This approximation can be greeted more readily as it involves integration rather than differentiation (as in the Cramér-Rao bound).

‡The complementary distribution function of a probability distribution function $P(x)$ is defined to be $\widetilde{P}(x) = 1 - P(x)$, the probability that a random variable exceeds x.

The reason for these rather obscure manipulations is now revealed: Because $\widetilde{P}(\cdot)$ is related to the probability distribution function of the absolute error, the right side of this equation is twice the mean-squared error $\mathcal{E}[\epsilon^2]$. The general Ziv-Zakai bound for the mean-squared estimation error of signal delay is thus expressed as

$$\mathcal{E}[\epsilon^2] \geq \frac{1}{L}\int_0^L \Delta \int_0^{L-\Delta} P_e(\tau, \Delta)\, d\tau\, d\Delta$$

In many cases, the optimum probability of error $P_e(\tau, \Delta)$ does not depend on τ, the time origin of the observations. This lack of dependence is equivalent to ignoring edge effects and simplifies calculation of the bound. Thus, the Ziv-Zakai bound for time-delay estimation relates the mean-squared estimation error for delay to the probability of error incurred by the optimal detector that is deciding whether a nonzero delay is present or not.

$$\begin{aligned}\mathcal{E}[\epsilon^2] &\geq \frac{1}{L}\int_0^L \Delta(L-\Delta)P_e(\Delta)\, d\Delta \\ &\geq \frac{L^2}{6}P_e(L) - \int_0^L \left(\frac{\Delta^2}{2} - \frac{\Delta^3}{3L}\right)\frac{dP_e}{d\Delta}\, d\Delta\end{aligned} \tag{6.7}$$

To apply this bound to time-delay estimates (unbiased or not), the optimum probability of error for the type of noise and the relative delay between the two signals must be determined. Substituting this expression into either integral yields the Ziv-Zakai bound.

The general behavior of this bound at parameter extremes can be evaluated in some cases. Note that the Cramér-Rao bound in this problem approaches infinity as either the noise variance grows or the observation interval shrinks to 0 (either forces the signal-to-noise ratio to approach 0). This result is unrealistic as the actual delay is bounded, lying between 0 and L. In this very noisy situation, one should ignore the observations and "guess" *any* reasonable value for the delay; the estimation error is smaller. The probability of error approaches $1/2$ in this situation no matter what the delay Δ may be. Considering the simplified form of the Ziv-Zakai bound, the integral in the second form is 0 in this extreme case.

$$\mathcal{E}[\epsilon^2] \geq \frac{L^2}{12}$$

The Ziv-Zakai bound is exactly the variance of a random variable uniformly distributed over $[0, L-1]$. The Ziv-Zakai bound thus predicts the size of mean-squared errors more accurately than does the Cramér-Rao bound.

Example [159]

Let the noise be Gaussian of variance σ_n^2 and the signal have energy E. The probability of error resulting from the likelihood ratio test is given by

$$P_e(\Delta) = Q\left(\left[\frac{E}{2\sigma_n^2}(1-\rho(\Delta))\right]^{1/2}\right)$$

The quantity $\rho(\Delta)$ is the normalized autocorrelation function of the signal evaluated at the delay Δ.

$$\rho(\Delta) = \frac{1}{E}\sum_l s(l)s(l-\Delta)$$

Evaluation of the Ziv-Zakai bound for a general signal is very difficult in this Gaussian noise case. Fortunately, the normalized autocorrelation function can be bounded by a relatively simple expression to yield a more manageable expression. The key quantity $1-\rho(\Delta)$ in the probability of error expression can be rewritten using Parseval's Theorem.

$$1-\rho(\Delta) = \frac{1}{2\pi E}\int_0^\pi 2|S(\omega)|^2[1-\cos(\omega\Delta)]\,d\omega$$

Using the inequality $1 - \cos x \le x^2/2$, $1-\rho(\Delta)$ is bounded from above by $\min\left\{\Delta^2\beta^2/2, 2\right\}$, where β is the root-mean-squared (RMS) signal bandwidth.

$$\beta^2 = \frac{\displaystyle\int_{-\pi}^{\pi}\omega^2|S(\omega)|^2\,d\omega}{\displaystyle\int_{-\pi}^{\pi}|S(\omega)|^2\,d\omega} \tag{6.8}$$

Because $Q(\cdot)$ is a decreasing function, we have $P_e(\Delta) \ge Q\big(\mu\min\{\Delta,\Delta^*\}\big)$, where μ is a combination of all of the constants involved in the argument of $Q(\cdot)$: $\mu = (E\beta^2/4\sigma_n^2)^{1/2}$. This quantity varies with the product of the signal-to-noise ratio E/σ_n^2 and the squared RMS bandwidth β^2. The parameter $\Delta^* = 2/\beta$ is known as the critical delay and is twice the reciprocal RMS bandwidth. We can use this lower bound for the probability of error in the Ziv-Zakai bound to produce a lower bound on the mean-squared estimation error. The integral in the first form of the bound yields the complicated, but computable result

$$\begin{aligned}\mathcal{E}[\epsilon^2] \ge{}& \frac{L^2}{6}Q\left(\mu\min\{L,\Delta^*\}\right) + \frac{1}{4\mu^2}P_{\chi_3^2}\left(\mu^2\min\{L^2,\Delta^{*2}\}\right)\\ &-\frac{2}{3\sqrt{2\pi}L\mu^3}\left[1-\left(1+\frac{\mu^2}{2}\min\{L^2,\Delta^{*2}\}\right)\exp\left\{-\mu^2\min\{L^2,\Delta^{*2}\}/2\right\}\right]\end{aligned}$$

The quantity $P_{\chi_3^2}(\cdot)$ is the probability distribution function of a χ^2 random variable having three degrees of freedom.† Thus, the threshold effects in this expression for the mean-squared estimation error depend on the relation between the critical delay and the signal duration. In most cases, the minimum equals the critical delay Δ^*, with the opposite choice possible for very low bandwidth signals.

The Ziv-Zakai bound and the Cramér-Rao bound for the time-delay estimation problem are shown in Fig. 6.2. Note how the Ziv-Zakai bound matches the Cramér-Rao bound only for large signal-to-noise ratios, where they both equal $1/4\mu^2 = \sigma_n^2/E\beta^2$. For smaller values, the former bound is much larger and provides a better indication of the size of the estimation errors. These errors are because of the "cycle skipping" phenomenon described earlier. The Ziv-Zakai bound describes them well, whereas the Cramér-Rao bound ignores them.

†This distribution function has the "closed-form" expression $P_{\chi_3^2}(x) = 1 - Q(\sqrt{x}) - \sqrt{x/2}\exp\{-x/2\}$.

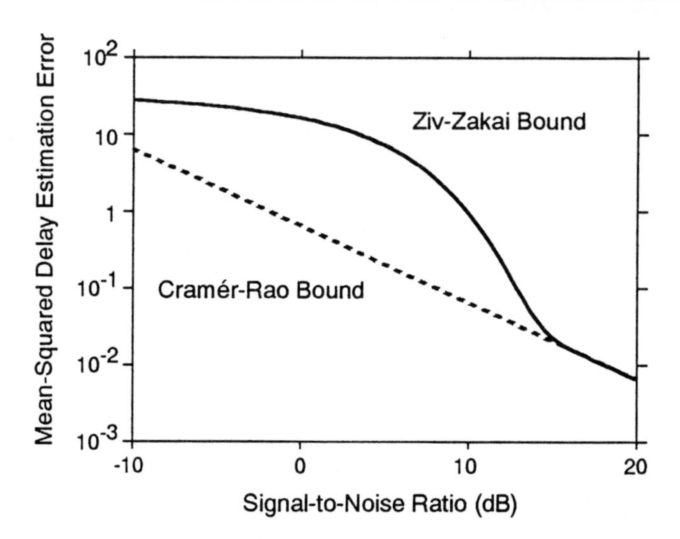

Figure 6.2 The Ziv-Zakai bound and the Cramér-Rao bound for the estimation of the time delay of a signal observed in the presence of Gaussian noise is shown as a function of the signal-to-noise ratio. For this plot, $L = 20$ and $\beta = 2\pi \times 0.2$. The Ziv-Zakai bound is much larger than the Cramér-Rao bound for signal-to-noise ratios less than 13 dB; the Ziv-Zakai bound can be as much as 30 times larger.

6.4 Linear Signal Waveform Estimation

When the details of a signal's waveform are unknown, describing the signal parametrically is usually unsatisfactory. We need techniques that estimate waveforms rather than numbers. For example, we may want to know the propagating signal's waveform contained in the noise-corrupted array output. Without some a priori information, this task is impossible; if neither the signal nor the noise is known, how can anyone discriminate one from the other? The key to waveform estimation is how much prior information we have about the signal and the noise, and how valid that information is. Given noisy observations of a signal throughout the interval $[L_i, L_f]$, the waveform estimation problem is to estimate accurately the value of the signal at some moment $L_f + l_e$. In most situations, the observation interval evolves with the passage of time while the estimation time is fixed relative to the occurrence of the most recent observation (in other words, l_e is a constant). Linear waveform estimation results when we apply a linear filter to the observations.

Waveform estimation problems are usually placed into one of three categories [4: 9–11] based on the value of l_e (see Fig. 6.3):

Interpolation. The interpolation or smoothing problem is to estimate the signal at some moment within the observation interval ($l_e < 0$). Observations are thus considered before and after the time at which the signal needs to be estimated. In practice, applying interpolation filtering means that the estimated signal waveform is produced some time *after* it occurred. Suppose we want to determine the direction

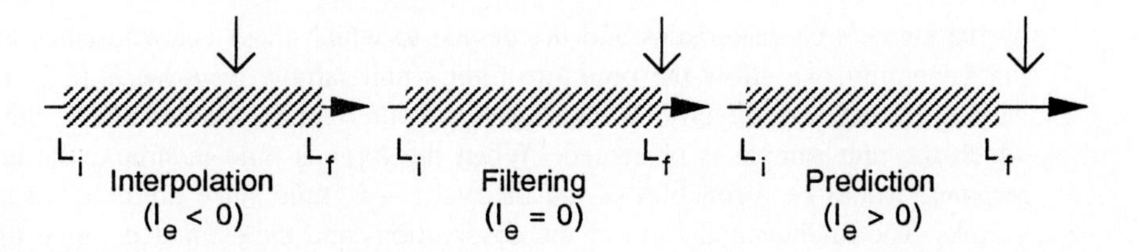

Figure 6.3 The three classical categories of linear signal waveform estimation are defined by the observation interval's relation to the time at which we want to estimate the signal value. As time evolves, so does the observation interval so that l_e, the interval between the last observation and the estimation time, is fixed.

of propagation of a signal when the signal's waveform is not well known. As the signal must be present in each sensor's output to determine direction, we can apply the interpolation problem to the array's output to enhance the signal. If this estimation delay can be tolerated, interpolation filtering, as we shall see, yields smaller estimation errors than the others.

Filtering. We estimate the signal at the end of the observation interval ($l_e = 0$). Thus, a waveform estimate is produced as soon as the signal is observed. The filtering problem arises when we want to remove noise (as much as possible) from noise-corrupted signal observations as they are obtained.

Prediction. Here, we attempt to predict the signal's value at some future time ($l_e > 0$). The signal's structure must be well known to enable us to predict what values the signal obtains. Prediction filters have obvious applications in sonar/radar tracking and stock market analysis. Of all the waveform estimation problems, this one produces the largest errors.

Waveform estimation algorithms are *not* defined by this categorization; each technique can be applied to each type of problem (in most cases). Instead, the algorithms are defined according to the signal model. Correctness of the signal model governs the utility of a given technique. Because the signal usually appears *linearly* in the expression for the observations (the noise is usually additive), *linear* waveform estimation methods—filters—are frequently employed.

6.4.1 General Considerations

In the context of *linear* waveform estimation, the signal as well as the noise is considered to be a stochastic sequence. Furthermore, the signal component $\tilde{s}$ in the observations is assumed to only be *related* to the signal s to be estimated and not necessarily equal to it: $y(l) = \tilde{s}(l) + n(l)$. For example, the observations may contain a filtered version of the signal when we require an estimate of the prefiltered waveform. In this situation, the signal filter is usually known. The noise and signal components are zero-mean random sequences statistically independent of each other. The optimum filter that provides the waveform estimate $\widehat{s}(l)$ can be time invariant (Wiener filters), time varying (Kalman filters), or data dependent (adaptive filters). Choosing an estimation strategy is determined by the signal's characteristics and the degree to which these characteristics are known. For generality, we allow the optimum filter's unit-sample response $h_\diamond(l,k)$ to be time varying: It depends directly on the values of l, the "time variable" and k, the "time" at which the unit sample is presented. When the filter is time invariant, the unit-sample response would be a function of the interval $l-k$, time since presentation of the unit sample. The fundamental form of the observations and the estimated signal in all linear waveform estimators is

$$y(l) = \tilde{s}(l) + n(l)$$

$$\widehat{s}(L_f + l_e) = \sum_{k=L_i}^{L_f} h_\diamond(L_f, k) y(k)$$

The estimate of the signal's value at $L_f + l_e$ is thus produced at time L_f in the filter's output. The duration of the filter's unit-sample response extends over the entire observation interval $[L_i, L_f]$.

The Orthogonality Principle that proved so useful in linear parameter estimation can be applied here. It states that the estimation error must be orthogonal to all linear transformations of the observations (see Eq. 6.1 {273}). For the waveform estimation problem, this requirement implies that

$$\mathcal{E}\left[\{s(L_f + l_e) - \widehat{s}(L_f + l_e)\} \sum_{k=L_i}^{L_f} h(L_f, k) y(k)\right] = 0 \quad \text{for all } h(\cdot,\cdot)$$

This expression implies that each observed value must be orthogonal to the estimation error at time $L_f + l_e$.

$$\mathcal{E}\left[\left\{s(L_f + l_e) - \sum_{j=L_i}^{L_f} h_\diamond(L_f, j) y(j)\right\} y(k)\right] = 0 \quad \text{for all } k \text{ in } [L_i, L_f]$$

Simplifying this expression, the fundamental equation that determines the unit-sample response of the linear minimum mean-squared error filter is

$$K_{s\tilde{s}}(L_f + l_e, k) = \sum_{j=L_i}^{L_f} K_y(j,k) h_\diamond(L_f, j) \quad \text{for all } k \text{ in } [L_i, L_f]$$

where $K_y(k,l)$ is the covariance function of the observations, equaling $\mathcal{E}[y(k)y(l)]$, and $K_{s\tilde{s}}(L_f + l_e, k)$ is the cross-covariance between the signal at $L_f + l_e$ and the signal-related component of the observation at k. When the signal and noise are uncorrelated, $K_y(k,l) = K_{\tilde{s}}(k,l) + K_n(k,l)$. Given these quantities, the preceding equation must then be solved for the unit-sample response of the optimum filter. This equation is known as the *generalized Wiener-Hopf equation.*

From the general theory of linear estimators, the mean-squared estimation error at index l equals the variance of the quantity being estimated minus the estimate's projection onto the signal.

$$\mathcal{E}[\epsilon^2(l)] = K_s(l,l) - \mathcal{E}[\widehat{s}(l)s(l)]$$

Expressing the signal estimate as a linear filtering operation on the observations, this expression becomes

$$\mathcal{E}[\epsilon^2(l)] = K_s(l,l) - \sum_{k=L_i}^{L_f} h_\diamond(L_f,k)K_{s\widehat{s}}(l,k)$$

Further reduction of this expression is usually problem dependent, as succeeding sections illustrate.

6.4.2 Wiener Filters

Wiener filters are the solutions of the linear minimum mean-squared waveform estimation problem for the special case in which the noise and the signal are *stationary* random sequences [68: 100–18];[152: 481–515];[162]. The covariance functions appearing in the generalized Wiener-Hopf equation thus depend on the difference of their arguments. Considering the form of this equation, one would expect the unit-sample response of the optimum filter to depend on its arguments in a similar fashion. This presumption is in fact valid, and Wiener filters are always time invariant.

$$\widehat{s}(L_f + l_e) = \sum_{k=L_i}^{L_f} h_\diamond(L_f - k)y(k)$$

We consider first the case in which the initial observation time L_i equals $-\infty$. The resulting filter uses all of the observations available at any moment.[†] The errors that result from using this filter are smaller than those obtained when the filter is constrained to use a finite number of observations (such as some number of recent samples). The choice of $L_i = -\infty$ corresponds to an infinite-duration impulse response (*IIR*) Wiener filter; in a succeeding section, L_i is finite and a finite-duration impulse response (*FIR*) Wiener filter results. The error characteristics of the *IIR* Wiener filter generally bound those of *FIR* Wiener filters because more observations are used. We write the generalized Wiener-Hopf equation for the *IIR* case as

$$K_{s\widehat{s}}(L_f + l_e - k) = \sum_{j=-\infty}^{L_f} K_y(j-k)\cdot h_\diamond(L_f - j) \quad \text{for all } k \text{ in } (-\infty, L_f]$$

Changing summation variables results in the somewhat simpler expression known as the Wiener-Hopf equation. It and the expression for the mean-squared estimation error are

[†]Presumably, observations have been continuously available since the beginning of the universe.

given by

$$K_{s\tilde{s}}(l+l_e) = \sum_{k=0}^{\infty} K_y(l-k)h_\diamond(k) \quad \text{for all } l \text{ in } [0,\infty)$$
$$\mathcal{E}[\epsilon^2] = K_s(0) - \sum_{k=0}^{L_f-L_i} h_\diamond(k)K_{s\tilde{s}}(l_e+k) \tag{6.9}$$

The first term in the error expression is the signal variance. The mean-squared error of the signal estimate cannot be greater than this quantity; this error results when the estimate always equals 0.

In many circumstances, we want to estimate the signal directly contained in observations: $y = s + n$. This situation leads to a somewhat simpler form for the Wiener-Hopf equation.

$$K_s(l+l_e) = \sum_{k=0}^{\infty} [K_s(l-k) + K_n(l-k)]\, h_\diamond(k) \quad \text{for all } l \text{ in } [0,\infty)$$

It is this form we solve, but the previous one is required in its solution.

Solving the Wiener-Hopf equation. The Wiener-Hopf equation at first glance appears to be a convolution integral, implying that the optimum filter's frequency response could be easily found. The constraining condition—the equation applies only for the variable l in the interval $[0,\infty)$—means, however, that Fourier techniques *cannot* be used for the general case. If the Fourier Transform of the left side of the Wiener-Hopf equation were evaluated only over the constraining interval, the covariance function on the left would be *implicitly* assumed 0 outside the interval, which is usually not the case. Simply stated but mathematically complicated, the covariance function of the signal outside this interval is not to be considered in the solution of the equation.

One set of circumstances does allow Fourier techniques. Let the Wiener filter be noncausal with $L_f = +\infty$. In this case, the Wiener-Hopf equation becomes

$$K_s(l) = \sum_{k=-\infty}^{\infty} K_y(l-k)h_\diamond(k) \quad \text{for all } l$$

As this equation must be valid for all values of l, a convolution sum emerges. The frequency response $H_\diamond(\omega)$ of the optimum filter is thus given by

$$H_\diamond(\omega) = \frac{\mathcal{S}_s(\omega)}{\mathcal{S}_s(\omega) + \mathcal{S}_n(\omega)}$$

where $\mathcal{S}_s(\omega)$ and $\mathcal{S}_n(\omega)$ are, respectively, the signal and the noise power spectra. Because this expression is real and even, the unit-sample response of the optimum filter is also real and even. The filter is therefore noncausal and usually has an infinite duration unit-sample response. This result is not often used in temporal signal processing but may find applications in spatial problems. Be that as it may, because this filter can use the entire set of observations to estimate the signal's value at any moment, it yields the smallest estimation error of *any* linear filter. Computing this error thus establishes a bound on how well any causal or *FIR* Wiener filter performs. The mean-squared estimation error of the noncausal Wiener filter can be expressed in the time domain or frequency domain.

$$\begin{aligned}\mathcal{E}[\epsilon^2] &= K_s(0) - \sum_{l=-\infty}^{\infty} h_\diamond(l) K_s(l) \\ &= \frac{1}{2\pi}\int_{-\pi}^{\pi} \frac{\mathcal{S}_s(\omega)\mathcal{S}_n(\omega)}{\mathcal{S}_s(\omega)+\mathcal{S}_n(\omega)}\,d\omega\end{aligned}$$

The causal solution to the Wiener-Hopf equation, the frequency response of the causal Wiener filter, is the product of two terms: the frequency response of a *whitening* filter and the frequency response of the signal estimation filter based on whitened observations [152: 482–93].

$$\boxed{H_\diamond(\omega) = \frac{1}{[\mathcal{S}_s+\mathcal{S}_n]^+(\omega)} \cdot \left[\frac{e^{+j\omega l_e}\mathcal{S}_s(\omega)}{[\mathcal{S}_s+\mathcal{S}_n]^{+*}(\omega)}\right]_+}$$

$[\mathcal{S}(\omega)]_+$ means the Fourier Transform of a covariance function's causal part, which corresponds to its values at nonnegative indices, and $\mathcal{S}^+(\omega)$ the stable, causal, and minimum-phase square root of $\mathcal{S}(\omega)$. Evaluation of this expression therefore involves both forms of causal-part extraction. This solution is clearly much more complicated than anticipated when we first gave the Wiener-Hopf equation. How to solve it is best seen by example, which we provide once we determine an expression for the mean-squared estimation error.

Error characteristics of Wiener filter output. Assuming that $\tilde{s}$ equals s, the expression for the mean-squared estimation error given in Eq. 6.9 {298}, can be simplified with the result

$$\mathcal{E}[\epsilon^2] = K_s(0) - \sum_{k=0}^{\infty} h_\diamond(k) K_s(l_e+k) \tag{6.10}$$

Applying Parseval's Theorem to the summation, this expression can also be written in the frequency domain as

$$\mathcal{E}[\epsilon^2] = K_s(0) - \frac{1}{2\pi}\int_{-\pi}^{\pi} \frac{1}{[\mathcal{S}_s+\mathcal{S}_n]^+(\omega)} \left[\frac{e^{+j\omega l_e}\mathcal{S}_s(\omega)}{[\mathcal{S}_s+\mathcal{S}_n]^{+*}(\omega)}\right]_+ \left[e^{-j\omega l_e}\mathcal{S}_s(\omega)\right]_+ d\omega$$

Noting that the first and third terms in the integral can be combined, the mean-squared error can also be written as

$$\mathcal{E}[\epsilon^2] = K_s(0) - \frac{1}{2\pi}\int_{-\pi}^{\pi} \left|\left[\frac{e^{+j\omega l_e}\mathcal{S}_s(\omega)}{[\mathcal{S}_s+\mathcal{S}_n]^{+*}(\omega)}\right]_+\right|^2 d\omega$$

The expression within the magnitude bars equals the frequency response of the second component of the Wiener filter's frequency response. Again using Parseval's Theorem to return to the time domain, the mean-squared error can be expressed directly either in terms of the filter's unit-sample response or in terms of signal and noise quantities by

$$\mathcal{E}[\epsilon^2] = K_s(0) - \sum_{k=0}^{\infty} K_{ss_w}^2(l_e + k)$$

where the latter quantity is the cross-covariance function between the signal and the signal after passage through the whitening filter.

Example

Let's estimate the value of $s(L_f + l_e)$ with a Wiener filter using the observations obtained up to and including time L_f. The additive noise in the observations is white, having variance $8/7$. The power spectrum of the signal is given by

$$\begin{aligned} \mathcal{S}_s(\omega) &= \frac{1}{5/4 - \cos\omega} \\ &= \frac{1}{1 - 0.5e^{-j\omega}} \cdot \frac{1}{1 - 0.5e^{+j\omega}} \end{aligned}$$

The variance of the signal equals the value of the covariance function (found by the inverse Fourier Transform of this expression) at the origin. In this case, the variance equals $4/3$; the signal-to-noise ratio of the observations, taken to be the ratio of their variances, equals $7/6$.

The power spectrum of the observations is the sum of the signal and noise power spectra.

$$\begin{aligned} \mathcal{S}_s(\omega) + \mathcal{S}_n(\omega) &= \frac{1}{1 - 0.5e^{-j\omega}} \frac{1}{1 - 0.5e^{+j\omega}} + \frac{8}{7} \\ &= \frac{16}{7} \frac{\left(1 - 0.25e^{-j\omega}\right)\left(1 - 0.25e^{+j\omega}\right)}{\left(1 - 0.5e^{-j\omega}\right)\left(1 - 0.5e^{+j\omega}\right)} \end{aligned}$$

The noncausal Wiener filter has the frequency response

$$\frac{\mathcal{S}_s(\omega)}{\mathcal{S}_s(\omega) + \mathcal{S}_n(\omega)} = \frac{7}{16} \frac{1}{\left(1 - 0.25e^{-j\omega}\right)\left(1 - 0.25e^{+j\omega}\right)}$$

The unit-sample response corresponding to this frequency response and the covariance function of the signal are found to be

$$h_\diamond(l) = \frac{7}{15}\left(\frac{1}{4}\right)^{|l|} \quad \text{and} \quad K_s(l) = \frac{4}{3}\left(\frac{1}{2}\right)^{|l|}$$

Using Eq. 6.10 {299}, we find that the mean-squared estimation error for the noncausal estimator equals $4/3 - 4/5 = 8/15$.

The convolutionally causal part of signal-plus-noise power spectrum consists of the first terms in the numerator and denominator of the signal-plus-noise power spectrum.

$$[\mathcal{S}_s + \mathcal{S}_n]^+(\omega) = \frac{4}{\sqrt{7}} \frac{1 - 0.25e^{-j\omega}}{1 - 0.5e^{-j\omega}}$$

The second term in the expression for the frequency response of the optimum filter is given by

$$\frac{e^{+j\omega l_e}\mathcal{S}_s(\omega)}{[\mathcal{S}_s+\mathcal{S}_n]^{+*}(\omega)}=\frac{\dfrac{e^{+j\omega l_e}}{\left(1-0.5e^{-j\omega}\right)\left(1-0.5e^{+j\omega}\right)}}{\dfrac{4}{\sqrt{7}}\dfrac{1-0.25e^{+j\omega}}{1-0.5e^{+j\omega}}}$$

$$=\frac{\sqrt{7}}{4}\frac{e^{+j\omega l_e}}{\left(1-0.5e^{-j\omega}\right)\left(1-0.25e^{+j\omega}\right)}$$

The additively causal part of this Fourier Transform is found by evaluating its partial fraction expansion.

$$\frac{\sqrt{7}}{4}\frac{e^{+j\omega l_e}}{\left(1-0.5e^{-j\omega}\right)\left(1-0.25e^{+j\omega}\right)}=\frac{e^{+j\omega l_e}}{2\sqrt{7}}\left[\frac{4}{1-0.5e^{-j\omega}}-\frac{2e^{+j\omega}}{1-0.25e^{+j\omega}}\right]$$

The simplest solution occurs when l_e equals zero: Estimate the signal value at the moment of the most recent observation. The first term on the right side of the preceding expression corresponds to the additively causal portion.

$$\left[\frac{\mathcal{S}_s(\omega)}{[\mathcal{S}_s+\mathcal{S}_n]^{+*}(\omega)}\right]_+=\frac{2}{\sqrt{7}}\frac{1}{1-0.5e^{-j\omega}}$$

The frequency response of the Wiener filter is found to be

$$H_\diamond(\omega)=\frac{\sqrt{7}}{4}\frac{1-0.5e^{-j\omega}}{1-0.25e^{-j\omega}}\cdot\frac{2}{\sqrt{7}}\frac{1}{1-0.5e^{-j\omega}}$$

$$=\frac{1}{2}\frac{1}{1-0.25e^{-j\omega}}$$

The Wiener filter has the form of a simple first-order filter with the pole arising from the whitening filter. The difference equation corresponding to this frequency response is

$$\widehat{s}(l)=\frac{1}{4}\widehat{s}(l-1)+\frac{1}{2}y(l)$$

The waveforms that result in this example are exemplified in Fig. 6.4.

To find the mean-squared estimation error, the cross-covariance between the signal and its whitened counterpart is required. This quantity equals the inverse transform of the Wiener filter's second component and thus equals $(2/\sqrt{7})\,(1/2)^l$, $l \geq 0$. The mean-squared estimation error is numerically equal to

$$\mathcal{E}[\epsilon^2]=\frac{4}{3}-\sum_{l=0}^{\infty}\left[\frac{2}{\sqrt{7}}\left(\frac{1}{2}\right)^l\right]^2$$

$$=\frac{20}{21}=0.95$$

which compares with the smallest possible value of 0.53 provided by the noncausal Wiener filter. Thus, little is lost by using the causal filter. The signal-to-noise ratio of the estimated signal is equal to $K_s(0)/\mathcal{E}[\epsilon^2]$. The causal filter yields a signal-to-noise ratio of 2.3, which

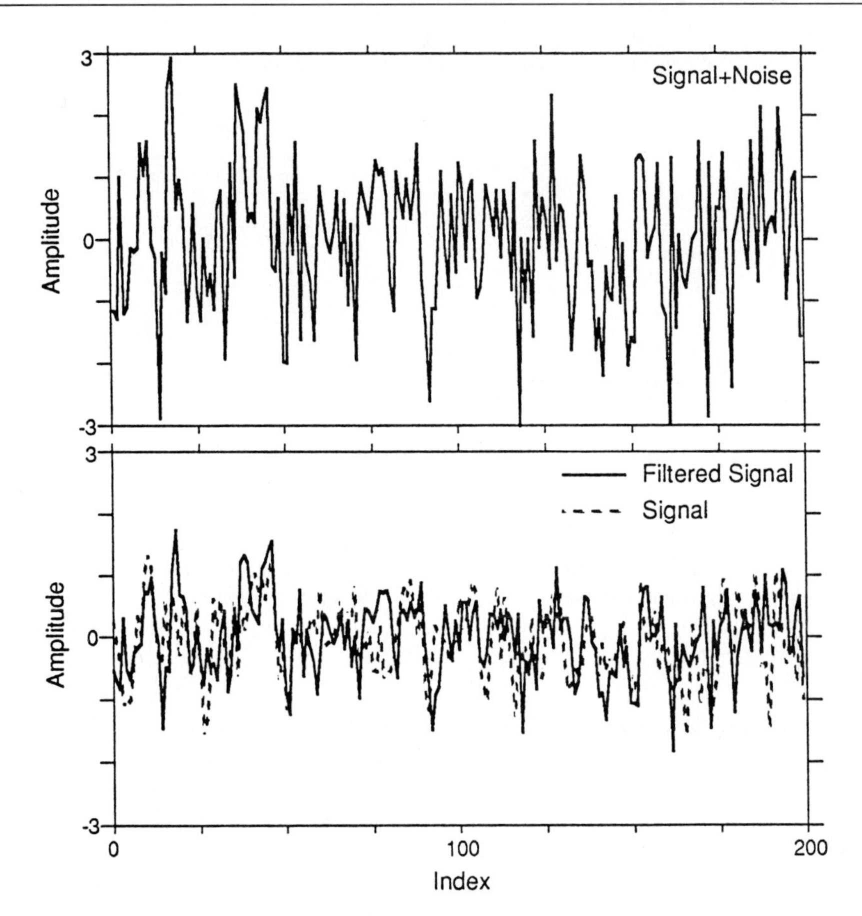

Figure 6.4 The upper panel displays observations having statistic characteristics corresponding to those given in the accompanying example. The output of the causal Wiener filter is shown in the bottom panel along with the actual signal, which is shown as a dashed line.

should be compared with the ratio of 1.17 in the observations. The ratio of the signal-to-noise ratios at the output and input of a signal processing operation is usually referred to as the *processing gain*. The best possible processing gain is 2.5 and equals 1.75 in the causal case. These rather modest gains are because of the close similarity between the power spectra of the signal and the noise. As the parameter of this signal's power spectrum is increased, the two become less similar, and the processing gain increases.

Now consider the case in which $l_e > 0$: We want to predict the signal's future value. The whitening filter portion of the solution does not depend on the value of l_e and is therefore identical to that just given. The second component of the Wiener filter's frequency response does depend on l_e and is given for positive values of l_e by

$$\left[\frac{e^{+j\omega l_e}\mathcal{S}_s(\omega)}{[\mathcal{S}_s+\mathcal{S}_n]^{+*}(\omega)}\right]_+ = \left[\frac{\frac{2}{\sqrt{7}}e^{+j\omega l_e}}{1-0.5e^{-j\omega}}\right]_+$$

The causal portion of this frequency response is found by shifting the unit-sample response to the *left* and retaining the positive-time portion. Because this frequency response has only one pole, this manipulation is expressed simply as a scaling.

$$\left[\frac{e^{+j\omega l_e}\mathcal{S}_s(\omega)}{[\mathcal{S}_s+\mathcal{S}_n]^{+*}(\omega)}\right]_+ = \frac{2}{\sqrt{7}}\frac{\left(\frac{1}{2}\right)^{l_e}}{1-0.5e^{-j\omega}}$$

The frequency response of the prediction filter is thus given by

$$H_\diamond(\omega) = \frac{1}{2}\frac{2^{-l_e}}{1-0.25e^{-j\omega}}$$

The optimum linear predictor is a *scaled* version of the signal estimator. The mean-squared error increases as the desired time of the predicted value exceeds the time of the last observation. In particular, the signal-to-noise ratio of the predicted value is given by

$$\frac{K_s(0)}{\mathcal{E}[\epsilon^2]} = \frac{1}{1-\frac{4}{7}\left(\frac{1}{2}\right)^{2l_e}}$$

The signal-to-noise ratio decreases rapidly as the prediction time extends into the future. This decrease is directly related to the reduced correlation between the signal and its future values in this example. This correlation is described by the absolute value of the signal's covariance function relative to its maximum value at the origin. As a covariance function broadens (corresponding to a lower frequency signal), the prediction error decreases. If a covariance function oscillates, the mean-squared prediction error varies in a similar fashion.

Finite-duration Wiener filters. Another useful formulation of Wiener filter theory is to constrain the filter's unit-sample response to have finite duration. To find this solution to the Wiener-Hopf equation, the values of L_f and L_i are chosen to be finite. Letting L represent the duration of the filter's unit-sample response ($L = L_f - L_i + 1$), the Wiener-Hopf equation becomes

$$K_{s\tilde{s}}(l+l_e) = \sum_{k=0}^{L-1} K_y(k-l)h_\diamond(k), \quad \text{for all } l \text{ in } [0, L-1]$$

This system of equations can be written in matrix form as $\mathbf{k}_{s\tilde{s}}(l_e) = \mathbf{K}_y\mathbf{h}_\diamond$.

$$\begin{bmatrix} K_{s\tilde{s}}(l_e) \\ K_{s\tilde{s}}(l_e+1) \\ \vdots \\ K_{s\tilde{s}}(l_e+L-1) \end{bmatrix} =$$

$$\begin{bmatrix} K_y(0) & K_y(1) & \cdots & K_y(L-1) \\ K_y(-1) & K_y(0) & \cdots & K_y(L-2) \\ \vdots & K_y(-1) & \ddots & \vdots \\ K_y(-L+1) & \cdots & K_y(-1) & K_y(0) \end{bmatrix} \cdot \begin{bmatrix} h_\diamond(0) \\ h_\diamond(1) \\ \vdots \\ h_\diamond(L-1) \end{bmatrix}$$

When the signal component of the observations equals the signal being estimated ($\tilde{s} = s$), the Wiener-Hopf equation becomes $\mathbf{k}_s(l_e) = \mathbf{K}_y\mathbf{h}_\diamond$. The $L \times L$ matrix $\mathbf{K}_y$ is the covariance matrix of the sequence of L observations. In the simple case of uncorrelated signal and noise components, this covariance matrix is the sum of those of the signal and the noise ($\mathbf{K}_y = \mathbf{K}_s + \mathbf{K}_n$). This matrix has an inverse in all but unusual circumstances with the result that the unit-sample response of the *FIR* Wiener filter is given by

$$\boxed{\mathbf{h}_\diamond = \mathbf{K}_y^{-1}\mathbf{k}_s(l_e)}$$

Because this covariance matrix is Toeplitz and Hermitian, its inverse can be efficiently computed using a variety of algorithms [105: 80–90]. The mean-squared error of the estimate is given by

$$\begin{aligned}\mathcal{E}[\epsilon^2] &= K_s(0) - \sum_{k=0}^{L-1} h_\diamond(k)K_s(l_e+k)\\ &= K_s(0) - \mathbf{k}_s^t(l_e)\mathbf{K}_y^{-1}\mathbf{k}_s(l_e)\end{aligned}$$

Linear prediction. One especially important variation of the *FIR* Wiener filter occurs in the unique situation in which no observation noise is present, and the signal generation model contains only poles [101, 104]. Thus, the signal $s(l)$ is generated by passing white noise $w(l)$ through a linear system given by the difference equation

$$s(l) = a_1 s(l-1) + a_2 s(l-2) + \cdots + a_p s(l-p) + w(l)$$

The coefficients $a_1, \ldots, a_p$ are unknown. This signal modeling approach is frequently used to estimate the signal's spectrum; this topic is addressed in §6.5.3 {334}.

As no noise is present in the observations, the filtered estimate of the signal ($l_e = 0$) equals $s(l)$ and the estimation error is exactly 0. The concern of linear prediction is not this trivial problem, but the so-called *one-step prediction problem* ($l_e = 1$): Predict the value of the signal at index l given values of $s(l-1), s(l-2), \ldots$. Thus, we seek a *FIR* Wiener filter predictor, which has the form

$$\widehat{s}(l) = h(0)s(l-1) + h(1)s(l-2) + \cdots + h(p-1)s(l-p)$$

Comparing the signal model equation to that for the Wiener filter predictor, we see that the model parameters $\{a_1, \ldots, a_p\}$ equal the Wiener filter's unit-sample response $h(\cdot)$ because the input $w(l)$ is uncorrelated from sample to sample. In linear prediction, the signal model parameters are used notationally to express the filter coefficients.

The Orthogonality Principle can be used to find the minimum mean-squared error predictor of the next signal value. By requiring orthogonality of the prediction error to each of the observations used in the estimate, the following set of equations results.

$$\begin{aligned}a_1K_s(0) + a_2K_s(1) + \cdots + a_pK_s(p-1) &= K_s(1)\\ a_1K_s(1) + a_2K_s(0) + \cdots + a_pK_s(p-2) &= K_s(2)\\ \vdots \qquad\qquad & \quad \vdots\\ a_1K_s(p-1) + a_2K_s(p-2) + \cdots + a_pK_s(0) &= K_s(p)\end{aligned}$$

In linear prediction, these are known as the *Yule-Walker equations*. Expressing them concisely in matrix form $\mathbf{K}_s\mathbf{a} = \mathbf{k}_s$, the solution is $\mathbf{a} = \mathbf{K}_s^{-1}\mathbf{k}_s$.

From the signal model equation, we see that the mean-squared prediction error $\mathcal{E}[\{s(l) - \widehat{s}(l)\}^2]$ equals the variance σ_w^2 of the white-noise input to the model. Computing the mean-squared estimation error according to Eq. 6.9 {298}, this variance is expressed by

$$\sigma_w^2 = K_s(0) - a_1 K_s(1) - \cdots - a_p K_s(p)$$

This result can be combined with the previous set of equations to yield a unified set of equations for the unknown parameters and the mean-squared error of the optimal linear predictive filter.

$$\begin{bmatrix} K_s(0) & K_s(1) & \cdots & K_s(p) \\ K_s(1) & K_s(0) & \cdots & K_s(p-1) \\ \vdots & K_s(1) & \ddots & \vdots \\ K_s(p) & \cdots & K_s(1) & K_s(0) \end{bmatrix} \cdot \begin{bmatrix} 1 \\ -a_1 \\ \vdots \\ -a_p \end{bmatrix} = \begin{bmatrix} \sigma_w^2 \\ 0 \\ \vdots \\ 0 \end{bmatrix} \tag{6.11}$$

To solve this set of equations for the model coefficients and the input-noise variance conceptually, we compute the preliminary result $\mathbf{K}_s^{-1}\boldsymbol{\delta}$. The first element of this vector equals the reciprocal of σ_w^2; normalizing $\mathbf{K}_s^{-1}\boldsymbol{\delta}$ so that its leading term is unity yields the coefficient vector $\mathbf{a}$. Levinson's algorithm can be used to solve these equations efficiently and simultaneously obtain the noise variance [105: 211–16].

6.4.3 Kalman Filters

Kalman filters are linear minimum mean-squared error waveform estimators that generalize Wiener filters to situations in which the observations have time-varying statistics [4];[58];[68: 269–301];[83]. On the surface, this problem may seem esoteric. Kalman filters, however, directly face the issue ducked by the Wiener filter: initiating a linear waveform estimator from the first observation. Wiener filters *tacitly* assume that observations have been forever available, an unrealistic requirement. More important, the structure of the Kalman filter provides direct insights into the nature of *MMSE* waveform estimators. For example, we should have noted previously that the Wiener filter's output does *not* have a power spectrum equal to that assumed for the signal. A simple illustration of this point is the noncausal Wiener filter; the power spectrum of its waveform estimate is

$$\mathcal{S}_{\hat{s}}(\omega) = \frac{\mathcal{S}_s^2(\omega)}{\mathcal{S}_s(\omega) + \mathcal{S}_n(\omega)}$$

This power spectrum tends toward the signal's power spectrum $\mathcal{S}_s(\omega)$ only when that of the noise tends to zero, not a very interesting situation. Why shouldn't the optimum filter make the power spectrum of its output equal that of the signal? This question and others are best answered in the context of the Kalman filter.

Rather than use the unit-sample response framework of linear waveform estimation developed in §6.4.1, Kalman filters are derived in terms of *state variable* characterizations of linear systems. This reexpression is, in fact, the key to gaining more insight into linear waveform estimators. The Kalman filter provides the basis for *tracking* systems that coalesce the outputs from several arrays to locate a moving source of propagating energy. Here, the concept of state becomes crucial to developing effective algorithms. Conversely, signals having zeros in their power spectra cannot be easily represented by state variable characterizations; the gain in insight more than compensates for the loss of generality.

State-variable modeling of observations. Signal production is modeled as passing white noise through a system characterized by the state vector $\mathbf{v}$. The *state equation*, which governs the behavior of the system, is given by

$$\mathbf{v}(l) = \mathbf{A}(l)\mathbf{v}(l-1) + \mathbf{u}(l)$$

The *observation equation* describes the signal $\mathbf{s}(l)$ as a linear combination of state vector components $\mathbf{C}(l)\mathbf{v}(l)$ observed in the presence of additive noise.

$$\mathbf{y}(l) = \mathbf{C}(l)\mathbf{v}(l) + \mathbf{n}(l)$$

Note that the observations and the signal can be vectors in this characterization; in this way, multiple, interdependent observations can be described and their estimates described as the result of a single, unified operation. In previous sections, the signal and noise models have been constrained to stationary random sequences. The explicit dependence of every aspect of the state and observation equations on the index l permits relaxation of that constraint. The noise vectors are assumed to be zero-mean, white Gaussian random vectors.

$$\mathcal{E}[\mathbf{u}(k)\mathbf{u}'(l)] = \mathbf{K}_u(l)\delta(k-l) \qquad \mathcal{E}[\mathbf{n}(k)\mathbf{n}'(l)] = \mathbf{K}_n(l)\delta(k-l)$$

Note the *components* of these noise vectors are allowed to be correlated in this framework. Additive colored noise in the observations can be described by enlarging the number of states in the state equation and defining the matrix $\mathbf{C}$ appropriately.

Kalman filter derivation. The minimum mean-squared error *linear* estimate of the signal is equivalent to estimating the state vector and then multiplying by the matrix $\mathbf{C}$: $\widehat{\mathbf{s}}(l) = \mathbf{C}(l)\widehat{\mathbf{v}}(l)$. We shall thus derive the optimal state estimate. Furthermore, this estimate is the *filtered* estimate: The estimate of the state at index l depends on all of the observations made up to and including l, commencing with $\mathbf{y}(0)$. We describe later how to extend this estimate to prediction problems.

Denote the state estimate by $\widehat{\mathbf{v}}(l|m)$, the notation intending to convey that the estimate at index l depends on *all* the observations made up to and including those made at index m $\{\mathbf{y}(m), \mathbf{y}(m-1), \ldots, \mathbf{y}(0)\}$. The filtered state estimate thus takes the form $\widehat{\mathbf{v}}(l|l)$, and we assume it to have a state variable characterization:

$$\widehat{\mathbf{v}}(l|l) = \mathbf{M}(l)\widehat{\mathbf{v}}(l-1|l-1) + \mathbf{G}(l)\mathbf{y}(l)$$

This formulation is verified by finding the matrices **M** and **G** that permit the Orthogonality Principle to be obeyed: The estimation error $\boldsymbol{\epsilon}(l|l) = \mathbf{v}(l) - \widehat{\mathbf{v}}(l|l)$ must be orthogonal to all linear transformations that can be applied to the observations. For instance, the waveform estimator is certainly a linear system acting on the observations; hence the estimator $\widehat{\mathbf{v}}(l-1|l-1)$ is a linear transformation of all of the observations obtained before index l and must be orthogonal to the estimation error.

$$\mathcal{E}\left[\{\mathbf{v}(l) - \widehat{\mathbf{v}}(l|l)\}'\widehat{\mathbf{v}}(l-1|l-1)\right] = 0$$

This property can be used to find the matrix **M**.[†]

$$\mathcal{E}\left[\{\mathbf{v}(l) - \mathbf{G}\mathbf{y}(l)\}'\,\widehat{\mathbf{v}}(l-1|l-1)\right] = \mathcal{E}\left[\{\mathbf{M}\widehat{\mathbf{v}}(l-1|l-1)\}'\,\widehat{\mathbf{v}}(l-1|l-1)\right]$$

Adding and subtracting the state $\mathbf{v}(l-1)$ from $\widehat{\mathbf{v}}(l-1|l-1)$ occurring in braces on the right side of this equation and applying the Orthogonality Principle results in

$$\mathcal{E}\left[\{\mathbf{v}(l) - \mathbf{G}\mathbf{y}(l)\}'\,\widehat{\mathbf{v}}(l-1|l-1)\right] = \mathcal{E}\left[\{\mathbf{M}\mathbf{v}(l-1)\}'\,\widehat{\mathbf{v}}(l-1|l-1)\right]$$

Using the observation equation to eliminate $\mathbf{y}(l)$ leaves

$$\mathcal{E}\left[\{\mathbf{v}(l) - \mathbf{G}\mathbf{C}\mathbf{v}(l) - \mathbf{G}\mathbf{n}(l)\}'\,\widehat{\mathbf{v}}(l-1|l-1)\right] = \mathcal{E}\left[\{\mathbf{M}\mathbf{v}(l-1)\}'\,\widehat{\mathbf{v}}(l-1|l-1)\right]$$

Because the noise $\mathbf{n}(l)$ is uncorrelated from index to index, it is uncorrelated with the state estimate $\widehat{\mathbf{v}}(l-1|l-1)$, which depends only on the noise values previous to index l. Thus, the expected value of the inner product between these two quantities is zero. Substituting the state equation into the left side of this equation and noting a similar simplification with respect to the input noise $\mathbf{u}(l)$ leaves

$$\mathcal{E}\left[\left\{(\mathbf{I} - \mathbf{G}\mathbf{C})\mathbf{A}\mathbf{v}(l-1)\right\}'\,\widehat{\mathbf{v}}(l-1|l-1)\right] = \mathcal{E}\left[\left\{\mathbf{M}\mathbf{v}(l-1)\right\}'\,\widehat{\mathbf{v}}(l-1|l-1)\right]$$

For this equality to be satisfied for all nonzero states and their estimates, the matrix **M** must satisfy

$$\mathbf{M} = [\mathbf{I} - \mathbf{G}\mathbf{C}]\mathbf{A}$$

The Kalman filter state estimator equation thus becomes

$$\boxed{\widehat{\mathbf{v}}(l|l) = \mathbf{A}(l)\widehat{\mathbf{v}}(l-1|l-1) + \mathbf{G}(l)\left[\mathbf{y}(l) - \mathbf{C}(l)\mathbf{A}(l)\widehat{\mathbf{v}}(l-1|l-1)\right]}$$

This rather complicated result hides an intriguing and useful interpretation. A critical aspect of this interpretation is the idea of the *one-step predictor* $\widehat{\mathbf{v}}(l|l-1)$: the estimate of the current state based on all previous observations. As no other observations are available, assume that the estimate is based solely on the estimate of the state at $l-1$: $\widehat{\mathbf{v}}(l-1|l-1)$. The predictor is therefore written

$$\widehat{\mathbf{v}}(l|l-1) = \mathbf{N}(l)\widehat{\mathbf{v}}(l-1|l-1)$$

[†] The dependence of the various matrices on time is notationally implicit in sequel. Consequently, reference to a matrix (such as **A**) denotes the matrix $\mathbf{A}(l)$ at index l.

where $\mathbf{N}$ is a matrix found by satisfying the Orthogonality Principle. This matrix is found to equal $\mathbf{A}(l)$ so that the one-step predictor is given by

$$\boxed{\widehat{\mathbf{v}}(l|l-1) = \mathbf{A}(l)\widehat{\mathbf{v}}(l-1|l-1)}$$

In the Wiener filter example of estimating a first-order signal observed in noise {300}, we found the one-step predictor to be a scaled version of the previous estimate. We now see that the Kalman filter generalizes that finding to higher-order signals.

The estimator equation can thus be expressed in terms of the one-step predictor.

$$\widehat{\mathbf{v}}(l|l) = \widehat{\mathbf{v}}(l|l-1) + \mathbf{G}[\mathbf{y}(l) - \mathbf{C}\widehat{\mathbf{v}}(l|l-1)]$$

The term $\boldsymbol{\nu}(l) = \mathbf{y}(l) - \mathbf{C}\widehat{\mathbf{v}}(l|l-1)$ serves as the input and represents that part of the current observations $\mathbf{y}(l)$ *not* predicted by the previous ones. This quantity is often referred to as the *innovations sequence*. This basic form of the estimator equation occurs in many linear estimation situations: The value of the one-step predictor is augmented by an input term equal to a matrix $\mathbf{G}$, the *Kalman gain matrix*, multiplying the innovations sequence.

To derive the equations determining the Kalman gain, the estimation errors $\boldsymbol{\epsilon}(l|l)$ and $\boldsymbol{\epsilon}(l|l-1)$ as well as the observation vector $\mathbf{y}(l)$ are given by

$$\begin{aligned}
\boldsymbol{\epsilon}(l|l-1) &= \mathbf{A}\boldsymbol{\epsilon}(l-1|l-1) + \mathbf{u}(l)\\
\boldsymbol{\epsilon}(l|l) &= \mathbf{A}\boldsymbol{\epsilon}(l-1|l-1) + \mathbf{u}(l) - \mathbf{GCA}\boldsymbol{\epsilon}(l-1|l-1) - \mathbf{G}[\mathbf{Cu}(l) + \mathbf{n}(l)]\\
\mathbf{y}(l) &= \mathbf{CA}[\boldsymbol{\epsilon}(l-1|l-1) + \widehat{\mathbf{v}}(l-1|l-1)] + \mathbf{Cu}(l) + \mathbf{n}(l)
\end{aligned}$$

Let $\mathbf{K}_\epsilon$ denote the covariance matrix of the state estimation error: $\mathbf{K}_\epsilon = \mathcal{E}[\boldsymbol{\epsilon}\boldsymbol{\epsilon}^t]$. The covariance matrix of the one-step prediction error is found from the first equation to be

$$\mathbf{K}_\epsilon(l|l-1) = \mathbf{A}\mathbf{K}_\epsilon(l-1|l-1)\mathbf{A}^t + \mathbf{K}_u$$

Because of the Orthogonality Principle, $\mathcal{E}\left[\boldsymbol{\epsilon}(l|l)\mathbf{y}^t(l)\right] = 0$. Using the last two equations, this constraint yields

$$\mathbf{K}_\epsilon(l|l-1)\mathbf{C}^t - \mathbf{G}[\mathbf{C}\mathbf{K}_\epsilon(l|l-1)\mathbf{C}^t + \mathbf{K}_n] = 0$$

We find the Kalman gain matrix to be

$$\boxed{\mathbf{G}(l) = \mathbf{K}_\epsilon(l|l-1)\mathbf{C}^t(l)\left[\mathbf{C}(l)\mathbf{K}_\epsilon(l|l-1)\mathbf{C}^t(l) + \mathbf{K}_n(l)\right]^{-1}}$$

Expressing the covariance matrix of the estimation error directly, a simple recursion for it emerges.

$$\mathbf{K}_\epsilon(l|l) = \mathbf{K}_\epsilon(l|l-1) - \mathbf{G}(l)\mathbf{C}(l)\mathbf{K}_\epsilon(l|l-1)$$

Thus, the Kalman gain matrix is found recursively from the covariance matrix of the estimation error and the covariance matrix of the one-step prediction error. This recursion amounts to a solution of the *Riccati equation*, the nonlinear difference equation governing the evolution of the gain matrix.

In summary, the Kalman filter equations are

Kalman Prediction Equations:

$$
\begin{aligned}
\widehat{\mathbf{s}}(l) &= \mathbf{C}(l)\widehat{\mathbf{v}}(l|l) \\
\widehat{\mathbf{v}}(l|l) &= \mathbf{A}(l)\widehat{\mathbf{v}}(l-1|l-1) + \mathbf{G}(l)\left[\mathbf{y}(l) - \mathbf{C}(l)\mathbf{A}(l)\widehat{\mathbf{v}}(l-1|l-1)\right]
\end{aligned} \qquad (6.12a)
$$

Kalman Gain Equations:

$$
\begin{aligned}
\mathbf{G}(l) &= \mathbf{K}_\epsilon(l|l-1)\mathbf{C}^t(l)\left[\mathbf{C}(l)\mathbf{K}_\epsilon(l|l-1)\mathbf{C}^t(l) + \mathbf{K}_n(l)\right]^{-1} \\
\mathbf{K}_\epsilon(l|l-1) &= \mathbf{A}(l)\mathbf{K}_\epsilon(l-1|l-1)\mathbf{A}^t(l) + \mathbf{K}_u(l) \\
\mathbf{K}_\epsilon(l|l) &= \left[\mathbf{I} - \mathbf{G}(l)\mathbf{C}(l)\right]\mathbf{K}_\epsilon(l|l-1)
\end{aligned} \qquad (6.12b)
$$

To perform these recursions, initial conditions must be specified. As observations are usually taken to begin at $l = 0$, values for $\widehat{\mathbf{v}}(-1|-1)$ and $\mathbf{K}_\epsilon(-1|-1)$ must somehow be established. The initial value $\widehat{\mathbf{v}}(-1|-1)$ equals the state estimate when no observations are available. Given no data, the minimum mean-squared error estimate of *any* quantity is the mean. As the noise terms in the state variable representation of the signal and the observations are assumed to have zero mean, the expected value of the state is 0, implying that the initial value of the state estimate should be 0. In this case, the variance of the estimation error equals that of the state. Thus, the initial condition of the Riccati equation solution is $\mathbf{K}_\epsilon(-1|-1) = \mathcal{E}[\mathbf{v}(-1)\mathbf{v}^t(-1)]$.

Example

The simplest example of a Kalman filter waveform estimator occurs when the signal generation model is first-order. Recalling this type of example from the previous section {300}, its state variable characterization is

$$v(l) = \frac{1}{2}v(l-1) + w(l) \quad \mathbf{K}_w = [1]$$

and the observation equation is

$$y(l) = v(l) + n(l) \quad \mathbf{K}_n = [8/7]$$

Thus, the quantities that determine the solution of the Kalman filter components are $\mathbf{A} = [1/2]$ and $\mathbf{C} = [1]$. The estimator equation is, therefore,

$$\widehat{s}(l) = \frac{1}{2}\widehat{s}(l-1) + G(l)[y(l) - \frac{1}{2}\widehat{s}(l-1)]$$

The covariance matrices are simply variances and the equations determining the Kalman gain are scalar equations.

$$
\begin{aligned}
G(l) &= \frac{\sigma_\epsilon^2(l|l-1)}{\sigma_\epsilon^2(l|l-1) + 8/7} \\
\sigma_\epsilon^2(l|l-1) &= \frac{1}{4}\sigma_\epsilon^2(l-1|l-1) + 1 \\
\sigma_\epsilon^2(l|l) &= \sigma_\epsilon^2(l|l-1) - G(l)\sigma_\epsilon^2(l|l-1)
\end{aligned}
$$

The initial value for the variance, $\sigma_\epsilon^2(-1|-1)$, equals the signal variance, which equals 4/3. The table shows the values of the quantities involved in the Kalman gain computation.

l	$\sigma_\epsilon^2(l\|l-1)$	$G(l)$	$\sigma_\epsilon^2(l\|l)$
0	1.3333	0.5385	0.6154
1	1.1538	0.5024	0.5742
2	1.1435	0.5001	0.5716
3	1.1429	0.5000	0.5714
⋮	⋮	⋮	⋮
∞	8/7	1/2	4/7

Note how the gain varies despite the fact that the signal model is stationary. This effect occurs because the filter is started with no prior observation values. After few iterations that correspond to this initial transient, the Kalman filter coefficients settle and the signal estimate is governed by the difference equation

$$\begin{aligned}\widehat{s}(l) &= \frac{1}{2}\widehat{s}(l-1) + \frac{1}{2}[y(l) - \frac{1}{2}\widehat{s}(l-1)] \\ &= \frac{1}{4}\widehat{s}(l-1) + \frac{1}{2}y(l)\end{aligned}$$

which is precisely the Wiener filter result. The Kalman filter has the advantage that there is an explicit starting procedure; the Wiener filter is the Kalman filter in "steady state" for otherwise stationary signals.

6.4.4 Dynamic Adaptive Filtering

The Wiener and Kalman filter approaches presume that the signal *and* the noise are statistically well characterized. The theoretical framework for each presumes that the signal has been produced by passing white noise through a *known* linear system. Kalman filter theory allows this system to be time varying, but the nature of the time variation must be known a priori. Each filter has a specific implementation and performance characteristics that can be determined before the observations are made. If the statistical characteristics of the signal and noise are known well, either of these waveform estimation techniques can be fruitfully used.

In many other situations, however, the signal or the observation noise are ill defined. They may vary in unpredictable ways or have poorly specified spectral or temporal characteristics. One typical problem is *interference*: Rather than just noise, the observations are corrupted by other signals that defy simple statistical characterization. Here, the waveform estimator must *learn* from the observations what is not known a priori and *adapt* its characteristics or structure as data become available. Rather than, for example, estimating a signal's power spectrum "off-line" and then incorporating the estimate into the filter's structure for subsequent observations, *dynamic adaptive filters* either take the imprecise knowledge into account directly or modify themselves while producing signal estimates [3, 36, 68, 74, 161].

Such filters can adjust two aspects of themselves: their structure and their coefficients. Consider a finite-duration (*FIR*) Wiener filter. The characteristics of this filter are its duration (L) and the values of the unit-sample response $\mathbf{h}$. The quantity L is a structural parameter because it determines the number of filter coefficients; the coefficients are determined by signal and noise characteristics. Structure adjustment is usually quite involved while changing coefficient values dynamically is much more plausible. For example, adjusting the filter coefficients from one sample to another by a presumably modest increment poses no fundamental difficulties (as shown in sequel). Modifying the number of coefficients, however, requires the ability to compute an initial value for a new coefficient, conceptually equivalent to "adjusting" the value from zero. Not only is this adjustment in reality a rather drastic change, but it also increases the computations required for the other coefficients. Consequently, we concentrate here on coefficient adaptations to changes in the observations.

The *FIR* Wiener filter consists of a finite set of coefficients that minimize the mean-squared estimation error. The closed form solution for this set is

$$\mathbf{h}_{\diamond} = \mathbf{K}_y^{-1}\mathbf{k}_{s\tilde{s}}(l_e)$$

where $\mathbf{K}_y$ is the covariance matrix of the observations (assumed to consist of signal plus noise) and $\mathbf{k}_{s\tilde{s}}(l_e)$ consists of L values starting at lag l_e of the cross-covariance function between the signal and its observed counterpart. For simplicity, only the one-step prediction problem ($l_e = 1$) is explicitly considered here. To determine a method of *adapting* the filter coefficients at each sample, two approaches have been used. The first one considered here, the least mean squares (*LMS*) algorithm, is to return to the linear minimum mean-squared error minimization problem and find iterative methods of solving it that can incorporate new observations "on the fly." The second approach, the recursive least squares (*RLS*) algorithm (considered in the next section), amounts to finding ways of updating the optimal Wiener solution at each index.

Least mean squares (LMS) algorithm. The *LMS* algorithm is intimately related to an optimization method known as the *method of steepest descent* [3: 46–52];[46]: Given a function to be minimized, the mean-squared error $\mathcal{E}[\epsilon^2]$ at index l, and the independent variables, the vector of filter coefficients $\mathbf{h}$, the minimum can be found iteratively by adjusting the filter coefficients according to the gradient of the mean-squared error.

$$\mathbf{h}(i+1) = \mathbf{h}(i) - \mu\,\nabla_h\,\mathcal{E}[\epsilon^2(l)]\,, \quad \mu > 0$$

The gradient "points" in the direction of the maximum rate of change of the mean-squared error for any set of coefficient values. The direction of the gradient's negative points toward the minimum. The often-used analogy is rolling a ball down the edge of a bowl. The direction the ball rolls at each instant corresponds to the locally maximum slope, the negative of the gradient at each spatial location. If the bowl is smooth and isn't too bumpy, the ball rolls to the global minimum; if bumpy, the ball may stop in a local minimum. Because of the characteristics of the mean-squared error, the bowl it forms is provably convex; hence, it has only one minimum. The method of steepest descent is therefore guaranteed to converge.

To compute this gradient, the mean-squared error must be expressed in terms of the filter coefficients. Letting the previous L observations be expressed by the observation vector $\mathbf{y} = \text{col}[y(l-1), \ldots, y(l-L)]$ with $\mathbf{y} = \tilde{\mathbf{s}} + \mathbf{n}$, the application of an arbitrary set of coefficients results in a mean-squared error given by

$$\begin{aligned}\mathcal{E}[\epsilon^2] &= \mathcal{E}\left[(s - \mathbf{h}'\mathbf{y})^2\right] \\ &= \sigma_s^2 - 2\mathbf{h}'\mathbf{k}_{s\tilde{s}} + \mathbf{h}'\mathbf{K}_y\mathbf{h}\end{aligned}$$

The gradient of this quantity is, therefore,

$$\nabla_h \mathcal{E}[\epsilon^2] = -2\mathbf{k}_{s\tilde{s}} + 2\mathbf{K}_y\mathbf{h}$$

and the method of steepest descent is characterized by the iterative equation for the filter coefficients given by[†]

$$\mathbf{h}(i+1) = \mathbf{h}(i) + 2\mu[\mathbf{k}_{s\tilde{s}} - \mathbf{K}_y\mathbf{h}(i)]$$

Here, the variable i denotes iteration number, which in general need not be related to index values: Filter adaptation could occur every once in a while rather than at every index. The term in the brackets can also be written as

$$\mathbf{k}_{s\tilde{s}} - \mathbf{K}_y\mathbf{h}(i) = \mathcal{E}[\epsilon(i)\mathbf{y}]$$

where $\epsilon(i) = s - \mathbf{h}'(i)\mathbf{y}$ is the estimation error.

The convergence of the method of steepest descent is controlled by the value of the *adaptation parameter* μ. The difference of the filter coefficients from the optimal set, $\mathbf{d}(i) = \mathbf{h}_\diamond - \mathbf{h}(i)$, is governed by an easily derived iterative equation.

$$\mathbf{d}(i+1) = [\mathbf{I} - 2\mu\mathbf{K}_y]\mathbf{d}(i)$$

This equation says that the error at each iteration is "proportional" to the error at the previous iteration; consequently, the method of steepest descent converges linearly. A closed form expression for the vector of coefficient differences is

$$\mathbf{d}(i) = [\mathbf{I} - 2\mu\mathbf{K}_y]^i\mathbf{d}(0)$$

Convergence is assured so long as the ith power of the matrix in brackets decreases to 0 as i becomes large. This condition is more easily expressed in terms of the eigenvalues of this matrix (see §B.5 {493}): A necessary and sufficient condition for convergence is that the magnitude of the eigenvalues of $\mathbf{I} - 2\mu\mathbf{K}_y$ must all be less than unity (see Prob. 7.16 {417}). Coupling this condition with the necessity that $\mu > 0$ results in the requirement for the method of steepest descent to converge, μ must be less than the reciprocal of the largest eigenvalue of $\mathbf{K}_y$.

$$\boxed{0 < \mu < \frac{1}{\max \lambda_{\mathbf{K}_y}}}$$

[†]The form of this equation is remarkably similar to a Kalman filter equation for estimating the filter coefficients. This similarity is no accident and reveals the fundamental importance of the ideas expressed by Kalman filter.

The "time constant" of the convergence is governed by the *smallest* eigenvalue and is greater than $1/(2\mu \min\{\lambda_{\mathbf{K}_y}\})$. The greatest rate of convergence is obtained with the smallest time constant; we should thus choose μ as near to its upper limit as possible. When the largest value of μ is selected, the rate of convergence of the method of steepest descent is proportional to the ratio of the largest and smallest eigenvalues of the observations' covariance matrix.

The preceding discussion described a numerical procedure for computing the coefficients of the *FIR* Wiener filter as an alternative to the closed-form expression that involves a matrix inverse. This procedure does *not* yield an adaptive filter because we must somehow calculate $\mathcal{E}[\epsilon(i)\mathbf{y}]$, but it does inspire an idea. Expressing the steepest descent iteration in terms of the estimation error yields

$$\mathbf{h}(i+1) = \mathbf{h}(i) + 2\mu\, \mathcal{E}[\epsilon(i)\mathbf{y}]$$

The *LMS* algorithm amounts to approximating the expected value by the product of the observation vector and the estimation error, a technique known as *stochastic approximation* [156]. The optimization iteration now coincides with the index of the observations and a desired signal ($i = l$). The equations governing the *LMS* algorithm are [3: 68–83];[68: 194–259]

$$\boxed{\begin{aligned} \widehat{s}(l) &= \sum_{k=0}^{L-1} h(l,k)y(l-k-1) = \mathbf{h}^t(l)\mathbf{y}(l-1) \\ \epsilon(l) &= s(l) - \widehat{s}(l) \\ \mathbf{h}(l+1) &= \mathbf{h}(l) + 2\mu\epsilon(l)\mathbf{y}(l-1) \end{aligned}}$$

Employing the *LMS* algorithm presumes that either the signal is known to some degree and we want to estimate some aspect of it or that we want to cancel it so that the error sequence $\epsilon(l)$ is "signal-free" [3: 87–98];[68: 7–31]. A typical example is one-step prediction where no observation noise is present: The filter predicts the signal value at index l, the error at time l is computed, and this error is used to adapt the filter coefficients that yield the *next* signal estimate. Because of the limited extent of the data used to change the filter coefficients, we can estimate well signals having slowly varying characteristics.

Because the filter coefficients depend on new observations and are continually updated via the estimate of the gradient, the *LMS* filter coefficients do *not* converge to a stable set of values even when the observations are stationary. Rather, after an initial transient response to the onset of the observations, the coefficients "settle" toward a set of values and vary randomly to some degree. Furthermore, the convergence bounds on the adaptation parameter must be sharpened to include the gradient estimate rather than its expected value [56]. For the *LMS* filter to converge, the adaptation parameter must satisfy [55, 56]

$$0 < \sum_j \frac{\mu\lambda_j}{1-2\mu\lambda_j} < 1 \quad \text{and} \quad 0 < \mu < \frac{1}{2\max_j\{\lambda_j\}}$$

where λ_j, $j = 1, \ldots, L$ denotes the eigenvalues of $\mathbf{K}_y$. Enforcing this condition requires clairvoyance because the covariance is assumed to be unknown. A more practical and more stringent bound is [55]

$$0 < \mu \leq \frac{1}{3\,\mathrm{tr}[\mathbf{K}_y]} \tag{6.13}$$

because the trace of the covariance matrix equals the total power in the observations, an easily estimated quantity in practice.

The variances of the filter's coefficients are also related to the adaptation parameter μ, but in the opposite way than the settling time. A small value of μ results in little change of the coefficients from one update to the next; the filter coefficients settle slowly but their long-term values do not vary greatly about the optimal values. Larger values of μ allow greater changes; more rapid settling occurs, but the long-term values have larger variations. Fig. 6.5 illustrates this point. The degree of the long-term variations in stationary environments is assessed by the so-called *misadjustment factor* M: the difference between the variances of the *LMS* estimate and the optimal, Wiener filter estimate normalized by the optimal value [68: 236–37].

$$M = \frac{\sigma_\epsilon^2 - \sigma_\epsilon^2(Wiener)}{\sigma_\epsilon^2(Wiener)}$$

The misadjustment factor M_{LMS} for the *LMS* algorithm is well approximated (for small values of the factor) by $\mu L \sigma_s^2$ [56];[68: 236–37].

On the basis of these simple considerations, one is tempted to consider changing the adaptation parameter in addition to the filter coefficients; conceptually, large values could be used initially and then slowly reduced to yield smaller coefficient variations about their optimal values. Given sufficient detail about the signal characteristics, such a scheme might be feasible. If this reduction of μ is tried in an ad hoc manner, however, the coefficients might settle about artificial values because they are not allowed to change.[†]

Recursive least squares (RLS) algorithm. Another approach in adaptive filtering is to approximate the solution for optimal set of filter coefficients *directly* without imposing the additional burden of approximating an optimization procedure as the *LMS* algorithm does [3: 111–21];[56];[68: 381–444]. This more direct approach results in filters having better convergence properties but at the expense of computational complexity.

Recalling the solution to the *FIR* Wiener filter, the filter coefficients are given by

$$\mathbf{h}_\diamond = \mathbf{K}_y^{-1}\mathbf{k}_{s\tilde{s}}$$

The data-dependent quantity in this expression is the covariance matrix of the observations. Not only is a recursive method of computing the inverse of this matrix sought, but also a recursive estimate of the matrix itself. To estimate the covariance matrix of size $L \times L$ from zero-mean data, the data are formed into a series of groups (frames) of duration L. The lth frame, denoted by the vector $\mathbf{y}(l)$, contains the most recent L

[†]Note that *any* set of vectors is a solution to $\mathbf{h}(l+1) = \mathbf{h}(l)$ when $\mu = 0$.

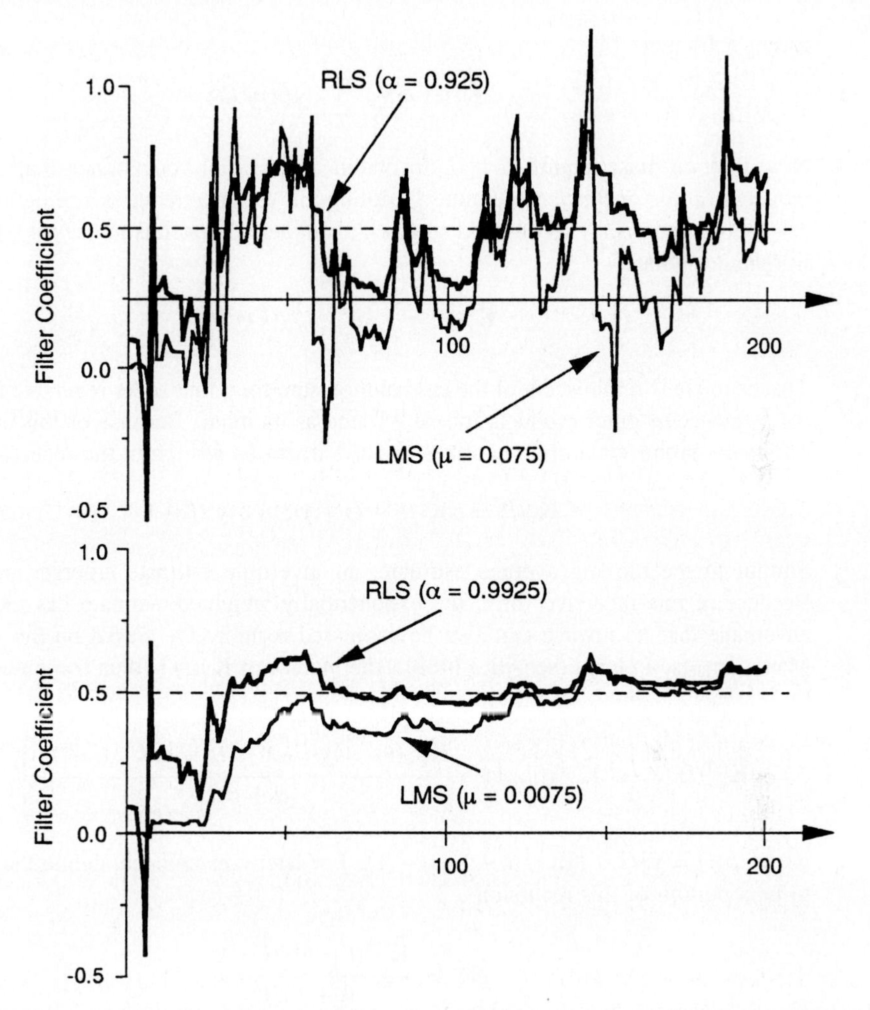

Figure 6.5 The filter coefficients of one-step predictors produced by first-order *LMS* and *RLS* algorithms are shown as a function of the number of observations. The coefficient's correct value is $1/2$. In the upper panel, both algorithms settle rapidly but vary greatly about their optimal values. In the lower, the algorithms' parameters were adjusted to reduce the variations of their estimates. Note that the settling time of the *RLS* filter coefficient is not strongly influenced by this parameter value change. In comparison, settling takes much longer in the *LMS* algorithm.

observations: $\mathbf{y}(l) = \text{col}[y(l), y(l-1), \ldots, y(l-L+1)]$. Two methods of estimating the covariance matrix are commonly used. Whichever is chosen, the estimate at index l must depend on *only* the observations obtained *before* that time.† The first method of estimating the covariance matrix is to average the outer products computed from the most

†This requirement is needed because we are focused on prediction problems.

recent S frames.

$$\widehat{\mathbf{K}}_y(l) = \frac{1}{S} \sum_{k=l-S}^{l-1} \mathbf{y}(k)\mathbf{y}^t(k)$$

Note that we must require $S \geq L$ to obtain an invertible covariance matrix from this *moving average* or *Bartlett* estimate. Updating this estimate requires storing all of the $S+L-1$ observations. A second and somewhat simpler approach is to use the *exponentially weighted* estimate†

$$\widehat{\mathbf{K}}_y(l) = \sum_{k=0}^{l-1} \alpha^{l-1-k}\mathbf{y}(k)\mathbf{y}^t(k)$$

This estimate is reminiscent of the convolution sum for a first-order recursive filter having the sequence of outer products for each frame as its input. Because of this analogy, this estimate's promised simplicity arises because it can be written in the recursive form

$$\widehat{\mathbf{K}}_y(l) = \alpha\widehat{\mathbf{K}}_y(l-1) + \mathbf{y}(l-1)\mathbf{y}^t(l-1)$$

Similar to the moving average estimate, an invertible estimate emerges once $l \geq S$. Because of this recursive form, the exponentially weighted estimate has the additional advantage that its inverse can also be expressed recursively. Based on the well-known Matrix Inverse Lemma (see §B.3 {489}), the inverse of $\widehat{\mathbf{K}}_y(l)$ is thus computed according to

$$\widehat{\mathbf{K}}_y^{-1}(l) = \frac{1}{\alpha}\left[\widehat{\mathbf{K}}_y^{-1}(l-1) - \frac{\widehat{\mathbf{K}}_y^{-1}(l-1)\mathbf{y}(l-1)\mathbf{y}^t(l-1)\widehat{\mathbf{K}}_y^{-1}(l-1)}{\alpha + \beta(l)}\right], \tag{6.14}$$

where $\beta(l) = \mathbf{y}^t(l-1)\widehat{\mathbf{K}}_y^{-1}(l-1)\mathbf{y}(l-1)$. For later convenience, define the vector $\mathbf{g}(l)$ to be a portion of this recursion

$$\mathbf{g}(l) = \frac{\widehat{\mathbf{K}}_y^{-1}(l-1)\mathbf{y}(l-1)}{\alpha + \beta(l)}$$

and we thus express the estimate's inverse by

$$\widehat{\mathbf{K}}_y^{-1}(l) = \frac{1}{\alpha}\left[\widehat{\mathbf{K}}_y^{-1}(l-1) - \mathbf{g}(l)\mathbf{y}^t(l-1)\widehat{\mathbf{K}}_y^{-1}(l-1)\right],$$

A similar recursive structure can be imposed on the second term in the expression for the *FIR* Wiener filter's unit-sample response that started all of this ($\mathbf{h}_\diamond = \mathbf{K}_y^{-1}\mathbf{k}_{s\tilde{s}}$). The vector $\mathbf{k}_{s\tilde{s}}$ arose from the cross-covariance of the to-be-estimated signal with the observations. This quantity can be approximated by the recursion

$$\widehat{\mathbf{k}}_{s\tilde{s}}(l) = \alpha\widehat{\mathbf{k}}_{s\tilde{s}}(l-1) + s(l)\mathbf{y}(l-1)$$

†Note that the frames used in both estimates overlap. If desired, nonoverlapping frames could be used with the advantage of a more statistically stable estimate. Observations would span a longer period, however.

The filter coefficients $\mathbf{h}(l+1)$ are given by $\widehat{\mathbf{K}}_y^{-1}(l)\widehat{\mathbf{k}}_{s\tilde{s}}(l)$, which on substitution of the recursions becomes

$$\mathbf{h}(l+1) = \mathbf{h}(l) + \mathbf{g}(l)\left[s(l) - \mathbf{h}^t(l)\mathbf{y}(l-1)\right]$$

The equations governing the *RLS* algorithm are therefore

$$\begin{aligned}
\widehat{s}(l) &= \mathbf{h}^t(l)\mathbf{y}(l-1) \\
\epsilon(l) &= s(l) - \widehat{s}(l) \\
\beta(l) &= \mathbf{y}^t(l-1)\widehat{\mathbf{K}}_y^{-1}(l-1)\mathbf{y}(l-1) \\
\mathbf{g}(l) &= \frac{\widehat{\mathbf{K}}_y^{-1}(l-1)\mathbf{y}(l-1)}{\alpha + \beta(l)} \\
\widehat{\mathbf{K}}_y^{-1}(l) &= \frac{1}{\alpha}\left[\widehat{\mathbf{K}}_y^{-1}(l-1) - \mathbf{g}(l)\mathbf{y}^t(l-1)\widehat{\mathbf{K}}_y^{-1}(l-1)\right] \\
\mathbf{h}(l+1) &= \mathbf{h}(l) + \mathbf{g}(l)\epsilon(l)
\end{aligned}$$

The difference equation for updating the inverse of the covariance matrix is numerically sensitive. This computational issue can be approached using so-called square-root algorithms that update the matrix square root of the covariance matrix [15]. Another approach based on numerical linear algebra is discussed in §7.5.3 {407}. The parameter α in this algorithm plays a similar role as μ in the *LMS* approach. Both parameters control the effective amount of data used in the averaging and hence the degree to which the algorithms can track variations in signal characteristics. The *RLS* algorithm *always* converges so long as $0 < \alpha < 1$; otherwise, the covariance estimate recursion is unstable. In addition, α controls the variance of the long-term estimate. The misadjustment factor of the *RLS* algorithm is approximately equal to $M_{\text{RLS}} = (1-\alpha)L$ [56]. Both procedures require initial conditions to begin their recursions. Usually, the filter coefficients are initially zero. The *RLS* algorithm requires an additional quantity: the covariance matrix of the observations. A typical choice for this initial covariance matrix estimate is a constant times the identity matrix, the constant equaling an estimate of the signal variance.

Example

Consider the first-order example given when we considered the Wiener filter {300} where no observation noise is present. The initial observation occurs at $l = 0$. Given that the statistical characteristics of the signal are known, the optimal one-step predictor is a constant times the previous observations. The adaptive algorithms, *LMS* and *RLS*, do not presume knowledge of signal characteristics *other* than filter duration. The optimal one-step predictor for a first-order signal is a zeroth-order system ($L = 1$). This type of consideration is an example of a structural constraint; usually, the duration of the adaptive filter is chosen longer than that expected to be required by the data. This choice must be made carefully as it directly influences the amount of computation in both algorithms. For simplicity, set $L = 1$.

The *LMS* algorithm requires a value for μ to ensure convergence; too large a choice leads to an unstable filter. In this simple, first-order example, the variance σ_y^2 of the observations is about 2.5 and μ must be less than 0.13.[†] The *RLS* algorithm can also use an estimate of the observations' variance to initialize the covariance matrix estimate. The choice of α is governed by the misadjustment factor. To compare fairly the two adaptive algorithms, we set the misadjustment factors equal. The values of the single filter coefficient are shown for each in Fig. 6.5 {315}. The *RLS* filter coefficient settles much more rapidly than the *LMS* one. The *LMS* algorithm must both "sense" the signal characteristics *and* search for the optimal filter; the *RLS* algorithm needs only to perform the latter chore.

The *RLS* algorithm, which generally produces better adaptive filters by most any measure, does require more computations than the *LMS* filter *in its present form*. Typically, the *RLS* algorithm converges quickly, taking about $2M$ steps to converge [68: 395]. This convergence rate is much greater than that provided by the *LMS* algorithm. The quickened convergence is bought with increased computational complexity: Computations for each *RLS* update are of order $O(L^2)$ and $O(L)$ for *LMS*. The computations of the covariance matrix's inverse can be reduced by exploiting the matrix's Toeplitz symmetries for both algorithms. Generically known as fast Kalman algorithms [3: 142–74];[68: 409–44];[100], they provide linear computational complexity for both algorithms.

6.5 Spectral Estimation

One of the commonest requirements in array signal processing systems, both explicitly and implicitly, is estimating a signal's Fourier spectrum. We have already seen that the array pattern amounts to a spatial spectrum computation (Eq. 4.4 {120}). By examining the spectrum estimation problem carefully, not only do criteria for selecting shading weights emerge but also novel array processing techniques that have superior performance characteristics. Toward these ends, this section summarizes techniques for estimating the power spectrum of a random signal or of a deterministic signal observed in the presence of noise. Because the covariance function and the power spectrum are Fourier Transform pairs, estimation of one is related to estimation of the other. Thus, the following discussion frequently alternates between these two quantities.

The *resolution* of any spectral estimate is the degree to which it can express spectral detail. For example, if the spectrum is smooth (like that arising from first-order signals), little resolution is demanded of the spectral analysis. Conversely, signals consisting of sinusoids that give rise to spectra having abrupt changes require much more resolution. The more usual criteria of assessing the quality of an estimate (bias, consistency, etc.) are subsumed under spectral resolution considerations.

Resolution generally depends on two factors: the variance of the estimate and the degree of spectral smoothing implicit in the spectral estimation algorithm. A large variance may well mean that differences in spectral amplitude at two frequencies cannot be easily distinguished because of large statistical variations. Although the various spectral estimation algorithms do have very different statistical properties, they all share the property that variance depends *directly* on the spectrum being estimated. Thus, to characterize statistical errors, a general understanding of the spectrum's shape and how it varies with frequency, what we term *spectral structure*, is required to specify criteria on variability.

[†]Because the system is first order, we use the constraint $\mu < 1/3\sigma_y^2$ (Eq. 6.13 {314}).

As more data are incorporated into the spectral estimate, the variance should decrease (the estimate is consistent); for consistent estimators, the variance is ultimately limited by the amount of data available. *Spectral smoothing* expresses the detail that can be gleaned from an estimate when the statistical errors are small. An algorithm's asymptotic resolution is determined entirely by the range of frequencies that influence the estimate at each frequency. In many algorithms, this range is dependent on spectral structure and on the variance we demand: Algorithms having high asymptotic resolution usually have large variances and *vice versa*. A kind of "Catch-22" emerges: We have to know the spectrum to understand an algorithm's characteristics, but we don't know the spectrum; that's why we're trying to estimate it. Understanding this dilemma is the key to discerning effective spectral estimation techniques.

The various algorithms used in spectral estimation fall into two broad categories: *nonparametric* and *parametric*. Some idea of the spectral resolution necessary to reveal the spectrum's structure is required not only to select between algorithms in the two categories but also to select an algorithm within the category.

Nonparametric. We require no strong preconceived notions about the nature of a signal that can be successfully subjected to nonparametric spectral analysis. Relatively straightforward application of the discrete Fourier Transform characterizes these algorithms. They are by far the best understood; examples are the periodogram and Bartlett's procedure.

Parametric. Parametric algorithms assume we know a great deal a priori about a spectrum's structure. They are based on an explicit model for signal generation with the model's only uncertainty being the values of a few parameters. Thus, we must know the model's structure, and given this structure, the spectral estimation problem becomes one of parameter estimation. If the spectrum does fall into the assumed class, the estimate is very accurate (high resolution is achieved). If not, the estimate does not just have poor resolution, it can be misleading.

6.5.1 Periodogram

Let $y(n)$ denote a zero-mean, stationary, stochastic sequence having covariance function $K_y(m) = \mathcal{E}[y(n)y(n+m)]$ and power spectrum $\mathcal{S}_y(\omega)$, which are related to each other as

$$\mathcal{S}_y(\omega) = \sum_{m=-\infty}^{+\infty} K_y(m) e^{-j\omega m}$$

$$K_y(m) = \frac{1}{2\pi} \int_{-\pi}^{\pi} \mathcal{S}_y(\omega) e^{+j\omega m}\, d\omega$$

Because the Fourier Transform is linear, a linear estimate of one is equivalent to some linear estimate of the other. This relationship does *not* imply that they share statistical characteristics, however. This subtle point is beautifully exemplified by the periodogram.

Covariance-based spectral estimation. We observe the signal over the finite interval $[0, D-1]$. The covariance function can be estimated by approximating the averaging implicit in its definition with

$$\widehat{K}_y(m) = \frac{1}{D} \sum_{n=0}^{D-|m|-1} y(n)y(n+m)\,, \quad 0 \le |m| \le D-1$$

The upper limit on this sum arises because of the data's finite extent: The signal $y(n+m)$ is provided for values of the index n ranging from $-m$ to $D-m-1$. Beyond lag $D-1$, this simple estimate provides no value. The expected value of this estimate is easily seen to be

$$\mathcal{E}[\widehat{K}_y(m)] = \left(1 - \frac{|m|}{D}\right) K_y(m)$$

Thus, this estimate is biased, but asymptotically unbiased. More important, bias differs for each lag, the smaller lags having much less bias than the larger ones. This effect is traced to the number of signal values that contribute to the estimate for a given lag: $D-m$ terms for lag m. Usually, we compute an average by summing a few terms, then dividing by the number of terms in the sum. The estimate of the covariance function given previously normalizes the sum by D no matter how many terms contribute to the sum; this choice leads to the bias. The variance of the estimate is given approximately by[†]

$$\mathcal{V}[\widehat{K}_y(m)] \approx \frac{1}{D} \sum_{n=-(D-m-1)}^{D-m-1} \left(1 - \frac{m+|n|}{D}\right) [K_y^2(n) + K_y(n+m)K_y(n-m)]$$

Consistent with the averaging interpretation just given, we find that the variance of the covariance function's estimate is bigger at large lags than at the small ones. Assuming the covariance function is square-summable, this summation remains finite as $D \to \infty$; hence the variance of the covariance estimate is proportional to $1/D$, and the estimate is consistent.

One way of estimating the power spectrum is to compute the Fourier Transform of the covariance function's estimate.

$$\widehat{\mathcal{S}}_y(\omega) = \sum_{m=-(D-1)}^{D-1} \widehat{K}_y(m) e^{-j\omega m}$$

This spectral estimate can be related directly to the observations by substituting the expression for the covariance estimate.

$$\widehat{\mathcal{S}}_y(\omega) = \frac{1}{D} \sum_{m=-(D-1)}^{D-1} \sum_{n=0}^{D-m-1} y(n)y(n+m)e^{-j\omega m}$$

[†]The reason for the approximation is the need for the *fourth* moment of the stochastic sequence y. This computation is dependent on the signal's joint amplitude distribution (see Prob. 6.25). This approximation, derived under a Gaussian assumption, also applies to other distributions [79: 171–89].

The summation limits result from the finite extent of the observations. We can express the availability of data directly by replacing the signal by the *windowed signal* $w_D^R(n)y(n)$, where $w_D^R(n)$ is the rectangular or "boxcar" window of duration D.

$$w_D^R(n) = \begin{cases} 1 & 0 \le n \le D-1 \\ 0 & \text{otherwise} \end{cases}$$

The limits on the summations now become infinite with the result

$$\widehat{\mathcal{S}}_y(\omega) = \frac{1}{D} \sum_{m=-\infty}^{\infty} \sum_{n=-\infty}^{\infty} w_D^R(n) w_D^R(n+m) y(n) y(n+m) e^{-j\omega m}$$

A simple manipulation of the summand yields in an insightful reexpression.

$$\widehat{\mathcal{S}}_y(\omega) = \frac{1}{D} \sum_{n=-\infty}^{\infty} w_D^R(n) y(n) e^{+j\omega n} \sum_{m=-\infty}^{\infty} w_D^R(l+m) y(l+m) e^{-j\omega(l+m)}$$

The latter sum does not depend on n; thus, we find the spectral estimate to be the product of the windowed signal's Fourier Transform and its conjugate. Defining $Y_D(\omega)$ to be the Fourier Transform of the windowed signal, this power spectrum estimate, termed the *periodogram*, is found to be [16];[105: 130–64];[118: 730–36]

$$\widehat{\mathcal{S}}_y(\omega) = \frac{1}{D} |Y_D(\omega)|^2$$

$$Y_D(\omega) = \sum_{n=-\infty}^{\infty} w_D^R(n) y(n) e^{-j\omega n}$$

The periodogram is, expectedly, biased because it is the Fourier Transform of a biased estimate of the covariance function. The expected value of the periodogram equals the Fourier Transform of the triangularly windowed covariance function.

$$\mathcal{E}[\widehat{\mathcal{S}}_y(\omega)] = \sum_{m=-(D-1)}^{D-1} \left(1 - \frac{|m|}{D}\right) K_y(m) e^{-j\omega m}$$

This window is colloquially termed a "rooftop" window; the literature terms it a *triangular* or *Bartlett* window.

$$w_{2D-1}^T(m) = 1 - \frac{|m|}{D}, \quad m = -D, \ldots, D$$

This lag-domain window is a consequence of the rectangular window we applied to the data. In the frequency domain, the periodogram's expected value is the convolution of the actual power density spectrum with the Bartlett window's Fourier transform.

$$\mathcal{E}[\widehat{\mathcal{S}}_y(\omega)] = \frac{1}{2\pi} \int_{-\pi}^{\pi} \mathcal{S}_y(\alpha) W_{2D-1}^T(\omega - \alpha)\, d\alpha$$

The Fourier Transform $W_{2D-1}^T(\omega)$ of the Bartlett window is known as the Féjer kernel.

$$W_D^T(\omega) = \frac{1}{D}\left(\frac{\sin \omega D/2}{\sin \omega/2}\right)^2$$

Window selection. These results derived for the rectangular window are easily generalized to other window shapes: Applying a window $w_D(n)$ to the data is equivalent to applying the window $w_{2D-1}^c(m)$ to the covariance function, where the covariance-domain window equals the autocorrelation function of the data window. The relationship of the data window $w_D(n)$ to the periodogram's expected value $\mathcal{E}[\widehat{S}_y(\omega)]$ is summarized as

$$\boxed{\begin{aligned} w_{2D-1}^c(m) &= \frac{1}{D}\sum_{n=-\infty}^{\infty} w_D(n)w_D(n+m) \\ W_{2D-1}^c(\omega) &= \sum_{m=-\infty}^{\infty} w_{2D-1}^c(m)e^{-j\omega m} \\ \mathcal{E}[\widehat{S}_y(\omega)] &= \frac{1}{2\pi}\int_{-\pi}^{\pi} S_y(\alpha)W_{2D-1}^c(\omega-\alpha)\,d\alpha \end{aligned}} \tag{6.15}$$

Because the actual power spectrum is convolved with the kernel $W_{2D-1}^c(\cdot)$, the periodogram is a *smoothed* estimate of the power spectrum. Only those aspects of the spectrum that vary over a range of frequencies *wider* than the width of the kernel are noticeable. This width can be defined much like a mainlobe as the distance between the smallest zeros surrounding the origin. For the Féjer kernel, this width is $4\pi/D$. The periodogram's bias thus has a very complicated structure, being neither additive nor multiplicative.

The relation of the periodogram's expected value to the convolution of the actual spectrum with the window's magnitude-squared Fourier Transform (Eq. 6.15) expresses the smoothing characteristics of periodogram-based spectral estimates. Using language gleaned from conventional beamforming, the Fourier Transform of a rectangular window $\sin(\omega D/2)/\sin(\omega/2)$ has a very narrow mainlobe and large ripples (sidelobes) elsewhere. The sidelobes are directly related to the abrupt beginning and termination of the window. We can use window shapes other than the rectangular to improve these smoothing characteristics.

Windows having reduced sidelobes have *tapered* ends: They slowly decrease to 0 near the ends. As expected, this reduction is compensated by an increase in mainlobe width. For example, a Bartlett window may be applied to a frame: The mainlobe of its Fourier Transform has twice the width of the rectangular window's transform, but the sidelobes' amplitudes are greatly reduced. Window taper also suppresses the data that are unfortunate enough to occur at a frame's edges; they tend to be ignored in the spectral estimate. Some well-known window functions along with their spectral smoothing functions (their Fourier Transforms) are shown in Fig. 6.6. A tabular display of their characteristics is provided in Table 6.1.

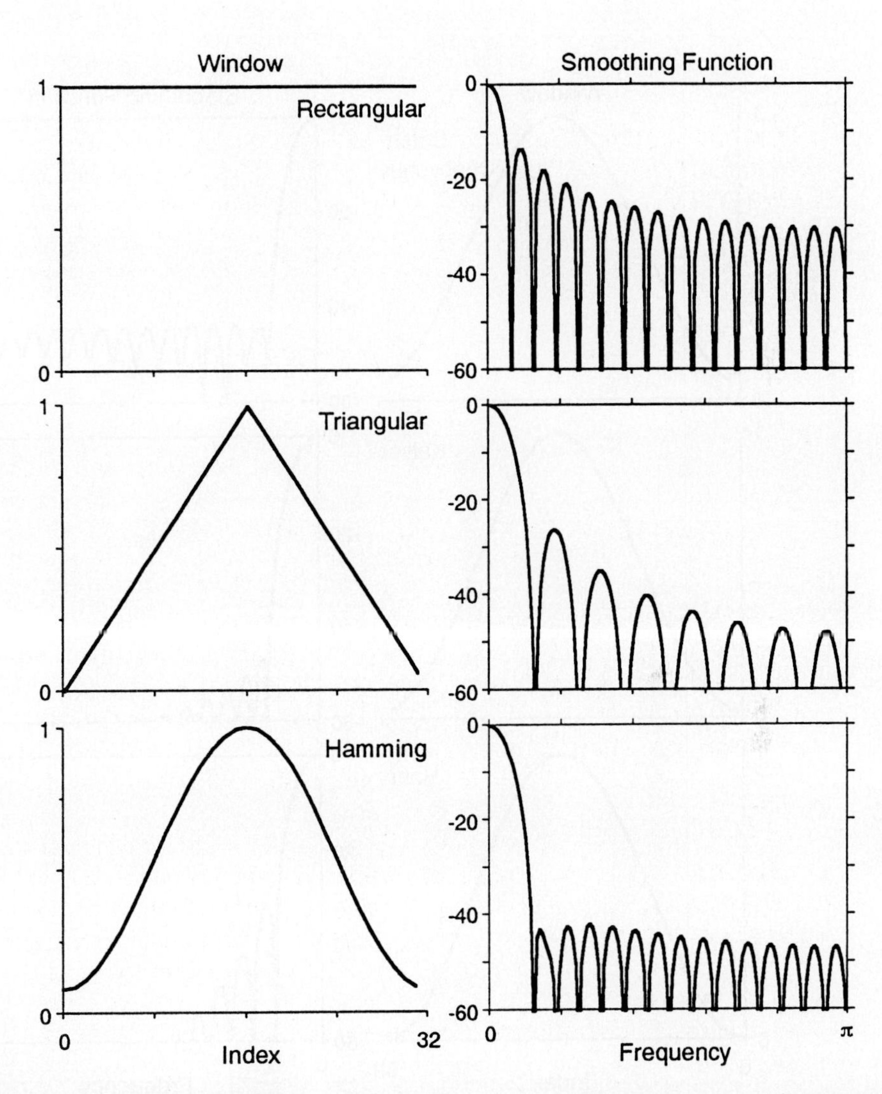

Figure 6.6 Window functions frequently used in spectral estimation are shown in the left column and their Fourier Transforms—smoothing functions—in the right.The spectral amplitudes are plotted in dB *re* the spectral amplitude at the origin. The window duration was $D = 32$. The latter quantities, when convolved with the true spectrum, yield the expected value of the Bartlett spectral estimate. The ideal of a narrow mainlobe and small sidelobes is never achieved as these examples show; a compromise is always required in practice.

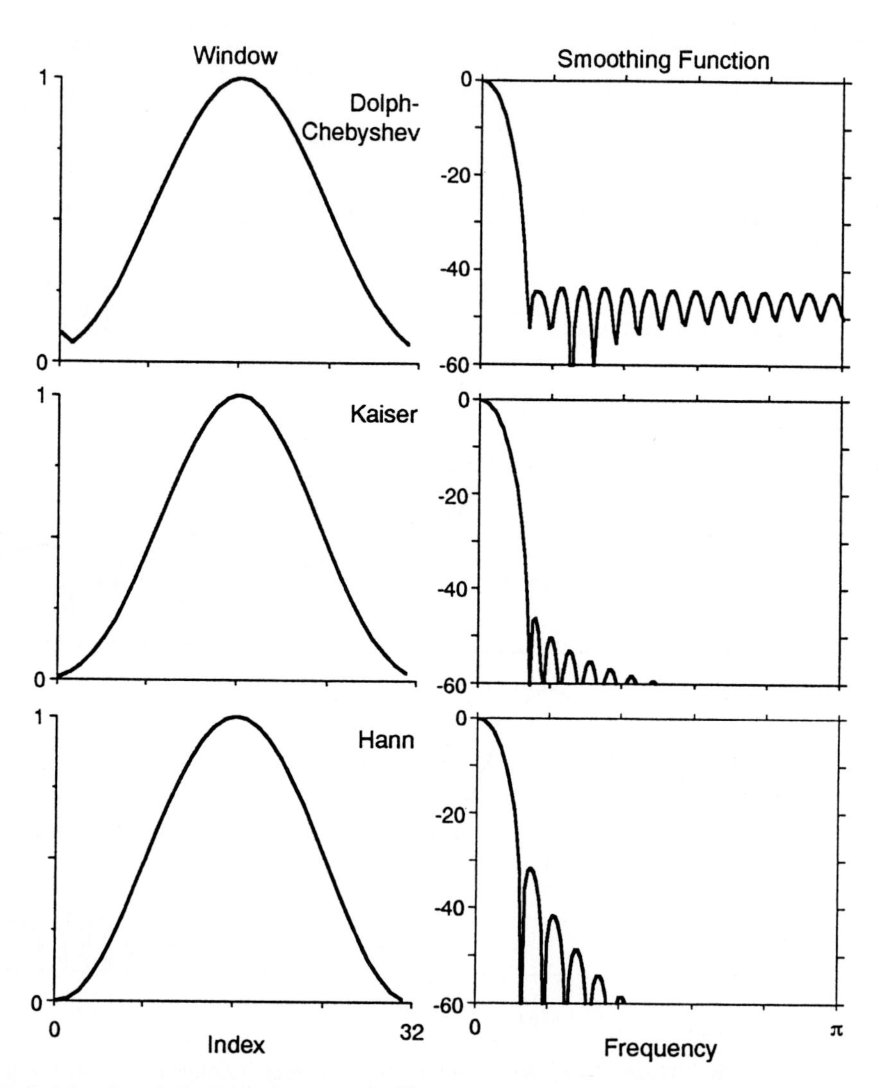

Figure 6.6 **(continued)** Window functions frequently used in spectral estimation are shown in the left column and their Fourier Transforms in the right. For those frequency ranges where no spectral values are shown, the smoothing functions are less than −60 dB. The parameters of the Dolph-Chebyshev and Kaiser windows are those used in Table 6.1.

Window	*Formula*[a]	*FWHM*[b]	*Highest Sidelobe Level* (dB)
Rectangular	1	$7.60/D$	-13
Triangular	$1 - \frac{\lvert n-D_{1/2}\rvert}{D_{1/2}}$	$11.2/D$	-27
Hamming	$(0.54 - 0.46\cos\frac{2\pi n}{D})$	$11.4/D$	-43
Dolph-Chebyshev[c]	$W(k) \propto \begin{cases} \cos\{D\cos^{-1}[\theta(k)]\}, & \lvert\theta(k)\rvert \le 1 \\ \cosh\{D\cosh^{-1}[\theta(k)]\}, & \lvert\theta(k)\rvert > 1 \end{cases}$ where $\theta(k) = \beta\cos\pi k/D$	$11.6/D$	-50
Kaiser[d]	$I_0\left[\pi\alpha\sqrt{1-\left(\frac{n-D_{1/2}}{D_{1/2}}\right)^2}\right]/I_0(\pi\alpha)$	$12.5/D$	-46
Hann	$\frac{1}{2}\left(1-\cos\frac{2\pi n}{D}\right)$	$12.6/D$	-32

[a]The formula for each window presumes that they are defined over the interval $0 \le n < D$ with $D_{1/2} = (D-1)/2$.

[b]Full-width half-maximum measure of mainlobe width.

[c]$W(k)$, $k = 0, \ldots, D-1$ is the discrete Fourier Transform of the window function. For the purposes of this formula, $\cosh^{-1}(x)$ is given by $\ln\left(\lvert x\rvert + \sqrt{x^2-1}\right)$. The parameter β equals $\cosh[(\cosh^{-1} 10^{\alpha})/D]$, where α is the number of decades the sidelobe heights lie below the mainlobe height. In this table, the numerical values were obtained with $\alpha = 2.5$.

[d]The parameter α controls the mainlobe width and sidelobe levels. The larger it is, the wider the mainlobe but smaller the sidelobes. The numbers presented here result from $\alpha = 2.0$.

TABLE 6.1 Characteristics of Commonly Used Windows [adapted from Harris].

In a general sense, the effect of windowing on spectral resolution is *independent* of the shape of the window. As more and more data are used, the ultimate resolution of periodogram-based spectral estimates is directly dependent on frame duration, with window choice changing the constant of proportionality via mainlobe width. All windows produce sidelobes of varying structure; sidelobe structure affects *spectral leakage*: the influence on the estimate at a particular frequency by remote portions of the spectrum.

Relation to conventional beamforming. Eq. 4.24 {172}, which we repeat here, describes conventional beamforming operations in the frequency domain.

$$Z(p,\omega) = \sum_{m=0}^{M-1} w_m Y_m(p,\omega) e^{j\omega p} e^{-j\omega\Delta_m}$$

Recall that $Y_m(p,\omega)$ denotes the short-time Fourier Transform of the mth sensor's output at time index p. For each temporal frequency ω, calculation of the array's frequency-domain output is a Fourier Transform with respect to space, with the sensor locations and assumed propagation direction combining to yield the sensor delays: $\Delta_m = \vec{k} \cdot \vec{x}_m$. Taking the periodogram discussion into consideration, array processing's penchant for regular linear arrays becomes clear: Because $\Delta_m = k_x d(m - M_{1/2})$ for regular linear arrays, *conventional beamforming is the periodogram-based spatial spectral estimate with* w_m *serving as the window.* The spectral smoothing function, the array pattern in beamforming terminology, determines an array's mainlobe-sidelobe structure. We can manipulate this structure by changing the array shading w_m. Note that the window's "duration" D now equals the number M of sensors in the array. Of the array patterns shown in Fig. 6.6, the Dolph-Chebyshev is usually considered the most interesting: Despite a wide mainlobe, the sidelobes have a constant, controllable height. We should not be surprised to learn that this window was developed for array processing applications in the mid-1940s [47].

For array geometries other than linear and regular, the fundamental relation between window shape and smoothing does not change: They are Fourier Transform pairs. Be that as it may, geometry does affect the smoothing function dramatically for two reasons: (1) The relation between spatial frequency and direction of propagation becomes complicated, and (2) sensor index may not appear linearly in the complex exponential. Figs. 3.28 {102}, 3.29 {102}, and 3.30 {105} illustrate these complexities for linear, unequally spaced arrays. We have learned how to deal with these complexities. Just as windowing in the time domain is related to windowing in the lag domain via an autocorrelation operation, the co-array emerges in array processing as the lag-domain window. The co-array's Fourier Transform is the spectral smoothing function for array processing's steered power responses.

Statistical characteristics. Despite the expected value of the periodogram being a seemingly reasonable approximation to a signal's power spectrum and despite the fact that the estimate of the covariance function is asymptotically unbiased and consistent, *the periodogram does not converge to the power spectrum* [79: 222–23];[93: 260–65]: The variance of the power spectrum does not tend to zero as the number of observations increases. For example, the periodogram's asymptotic variance, for Gaussian signals, tends at any frequency toward the square of the power density spectrum (see Prob. 6.25).

$$\lim_{D\to\infty} \mathcal{V}[\widehat{\mathcal{S}}_y(\omega)] \propto \mathcal{S}_y^2(\omega)$$

Because the asymptotic variance is not 0, the periodogram is *not* a consistent estimate of the power spectrum. In estimation theory, the periodogram is perhaps the most famous example of an inconsistent, yet asymptotically unbiased estimator. This lack of

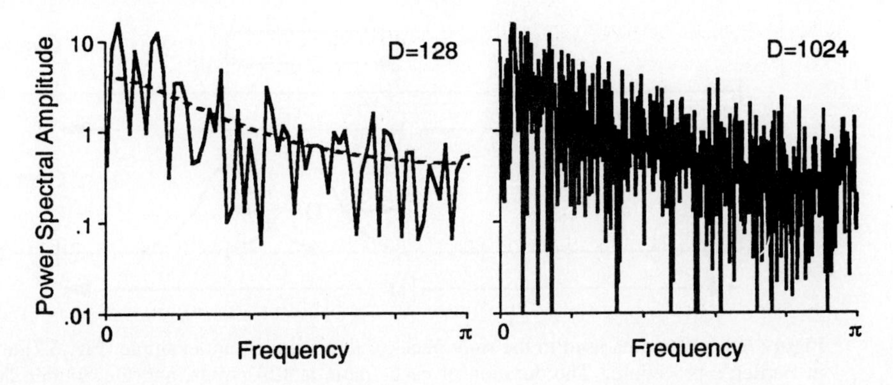

Figure 6.7 Periodograms were computed from the example signal used in the Wiener and adaptive filtering sections of this chapter. A rectangular window was applied in each case with the spectrum in the left panel computed from 128 data points and from 1,024 points for the one on the right. Note how the estimate's variability does not stabilize as more data are incorporated.

convergence is illustrated in Fig. 6.7.

Why shouldn't the periodogram converge? The reason lies in the expression of the power spectrum as an integral transformation of the estimated covariance function. Recall that the variances of the covariance function's values at higher lags were large. These terms are added in the Fourier sum, and their total variability is never compensated by the normalization factor. Realizing this flaw in the periodogram is the key to finding alternate spectral estimates that do converge.

Bartlett's procedure. To obtain a consistent spectral estimate from the empirical covariance function, either higher lags must not be used in the computations or more data should contribute toward their average. With the latter idea in mind, from the total of N observations, frame the data into L sections of equal duration D ($N = LD$). The *Bartlett procedure* forms a power spectrum estimate by averaging the spectra computed from each frame [11];[105: 153–58];[160].†

$$\widehat{\mathcal{S}}_y(\omega) = \frac{1}{L}\sum_{l=0}^{L-1}\widehat{\mathcal{S}}_y^{(l)}(\omega)$$

$$\widehat{\mathcal{S}}_y^{(l)}(\omega) = \frac{1}{D}\left|Y_D^{(l)}(\omega)\right|^2$$

$$Y_D^{(l)}(\omega) = \sum_{n=0}^{D-1} w_D(n)y(lD+n)e^{-j\omega n}$$

†The quantity $Y_D^{(l)}(\omega)$ corresponds exactly to the short-time Fourier Transform $Y(lD,\omega)e^{jlD\omega}$ defined previously (Eq. 4.25 {173}). The more streamlined current notation is employed to simplify the equations for Bartlett's procedure. Note that the linear phase shift $e^{jlD\omega}$ owing to the signal's nonzero time origin is removed by the magnitude operation of the second equation.

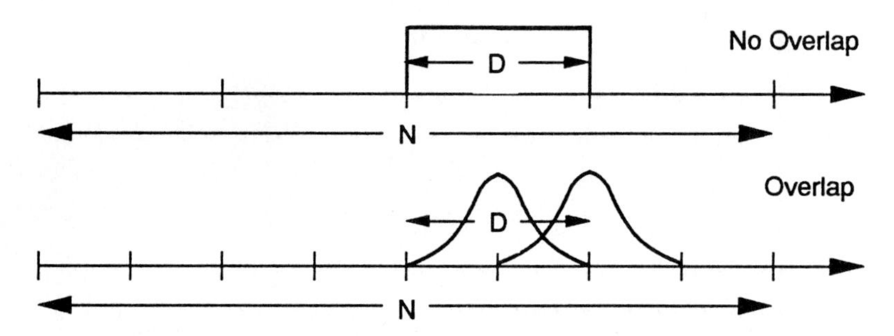

Figure 6.8 The data used in the right panel of the periodogram example (Fig. 6.7) are now used in Bartlett's procedure. The duration of each frame is 128 points, and the Fourier Transform of each frame has the same length. Eight periodograms are averaged to produce this spectrum; thus, a total of 1,024 data points were used in this computation and in the last portion of the periodogram example. The marked decrease in statistical variability is evident. Also note that the variability has roughly the same amplitude at each frequency. This effect is because of the relation of the spectral estimate's variance to the square of the actual spectral amplitude and to the logarithmic vertical scale.

In this case, periodograms are computed from each frame; conceptually, any spectral estimate could be averaged in this way. Assuming that the frames are mutually independent, the variance of the Bartlett estimate equals the sum of the individual variances divided by L.

$$\mathcal{V}[\widehat{\mathcal{S}}_y(\omega)] \propto \frac{1}{L}\mathcal{S}_y^2(\omega) \tag{6.16}$$

Thus, as $N \to \infty$, maintaining fixed duration frames implies that $L \to \infty$ so that the Bartlett spectral estimate is consistent. The nonconvergent periodogram example is reexamined in Fig. 6.8 with the Bartlett procedure.

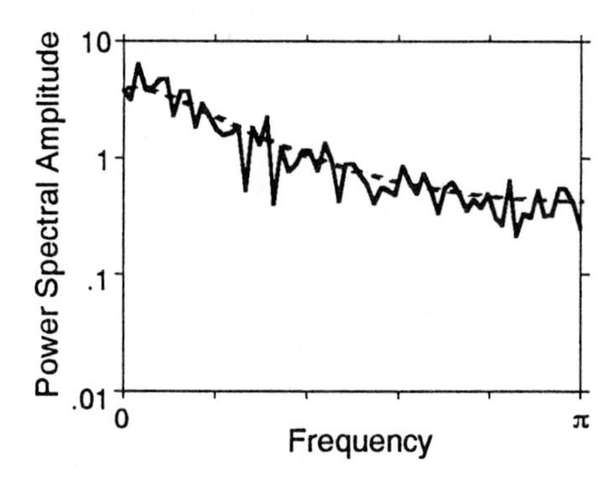

Figure 6.9 In Bartlett's spectral estimation procedure, the observations are subdivided into frames of length D. The periodogram is computed over each frame, and these spectra are averaged to produce a spectral estimate. The frames do not usually overlap, as shown in the upper portion, where a rectangular window is used. A 50% overlap or, said another way, 50% stride, is typical when one of the gently tapered windows (such as the Hann window shown here) is used.

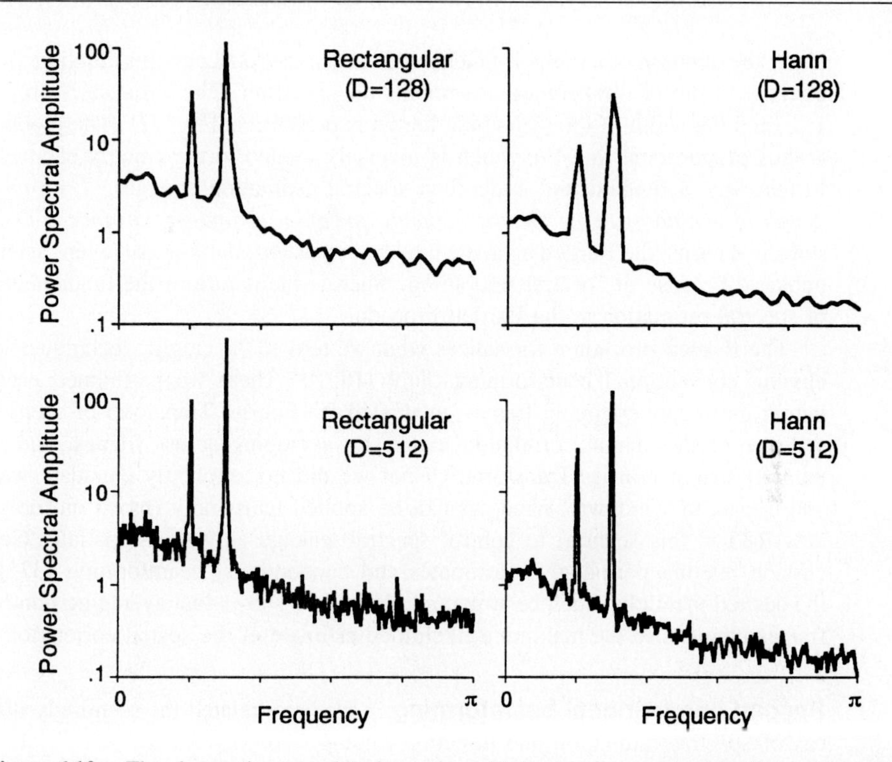

Figure 6.10 The observations consist of two sinusoids having frequencies $2\pi \times 0.1$ and $2\pi \times 0.15$ and additive noise having first-order lowpass spectral characteristics. The power of the higher frequency sinusoid is four times that of the lower. This additive noise is the same as that presented in the adaptive filtering examples. The number of observations available equals 8,192. The left column portrays the Bartlett spectral estimate that uses a rectangular data window with no overlap, the right column a Hann window with a 50% overlap. The top row uses 128-point component periodograms, the bottom 512. Because of the smaller energy of the Hann window, the spectra produced by the Hann window are scaled relative to the ones produced by the rectangular window (both windows had unity height).

When windows other than the rectangular are used, data aligned with the ends of our somewhat arbitrarily selected frames do not contribute to the spectral estimate. To counteract this arbitrary stress on portions of the data, frames are overlapped (Fig. 6.9): The separation between successive windows, the stride, is less than the window's duration. A 50% overlap is commonly used. With overlap, many more frames are obtained from a given set of data. This increased number does *not* mean that the variance of the estimate is reduced proportionally. The variance of the spectra resulting from Bartlett's procedure is inversely proportional to the number of *statistically independent* frames; overlapping frames introduces a dependence between them. The number of statistically independent frames equals the number of nonoverlapping frames. The effect of window shape and overlapping frames is illustrated in Fig. 6.10.

Consistency in this spectral estimate has been obtained at the expense of increased bias. The duration of a frame is much smaller than the total, and this duration increases the effective width of the kernel that smooths the spectrum. The variance is proportional to $1/L$, and the width of the smoothing kernel is proportional to $1/D$. The product of these is thus proportional to $1/N$, which is inversely related to the amount of available data. In this way, a fundamental tradeoff in spectral estimation emerges: *The product of the degree of smoothing of the spectral estimate and of the estimate's variance is a constant.* In statistical terms, the Bartlett estimate may be consistent, but it is *not* (even) asymptotically unbiased because of spectral smoothing. Such is the nature of the fundamental tradeoff of spectral estimation in the Bartlett procedure.

The Bartlett procedure formalizes what we termed "averaging techniques" when discussing conventional beamforming (§4.9 {182}). There we partitioned each sensor's output into nonoverlapping frames, evaluated the Fourier Transform of each, formed an estimate of the spatial correlation matrix by averaging across frames, and then computed a spatial Fourier Transform. What we did not explicitly elucidate was the role and impact of windows. Windows can be applied temporally (based on the procedures described in this section) to control spectral leakage as well as spatially (see the discussion relating periodogram estimates and conventional beamforming {325}) to form the desired mainlobe-sidelobe structure. Prob. 6.27 shows that averaging nonoverlapping frames amounts to the maximum likelihood estimate of the spatial correlation matrix.

Beyond conventional beamforming. We have related the seemingly disparate areas of spectral estimation and direction finding. So why should this chapter (and this book, for that matter) continue? Because of the direct effect of window duration (aperture in array processing terminology) on mainlobe width, we seem to have reached a "resolution wall"; no enhancements, other than the relatively minor contribution of shading choice, improves resolution. Here is where modern spectral estimation algorithms come to the forefront. The remainder of this section details algorithms where the smoothing function depends on the spectrum being estimated. Stated another way, window shape varies with frequency according to spectral structure, enabling significant resolution increases to occur. We defer to the next chapter tailoring these algorithms to the array processing problem. The lesson is that these spectral estimation algorithms are worth understanding on their own merits before we apply them to our view of an "important" problem.

6.5.2 Minimum Variance Spectral Estimation

Bartlett's procedure yields a consistent spectral estimate by averaging periodograms computed over short frames of the observations. An alternate approach is to compute the covariance function over a few lags, but compute it over the entire data set. After applying any tapered window to this estimate, computing the Fourier Transform results in a spectral estimate that does not differ greatly from that yielded by Bartlett's procedure. A more interesting use of the covariance estimate is the so-called *minimum variance* spectral estimate [28];[29];[105: 350–57], which uses the covariance function in the solution of a somewhat artificial optimization problem. The resulting estimate is a *nonlinear* function of the data; hence its characteristics are not readily analyzed. Suffice it to say that its spectral resolution is greatly superior to Bartlett's procedure, but much more computation is needed to produce the estimate.

Similar to the *FIR* Wiener filter problem, assume that the observations are sent through a finite-duration filter that passes a known signal with unity gain while minimizing the contribution to the filter's output of all other components in the observations. Let the known signal be denoted by the column matrix $\mathbf{s}$ and the filter's unit-sample response by $\mathbf{h}$. Require the filter's output to the signal component $\mathbf{h}'\mathbf{s}$ to be unity while minimizing the mean-squared value of the entire output $\mathcal{E}[|\mathbf{h}'\mathbf{y}|^2] = \mathbf{h}'\mathbf{K}_y\mathbf{h}$. The minimum variance approach imposes the constrained optimization problem [80, 96]

$$\min_{\mathbf{h}} \mathbf{h}'\mathbf{K}_y\mathbf{h} \quad \text{subject to } \mathbf{h}'\mathbf{s} = 1$$

This kind of constrained optimization problem is easily solved using Lagrange multipliers (see §C.2 {502}). The filter that yields the minimum variance† signal estimate has unit-sample response

$$\boxed{\mathbf{h}_\diamond = \frac{\mathbf{K}_y^{-1}\mathbf{s}}{\mathbf{s}'\mathbf{K}_y^{-1}\mathbf{s}}}$$

To apply this somewhat arbitrary problem and its solution to spectral estimation, first note that the solution depends on *only* the observations' covariance matrix over the few lags that correspond to the assumed filter duration. Given an estimate of this covariance matrix, we can approximate the optimal filter by using the estimate in the formula given earlier. To focus this reasoning further, let the "signal" be a complex exponential corresponding to the frequency at which we want to estimate the spectrum: $\mathbf{s} = \mathbf{e}(\omega) = \text{col}\left[1, e^{j\omega}, e^{j2\omega}, \ldots, e^{j(M-1)\omega}\right]$. Thus, the length-$M$ filter passes this signal with unity gain—performs a spectral analysis—while minimizing the mean-squared value of the filter's output (asking for the maximum resolution possible by rejecting the influence of all other spectral components). In this way, a somewhat artificial optimal filtering problem has been recast as a potentially interesting spectral estimation procedure. The power spectrum would correspond to the filter's output power; this quantity equals the mean-squared value of the filtered observations. Substituting the optimal unit-sample response into the expression for the output's mean-squared value results in the minimum variance spectrum.

$$\mathcal{S}_y^{\text{MV}}(\omega) = \mathcal{E}\left[\left|\mathbf{h}_\diamond'\mathbf{y}\right|^2\right] = \left[\mathbf{e}'(\omega)\mathbf{K}_y^{-1}\mathbf{e}(\omega)\right]^{-1} \tag{6.17}$$

The minimum variance spectral estimate is obtained by replacing the covariance matrix

† The observations are assumed to be zero mean. Thus, the mean-squared value and the variance are equal. Historically, even in nonzero mean cases the solution to the "minimum variance" problem is taken to be this minimum mean-squared value solution. So much for the name of an algorithm suggesting what aspects of the data are considered important or how the result was derived.

by its estimate.

$$
\boxed{\begin{aligned}
\widehat{\mathcal{S}}_y^{\mathrm{MV}}(\omega) &= \left[\mathbf{e}'(\omega)\widehat{\mathbf{K}}_y^{-1}\mathbf{e}(\omega)\right]^{-1} \\
\widehat{K}_{i,j}^{y} &= \frac{1}{D}\sum_{n=0}^{D-|i-j|-1} y(n)y(n+|i-j|)\,, \quad 0 \le i, j < M
\end{aligned}}
\tag{6.18}
$$

This spectral estimate is not a simple Fourier Transform of the covariance estimate: The covariance matrix estimate is formed from length-D observations, its inverse computed and used in a quadratic form that is approximately equivalent to the Fourier Transform of a matrix, and the reciprocal evaluated. Conceptually, the quadratic form is the kernel's Fourier Transform. The Fourier Transform of a column matrix is concisely expressed as $\mathbf{e}'(\omega)\mathbf{y}$; the combination of premultiplying and postmultiplying a matrix by $\mathbf{e}$ amounts to a Fourier Transform (see Prob. 6.28). Fast Fourier Transform algorithms can be used to transform a column matrix, but not a square one unless the matrix has many symmetries. Although a covariance matrix may be Toeplitz as well as Hermitian, its inverse is only Hermitian, thereby obviating fast algorithms. Note that the number M of covariance function lags used in the estimate is usually much smaller than the number of observations D. In contrast to Bartlett's procedure, where many covariance estimates computed over short frames are averaged to produce a covariance estimate, here the entire set of observations is used to produce estimates for a similar number of lags. A Bartlett-procedure-like variation of estimating the covariance matrix is derived in Prob. 6.27. Another consequence of $M \ll D$ is the number of frequencies at which the quadratic form can be evaluated to reflect the algorithm's resolution accurately. In most cases, the width of the smoothing function is comparable with that of the Féjer kernel implicit in rectangular data windows of length D. Despite the comparatively small size of the matrix, the quadratic form needs evaluation at *many* frequencies to capture the minimum variance algorithm's resolution. A computational note: The inverse of this matrix need not be computed directly. The minimum variance estimate is intimately related to the so-called autoregressive (AR) spectral estimate described in the next section (Eq. 6.21 {336}). This relationship not only expresses the deep relationship between two very different spectral estimators but also provides a computationally expedient method of calculating the minimum variance estimate.

The minimum variance spectrum is related to the signal's power spectrum similarly to the periodogram: The minimum variance spectrum is a smoothed version of the true spectrum. To demonstrate this property, any spectral estimate $\widehat{\mathcal{S}}_y(\omega)$ given by the mean-squared value of weighted observations is related to the covariance function by

$$
\mathcal{E}\left[\widehat{\mathcal{S}}_y(\omega)\right] = \mathcal{E}\left[\left|\mathbf{h}'\mathbf{y}\right|^2\right] = \sum_{k,l} h(k)K_y(k-l)h(l)
$$

Using Parseval's Theorem, this expression for the spectrum at frequency ω_0 is given by

$$
\mathcal{E}\left[\widehat{\mathcal{S}}_y(\omega_0)\right] = \frac{1}{2\pi}\int_{-\pi}^{\pi} |H(\omega_0,\omega)|^2\, \mathcal{S}_y(\omega)\, d\omega
$$

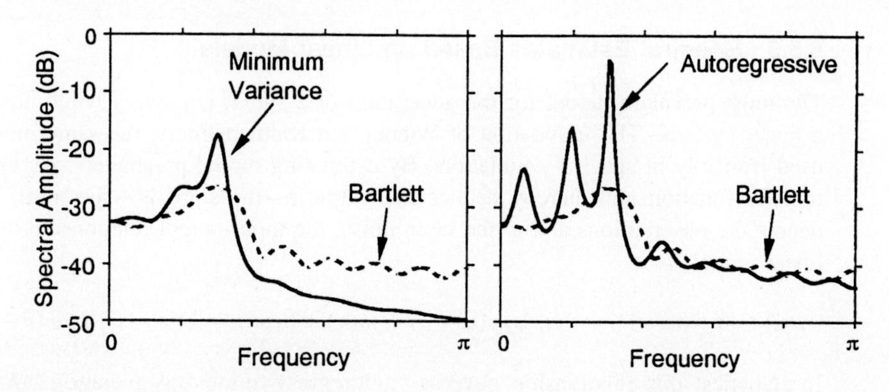

Figure 6.11 The plots depict minimum variance and *AR* spectral estimates computed from the same data ($D = 8,192$) used in Fig. 6.10 {329}. Only the *AR* spectrum has a peak at the frequency of the smaller sinusoid ($\omega = 2\pi \times 0.1$); however, that peak is mitigated by the presence of a spurious peak located at a lower frequency. The size M of the covariance matrix is 16.

$H(\omega_0, \omega)$ is the Fourier Transform of the filter's unit-sample response $\mathbf{h}$; for the minimum variance spectral estimate, this smoothing function is given by

$$H^{\mathrm{MV}}(\omega_0, \omega) = \frac{\mathbf{e}'(\omega)\mathbf{K}_y^{-1}\mathbf{e}(\omega_0)}{\mathbf{e}'(\omega_0)\mathbf{K}_y^{-1}\mathbf{e}(\omega_0)}$$

Because the minimum variance smoothing function is itself dependent on the data's characteristics, it changes form according to the shape of the signal's spectrum near the frequency of interest. In this way, the smoothing function depends directly on the value of the analysis frequency (ω_0), not on the difference between ω_0 and ω. The smoothing kernel is dependent on the structure of the spectrum; *its resolution is therefore data dependent*.

When the observations are Gaussian, the variance of the minimum variance spectral estimate is given by [30]

$$\mathcal{V}[\widehat{\mathcal{S}}_y^{\mathrm{MV}}(\omega)] = \frac{1}{D - M + 1}\,\mathcal{E}[\widehat{\mathcal{S}}_y^{\mathrm{MV}}(\omega)]$$

Thus, for fixed-size covariance matrices, the minimum variance spectral estimate is consistent. The estimate's expected value is proportional to the true minimum variance spectrum.

$$\mathcal{E}[\widehat{\mathcal{S}}_y^{\mathrm{MV}}(\omega)] = \frac{D - M + 1}{D}\,\frac{1}{2\pi}\int_{-\pi}^{\pi} \left|H^{\mathrm{MV}}(\omega_0, \omega)\right|^2 \mathcal{S}_y(\omega)\, d\omega$$

The minimum variance spectral estimate is therefore asymptotically unbiased and consistent. For a slightly larger variance than Bartlett's procedure, the minimum variance estimator provides much better resolution. Fig. 6.11 demonstrates this well.

6.5.3 Spectral Estimates Based on Linear Models

The most prevalent model for the generation of a signal is passing white noise through a linear system. The foundation of Wiener and Kalman filters, the same model can be used fruitfully in spectral estimation. By estimating model parameters, we can find the transfer function and thereby deduce the output's—the signal's—spectrum. Letting y denote the observations and w the input noise, the input-output relationship of a rational linear model is

$$y(n) = a_1 y(n-1) + \cdots + a_p y(n-p) + w(n) + b_1 w(n-1) + \cdots + b_q w(n-q) \quad (6.19)$$

In statistics, this relationship governs "autoregressive-moving average" (*ARMA*) models [105: 172–86]. The word "autoregressive" refers to models where the signal depends on itself in a linear fashion. "Moving average" refers to a linear combination of white-noise values. Thus, the model's autoregressive portion refers to the recursive aspect related to the poles and the moving average to the zeros of some linear system's transfer function. Spectral estimation based on this linear model for the signal amounts to estimation of the model's parameters. Given these values, the *ARMA*(p, q) spectral estimate has the form

$$\boxed{\widehat{\mathcal{S}}_y^{\text{ARMA}}(\omega) = \hat{\sigma}_w^2 \frac{\left|1 + \hat{b}_1 e^{-j\omega} + \cdots + \hat{b}_q e^{-j\omega q}\right|^2}{\left|1 - \hat{a}_1 e^{-j\omega} - \cdots - \hat{a}_p e^{-j\omega p}\right|^2}}$$

AR spectral estimation. The *AR* method is by far the most popular *parametric* or *model-based* spectral estimators [105: 189–274]. Translating into signal processing terms, an autoregressive signal is produced at the output of an all-pole linear filter driven by white noise.

$$y(n) = a_1 y(n-1) + \cdots + a_p y(n-p) + w(n)$$

The parameter p denotes the model's *order*. One method of estimating the parameters of the *AR* spectral estimate was given in a previous section (Eq. 6.11 {305}). The *linear predictive* solution is found by finding the minimum mean-squared error one-step predictor.†

$$\frac{1}{\hat{\sigma}_w^2} \begin{bmatrix} 1 \\ -\hat{a}_1 \\ \vdots \\ -\hat{a}_p \end{bmatrix} = \left(\widehat{\mathbf{K}}_y\right)^{-1} \begin{bmatrix} 1 \\ 0 \\ \vdots \\ 0 \end{bmatrix}$$

The variance estimate $\hat{\sigma}_w^2$ is equal to the reciprocal of the first component of $\widehat{\mathbf{K}}_y^{-1}\boldsymbol{\delta}$, where $\boldsymbol{\delta} = \text{col}[1, 0, \ldots, 0]$. Because of the way the input term $w(n)$ enters the difference

†Note the care used in denoting the inverse of the covariance matrix estimate rather than the estimate of the inverse. Directly estimating the inverse is touchy at best. We have discussed several methods of estimating the covariance matrix ({316} and {332}). The linear predictive literature has developed even more [101].

equation for the signal, the mean-squared prediction error equals $\hat{\sigma}_w^2$. The *AR* spectral estimate has the form

$$\boxed{\widehat{\mathcal{S}}_y^{\text{AR}}(\omega) = \frac{\boldsymbol{\delta}^t \widehat{\mathbf{K}}_y^{-1} \boldsymbol{\delta}}{\left|\boldsymbol{\delta}^t \widehat{\mathbf{K}}_y^{-1} \mathbf{e}(\omega)\right|^2} = \frac{\hat{\sigma}_w^2}{\left|1 - \hat{a}_1 e^{-j\omega} - \cdots - \hat{a}_p e^{-j\omega p}\right|^2}} \tag{6.20}$$

An example of this estimate is shown in Fig. 6.11 {333}. Another popular technique for estimating these parameters is Burg's method [105: 213–15].

The *AR* estimator is also known as the *maximum entropy* spectral estimate [24, 78]. How this name arose is an interesting story that recalls the relation between the power spectrum and the covariance function. The asymptotic resolution of the periodogram-based spectral estimates is directly related to how many lags of the covariance function can be estimated accurately: More lags imply higher resolution. Suppose $p+1$ lags can be well estimated. Can the higher lags be accurately *extrapolated* from the well-known ones in such a way that the spectral resolution is increased? Seemingly, this approach is trying to "get something for nothing." Not any extrapolation, however, can result in a covariance function; the Fourier Transform of any potential covariance function must be real and nonnegative. Imposing this restriction greatly reduces the number of possible extrapolations, but an infinite number remain. From statistical arguments, one can show that the spectral estimate that coincides with the known values of the covariance function and corresponds to the most probable extrapolation of the covariance function maximizes the *entropy* $\mathcal{H}[\mathcal{S}]$ of the power spectrum $\mathcal{S}$.

$$\mathcal{H}[\mathcal{S}] = \frac{1}{2\pi} \int_{-\pi}^{\pi} \ln \mathcal{S}(\omega)\, d\omega$$

This framework defines a constrained maximization problem, the solution to which is an *AR* spectral estimate of the form given earlier! Thus, the *AR*, model-based technique holds the promise of greater resolution than Bartlett's procedure.

By noting that the squared magnitude of a sequence's Fourier Transform equals the Fourier Transform of its autocorrelation, we find an alternative to the denominator expression in the *AR* spectral estimate.

$$\left|1 - \hat{a}_1 e^{-j\omega} - \cdots - \hat{a}_p e^{-j\omega p}\right|^2 = \sum_{m=-p}^{p} R_{\hat{a}}^{\text{AR}}(m) e^{-j\omega m}$$

$$R_{\hat{a}}^{\text{AR}}(|m|) = \sum_{k=0}^{p-m} \hat{a}_k \hat{a}_{k+m}\, , \, m \geq 0$$

For the purposes of this reexpression, $\hat{a}_0 = -1$. This reexpression is not particularly useful as it stands: It provides no computational advantages. Its value lies in its relation to a similar expression for the minimum variance spectral estimate [111]. Because the solution for the minimum variance estimator (Eq. 6.18 {332}) also contains the inverse of the covariance matrix, the minimum variance spectral estimate can be related to the

AR parameter estimates.

$$\widehat{S}_y^{\text{MV}}(\omega) = \frac{\hat{\sigma}_w^2}{\sum_{m=-p}^{p} R_{\hat{a}}^{\text{MV}}(m) e^{-j\omega m}}$$
$$R_{\hat{a}}^{\text{MV}}(|m|) = \sum_{k=0}^{p-m} (p+1-m-2k)\hat{a}_k \hat{a}_{k+m}\ , m \geq 0 \tag{6.21}$$

The correlation function resembles an expression for a windowed version of the *AR* parameter estimates' autocorrelation. It does point out one of the ways that minimum variance and *AR* spectral estimates can be related. More important, because efficient methods exist for computing the *AR* parameter estimates, those algorithms can be used to compute the minimum variance estimate as well.

A critical issue in *AR* spectral estimation is value of the model order p. As model order is increased, more parameters become available to model the signal's spectrum, suggesting that mean-squared prediction error decreases. Using larger values of p, however, implies accurate knowledge of the covariance function at larger lags, which we know are less reliable statistically. Thus, a parsimonious criterion for selecting a value for p is required. Two techniques are most well-known: the *A* information criterion (*AIC*) [2] and the minimum description length (*MDL*) criterion [130, 138]. Each of these criteria seeks the model order that minimizes the sum of two terms: The first is the negative logarithm of the likelihood function, where the *AR* parameters are evaluated at their maximum likelihood estimates, and the second is a constant times the number of parameters $p+1$. The criterion functions differ only in the constant used in the second term [101];[105: 230–31].[†]

$$AIC(p) = -\ln \Lambda(\hat{\sigma}_w^2, \hat{a}_1, \ldots, \hat{a}_p) + p$$
$$MDL(p) = -\ln \Lambda(\hat{\sigma}_w^2, \hat{a}_1, \ldots, \hat{a}_p) + \frac{1}{2} p \ln D$$

As p increases, the first term decreases while the second increases, thereby guaranteeing a unique minimum. For the *AR* parameter estimation problem, the negative logarithm of the likelihood function is proportional to the logarithm of the prediction error. Minimizing either of these criteria selects a model order best fitting the observations in terms of spectral variation with frequency and of data length [101].

$$AIC(p) = \ln \hat{\sigma}_w^2(p) + \frac{2p}{D_e}$$
$$MDL(p) = \ln \hat{\sigma}_w^2(p) + \frac{p \ln D}{D_e}$$

[†]Because we are only interested in the behavior of these criterion functions with variations in p, additive terms not depending on p are dropped as they do not affect the location of the minimum. These criterion functions have been so manipulated to their simplest form.

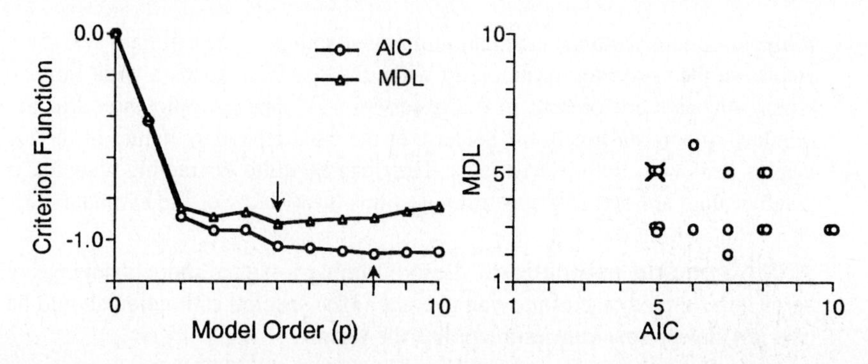

Figure 6.12 In the left panel, example *AIC* and *MDL* criterion functions are shown for the data used in Fig. 6.11. The observations thus consist of two sinusoids and additive first-order lowpass noise. Each frame has a duration of 512 samples and was viewed through a Hann window. This sum of three autoregressive sequences does *not* have an all-pole spectrum but serves to illustrate the difficulties in applying model-based criteria in the real world. The "correct" model order is five; the example shows *MDL* correctly judging the model order while *AIC*'s estimate is high (eight). The right panel summarizes the two algorithm's estimates for 16 different realizations of the same prototypical signal. The "correct" choice of five is indicated by the X. Although neither algorithm performed well, *AIC*'s estimates are consistently higher than *MDL*'s, which theory predicts.

The quantity D_e denotes the effective number of observations used in the analysis because of data windowing. It is expressed by the ratio of the energy when the window is applied to that obtained when a rectangular window is used. For example, $D_e = 0.374D$ for the Hann window. These criterion functions appear to differ little but they do have important differences (Fig. 6.12). The *AIC* estimate is inconsistent [84]: The estimate is usually too large no matter how many observations are available. In contrast, the *MDL* estimate is always consistent [138]. Thus, the *MDL* criterion is more accurate when a "large"‡ number of observations is available; with fewer observations, either model order method can be used to advantage.

Statistical characteristics of the *AR* spectral estimate are quite different from those of the others described here [8, 26]. Although the probability distribution of the *AR* spectral estimate is not known, the estimate's confidence interval has a F distribution having $(2, D-p)$ degrees of freedom (for a description of the F distribution, see §A.1.7 {479}). This result means that the spectral estimate can demonstrate fairly large deviations from the true spectrum, with deviations roughly proportional to the true value.

Using matrix notation, we can express the *AR* spectral estimate by

$$\widehat{\mathcal{S}}_y^{\text{AR}}(\omega) = \hat{\sigma}_w^2 \left|\mathbf{e}'(\omega)\widehat{\mathbf{K}}_y^{-1}\boldsymbol{\delta}\right|^{-2}$$

Interestingly, this formula *cannot* be expressed as the expected value of a weighted combination of the observations as can the minimum variance estimate (Eq. 6.17 {331}). The *AR* estimate does have data-dependent spectral resolution, which tends to be superior

‡What is left unsaid is how large is "large." Despite inconsistency, the larger value yielded by the *AIC* technique is the correct way to err: better too many parameters than not enough.

to the minimum variance estimate in many situations. The estimate also tends to produce ripples in the spectrum, particularly when the spectrum spans a wide range of values (as when sinusoids are present in the observations). These ripples are akin to sidelobes in window-smoothing functions; because of the data-dependent nature of the estimate, these ripples vary with the observations. They can be quite distracting, causing us to wonder whether they are actually spectral variations or artifacts of the estimate (Fig. 6.11).

ARMA spectral estimation. Before turning to signal models having both poles and zeros, the simpler case of moving average (*MA*) spectral estimation should be mentioned. The *MA* power spectrum estimate has the form

$$\widehat{\mathcal{S}}_y(\omega) = \hat{\sigma}_w^2 \left|1 + \hat{b}_1 e^{-j\omega} + \cdots + \hat{b}_q e^{-j\omega q}\right|^2$$

The model parameters need not be found to compute this estimate: Given the first $q+1$ lags of the covariance function, the *MA* estimate amounts to computing the Fourier Transform of these covariance function values. This estimate is hardly novel and is not frequently referenced. It amounts to applying a much narrower rectangular window than the observation length to the estimated covariance function.

In reality, models for the observations frequently require *both* poles and zeros. For example, consider the simple case in which the observations consist of a sum of signals, each of which has a first-order spectrum. The power spectrum of the sum has *both* poles and zeros, whereas neither component has zeros.

$$\frac{1}{\left(1-a_1e^{-j\omega}\right)\left(1-a_1e^{+j\omega}\right)} + \frac{1}{\left(1-a_2e^{-j\omega}\right)\left(1-a_2e^{+j\omega}\right)} =$$
$$\frac{\left(1-a_1e^{-j\omega}\right)\left(1-a_1e^{+j\omega}\right) + \left(1-a_2e^{-j\omega}\right)\left(1-a_2e^{+j\omega}\right)}{\left(1-a_1e^{-j\omega}\right)\left(1-a_1e^{+j\omega}\right)\left(1-a_2e^{-j\omega}\right)\left(1-a_2e^{+j\omega}\right)}$$

An *ARMA* model for observations is quite realistic and can encompass a much wider variety of situations than *AR* ones. Unfortunately, the equations that govern the *ARMA* model's parameters are nonlinear. In contrast, the equations governing the *AR* parameters are linear, and the only computational issue is how quickly they can be solved. The issue in *ARMA* spectral estimation is *how* to solve them. We can derive a set of equations that determine the *ARMA* parameters by considering the difference equation that governs the observations' covariance function. Using the difference equation for the observations (Eq. 6.19 {334}), we find their covariance function obeys

$$K_y(m) = \begin{cases} \displaystyle\sum_{k=1}^{p} a_k K_y(m-k) + K_{wy}(m) + \sum_{k=1}^{q} b_k K_{wy}(m-k), & m = 0, \ldots, q \\ \displaystyle\sum_{k=1}^{p} a_k K_y(m-k), & m > q \end{cases}$$

$K_{wy}(m-k) = \mathcal{E}[w(l-k)y(l-m)]$ is the cross-covariance function between the observations and the model's input. Because the process $w(\cdot)$ is white, this cross-covariance function equals the model's unit-sample response times the variance of the white-noise input.

$$K_{wy}(m-k) = \mathcal{E}\left[\sum_{n=-\infty}^{l-m} w(l-k)\cdot w(n)h(l-m-n)\right]$$
$$= \sigma_w^2 h(k-m)$$

The equations that must be solved for the *ARMA* parameters become

$$K_y(m) = \begin{cases} \sum_{k=1}^{p} a_k K_y(m-k) + \sigma_w^2 h(-m) + \sigma_w^2 \sum_{k=1}^{q} b_k h(k-m), & m = 0,\ldots,q \\ \sum_{k=1}^{p} a_k K_y(m-k), & m > q \end{cases}$$

The unit-sample response is, of course, dependent on the values of the parameters a_k and b_k. The multiplication of the unit-sample response and the coefficients b_k is thus nonlinear. Much active research is focused on efficient methods of solving these equations [89: Chap. 10];[105: Chap. 10]. One method is derived in Prob. 6.29.

Summary

Estimation theory works hand in hand with detection theory not only in array processing problems, but in all statistical signal processing problems. Good references in modern estimation theory are Poor and van Trees. The authors have not presented material in this chapter, however, that makes this point clearly for array processing applications; that is the subject of the next chapter. The range of problems under estimation theory's purview—we have covered parameter estimation, signal parameter estimation, waveform estimation, and spectrum estimation—encompasses virtually all array processing issues. The techniques described here can all be fruitfully applied to the array output to help determine a propagating signal's characteristics or assess the background noise. The next two chapters go many steps further. As promised, they contain direction-finding algorithms based on data-adaptive spectrum estimation techniques. There, we exploit model-order determination ideas to estimate the number of propagating signals in the array's field and Kalman filters track moving sources of propagating energy by merging the outputs from several arrays.

Beyond what is presented here, several other topics in estimation theory have not been discussed to keep this chapter at a manageable size. *Robust estimation theory* has the goal of providing us parameter or signal estimates when we have only partially complete information. For example, we might want to know a signal's amplitude when we only have a rough idea of its waveform. A recent survey of robust estimation procedures is found in Kassam and Poor [85]. The next chapter explores robust array processing algorithms as constrained optimization problems. Another interesting topic is *probability density function estimation* [143]. Because we based many of our algorithms on the Gaussian assumption, we would be wise to at least check its veracity. Estimation algorithms tailored to non-Gaussian observations are few; robust procedures are frequently invoked to deal with such data.

Despite the brevity of the coverage given detection theory and estimation theory, they complement propagating wave physics as the mathematical heart of array signal processing; practitioners and researchers should monitor developments in estimation theory through journals such as *IEEE Transactions on Information Theory*, the *Annals of Statistics*, and the *Journal of Time Series Analysis*.

Problems

6.1 Estimates of identical parameters are heavily dependent on the assumed underlying probability densities. To understand this sensitivity better, consider the following variety of problems, each of which asks for estimates of quantities related to variance. Determine the bias and consistency in each case.

(a) Compute the maximum a posteriori and maximum likelihood estimates of ξ based on L statistically independent observations of a Maxwellian random variable y.

$$p_{y|\xi}(y|\xi) = \sqrt{\frac{2}{\pi}}\xi^{-3/2}y^2e^{-\frac{1}{2}y^2/\xi} \qquad y > 0, \xi > 0$$

$$p_\xi(\xi) = \lambda e^{-\lambda\xi}, \qquad \xi > 0$$

(b) Find the maximum a posteriori estimate of the variance σ^2 from L statistically independent observations having the exponential density

$$p_y(y) = \frac{1}{\sqrt{\sigma^2}}e^{-y/\sqrt{\sigma^2}} \qquad y > 0$$

where the variance is uniformly distributed over the interval $[0, \sigma^2_{\text{max}})$.

(c) Find the maximum likelihood estimate of the variance of L identically distributed, but dependent Gaussian random variables. Here, the covariance matrix is written $\mathbf{K}_y = \sigma^2\widetilde{\mathbf{K}}_y$, where the normalized covariance matrix has trace $\text{tr}[\widetilde{\mathbf{K}}_y] = L$. Assume the random variables have zero mean.

6.2 Imagine yourself idly standing on the corner in a large city when you note the serial number of a passing beer truck. Because you are idle, you wish to estimate (guess may be more accurate here) how many beer trucks the city has from this single observation.

(a) Making appropriate assumptions, the beer truck's number is drawn from a uniform probability density ranging between zero and some unknown upper limit, find the maximum likelihood estimate of the upper limit.

(b) Show that this estimate is biased.

(c) In one of your extraordinarily idle moments, you observe throughout the city L beer trucks. Assuming them to be independent observations, now what is the maximum likelihood estimate of the total?

(d) Is this estimate of ξ biased? asymptotically biased? consistent?

6.3 We make L observations $y_1, \ldots, y_L$ of a parameter ξ corrupted by additive noise ($y_l = \xi + n_l$). The parameter ξ is a Gaussian random variable [$\xi \sim \mathcal{N}(0, \sigma_\xi^2)$] and n_l are statistically independent Gaussian random variables [$n_l \sim \mathcal{N}(0, \sigma_n^2)$].

(a) Find the *MMSE* estimate of ξ.

(b) Find the maximum a posteriori estimate of ξ.

(c) Compute the resulting mean-squared error for each estimate.

(d) Consider an alternate procedure based on the same observations y_l. Using the *MMSE* criterion, we estimate ξ immediately after each observation. This procedure yields the sequence of estimates $\widehat{\xi}_1(y_1), \widehat{\xi}_2(y_1, y_2), \ldots, \widehat{\xi}_L(y_1, \ldots, y_L)$. Express $\widehat{\xi}_l$ as a function of $\widehat{\xi}_{l-1}$, σ_{l-1}^2, and y_l. Here, σ_l^2 denotes the variance of the estimation error of the lth estimate. Show that

$$\frac{1}{\sigma_l^2} = \frac{1}{\sigma_\xi^2} + \frac{l}{\sigma_n^2}$$

6.4 Although the maximum likelihood estimation procedure was not clearly defined until early in the 20th century, Gauss showed in 1805 that the Gaussian density† was the *sole* density for which the maximum likelihood estimate of the mean equaled the sample average. Let $\{y_0, \ldots, y_{L-1}\}$ be a sequence of statistically independent, identically distributed random variables.

(a) What equation defines the maximum likelihood estimate $\widehat{m}_{\text{ML}}$ of the mean m when the common probability density function of the data has the form $p(y - m)$?

(b) The sample average is, of course, $\sum_l y_l/L$. Show that it minimizes the mean-squared error $\sum_l [y_l - m]^2$.

(c) Equating the sample average to $\widehat{m}_{\text{ML}}$, combine this equation with the maximum likelihood equation to show that the Gaussian density uniquely satisfies the equations.

Note: Because both equations equal 0, they can be equated. Use the fact that they must hold for *all* L to derive the result. Gauss thus showed that mean-squared error and the Gaussian density were closely linked, presaging ideas from modern robust estimation theory.

6.5 In an example {276}, we derived the maximum likelihood estimate of the mean and variance of a Gaussian random vector. You might wonder why we chose to estimate the variance σ^2 rather than the standard deviation σ. Using the same assumptions provided in the example, let's explore the consequences of estimating a *function* of a parameter [152: Probs. 2.4.9, 2.4.10].

(a) Assuming that the mean is known, find the maximum likelihood estimates of first the variance, then the standard deviation.

(b) Are these estimates biased?

(c) Describe how these two estimates are related. Assuming that $f(\cdot)$ is a monotonic function, how are $\widehat{\xi}_{\text{ML}}$ and $\widehat{f(\xi)}_{\text{ML}}$ related in general?

These results suggest a general question. Consider the problem of estimating some function of a parameter ξ, say $f_1(\xi)$. The observed quantity is y and the conditional density $p_{y|\xi}(y|\xi)$ is known. Assume that ξ is a nonrandom parameter.

(d) What are the conditions for an efficient estimate $\widehat{f_1(\xi)}$ to exist?

(e) What is the lower bound on the variance of the error of any unbiased estimate of $f_1(\xi)$?

(f) Assume an efficient estimate of $f_1(\xi)$ exists; when can an efficient estimate of some other function $f_2(\xi)$ exist?

† It wasn't called the Gaussian density in 1805; this result is one of the reasons why it is.

6.6 Let the observations $y(l)$ consist of statistically independent, identically distributed Gaussian random variables having zero mean but unknown variance. We wish to estimate σ^2, their variance.

(a) Find the maximum likelihood estimate $\widehat{\sigma^2}_{\text{ML}}$ and compute the resulting mean-squared error.

(b) Show that this estimate is efficient.

(c) Consider a new estimate $\widehat{\sigma^2}_{\text{NEW}}$ given by $\widehat{\sigma^2}_{\text{NEW}} = \alpha\widehat{\sigma^2}_{\text{ML}}$, where α is a constant. Find the value of α that minimizes the mean-squared error for $\widehat{\sigma^2}_{\text{NEW}}$. Show that the mean-squared error of $\widehat{\sigma^2}_{\text{NEW}}$ is less than that of $\widehat{\sigma^2}_{\text{ML}}$. Is this result compatible with part b?

6.7 Let the observations be of the form $\mathbf{y} = \mathbf{H}\boldsymbol{\xi} + \mathbf{n}$ where $\boldsymbol{\xi}$ and $\mathbf{n}$ are statistically independent Gaussian random vectors.

$$\boldsymbol{\xi} \sim \mathcal{N}(\mathbf{0}, \mathbf{K}_\xi) \qquad \mathbf{n} \sim \mathcal{N}(\mathbf{0}, \; \mathbf{K}_n)$$

The vector $\boldsymbol{\xi}$ has dimension M; the vectors $\mathbf{y}$ and $\mathbf{n}$ have dimension N.

(a) Derive the minimum mean-squared error estimate of $\boldsymbol{\xi}$, $\widehat{\boldsymbol{\xi}}_{\text{MMSE}}$, from the relationship $\widehat{\boldsymbol{\xi}}_{\text{MMSE}} = \mathcal{E}[\boldsymbol{\xi}|\mathbf{y}]$.

(b) Show that this estimate and the optimum linear estimate $\widehat{\boldsymbol{\xi}}_{\text{LIN}}$ derived by the Orthogonality Principle are equal.

(c) Find an expression for the mean-squared error when these estimates are used.

6.8 To illustrate the power of importance sampling, let's consider a somewhat naïve example. Let z have a zero-mean Laplacian distribution {261}; we want to employ importance sampling techniques to estimate $\Pr[z > \gamma]$ (despite the fact that we can calculate it easily). Let the density for $\tilde{z}$ be Laplacian having mean γ.

(a) Find the weight c_l that must be applied to each decision based on the variable $\tilde{z}$.

(b) Find the importance sampling gain. Show that this gain means that a *fixed* number of simulations are needed to achieve a given percentage estimation error (as defined by the coefficient of variation). Express this number as a function of the criterion value for the coefficient of variation.

(c) Now assume that the density for $\tilde{z}$ is Laplacian, but with mean m. Optimize m by finding the value that maximizes the importance sampling gain.

6.9 Suppose we consider an estimate of the parameter ξ having the form $\widehat{\xi} = \mathcal{L}(\mathbf{y}) + C$, where $\mathbf{y}$ denotes the vector of the observables and $\mathcal{L}(\cdot)$ is a linear operator. The quantity C is a constant. This estimate is *not* a linear function of the observables unless $C = 0$. We are interested in finding applications for which it is advantageous to allow $C \neq 0$. Estimates of this form we term "quasi-linear."

(a) Show that the optimum (minimum mean-squared error) quasi-linear estimate satisfies

$$\mathcal{E}[\langle \mathcal{L}_\diamond(\mathbf{y}) + C_\diamond - \xi, \mathcal{L}(\mathbf{y}) + C\rangle] = 0, \quad \text{for all } \mathcal{L}(\cdot) \text{ and } C$$

where $\widehat{\xi}_{\text{QLIN}} = \mathcal{L}_\diamond(\mathbf{y}) + C_\diamond$.

(b) Find a general expression for the mean-squared error incurred by the optimum quasi-linear estimate.

(c) Such estimates yield a smaller mean-squared error when the parameter ξ has a nonzero mean. Let ξ be a scalar parameter with mean m. The observables comprise a vector $\mathbf{y}$ having components given by $y_l = \xi + n_l, l = 1, \ldots, N$ where n_l are statistically independent Gaussian random variables $[n_l \sim \mathcal{N}(0, \sigma_n^2)]$ independent of ξ. Compute expressions for $\widehat{\xi}_{\text{QLIN}}$ and $\widehat{\xi}_{\text{LIN}}$. Verify that $\widehat{\xi}_{\text{QLIN}}$ yields a smaller mean-squared error when $m \neq 0$.

6.10 On page 288, we questioned the existence of an efficient estimator for signal parameters. We found in the succeeding example that an unbiased efficient estimator exists for the signal amplitude. Can a nonlinearly represented parameter, such as time delay, have an efficient estimator?

(a) Simplify the condition for the existence of an efficient estimator by assuming it to be unbiased. Note carefully the dimensions of the matrices involved.

(b) Show that the only solution in this case occurs when the signal depends "linearly" on the parameter vector.

6.11 In the "classic" radar problem, not only is the time of arrival of the radar pulse unknown but also the amplitude. In this problem, we seek methods of simultaneously estimating these parameters. The received signal $y(l)$ is of the form

$$y(l) = \xi_1 s(l - \xi_2) + n(l)$$

where ξ_1 is Gaussian with zero mean and variance σ_1^2 and ξ_2 is uniformly distributed over the observation interval. Find the receiver that computes the maximum a posteriori estimates of ξ_1 and ξ_2 jointly. Draw a block diagram of this receiver and interpret its structure.

6.12 We stated without derivation the Cramér-Rao bound for estimates of signal delay (Eq. 6.6 {289}).

(a) The parameter ξ is the delay of the signal $s(\cdot)$ observed in additive, white Gaussian noise: $y(l) = s(l - \xi) + n(l), l = 0, \ldots, L - 1$. Derive the Cramér-Rao bound for this problem.

(b) On page 293, this bound is claimed to be given by $\sigma_n^2/E\beta^2$, where β^2 is the mean-squared bandwidth {293}. Derive this result from your general formula. Does the bound make sense for all values of signal-to-noise ratio E/σ_n^2?

(c) Using optimal detection theory, derive the expression {292} for the probability of error incurred when trying to distinguish between a delay of τ and a delay of $\tau + \Delta$. Consistent with the problem posed for the Cramér-Rao bound, assume the delayed signals are observed in additive, white Gaussian noise.

6.13 In many analyses of array processing algorithms, the signal as well as the noise are modeled as Gaussian processes. The authors use a deterministic signal model throughout this book. Rather than discussing the merits of each model here, let's explore what differences the Cramér-Rao bound has. Assume that the signal contains unknown parameters $\boldsymbol{\xi}$, that it is statistically independent of the noise, and that the noise covariance matrix is known.

(a) What forms do the conditional densities of the observations take under the two assumptions? What are the two covariance matrices?

(b) Assuming the stochastic signal model, show that each element of the Fisher information matrix has the form

$$F_{ij} = \frac{1}{2} \operatorname{tr}\left[\mathbf{K}^{-1} \frac{\partial \mathbf{K}}{\partial \xi_i} \mathbf{K}^{-1} \frac{\partial \mathbf{K}}{\partial \xi_j}\right]$$

where $\mathbf{K}$ denotes the covariance matrix of the observations. Make this expression more complex by assuming the noise component has no unknown parameters.

(c) Compare the stochastic and deterministic bounds, the latter is given by Eq. 6.5 {287}, when the unknown signal parameters are amplitude and delay. Assume the noise covariance matrix equals $\sigma_n^2\mathbf{I}$. Do these bounds have similar dependence on signal-to-noise ratio?

6.14 We derived the Ziv-Zakai bound for a two-sensor "array." Let's extend that result to an M-sensor linear array with equal intersensor spacing d. One signal and additive white Gaussian noise compose the measured field. Because the sensors are equally spaced, one parameter, the delay-per-sensor Δ, characterizes the direction of propagation. We need to determine how array parameters M and d affect how well we can estimate Δ.

(a) Show the Ziv-Zakai bound given by Eq. 6.7 {292} applies but with $P_e(\Delta)$ interpreted for the M-sensor case.

(b) Find $P_e(\Delta)$.

(c) Using the same signal as in the example {292}, how do array parameters enter into the final bound on the delay estimate?

(d) Plot the bound for a 10-sensor array. Translate the lower mean-squared estimation error into a signal-to-noise ratio gain produced by the array.

6.15 The *Eckhart filter* is an optimum linear filter that maximizes the signal-to-noise ratio of its output [50]. To find the unit-sample response of the *FIR* Eckhart filter, consider observations of the form $\mathbf{y} = \mathbf{s}+\mathbf{n}$ where the covariance matrix of the noise is known. The signal-to-noise ratio is computed according to $\mathcal{E}[\|\mathbf{h}'\mathbf{s}\|^2]/\mathcal{E}[\|\mathbf{h}'\mathbf{n}\|^2]$, where $\mathbf{h}$ is the desired unit-sample response.

(a) Assuming the signal is nonrandom, find the Eckhart filter's unit-sample response.

(b) What is the signal-to-noise ratio produced by the Eckhart filter? How does it compare with that produced by the corresponding Wiener filter?

(c) Now assume the signal is random, having covariance matrix $\mathbf{K}_s$. Characterize the Eckhart filter.

6.16 Consider the interpolation problem where the number of observations L is odd and that we want to estimate the signal value located in the middle of the frame: $l_e = -(L-1)/2$. Assume the observations have the form $y(l) = s(l) + n(l)$ and that the covariance functions of the signal and the noise are known.

(a) What equations determine the unit-sample response of the *FIR* Wiener filter?

(b) Show that the solution to this problem has linear phase: The unit-sample response is symmetric about $(L-1)/2$.

6.17 Accurately estimating the parameters of an *AR* model using linear prediction is contingent on zero-mean observations: We must use covariance rather than correlation matrix estimates to calculate these parameters. To illustrate this point, consider a sequence of observations $y(l)$ having mean m. Rather than subtracting the mean, you calculate the parameters presuming zero-mean data.

(a) How is the "covariance matrix" calculated by ignoring the mean related to the actual one? Relate the inverse of this matrix to the inverse of the true covariance matrix.

(b) Find an expression relating the true *AR* parameters to the computed ones. Show that the quantity $1/m^2 + \mathbf{1}'\mathbf{K}^{-1}\mathbf{1}$ controls the deviation of the calculated parameters from the true ones.

6.18 Because of the special structure of the equation $\mathbf{K}_y\mathbf{a} = \sigma_w^2\boldsymbol{\delta}$ that defines the linear prediction parameters, the *AR* power spectrum can be directly related to the minimum variance power spectrum [25];[105: 354–56].

(a) Let $\mathbf{a}_r = \text{col}[1, -a_1^r, \ldots, -a_r^r], r = 0, \ldots, p$, denote the solution to the rth-order linear prediction problem. Form the matrix $\mathbf{A}$ having these vectors as its columns.

$$\mathbf{A} = \begin{bmatrix} 1 & 0 & \cdots & 0 & 0 \\ -a_1^p & 1 & \cdots & 0 & 0 \\ -a_2^p & -a_1^{p-1} & \cdots & 0 & 0 \\ \vdots & \vdots & \ddots & 1 & 0 \\ -a_p^p & -a_{p-1}^{p-1} & \cdots & -a_1^1 & 1 \end{bmatrix}$$

Show that $\mathbf{K}_y \mathbf{A} = \mathbf{U}$, where $\mathbf{U}$ is an upper triangular matrix. Here, $\mathbf{K}_y$ is a $(p+1) \times (p+1)$ matrix.

(b) Show that $\mathbf{A}'\mathbf{K}_y\mathbf{A} = \mathbf{D}$, where $\mathbf{D}$ is a diagonal matrix. What are its diagonal elements?

(c) Show that $\mathbf{K}_y^{-1} = \mathbf{A}\mathbf{D}^{-1}\mathbf{A}'$.

(d) Let's put these matrix properties of linear prediction to use. Using the result of part c and the formula $\mathbf{e}'(\omega)\mathbf{K}_y^{-1}\mathbf{e}(\omega)$ for the reciprocal of the minimum variance spectrum, relate the spectra produced by the two techniques.

(e) From this relationship, predict which spectral estimation technique would have greater resolution.

6.19 Additive noise in observations cannot be taken lightly when dealing with model-based techniques. Consider an *AR* signal observed in additive white noise.

(a) Show that the observations are *not* well described as an *AR* process. How would you characterize the observations?

(b) Find an expression relating the signal's *AR* parameters to those calculated from the observations by the linear prediction technique.

6.20 The signal has a power spectrum given by

$$\mathcal{S}_s(\omega) = \frac{17}{20} \cdot \frac{1 + \frac{8}{17}\cos\omega}{1 - \frac{4}{5}\cos\omega}$$

This signal is observed in additive white noise having variance equaling 6.

(a) Find the unit-sample response of the noncausal Wiener filter.

(b) Find the difference equation governing the causal Wiener filter ($l_e = 0$).

(c) Compare this equation with that of the Kalman filter.

(d) Calculate the signal processing gain of Wiener filter.

6.21 The state variable formulation of the Kalman filter allows its use in a variety of problems in a unified manner.

(a) Find a state variable characterization of a second-order, all-pole system governed by the input-output equation

$$y(l) = a_1 y(l-1) + a_2 y(l-2) + \sigma u(l)$$

where $u(l)$ is white Gaussian noise having unit variance.†

†Note that the answer to this part is not unique.

(b) Now consider an array of M sensors, each of which is producing a signal governed by the difference equation given previously corrupted by additive noise. The model input $u(\cdot)$ for each sensor's output is common. The observation noise at each sensor is white, Gaussian, has variance σ_n^2, and is statistically independent of the observation noise at the other sensors. Find the state variable characterization for the total array output. Describe the state transition matrix's structure.

(c) The sensor outputs $\mathbf{y}(l)$ are coalesced into an array output $z(l)$ according to the transformation $z(l) = \mathbf{w}^t\,\mathbf{y}(l)$, where $\mathbf{w}$ denotes the array shading. Derive the Kalman filter for estimating the signal component in the array output.

6.22 On page 313, the *settling time* of *LMS* filters was related to the eigenvalues of the observations' covariance matrix. A little analysis coupled with engineering knowledge makes these claims clear. Let's discuss the vector difference equation $\mathbf{v}(l) = \mathbf{A}\mathbf{v}(l-1)$, with $\mathbf{A}$ a symmetric matrix.

(a) Because this vector difference equation is first order, it represents a set of coupled first-order difference equations. Thus, some homogeneous M-dimensional linear system is described. Show that the homogeneous response can be expressed as a weighted linear combination of the eigenvectors of $\mathbf{A}$.

(b) Use the matrix's eigen decomposition to show that the norm of $\mathbf{v}$ remains bounded for any initial condition if $\max|\lambda_{\mathbf{A}}| < 1$. Letting $\mathbf{A} = \mathbf{I} - 2\mu\mathbf{K}_y$, show that this bound implies the convergence condition $0 < \mu < 1/\max\lambda_{\mathbf{K}_y}$ for the *LMS* algorithm.

(c) The rate at which a linear system reaches steady state is determined by the mode having the largest time constant. Use this analogy to show that the decay rate of our difference equation is determined by the eigenvalue of $\mathbf{A}$ having the largest magnitude and that the corresponding time constant equals

$$\tau = -\frac{1}{\ln\max|\lambda_{\mathbf{A}}|}$$

(d) Find the convergence rate of the *LMS* adaptive filter that reaches steady state smoothly (no oscillatory adaptation).

6.23 Consider a second-order *AR* process that we wish to predict using the *LMS* algorithm. As an alternative to the linear prediction technique, such an approach might be to cope with time-varying *AR* signals where covariance matrix calculation would be suspect. Let the *LMS* filter also be second order.

(a) From the *AR* parameters, find the ratio of the largest and smallest eigenvalues of the observations' covariance matrix. Assume that the parameter a_1 is positive. For a fixed value of a_2, plot this ratio.

(b) Under what conditions will *LMS* settling be the slowest?

(c) Simulate this *LMS* filter to demonstrate the effect of eigenvalue spread on settling time for a fixed value of μ.

6.24 *Adaptive line enhancement* illustrates well an important application of dynamic adaptive filters. Assume the observations are known to contain a signal (as well as ever-present noise) having period P. This signal is not known more precisely than this: It could be a sinusoid of arbitrary amplitude and phase, or a distorted sinusoid having unknown harmonic structure. We want to design a *FIR* dynamic adaptive filter that enhances the periodic signal (which has a line spectrum) and thereby reduce the observation noise.

(a) The crucial part of the design is the selection of the error signal, whose energy is minimized by the filter. Assuming the noise is white, how would the error signal be chosen to enhance signals of period P? How long should the adaptive filter be in this application? Draw a block diagram of the resulting line enhancer.

(b) The settling time and mean-squared error are governed by the characteristics of the observations' covariance matrix. In this problem, however, we should not assume that the signal component is random. As $\mathbf{y} = \mathbf{s} + \mathbf{n}$, what is $\mathbf{K}_y$ when the noise has zero mean? Determine the smallest and largest eigenvalues of this matrix. Express the smallest settling time in terms of the signal-to-noise ratio.

6.25 The covariance function estimate described on pages 320ff was biased (the authors claim) because the number of terms used at each lag varied without a corresponding variation in the normalization. Let's explore that claim closely. Assume that we now estimate the covariance function according to

$$\widehat{K}_y(m) = \frac{1}{D - |m|} \sum_{n=0}^{D-|m|-1} y(n)y(n+m)\,, \quad 0 \leq |m| \leq D-1$$

(a) Find the expected value of this revised estimator, and show that it is indeed unbiased.

(b) To derive the variance of this estimate, we need the fourth moment of the observations, which is conveniently given in the appendix (§A.1.6 {478}). Derive the covariance estimate's variance and determine whether it is consistent or not.

(c) Evaluate the expected value and variance of the spectral estimate corresponding to this covariance estimate.

(d) Does spectral estimate consistency become a reality with this new estimation procedure?

6.26 Although we have determined expressions for the expected value and variance of the Bartlett spectral estimate, we also need to determine the spectrum's probability distribution that has these moments. Assume that the signal is Gaussian, has duration N, and has been partitioned into L nonoverlapping frames of duration D.

(a) Assuming a rectangular window, what is the probability distribution at each nonzero frequency (other than π) of each component periodogram?

(b) Continuing part a, characterize the probability distribution of the Bartlett spectral estimate.

(c) Why are the frequencies 0 and π special? How are their probability distributions different?

(d) Now suppose that an arbitrary window shape is applied to each frame; how does the probability distribution of the spectral estimate at intermediate frequencies change?

6.27 Rather than the covariance matrix estimate given in Eq. 6.18 {332}, we can use a Bartlett-procedure-like maximum likelihood estimate. Imagine the length N observations segmented into L frames each having duration D. The lth frame is formed into the column matrix $\mathbf{y}_l$ as $\text{col}[y(lD), y(lD+1), \ldots, y(lD+D-1)]$, $l = 0, \ldots, L-1$.

(a) Assuming the frames are statistically independent and are each distributed as a zero-mean Gaussian random vector having covariance matrix $\mathbf{K}_y$, find the log-likelihood function.

(b) By evaluating the gradient of this expression with respect to the matrix $\mathbf{K}$, find the maximum likelihood estimate of the covariance matrix $\widehat{\mathbf{K}}^{y}_{\text{ML}}$.

(c) How does this estimate differ from that provided in Eq. 6.18?

Note: The matrix gradient identities provided in §B.3 {491} should prove useful.

6.28 The rather interesting form of the minimum variance spectral estimate (Eq. 6.18 {332}) has a periodogram counterpart.

(a) Show that the quadratic form $\mathbf{e}'(\omega)\widehat{\mathbf{K}}_y\mathbf{e}(\omega)$ equals the periodogram.

(b) Show that the Bartlett spectral estimate can be written as $\mathbf{e}'(\omega)\widehat{\mathbf{K}}^y_{\text{ML}}\mathbf{e}(\omega)$, where $\widehat{\mathbf{K}}^y_{\text{ML}}$ is the subject of Prob. 6.27.

6.29 The equations governing the covariance function of an *ARMA* process {339} have one interesting aspect: For $m > q$, $K_y(m) = \sum_{k=1}^{p} a_k K_y(m-k)$.

(a) Based on this equation, develop an algorithm for determining the *AR* parameters of an *ARMA* process. The resulting equations are known as the *higher-order Yule-Walker equations*.†

(b) Show how to estimate the *MA* parameters by inverse filtering the observations with a *FIR* filter derived from the *AR* parameter estimates.

6.30 To illustrate the importance of *ARMA* models, consider a simple, but ubiquitous variant of the *AR* signal modeling [106]. We observe an *AR* signal $s(l)$ (order p) in the presence of statistically independent white noise $n(l)$ having variance σ_n^2.

(a) Express the power spectrum of the observations in terms of the signal's linear predictive parameters. Use this result to derive an equivalent *ARMA* model for the observations $y(l)$. What are the orders of this model?

The zeros introduced by the additive noise are undistinguished: Their locations depend entirely on the pole locations and on the noise variance σ_n^2. We seek an algorithm that might calculate the *AR* parameters despite the presence of additive noise.

(b) Express the covariance function of the observations in terms of the signal's covariance function and the noise variance. Find a difference equation obeyed by this covariance function.

(c) Develop two ideas for using this difference equation to solve for the *AR* parameters. One should be based on the higher-order Yule-Walker equations (Prob. 6.29); the other, which might be considered more interesting, on how this equation compares with that governing the signal observed *without* additive noise.

6.31 Random variables can be generated quite easily if the probability *distribution* function is "nice." Let X be a random variable having distribution function $P_X(\cdot)$.

(a) Show that the random variable $U = P_X(X)$ is uniformly distributed over $(0, 1)$.

(b) Based on this result, how would you generate a random variable having a specific density with a uniform random variable generator, which is commonly supplied with most computer and calculator systems?

(c) How would you generate random variables having the hyperbolic secant density $p_X(x) = 1/2\text{sech}(\pi x/2)$?

(d) Why is the Gaussian not in the class of "nice" probability distribution functions? Despite this fact, the Gaussian and other similarly unfriendly random variables can be generated using tabulated rather than analytic forms for the distribution function.

---

†The estimates calculated this way often have much larger variances than those computed from the usual Yule-Walker equations [132].

Chapter 7

Adaptive Array Processing

Conventional delay-and-sum beamforming (and its equivalent variants discussed in Chap. 4) makes specific assumptions on the form of the propagating signal. For example, the relation between direction of propagation and delay per sensor must be known. After compensating for these delays, this algorithm weights and sums the sensors' outputs to form a signal estimate. The parameters of the delay-and-sum algorithm do not depend on signal characteristics (such as temporal spectrum) or on noise characteristics. Cases do exist for which this approach provides an optimal answer; in other situations, however, this method is inadequate when compared with algorithms specifically tailored to the signal and noise situation at hand. For example, if the number of propagating signals is known, special-purpose algorithms can be derived that have superior performance characteristics than otherwise possible. By modifying conventional beamforming algorithms to incorporate spatial correlation, we can achieve optimal detection performance. Although similar results are presented for estimating signals' directions of propagation, such algorithms presume that critical signal and noise characteristics can be measured a priori. Most array processing problems are more complicated than that, with electromagnetic noise fields varying with time of day being one of many counterexamples.

Armed with techniques derived from optimal detection and estimation theory, we seek in this chapter to derive signal processing algorithms for the outputs of an array's sensors that *adapt* their computations to the characteristics of the observations. To the unaccustomed, the literature uses the nomenclature "adaptive" inconsistently: The term can mean any algorithm whose characteristics depend on the data it receives or refer to only those algorithms that update themselves as each observation is acquired.† Here, we intend "adaptive" to mean the former while denoting the latter as "dynamic adaptive" algorithms. Generally, adaptive methods modify sensor output weights and delays according to several aspects of the observations: signal frequency content, spatial and temporal noise characteristics, and signal number. If enough observations are available, these algorithms have the capability of yielding signal processing performances

†In linear filtering theory, an example of the former would be the Wiener filter and of the latter the *LMS* algorithm.

far exceeding those of conventional techniques, but not as good as those provided by special-purpose algorithms tailored to specific signal and noise situations. Adaptive algorithms make no more assumptions than the conventional methods, but because they are extracting superior performance by a careful examination of the observations, they tend to be much more sensitive to these assumptions, many of which are subtle. For example, accuracy of sensor locations and the transfer characteristics of the sensor electronics (amplitude and phase) partially determine the accuracy of the transformation from direction of propagation to delay per sensor. Conventional techniques are generally less sensitive to sensor calibration errors than the adaptive ones. Use of adaptive techniques is like an armored knight capable of wielding a variety of weaponry: The dragon is often slain, but impaired vision can lead to choosing the wrong weapon (e.g., attacking a mouse with a broadsword).

Many of the adaptive techniques are based on the so-called high-resolution spectral estimation techniques—minimum variance and linear prediction—and on dynamic adaptive filtering methods discussed in the previous chapter. Methods inspired by the array processing problem have also been developed that are special-purpose variations of these approaches. Fundamental to the spectral techniques is the relationship between a signal's direction of propagation and its spatial spectrum. The filtering techniques find application when the directions of sources of unwanted propagating energy—jammers—are known. These techniques attempt to eliminate these extraneous signals so that other signals propagating in the medium can be made more apparent. Not unexpectedly, both types of adaptive signal processing algorithms succeed when the variety of signals considered at one time is small; of specific utility in adaptive processing are narrowband signals, sinusoids in particular. We accommodate wider bandwidth signals by applying a sequence of narrowband algorithms or by using specifically designed wideband methods [94]. Thus, much of what follows is based on a narrowband signal assumption: Early processing stages for each sensor contain a bandpass filter or temporal spectral analysis.

7.1 Signal Parameter Estimation

Accurate estimates of the parameters that help characterize each signal, its amplitude, and its direction of propagation compose the common goal of virtually all array signal processing algorithms. One obvious approach is to use estimation techniques, such as the maximum likelihood method, to find the signal parameters. Unfortunately, this simple approach yields algorithms that are quite complex, but they do provide insights into the problem at hand and set the stage for adaptive algorithms.

7.1.1 Maximum Likelihood Estimation

Let the vector $\mathbf{y}$ denote a single observation from the array, be it in the time domain or in the frequency domain. Consider the case in which a single deterministic signal is present in each sensor's output along with ever-present additive noise: $\mathbf{y} = A\mathbf{s}(\vec{\zeta}^{\,o}) + \mathbf{n}$. The normalized signal waveform is denoted by $\mathbf{s}(\vec{\zeta}^{\,o})$, where, as is typical in array processing problems, the signal's direction of propagation $\vec{\zeta}^{\,o}$ and the signal's amplitude A are unknown. Note that the delay vector has real components, whereas the observation vector, the signal, and the signal amplitude may be complex. Assuming Gaussian noise with independent real and imaginary parts and known covariance matrix $\mathbf{K}_n$, we find the maximum likelihood estimate of A and $\vec{\zeta}^o$ as the solution of (see §6.3.2 {286})

$$\min_{A,\vec{\zeta}^o} \left[\mathbf{y} - A\mathbf{s}(\vec{\zeta}^o)\right]' \mathbf{K}_n^{-1} \left[\mathbf{y} - A\mathbf{s}(\vec{\zeta}^o)\right]$$

As described in appendix C {501}, we find the minimum by evaluating this quantity's derivative with respect to each parameter, setting the results to zero, and solving.†

$$A: \quad -\mathbf{s}'(\vec{\zeta}^o)\mathbf{K}_n^{-1}\left(\mathbf{y} - A\mathbf{s}(\vec{\zeta}^o)\right) = 0$$

$$\vec{\zeta}^o: \quad \mathrm{Re}\left[A\mathbf{y}'\mathbf{K}_n^{-1}\dot{\mathbf{s}}(\vec{\zeta}^o) - |A|^2\mathbf{s}(\vec{\zeta}^o)\mathbf{K}_n^{-1}\dot{\mathbf{s}}(\vec{\zeta}^o)\right] = 0$$

In the second equation, $\dot{\mathbf{s}}(\vec{\zeta}^o)$ denotes the formal derivative of the signal vector with respect to direction of propagation. We easily find the maximum likelihood estimate of the signal's amplitude from the first equation.

$$\boxed{\widehat{A}_{\mathrm{ML}} = \frac{\mathbf{s}'(\vec{\zeta}^o)\mathbf{K}_n^{-1}\mathbf{y}}{\mathbf{s}'(\vec{\zeta}^o)\mathbf{K}_n^{-1}\mathbf{s}(\vec{\zeta}^o)}} \tag{7.1}$$

This solution depends not only on the signal but also on the direction-of-propagation vector $\vec{\zeta}^o$ that is not yet known: *The maximum likelihood estimates of amplitude and direction are thereby coupled.* Unfortunately, determining the maximum likelihood estimate for $\vec{\zeta}^o$ is at best analytically tedious, more often than not requiring numeric solution. This circumstance is best illustrated by the simplest possible example: the linear array.

Example

Consider a linear array of equally spaced sensors (spacing d) receiving a narrowband signal. The direction of propagation is represented by the single angular variable ϕ, the angle of propagation with respect to broadside. The signal vector thus has the form

$$\mathbf{s}(\phi) = \begin{bmatrix} \exp\{+j\pi\frac{d}{\lambda/2}\sin\phi \cdot M_{1/2}\} \\ \vdots \\ \exp\{-j\pi\frac{d}{\lambda/2}\sin\phi \cdot (m - M_{1/2})\} \\ \vdots \\ \exp\{-j\pi\frac{d}{\lambda/2}\sin\phi \cdot M_{1/2}\} \end{bmatrix} \tag{7.2}$$

The amplitude estimate $\widehat{A}_{\mathrm{ML}}$ has a closed form solution dependent on the angle's estimate. To estimate ϕ, we must determine the derivative $\dot{\mathbf{s}}(\phi)$, substitute it into the estimation

†To minimize with respect to amplitude, we evaluate the derivative with respect to A^* and set it equal to 0 (see the discussion surrounding Eq. C.1 {502}, for details).

equation, and solve. As we shall see, the last step causes analytic difficulty. Proceeding anyway, the derivative equals

$$\dot{\mathbf{s}}(\phi) = -j\pi\frac{d}{\lambda/2}\cos\phi \begin{bmatrix} -M_{1/2}\exp\{+j\pi\frac{d}{\lambda/2}\sin\phi\cdot M_{1/2}\} \\ \vdots \\ (m-M_{1/2})\exp\{-j\pi\frac{d}{\lambda/2}\sin\phi\cdot(m-M_{1/2})\} \\ \vdots \\ M_{1/2}\exp\{-j\pi\frac{d}{\lambda/2}\sin\phi\cdot M_{1/2}\} \end{bmatrix}$$

The common factor of $\cos\phi$ means that $\phi = \pm\pi/2$ are *always* possible *data-independent* solutions. Several other solutions exist, meaning that we must test each to determine whether it maximizes the likelihood function or not. Searching this way is not simple; the diligent can search for analytic solutions (see Prob. 7.1).

We are led to an alternative interpretation of the maximum likelihood solution when we consider the special case of spatially white noise: $\mathbf{K}_n = \sigma_n^2\mathbf{I}$. In this case, $\mathbf{s}'\mathbf{K}_n^{-1}\mathbf{s} = M/\sigma_n^2$, the amplitude estimate equals $\mathbf{s}'\mathbf{y}/M$ and the quantity $\mathbf{s}'\mathbf{K}_n^{-1}\dot{\mathbf{s}}$ equals 0. $\widehat{A}_{\mathrm{ML}}$ thus equals the value of the constant-shading array pattern in the direction $\widehat{\phi}_{\mathrm{ML}}$. This angle estimate is that which yields the maximum real part for the array pattern. Thus, in the spatially white-noise, single-signal case, *conventional beamforming yields maximum likelihood estimates*. In this sense, constant-shading conventional beamforming is adapted to this special signal and noise circumstance.

Reconsider the original optimization problem by expanding the quadratic form on page 351 and expressing it as a maximization problem.

$$\max_{A,\vec{\zeta}^o}\left\{2\,\mathrm{Re}[A\mathbf{s}(\vec{\zeta}^o)\mathbf{K}_n^{-1}\mathbf{y}] - |A|^2\mathbf{s}'(\vec{\zeta}^o)\mathbf{K}_n^{-1}\mathbf{s}(\vec{\zeta}^o)\right\}$$

The first term represents a conventional beamforming calculation in which the "shading" sequence has become the matrix $\mathbf{K}_n^{-1}$. We saw this generalization occur when we applied optimum colored-noise detection to the array processing problem {252}. Once we compute the array pattern in this enlightened way, the maximum likelihood estimate is found by maximizing the array pattern once the array pattern's variation to the signal $(\mathbf{s}'\mathbf{K}_n^{-1}\mathbf{s})$ is subtracted. In the spatially white-noise case, this variation is constant and can be ignored. Thus, as we have argued before, conventional beamforming expresses the correct idea for good signal processing operations but not the correct form.

Several important observations can be deduced from the maximum likelihood estimate's properties. To determine if a signal is present at all, a detection threshold can be applied quite easily to the squared magnitude of the amplitude estimate. Note that this estimate as well as the direction of propagation estimate requires knowledge of the noise-covariance matrix. If the application is such that the noise statistics can be measured without "corruption" by a signal, then this matrix can be estimated in the obvious way. One application fulfilling this requirement is the radar problem, in which the presence of a return from the target is under the control of the transmitter, which can easily coordinate with the receiver. In such problems, the maximum likelihood estimate of the noise-covariance matrix can be computed and used to advantage. In other problems, we may not be able to predict the signal's presence a priori or the time at which we measure noise characteristics may be distant from the observation time. The maximum likelihood approach to signal parameter estimation makes little sense in such cases.

Another issue is our assumption that only one signal is present in the observations [139, 158]. Presuming that the number of signals can be estimated,† the computations required by the maximum likelihood estimate grow more rapidly than proportionally. For example, the amplitude estimates in the two-signal case are found by

$$\widehat{A}_{1,\mathrm{ML}}\mathbf{s}'(\vec{\zeta}_1^o)\mathbf{K}_n^{-1}\mathbf{s}(\vec{\zeta}_1^o) + \widehat{A}_{2,\mathrm{ML}}\mathbf{s}'(\vec{\zeta}_1^o)\mathbf{K}_n^{-1}\mathbf{s}(\vec{\zeta}_2^o) = \mathbf{s}'(\vec{\zeta}_1^o)\mathbf{K}_n^{-1}\mathbf{y}$$
$$\widehat{A}_{1,\mathrm{ML}}\mathbf{s}'(\vec{\zeta}_2^o)\mathbf{K}_n^{-1}\mathbf{s}(\vec{\zeta}_1^o) + \widehat{A}_{2,\mathrm{ML}}\mathbf{s}'(\vec{\zeta}_2^o)\mathbf{K}_n^{-1}\mathbf{s}(\vec{\zeta}_2^o) = \mathbf{s}'(\vec{\zeta}_2^o)\mathbf{K}_n^{-1}\mathbf{y}$$

We are forced to solve a set of linear equations to find the amplitude estimates: *The amplitudes must be estimated jointly.* Realizing that two direction of propagation vectors now need to be found, a simple solution no longer emerges: We must numerically maximize a function of two variables.‡ The usual approach is to assume that the single-signal estimate approximates the optimal one: Apportion array pattern maxima (computed with the inverse noise covariance matrix shading) to individual signals. This approach ignores cross-terms between the signals; the term $\mathbf{s}'(\vec{\zeta}_2^o)\mathbf{K}_n^{-1}\mathbf{s}(\vec{\zeta}_1^o)$ in the joint amplitude estimation equations illustrates the coupling. Thus, in the two-signal case, the direction of propagation vectors that individually maximize the likelihood function yield *approximate* signal parameter estimates. As shown in later sections, the simplicity of this approximation is countered by an increase in bias, which in many cases is not asymptotically zero. Alternative, computationally simpler methods must be found that permit beamforming and direction of propagation estimation without a priori knowledge of the noise covariance matrix and cope well with multiple signals.

7.1.2 Least-Squares Fitting of Spatial Correlation Matrix

The complexities imposed by exact maximum likelihood estimates can be mitigated by changing the problem formulation. Rather than use the observations directly, a particularly simple, but powerful, adaptive signal processing technique exploits the structure of the spatiospectral correlation matrix. As shown by Eq. 2.21 {54}, the deterministic signal case yields $\mathbf{R} = \mathbf{SCS}' + \mathbf{K}_N$. For expository simplicity, suppose two narrowband signals are observed by a linear array (constant spacing d) in additive (spatially) white noise. The signal vector thus has the form of an ideal signal propagation vector $\mathbf{e}(\phi)$ given by Eq. 7.2 {351}. The spatiospectral correlation matrix corresponding to this signal and noise situation has the structure

$$\mathbf{R} = \begin{bmatrix} \mathbf{e}(\phi_0) & \mathbf{e}(\phi_1) \end{bmatrix} \cdot \begin{bmatrix} A_0^2 & cA_0A_1 \\ c^*A_0A_1 & A_1^2 \end{bmatrix} \cdot \begin{bmatrix} \mathbf{e}'(\phi_0) \\ \mathbf{e}'(\phi_1) \end{bmatrix} + \sigma_N^2\mathbf{I}$$

†Methods for determining the number of signals present are discussed in §7.3.5 {390}.

‡More variables must be scanned if the array has something other than linear geometry or if the source lies in the array's near field.

The parameter c is the normalized complex coherence between the two signals. Despite the complexity of this expression, note how a relatively few parameters, ϕ_0, ϕ_1, A_0, A_1, c, and σ_N^2, *completely* characterize the matrix.

The parametric approach simply finds the matrix having the form given earlier that best "fits" a measured spatiospectral correlation matrix $\widehat{\mathbf{R}}$ [99]. Denoting this set of parameters by the vector $\boldsymbol{\xi}$, the so-called *parametric method* extracts signal and noise parameters by finding the solution to

$$\boxed{\min_{\boldsymbol{\xi}} \left\| \widehat{\mathbf{R}} - \mathbf{R}(\boldsymbol{\xi}) \right\|^2}$$

The number of parameters sought—for N_s signals, there are N_s propagation directions, N_s amplitudes, and no more than $N_s(N_s-1)/2$ complex intersignal coherences as well as the noise variance (a total of 7 real-valued parameters in the two-signal case)—must not be greater than the number of quantities used in the fitting procedure. Here, the number of real quantities provided by the observations equals M^2: The real and imaginary parts of the intersensor correlations contained in the measured spatiospectral correlation matrix. So long as $N_s + 1 \leq M$, a unique solution to the fitting problem is guaranteed to exist. Numerical methods are used to solve this fitting problem because of its complexity. A prerequisite for this method is knowledge of the number of signals N_s; §7.3.5 {390} summarizes techniques for determining this number.

Example

To provide some basis for comparing this method with those that follow, the various methods are tested on a common set of observations. Two noncoherent signals are present in the acoustic field having directions of propagation of 10° and −20°. The amplitude of the first is unity and the second 0.5. The noise variance equals 1. The array consists of ten linearly placed sensors with half-wavelength spacing ($d = \lambda/2$). The results of applying the parametric method are shown in Table 7.1. Note that the maximum likelihood estimate of the correlation matrix employed in this example is not invertible in the $L = 5$ case.

Parameter	Actual Value	Estimated Value	
		$L = 5$	$L = 50$
σ_N^2	1.0	0.98 ± 0.15	0.99 ± 0.05
A_0^2	1.0	0.99 ± 0.21	1.00 ± 0.07
ϕ_0	20°	$20.0° \pm 0.7$	$20.0° \pm 0.2$
A_1^2	0.25	0.25 ± 0.11	0.26 ± 0.04
ϕ_1	−10°	$-10.2° \pm 2.1$	$-10.0° \pm 0.5$

Estimates are shown for two values of the time-bandwidth product L; one is less than the number of sensors in the array and the other greater ($M = 10$). The simulations employ a zero-intersignal coherence model over 1,000 trials. Errors are expressed as standard deviations. D. A. Linebarger, University of Texas at Dallas, provided these simulation results.

TABLE 7.1 Parametric Method Example.

Because the measured spatial correlation matrix converges to its ideal form as the time-bandwidth product increases, the signal and noise parameter estimates are consistent *if* the signal and noise models are accurate. If the assumed models are not quite correct (for example, the noise is not spatially white), all of the parameter estimates are affected to some degree. This characteristic is typical of all parametric methods: If the presumed form of the observations is correct, it provides a powerful clue to signal and noise characteristics; if the presumptions are wrong, not only is the clue wrong and the resulting conclusions in error, we have no indication that errors are present.

7.2 Constrained Optimization Methods

Many adaptive array processing techniques can be derived by solving a constrained mean-squared optimization problem [20, 80, 96]. Let $\mathbf{y}$ denote the vector of observations, lying in either frequency or time domains, obtained from the entire array. Adaptive algorithms based on optimization problem solutions consist of applying a *weight vector* $\mathbf{w}$ to the observation vector with the intent of minimizing the mean-squared value of the weighted observations $\mathcal{E}[|\mathbf{w}'\mathbf{y}|^2]$ subject to the side constraint that $\mathbf{Cw} = \mathbf{c}$. $\mathbf{C}$ is the *constraint matrix* and $\mathbf{c}$ is a column matrix of constraining values. For frequency-domain problems, the weight vector is complex valued and comprises a set of amplitude weights (acting as a spectral window) and phase shifts (to compensate for delay-per-sensor phase effects). An example of a frequency-domain constraint would be requiring the weight vector to either accentuate signals propagating from some direction ($\mathbf{s}'\mathbf{w} = 1$) or suppress jammers ($\mathbf{s}'\mathbf{w} = 0$). Thus, the constraint matrix would be formed from the set of ideal signals expressing various propagation directions and the constraining values would be chosen appropriately. Solving the optimization problem then amounts to finding the set of weights that results in the lowest-power array output subject to the directional constraints; minimizing power presumably reduces the deleterious effects of noise and unwanted signals. In time-domain problems, we would want the optimal weight vector to reinforce any signal propagating across the array with a specific set of delays. The constraint might be to focus on a specific signal waveform while suppressing others. In either case, adaptive array processing algorithms vary the shading according to signal and noise characteristics present at the time the observations are obtained. As $\mathcal{E}[|\mathbf{w}'\mathbf{y}|^2] = \mathbf{w}'\mathbf{R}\mathbf{w}$, we formally state the general optimization problem for adaptive array processing as

$$\boxed{\min_{\mathbf{w}} \mathbf{w}'\mathbf{R}\mathbf{w} \quad \text{subject to } \mathbf{Cw} = \mathbf{c}}$$

As shown in appendix C, the solution to this problem is easily derived using Lagrange multipliers. There, we find the solution vector $\mathbf{w}_\diamond$ for the general constrained adaptive array processor to be

$$\boxed{\mathbf{w}_\diamond = \mathbf{R}^{-1}\mathbf{C}'\left(\mathbf{C}\mathbf{R}^{-1}\mathbf{C}'\right)^{-1}\mathbf{c}}$$

with array output power

$$\boxed{\mathcal{P} = \mathbf{w}_\diamond'\mathbf{R}\mathbf{w}_\diamond = \mathbf{c}'\left(\mathbf{C}\mathbf{R}^{-1}\mathbf{C}'\right)^{-1}\mathbf{c}}$$

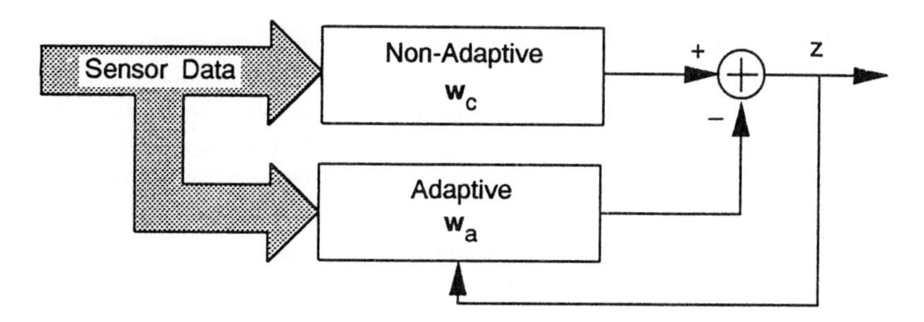

Figure 7.1 The solution to the constrained adaptive beamformer can be decomposed into two orthogonal components: a completely nonadaptive component having a weight vector given by $\mathbf{w}_c = \mathbf{C}'(\mathbf{C}\mathbf{C}')^{-1}\mathbf{c}$ and the adaptive component that becomes nonzero whenever the signal field received by the array is not white. This decomposition forms the foundation of the generalized sidelobe canceller.

These results lie at the heart of many important adaptive beamforming techniques.

The projection matrix {497} associated with the constraint equation is $\mathbf{P_C} = \mathbf{C}'(\mathbf{C}\mathbf{C}')^{-1}\mathbf{C}$.† This matrix serves to decompose the optimum weight vector into two *orthogonal* components. This decomposition of the constrained adaptive beamformer is illustrated in Fig. 7.1. The projection of the optimum weight vector onto the constraints is $\mathbf{w}_c = \mathbf{P_C}\mathbf{w}_\diamond = \mathbf{C}'(\mathbf{C}\mathbf{C}')^{-1}\mathbf{c}$. Note that this vector does not depend on the spatial correlation matrix and therefore describes the *nonadaptive component* of the solution. This component describes the entire solution when the spatial correlation matrix equals an identity matrix and frequently describes a conventional beamformer. The remaining component does depend on the observations and represents the purely adaptive component of array processing algorithms based on constrained optimization. Thus, these algorithms differ from conventional ones whenever *any* signal is present, and because of the constraints and the orthogonal decomposition, yield better performance. This decomposition of the optimal weight vector motivates the *generalized sidelobe canceller* described in §7.2.4 {369}.

Example [57]

An example application of an adaptive beamformer based on constrained optimization is the wideband system depicted in Fig. 7.2. The array is steered in a prescribed propagation direction according to the set of delays expressed by the column matrix **1**. The output of each sensor is then passed through a *FIR* filter having N coefficients. This time-domain filtering is intended to provide some rejection of interference not lying in the proper temporal frequency region. Rather than specifying identical time-domain filters, we allow each coefficient to adapt separately with constraints employed to force a specified temporal frequency response on the array output. As shown in the schematic of the system, the weight vector has MN components and the observation vector $\mathbf{y}$ has the concatenated form $\mathbf{y} = \text{col}[\mathbf{y}_0, \ldots, \mathbf{y}_{M-1}]$. The matrix $\mathbf{R}$ is thus a spatiotemporal correlation matrix. To determine the constraint matrix and the constraining values, assume that the array is "pointed" toward the signal of interest—the delays have been selected according to the assumed propagation direction—so that the propagating signal of interest has a constant wavefront when it enters the *FIR* filters. In the ideal situation in which other signals and noise are absent, the *FIR* filters appear to

†So long as the constraints are linearly independent (not redundant), the matrix $\mathbf{C}\mathbf{C}'$ is invertible.

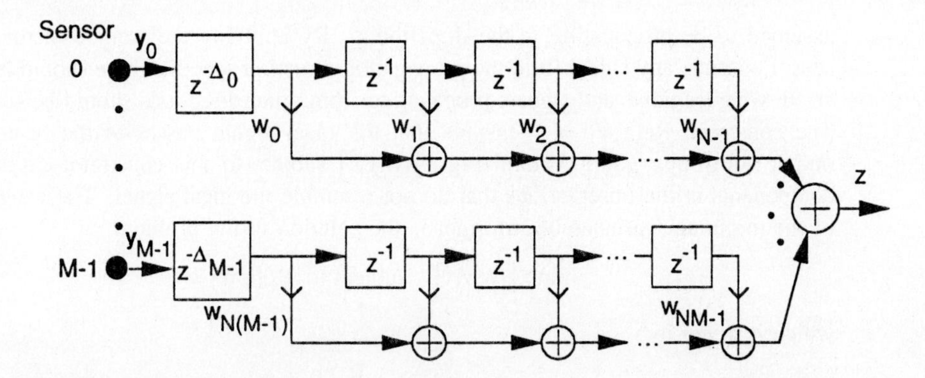

Figure 7.2 Once each sensor's output has been delayed by an amount appropriate for the assumed propagation direction, each output is passed through a *FIR* filter whose coefficients are selected individually. The sum of the weights applied to the nth tap on each filter is constrained to equal the value of the nth value of a predesigned unit sample response.

be driven by a common input. The resulting *FIR* filter describing the relation between array input and output has tap values equal to the sum of component-filter tap values occurring at the temporal delay (in other words, the same snapshot time). Because the nth snapshot occurs every n entries in $\mathbf{y}$, the nth row of the constraint matrix selects the corresponding entries from the shading vector to impose the *FIR* filter constraint.

$$\mathbf{C} = \overbrace{\begin{bmatrix} \underbrace{1\ 0\ 0\cdots0}_{N} & \underbrace{1\ 0\ 0\cdots0}_{N} & \cdots \\ 0\ 1\ 0\cdots0 & 0\ 1\ 0\cdots0 & \cdots \\ \vdots & \vdots & \\ 0\ 0\cdots0\ 1 & 0\ 0\cdots0\ 1 & \cdots \end{bmatrix}}^{MN} \Bigg\} N$$

The constraint value c_n equals the nth term in the unit sample response of a predesigned *FIR* filter.

We find that the nonadaptive portion of this system, expressed by $\mathbf{P_C w_\diamond}$, applies the predesigned filter to each sensor's output: Each filter is the same.

$$\mathbf{P_C w_\diamond} = \mathbf{C}'(\mathbf{CC}')^{-1}\mathbf{c} = \frac{1}{M}\,\mathrm{col}[\mathbf{c},\ldots,\mathbf{c}]$$

The adaptive portion "kicks in" when nonwhite (spatially and temporally) noise occurs or when signals propagating from other than the assumed propagation direction arise.

7.2.1 Minimum Variance Beamforming

As one example of a beamforming technique that can be obtained as the solution of a constrained optimization problem, the *minimum variance* method is perhaps the most straightforward [35: 45–52];[80];[96]. Let $\mathbf{e}(\vec{\zeta}\,)$ represent an ideal, unit-amplitude signal assumed to be propagating in the direction $\vec{\zeta}$. By applying the weight vector $\mathbf{w}$ to the sensor outputs, any signal from the propagation direction specified by $\mathbf{e}$ should be emphasized, whereas noise and signals propagating from other directions should be suppressed. The constraint $\text{Re}[\mathbf{e}'\mathbf{w}] = 1$ ensures that the ideal signal passes to the beamformer's output with unity gain. Minimizing $\mathcal{E}\left[|\mathbf{w}'\mathbf{y}|^2\right]$ subject to this constraint suppresses all components in the observations that do not resemble the ideal signal. The weight vector of the minimum variance beamformer is the solution to the problem

$$\min_{\mathbf{w}} \mathcal{E}\left[|\mathbf{w}'\mathbf{y}|^2\right] \quad \text{subject to} \quad \text{Re}[\mathbf{e}'\mathbf{w}] = 1$$

which is given by

$$\boxed{\mathbf{w}_\diamond = \frac{\mathbf{R}^{-1}\mathbf{e}}{\mathbf{e}'\mathbf{R}^{-1}\mathbf{e}}} \tag{7.3}$$

Note that the optimum weight vector not only depends on the spatial correlation matrix of the observations but also on the assumed propagation direction $\vec{\zeta}$. Thus, as various directions are scanned, the weights change; they adapt to the signal and noise components in the observations. The nonadaptive component $\mathbf{w}_c$ equals $\mathbf{e}(\mathbf{e}'\mathbf{e})^{-1} \cdot 1 = \mathbf{e}/M$. The quantity $\mathcal{P}(\vec{\zeta}\,) = \mathbf{w}_\diamond'\mathbf{R}\mathbf{w}_\diamond$ is the power in the beamformer's output in the assumed propagation direction. The output power of the minimum variance beamformer, which expresses its steered response, has the intriguing formula [28, 31]

$$\boxed{\mathcal{P}^{\text{MV}}(\vec{\zeta}\,) = \left[\mathbf{e}'(\vec{\zeta}\,)\mathbf{R}^{-1}\mathbf{e}(\vec{\zeta}\,)\right]^{-1}}$$

The reader might well wonder why this technique is termed "minimum variance" beamforming; a variance appears nowhere in the derivation. The beamformer output, $\mathbf{w}'\mathbf{y}$, contains a sum of signal and noise components, of which the noise is guaranteed to have a probabilistic model (at least implicitly). The signal(s), depending on the nature of the source(s) of the propagating energy, *may or may not* have a probabilistic foundation. If random, the mean of the signal component can be assumed zero, and the resulting beamformer output also has zero mean. In this case, its mean-squared value equals its variance; hence the name used here: minimum variance. If not random, the mean-squared value does not equal the variance, and the technique is inappropriately named. The terminology is further confused by references to the minimum variance technique in the literature as the maximum likelihood method (*MLM*) or Capon's method [28, 31]. In the radar community, this algorithm is termed an Applebaum or Howells-Applebaum array [5, 75]. Early sonar papers also refer to this technique as "adaptive beamforming" as it is one of the earliest adaptive methods; since then, many others have been developed, and it can no longer claim to be the sole adaptive algorithm. The etymology of these names can be traced to imprecision in signal modeling and misuse of statistical terminology. When Capon considered the maximum likelihood estimate of the signal amplitude described in §7.1.1 {350}, he noted its similarity to the formula for the minimum variance beamformer output.

$$\widehat{A}_{\text{ML}} = \left(\frac{\mathbf{K}_n^{-1}\mathbf{e}}{\mathbf{e}'\mathbf{K}_n^{-1}\mathbf{e}}\right)' \cdot \mathbf{y} \quad \text{vis-à-vis} \quad \mathbf{w}_\diamond'\mathbf{y} = \left(\frac{\mathbf{R}^{-1}\mathbf{e}}{\mathbf{e}'\mathbf{R}^{-1}\mathbf{e}}\right)' \cdot \mathbf{y}$$

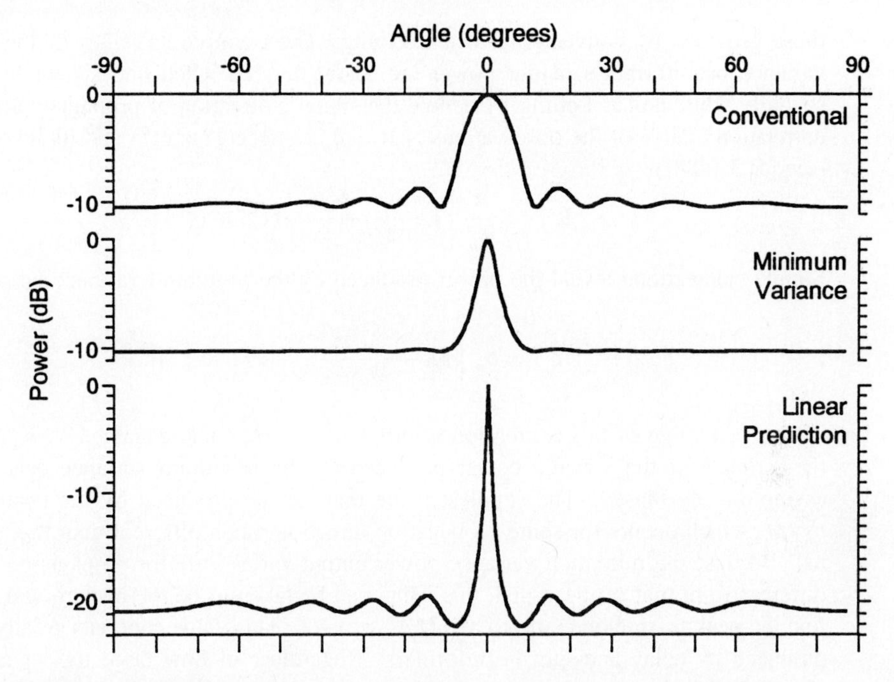

Figure 7.3 The steered responses that result when the exact spatial correlation matrix is used in array processing computations are shown. The noise is white and has unity variance. A single, unit-amplitude signal propagates from broadside to the regular linear array comprised of ten sensors ($d = \lambda/2$). The conventional beamformer with a rectangular window is used in the top panel, the minimum variance in the middle panel, and linear prediction in the bottom.

The *noise covariance matrix* appears in maximum likelihood amplitude estimate whereas the *observations' correlation matrix* appears in the minimum variance output. In Capon's seismic problem, the background "noise" was highly variable (nonstationary), and the noise covariance matrix could not be meaningfully estimated. He suggested using the correlation matrix (no signal component subtracted) rather than the covariance matrix because, from an "engineering" viewpoint, they are closely related [28]. The minimum variance weight vector is not the maximum likelihood solution of any known estimation problem. The derivation of the minimum variance technique given here, which avoids probabilistic signal models and maximum likelihood solutions, is based on the one given by Lacoss; even he grudgingly uses the name "maximum likelihood" in his paper. The origins of the terminology are indeed confusing and might even support a feeling that the method is ad hoc. Although such judgments have some foundation, the minimum variance beamformer has important characteristics described in sequel that make it one of the most important adaptive algorithms in applications [119].

The steered responses produced by the conventional and minimum variance algorithms are compared in Fig. 7.3. The minimum variance technique yields a narrower mainlobe and its "sidelobes," more accurately termed its *background level*, are smoother than those provided by conventional beamforming. The asymptotic values of the minimum variance beamformer's output power are easily derived when one signal is present in spatially white noise. Letting $\vec{\zeta}^o$ denote the signal's direction of propagation, the spatial correlation matrix of the observations is $\mathbf{R} = \sigma_n^2\mathbf{I} + A^2\mathbf{e}(\vec{\zeta}^o)\mathbf{e}'(\vec{\zeta}^o)$ and its inverse equals (see §B.3 {488})

$$\mathbf{R}^{-1} = \frac{1}{\sigma_n^2}\left[\mathbf{I} - \frac{A^2}{MA^2+\sigma_n^2}\mathbf{e}(\vec{\zeta}^o)\mathbf{e}'(\vec{\zeta}^o)\right]$$

Simple calculations reveal the power produced by the minimum variance beamformer.

$$\mathcal{P}^{\text{MV}}(\vec{\zeta}\,) = \sigma_n^2\left[M - \frac{A^2}{MA^2+\sigma_n^2}|\mathbf{e}'(\vec{\zeta}\,)\mathbf{e}(\vec{\zeta}^o)|^2\right]^{-1}$$

The largest value of this expression occurs when $\vec{\zeta} = \vec{\zeta}^o$ and equals $A^2 + \sigma_n^2/M$. Thus, the estimate of the signal's power produced by the minimum variance beamformer is asymptotically biased. The smallest value that can be produced by the beamformer is σ_n^2/M, which occurs for some propagation direction much different than that of the signal. Because the minimum variance power output varies little for propagation directions different from that of the signal, this value can be taken to be the background level. We find the peak-to-sidelobe ratio to be $(MA^2/\sigma_n^2)+1$. This value contrasts greatly with that produced by delay-and-sum beamformers: Regardless of how large the signal-to-noise ratio may be, they yield a constant peak-to-sidelobe ratio. The variation of this ratio with the signal-to-noise ratio in the beamformer's output is characteristic of adaptive beamformers: The beamformer's characteristics vary with signal and noise properties.

Further insights into the adaptive nature of the minimum variance technique are obtained by considering a two-signal example (Fig. 7.4). What reveals the minimum variance method's adaptive character is how its beampattern varies with signal and noise properties (through the spatial correlation matrix) and with the presumed direction of propagation (through **e**). Although the mainlobe has width comparable to that of conventional beamforming, the sidelobe structure differs dramatically (Fig. 7.4). When the presumed propagation direction corresponds to that of one of the signals, the minimum variance weight vector produces a beampattern having a null in the other signal's direction. If the presumed direction corresponds to no signal, the beampattern has nulls placed in each of the signal propagation directions. In contrast, the nulls of conventional beampatterns do not depend on the presence of other signals and are located at fixed directions relative to the assumed propagation direction. Because of this change in the weight vector with presumed direction of propagation, signals arriving from other directions have less influence, allowing the presumed signal's power to be accurately computed. Said another way, the weight vector adapts to the signal and noise environment. Note, however, that for any presumed propagation direction, we use the weight vector provided by the minimum variance algorithm in the same manner as in the conventional beamformer. The core signal processing idea of weighting sensor outputs in the frequency domain to reinforce maximally the signal remains; the minimum variance technique computes a weight vector a posteriori from the available observations, whereas conventional beamforming's computes it a priori.

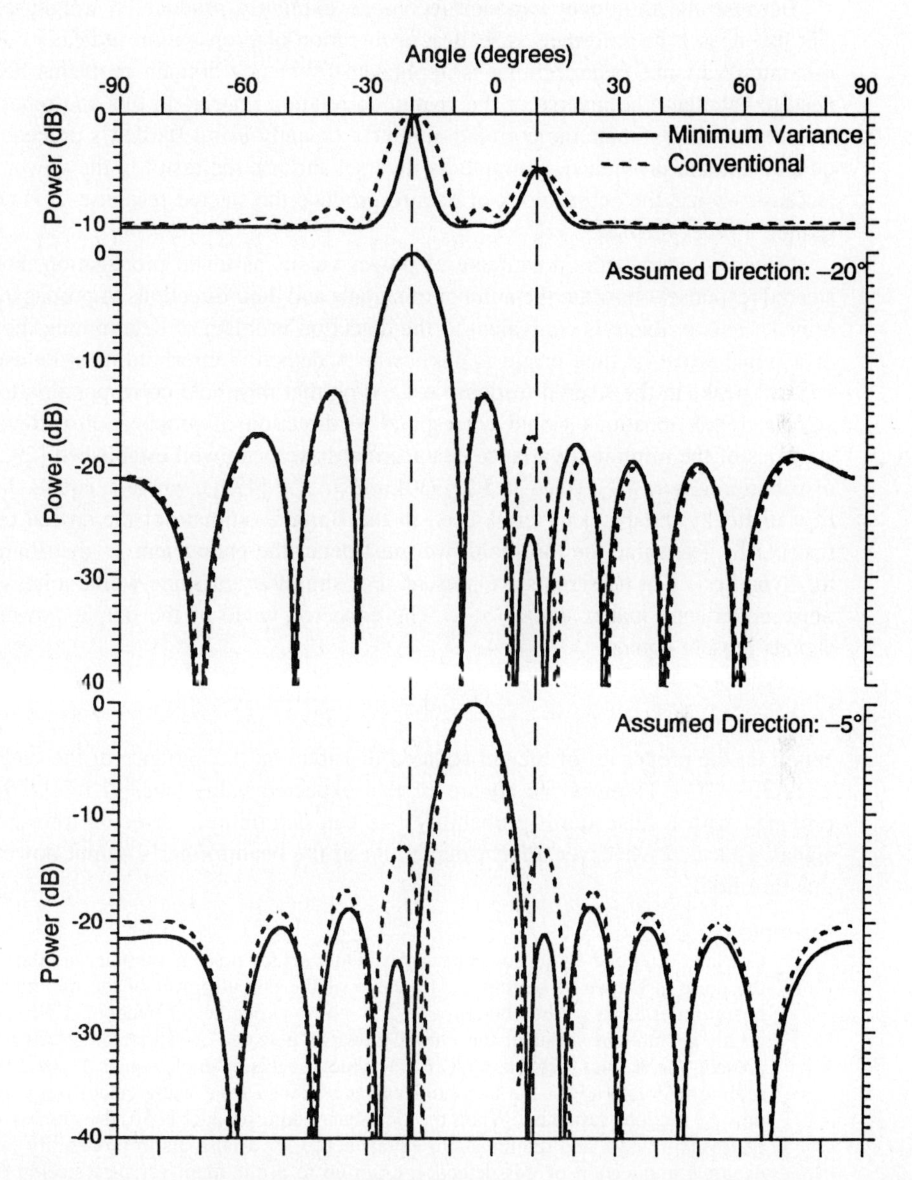

Figure 7.4 The top panel shows the steered responses of minimum variance and conventional beamformers for our example two-signal case. The center panel shows the minimum variance algorithm's beampattern corresponding to the -20° signal direction. Conventional beamforming's beampattern is shown for reference. A similar comparison for the 5° assumed propagation direction (which corresponds to no signal) is shown in the bottom panel. Note how the minimum variance method tends to null signals *not* propagating in the assumed propagation direction.

Because the minimum variance technique explicitly produces a weight vector, we can use it as a beamformer as well as a direction-of-propagation indicator. Primarily, minimum variance beamforming is applied to frequency domain problems because we need to calculate the inverse of the spatial correlation matrix. In this and other adaptive techniques, we estimate the correlation matrix (usually using Bartlett's procedure in frequency domain problems), compute its inverse, and use the result in the power estimate. Because we use the entire block of data to produce the steered response, it is commonly termed a *block method.*†

Once provided with an estimate of power versus assumed propagation direction—a steered response—how are the number of signals and their directions of propagation determined? This problem is equivalent to the detection problem of determining the presence of a signal when its time origin is unknown. A detection threshold must be established, with all peaks in the steered response exceeding that threshold corresponding to possible signals. Peak locations would correspond to direction of propagation estimates. The statistics of the minimum variance beamformer have been well established [29, 30]: The distribution is $\chi^2_{2(L-M+1)}$ (see §6.5.2 {330} and §A.1.7 {479}), where L equals the number of statistically independent terms used in the Bartlett estimate of the spatial correlation matrix. To determine the threshold, we must detail the parameters χ^2 distribution under the hypothesis that no signals are present. For simplicity, assume white noise so that the noise covariance matrix equals $\sigma_n^2\mathbf{I}$. The expected value of the output power with no signals present equals

$$\mathcal{E}[\mathcal{P}^{\mathrm{MV}}] = \frac{L-M+1}{L}\frac{\sigma_n^2}{M} \tag{7.4}$$

Based on the properties of the chi-squared distribution, the variance of the output power is $1/(L-M+1)$ times the square of this expected value (see §A.1.7 {479}). Once provided with a false-alarm probability, we can determine a detection threshold γ. A signal is located whenever a *local* maximum of the beamformer's output power exceeds this threshold.

Example

Consider a regular 10-sensor linear array. Only sensor noise is present, and the variance of the noise is known to be unity. The value of the output power of the minimum variance beamformer has a χ^2 distribution with $2(L-M+1)$ degrees of freedom. Rather than being a unit χ^2 random variable, the output power is a scaled χ^2 variable. When no signal is present, the scaling factor is $\sigma_n^2/(2LM)$, which in this example equals $1/20L$. We use this scaling factor to adjust the threshold values obtained from using calculations that employ unit χ^2 random variables. When the time-bandwidth product is 50, the number of degrees of freedom is 82, and the threshold equals 0.115 for a false-alarm probability of 10^{-2}. An example application of this detection criterion to a minimum variance steered response is shown in Fig. 7.5. Both signals are easily located and no false alarms are present in this example.

As expressed by Eq. 4.18 {141} and Prob. 4.16 {194}, the optimal array gain occurs when $\mathbf{w} \propto \mathbf{K}_n^{-1}\mathbf{e}$ and equals $\mathrm{tr}[\mathbf{K}_n]\mathbf{e}'\mathbf{K}_n^{-1}\mathbf{e}/M$. The minimum variance solution for the

†Alternatives to block methods incorporate the observations as they arrive. Such approaches are termed dynamic adaptive methods; they are discussed in §7.5 {402}.

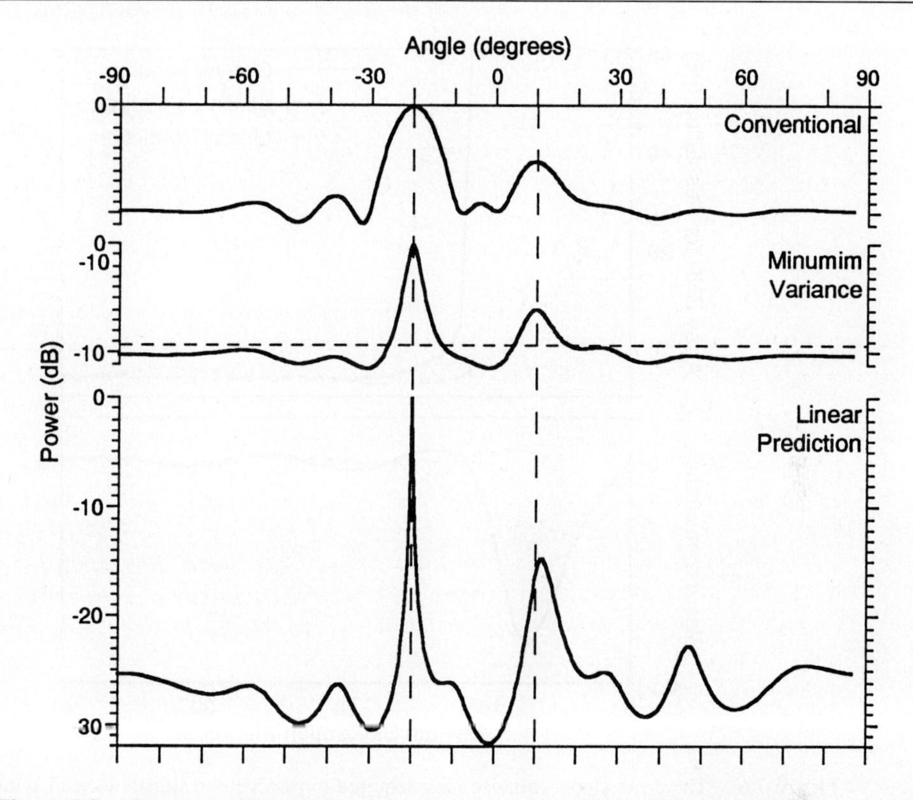

Figure 7.5 Steered responses computed using the conventional, minimum variance, and linear prediction techniques are shown for a regular 10-sensor linear array. The time-bandwidth product of the correlation matrix estimate is 50 and the noise variance unity. The indicated detection threshold corresponds to a false-alarm probability of 10^{-2}. Two signals are present in the observations, one having a bearing of 10° and the other -20°. The signal propagating from a bearing of 10° has $1/4$ the power of the other.

weight vector (Eq. 7.3 {358}) greatly resembles this expression. They differ because the optimal gain depends on the noise covariance matrix while the minimum variance gain depends on the field's correlation matrix. Surprisingly, the minimum variance solution *does* provide optimal array gain when only one signal is present, and the assumed propagation direction coincides with the signal's propagation direction (see Prob. 7.7). Thus in terms of signal detectability in additive Gaussian noise, the minimum variance method is optimal.

Another criterion used to evaluate a beamformer's performance is *resolution*: the ability to distinguish two signals having nearly equal directions of propagation.† Because of the adaptive characteristics of minimum variance beamforming, it affords a far

†On page 142, resolution is defined in terms of mainlobe width. For conventional methods, this definition gives some insight into the minimum resolvable directions of propagation for two sources. For adaptive methods, single-source and multiple-source steered responses are quite different; the multiple-source resolution criterion has become the standard.

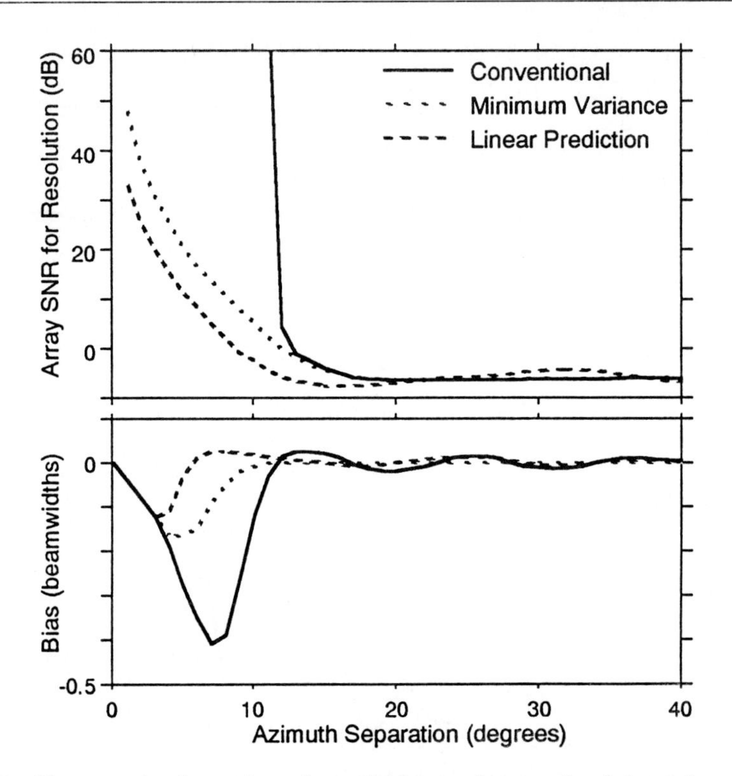

Figure 7.6 The array signal-to-noise ratio required to resolve two signals located symmetrically about a 10-element array's broadside is shown in the top panel as a function of the signals' bearing separation. These so-called *resolution curves* are shown for the minimum variance, linear predictive, and the conventional delay-and-sum beamformer. The bottom panel depicts the bias of the direction of propagation estimate for one signal. This bias is expressed in beamwidths, which equals width of the conventional beamformer's mainlobe (about 11° in this case). A negative-valued bias is indicative of a peak location being closer to broadside than the actual direction of propagation. The intersignal coherence equals 0 in this example.

superior resolution than conventional delay-and-sum methods. Direct analytic derivation of an adaptive technique's resolution, however, has defied all attempts to date. The extant results concern *asymptotic* resolution: that attainable if we had an infinite amount of data [37, 44]. These asymptotic results are heavily dependent on array geometry. The output signal-to-noise ratios required to resolve two equi-amplitude signals with a linear array for conventional and minimum variance beamformers are shown in Fig. 7.6. Coupling these results with rules of thumb concerning how much data are sufficient, an accurate indication of the algorithm's resolution emerges [44]. For these asymptotic results to hold for the minimum variance algorithm, the time-bandwidth product of the spatial correlation matrix estimate must be much greater than the number of sensors in the array ($L \gg M$). When insufficient data do not permit applying these results, the resolution characteristics are not known other than by simulation.

Once we resolve two signals, the next concern is the accuracy of the direction of propagation estimates: Do the estimates contain any bias? As with resolution, analytic predictions of the bias are rare. Asymptotic bias has been calculated, and the results are shown in Fig. 7.6 for a regular linear array [44]. The results indicate the superior accuracy of estimates provided by the minimum variance technique when compared with delay-and-sum techniques. For the minimum variance beamformer, this accuracy increases (the bias is reduced) as signal-to-noise ratio increases; for delay-and-sum beamforming, it does not change. This gratifying characteristic results from the adaptive nature of the minimum variance technique.

7.2.2 Linear Predictive Method

Linear prediction, a parametric, model-based, signal modeling, and spectrum estimation approach, can easily be extended to the array processing arena. We express the output of a selected sensor (the m_0^{th}, say) as a weighted linear combination of the outputs of other sensors. Because signal time delay becomes a complex-valued amplitude scaling in the frequency domain, linear prediction finds more direct application there than in the time-domain. Defining $Y_m^{(N)}(\omega)$ to be the Fourier Transform of an arbitrarily selected N-point section of the output of the mth sensor, the linear predictive framework is to express $Y_{m_0}^{(N)}(\omega)$ as

$$Y_{m_0}^{(N)}(\omega) = -\sum_{m \neq m_0} w_m^* Y_m^{(N)}(\omega)$$

where $\{w_m\}$ is a set of complex-valued weights. These weights become the parameters sought by the linear prediction method. If each of the sensor outputs contained a single propagating signal so that $Y_m^{(N)}(\omega) = \exp\{-j\omega\Delta_m\}S(\omega)$, weights equal to $w_m = -\exp\{+j\omega(\Delta_{m_0} - \Delta_m)\}/(M-1)$ would satisfy the linear predictive relationship.

Letting $\mathbf{y}$ denote the vector $\text{col}[Y_0^{(N)}(\omega), \ldots, Y_{M-1}^{(N)}(\omega)]$, we can easily determine the optimal linear predictive weight vector using the constrained optimization framework: Minimize $\mathcal{E}[|\mathbf{w}'\mathbf{y}|^2]$ (the mean-squared prediction error) subject to the constraint that the weight vector's m_0^{th} component equals unity. This constraint is expressed as $\boldsymbol{\delta}_{m_0}'\mathbf{w} = 1$, where $\boldsymbol{\delta}_{m_0}$ is a column vector whose sole nonzero entry equals 1 in the m_0^{th} position. The optimal weight vector for the linear predictive beamformer is [80]

$$\boxed{\mathbf{w}_\diamond = \frac{\mathbf{R}^{-1}\boldsymbol{\delta}_{m_0}}{\boldsymbol{\delta}_{m_0}'\mathbf{R}^{-1}\boldsymbol{\delta}_{m_0}}}$$

These weights have a much different interpretation than in minimum variance beamforming. Rather than expressing a set of amplitudes and phases that coalesces the sensor outputs into a unified expression for a signal propagating from a specific direction, the linear predictive weights express a *model* for the measured field. The weights do not depend on a chosen direction of propagation; they express simultaneously *all* of the propagating signals. *This inability to focus on a particular direction of propagation means that linear prediction cannot be used as a beamformer*: The algorithm cannot provide an output waveform and can only be used to indicate directions of propagation. To extract directions of propagation from the weights, we rely on the spatial spectrum interpretation of narrowband array processing. Applying the concepts of *AR* spectral analysis (see Eq. 6.20 {335}), we find the linear predictive steered response to be

$$\boxed{\mathcal{P}^{\mathrm{LP}}(\vec{\zeta}\,) = \frac{\boldsymbol{\delta}'_{m_0}\mathbf{R}^{-1}\boldsymbol{\delta}_{m_0}}{\left|\boldsymbol{\delta}'_{m_0}\mathbf{R}^{-1}\mathbf{e}(\vec{\zeta}\,)\right|^2}} \tag{7.5}$$

"Significantly large" peaks of this quantity correspond to the signals' directions of propagation.

Fig. 7.3 {359} depicts an example linear predictive steered response using an exact spatiospectral correlation matrix. Although the linear predictive method provides the narrowest signal-related peak among the algorithms used there, its "sidelobes" contain significant ripple, a phenomenon we noted examining the linear predictive spectral estimator (Fig. 6.11 {333}). Using the expression for the exact correlation matrix and making the simplifying assumption $MA^2 \gg \sigma_n^2$, the peak's amplitude is found to be $M(M-1)A^4/\sigma_n^2$ and the nominal background level σ_n^2. In comparison with both conventional and minimum variance beamforming, the linear predictive technique provides a peak-to-background ratio (roughly) proportional to M^2A^4/σ_n^4, whereas the others provide a ratio equal to the square-root of this value. Thus, the way we computed the linear predictive spectrum has artificially introduced greater disparity between the peak and background values; we could square other algorithms' responses and enhance their ratios too! To level the playing field, this book's convention is to force responses to produce peak values proportional to the signal's power. We must therefore apply a square-root to the linear predictive steered response to reflect its signal processing gains more accurately.

One difficulty with the linear predictive technique is that it only applies approximately to interesting array processing problems. Two commonly occurring situations cause this modeling inaccuracy: additive noise in the observations and multiple propagating signals. Although each signal may be *AR*, the presence of additive noise leads to the introduction of zeros that forces an *ARMA* model on the observations (see Prob. 6.30). Further, the superposition of two or more *AR* signals is not *AR*; it's also *ARMA* [71] (see Prob. 7.11). As we have seen {338}, evaluation of the *ARMA* model's coefficients is complicated. For computational simplicity, only the linear predictive (all-pole) model is used with the understanding that modeling errors become apparent once the signal-to-noise ratio becomes too small.

Another issue is the selection of m_0: the reference sensor whose output is predicted by merging all other sensors' outputs. The choice of m_0 can have dramatic effects on the spatial spectrum. No one choice yields uniformly "best" results (see Prob. 7.12). For linear arrays, two choices are evident: a sensor located at either end or one (if present) located at the array's phase center. In this case, an end reference is usually chosen. For other array geometries, an "end" cannot be defined; thus, a sensor near the phase center is usually selected as the reference.

The resolution afforded by the linear predictive method exceeds that of minimum variance beamforming. Fig. 7.6 {364} contains a plot of the resolution curves for the

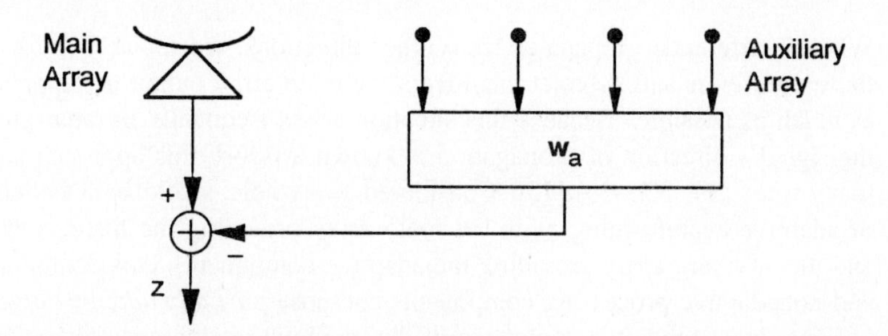

Figure 7.7 In the sidelobe canceller, we focus the main array's sensors in a particular direction. The auxiliary elements then search for signals propagating from other directions that lie in the main array's sidelobes. We use these auxiliary elements to estimate these interfering signals and subtract them from the main array's output.

linear predictive and minimum variance algorithms. The linear predictive algorithm requires an approximately 5–10 dB smaller signal-to-noise ratio than minimum variance beamforming to resolve two equiamplitude signals. Peak locations are unbiased when only one signal is present; in the case of two signals, the bias can be larger than that of minimum variance beamforming (again see Fig. 7.6).

This increased resolution performance is mitigated by the inability to define precisely a detection statistic. When we use the estimated spatial correlation matrix (estimated by Bartlett's procedure as in the minimum variance technique), the distribution of the linear predictive steered response is related to an F distribution [26]. This distribution does *not* have an exponentially decreasing tail; instead, it decays in an algebraic fashion (see §A.1.7 {479}). This dependence distinguishes linear predictive responses from minimum variance or conventional ones: The more slowly decreasing tail means that linear predictive responses have much larger variability and much more data are required for comparable variability. This large variability does not prevent the computation of a detection statistic; however, a signal's presence not only results in a large peak near the proper direction of propagation but also produces ripples in the steered response at other directions similar to sidelobes produced by conventional methods. Consequently, even far from those directions corresponding to propagating signals the null distribution does not adequately characterize the observed steered response; computations of detection thresholds based on the noise-only assumptions yield false-alarm probabilities larger than intended. Fig. 7.5 {363} illustrates a simple linear predictive steered response that demonstrates this problem.

7.2.3 Sidelobe Cancellers

A variation of the linear predictive approach to array signal processing is the so-called *sidelobe canceller*. As depicted in Fig. 7.7, we augment a main array with *auxiliary* sensors to cancel directional interference (e.g., jammers) that are located in the sidelobes of the main array's steered response. We use some sort of processing strategy, usually conventional, to focus the main array on a source having a particular direction of propagation.

When interference propagates from other directions, we focus the auxiliary sensors on these interferers and subtract them from the main array output to suppress the interferers as much as possible. Because this situation arises frequently in radar problems in which the signal's direction of propagation is known a priori, this approach has been used for many years [35: 336–39];[75]. Considered as a whole, sidelobe cancellers provide a type of adaptive beamforming, with the main array providing the fixed, nonadaptive portion and the auxiliary array providing the adaptive component. This combination of adaptive and nonadaptive processing components compose *partially adaptive algorithms.*

The linear predictive technique of the previous section exemplifies one kind of sidelobe canceller; identical main array and auxiliary array elements are used, and the signal as well as interferers are present in all sensors. In this case, careful distinction is made between the two types of elements. Denoting the output of the main array by z_{m_0}, we express the output of the entire array by

$$z = z_{m_0} + \sum_m w_m^* y_m$$

Letting the observation vector $\mathbf{y}$ be $\text{col}[z_{m_0}, y_1, y_2, \ldots]$, $\mathbf{w}'\mathbf{y}$ succinctly expresses the array output, where the weight vector is $\text{col}[1, w_1, w_2, \ldots]$. Because the auxiliary array contains no signal, we should adjust the weights to minimize array output power. The familiar optimization problem emerges: Minimize $\mathcal{E}[|\mathbf{w}'\mathbf{y}|^2]$ subject to the constraint that $\boldsymbol{\delta}'\mathbf{w} = 1$ (here $\boldsymbol{\delta} = \text{col}[1, 0, 0, \ldots]$). The solution to this problem is also familiar.

$$\mathbf{w}_\diamond \propto \mathbf{R}^{-1}\boldsymbol{\delta}$$

The proportionality constant ensures that the leading term in the weight vector is unity.

We obtain deeper insight into this result by examining the structure of the spatial correlation matrix when one "sensor"—the main array—has rather different characteristics from the others. Because the optimum weight vector is the solution of the matrix equation $\mathbf{R}\mathbf{w}_\diamond \propto \boldsymbol{\delta}$, expand this equation into main array and auxiliary components.

$$\begin{bmatrix} R_{m_0,m_0} & \mathbf{r}'_{m_0,a} \\ \mathbf{r}_{m_0,a} & \mathbf{R}_a \end{bmatrix} \begin{bmatrix} 1 \\ \mathbf{w}_a \end{bmatrix} \propto \begin{bmatrix} 1 \\ \mathbf{0} \end{bmatrix}$$

Here, R_{m_0,m_0} represents the mean-squared value of the main array output and $\mathbf{r}_{m_0,a}$ the vector of cross-correlations between the output of each auxiliary array element and the main array output: $\mathbf{r}_{m_0,a} = \text{col}\big[\mathcal{E}[z_{m_0}^* y_1], \mathcal{E}[z_{m_0}^* y_2], \ldots\big]$. The weights applied to the outputs of the auxiliary array's elements compose the vector $\mathbf{w}_a$. From the homogeneous portion $\mathbf{r}_{m_0,a} + \mathbf{R}_a\mathbf{w}_a = \mathbf{0}$ of our matrix equation, we find the auxiliary array's weight vector, which corresponds to the sidelobe canceller's adaptive component, to be

$$\boxed{\mathbf{w}_a = -\mathbf{R}_a^{-1}\mathbf{r}_{m_0,a}}$$

The correlation matrix $\mathbf{R}_a$, as well as the vector $\mathbf{r}_{m_0,a}$ of cross correlations between auxiliary array elements and the main array, can be estimated with any of several techniques; among them are the exponential weighting and moving average methods {316}.

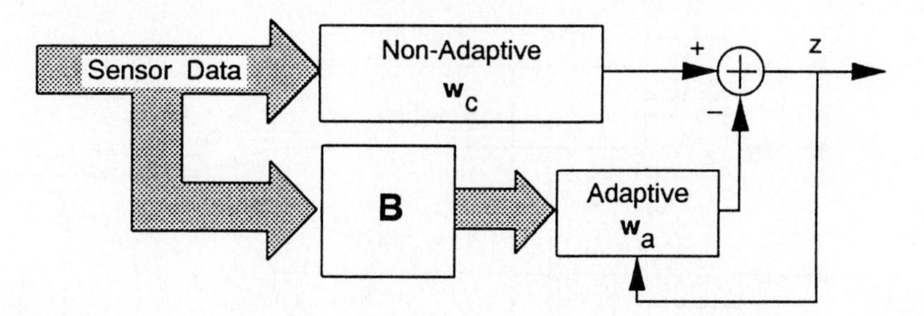

Figure 7.8 A nonadaptive portion having weight vector $\mathbf{w}_c$ and an adaptive portion having two parts compose the generalized sidelobe canceller. The blocking matrix **B** prevents the adaptive portion from cancelling any signal propagating in the assumed propagation direction. The weight vector $\mathbf{w}_a$ adapts to other signals with the intent of deleting them from the nonadaptive portion should they "leak" through the sidelobes.

7.2.4 Generalized Sidelobe Cancellers

A generalization of the sidelobe canceller structure shown in Fig. 7.8 has no main array, but has a main beam steered in some assumed propagation direction by a conventional beamformer according to the weight vector $\mathbf{w}_c$ [22, 64]. Design of this weight vector is based on the desired mainlobe width and sidelobe height. The adaptive portion wants to remove any other signals from appearing in the output. Because this portion has the same input as the conventional beamformer, care must be taken not to remove the desired signal. Preventing this unwanted signal suppression underlies the insertion of the matrix **B**, the *blocking matrix*. This matrix takes into account the assumed propagation direction and prevents any signal propagating from that direction from passing through. The succeeding adaptive section, which estimates signals that do pass through by considering the total output, can then freely adjust the weight vector $\mathbf{w}_a$ to emphasize these signals so that they can be subtracted from the main beam's output. Fig. 7.9 depicts one simple realization of the generalized sidelobe canceller. Explicit separation of the beamsteering delays from the amplitude shading demonstrates that the blocking matrix must cancel any input component having a constant wavefront. Thus, for unit-amplitude shading, the matrix **B** must have rows whose elements sum to 0. After removing the constant-wavefront signals, the adaptive section emphasizes the remainder based on what's in the output and how the conventional beamformer shades the desired signal.

The general mathematical framework for the generalized sidelobe canceller relies on *unconstrained* optimization. The canceller's output is given by $\mathbf{w}_c'\mathbf{y} - \mathbf{w}_a'\mathbf{B}\mathbf{y}$, which reveals that the overall weight vector $\mathbf{w}$ is $\mathbf{w}_c - \mathbf{B}'\mathbf{w}_a$. We choose the conventional beamformer's weight vector a priori and design the blocking matrix and the adaptive weight vector to minimize output power. Solving the unconstrained optimization problem for the generalized sidelobe canceller yields $\mathbf{w}_a$.

$$\boxed{\min_{\mathbf{w}_a} \left(\mathbf{w}_c - \mathbf{B}'\mathbf{w}_a\right)' \mathbf{R} \left(\mathbf{w}_c - \mathbf{B}'\mathbf{w}_a\right) \Longrightarrow \mathbf{w}_a = (\mathbf{B}\mathbf{R}\mathbf{B}')^{-1}\mathbf{B}\mathbf{R}\mathbf{w}_c} \tag{7.6}$$

To find **B**, demand that this beamformer have the same solution as that of adaptive

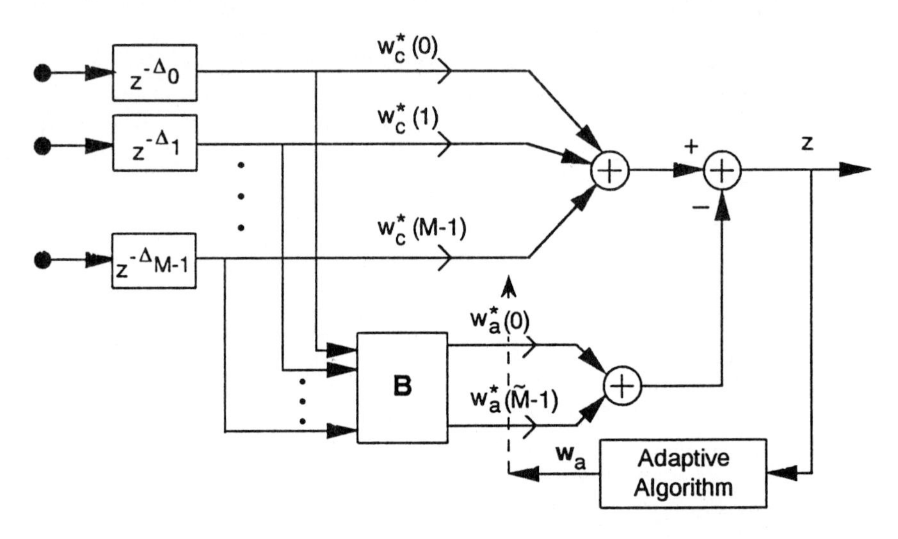

Figure 7.9 A particularly simple structure for the generalized sidelobe canceller has the delays imposed at each sensor separated from the main beam's amplitude shading. Signals in the adaptive section thus appear to have constant wavefronts. Simple blocking matrices can then be derived and the adaptive weights chosen to remove all other propagating signals.

algorithms derived by constrained optimization problem in §7.2 {355}. There, the problem is to minimize $\mathbf{w}'\mathbf{R}\mathbf{w}$ subject to the linear constraint $\mathbf{C}\mathbf{w} = \mathbf{c}$ on the weight vector $\mathbf{w}$. Equating its solution $\mathbf{R}^{-1}\mathbf{C}'\left(\mathbf{C}\mathbf{R}^{-1}\mathbf{C}'\right)^{-1}\mathbf{c}$ to ours for $\mathbf{w}_c - \mathbf{B}'\mathbf{w}_a$ yields

$$\left[\mathbf{I} - \mathbf{B}'(\mathbf{B}\mathbf{R}\mathbf{B}')^{-1}\mathbf{B}\mathbf{R}\right]\mathbf{w}_c = \mathbf{R}^{-1}\mathbf{C}'(\mathbf{C}\mathbf{R}^{-1}\mathbf{C}')^{-1}\mathbf{c}$$

Note that we have not presumed the blocking matrix to be invertible. To derive this matrix's properties, this equation's structure must be carefully examined. Multiplying each side by $\mathbf{B}\mathbf{R}$ yields

$$\mathbf{B}\mathbf{C}'(\mathbf{C}\mathbf{R}^{-1}\mathbf{C}')^{-1}\mathbf{c} = \mathbf{0}$$

If $\mathbf{B}$ were invertible, this equation would not make sense, implying that the generalized sidelobe canceller could not mimic *any* adaptive beamforming algorithm found as the solution of a constrained optimization problem. Wisely pursuing noninvertible solutions, we demand that $\mathbf{B}$'s null space must contain the vector $\mathbf{C}'(\mathbf{C}\mathbf{R}^{-1}\mathbf{C}')^{-1}\mathbf{c}$ but not depend on the characteristics of the data; the weight vector $\mathbf{w}_a$ should account for signal and noise characteristics. To satisfy these properties, the blocking matrix in the generalized sidelobe canceller must be such that

$$\boxed{\mathbf{B}\mathbf{C}' = \mathbf{0}}$$

As the unadapted weight vector $\mathbf{w}_c$ equals $\mathbf{C}'(\mathbf{C}\mathbf{C}')^{-1}\mathbf{c}$, this requirement means that each row of the blocking matrix must be orthogonal to the unadapted weight vector: $\mathbf{B}\mathbf{w}_c = \mathbf{0}$. Because the constraint matrix $\mathbf{C}$ is rectangular, wherein the number of rows equals the number of constraints $\dim(\mathbf{c})$, *the dimension of* $\mathbf{B}$*'s null space must equal the number of constraints.* This null space provides the blocking property the matrix must possess. The blocking matrix is most conveniently taken to be a rectangular matrix having size $[\dim(\mathbf{y}) - \dim(\mathbf{c})] \times \dim(\mathbf{y})$ where the rows are linearly independent while satisfying $\mathbf{BC}' = \mathbf{0}$.

Example [64]

As an example application of the generalized sidelobe canceller's structure, let's return to the minimum variance narrowband beamformer described in §7.2.1 {357}. This frequency-domain adaptive beamformer has only one constraint with the constraint matrix equaling $\mathbf{e}'$, the conjugate transpose of an ideal, unit-amplitude sinusoid propagating in the assumed propagation direction. As shown previously {358}, the nonadaptive weight vector equals $\mathbf{e}/M$. The blocking matrix has dimensions $(M-1) \times M$ where each row must be linearly independent and orthogonal to $\mathbf{e} = \text{col}[e_0, \ldots, e_{M-1}]$. By explicitly incorporating the phase shifts corresponding to the assumed propagation direction as shown in Fig. 7.9, simple structures for the blocking matrix result that do *not* depend on the assumed propagation direction. Many choices for the blocking matrix are possible; one of those having linearly independent rows is the $(M-1) \times M$ matrix

$$\mathbf{B} = \begin{bmatrix} e_1 & -e_0 & 0 & \cdots & 0 \\ 0 & e_2 & -e_1 & 0 & \cdots \\ \vdots & \ddots & \ddots & \ddots & \\ 0 & \cdots & 0 & e_{M-1} & -e_{M-2} \end{bmatrix}$$

This blocking matrix explicitly contains the assumed propagation direction and fits into the structure shown in Fig. 7.8. To achieve the simpler structure shown in Fig. 7.9, where the blocking matrix does not depend on the assumed propagation direction, the rows of $\mathbf{B}$ must be orthogonal to the vector equaling the magnitude of the nonadaptive steering vector: $\mathbf{B}\,\text{col}\big[|e_0|, \ldots, |e_{M-1}|\big] = \mathbf{0}$. Thus, for the minimum variance beamformer, $|e_m| = 1$, and the blocking matrix has the form

$$\mathbf{B} = \begin{bmatrix} 1 & -1 & 0 & \cdots & 0 \\ 0 & 1 & -1 & 0 & \cdots \\ \vdots & \ddots & \ddots & \ddots & \\ 0 & \cdots & 0 & 1 & -1 \end{bmatrix}$$

7.2.5 Pisarenko Harmonic Decomposition

Many direction of propagation estimation algorithms are based on the idea of *null steering*: Find the set of weights to apply to the sensor outputs that minimizes output power without placing signal-dependent constraints. These algorithms operate by optimally reducing the impact of signal and noise on the array's output. Null-steering algorithms are based on pessimism: By examining where signals aren't located, we know by an elimination process where they are. This philosophy contrasts with the optimism of the minimum variance algorithm, which tries to enhance a signal's presence. Null-steering algorithms find a set of weights that leaves the weighted sum of sensor outputs with as little signal and noise power as possible. One null-steering solution preferred by extreme pessimists would be to apply zero-valued weights; to avoid this trivial and useless solution, we impose a constraint on the weights. One technique based on this approach is the *Pisarenko harmonic decomposition algorithm* [121]. Named for a Russian mathematician, the optimization problem is to minimize the mean-squared value of the array output while constraining the norm of the weight vector to be unity.

$$\boxed{\min_{\mathbf{w}} \mathbf{w}'\mathbf{R}\mathbf{w} \quad \text{subject to } \mathbf{w}'\mathbf{A}\mathbf{w} = 1}$$

The matrix $\mathbf{A}$, a positive-definite symmetric matrix, serves to balance the relative importance of portions of the weight vector over others. Evaluating the gradient of the objective function provided by the Lagrange multiplier technique (described in §C.2 {502}), the weight vector must satisfy

$$\mathbf{R}\mathbf{w} = -\lambda \mathbf{A}\mathbf{w}$$

Thus, $\mathbf{w}$ must be a generalized eigenvector of the spatial correlation matrix (see §B.5 {493}). Of the M possible eigenvectors, the one that minimizes the mean-squared array output (the quadratic form of the optimization problem) is the one corresponding to the smallest eigenvalue $\min\{\lambda_i, \mathbf{v}\}_{(\mathbf{R},\mathbf{A})}$, if the smallest is unique. If the smallest eigenvector has multiplicity, *any* unit-norm vector expressible as a linear combination of the eigenvectors associated with the smallest eigenvalue forms a possible weighting vector (the solution is not unique).†

By selecting the eigenvector corresponding to the smallest eigenvalue, we obtain a weight vector having the least in common with the signal propagation vectors: This eigenvector is approximately orthogonal to all signal vectors. Taking $\mathbf{A}$ to equal the identity matrix, which places equal importance on the weight vector's components, we find that the product $|\left(\mathbf{v}_R^{\min}\right)' \mathbf{e}(\vec{\zeta}\,)|$ is small whenever the assumed propagation direction corresponds to a signal's direction of propagation [27, 102]. A "peak-picking" approach to determining signal directions, as used in other algorithms, results by considering the reciprocal of this quantity. Although signal directions are associated with peaks in Pisarenko's method, extraneous peaks are also present, making the selection of detection thresholds difficult (see Fig. 7.10). A peak's amplitude has little to do with a signal's amplitude; peaks are merely indicators of possible signal propagation directions and do not convey amplitude information.

We can estimate the signal amplitudes if the number of signals present in the array's field are known‡ and the signals are not coherent. Letting $\mathbf{e}_i$ denote the ideal (unit-amplitude) signal vector propagating from one of the directions just found, the quantity

†The eigenvectors corresponding to eigenvalues having multiplicity are also not unique. Be that as it may, a vector space having dimension equal to the eigenvalue's multiplicity can be defined. The eigenvectors as well as the weight vector are contained in this space. The upshot of this detail is that the lack of uniqueness does not matter. Most numerical routines produce eigenvectors for these eigenvalues; any of them can be used as a weight vector.

‡See §7.3.5 {390} for a discussion of techniques to determine the number of sources.

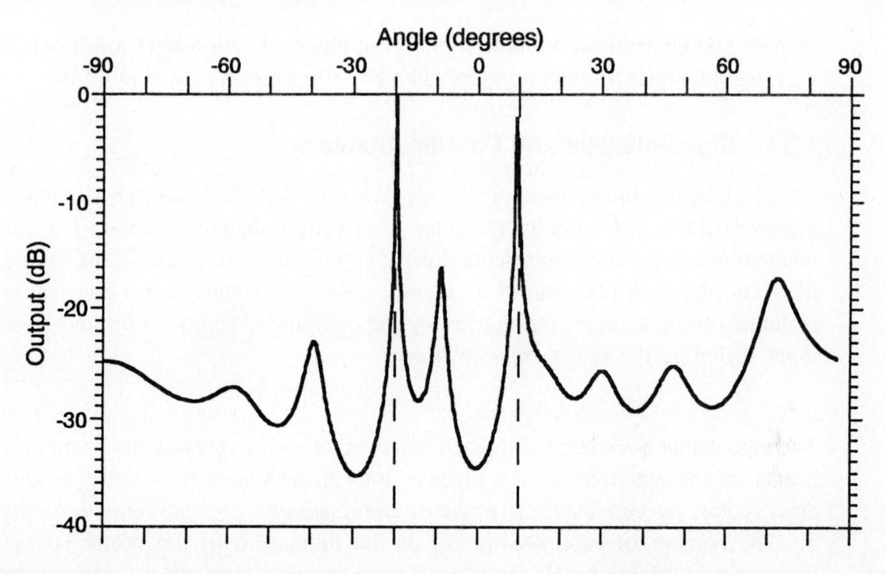

Figure 7.10 The quantity $|\left(\mathbf{v}_R^{\min}\right)' \mathbf{e}(\vec{\zeta}\,)|^{-1}$ is shown for the same data used in Fig. 7.5 {363}. Although peaks corresponding to signal directions are evident, other peaks are equally prominent. The vertical scale is expressed in decibels.

$\mathbf{e}_i'\mathbf{R}\mathbf{e}_i$ is

$$\mathbf{e}_i'\mathbf{R}\mathbf{e}_i = \mathbf{e}_i'\mathbf{K}_n\mathbf{e}_i + A_i^2M^2 + \sum_{j\neq i} A_j^2 \left|\mathbf{e}_i'\mathbf{e}_j\right|^2$$

Assuming the noise covariance matrix is known, we have an easily solved set of linear equations for the signals' squared amplitudes. Because the peak-picking technique does not always yield dramatic indications of the signals' directions of propagation, more peaks than the estimated number of signals are usually selected. With this larger than necessary set of equations to solve for the signals' squared amplitudes, no signal is ignored while the erroneous directions yield small amplitude values. The idea is to use the estimated number of signals *after* candidate signals have been selected, not before.

7.3 Eigenanalysis Methods

From rather specialized applications like the development of the matched filter structure for optimum detectors to far-reaching concepts like Fourier analysis in linear system theory, *eigenanalysis*—the determination of the natural coordinate system for a given quantity—occurs frequently in the development of signal processing algorithms. In abstract mathematics, eigenanalysis is equated with spectral analysis: A linear operator's spectrum, for example, is defined to be its set of eigenvalues and its "natural" basis the set of normalized eigenvectors. In engineering, spectral analysis connotes Fourier analysis; any connection between Fourier analysis and eigenanalysis is downplayed. In the authors' humble and contrasting opinion, reinforcing this connection keys a full understanding of why eigenanalysis yields important adaptive array processing algorithms.

7.3.1 Eigenanalysis and Fourier Analysis

Let $\mathcal{L}[\cdot]$ denote a linear operator: $\mathcal{L}[a_1x_1+a_2x_2] = a_1\,\mathcal{L}[x_1]+a_2\,\mathcal{L}[x_2]$, where x represents a vector (in the sense of a linear vector space) and a denotes a scalar. This rather general mathematical operation represents diverse signal processing operations such as the linear filter and the multiplication of a column matrix by a matrix. An *eigenvector* v of $\mathcal{L}[\cdot]$ is defined to be a vector reproduced by the operator: The operator thus "gives back" the input scaled by the scalar *eigenvalue* λ.

$$\mathcal{L}[v] = \lambda v, \quad v \neq 0$$

The eigenvalue associated with each eigenvector thus expresses the operator's gain when it acts on the eigenvector. An eigenvector can be scaled by a constant and still be an eigenvector; by convention, *we choose the norm of the eigenvector to be unity*: $\|v\| = 1$. The number of eigenvectors equals the dimension of the vector space defined by the operator's domain. We equate a linear operator's *spectrum* to the collection of its eigenvalues. In those cases in which the spectrum contains a finite number of values, eigenvalue indexing denotes an ordering: The largest eigenvalue is denoted by λ_1 and the smallest by λ_M, M being the number of eigenvectors. The *eigensystem* $(\lambda_i, v_i)_{\mathcal{L}}$ tersely refers to the collection of eigenvectors and associated eigenvalues.

We can generalize the concept of an eigensystem by considering those vectors that $\mathcal{L}[\cdot]$ gives back not just proportionally, but as another linear operator $\mathcal{L}_o[\cdot]$ acting on the vector.

$$\mathcal{L}[v] = \lambda\,\mathcal{L}_o[v], \quad v \neq 0$$

In this situation, the generalized eigenvectors of $\mathcal{L}[\cdot]$ are those that appear to be outputs of a second linear system. The second linear system $\mathcal{L}_o[\cdot]$ represents a constraining system, forcing us to view eigenvectors as being proportional outputs of *both* operators. Thus, we denote the linear operator pair defining the generalized eigensystem as $(\mathcal{L}, \mathcal{L}_o)$. When the usual eigensystem of $\mathcal{L}[\cdot]$ contains M elements, the number in $(\mathcal{L}, \mathcal{L}_o)$'s eigensystem equals M if and only if the constraining system has no zero eigenvalues.[†]

The usual engineering example of a linear operator is the linear, time-invariant system. Here, signals are vectors and the convolution sum defines the linear transformation. The eigenvectors of the convolution sum are the complex exponential signals $\exp\{j\omega n\}$ and we find the eigenvalue associated with each frequency to be the system's transfer function evaluated at that frequency.

$$\mathcal{L}\left[e^{j\omega n}\right] = \sum_{k=-\infty}^{\infty} h(n-k)e^{j\omega k} = H(e^{j\omega})e^{j\omega n}$$

[†]To demonstrate the plausibility of this nonzero-eigenvalue property, consider the case in which the null spaces of $\mathcal{L}[\cdot]$ and $\mathcal{L}_o[\cdot]$ are disjoint: $\mathcal{L}_o[v] = 0 \Longrightarrow \mathcal{L}[v] \neq 0$ and vice versa. Clearly, no nonzero vector satisfies $\mathcal{L}[v] = \lambda\,\mathcal{L}_o[v]$ and the eigensystem is empty. Another extreme occurs when they are projection operators (see §B.2 {488}) having the same null space; in this case, an infinite number of generalized eigenvectors exist. We concentrate in sequel on the "nice" case: $\mathcal{L}_o[\cdot]$ has no null space.

Note that the number of eigenvectors is uncountably infinite: Any value of ω serves. Thus, the linear filter's spectrum is equivalent to the Fourier spectrum.

The utility of eigenanalysis rests not only on the defining input-reproducing property, but also on the orthogonality of the eigenvectors and their completeness. *When* these properties hold, they permit any vector x to be represented by a linear eigenvector expansion, thereby allowing eigenvectors to define a coordinate system.

$$x = \sum_i a_i v_i \quad \text{with} \quad a_i = \langle x, v_i \rangle$$

This coordinate system is tailored to the linear operator defining the eigensystem; it allows the associated linear operator's output to have an easily calculated representation.

$$\mathcal{L}[x] = \sum_i \lambda_i a_i v_i$$

A similar expression underlies the widespread use of the Fourier Transform in linear system theory. Thus, in the context of a given linear operator, the set of orthonormal eigenvectors constitutes the "natural" coordinate system *when it exists.* The eigenvectors of a linear operator do not necessarily form an orthogonal set. Fortunately, for most interesting linear operators and for all of those discussed here, they do.

In many adaptive array processing algorithms, the key quantity is the spatial correlation matrix $\mathbf{R}$. Its ubiquity in beamforming algorithms forces consideration of its natural coordinate system and what can be learned from the coordinate system's structure. Correlation matrices all have Hermitian symmetry ($\mathbf{R}' = \mathbf{R}$) and are (usually)[†] positive definite ($\mathbf{x}'\mathbf{R}\mathbf{x} > 0, \forall\, \mathbf{x} \neq \mathbf{0}$). As shown in §B.5 {493}, the eigenvalues of such matrices are always positive numbers, and the eigenvectors are orthogonal with respect to the matrix *and* the identity matrix (see Prob. 7.15).

$$\mathbf{v}_i'\mathbf{R}\mathbf{v}_j = \lambda_i \delta_{ij} \quad \text{and} \quad \mathbf{v}_i'\mathbf{v}_j = \delta_{ij}$$

Consequently, the eigenvectors form an orthonormal set, and we can freely use the eigenvectors as our coordinates. For example, positive-definite Hermitian matrices as well as their inverses can be expressed in terms of their eigensystems—their spectra—by

$$\mathbf{R} = \sum_{i=1}^{M} \lambda_i \mathbf{v}_i \mathbf{v}_i' \quad \text{and} \quad \mathbf{R}^{-1} = \sum_{i=1}^{M} \lambda_i^{-1} \mathbf{v}_i \mathbf{v}_i' \tag{7.7}$$

Eigenanalysis allows us to interpret the quadratic form, clearly one of the important quantities arising in array processing algorithms. The ratio of a quadratic form and the norm of the vector used in the quadratic form is called a *Rayleigh quotient* (see §B.5 {495}).

$$q(\mathbf{x}) \equiv \frac{\mathbf{x}'\mathbf{R}\mathbf{x}}{\mathbf{x}'\mathbf{x}}$$

[†]In degenerate cases, correlation matrices are only nonnegative definite: There exists some $\mathbf{x}$ such that $\mathbf{x}'\mathbf{R}\mathbf{x} = 0$. This situation can only occur when some element on the matrix's diagonal equals zero, implying that component of the random vector defining $\mathbf{R}$ has zero mean-squared value. This situation can occur when the Fourier spectrum equals zero for either distinct frequencies or a frequency range.

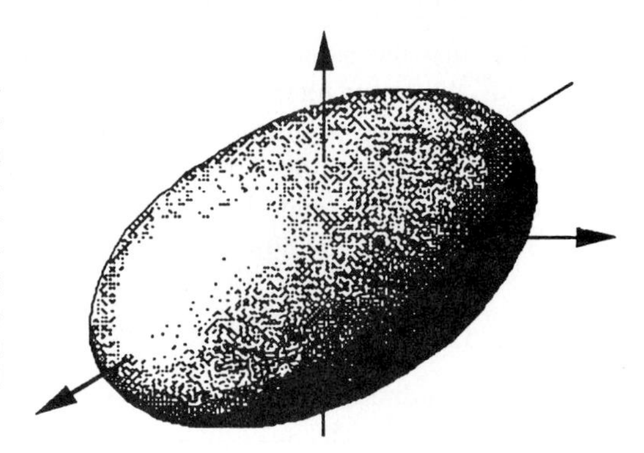

Figure 7.11 Depicted is the constant-value surface of the Rayleigh quotient in three-dimensional space. All such surfaces resemble rugby balls, more accurately termed multidimensional ellipsoids. The larger values correspond to directions aligned with eigenvectors having the larger eigenvalues, the smaller values with those having smaller eigenvalues. In this example, the eigenvalues are 1,1, and 2.

Because this quotient is invariant with respect to scaling the vector $\mathbf{x}$, we can take the norm of $\mathbf{x}$ to be unity and consider the thus constrained properties of the quadratic form alone. Fig. 7.11 shows the surface generated by the Rayleigh quotient as the direction of the vector $\mathbf{x}$ varies. By representing $\mathbf{x}$ in terms of the eigenvectors of $\mathbf{R}$ ($\mathbf{x} = \sum a_i \mathbf{v}_i$), the quotient can be expressed as $\mathbf{x}'\mathbf{R}\mathbf{x} = \sum \lambda_i a_i^2$. As $\mathbf{R}$ is a positive-definite matrix, all of its eigenvalues are positive. As shown in §B.5, the vector minimizing (maximizing) the Rayleigh quotient corresponds to the eigenvector having the smallest (largest) eigenvalue.

$$\min_{\mathbf{x}} \frac{\mathbf{x}'\mathbf{R}\mathbf{x}}{\mathbf{x}'\mathbf{x}} = \lambda_M \quad \text{and} \quad \max_{\mathbf{x}} \frac{\mathbf{x}'\mathbf{R}\mathbf{x}}{\mathbf{x}'\mathbf{x}} = \lambda_1$$

These extrema occur when $\mathbf{x} = \mathbf{v}_M$ and $\mathbf{x} = \mathbf{v}_1$, respectively.

We take a correlation matrix's Fourier spectrum to be the value of the quadratic form $\mathbf{e}'\mathbf{R}\mathbf{e}$, where, for our purposes, the Fourier vector $\mathbf{e}$ is taken to be an ideal, unit-amplitude signal vector {54}.† Note that $\|\mathbf{e}\|^2 = M$. In terms of the matrix's eigenspectrum, the Fourier spectrum is expressed by

$$\mathbf{e}'\mathbf{R}\mathbf{e} = \sum_{i=1}^{M} \lambda_i |\mathbf{e}'\mathbf{v}_i|^2$$

Because Fourier vectors $\mathbf{e}$ are parameterized by a frequency variable, the Fourier spectrum assumes a continuum of values. Because of the properties just derived for the Rayleigh quotient, *the Fourier spectral values must lie within the interval* $[M\lambda_M, M\lambda_1]$; see Fig. 7.12 for an example. As frequency is varied, if a Fourier vector happens to correspond to one of $\mathbf{R}$'s eigenvectors (say the ith), the Fourier spectrum equals $M\lambda_i$. A correlation matrix's eigenvectors usually do *not* correspond to Fourier vectors, however. For example, with a regular linear array, the spatiospectral correlation matrix is Toeplitz as well as Hermitian and ideal signal vectors equal the usual *DFT* vectors. As the size of a Toeplitz-Hermitian matrix grows, its eigenvectors *approach* the *DFT* vectors [62]. Circulant matrices are the only finite-dimensional Toeplitz-Hermitian matrices having the

†This seemingly ad hoc definition is based on the reasoning $\mathbf{e}'\mathbf{R}\mathbf{e} = \mathbf{e}'\,\mathcal{E}[\mathbf{x}\mathbf{x}']\mathbf{e} = \mathcal{E}\big[|\mathbf{e}'\mathbf{x}|^2\big]$ for some random vector $\mathbf{x}$. The inner product of $\mathbf{e}$ with any vector amounts to computing the vector's Fourier Transform. Thus, the quadratic form $\mathbf{e}'\mathbf{R}\mathbf{e}$ amounts to the power spectrum for the correlation matrix $\mathbf{R}$.

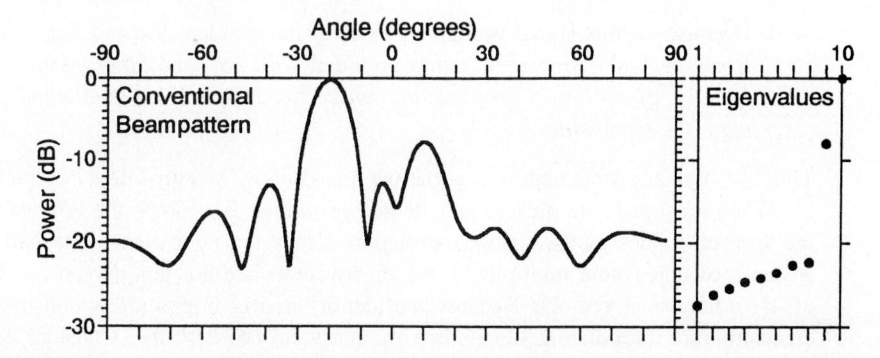

Figure 7.12 A conventional beamformer's steered response, which equals the quadratic form $\mathbf{e}'\mathbf{Re}$, is depicted along with the correlation matrix's eigenvalues. Note that the range spanned by the Fourier spectrum matches closely the range of eigenvalues. Furthermore, the fraction of large eigenvalues roughly corresponds to the fraction of the Fourier spectrum having large values.

DFT vectors as their eigenvectors (see §B.2 {488}). This special case occurs rarely in array signal processing applications and thus Fourier analysis does *not* correspond directly to spectral (eigen) analysis. Consequently, signal vectors cannot form the natural coordinates of a spatial correlation matrix (except in special circumstances; see Prob. 7.17) even though the matrix is composed of them. Resolving this tension initiates a deeper understanding of array processing algorithms, resulting in new ones.

7.3.2 Noise and Signal-Plus-Noise Subspaces

Using eigenanalysis to develop algorithms hinges on the relation between the eigenvectors of spatial correlation matrices and signal vectors. The spatial correlation matrix, temporal or spectral, has the form $\mathbf{R} = \mathbf{K}_n + \mathbf{SCS}'$, where $\mathbf{C}$ is the intersignal coherence matrix {54}. To employ eigenanalysis fruitfully in developing array signal processing algorithms, we must understand the general constraints imposed by the noise and the signals on the spatial correlation matrix's eigenstructure. Because analytic expressions for the eigenvalues of spatial correlation matrices are at best complex in the simpler cases and impossible in the remainder, we are forced to accept qualitative constraints that guide rather than instruct. We have to consider many special cases: spatially white versus colored noise, the number of signals, intersignal coherence, and empirical versus ideal matrices. Because the description that follows winds its way among these waystations, the authors feel obliged to give away now what's at the end of the path.

- By *sorting* the eigenvalues, the associated eigenvectors can be subdivided into two groups that each form a basis: One group spans signal vectors and noise components indistinguishable from signals, and the other spans the remaining noise components.
- The largest N_s eigenvalues, where N_s denotes the number of signals, defines the first group.

- Because distinct signal vectors are linearly independent, we can find signal vectors from the first eigenvector group provided with *only* the array geometry. Consequently, *directions of propagation can be found by sorting the spatial correlation matrix's eigenvalues.*

This last dramatic statement makes the trip that follows worthwhile.

When no signals are present and the noise is white, $\mathbf{K}_n = \sigma_n^2 \mathbf{I}$, the Fourier values and the spectral values of the spatial correlation matrix are equivalent: They all equal σ_n^2. Because of eigenvalue multiplicity, the eigenvectors are not unique and can be *any* set of M orthonormal vectors. Because the identity matrix imposes few constraints on the eigenstructure, introducing signals is particularly easy to describe. Consider the simplest case in which a single signal propagates toward the array. The spatial correlation matrix is then given by

$$\mathbf{R} = \sigma_n^2 \mathbf{I} + A^2 \mathbf{e}\mathbf{e}'$$

where $\mathbf{e}$ represents a unit-amplitude propagating signal. The eigenvector corresponding to the largest eigenvalue is simply the signal vector scaled by $\sqrt{M}$; the associated eigenvalue equals $\sigma_n^2 + MA^2$ (see Prob. 7.17). The remaining eigenvalues equal σ_n^2 and the corresponding eigenvectors comprise a nonunique orthonormal set, each element of which must be orthogonal to the signal. Note what has happened here: Using eigenanalysis, the signal vector (and equivalently the direction of propagation) as well as its amplitude can now be computed *exactly*. We simply perform an eigenanalysis and concentrate on the largest eigenvalue and its associated eigenvector. The eigenvector is proportional to the signal and subtracting the small eigenvalue from the largest yields the amplitude. Consequently, we have a primitive method of determining the signal vector without computing a steered response or without knowing the array geometry. To determine the direction of propagation, we must know the geometry; conversely, if we know the direction of propagation, we can find the geometry (see Prob. 7.24). Unfortunately, only one signal can be present, the noise must be white, and the exact spatial correlation matrix of the observations must be (magically) available. Despite these severe restrictions, this result's simplicity and power lead us to consider more realistic circumstances with the eigenanalysis idea in mind.

Complicating the situation first by considering more than one signal, we must face the possibility of intersignal coherence. When two signals are present, *no more* than two eigenvalues of $\mathbf{R}$ are related to the signals (see Prob. 7.17). The coherent signal case is sufficiently complicated to warrant a separate discussion in §7.3.4 {385}. Turning to the incoherent case first so that intersignal coherence matrix is diag$[A_1^2, A_2^2]$, the spatial correlation matrix in the white-noise case has the form

$$\mathbf{R} = \sigma_n^2 \mathbf{I} + A_1^2 \mathbf{e}(\vec{\zeta}_1^o)\mathbf{e}'(\vec{\zeta}_1^o) + A_2^2 \mathbf{e}(\vec{\zeta}_2^o)\mathbf{e}'(\vec{\zeta}_2^o)$$

The ideal signal vectors are linearly independent so long as the directions of propagation differ, and the array does not introduce direction ambiguities. Two distinguished, signal-related eigenvalues appear, while the remaining ones remain at their noise-only values. Just as in the one-signal case, signal presence has subdivided the spatial correlation matrix's eigenvalues into two distinct sets: one characterized by signal and noise parameters, the other only by the noise. As with the one signal case, the eigenvectors corresponding to the smallest $M-2$ eigenvalues form a *noise subspace* in which signal vectors have no components. The remaining eigenvectors, which correspond to the larger eigenvalues, do *not* equal the signal vectors: Eigenvectors of Hermitian matrices *must* be orthogonal, and signal vectors are often not orthogonal (see Prob. 7.17). These eigenvectors, however, do form a basis that expresses the signal vectors. Mathematically, if $\mathbf{V}_{s+n}$ denotes the $M\times 2$ matrix having columns equal to the two largest eigenvectors, $\mathbf{V}_{s+n} = [\mathbf{v}_1\ \mathbf{v}_2]$, the matrix of signal vectors $\mathbf{S}$ must be related to $\mathbf{V}_{s+n}$ by an invertible 2×2 matrix $\mathbf{T}$.

$$\mathbf{V}_{s+n}\mathbf{T} = \mathbf{S}$$

As signal vectors are linearly independent, *no other signals can be represented by this eigenvector basis*. Hence, each signal vector can be expressed as a linear combination of these eigenvectors and *vice versa*. The noise portion of this subspace contributes σ_n^2 to the eigenvalues but does not affect the eigenvectors in this white-noise situation. Thus, the signals entirely determine the two largest eigenvectors but not the associated eigenvalues. The eigenvectors corresponding to the two largest eigenvalues define a *signal-plus-noise* subspace.

This structure holds for N_s signals so long as $N_s < M$. The N_s largest eigenvectors define a signal-plus-noise subspace; we can form these eigenvectors into a $M\times N_s$ matrix $\mathbf{V}_{s+n}$ and find the signal vectors via the $N_s\times N_s$ matrix $\mathbf{T}$: $\mathbf{V}_{s+n}\mathbf{T} = \mathbf{S}$. The transformation matrix is *not* known a priori; its components are determined by the signal vectors and amplitudes. We thus conclude that this matrix cannot be found in practice, we must seek alternative methods of determining $\mathbf{S}$ from $\mathbf{V}_{s+n}$. The noise only affects the signal-plus-noise subspace eigenvalues by adding the noise variance to them. In the noise subspace (dimension $M - N_s$), no signal components are present and the eigenvalues all equal the noise variance. When we develop algorithms, computing the matrix of signal vectors will be based on the characteristics of the noise subspace: Effective algorithms result when we look where the signals aren't, not where they are. As the special cases illustrate, we require *no* knowledge of the signals to define these subspaces: We can partition into the two subspaces according to the grouping of eigenvalues.

When no pair of signals is perfectly coherent and the additive noise is white, the ideal spatial correlation matrix's eigensystem defines well-formed signal-plus-noise and noise subspaces. We must estimate the correlation matrix from the array data, however. This estimate does not equal the ideal and, in particular, the estimate's noise component does not equal an identity matrix. This component's eigenvalues are no longer identical as in the ideal case, and the noise covariance matrix tends to induce "preferential" directions on its eigenstructure: Some eigenvalues are larger than others and adding signals affects this structure in unpredictable ways.†

A similar situation occurs when the noise is not spatially white. Even in the ideal case, the noise spatial correlation matrix no longer equals an identity matrix, and the

†We have in mind the preferential directions induced on quadratic forms, such as those illustrated by Fig. 7.11. The coordinates equal the eigenvectors that, because they are derived from an estimated matrix, point in directions unrelated to the signals. Including signals adjusts the eigenstructure in complicated ways because now *both* the signals and the noise have preferential directions.

eigenvalues' exact form is open to question. Mathematically, we can explore this situation with generalized eigenanalysis. Assume the correlation matrix has the form $\mathbf{SCS}' + \sigma_n^2 \widetilde{\mathbf{K}}_n$ where $\widetilde{\mathbf{K}}_n$ is the invertible noise covariance matrix normalized to have trace equal to M. Consider the eigenvalue problem

$$\mathbf{Rv} = \lambda \widetilde{\mathbf{K}}_n \mathbf{v}$$

Because the signal-related term has rank equal to the number of signals N_s, it has a null space, defined by $(\mathbf{SCS}')\mathbf{v} = \mathbf{0}$, of dimension $M - N_s$. Such vectors form generalized eigenvectors with eigenvalues equal to σ_n^2. Because these vectors are unrelated to signals, they define a noise subspace and the remaining eigenvectors define a signal-plus-noise subspace. As this signal term is nonnegative definite, the eigenvalues corresponding to the signal-plus-noise subspace must exceed σ_n^2. Consequently, without having a preconceived notion of the signals, we can find the two subspaces by sorting the eigenvalues. Because of the properties of generalized eigenvectors (see §B.1 {495}), these subspaces are orthogonal with respect to $\widetilde{\mathbf{K}}_n$. The matrix of signal vectors can be found from the eigenvectors composing the signal-plus-noise subspace by defining the matrix $\mathbf{V}_{s+n} = \widetilde{\mathbf{K}}_n[\mathbf{v}_1 \cdots \mathbf{v}_{N_s}]$. With this matrix, the relationship expressed previously also applies: $\mathbf{V}_{s+n}\mathbf{T} = \mathbf{S}$. Generalized eigenanalysis thus allows the two subspaces to be defined.

From a more practical standpoint, not only do we not know the matrix $\mathbf{T}$, we often don't know what the noise covariance matrix $\widetilde{\mathbf{K}}_n$ is. We usually only have general knowledge of its characteristics through physical considerations. Without an exact noise covariance matrix, generalized eigenanalysis cannot be used to define signal-plus-noise and noise subspaces clearly. All is not lost; because of the relationship between Fourier and eigenanalyses, we can exploit the Fourier properties of colored noise to assess the noise covariance matrix's eigenvalues qualitatively. When nonwhite noise arises from signals propagating toward the array from all directions, its Fourier spectral characteristics depend on array geometry and physical propagation constraints. These dependencies are illustrated by spherically and cylindrically symmetric propagation modes. We have seen that portions of the Fourier spectrum do not correspond to any propagating waves, resulting in the so-called *invisible regions* (see earlier discussion {91}). Despite their "invisibility," which is a physical constraint, the existence of such spectral regions influences the eigenvalues of the spatial correlation matrix, a mathematical constraint. Because of the relation between Fourier analysis and spectral (eigen) analysis, the smallest value attainable by $\mathbf{e}'\mathbf{Re}$ is $M\lambda_M$ whether $\mathbf{e}$ is visible or not. The lack of propagating energy in invisible regions forces the smaller eigenvalues to correspond to these regions. The smallest eigenvalue must be nonzero as spatial correlation matrices are positive definite. The larger Fourier values, which correspond to visible regions, are thus related to propagating energy.

Within frequency regions that do correspond to propagating energy, the Fourier spectrum's variations affect the spread of the spatial correlation matrix's larger eigenvalues. If the noise Fourier spectrum is relatively constant, the corresponding eigenvalues are nearly equal. Thus, the large eigenvalues of spherically symmetric isotropic noise are approximately constant (Fig. 7.13). In the cylindrical propagation mode, the power spectrum is far from constant, and the eigenvalues corresponding to propagating energy show a similar spread.

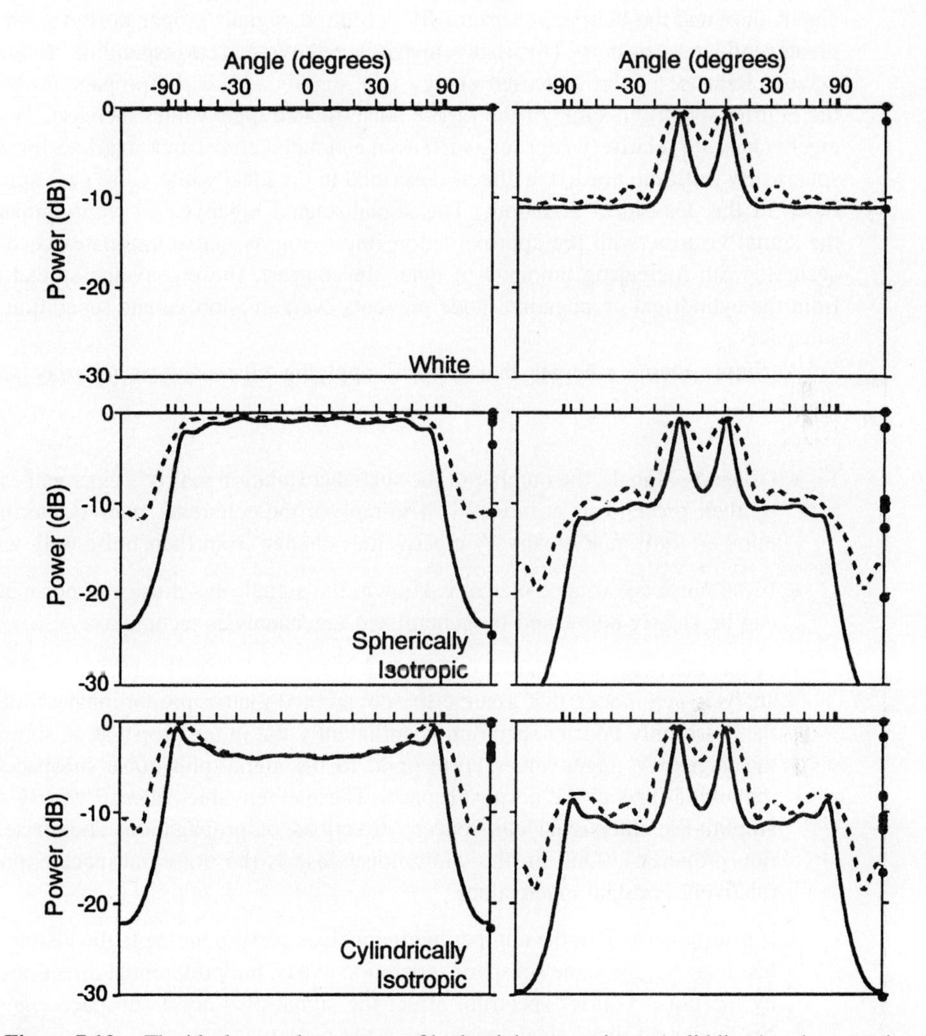

Figure 7.13 The ideal steered responses of both minimum variance (solid lines) and conventional (dashed) algorithms obtained under various noise and signal conditions are shown. Three noise models, one in each row, are used: spatially white noise in the top row, spherically isotropic in the middle, and cylindrically isotropic in the bottom. The left column displays responses when only noise is present, whereas the right shows those that result when two signals plus noise are present. The signals have equal energy and are propagating from 10° and −10°. In each case, noise and signal have equal energy. The eigenvalues of spatial spectral correlation matrices for each plot are shown on the right vertical scale of each panel, with the largest eigenvalue defining the vertical scale's reference. The separation between adjacent sensors of the 10-element linear array equals $3/8\lambda$.

A signal's presence does not change the relationship between the spatial correlation eigenvalues and the Fourier spectrum. By definition, signals propagate and should exert a greater influence on noise covariance matrix's eigenvalues corresponding to propagating waves. Because of the increased energy that signals add to the propagating portion of the Fourier spectrum, some of the *larger* noise-related eigenvalues increase. When these eigenvalues are relatively constant, such as in estimated correlation matrices for white and spherically isotropic noise, the effects described in the ideal white-noise case approximate those in this less ideal situation: The signal-related eigenvectors approximately span the signal vectors, with the approximation improving as matrix estimates become more accurate with increasing amounts of data. In contrast, the eigenvalue spread resulting from the cylindrical propagation mode prevents even an approximate separation into two subspaces.

A simple theme emerges that governs applying eigenanalysis to array processing problems.

- Given N_s signals, the number of the spatial correlation matrix's eigenvalues affected by their presence is at most N_s. The rank of the coherence matrix determines this number: Full rank means N_s eigenvalues change from their noise-only values.
- If the noise covariance matrix is known, the signal-plus-noise and noise subspaces can be clearly delineated by generalized eigenanalysis techniques.
- When the noise covariance matrix is not known, we fall back on simpler eigenanalysis techniques that assume the covariance matrix proportional to identity. If the noise-only Fourier spectrum is sufficiently flat in the propagating signal region, the largest N_s eigenvalues correspond to the signal-plus-noise subspace and the remaining ones to the noise subspace. These eigenvalues depend greatly on signal amplitudes, intersignal coherences, directions of propagation, and noise correlation properties. Only in the white-noise case is the noise subspace expressed by relatively constant eigenvalues.
- If insufficiently flat, the number of eigenvalues corresponding to the visible region is less than N_s, the same subspace separation exists, but preferential directions induced by the noise Fourier spectrum affect the signal-plus-noise subspace eigenvectors. No longer do these eigenvectors span the signals.

7.3.3 Eigenanalysis and Adaptive Methods

Most signal processing algorithms, adaptive or not, merge data lying in signal-plus-noise and noise subspaces. Consider the minimum variance algorithm as one example. As stated earlier by Eq. 7.7 {375}, the eigenvectors of a spatial correlation matrix and its inverse are identical while the corresponding eigenvalues are reciprocals of each other. Interestingly, when we use the generalized eigensystem, the *same* expression holds for the inverse, but not for the matrix {495}. Expressed in terms of the eigensystem, the output power has the form

$$\mathcal{P}^{MV}(\vec{\xi}\,) = \left[\sum_i \lambda_i^{-1} |\mathbf{e}'(\vec{\xi}\,)\mathbf{v}_i|^2\right]^{-1}$$

Defining $a_i(\vec{\xi}\,) = \mathbf{e}'(\vec{\xi}\,)\mathbf{v}_i$ to be the inner product (projection) of the ideal signal vector onto the ith eigenvector,

$$\mathcal{P}^{MV}(\vec{\xi}\,) = \left[\sum_i \lambda_i^{-1} |a_i(\vec{\xi}\,)|^2\right]^{-1}$$

Note that we need not consider both signal-plus-noise and noise subspaces to produce, for example, estimates of directions of propagation; *either* suffices to specify the signal vectors. The signal-plus-noise subspace informs the knowledgeable user where the signals are, the noise subspace where they aren't. Because these natural coordinate systems represent the same information so differently, combining them usually results in inferior performance when compared with considering only one. Algorithms in the latter category are given the general name of *eigenanalysis* algorithms. The key for developing an eigenanalysis algorithm for determining direction of propagation is to select the subspace that maximizes the output power *only* in the directions of propagation while not enhancing non–signal-related directions.

The larger eigenvalues, which by the indexing convention have the largest indices {493}, correspond to the signal-plus-noise subspace and the smaller ones to the noise subspace. Assuming N_s signals, we partition the bracketed term into these subspaces [14, 81, 129, 136].

$$\mathcal{P}^{MV}(\vec{\xi}\,) = \Bigg[\underbrace{\sum_{i=1}^{N_s} \lambda_i^{-1} |a_i(\vec{\xi}\,)|^2}_{\text{signal-plus-noise subspace}} + \underbrace{\sum_{i=N_s+1}^{M} \lambda_i^{-1} |a_i(\vec{\xi}\,)|^2}_{\text{noise subspace}}\Bigg]^{-1}$$

When $\mathbf{e}(\vec{\xi}\,)$ represents an actual propagating signal, the coefficients $\{a_i\}$ are large in the signal-plus-noise subspace and small in the noise subspace. In the ideal situation of an identity noise covariance matrix, the *only* nonzero coefficients are those contained in the signal-plus-noise subspace. The squared magnitudes of the coefficients are weighted by the reciprocal eigenvalues, which are larger in the signal-plus-noise subspace. In this way, the larger coefficient values are reduced in the sum. Peaks in output power obtain when the sum becomes *small* as the assumed propagation direction varies. Ironically, the signal-related terms rather than the noise terms restrain the sum from achieving as small a value as possible. By dropping signal-related terms and considering only those in the noise subspace, we obtain a more dramatic indication of a signal's direction of propagation: The eigenvalue weighting is constant, but the signal-plus-noise subspace coefficients increase in value just when we would like them to decrease. Thus, because the sum's reciprocal is important, we want to consider where the signals aren't—the noise subspace—rather than where they are. This approach is indicative of most, if not all, of the eigenanalysis methods applied to adaptive beamforming.

The so-called *eigenvector method* [81] consists of using the truncated eigenvector expansion for the inverse of the estimated spatial correlation matrix in adaptive beamforming methods.

$$\boxed{\widehat{\mathbf{R}}_{\mathrm{EV}}^{-1} = \sum_{i=N_s+1}^{M} \lambda_i^{-1} \mathbf{v}_i \mathbf{v}_i'} \tag{7.8a}$$

The multiple signal classification (*MUSIC*) algorithm [135, 136] also truncates the eigenexpansion and, in addition, sets the noise subspace eigenvalues to unity.

$$\boxed{\widehat{\mathbf{R}}_{\mathrm{MUSIC}}^{-1} = \sum_{i=N_s+1}^{M} \mathbf{v}_i \mathbf{v}_i'} \tag{7.8b}$$

As identical eigenvalues correspond to white noise, this modification of the noise subspace's eigenvalues amounts to whitening the noise subspace. Thus, in the context of the minimum variance algorithm, two eigenanalysis-based variants of the steered response emerge.

$$\mathcal{P}^{\mathrm{EV}} = \left[\mathbf{e}'\widehat{\mathbf{R}}_{\mathrm{EV}}^{-1}\mathbf{e}\right]^{-1} \qquad \mathcal{P}^{\mathrm{MUSIC}} = \left[\mathbf{e}'\widehat{\mathbf{R}}_{\mathrm{MUSIC}}^{-1}\mathbf{e}\right]^{-1}$$

Using either of these quantities in adaptive methods can result in a more dramatic indication of the signals' propagation directions. Fig. 7.14 demonstrates the effectiveness of truncating the eigensystem of the spatial correlation matrices.

Truncated eigenexpansions can be used with both the minimum variance and linear predictive methods [14, 80, 81, 95] as shown in Fig. 7.15. Interestingly, the eigenanalysis-based modifications of these adaptive methods appear to "lift" the signals out of the noise, leaving the noise portions of the steered responses unchanged. By calculating the amplitude of the eigenanalysis-modified minimum variance steered response, this enhancement can be interpreted as an effective increase in the number of sensors in the array and a similar increase in effective aperture [81]. The effective number M_{eff} of sensors for either eigenanalysis method in the single-signal case is

$$M_{eff} = \frac{M}{\dfrac{\sigma_n^2}{A^2}\dfrac{\gamma^2}{M} + \dfrac{M}{L}}$$

where γ^2 is a variance-like expression that measures the spread of the eigenvalues in the noise covariance component in the spatial correlation matrix's estimate. This spread usually decreases with increasing time-bandwidth product L. When only white noise is present, this spread tends toward zero with increasing L as the ideal noise covariance matrix has equal eigenvalues. At this limit, the effective number of sensors equals the time-bandwidth product. As shown in the upper portion of Fig. 7.15, the ratio of peak values to background noise levels is consistent with this prediction when either eigenanalysis method is used. Thus, only the amount of available data limits the resolution

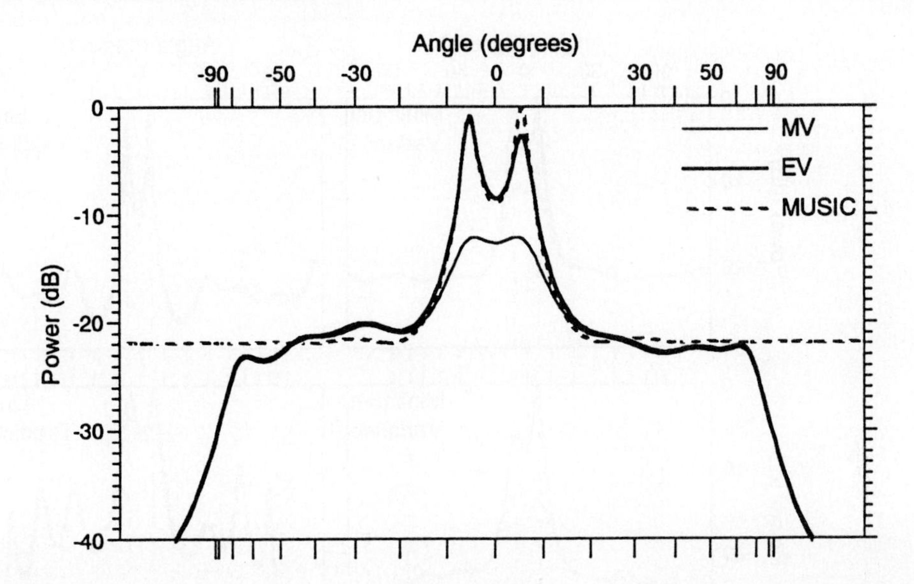

Figure 7.14 The eigenvector (*EV*) and *MUSIC* variations of truncating the eigenexpansions of the spatial correlation matrix's inverse are shown in the context of the minimum variance algorithm. Two signals are propagating from -5° and 5° in the presence of spherically isotropic noise. The signal-to-noise ratio for each signal is 0 dB, and the time-bandwidth product of the spatial correlation matrix's estimate equals 100. The minimum variance algorithm alone does not have sufficient resolving power to distinguish the two signals. Either of the eigenanalysis-based methods can be at about the same level of performance. The greatest difference between the methods occurs in the invisible region; the eigenvector method yields the same results as minimum variance while *MUSIC* yields a flattened (whitened) result.

capabilities of eigenanalysis-based methods in the white-noise case [87]. When nonwhite noise is present, the eigenanalysis methods cannot provide arbitrary resolution. Because the underlying noise correlation matrix has unequal eigenvalues (Fig. 7.13 {381}), the limiting value of γ^2 equals some nonzero value. Once the ratio M/L becomes smaller than $\sigma_n^2\gamma^2/A^2M$, the point of diminishing return is reached: Further increases in time-bandwidth product do not increase M_{eff}, and no further increases in spatial resolution occur. An illustration of this effect is shown in Fig. 7.16. Without employing eigenanalysis, the asymptotic resolution of adaptive methods is always limited. The locations as well as the amplitudes of the peaks yielded by the eigenanalysis methods are affected by background noise characteristics. If white, the peak locations are asymptotically unbiased estimates of directions of propagation; if not white, peak locations are frequently biased. Kaveh and Barabell [86] provide detailed statistical analyses of the *MUSIC* method applied to both minimum variance and linear prediction techniques.

7.3.4 Signal Coherence

Signal coherence adversely affects adaptive algorithms. Eigenanalysis best portrays the difficulties and allows us to assess how well signal processing countermeasures work.

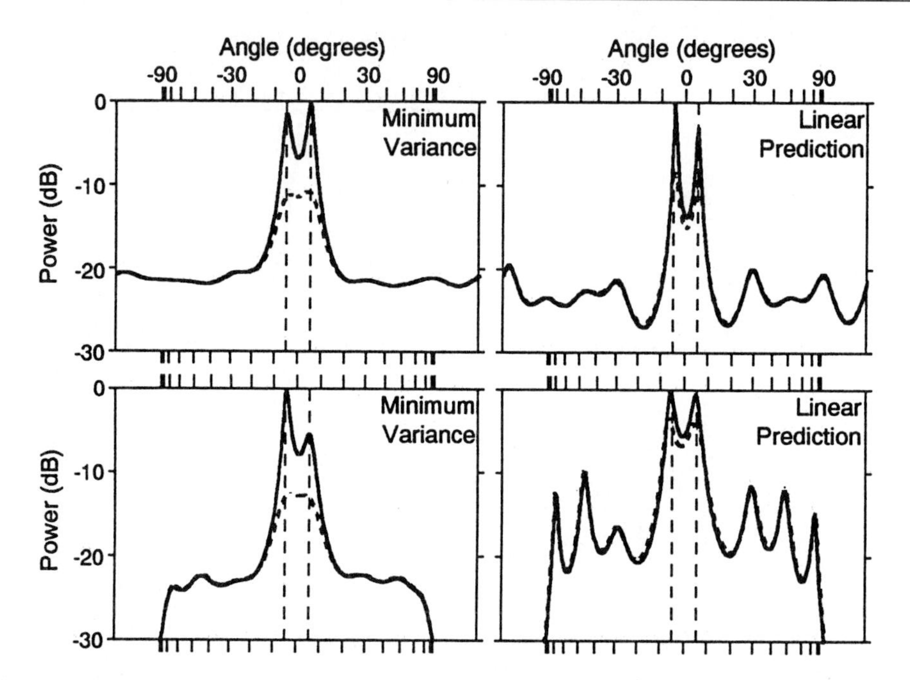

Figure 7.15 The steered responses resulting from employing the eigenanalysis methods (solid lines) are compared with those produced by the usual adaptive methods (dashed lines) using computer simulations. The left column displays the application of the minimum variance method and the right linear prediction. In the top row, the noise is spatially white, whereas in the bottom spherically isotropic noise is present. The array contains ten sensors in a regular linear geometry (separation of $3/8\lambda$). Two equistrength sources are present, one located at 5° and the other at -5°, with a signal-to-noise ratio at each sensor of 0 dB. The reference for each panel is the peak value of the steered response derived with the eigenvector method. The time-bandwidth product of the spatial correlation function estimate equals 100.

When the intersignal coherence matrix is singular in the two-signal case (the two signals are *perfectly* coherent), one rather than two eigenvalues differs from σ_n^2. This matrix causes the signal term in the spatial correlation matrix to resemble a *single* outer product, which is not composed of an ideal signal vector. For example, the intersignal coherence matrix $\mathbf{C} = \begin{bmatrix} A^2 & -A^2 \\ -A^2 & A^2 \end{bmatrix}$ is singular and results in a spatial correlation matrix of the form

$$\mathbf{R} = \sigma_n^2 \mathbf{I} + A^2 \left[\mathbf{e}(\vec{\zeta}_1^o) - \mathbf{e}(\vec{\zeta}_2^o)\right]\left[\mathbf{e}(\vec{\zeta}_1^o) - \mathbf{e}(\vec{\zeta}_2^o)\right]'$$

$M-1$ eigenvalues equal σ_n^2 with the sole distinguished one equaling $MA^2 + \sigma_n^2$. The eigenvector corresponding to the largest eigenvalue equals the difference of the two signals. Because signal vectors are linearly independent, a linear combination of signals cannot be a signal. Thus, this eigenvector is related to the signal vectors, but neither signal cannot be obtained from it without knowing the other; stated another way, the magic matrix $\mathbf{T}$ does not exist. We see that *perfect* coherence introduces a degeneracy

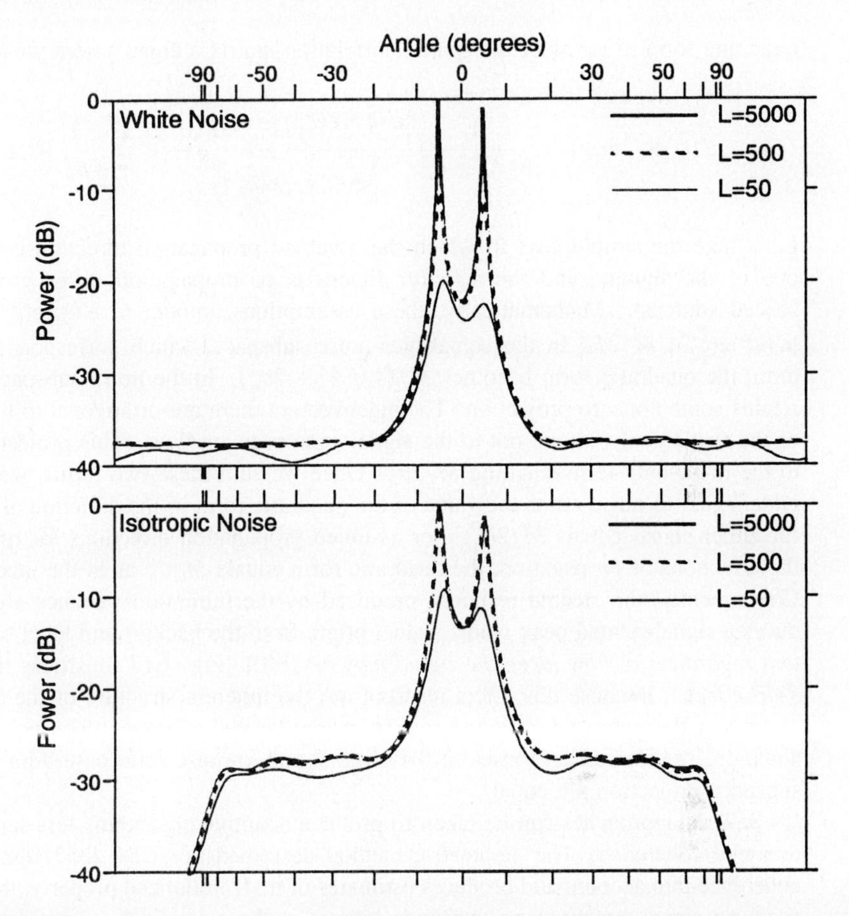

Figure 7.16 Steered responses resulting from using the eigenvector method in conjunction with the minimum variance algorithm are shown as the time-bandwidth product varies ($L = 50, 500, 5000$). The array and signal properties are the same as those in Fig. 7.15. The noise is spatially white in the upper panel and is spherically isotropic in the lower. Note how the factor-of-ten change in the time-bandwidth product results in a 10-dB change in the peak of the response when white noise is present (upper panel). In contrast, because of the isotropic noise, no such concomitant change is seen.

and eigenvalue grouping no longer corresponds to a signal-plus-noise and noise subspace segmentation.†

This degeneracy affects adaptive algorithms in dramatic fashion. Consider the minimum variance algorithm. In our perfectly coherent example, expressing its defining

†Only when the coherence matrix is singular does this degenerate situation occur. Non-diagonal, full-rank matrices yield full-rank signal-plus-noise subspaces, but still reduce the signal processing capabilities of adaptive and eigenanalysis methods.

quadratic form in terms of the spatial correlation matrix's eigensystem yields

$$\mathbf{e}'(\vec{\zeta})\mathbf{R}^{-1}\mathbf{e}(\vec{\zeta}) = \frac{1}{MA^2+\sigma_n^2}\frac{\left|\mathbf{e}'(\vec{\zeta})[\mathbf{e}(\vec{\zeta}_1^o)-\mathbf{e}(\vec{\zeta}_2^o)]\right|^2}{\left\|\mathbf{e}(\vec{\zeta}_1^o)-\mathbf{e}(\vec{\zeta}_2^o)\right\|^2} + \sum_{i=2}^{M}\frac{1}{\sigma_n^2}\left|\mathbf{e}'(\vec{\zeta})\mathbf{v}_i\right|^2$$

Let's take the simple case in which the assumed propagation direction corresponds to one of the signals, and the signals' directions of propagation differ greatly (widely spaced sources). Mathematically, these assumptions amount to $\mathbf{e}'(\vec{\zeta})\mathbf{e}(\vec{\zeta}_1^o) = M$ and $|\mathbf{e}'(\vec{\zeta}_1^o)\mathbf{e}(\vec{\zeta}_2^o)| \ll M$. In the signal-plus-noise subspace, which corresponds to the first term, the quadratic form becomes $M/(4MA^2+2\sigma_n^2)$. In the noise subspace, the signal retains some nonzero projection: The eigenvectors there are orthogonal to the difference of the two signal vectors, not to the signal vectors themselves. This projection amounts to the noise-only term equaling $M/2\sigma_n^2$. Once we sum these two terms, we see that for large signal-to-noise ratios the value of the quadratic form in the direction of propagation for either signal equals $M/2\sigma_n^2$. For assumed propagation directions far removed from the directions of propagation, the quadratic form equals M/σ_n^2 as in the incoherent case. Consequently, the steered response produced by the minimum variance algorithm produces a signal-related peak whose value differs from the background level by a factor of two *regardless of how large the signal may be* [140]. Fig. 7.17 illustrates this so-called *3-dB effect*. Because this effect results from the inherent structure of the eigensystem, restricting the inverse correlation matrix to the "noise" subspace does not improve the situation; the steered responses in the large signal-to-noise ratio case with and without subspace projection are equal.

Several approaches can be taken to produce adaptive algorithms less sensitive to intersignal coherence. The parametric method described in §7.1.2 {353} explicitly takes coherence into account and produces estimates of it. If initialized properly, this algorithm produces signal direction and amplitude estimates whose precision is unaffected by coherence [99]. Unfortunately, such explicit algorithms are few; the most common approach is to reduce the measured coherence using signal processing means, but at the cost of reduced direction-finding capability [54, 141]. For example, in §4.9.2 {186}, we found that subarray averaging reduces intersignal coherence. Equipped with decompositions for the spatial correlation matrix, we can justify that claim and employ eigenanalysis as well as other adaptive methods to the coherent source problem.

The subarray averaging approach applies only to the regular linear array. As shown in Fig. 4.35 {188}, an array of M sensors is decomposed into $M-M_s+1$ overlapping subarrays each having M_s sensors. A correlation matrix is formed by averaging the matrices computed for each subarray.

$$\overline{\mathbf{R}} = \frac{1}{M-M_s+1}\sum_{i=0}^{M-M_s}\mathbf{R}_i$$

When spatially white noise is present, each subarray's correlation matrix has the form $\mathbf{R}_i = \sigma_n^2\mathbf{I} + \mathbf{S}_i\mathbf{C}_i\mathbf{S}_i'$. Because the same signals are present in each subarray, the same

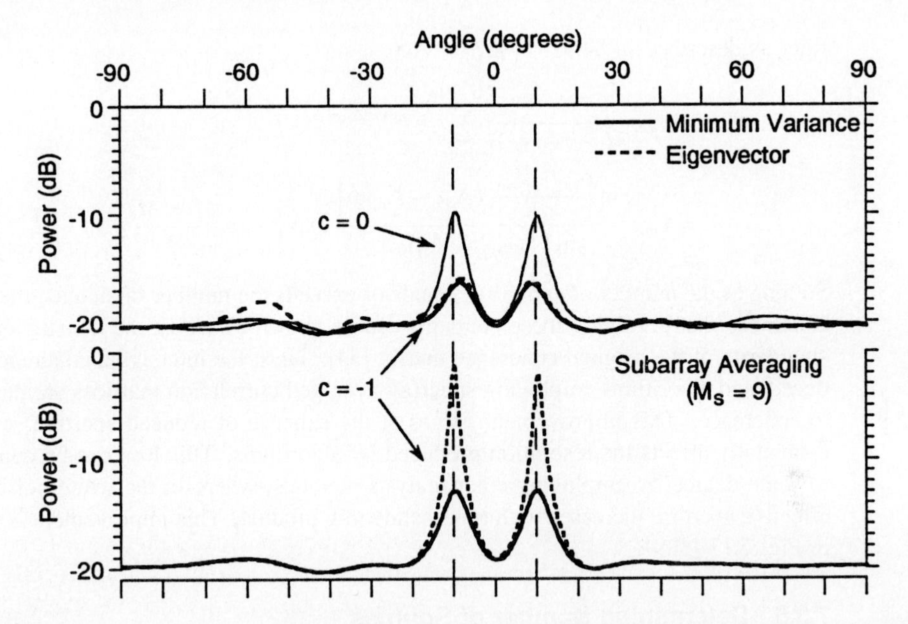

Figure 7.17 The effect of intersignal coherence is illustrated by the steered response computed with the minimum variance technique and drawn by the thick solid line in the upper panel. The thinner solid line shows the steered response obtained under zero signal coherence conditions. Applying the eigenvector method (dashed line) does not enhance signal-related peaks. In fact, background "peaks" seem slightly enhanced. The signals have equal amplitude and are propagating from $\pm 10°$. The linear array consisted of ten sensors with intersensor spacing $\lambda/2$. The bottom panel illustrates the effect of subarray averaging. With only two subarrays ($M_s = 9$), the effects of coherence are reduced when the minimum variance (solid line) and eigenvector (dashed line) methods employ the averaged correlation matrix. The time-bandwidth product of the matrix estimates in all cases is 100.

intersignal coherence matrix applies to each subarray: $\mathbf{C}_i = \mathbf{C}$. The subarrays' signal matrices are related by the propagation delay from one array to another. Because of the regular linear array's geometry, this relationship is expressed by $\mathbf{S}_i = \mathbf{S}_0\mathbf{D}^{i-1}$ where $\mathbf{D} = \text{diag}[e^{-j\omega\Delta_1}, \ldots, e^{-j\omega\Delta_{N_s}}]$ is a diagonal matrix expressing the intersensor propagation delays for the N_s signals. Thus, the subarray averaged correlation matrix has the equivalent form

$$\overline{\mathbf{R}} = \sigma_n^2\mathbf{I} + \frac{1}{M - M_s + 1}\sum_{i=0}^{M-M_s} \mathbf{S}_0\mathbf{D}^i\mathbf{C}\mathbf{D}'^i\mathbf{S}_0'$$

which indicates that the intersignal coherence matrix resulting from subarray averaging becomes $\overline{\mathbf{C}} = \left(\sum_{i=0}^{M-M_s} \mathbf{D}^i\mathbf{C}\mathbf{D}'^i\right)/(M - M_s + 1)$. This averaging operation reduces off-diagonal entries by a quantity corresponding to the frequency response of a "boxcar" *FIR* filter as described in §4.9.2 {186}.

$$\overline{c}_{mn} = c_{mn} \frac{1}{M - M_s + 1} \sum_{i=0}^{M-M_s} e^{-j\omega i(\Delta_m - \Delta_n)}$$

$$= c_{mn} \frac{\sin\left[\frac{(M-M_s+1)}{2}(\Delta_m - \Delta_n)\omega\right]}{\sin\left[\frac{1}{2}(\Delta_m - \Delta_n)\omega\right]} \exp\left\{-j\omega \frac{M - M_s}{2}(\Delta_m - \Delta_n)\right\}$$

So long as the number of subarrays equals or exceeds the number of signals, the averaged coherence matrix having these elements can be shown to *always* be nonsingular despite singularity of the original coherence matrix [54]. Thus, the intersignal coherences are reduced, and algorithms employing subarray-averaged correlation matrices are desensitized to coherence. This improvement comes at the expense of reduced aperture, which fundamentally affects the resolution produced by algorithms. This loss can be compensated to some degree by employing eigenanalysis methods, where in the white-noise case the effective aperture increases with time-bandwidth product. This improvement is portrayed in Fig. 7.17 {389}.

7.3.5 Determining Number of Sources

The eigenanalysis methods critically depend on having a value for N_s, the number of signals present in the sensor outputs. Without a value for this parameter, these methods are helpless. From the discussion of the eigenstructure of the spatial correlation matrix in §7.3.2 {377}, the number of signals is equivalent to the number of eigenvalues that differ from the eigenvalues of the noise-only spatial correlation matrix $\mathbf{K}_n$. In many cases, not only is the ideal $\mathbf{K}_n$ not known, but the time-bandwidth product of its estimate must agree with that of the spatial correlation matrix estimate used in the beamforming computations. Instead of comparing signal-plus-noise eigenvalues with noise-only eigenvalues in an ad hoc fashion, we determine the number of signals with statistical tests based on the observations' spatial correlation matrix.

Estimating N_s is equivalent to the model order determination problem described in §6.5.3 {334} [157]. To use the *AIC* and *MDL* criteria, we must first establish a parametric model. Here the "model" is the ideal spatial correlation matrix. We assume the observation vectors to be independent, zero-mean, Gaussian random vectors having a covariance matrix equal to the spatial correlation matrix.† These observation vectors consist of white noise and an independent signal component having a covariance matrix of rank N_s, which mimics the $\mathbf{SCS}'$ term we have been using. The joint distribution of the L complex-valued observation vectors $\mathbf{y}_1, \ldots, \mathbf{y}_L$ is (see §A.3 {478})

$$p(\mathbf{y}_1, \ldots, \mathbf{y}_L \mid \boldsymbol{\xi}) = \prod_{l=1}^{L} \frac{1}{\det[\pi \mathbf{K}(\boldsymbol{\xi})]} \exp\left\{-\mathbf{y}_l' \mathbf{K}^{-1}(\boldsymbol{\xi}) \mathbf{y}_l\right\}$$

†For the purposes of determining the number of signals, the signals as well as the noise are assumed Gaussian. Thus, they do not appear as the mean of the probability density, but instead in the covariance matrix $\mathbf{K}$. This departure from the usual model is made for analytic expedience: When you have a solution to a problem close to yours, modify your problem to become the other and then see how well the solution works.

Using matrix properties given in §B.3 {491}, the argument of the exponential can be written as

$$\sum_{l=1}^{L} \mathbf{y}_l' \mathbf{K}^{-1}(\boldsymbol{\xi}) \mathbf{y}_l = L \,\mathrm{tr}[\mathbf{K}^{-1}(\boldsymbol{\xi}) \widehat{\mathbf{K}}]$$

where $\widehat{\mathbf{K}} = \sum_{l=1}^{L} \mathbf{y}_l \mathbf{y}_l' / L$ represents the Bartlett estimate of the spatial correlation matrix. The various parameters of the model—the eigenvalues, the eigenvectors, and the variance—are represented by the vector $\boldsymbol{\xi}$. The number of parameters in this model would seem to be N_s (the number of eigenvalues) plus $2N_s M$ (the real and imaginary parts of N_s length-M eigenvectors) plus 1 (the noise variance). Note that the eigenvectors must be mutually orthogonal and have unit norm. The orthogonality constraint reduces the initial estimate by $N_s(N_s - 1)$, and the common norm constraint reduces it by an additional N_s. Thus the actual "model order" equals $N_s(2M - N_s + 1) + 1$.

To compute the *AIC* and *MDL* criteria, the maximum likelihood estimates of the parameters must be found. From any text in multivariate statistics [110], the maximum likelihood estimates of the N_s largest eigenvalues and their associated eigenvectors are simply the N_s largest components in the eigenanalysis of $\widehat{\mathbf{K}}$, which can be computed using numerical routines. The estimate of the noise variance equals the average value of the $M - N_s$ smallest eigenvalues.

$$\hat{\sigma}^2_{\mathrm{ML}} = \frac{1}{M - N_s} \sum_{i=N_s+1}^{M} \hat{\lambda}_i$$

Substituting these estimates for the parameter values into the logarithm of the likelihood function yields the data-dependent component for the *AIC* and *MDL* criteria.

$$-\ln p(\mathbf{y}_1, \ldots, \mathbf{y}_L \mid \widehat{\boldsymbol{\xi}}_{\mathrm{ML}}) = -L(M - N_s) \ln \left[\frac{\left(\prod_{i=N_s+1}^{M} \hat{\lambda}_i \right)^{1/(M-N_s)}}{\dfrac{1}{M - N_s} \sum_{i=N_s+1}^{M} \hat{\lambda}_i} \right]$$

This rather curious term in the brackets is the ratio of the geometric and arithmetic means of the $M - N_s$ smallest eigenvalues of the estimated spatial correlation matrix. These eigenvalues correspond to the noise subspace for each presumed value of N_s. If N_s is accurate, these eigenvalues should be nearly equal because of the white-noise assumption. If precisely equal, the geometric and arithmetic means are equal. When N_s is chosen too small so that the eigenvalues are unequal, the arithmetic mean always exceeds the geometric mean and the log-likelihood function increases dramatically. The criterion

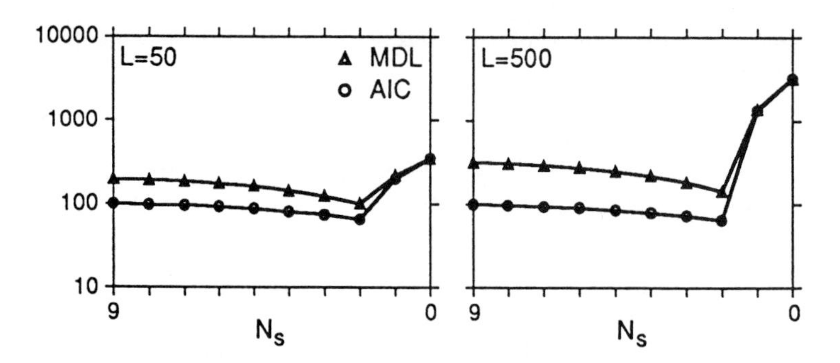

Figure 7.18 The *AIC* and *MDL* criterion functions are plotted against the presumed number of sources for two examples. The *MDL* criterion function is represented by the open triangles in both panels, the *AIC* by the open circles. The spatial correlation matrices are those used in the upper panel of Fig. 7.16 having a time-bandwidth product of 50 (left panel here) and 500 (right panel). The presumed number of sources decreases along the horizontal axis because the criteria are computed with an increasing number of eigenvalues starting with the smallest. In these examples, both criteria yield the correct estimate of two sources.

functions for determining the number of signals from the eigenvalues are [157][†]

$$
\boxed{\begin{aligned}
\text{AIC}(N_s) &= -\ln p(\mathbf{y}_1, \ldots, \mathbf{y}_L \mid \widehat{\xi}_{\text{ML}}) + N_s(2M - N_s + 1) \\
\text{MDL}(N_s) &= -\ln p(\mathbf{y}_1, \ldots, \mathbf{y}_L \mid \widehat{\xi}_{\text{ML}}) + \frac{1}{2} N_s(2M - N_s + 1) \ln L
\end{aligned}}
$$

These criterion functions are shown in Fig. 7.18 for two of the spatial correlation matrices used in Fig. 7.16 {387}. The value of N_s minimizing the selected criterion function becomes the estimate of the number of signals present in the spatially white noise field. As might be expected, the minimum of either criterion becomes more pronounced as the time-bandwidth product increases. Because of the noise, the maximum number of sources that can be objectively estimated with these techniques is $M-1$, the number of sensors minus one. When coherent signals are present, this maximum number may be smaller.

The *MDL* criterion yields consistent estimates of the number of signals while the *AIC* criterion does not {337}. The *AIC* criterion usually yields a slightly larger value for the number of signals. This bias affects the eigenanalysis methods little, whereas too low an estimate does not yield the predicted resolution increases [81]. The value of N_s minimizing either criterion provides an estimate of the number of signals for the eigenanalysis methods. Because this estimate is based on the spatial correlation matrix's eigenvalues rather than on a signal model, an estimate emerges that does not

[†]Two details about these formulas need clarification. The first concerns the missing "+1" from the model-order term: This term is not computed because we drop any term not depending on the data or on N_s. As for the second, the scholarly reader will find this expression differing from that given by Wax and Kailath [157: Eqs. 16 and 17]. They informed the authors that typographical errors slipped into their paper and that the formula presented here is more correct.

require knowledge of signal characteristics (directions of propagation, amplitudes, and coherences).

7.4 Robust Adaptive Array Processing

Adaptive array processing algorithms extract so much information from the observations that they tend to be sensitive to slight deviations in fundamental assumptions when reality departs from model. For example, the assumed form for the ideal signal vector $\mathbf{e}$ used so frequently in adaptive algorithms implies that sensor positions, sensor gains, and sensor phase shifts are all known *perfectly*. Realistically, all of these are in error to some degree, and, as a consequence, the adaptive methods can lose much of their performance capabilities [35: 344–61]: They are not robust. Conventional beamforming algorithms do not rely heavily on such precision; as we shall see, they are robust but lack the signal processing capabilities of the adaptive methods. All is not lost; recognizing how such imperfections manifest themselves leads to adaptive algorithms that cope with small deviations from the nominal without suffering large performance losses.

7.4.1 Effects of Signal Modeling Errors

The ideal form of the signal represented by the vector $\mathbf{e}$ implies we have perfectly surveyed the sensor locations, and that identical transducers and amplifiers lead from the sensors to the digitizers. As this utopia is at best costly, if not impossible to achieve, we must probe the effects of signal modeling errors on adaptive algorithms [39, 53, 77, 155]. Assuming that the ideal gain g_m of the mth sensor is unity, the true gain for frequency-domain beamforming can be expressed as $g_m = 1 + \delta g_m$, where δg_m represents an unknown *complex-valued* quantity that summarizes errors in sensor locations and in sensor gains (amplitude and phase). Note that position errors lead to phase changes dependent on the signal's direction of propagation; the gain perturbations δg_m also depend on the signals present in the propagating field and the assumed propagation direction. These gain perturbations apply to propagating signals, deviating the observations from the ideal signal assumed by the signal processing algorithms. Instead of $\mathbf{e}$, the observed signal is $\mathbf{Ge}$, where $\mathbf{G} = \mathbf{I} + \text{diag}[\delta g_0, \ldots, \delta g_{M-1}]$. Assuming that the spatial correlation matrix has the form $\mathbf{R} = \sigma_n^2 \mathbf{I} + A^2 \mathbf{G}\mathbf{e}\mathbf{e}'\mathbf{G}'$, the minimum variance beamformer using the ideal signal model would produce a signal power estimate in the direction of propagation equaling

$$(\mathbf{e}'\mathbf{R}^{-1}\mathbf{e})^{-1} = \frac{A^2 + \dfrac{\sigma_n^2}{M} + \dfrac{2A^2}{M}\sum_m \text{Re}[\delta g_m] + \dfrac{A^2}{M}\sum_m |\delta g_m|^2}{1 + \dfrac{A^2}{\sigma_n^2}\sum_m |\delta g_m|^2 - \dfrac{A^2}{M\sigma_n^2}\left|\sum_m \delta g_m\right|^2}$$

Assuming that the sum of the deviations tends to be small (but not the sum of their magnitudes), this result simplifies to

$$(\mathbf{e}'\mathbf{R}^{-1}\mathbf{e})^{-1} = \left(A^2 + \frac{\sigma_n^2}{M}\right) \cdot \left(\frac{1 + \dfrac{A^2/\sigma_n^2}{1 + MA^2/\sigma_n^2} \sum_m |\delta g_m|^2}{1 + \dfrac{A^2}{\sigma_n^2} \sum_m |\delta g_m|^2} \right) \tag{7.9}$$

For small or moderate signal-to-noise ratios A^2/σ_n^2, the factor on the right nearly equals 1: Gain errors do not perturb the signal processing algorithm in this regime. As the signal-to-noise ratio *increases*, however, this factor becomes proportional to $1/A^2$, and the signal is actually suppressed.

$$\lim_{A^2/\sigma_n^2 \to \infty} (\mathbf{e}'\mathbf{R}^{-1}\mathbf{e})^{-1} = \frac{\sigma_n^2}{M} \left(1 + \frac{1}{\sum_m |\delta g_m|^2 / M}\right), \quad \sum_m |\delta g_m|^2 \neq 0$$

The power estimate thus produced does not depend on the signal's power, but on the noise variance and on the mean-squared perturbation. Because adaptive algorithms designed by constrained optimization methods reject signals that do not resemble the assumed ideal, they may, as just illustrated, actually reject observed signals when gain errors are not taken into account. Similar calculations for conventional beamforming yield a power output equal to

$$\mathbf{e}'\mathbf{R}\mathbf{e} = A^2 + \sigma_n^2/M + 2\sum_m \mathrm{Re}[\delta g_m]/M + \sum_m |\delta g_m|^2/M^2$$

In contrast to the minimum variance method, conventional beamformers are not extraordinarily sensitive to gain perturbations: As signal power increases, sensor errors have lesser effect. Conventional beamformers are *robust* to gain errors. Fig. 7.19 illustrates the differences in the methods' sensitivities to perturbations.

7.4.2 Amplitude and Derivative Constraints

Adaptive methods tend to produce more precise indications of source locations as signal-to-noise ratio increases. This effect is illustrated in Fig. 7.20 for the minimum variance method. Consequently, signal model inexactitudes may mean that the algorithm misses signals because it concentrates so precisely on ideal signals. Creating a robust alternative would require a controlled loss of resolution by widening each signal's steered response [145, 147]. One obvious approach is to constrain the response amplitude in directions symmetrically placed about the assumed propagation direction. For example, a two-point constraint would force unity gain in directions close to the assumed propagation direction with $\mathbf{e}'(\vec{\zeta} + \delta\vec{\zeta})\mathbf{w} = 1$ and $\mathbf{e}'(\vec{\zeta} - \delta\vec{\zeta})\mathbf{w} = 1$; a three-point constraint would force unity gain in the assumed propagation direction ($\mathbf{e}'(\vec{\zeta})\mathbf{w} = 1$) in addition to these nearby directions. This idea yields multiple linear constraints on the weight vector and can be considered a special case of the constrained optimization approach described

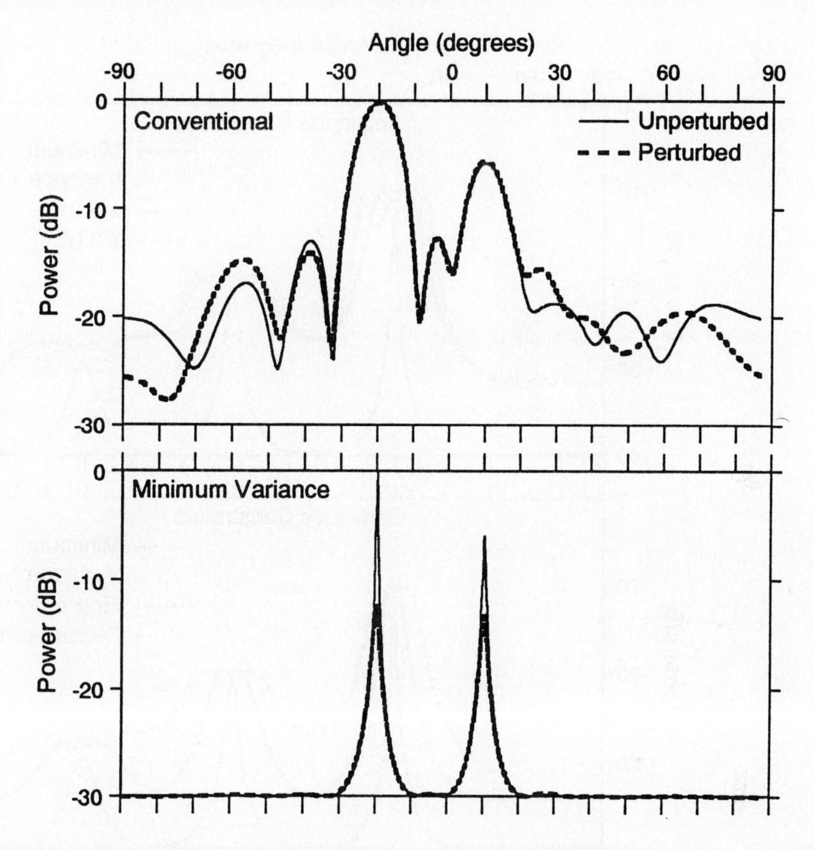

Figure 7.19 Steered responses are shown for the conventional and minimum variance beamformers for the same 10-element array and the same data. In each plot, the solid line represents the response to two signals observed in additive sensor noise with no errors in sensor gain or sensor position. The signal-to-noise ratio is 20 dB for the larger signal, 14 dB for the other. The dashed line denotes the response when the sensor gain is augmented by a complex-valued, unaccounted error simulated as a complex Gaussian random variable having independent components and independently chosen for each sensor. The standard deviation of the error equals 0.15.

in §7.2 {355}. The constraint matrix and vector of constraining values for the three-point example are

$$\mathbf{C} = \begin{bmatrix} \mathbf{e}'(\vec{\zeta} - \delta\vec{\zeta}\,) \\ \mathbf{e}'(\vec{\zeta}\,) \\ \mathbf{e}'(\vec{\zeta} + \delta\vec{\zeta}\,) \end{bmatrix} \qquad \mathbf{c} = \begin{bmatrix} 1 \\ 1 \\ 1 \end{bmatrix}$$

Use of the three-point constraint is illustrated in Fig. 7.20. If we choose too great a spacing between the constraint directions, individual peaks in the constraint directions appear. Chosen properly, amplitude constraints flatten the top of the signal peaks as desired. This loss of resolution is also accompanied by a reduction in array gain; note how we can no longer distinguish the weaker signal on the right in Fig. 7.20.

We can also improve robustness but reduce resolution by forcing derivatives of the

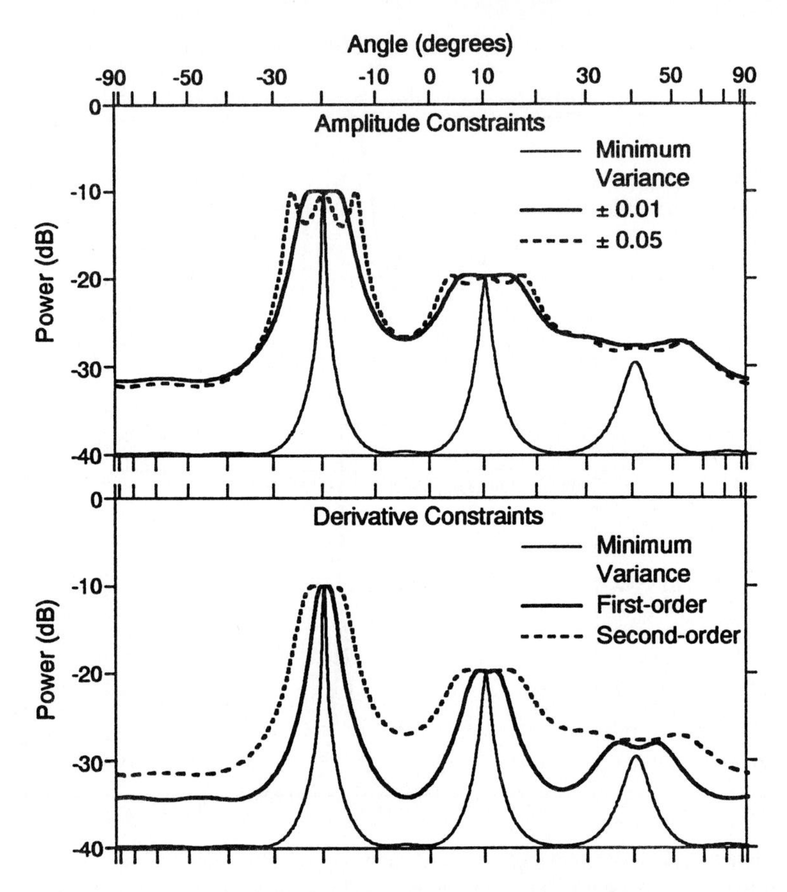

Figure 7.20 In each panel, three signals are present with the signal power decreasing by a factor of ten from left to right. Note that the width of the left peak is dramatically narrower than the rightmost. In the upper panel, the steered responses resulting from three-point amplitude constraints are shown. The constraints are placed in the assumed propagation direction and symmetrically on each side. The spacing between the constraint directions is expressed in units of digital frequency, in which a range of unity corresponds to sweeping incidence angle by 180°. First- and second-order derivative constraints are used in the lower panel on the same data.

algorithm's steered response to be zero in the assumed propagation direction [6, 52, 145]. With this kind of constraint, leading terms in the response's Taylor series taken about each assumed propagation direction can be set to zero, thereby rendering the algorithm insensitive to signal perturbations. Assuming that the ideal, unit-amplitude signal vector $\mathbf{e}$ depends on azimuth θ and elevation ϕ, its formal Taylor series taken about an assumed propagation direction (θ_0, ϕ_0) is

$$\mathbf{e}(\theta, \phi) = \sum_{n=0}^{\infty} \frac{1}{n!} \left(\delta\theta \frac{\partial}{\partial\theta} + \delta\phi \frac{\partial}{\partial\phi} \right)^n \mathbf{e}(\theta_0, \phi_0)$$

We zero the linear term by enforcing the pair of linear constraints

$$\frac{\partial \mathbf{e}'}{\partial \theta}\mathbf{w} = 0 \qquad \frac{\partial \mathbf{e}'}{\partial \phi}\mathbf{w} = 0$$

We similarly cancel the second-order term by forcing the second derivatives and the mixed partial to be zero. Constraints higher than second order are seldom used in practice. Some array geometries such as the linear array may not allow resolution in azimuth or elevation. In such cases, only the pertinent derivatives need be constrained.

Example

For simplicity, we consider the second-order derivative constraint for a linear array. The constrained optimization problem that yields the adaptive derivative-constrained beamformer is

$$\min_{\mathbf{w}} \mathbf{w}'\mathbf{R}\mathbf{w} \text{ subject to } \mathbf{e}'\mathbf{w} = 1;\ \frac{\partial \mathbf{e}'}{\partial \phi}\mathbf{w} = 0;\ \frac{\partial^2 \mathbf{e}'}{\partial \phi^2}\mathbf{w} = 0$$

With this formulation, we (again) have a special case of the constrained optimization problem described in §7.2 {355}. We find that the constraint matrix $\mathbf{C}$ and the vector $\mathbf{c}$ of constraining values are

$$\mathbf{C} = \begin{bmatrix} \mathbf{e}' \\ \dfrac{\partial \mathbf{e}'}{\partial \phi} \\ \dfrac{\partial^2 \mathbf{e}'}{\partial \phi^2} \end{bmatrix} \qquad \mathbf{c} = \begin{bmatrix} 1 \\ 0 \\ 0 \end{bmatrix}$$

The upper-left entry of the matrix $\left(\mathbf{C}\mathbf{R}^{-1}\mathbf{C}'\right)^{-1}$ expresses the array's output power $\mathbf{w}'\mathbf{R}\mathbf{w}$ in the propagation direction specified by $\mathbf{e}$. This entry can be selected by using the constraint vector:

$$\mathbf{w}'\mathbf{R}\mathbf{w} = \mathbf{c}'\left(\mathbf{C}\mathbf{R}^{-1}\mathbf{C}'\right)^{-1}\mathbf{c}$$

The results of employing derivative constraints are illustrated in Fig. 7.20.

The greater the number of derivatives constrained, the wider the signal-related peaks and the smaller the array gain becomes. Derivative constraints cannot yield multimodal signal peaks as can too widely placed amplitude constraints. Properly placed amplitude constraints can yield finer control of asymptotic (large signal-to-noise ratio) peak widths, however. Using either approach, we can scan assumed propagation directions with some assurance that we won't miss signals.

Although amplitude and derivative constraints can be placed with respect to physical propagation directions, from a signal processing viewpoint a more convenient choice would be to place constraints on the intersensor delays: The calculations become simpler and take into account the inherent variability of angular resolution with propagation direction. Consider the regular linear array. The components of the ideal signal vector can be expressed in either of the forms

$$e_m(\phi) = e^{+j\pi \frac{d}{\lambda/2}\sin\phi\cdot(m-M_{1/2})}$$

$$e_m(\Delta) = e^{+j\Delta\cdot(m-M_{1/2})}, \quad \Delta = \pi\frac{d}{\lambda/2}\sin\phi$$

In the former expression, the width of a single signal's response varies with the incidence angle ϕ. This variation of the response's mainlobe is an inherent property of the regular linear array. To effect controlled broadening via constraints, the *relative angular* placement of amplitude constraints would need to vary with the assumed propagation direction. In contrast, expressing the ideal signal vector in terms of delay results in a mainlobe width varying with propagation direction as usual but with amplitude constraints placed at *fixed* incremental delays. If derivative constraints are employed, the nonlinear expression of angle in the exponent of the signal vector means that the vector's derivatives have a multiplicative term dependent on the assumed propagation direction. Again, using delays results in simpler expressions from both mathematical and computational viewpoints. The amplitude constraints exemplified in Fig. 7.20 were spaced according to delay.

This simplicity is illustrated when multiple constraints are used in the context of the generalized sidelobe canceller structure discussed in §7.2.4 {369} [23]. The blocking matrix $\mathbf{B}$ is directly related to the constraints placed on the weight vector. As shown there, the null space of this matrix must contain the vector $\mathbf{C}'\left(\mathbf{C}\mathbf{R}^{-1}\mathbf{C}'\right)^{-1}\mathbf{c}$. One way of satisfying this requirement has the rows of $\mathbf{B}$ orthogonal to the columns of $\mathbf{C}'$.† In this way, $\mathbf{B}\mathbf{C}' = \mathbf{0}$, thereby satisfying the null-space requirement. In amplitude-constrained robust beamforming, the columns of $\mathbf{C}'$ are ideal signal vectors centered about the assumed propagation direction. By first removing the delay per sensor as shown in Fig. 7.9 {370}, we can place constraints symmetrically about the propagation direction. In the two-point constraint, the columns of the constraint matrix consist of $\mathrm{col}[e_0(-\delta\Delta), \ldots, e_{M-1}(-\delta\Delta)]$ and $\mathrm{col}[e_0(+\delta\Delta), \ldots, e_{M-1}(+\delta\Delta)]$. To be orthogonal to two linearly independent vectors, each row must contain at least three nonzero elements. To simplify finding the rows of the blocking matrix, require that each be a shifted version of the other. Simple calculations yield an $(M-2) \times M$ blocking matrix of the form

$$\mathbf{B} = \begin{bmatrix} 1 & -2\cos\delta\Delta & 1 & 0 & \cdots & 0 \\ 0 & 1 & -2\cos\delta\Delta & 1 & 0 & \cdots \\ \vdots & \ddots & \ddots & \ddots & & \\ 0 & \cdots & 0 & 1 & -2\cos\delta\Delta & 1 \end{bmatrix}$$

Similar methods can be used on derivative-constrained robust beamformers.

7.4.3 Robust Constrained Optimization

As described in the previous section, adaptive beamforming algorithms based on constrained optimization techniques can be made more robust by including additional constraints that make the algorithm less "picky." This *indirect* approach is based on the intuitive idea of reducing the algorithm's spatial resolution so that imprecisely modeled signals won't be suppressed so readily. Robust adaptive algorithms can be derived *directly* by solving a variety of constrained optimization problems that *explicitly* acknowledge the

---

†Recall that the number of rows in the blocking matrix equals the number of sensors *minus* the number of linear constraints.

existence of unpredictable errors in the sensors' outputs [40]. The most accessible example of an optimization problem that yields a robust adaptive array signal processing algorithm is

$$\boxed{\min_{\mathbf{w}} \mathbf{w}'\mathbf{R}\mathbf{w} \quad \text{subject to } (\mathbf{e}+\boldsymbol{\delta})'\mathbf{w} = 1 \text{ and } \|\boldsymbol{\delta}\|^2 \le \epsilon^2}$$

The vector $\boldsymbol{\delta}$ describes unforetold errors in the ideal signal vector. The constraints insist that whatever the "true" ideal signal may be, it must be passed by the weight vector with unity gain. The second constraint notes that the root-mean-squared error is bounded by ϵ. The latter constraint is the only one placed on the error vector; we assume no further information about it.

A constrained optimization problem similar to this is solved in §C.2.2 {505}. Both $\mathbf{w}$ and $\boldsymbol{\delta}$ are considered as unknowns. Forming the Lagrangian and computing its gradient with respect to the conjugate of each vector yields the pair of equations

$$\mathbf{R}\mathbf{w}_\diamond + \lambda_1(\mathbf{e}+\boldsymbol{\delta}) = 0$$
$$\lambda_1\mathbf{w}_\diamond + \lambda_2\boldsymbol{\delta} = 0$$

where λ_1 and λ_2 represent the Lagrange multipliers for the two constraints. As discussed in the appendix, the inequality constraint and its Lagrange multiplier λ_2 have special properties: Either λ_2 is zero or $\|\boldsymbol{\delta}\|^2 = \epsilon^2$ with $\lambda_2 \neq 0$. For λ_2 to be 0, the second equation given previously suggests that the weight vector would necessarily be 0. Because this solution lacks interest and applicability, the latter property must hold, which requires $\mathbf{w}$ and $\boldsymbol{\delta}$ to be proportional quantities.

$$\boldsymbol{\delta} = -\frac{\lambda_1}{\lambda_2}\mathbf{w}_\diamond \Longrightarrow \|\mathbf{w}_\diamond\|^2 = \left(\frac{\lambda_2}{\lambda_1}\right)^2 \epsilon^2$$

The inequality constraint on the error vector's length has thus indirectly placed a constraint on the weight vector's length. The (incomplete) solution for the weight vector in the robust adaptive beamformer is, assuming the inverse exists,

$$\boxed{\mathbf{w}_\diamond = -\lambda_1 \left(\mathbf{R} - \frac{\lambda_1^2}{\lambda_2}\mathbf{I}\right)^{-1} \mathbf{e}}$$

Unfortunately, explicitly computing the Lagrange multipliers is difficult. Multiple solutions are possible; each must be tested in the expression for the weight vector to determine which set minimizes $\mathbf{w}'\mathbf{R}\mathbf{w}$. Alternative to this analytic approach, we can determine the optimal weight vector with dynamic adaptive algorithms; see §7.5.4 {409}.

Many constrained optimization problems have solutions similar to that just given. Interesting in their own right, these problems allow simpler computations—analytically or iteratively—for the unknown parameters. One of these problem statements is not overtly robust: It does not explicitly express uncertainty in the ideal signal.

$$\min_{\mathbf{w}} \mathbf{w}'\mathbf{R}\mathbf{w} \quad \text{subject to } \mathbf{e}'\mathbf{w} = 1 \text{ and } \|\mathbf{w}\|^2 \le \beta^2$$

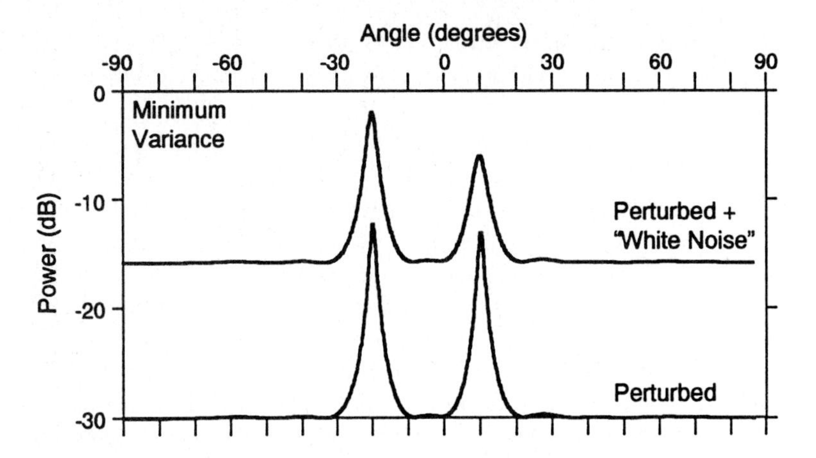

Figure 7.21 The minimum variance steered response produced by a perturbed linear array (same data as used in Fig. 7.19 {395}) is shown. The correlation matrix formed from these data is regularized by adding a multiple of the identity matrix, a procedure equivalent to adding spatially white noise to the data. The diagonal values of this added matrix equaled the power of the smaller signal at each sensor. Note how the signal power estimates improve with regularization, but that the background noise level increases. If the signal amplitudes were perfectly restored, they would have powers of 0 dB and −6 dB.

By placing a limit on the weight vector's norm, however, a robustlike solution emerges [103]. Forming the Lagrangian in the usual way, we find the gradient with respect to the conjugate of the weight vector to be

$$\mathbf{R}\mathbf{w}_\diamond + \lambda_1 \mathbf{e} + \lambda_2 \mathbf{w}_\diamond = \mathbf{0}$$

where λ_2 is the Lagrange multiplier corresponding to the inequality constraint. As just described, this multiplier can be zero or nonzero, the latter instance requiring that the constraint be met with equality. If $\lambda_2 = 0$, the usual minimum variance solution results. If nonzero, the solution is

$$\mathbf{w}_\diamond = -\lambda_1 \left(\mathbf{R} + \lambda_2 \mathbf{I}\right)^{-1} \mathbf{e}$$

The norm of the weight vector is thus held constant by adding an appropriately scaled identity matrix to the spatial correlation matrix.

While noting that these multipliers do not have the same values as in the previous problem, the suggestive similarity of the solution's *form* cannot be ignored: In both cases, the solution amounts to adding a scaled identity matrix to the spatial correlation matrix. The "size" of this modification depends on the Lagrange multipliers, whose values almost surely change with assumed propagation direction. We can develop a suboptimal solution akin to the optimal, however, by adding a fixed identity matrix to the estimated spatial correlation matrix. Fig. 7.21 displays the result of applying this simple approach. Although the signals are less distinct, previously erroneous signal power estimates are much more accurate. The exact solution to the robust constrained optimization problem (if it could be computed) would be expected to yield superior performance. Essentially, these various solutions suggest that adding spatially white noise† to the observations desensitizes signal processing algorithms to signal modeling misconceptions. We artificially add this noise through the identity matrix rather than physically by adding white-noise sources to the sensors' outputs. This rather counterintuitive result was implicitly present in the previous discussion {394}. Adding an identity matrix so as to desensitize an algorithm to errors is known as *regularization* in numerical linear algebra [97] or *diagonal loading* in array processing [32].‡

The sensitivity to sensor-gain modeling errors increases with signal-to-noise ratio; our current results can be viewed as reducing this sensitivity by taking the expedient of reducing (artificially) the signal-to-noise ratio by adding white noise. This view allows us to specify the artificial white-noise variance. Inspecting Eq. 7.9 {394}, "large" signal-to-noise ratios of modeling-sensitive algorithms occur whenever

$$\frac{A^2}{\sigma_n^2}\sum_m |\delta g_m|^2 \gg 1$$

Decomposing the noise variance into artificial and physical components, $\sigma_n^2 = \sigma_A^2 + \sigma_P^2$, we want to prevent large signal-to-noise ratios while maintaining signal integrity ($A^2 > \sigma_n^2/M$). These requirements amount to satisfying the inequalities

$$A^2\sum_m |\delta g_m|^2 < \sigma_A^2 + \sigma_P^2 < MA^2$$

The robust constrained optimization problem sets the squared norm of $\boldsymbol{\delta}$, which equals $\sum_m |\delta g_m|^2$, to ϵ^2. In this way, we can relate the artificial noise contribution to problem parameters: The signal-to-noise-ratio of each sensor's ideal output should lie in the range

$$\boxed{\frac{1}{M} < \frac{A^2}{\sigma_A^2 + \sigma_P^2} < \frac{1}{\epsilon^2}}$$

to yield an adaptive algorithm insensitive to unknown sensor position or gain perturbations while maintaining sensitivity to signals that are present. If more than one signal is present, we could vary the amount of artificial noise for each signal's assumed propagation direction.

Example

To illustrate how to select the size of identity matrix added to the spatial correlation matrix to achieve robustness, we must choose a value for the amount of uncertainty ϵ^2. This quantity represents the strength of perturbations on the ideal signal model vector $\mathbf{e}$, which has norm-squared M. Thus, ϵ^2 must be some fraction of M: The larger the fraction, the

†Temporal white noise is added if the system is wideband.

‡Regularization is also used in *high* signal-to-noise ratio conditions even when no sensor errors are present. Without regularization under these conditions, the ratio of the largest and smallest eigenvalues of the spatial correlation matrix becomes quite large, indicating numerical problems in the matrix inversion process. Regularizing the matrix increases the smallest eigenvalue, thereby reducing numerical difficulties.

greater the expected modeling errors. Choosing this fraction to be 1/10 in this example, we find that

$$1 < \frac{MA^2}{\sigma_A^2 + \sigma_P^2} < 10$$

Solving for σ_A^2, the scaling of the desensistizing identity matrix, we have

$$MA^2 - \sigma_P^2 > \sigma_A^2 > \frac{1}{10}MA^2 - \sigma_P^2$$

The upper bound keeps us from adding too much white noise, which would hide the signal, whereas the lower bound defines how much noise we need to desensitize the algorithm. The ambient noise background level σ_P^2 can be estimated from the steered response. Once this value is known, we can proceed to the signal amplitude. Threshold signals occur at 0 dB signal-to-noise ratio at a sensor: $A^2 = \sigma_P^2$. For these signals,

$$(M-1)\sigma_P^2 > \sigma_A^2 > \frac{M-10}{10}\sigma_P^2$$

At the other extreme, $MA^2 >> \sigma_P^2$ defines strong signals, which yields the range

$$MA^2 > \sigma_A^2 > \frac{1}{10}MA^2$$

Thus, the value must be adjusted to signal strength in large signal-to-noise situations and to background noise levels otherwise. Usually, small values are chosen first, then increased to enhance the stronger signals. In either case, our 10% error criterion allows a decade of values for σ_A^2 to satisfy robustness and signal processing constraints.

7.5 Dynamic Adaptive Methods

The methods described in the previous sections of this chapter require *direct* estimation of the spatial correlation matrix, with either the matrix's inverse or its eigenvalues and eigenvectors needed to determine the steered response and the shading. Hence, they are referred to as *block methods*: Only after a block of data is obtained and the correlation matrix estimate computed from this block, using Bartlett's procedure for example, can the array weights be computed. A different approach is *dynamic adaptation*: Array weights are recomputed as often as the sections used in Bartlett's procedure become available.

7.5.1 LMS Algorithm

The most well-known algorithm for updating the coefficients of a *FIR* filter on a "sample-by-sample" basis is the *LMS* algorithm {311}. With the notation translated to the present array processing problem, the equation for the shading vector at index $l+1$ is

$$\mathbf{w}(l+1) = \mathbf{w}(l) - \mu\, \nabla(|\epsilon(l)|^2)$$

where $\epsilon(l)$ is the estimation error given by the difference between the array output and its desired value $d(l)$: $\epsilon(l) = \mathbf{w}'(l)\mathbf{y}(l) - d(l)$. The indicated gradient arises from applying the method of steepest descent to the problem of minimizing the mean-squared estimation error with respect to the weight vector. In this method, the gradient represents the direction in which the mean-squared error increases most rapidly. When the weight vector is complex, the gradient must be computed with respect to the complex conjugate of $\mathbf{w}$ {502}. The required gradient $\nabla_{\mathbf{w}^*}(|\epsilon(l)|^2)$ equals $-\big[d(l) - \mathbf{w}'(l)\mathbf{y}(l)\big]^*\mathbf{y}(l)$; the update equation for the optimum weight vector in the *LMS* algorithm becomes

$$\boxed{\mathbf{w}(l+1) = \mathbf{w}(l) + \mu[d(l) - \mathbf{w}'(l)\mathbf{y}(l)]^*\mathbf{y}(l)}$$

Although mathematically this algorithm converges to the "desired" result, a practical method for specifying the desired signal $d(l)$ in the array output must be called into question. Cynically, if the desired output is known, why perform any array processing at all? More specifically, the desired output is a constant, implying that the sensor delays corresponding to some direction of propagation have been cancelled by the weight vector. Note that nowhere is the assumed propagation direction specified; even if a value for the desired signal $d(l)$ could be specified and the *LMS* algorithm converged, the resulting weight vector could not be related to the signal's direction of propagation. Consequently, the *LMS* algorithm cannot be used directly, but the basic structure of its update equation for the weight vector underlies virtually all dynamic adaptation algorithms. Realistic algorithms differ from this seminal *LMS* algorithm by their explicit inclusion of an assumed propagation direction and of a quantity to be minimized that does not involve the estimation error.

7.5.2 Frost's Adaptive Algorithm

The most well-known dynamic adaptation algorithm for array processing was developed by Frost to find a time-adaptive algorithm based on the *LMS* approach. His idea was based on the constrained optimization method described earlier (§7.2 {355}). In that method, we minimized the power in the array's output with respect to the weight vector subject to a set of constraints. The analytic method of solving this constrained optimization problem is the *unconstrained* optimization of the Lagrangian. Frost's idea was to use the Lagrangian instead of the mean-squared estimation error as the quantity minimized by the *LMS* algorithm.

Formally, the basic constrained optimization problem of array signal processing is

$$\min_{\mathbf{w}} \mathbf{w}'\mathbf{R}\mathbf{w} \quad \text{subject to } \mathbf{C}\mathbf{w} = \mathbf{c}$$

The $K \times M$ matrix $\mathbf{C}$ describes a set of K linearly independent constraints on the M-dimensional weight vector; $\mathbf{c}$ is the K-dimensional vector of constraining values. The Lagrangian associated with this problem equals $\mathbf{w}'\mathbf{R}\mathbf{w} + \boldsymbol{\lambda}'(\mathbf{C}\mathbf{w} - \mathbf{c}) + \boldsymbol{\lambda}^t(\mathbf{C}^*\mathbf{w}^* - \mathbf{c}^*)$. The update equation for the weight vector of the Frost algorithm has the general form

$$\mathbf{w}(l+1) = \mathbf{w}(l) - \mu\,\nabla_{\mathbf{w}^*}\big[\mathbf{w}'\mathbf{R}\mathbf{w} + \boldsymbol{\lambda}'(\mathbf{C}\mathbf{w} - \mathbf{c}) + \boldsymbol{\lambda}^t(\mathbf{C}^*\mathbf{w}^* - \mathbf{c}^*)\big]$$

We find the gradient to be $\mathbf{R}\mathbf{w} + \mathbf{C}'\boldsymbol{\lambda}$. To determine the Lagrange multiplier, require that the updated weight vector $\mathbf{w}(l+1)$ satisfy the constraint $\mathbf{C}\mathbf{w}(l+1) = \mathbf{c}$. Imposing this

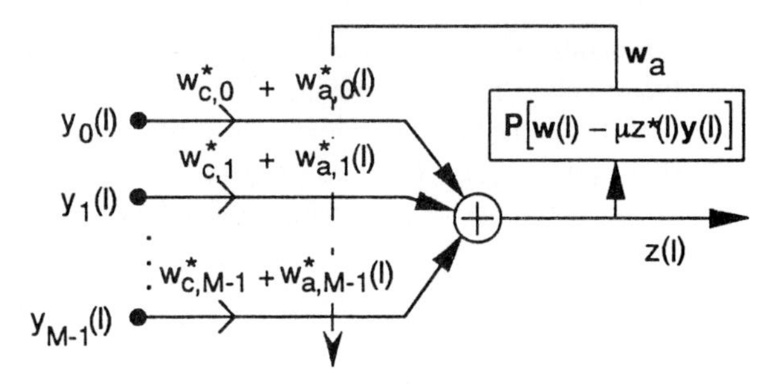

Figure 7.22 The weight vector in the Frost algorithm can be decomposed into two portions: a nominal one $\mathbf{w}_c$ and a data-dependent one $\mathbf{w}_a$. Note the similarity of this structure to that of the generalized sidelobe canceller.

constraint the update equation results in

$$\mathbf{w}(l+1) = \mathbf{w}(l) - \mu[\mathbf{I} - \mathbf{C}'(\mathbf{C}\mathbf{C}')^{-1}\mathbf{C}]\mathbf{R}\mathbf{w}(l) + \mathbf{C}'(\mathbf{C}\mathbf{C}')^{-1}[\mathbf{c} - \mathbf{C}\mathbf{w}(l)]$$

The last term in this equation, the constraint equation on the weight vector $\mathbf{w}(l)$, is 0 when the quantities involved are known exactly. Because these are approximated in the final form of the equation, leaving this term as is provides some measure of numerical stability to the algorithm. Defining $\mathbf{w}_c$ to be $\mathbf{C}'(\mathbf{C}\mathbf{C}')^{-1}\mathbf{c}$, the nonadaptive component of the weight vector, and $\mathbf{P}$ to be $\mathbf{I} - \mathbf{C}'(\mathbf{C}\mathbf{C}')^{-1}\mathbf{C}$, the projection matrix associated with the data-dependent component of the weight vector,† we have a concise expression for the update equation.

$$\mathbf{w}(l+1) = \mathbf{w}_c + \mathbf{P}[\mathbf{w}(l) - \mu\mathbf{R}\mathbf{w}(l)]$$

The unknown quantity here is the spatial correlation matrix, not the error signal as in the original *LMS* algorithm. Typical of dynamic algorithms, we replace this matrix by some estimate $\widehat{\mathbf{R}}$ of it, the simplest being the outer product of the array output vector with itself at the lth index: $\mathbf{y}(l)\mathbf{y}'(l)$. The equations summarizing Frost's adaptive beamforming algorithm are

$$\boxed{\begin{aligned} \mathbf{P} &= \mathbf{I} - \mathbf{C}'(\mathbf{C}\mathbf{C}')^{-1}\mathbf{C} \\ \mathbf{w}_c &= \mathbf{C}'(\mathbf{C}\mathbf{C}')^{-1}\mathbf{c} \\ z(l) &= \mathbf{w}'(l)\mathbf{y}(l) \\ \mathbf{w}(l+1) &= \mathbf{w}_c + \mathbf{P}[\mathbf{w}(l) - \mu z^*(l)\mathbf{y}(l)] \end{aligned}}$$

The components of these equations can be interpreted in light of the generalized sidelobe canceller's structure (Fig. 7.22). The vector $\mathbf{w}_c$, the nonadaptive portion of the constrained optimization beamformer, does not depend on the data; it can be precomputed as indicated and, more important, the adaptation should *not* let the data affect it. The remaining portion of the update equation describes the data-dependent (adaptive) component of the optimal solution *orthogonal* to the data-independent nominal solution.

We can examine the convergence of Frost's algorithm by considering the expected value of the update equation.

$$\mathcal{E}[\mathbf{w}(l+1)] = \mathbf{w}_c + \mathbf{P}\,(\mathbf{I} - \mu\mathbf{R})\,\mathcal{E}[\mathbf{w}(l)]$$

---

†The nonadaptive weight vector and the associated projection matrix are described on page 356.

Assuming that a steady-state solution exists, the expected value of the weight vector equals the solution $\mathbf{R}^{-1}\mathbf{C}'(\mathbf{C}\mathbf{R}^{-1}\mathbf{C}')^{-1}\mathbf{c}$ of the constrained optimization problem given in §7.2 {355} *regardless* of the value of μ. For a steady-state solution to exist, the gain parameter μ must be sufficiently small; consideration of the dynamics of the equation for the expected weight vector indicate that

$$0 < \mu < \frac{1}{\max \lambda_{\mathbf{PRP}}}$$

Using the criterion we developed for time series that takes the estimation of $\mathbf{R}$ into account (Eq. 6.13 {314}), the adaptation parameter must satisfy

$$0 < \mu < \frac{1}{3\,\mathrm{tr}[\mathbf{PRP}]}$$

The "time-constant" of the convergence is $1/(\mu \min \lambda_{\mathbf{PRP}})$, where $\min \lambda_{\mathbf{PRP}}$ is the smallest *nonzero* eigenvalue of **PRP**.†

Example

Let's apply the Frost dynamic adaptive algorithm to the problem defining the minimum variance block-adaptive technique. Thus, $\mathbf{C} = \mathbf{e}'$—an ideal, unit-amplitude signal vector corresponding to some assumed propagation direction—and $\mathbf{c} = 1$. The update equation for the weight vector becomes

$$\mathbf{w}(l+1) = \frac{\mathbf{e}}{M} + \left(\mathbf{I} - \frac{\mathbf{e}\mathbf{e}'}{M}\right)\left[\mathbf{w}(l) - \mu z^*(l)\mathbf{y}(l)\right]$$

where M is the number of sensors in the array.

Convergence of this equation is determined by the product **PRP**. In this example, $\mathbf{P} = \mathbf{I} - \mathbf{e}\mathbf{e}'/M$, which, as promised, has one zero-valued eigenvalue with the associated eigenvector equaling $\mathbf{e}$. The product **PRP** thus consists only of those components orthogonal to the ideal signal vector $\mathbf{e}$. Thus, this matrix's smallest nonzero eigenvalue equals the smallest eigenvalue of the noise-only correlation matrix and the smallest its trace can be is the trace of the noise-only correlation matrix. To choose μ, it must be less than $1/3M\sigma_n^2$. Values considerably less than this value may be required in practice. The power of the array output at each iteration is shown in Fig. 7.23 when two signals are present in the array's field.

Considering Fig. 7.23, we see that the exponential decay/increase characteristic of the *LMS* algorithm in time series applications is not always present. Note how the responses in the bottom row decay as expected initially, but when a signal arises in the assumed propagation direction, the power jumps instantly to the proper value. This type of behavior is not evident when we use the *LMS* algorithm on time series (see Fig. 6.5 {315}). Constraints make the difference here; because of them, a fixed weight vector $\mathbf{w}_c$ *orthogonal* to the adaptive component persists in the weight vector's update equation. Because of the fixed component, we should expect output changes as rapid as those of the input plus exponentially changing ones because of the adaptive component.

†The product **PRP** has K identically zero eigenvalues where K is the number of constraints (the dimension of $\mathbf{c}$). This property can be traced to constraint-matrix properties and to $\mathbf{P}$ being a projection matrix.

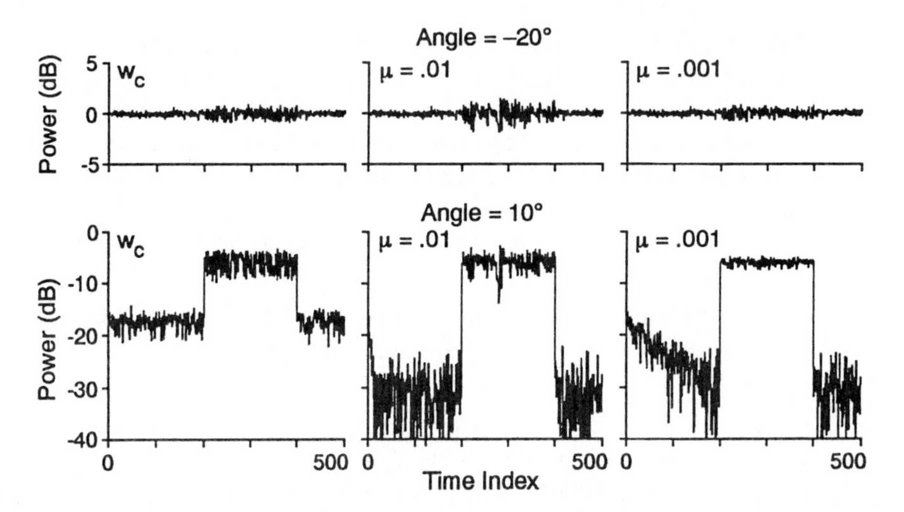

Figure 7.23 The regular linear array consists of ten sensors. The observations contain two signals, one propagating from -20° and the other from 10°, and additive noise. The amplitude of the second is one-half that of the first. The signal characteristics are the same as those used in constructing Fig. 7.4 {361}; the noise has one-tenth the power here. The second signal is present from observations 200 through 400 while the first was present throughout. The power estimate $|z(l)|^2$ on a decibel scale (reference value of unity) is plotted during the observation interval. The top row shows the output power when the assumed propagation direction corresponds to the first signal, the bottom row when the beam is steered toward the second. The plot shown in the left column corresponds to a nonadaptive beamformer having a weight vector equaling $\mathbf{e}/M$, the data-independent component of the weight vector given in the example. The center and right columns demonstrate the dynamic adaptive beamformer's outputs for two values of the adaptation parameter μ.

To demonstrate these effects, the output power in the nonadapted response is shown in the left column of Fig. 7.23. The beampattern corresponding to this weight vector is the conventional one shown in the center panel of Fig. 7.4 {361}. From this beampattern, we see that the constant weight vector reduces the signal that's not propagating in the assumed direction by about 17 dB. The beampattern corresponding to the entire weight vector is also shown there; the adaptive beamformer reduces this signal another 11 dB. The Frost algorithm adapts the weight vector so that it passes from the first spatial response to the second, thereby reducing the other signal's effect. The output power plotted in the remaining columns of Fig. 7.23 demonstrate little change as the other signal is greatly reduced by the fixed component; the expected exponential change in the output power *is* there, but is masked by noise because the fixed component reduces the second signal so much. The adaptation is more visible in the initial portions of the center and right panels in the bottom row. The larger the value of μ, the faster the adaptation.

The concomitant variability of the output power, however, does *not* follow the usual *LMS* algorithm tradeoff (decreasing μ yields slower adaptation rates and smaller variabilities). Note the initial portions of the center and right entries in the top row of Fig. 7.23; despite a factor-of-ten difference in the adaptation parameter, the variability is virtually identical, differing by less than 1% during the first 200 samples. During the time the second signal is present, the outputs obviously differ. Again, the constraints underlie this behavior. The adaptive component changes little during the initial segment and has greater effect while second signal is present; the variability related to the constant weight vector is, of course, fixed. Generally speaking, less variability requires a smaller value of the adaptation parameter μ.

In summary, you cannot expect the predicted effects of a smaller μ—slower adaptation rates and less variability—to always be as dramatic as expected. If there are changes in signals having directions of propagation equal to the assumed propagation direction, the weight adapts rapidly. The constraint is instantaneous, applying at each iteration of the update equation; any change in the signal gain in the assumed propagation direction can be quickly accommodated. In contrast, signals arising from other directions demonstrate the *LMS* tradeoff. Their suppression is an indirect consequence of the constraint and, as shown in the examples, adaptation to their changes occurs with the predicted time constant. Furthermore, as $\min \lambda_{\mathbf{PRP}}$ is related to the noise variance and $\max \lambda_{\mathbf{PRP}}$ to the total signal power, the adaptation rate depends roughly on the signal-to-noise ratio: The smaller it is, the slower the adaptation.

7.5.3 Matrix Update Algorithms

In contrast to Frost's algorithm, which uses the method of steepest descent to minimize the constrained quadratic form $\mathbf{w}'\mathbf{R}\mathbf{w}$, *update* methods directly calculate the minimum power solution $\mathbf{c}'(\mathbf{C}\mathbf{R}^{-1}\mathbf{C}')^{-1}\mathbf{c}$ of constrained optimization methods by modifying the current solution to incorporate the latest observations. As we discussed in §6.4.4 {314}, the *RLS* algorithm bypasses numerical minimization by the method of steepest descent and recursively updates the inverse of the correlation matrix. Not only does this approach speed convergence (at the expense of more computations), but it allows us to apply eigenanalysis methods. The following describes the dynamic update methods for adaptive methods; a dynamic update method for the eigensystem of a correlation matrix is described in DeGroat and Roberts.

Given the lth observation vector $\mathbf{y}(l)$, we (conceptually) update the spatial correlation matrix estimate $\widehat{\mathbf{R}}(l-1)$ by the difference equation

$$\widehat{\mathbf{R}}(l) = \alpha\widehat{\mathbf{R}}(l-1) + \mathbf{y}(l)\mathbf{y}'(l), \quad 0 < \alpha < 1$$

From this update equation, we can obtain a recursive equation for the inverse correlation matrix so necessary for adaptive methods based on constrained optimization. The *RLS* approach to updating the inverse matrix of $\widehat{\mathbf{R}}(l)$ results from applying the matrix inverse formula {489} and Eq. 6.14 {316}:

$$\widehat{\mathbf{R}}^{-1}(l) = \frac{1}{\alpha}\left[\widehat{\mathbf{R}}^{-1}(l-1) - \frac{\mathbf{x}(l)\mathbf{x}'(l)}{\alpha + \beta(l)}\right]$$

where $\mathbf{x}(l) = \widehat{\mathbf{R}}^{-1}(l-1)\mathbf{y}(l)$ and $\beta(l) = \mathbf{y}'(l)\widehat{\mathbf{R}}^{-1}(l-1)\mathbf{y}(l)$. To compute the array output power, we need to compute $\mathbf{C}\mathbf{R}^{-1}\mathbf{C}'$; using the iterative equation for the inverse, we find that

$$\mathbf{C}\mathbf{R}^{-1}(l)\mathbf{C}' = \frac{1}{\alpha}\left[\mathbf{C}\mathbf{R}^{-1}(l-1)\mathbf{C}' - \frac{\mathbf{C}\mathbf{x}(l)\mathbf{x}'(l)\mathbf{C}'}{\alpha+\beta(l)}\right]$$

Unfortunately, two problems arise. First of all, no matter what algorithm we may be considering, we need to compute $\mathbf{x}(l)$, which contains the spatial correlation matrix's inverse explicitly. Thus, no simple update equation for the power estimate of *any* constrained optimization algorithm can be found this way; we would be forced to carry along the iterative equation for $\widehat{\mathbf{R}}^{-1}$. Second, Schreiber points out that the iterative equation for the inverse matrix is numerically unstable. This instability is traced to the weighting by $1/\alpha$: This quantity exceeds unity and magnifies numeric errors.† Apparently, our first attempt at a more direct algorithm has encountered problems.

Rather than basing algorithms on update equations for the correlation matrix, we have better success considering its *Cholesky factors* [137]. As shown in the appendix {492}, every positive definite Hermitian matrix can be expressed as $\mathbf{L}\mathbf{L}'$ where $\mathbf{L}$ is a lower triangular matrix.‡ The recursive equation for the correlation matrix $\widehat{\mathbf{R}}$ now becomes

$$\mathbf{L}(l)\mathbf{L}'(l) = \alpha\mathbf{L}(l-1)\mathbf{L}'(l-1) + \mathbf{y}(l)\mathbf{y}'(l)$$

To find $\mathbf{L}(l)$, consider the temporary $M{\times}M{+}1$ matrix $\mathbf{T}(l) = [\mathbf{y}(l) \quad \alpha^{1/2}\mathbf{L}(l-1)]$, which allows us to express the update equation concisely as $\mathbf{L}(l)\mathbf{L}'(l) = \mathbf{T}(l)\mathbf{T}'(l)$. A unitary matrix $\mathbf{Q}$ exists that can zero the first column of $\mathbf{T}(l)$: $\mathbf{T}(l)\mathbf{Q} = [\mathbf{0} \quad \mathbf{L}(l)]$. The matrix $\mathbf{Q}$ is formed from a sequence of M Givens rotations {490} that leaves the remaining columns to form a lower triangular matrix.¶ The surprising emergence of $\mathbf{L}(l)$ results from the fact that $\mathbf{T}(l)\mathbf{Q}\,[\mathbf{T}(l)\mathbf{Q}]' = \mathbf{L}(l)\mathbf{L}'(l)$.

Armed with Cholesky factor updates, we can find the array output power produced by any constrained optimization method. The product $\mathbf{C}\widehat{\mathbf{R}}^{-1}(l)\mathbf{C}'$ now becomes $\mathbf{C}\mathbf{L}'^{-1}(l)\mathbf{L}^{-1}(l)\mathbf{C}'$. Rather than computing $\mathbf{L}$'s inverse and multiplying it times the constraint matrix, a better approach in some cases is to solve the matrix equation $\mathbf{L}(l)\mathbf{A}(l) = \mathbf{C}'$ and express the constrained algorithm's output power as $\mathbf{c}'(\mathbf{A}'\mathbf{A})^{-1}\mathbf{c}$. As detailed in the following example, we can easily solve this matrix equation because of $\mathbf{L}$'s lower triangular structure. In summary, an algorithm for dynamically updating the output power of algorithms based on constrained optimization has the steps

$$\mathbf{T}(l) = \left[\mathbf{y}(l) \quad \alpha^{1/2}\mathbf{L}(l-1)\right]$$

$$\left[\mathbf{0} \quad \mathbf{L}(l)\right] = \mathbf{T}(l)\mathbf{Q} \quad (M \text{ Given's rotations})$$

$$\text{Solve} \quad \mathbf{L}(l)\mathbf{A}(l) = \mathbf{C}'$$

$$\mathbf{c}'[\mathbf{C}\widehat{\mathbf{R}}^{-1}(l)\mathbf{C}']^{-1}\mathbf{c} = \mathbf{c}'[\mathbf{A}'(l)\mathbf{A}(l)]^{-1}\mathbf{c}$$

---

†The way in which Frost imposed the optimization problem's constraint mitigates numerical errors.

‡The appendix writes the Cholesky factorization as $\mathbf{L}\mathbf{D}\mathbf{L}'$ where $\mathbf{D}$ is diagonal. Our expression for the factorization merely merges $\mathbf{D}^{1/2}$ with the lower triangular matrix, which forces its diagonal elements L_{ii} to equal $D_{ii}^{1/2}$. In short, nothing fundamental is changed by ignoring the diagonal matrix.

¶Each Given's rotation zeros a first-column element by rotating with respect to that element and the diagonal element of $\mathbf{L}$ in the same row. The unitary matrix multiplication performing this chore can thus zero the first column while retaining the lower triangular form of the remaining columns.

Note that because $\mathbf{C}$ need not be square, neither is $\mathbf{A}(l)$.

Example

Consider the minimum variance algorithm where $\mathbf{C} = \mathbf{e}'$ and $\mathbf{c} = 1$. Thus, the sought-after steered response becomes $\left(\mathbf{e}'\widehat{\mathbf{R}}^{-1}\mathbf{e}\right)^{-1}$. Updating the Cholesky factor $\mathbf{L}(l)$ does not depend on the array processing algorithm. The details of applying Givens rotations are described in §B.3 {490}. We must be careful to apply the rotations in the correct order so as to produce a column of zeros. The first rotation matrix $\mathbf{Q}_1$ zeros the first element of the column by rotating with respect to the first two columns. The next rotation matrix $\mathbf{Q}_2$ zeros the second element of the first transformation's first column by rotating with respect to the first and third columns. Because of the lower triangular form of the second through Mth columns of $\mathbf{T}(l)$, successive rotations do not affect previously zeroed elements. The transformation matrix $\mathbf{Q}$ that zeros the first column equals the product $\mathbf{Q}_1\mathbf{Q}_2\cdots$.

Assuming we have the updated Cholesky factor $\mathbf{L}(l)$, we need to solve the matrix equation $\mathbf{L}(l)\mathbf{A}(l) = \mathbf{e}$. The matrix $\mathbf{A}(l)$ is thus a column matrix. How to solve this equation depends on the matrix dimensions and the number of assumed propagation directions. The dimension of the Cholesky factor always equals the number of array elements M. If the number of assumed propagation directions is comparable with or smaller than M, we should not compute $\mathbf{L}^{-1}(l)$ directly; by expressing the matrix equation in detail, we find that solving for $\mathbf{A}(l)$ requires fewer computations.

$$\begin{bmatrix} L_{11} & 0 & 0 & \cdots & \\ L_{12} & L_{22} & 0 & \cdots & \\ \vdots & \vdots & \ddots & & \\ L_{1M} & L_{2M} & & L_{MM} \end{bmatrix} \begin{bmatrix} A_1 \\ A_2 \\ \vdots \\ A_M \end{bmatrix} = \begin{bmatrix} e_1 \\ e_2 \\ \vdots \\ e_M \end{bmatrix}$$

Clearly, $A_1 = e_1/L_{11}$. The second equation has the form $L_{12}A_1 + L_{22}A_2 = e_2$; because A_1 is known, we can solve for A_2, and so on. The typical case has M much smaller than the number of assumed propagation directions, making direct evaluation the inverse matrix computationally more attractive. The lower triangular form makes computing $\mathbf{L}^{-1}$ simple: The inverse also has a lower triangular form with the diagonal equaling the reciprocal of $\mathbf{L}$'s diagonal elements and the subdiagonal entries found by solving simple linear equations. For each assumed propagation direction we thus solve for $\mathbf{A}(l)$ directly or compute $\mathbf{A}(l) = \mathbf{L}^{-1}(l)\mathbf{C}'$ and update the array output power by summing the squares of $\mathbf{A}$'s components.

7.5.4 Dynamic Generalized Sidelobe Cancellers

By merging the ideas contained in the generalized sidelobe canceller and the *LMS* algorithm, many interesting dynamic adaptive algorithms can be derived. The dynamic adaptive structure is shown in Fig. 7.24. We begin by choosing a weight vector $\mathbf{w}_c$ for the nonadaptive section of the beamformer to yield a nominal beampattern satisfying our concept of the ideal tradeoff between mainlobe width and sidelobe height and structure. The blocking matrix $\mathbf{B}$ must reject signals propagating in the assumed propagation direction. It can also incorporate additional spatial constraints, both amplitude and derivative, that yield robust beamforming algorithms as discussed in §7.4 {393}. We dynamically update the weight vector $\mathbf{w}_a$ for the adaptive section so that any signals common to the two sections' outputs are cancelled. This condition allows *direct* use of the *LMS* algorithm: The desired signal is zero in the generalized sidelobe canceller structure. The

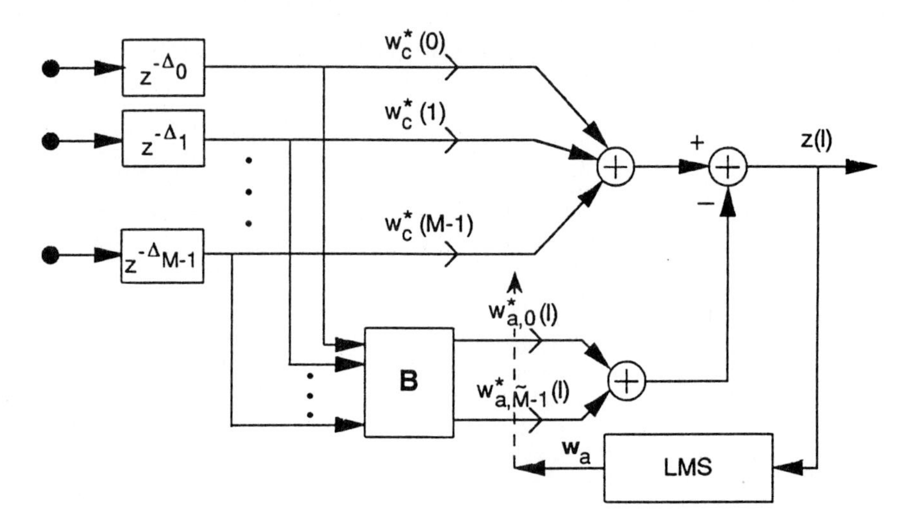

Figure 7.24 Many dynamically adapted beamformers can be structured with a nominal, non-adaptive section and an adaptive section in which the weights are based on an unconstrained optimization problem. The blocking matrix can be used to enforce a variety of constraints, robust ones especially.

equations describing the dynamic generalized sidelobe canceller are

$$\boxed{\begin{aligned} \mathbf{y}_B(l) &= \mathbf{B}\mathbf{y}(l) \\ z_c(l) &= \mathbf{w}'_c\mathbf{y}(l) \\ z_a(l) &= \mathbf{w}'_a(l)\mathbf{y}_B(l) \\ z(l) &= z_c(l) - z_a(l) \\ \mathbf{w}_a(l+1) &= \mathbf{w}_a(l) + \mu z^*(l)\mathbf{y}_B(l) \end{aligned}}$$

The adaptation parameter μ must be set according to the eigenstructure of the matrix $\mathbf{BRB}'$.

Robust beamformers based on adding artificial white noise can also be computed using the dynamic generalized sidelobe canceller [77]. The optimization problem describing the generalized sidelobe canceller {369} can be recast for robust adaptive beamformers as

$$\min_{\mathbf{w}_a} \left(\mathbf{w}_c - \mathbf{B}'\mathbf{w}_a\right)' \left(\mathbf{R} + \sigma_A^2\mathbf{I}\right) \left(\mathbf{w}_c - \mathbf{B}'\mathbf{w}_a\right)$$

By computing the gradient with respect to the conjugate of the adapted weight vector, the robust formulation for the generalized sidelobe canceller requires modification of only the dynamic equation for the adapted weight vector $\mathbf{w}_a$.

$$\mathbf{w}_a(l+1) = \left(\mathbf{I} - \sigma_A^2\mathbf{BB}'\right)\mathbf{w}_a(l) + \mu z^*(l)\mathbf{y}_B(l)$$

We can determine the value of the artificial white noise's variance σ_A^2 according to the signal's amplitude and the expected mean-squared perturbations. As these are usually not known precisely, σ_A^2 can only be set approximately.

An alternative technique [40, 103], which does *not* require σ_A^2 to be specified explicitly, is based on the second robust constrained optimization problem described in §7.4.3 {399}. There, a constraint is placed on the norm of overall system weight vector $\mathbf{w}$, here given by $\mathbf{w}_c - \mathbf{B}'\mathbf{w}_a$. As the weight vector $\mathbf{w}_c$ is not adapted, the norm constraint applies only to the adapted weight vector. Because these component terms are orthogonal, the norm constraint $\|\mathbf{w}\|^2 \leq \beta^2$ becomes

$$\|\mathbf{B}'\mathbf{w}_a\|^2 \leq \beta^2 - \|\mathbf{w}_c\|^2$$

Once we choose the fixed weight vector $\mathbf{w}_c$, the constraining value β^2 placed on the adapted weight vector can be computed. We can also derive an update equation for $\widetilde{\mathbf{w}}_a = \mathbf{B}'\mathbf{w}_a$ by multiplying the last equation for the generalized sidelobe canceller by $\mathbf{B}'$.

$$\widetilde{\mathbf{w}}_a(l+1) = \widetilde{\mathbf{w}}_a(l) + \mu z^*(l)\mathbf{B}'\mathbf{y}_B(l)$$

To become a robust algorithm, we force this weight vector to satisfy its norm constraint at each update. The robust formulation of the generalized sidelobe canceller, which is based on a norm constraint placed on the weight vector, is governed by the equations

$$\boxed{\begin{aligned}
\mathbf{y}_B(l) &= \mathbf{B}\mathbf{y}(l)\\
\mathbf{v}(l) &= \widetilde{\mathbf{w}}_a(l) + \mu z^*(l)\mathbf{B}'\mathbf{y}_B(l)\\
\widetilde{\mathbf{w}}_a(l+1) &= \begin{cases} \mathbf{v}(l), & \|\mathbf{v}(l)\|^2 \leq \beta^2 - \|\mathbf{w}_c\|^2 \\ \sqrt{\beta^2 - \|\mathbf{w}_c\|^2}\dfrac{\mathbf{v}(l)}{\|\mathbf{v}(l)\|}, & \|\mathbf{v}(l)\|^2 > \beta^2 - \|\mathbf{w}_c\|^2 \end{cases}\\
z(l) &= [\mathbf{w}_c - \widetilde{\mathbf{w}}_a(l)]'\mathbf{y}(l)
\end{aligned}}$$

The motivation for this approach is relatively simple. When the assumed propagation direction does not coincide with a signal's direction of propagation, all adaptive beamformers based on constrained optimization produce weight vectors that equal the quiescent value $\mathbf{w}_c$.† When the assumed propagation direction does correspond to a direction of propagation, an additional component, the adaptive component, becomes apparent. Because this component is orthogonal to the quiescent one, the adaptive component's norm must increase. Robustness constraints limit the maximum value of the total weight vector's norm; because the fixed component cannot change, only the adaptive weight vector's norm need be limited. The equations given earlier simply scale the adaptive component without changing its direction: It must remain orthogonal to the fixed component. Conceptually, clamping the norm of the weight vector leaves the noise-only portion of the beampattern unchanged while dynamically increasing the artificial

†One consequence of this observation is that β^2 had better be larger than the norm of the quiescent weight vector. Otherwise, the algorithm would try to be robust when no signals were present.

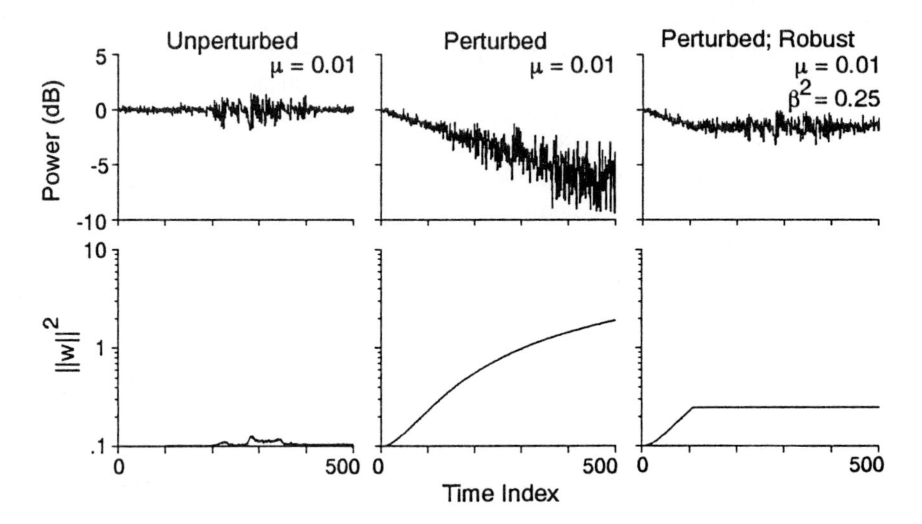

Figure 7.25 The array and signal fields are the same as in Fig. 7.23. The top row portrays the output power computed by dynamic adaptive methods when the assumed propagation direction equals the bearing (-20°) of the largest amplitude signal. In all cases, the adaptation parameter μ was equal to 0.01. The bottom row displays the squared norm of the optimal weight vector $\mathbf{w}$ computed by the adaptive algorithm. For the results shown in the left column in which the Frost algorithm is used, no sensor errors are present; these data are identical to those used in Fig. 7.23. For the center and right columns of panels, the sensor errors with a standard deviation of 0.15 are present as they are in Fig. 7.19 {395}. Output of the Frost algorithm is shown in the center column while the robust variation of this algorithm is shown on the right; β^2 equaled 0.25.

white-noise component in the spatial correlation matrix as the signal becomes larger. As required, the signal-to-noise ratio remains in the proper range for a robust beamformer.

Example

For the narrowband example portrayed in Fig. 7.23 {406}, the effects of sensor gain and phase errors are apparent in Fig. 7.25. In this example, the squared norm of the weight vector did not greatly exceed that (0.1) of the quiescent weight vector $\mathbf{w}_c$. When sensor errors are not taken into account, the signal power estimated by the Frost algorithm continually diminishes, falling to more than 5 dB below the true value. Note how the weight vector's norm steadily increases; the robust Frost algorithm squelches this increase, preventing such large losses from occurring. The value of β^2 used in this example was selected arbitrarily but mindful of the quiescent weight vector's norm (0.1).

Summary

Known algorithms that effectively deal with sensor outputs to yield direction of propagation and propagating waveform estimates are based on delay-and-sum beamformers, the basic structure of which also encompasses adaptive algorithms. By modifying their parameters to measured field characteristics rather than retaining fixed values, the adaptive algorithms described in this chapter yield signal processing performances superior to the conventional ones described in Chap. 4 [35, 69, 76, 108, 115]. For example, adaptive algorithms provide superior angular resolution, which can increase with signal-to-noise ratio or time-bandwidth product (depending on the algorithm). We have found, however, that adaptive algorithms begin with a well-designed conventional beamformer, then adapt their characteristics to this nominal framework. The generalized sidelobe canceller structure clearly indicates this underlying decomposition, which forms the basis for both block and dynamic adaptive methods. Consequently, despite the current concentration on adaptive algorithms, window design and sampling effects must be understood and incorporated into algorithm design to develop effective adaptive array processing algorithms.

Eigenanalysis algorithms depart from this picture. They rely on different ideas from constrained optimization algorithms; the problems discuss some of the more recent algorithms. Although providing superior direction of propagation algorithms, they cannot be used as for beamformers: They do not provide waveform estimates. Their superior resolution, however, makes them useful in tracking algorithms, which coalesce directional information from several arrays.

Adaptive algorithms do have a negative side: They are sensitive to modeling assumptions. Sensor position errors, uncertainties in sensor filtering characteristics, and wavefront imperfections can bring an adaptive algorithm "to its knees." Because modeling assumptions for algorithm development can never describe the physical application precisely, understanding how to incorporate uncertainty into design becomes an essential tool. As we have seen, *robust* algorithms can be developed that explicitly incorporate uncertainties into their design but at some performance sacrifice. The knowledgeable designer must trade this loss against robustness in developing effective algorithms.

Problems

7.1 Consider a regular linear array. The field is known to contain a single narrowband signal plus spatially white Gaussian noise. Sensor outputs are sampled and a length D snapshot collected. The snapshot encompasses many cycles of the sinusoid.

(a) Find the maximum likelihood estimate for the signal's amplitude. Interpret this result in terms of standard signal processing operations when the sampled noise is temporally white.

(b) Show that the maximum likelihood estimate of the signal's direction of propagation corresponds to maximizing the magnitude of the estimated amplitude. Interpret this result in terms of array processing algorithms when the sampled noise is temporally white.

(c) An alternative estimator uses the analytic approach to find the direction of propagation rather than search. How many potential solutions must be tested before the optimal one is found?

7.2 The maximum likelihood amplitude estimate given in Eq. 7.1 {351} greatly resembles the sufficient statistic of the optimal detector (Eq. 5.7 {228}).

(a) Show that the optimal detector consists of maximizing the sufficient statistic with respect to direction of propagation and comparing the result to a threshold.

(b) Assuming B beams (look directions), under what conditions are the corresponding sufficient statistics statistically independent? What is the detection threshold in this case? Assume spatially white noise.

7.3 In Prob. 6.14 {344}, we derived the Ziv-Zakai bound for a linear array. This bound can be interpreted as expressing the ultimate resolution provided by any array processing algorithm. Here, we express the improved resolution provided by adaptive beamforming methods in a very different way; rather than mean-squared estimation error, resolution expresses the ability to distinguish two closely spaced sources.

(a) How would you translate a mean-squared error result into a resolution-like result?

(b) Using your technique, compare the Ziv-Zakai bound with the resolution curves presented in Fig. 7.6 {364}. Interpret your results.

7.4 When we have a rectangular array geometry, we need to find both the elevation and azimuth angles to determine the direction of propagation. Assuming the source to be narrowband and the array to have an equal number of sensors along each side in the (x, y) plane, we seek the maximum likelihood estimate of the signal's direction of propagation.

(a) Express the normalized signal vector in terms of elevation and azimuth angles and in terms of the intersensor delays Δ_x and Δ_y. Which mode of expression do you think yields the simplest estimate?

(b) Assuming we use intersensor delays, find analytic expressions that the maximum likelihood estimates of Δ_x and Δ_y must satisfy. Note any coupling between estimation equations.

7.5 We defined the projection matrix $\mathbf{P_C}$ for algorithms derived by constrained optimization. Because this matrix and its properties play an important role in developing algorithms, we need to appreciate its properties.

(a) Show that $\mathbf{P_C} = \mathbf{C}'(\mathbf{CC}')^{-1}\mathbf{C}$ is indeed a projection matrix.

(b) Given that $\dim[\mathbf{c}] = K$, what is the dimension of $\mathbf{P_C}$? Assuming $K < M$, what is the smallest dimension that its null space can be? the largest?

(c) Using pseudoinverse theory, show that, as claimed, the matrix $\mathbf{CC}'$ is invertible if the constraint matrix has linearly independent rows.

(d) Show that the optimal weight vector's projection onto the constraints does not depend on the observations. How many constraints can be fruitfully placed on a M-sensor array?

7.6 The example described on page 356 details how to apply constrained optimization techniques to a wideband beamformer. This result needs further exploration.

(a) Show that the adaptive portion of the beamformer indeed equals 0 when the noise is spatially and temporally white and no signal is present.

(b) When the noise is not temporally white (as is often the case) but remains spatially white, find the adaptive weight vector. Do the component *FIR* filters differ in this case? Again assume no signal is present.

(c) Find the optimal filter coefficient constraint vector that maximizes the output signal-to-noise ratio when the noise is spatially, but not temporally, white.

7.7 The minimum variance method achieves optimal array gain under certain conditions. Let's explore what these might be to better appreciate the algorithm.

(a) Assume one signal is present in the observations. Using the minimum variance weight vector, show that we obtain maximum array gain when the algorithm's assumed propagation direction equals the signal's direction of propagation.

(b) Now consider the two-signal case. Assuming incoherent signals, calculate the array gain when the array is steered toward one of the signals.

(c) Mathematically, when is this array gain optimal? Provide a physical interpretation for this condition.

7.8 We can illustrate the properties of the minimum variance beamformer by considering the spatially white noise case: $\mathbf{K}_n = \sigma^2\mathbf{I}$.

(a) Show that the optimal weight vector $\mathbf{w}_\diamond$ is proportional to $\mathbf{e}$. What is the constant of proportionality, and how does it vary with signal and array parameters? Take the single signal case as your prototype.

(b) Discuss whether this result means that conventional beamforming is optimal in the spatially white noise case. Under what conditions is it optimal?

(c) Derive the expression given for the output power produced by the minimum variance steered response and show that the peak-to-sidelobe ratio equals $(MA^2/\sigma_n^2) + 1$. Derive this ratio for conventional beamforming and contrast the results.

7.9 We can easily extend the minimum variance technique to wideband situations. Define the observation vector $\mathbf{y}$ to be the concatenation of frequency domain sensor outputs at various temporal frequencies.

$$\mathbf{y} = \text{col}\big[\mathbf{Y}(\omega_1) \cdots \mathbf{Y}(\omega_K)\big]$$

(a) Determine an expression for the ideal signal vector $\mathbf{e}$. Note that signal amplitude can vary with frequency; denote the signal amplitude at frequency ω_k as A_k.

(b) If one signal is present in the field, find the steered response for this wideband problem. As usual in spectral domain solutions, assume that the noise components at different frequencies are uncorrelated. Express in words the nature of the wideband minimum variance solution, comparing it to the conventional beamforming solution.

(c) Assume that the noise is spatially white, but having a variance that depends on frequency. Find the peak-to-background ratio.

7.10 We need to investigate the details of the steered response produced by the linear prediction method (Eq. 7.5 {366}) [80]. Assume the noise to be spatially white and that the observations contain only one signal. Take the predicted sensor to be the first ($m = 0$).

(a) Find the exact expression for the steered response in the signal's direction of propagation. Simplify your result to the situation in which $MA^2 \gg \sigma_n^2$.

(b) Contrast the simplified formulas for the linear predictive and minimum variance steered responses. You should find that the linear prediction algorithm produces squared amplitude values. Compare the two algorithms on a leveled playing field.

7.11 Consider the narrowband array processing problem of determining the directions of propagation of signals propagating toward a regular linear array. We need to probe the ability of linear prediction to model accurately the measured field.

(a) Show that the frequency-domain sensor outputs resulting from a single signal propagating toward the array can be modeled by a linear predictive (*AR*) model. Find the ideal weights and how they vary with m_0, the sensor whose output is predicted by a linear combination of the others.

(b) Show that the superposition of two propagating waves is *not* described by the linear predictive model. What is a more appropriate model?

(c) Now assume that spatially white noise in addition to a single propagating signal constitute the measured field. Show that sensor outputs are also not described by a linear predictive model. Do the weights calculated by the linear predictive method yield the correct values anyway?

(d) Find the direction corresponding to the largest peak in the linear predictive steered response as a function of the noise variance.

7.12 A simple example illustrates the problems in choosing the distinguished sensor m_0 for the linear predictive method. Assume a regular linear array toward which signals propagate in the presence of spatially white noise.

(a) Find an expression for the mean-squared prediction error. We can use this quantity to assess which sensor to choose: The one yielding the smallest prediction error should provide the best fit to the observations.

(b) Compute this error as a function of direction of propagation and m_0 for the single signal case. Show that in this case selecting any sensor as the reference minimizes the prediction error.

(c) Assume two equiamplitude, incoherent sources are present in the field. Show that minimum prediction error varies with choice of reference sensor. Do you think one choice for m_0 minimizes the prediction error for all possible directions of propagation? Assuming we desire to develop an optimal algorithm for applications (a noble goal), interpret your answer.

7.13 When designing an auxiliary array as a sidelobe canceller for a main antenna, we first need to determine the number of elements in the array. Assume that in addition to the source located in a known position N_s jammers can be present in the field, each of which is weighted by some unknown sidelobe gain.

(a) Show that when $N_s = 1$ and the jammer and source are incoherent, one auxiliary array element suffices to cancel the jammer.

(b) When the jammer and source are coherent, show that the jammer cannot be cancelled. How is the degree of suppression, as expressed by the source-to-jammer ratio, related to jammer power?

(c) Assuming that $N_s > 1$, show that even in the case of incoherent jammers a sidelobe canceller cannot perform perfectly as it can in the case $N_s = 1$.

Note: Answering the last part should *not* be based on a detailed solution of the sidelobe canceller equations, but instead on the capabilities of linear prediction to model two or more signals, jammers in this case (Prob. 7.11).

7.14 More than a few details concerning the generalized sidelobe canceller need exploration. Rather than equating the generalized sidelobe canceller with algorithms derived via constrained optimization, we design the blocking matrix according to our own criteria. For example, to suppress jammers, we may want to remove all signals arising from a given set of directions. This suppression is best accomplished by predesigning the fixed shading vector $\mathbf{w}_c$ to cancel these signals. After going to all this design work, we certainly don't want the adaptive section to let them back in!

(a) Demonstrate that the solution expressed for $\mathbf{w}_a$ in Eq. 7.6 {369} is indeed correct.

(b) Assuming that two jammer-suppression constraints are imposed on the blocking matrix, what is the dimension of $\mathbf{w}_a$?

(c) Assume we have a regular linear array with M sensors. The rows of the blocking matrix must be orthogonal to each jamming signal. Characterize a prototypical row of $\mathbf{B}$ as a linear *FIR* filter. What criterion would govern the filter's design?

(d) Using the special nature of the linear array's output, show that N_s jammers can be cancelled exactly so long as M exceeds N_s.

7.15 The definition of an inner product can be generalized to include positive definite, Hermitian matrices **A**:

$$\langle \mathbf{x}, \mathbf{y} \rangle_{\mathbf{A}} = \mathbf{x}'\mathbf{A}\mathbf{y}$$

The usual inner product $\langle \mathbf{x}, \mathbf{y} \rangle = \mathbf{x}'\mathbf{y}$ thus occurs when the matrix equals the identity. Let $\{\lambda_i, \mathbf{v}_i\}$ denote the eigensystem of **A**.

(a) Why do we require **A** to be positive definite and Hermitian?

(b) Show that the eigenvectors of **A** are orthogonal with respect to *both* inner products: $\langle \mathbf{v}_i, \mathbf{v}_j \rangle_{\mathbf{A}} = \delta_{ij}$ and $\langle \mathbf{v}_i, \mathbf{v}_j \rangle = \delta_{ij}$. Show this first when the associated eigenvalues are distinct, then consider the case when the eigenvalues have multiplicity.

(c) Because **A** is positive definite, its inverse exists. Show that the eigensystem of the inverse matrix equals $\{1/\lambda_i, \mathbf{v}_i\}$.

(d) Show that the matrix **A** can be represented as

$$\mathbf{A} = \sum_{i=1}^{M} \lambda_i \mathbf{v}_i \mathbf{v}_i'$$

7.16 We frequently employ matrix eigenanalysis to explore the convergence of dynamic adaptive methods. Consider the vector difference equation

$$\mathbf{z}(l) = \mathbf{A}\mathbf{z}(l-1) + \mathbf{y}(l)$$

(a) Express the output **z** and the input **y** in **A**'s eigensystem, thereby allowing the difference equation to be expressed as an uncoupled set of first-order difference equations.

(b) Find the criterion that guarantees stability of the set of uncoupled equations.

(c) How does this criterion apply to the original difference equation?

7.17 Consider the spatial correlation matrix resulting from a single signal propagating in spatially white noise.

$$\mathbf{R} = \sigma_n^2 \mathbf{I} + A^2 \mathbf{e}\mathbf{e}'$$

(a) Find the eigensystem for this matrix. Note that the eigenvectors are not unique; give an algorithm for finding the eigensystem. Describe this eigensystem's associated rugby ball (Rayleigh quotient).

(b) Consider the two-signal case $\mathbf{R} = \sigma_n^2 \mathbf{I} + A_1^2 \mathbf{e}_1 \mathbf{e}_1' + A_2^2 \mathbf{e}_2 \mathbf{e}_2'$, $\mathbf{e}_1 \neq \mathbf{e}_2$. What is the intersignal coherence matrix **C** in this case? Can two ideal signal vectors for a regular linear array be orthogonal?

(c) Find the eigensystem for the two-signal case. Under what conditions are $\mathbf{e}_1$ and $\mathbf{e}_2$ eigenvectors? Hint: Consider eigenvectors of the form $a_1\mathbf{e}_1 + a_2\mathbf{e}_2$.

(d) Find the matrix **T** that transforms the eigenvectors corresponding to the two largest eigenvalues into the signal vectors: $\mathbf{V}\mathbf{T} = \mathbf{S}$ where $\mathbf{V} = [\mathbf{v}_1\ \mathbf{v}_2]$.

(e) Find the matrix that similarly transforms a column matrix of the eigenvalues into the column matrix of squared amplitude values.

7.18 *Forward-backward smoothing* is frequently employed in linear prediction techniques to improve correlation matrix estimates [149]. Because this technique is not specific to the algorithm, its estimate can be incorporated into other algorithms as well.

(a) Consider the reversed and conjugated observation vector $\mathbf{Jy}^*$ where $\mathbf{J}$ denotes the exchange matrix (see §B.2 {487}). Show that this vector's correlation matrix $\mathbf{R}_e$ equals the correlation matrix for $\mathbf{y}$ indexed in "reverse" and conjugated.

(b) Relate the eigensystem of $\mathbf{R}_e$ to that for $\mathbf{R}$.

(c) When the spatial correlation matrix is Toeplitz as well as Hermitian, show that its eigenvectors have the property $\mathbf{v} = \mathbf{Jv}^*$. What array geometry yields correlation matrices having this structure?

(d) The forward-backward smoothed estimate of the spatial correlation matrix amounts to averaging the estimated correlation matrices of the observations and of the reversed-conjugated observations.

$$\widehat{\mathbf{R}} = \frac{1}{2L}\sum_{l=1}^{L} \mathbf{y}_l\mathbf{y}_l' + \mathbf{Jy}_l^*\,(\mathbf{Jy}_l^*)'$$

Find the expected value of this estimate and the associated eigensystem.

7.19 Eigenanalysis-based techniques have often been said to provide "super-resolution" because they allow two sources to be distinguished at much closer angular locations than do their noneigen counterparts (for example, *MUSIC* provides much better resolution than the minimum variance algorithm). Care must be taken in making such brash statements as this problem demonstrates [88]. Consider the *MUSIC*-like variant of conventional beamforming in which the spatial correlation matrix's N_s largest eigenvalues are retained and a Bartlett spectrum produced. Because eigenanalysis techniques are usually given names, we call this algorithm the *MB* algorithm.

$$\mathbf{R}_{\text{MB}} = \sum_{i=1}^{N_s} \mathbf{v}_i\mathbf{v}_i' \qquad \mathcal{P}^{\text{MB}}(\vec{\zeta}\,) = \mathbf{e}'(\vec{\zeta}\,)\mathbf{R}_{\text{MB}}\mathbf{e}(\vec{\zeta}\,)$$

(a) Assuming that two incoherent sources located symmetrically about the broadside of a regular linear array are observed in the presence of spatially white noise, find an analytic expression for $\mathcal{P}^{\text{MB}}$. For simplicity, express this steered response in terms of delay rather than incidence angle.

(b) Show that in most interesting cases this technique provides no more resolution than its noneigen counterpart. (Who said eigenanalysis techniques are always superior?)

(c) Show that the steered response $\mathcal{P}^{\text{MUSIC}}(\vec{\zeta}\,)$ produced by the *MUSIC* algorithm equals $1/(M - \mathcal{P}^{\text{MB}}(\vec{\zeta}\,))$. Noting that $0 \le \mathcal{P}^{\text{MB}}(\vec{\zeta}\,) \le M$, we see that the two algorithms are related by a monotonic transformation. Would you now conclude that our new technique has poorer resolution than that provided by the *MUSIC* algorithm?

7.20 In §7.3.4 {385}, we discovered that perfect coherence induces the "3-dB" effect: The peak-to-background ratios for adaptive beamforming and direction-finding algorithms equals 2 no matter how large the signal amplitude might be. The argument focused on one specific value of the intersignal coherence, however; let's explore a more general scenario.

(a) The derived result hinged on each signal's projection into the "noise" subspace equaling $M/2$; demonstrate the validity of this result.

(b) Assume that two equiamplitude signals are perfectly coherent ($|c| = 1$) but with some phase. Show that the 3-db effect remains for the minimum variance algorithm.

(c) What effect do coherent signals have on the conventional beamforming algorithm? Assume unit shading.

7.21 In addition to spatial averaging, forward-backward smoothing (Prob. 7.18) can also be employed to reduce intersignal coherence in the spatial correlation matrix.

(a) Assuming the ideal spatial correlation matrix produced by a regular linear array has the form $\mathbf{R} = \sigma_n^2\mathbf{I}+\mathbf{SCS}'$, show that the average of $\mathbf{R}$ and $\mathbf{JR}^*\mathbf{J}'$ has the form $\sigma_n^2\mathbf{I}+\mathbf{S}\widetilde{\mathbf{C}}\mathbf{S}'$. Summarize the effects of forward-backward smoothing on intersignal coherence.

(b) Show that spatial averaging and forward-backward smoothing can be applied in either order and produce the same correlation matrix.

7.22 The so-called "estimation of signal parameters via rotational invariance techniques" (*ESPRIT*) algorithm is a novel application of eigenanalysis techniques [131]. Here, we must partition the array into two subarrays, each of which has the same geometry and are separated by a known amount D (see Fig. 7.26). Defining $\mathbf{Z}_1$ and $\mathbf{Z}_2$ to be the subarrays' frequency-domain outputs, we note that their signal portions are related by $\mathbf{S}^{(2)} = \mathbf{S}^{(1)}\mathbf{\Phi}$ where $\mathbf{\Phi}$ is a diagonal matrix whose elements correspond to the intersubarray propagation delays and $\mathbf{S}^{(i)}$ denotes the ith array's signal matrix $[\mathbf{S}_1^{(i)} \cdots \mathbf{S}_{N_s}^{(i)}]$.

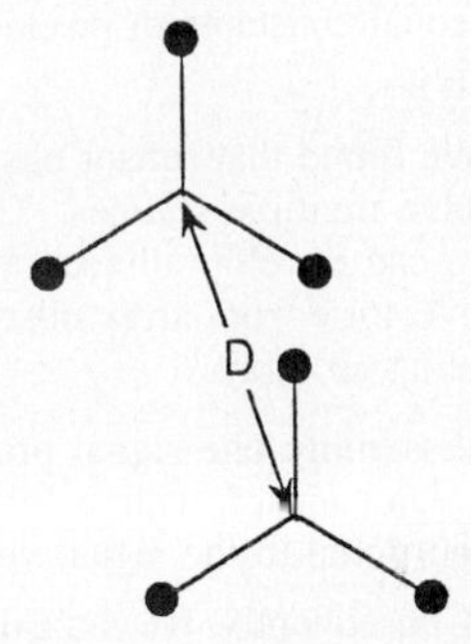

Figure 7.26 To use the *ESPRIT* algorithm, the array geometry must be such that two identical subarrays can be defined. Here, two triangular subarrays comprise the array. The separation D between each subarray's coordinate system must be known.

(a) Show that all array geometries compatible with the *ESPRIT* algorithm must have some co-array value equaling $M/2$. Show that the regular linear array having an even number of sensors has this property. Which circular arrays have it?

(b) Assuming N_s signals, find the matrix $\mathbf{\Phi}$. The algorithm's key point is that this matrix contains *all* the information necessary to determine the directions of propagation if $N_s \leq M/2$. Prove this fact.

(c) Assuming spatially white noise, find an analytic expression for the entire array's spatial correlation matrix in terms of $\mathbf{S}^{(i)}$ and $\mathbf{\Phi}$. Assume that $\mathbf{Z} = \text{col}[\mathbf{Z}_1, \mathbf{Z}_2]$.

(d) The matrix $\mathbf{V}_{s+n}$ of eigenvectors forming the signal-plus-noise subspace can be partitioned according to the subarrays: $\mathbf{V}_{s+n} = \text{col}[\mathbf{V}_{s+n}^1, \mathbf{V}_{s+n}^2]$. Form the $M/2 \times 2N_s$ matrix $\mathbf{V}$ by placing the subarray components side by side: $\mathbf{V} = [\mathbf{V}_{s+n}^1 \mathbf{V}_{s+n}^2]$. Show that this matrix has rank N_s, which implies that its *null space* defined by all vectors $\mathbf{x}$ satisfying $\mathbf{Vx} = \mathbf{0}$ has dimension N_s.

(e) Let the $2N_s \times N_s$ matrix $\mathbf{X} = \text{col}[\mathbf{X}_1, \mathbf{X}_2]$ contain the basis vectors of this null space. From $\mathbf{VX} = \mathbf{0}$, show that the matrix $-\mathbf{X}_1\mathbf{X}_2^{-1}$ has eigenvalues equal to the diagonal elements of $\mathbf{\Phi}$. How does this result enable us to find the directions of propagation? Note: To answer part e, you will need the fact, which you should show, that, for any square matrix $\mathbf{A}$ and invertible matrix $\mathbf{B}$, the matrices $\mathbf{A}$ and $\mathbf{BAB}^{-1}$ share the same eigenvalues (but not necessarily the same eigenvectors).

Thus, from a rather complicated, but well-founded, sequence of eigenanalyses, directions of propagation can be found directly without appeal to searching peaks in some steered response.

7.23 Eigenanalysis of the estimated spatial correlation matrix yields benefits other than providing new algorithms. To calculate the steered response for the minimum variance algorithm, we need to calculate $\mathbf{e}'\widehat{\mathbf{R}}^{-1}\mathbf{e}$. We have argued that such expressions amount to calculating the Fourier spectrum of a matrix; we would like to use the *FFT*. We can if we have a regular linear array geometry.

(a) How would you use the *FFT* to calculate directly the Fourier spectrum of $\widehat{\mathbf{R}}^{-1}$ found from a regular linear array? What is the complexity of the calculation? Note: The adaptive algorithms provide such high resolution that many more spectral points than the matrix's dimension need to be computed. For example, the authors use 1,024-point transforms for 10-sensor array data. Assume the transform length equals $100M$.

(b) Consider the eigenexpansion for the spatial correlation matrix's inverse (Eq. 7.8a {384}). Based on this expression, how would the Fourier spectrum be computed? Compare the complexity of the two spectral methods.

(c) The complexity of evaluating the eigensystem of a Hermitian matrix is proportional to $5M^3$. Over what ranges of M is one *FFT* algorithm faster than the other? Assume equal constants of proportionality in the *FFT* and eigenanalysis algorithmic complexities.

7.24 We have found that sensor positioning errors can degrade the ability of adaptive algorithms to resolve multiple sources. The *calibration* problem is to locate the sensors accurately so that we can develop an accurate signal model; in other words, we want calculate the ideal signal vector $\mathbf{e}$ from array observations. Surprisingly, eigenvector techniques offer a way of calibrating an array.

(a) Assuming one signal propagating in a white-noise background, show that the eigenvector corresponding to the largest eigenvalue of the ideal correlation matrix is proportional to the signal vector.

(b) Consequently, we can calibrate the array without knowing the sensor positions a priori. Detail how to generate a signal model vector from the eigenvector corresponding to the largest eigenvalue.

(c) In less ideal situations, we must estimate the spatial correlation matrix and propagating signals other than the calibration signal may be present in the observations. Describe how the eigenvector corresponding to the largest eigenvalue deviates from the actual signal when an unwanted signal corrupts the observations. Detail the nature of the error as a function of a single corrupting signal's direction of propagation and strength relative to the calibration signal.

7.25 Sensor gain and position errors confound adaptive methods that are unaware of their presence. Are eigenanalysis methods affected similarly? Assume that the true signal vector has the form $\mathbf{Ge}$, where $\mathbf{G} = \mathbf{I} + \text{diag}[\delta g_0, \ldots, \delta g_{M-1}]$ with δg_m representing a complex-valued perturbation and $\mathbf{e}$ equaling the assumed signal vector.

(a) Find the eigensystem of the spatial correlation matrix corresponding to the inaccurately known array. Assume spatially white noise.

(b) Calculate the projection of $\mathbf{e}$ onto this matrix's noise subspace. Will the eigenvector method produce an accentuated peak in the assumed propagation direction? Relate the amplitude of this peak to signal-to-noise ratio and mean-squared perturbation.

7.26 For analytic simplicity, let's discuss a two-point amplitude-constrained minimum variance algorithm for the single-signal, white-noise case. The constraints are placed symmetrically about the assumed propagation direction.

(a) Analytically express the algorithm's steered response in terms of the inverse correlation matrix and the two signal vectors that define the constraint directions.

(b) Consider enhancing the resolution of this method by employing eigenanalysis. Describe the effects of using the *MUSIC* method when the correction matrix is known exactly.

7.27 Applying derivative constraints can have the same affect as amplitude constraints [145]. To analyze this situation, consider approximating three-point amplitude constraints placed symmetrically about the assumed propagation direction by first- and second-derivative constraints.

(a) Assume that an invertible linear transformation $\mathbf{A}$ is applied to the constraint matrix in an adaptive algorithm derived by constrained optimization. The constraint vector is left untouched. Show that the resulting algorithm's output power equals that computed when the original constraint matrix is used in conjunction with the constraint vector $\mathbf{A}'\mathbf{c}$.

(b) Employing Taylor series approximations, represent amplitude constraints by a linear combination of derivative constraints. Find the matrix $\mathbf{A}$ that converts derivative constraints into amplitude constraints. Calculate the determinant of this matrix to assess its invertibility.

(c) Show that output power produced by a derivative-constrained system well approximates that produced by an amplitude-constrained system.

7.28 Implementing amplitude constraints within the context of the generalized sidelobe canceller structure is described on page 398ff. Let's explore how derivative constraints can be implemented. Assume that constraints are imposed on the signal model's derivatives with respect to intersensor delay rather than incidence angle.

(a) From the constraint matrix defining first- and second-derivative constraints, what conditions must the rows of the blocking matrix satisfy?

(b) Find a blocking matrix satisfying these conditions that has identical, but shifted, rows. Use linear filtering ideas to derive the matrix's elements.

7.29 Amplitude constraints can be placed on arrays having other than linear geometries. Consider a square array lying in the (x, y) plane and having M^2 sensors. Placing constraints in terms of elevation and azimuth is, at best, tedious analytically; we simplify the problem by considering the delays Δ_x and Δ_y instead.

(a) Geometrically, we can select the delays at which to impose constraints by sketching their locations in the (Δ_x, Δ_y) plane. Assuming only symmetrically placed constraints, find five- and nine- point constraint "geometries." Find the associated constraint matrices.

(b) Assuming a generalized sidelobe canceller enforces the five-point amplitude constraint, find the algorithm's blocking matrix.

7.30 In §7.4.3 {398}, we derived an expression for the shading vector derived from constrained optimization algorithms that employ uncertainty constraints on the signal model.

$$\mathbf{w}_\diamond = -\lambda_1 \left(\mathbf{R} - \frac{\lambda_1^2}{\lambda_2}\mathbf{I}\right)^{-1} \mathbf{e}$$

We did not fully explore, however, the difficulties of explicitly finding the Lagrange multipliers associated with the problem.

(a) Show that $\mathbf{w}_\diamond'\mathbf{R}\mathbf{w}_\diamond = -\lambda_1$, thereby indicating that this multiplier is negative.

(b) Find two simultaneous equations that can be solved for the Lagrange multipliers. What are the order and nature of these equations?

(c) Assuming that the spatial correlation matrix has the simplistic form $\mathbf{R} = \sigma^2\mathbf{I}$, find the equations governing the Lagrange multipliers.

7.31 Beamforming robustness can be expressed in a quite general way, allowing constraints to be placed on any number of potentially faulty conditions but resulting in the same constrained optimization problem [53]. Define the mean-squared beamforming error to be the weighted squared error of the beamformer's constraints.

$$\epsilon^2 = \frac{1}{\|\mathbf{c}\|^2} \cdot \|\mathbf{c} - \mathbf{C}\mathbf{w}\|^2$$

Rather than the usual norm, we select a norm that reflects unaccounted variations in a set of parameters expressed by the column matrix $\boldsymbol{\xi}$ whose nominal (modeled) values are $\boldsymbol{\xi}_0$.

$$\|\mathbf{x}\|^2 = \underset{\boldsymbol{\xi} \in \Re_{\xi_0}}{\int \cdots \int} W(\boldsymbol{\xi})\mathbf{x}'\mathbf{x}\, d\boldsymbol{\xi}$$

$\Re_{\xi_0}$ is a rectangle describing potential variations of K parameters about their nominal values: $\Re_{\xi_0} = \{\boldsymbol{\xi} : \xi_1 \in (\xi_1^0 - \delta\xi_1, \xi_1^0 + \delta\xi_1), \ldots, \xi_K \in (\xi_K^0 - \delta\xi_K, \xi_K^0 + \delta\xi_K)\}$. $W(\boldsymbol{\xi})$ denotes a weighting function used to express the relative severity of parameter variations. For example, we could explicitly express mainlobe broadening indirectly expressed by amplitude or derivative constraints (§7.4.2 {394}). For a linear array, ξ would equal azimuth (or delay), and the weighting function would equal 1 over the desired angular range. The usual minimum variance constraints would be selected: $\mathbf{C} = \mathbf{e}'(\xi)$, $\mathbf{c} = 1$. Rather than constraining the processing gain for a set of directions as in amplitude constrained beamforming, the error ϵ^2 represents the total mean-squared error over the angular range.

(a) Show that the mean-squared error in the general case can be expressed much more concisely as $\epsilon^2 = \mathbf{w}'\mathbf{Q}\mathbf{w} - \mathbf{q}'\mathbf{w} - \mathbf{w}'\mathbf{q} + 1$ where

$$\mathbf{Q} = \underset{\boldsymbol{\xi} \in \Re_{\xi_0}}{\int \cdots \int} W(\boldsymbol{\xi})\mathbf{C}'\mathbf{C}\, d\boldsymbol{\xi}$$

$$\mathbf{q} = \underset{\boldsymbol{\xi} \in \Re_{\xi_0}}{\int \cdots \int} W(\boldsymbol{\xi})\mathbf{c}\, d\boldsymbol{\xi}$$

(b) Show that the mean-squared error can be written as $(\mathbf{w} - \mathbf{w}_\diamond)'\mathbf{Q}(\mathbf{w} - \mathbf{w}_\diamond) + \epsilon_\diamond^2$ where $\mathbf{w}_\diamond$ represents the weight vector achieving the minimum mean-squared error $\epsilon_\diamond^2$. Find an expression for this minimum in terms of $\mathbf{Q}$ and $\mathbf{q}$.

(c) Modeling inaccuracies would result in the mean-squared error increasing uncontrollably above its optimum value. A robust beamformer results by finding the weight vector that minimizes array output power while placing an upper limit on the mean-squared constraint error. Find the optimum weight vector that solves the constrained optimization problem

$$\min_{\mathbf{w}} \mathbf{w}'\mathbf{R}\mathbf{w} \quad \text{subject to} \quad (\mathbf{w} - \mathbf{w}_\diamond)'\mathbf{Q}(\mathbf{w} - \mathbf{w}_\diamond) \leq \beta^2 \,,\, \beta^2 > \epsilon_\diamond^2$$

Note: You may not be able to calculate the Lagrange multiplier; leave your answer in terms of it.

(d) Contrast this robust beamformer's weight vector with that obtained by constraining the norm of signal modeling errors.

7.32 Frost's algorithm is quite general, applying to any algorithm based on constrained optimization. The only tricky aspect of its various applications is the selection of the adaptation parameter μ. Let's explore this problem-dependent aspect with the linear prediction method.

(a) Detail the equations for Frost's algorithm that produce the shading vector for the linear prediction method. What is the projection matrix $\mathbf{P}$?

(b) Determine how μ should be selected.

7.33 A simple dynamic adaptive algorithm that results from discretizing a servoloop-like differential equation is frequently employed in high-frequency applications (e.g., radar) [35: Chap. 2];[63]. Part of the method's design requires it to yield a shading having a pre-designed weight vector $\mathbf{w}_d$ when no signal is present. The weighted output to the ideal signal $\mathbf{e}$ defining an assumed propagation direction becomes $\mathbf{W}_d\mathbf{e}$, where $\mathbf{W}_d = \text{diag}[w_0^d, \ldots, w_{M-1}^d]$. The difference equation defining the algorithm is

$$\mathbf{w}(l+1) = \mathbf{w}(l) + \alpha\left[\beta \mathbf{W}_d \mathbf{e} - z^* \mathbf{y}\right]$$

(a) Assuming that a steady-state solution exists, show that the algorithm's shading vector becomes proportional to that produced by the minimum variance technique.

(b) Show that when only spatially white noise is present the weight vector approaches a constant times the predesigned shading.

(c) Determine the stability criterion for this system.

(d) Contrast this algorithm with Frost's, taking in both cases $\mathbf{w}_c = \mathbf{e}/M$. What are the essential differences between the two algorithms?

7.34 When selecting the upper limit for the weight vector's norm in the robust Frost algorithm, the ideal weight vector as well as the quiescent weight vector should be considered.

(a) Derive an expression for $\|\mathbf{w}\|^2$ resulting from the minimum variance algorithm.

(b) Assuming a single signal is observed in the presence of spatially white noise, find the maximum norm of the weight vector when the assumed propagation direction corresponds to a signal's direction of propagation and when it does not correspond to a signal. How does this result affect the designer's choice for β^2?

Chapter 8

Tracking

Array processing algorithms developed in previous chapters apply to those unfortunately rare situations in which the source of propagating energy moves little relative to the array. When the source does move, the resulting propagating signal characteristics usually violate assumptions made by processing algorithms. Block methods, conventional beamforming and minimum variance to name a couple, assume signal models in which the direction of propagation remains constant throughout the averaging process. Dynamic adaptive methods can be used when the source moves, but they do not associate source locations with *particular* sources. For example, two sources may maneuver such that their positions relative to the array temporarily overlap. From the algorithm's viewpoint, two sources are present at first, then one while they overlap, then two reappear. How do we associate reemerging source locations with previous ones? To complicate matters further, arrays move: How do we take into account array motion in locating sources with respect to some fixed coordinate system?

Tracking algorithms intend to fulfill one of the ultimate goals of array processing algorithms: Determine the location and motion of identified sources. They extend the algorithms described in previous chapters in several ways.

- Explicit inclusion of source motion dynamics and array motion into the signal processing algorithm.
- In addition to source location, production of source motion estimates, usually in the form of source velocity, while following the source over long periods.
- Rejection of false alarms and gliding past momentarily missing measurements.

Two basic tracking methods comprise this chapter. The first, *single-array tracking*, represents a logical extension of the conventional and adaptive algorithms discussed in previous chapters. Combining an array's measurement of source position,[†] as obtained from a steered response, with constraints imposed by *array platform motion* and *source*

[†]"Position" is not meant to be taken literally here. Rather, position-related measurements such as angle or range are intended.

motion modeling creates a *tracking algorithm*. To be viable, tracking algorithms must yield smaller variances for source position than can the array processing algorithm: They must yield a signal processing gain. For example, incorporating source motion gives us the ability to distinguish between sources and spurious array measurements. False-alarms and missing measurements, which are equivalent to noise, *clutter*, for tracking algorithms, can be so labeled because source motion models restrict how measurements evolve over time.

The second method, *multiarray tracking*, opens a new set of array processing problems that require new signal processing ideas. The same set of sources cannot be assumed present in every array's field. Furthermore, even when each array produces a measurement related to a specific source, we must somehow select that particular measurement from the arrays' other measurements. This association between data and source, the so-called *data association problem*, cannot be made unambiguously from a single snapshot's measurement. We must involve the tracker's prediction of future source motion to assess the quality of individual location-motion estimates. Assignment of source identification numbers to array outputs opens the door to new kinds of algorithms, some of which can reasonably be based on higher level (symbolic) signal processing ideas.

8.1 Source Motion Models

Using a two-dimensional problem as our ever-present example, denote a source's position by the tuple $\vec{x}^o(t) = \left(\vec{x}^o_x(t), \vec{x}^o_y(t)\right)$ and its velocity by $\dot{\vec{x}}^o(t) = \left(\dot{\vec{x}}^o_x(t), \dot{\vec{x}}^o_y(t)\right)$ *in continuous time*. The array moves with velocity $\left(\dot{x}_x(t), \dot{x}_y(t)\right)$. Because the array can move, we cannot reasonably choose the coordinate system's origin to coincide with that of the array. Instead, we choose a fixed coordinate system, usually with its origin and the array's coinciding at $t = 0$. We define a continuous-time *position state vector* $\mathbf{v}(t) = \text{col}[\vec{x}^o_x(t) - x_x(t), \dot{\vec{x}}^o_x(t) - \dot{x}_x(t), \vec{x}^o_y(t) - x_y(t), \dot{\vec{x}}^o_y(t) - \dot{x}_y(t)]$ to express the source's motion relative to the array. The relation between position, velocity, and acceleration lead to the state equation

$$\dot{\mathbf{v}}(t) = \begin{bmatrix} 0 & 1 & 0 & 0 \\ 0 & 0 & 0 & 0 \\ 0 & 0 & 0 & 1 \\ 0 & 0 & 0 & 0 \end{bmatrix} \mathbf{v}(t) + \begin{bmatrix} 0 \\ a_x(t) \\ 0 \\ a_y(t) \end{bmatrix}$$

Usually, motions in the Cartesian directions are assumed decoupled, meaning that the accelerations $a_x(t)$ and $a_y(t)$ are independent inputs. We express this state equation succinctly as $\dot{\mathbf{v}}(t) = \widetilde{\mathbf{A}}\mathbf{v}(t) + \widetilde{\mathbf{u}}(t)$. We need to convert this physically based continuous-time model into a discrete-time one that corresponds to sampling the source's motion when snapshots are taken. Assume each snapshot spans time T. From the theory of state variable descriptions for stationary (time-invariant) systems the state vector can be directly related to the input as

$$\mathbf{v}(t) = \int_0^t e^{\widetilde{\mathbf{A}}(t-\tau)} \widetilde{\mathbf{u}}(\tau)\, d\tau$$

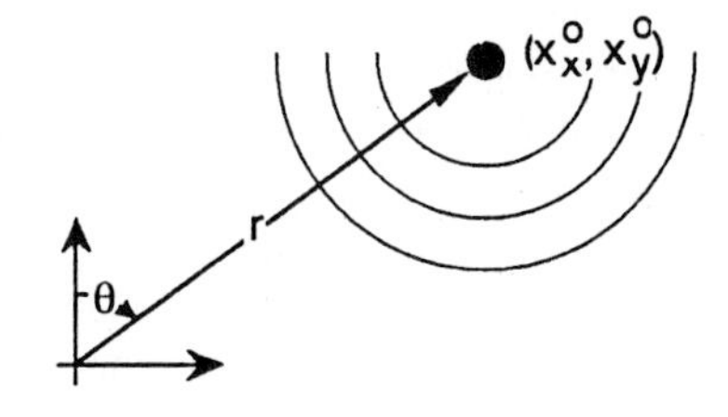

Figure 8.1 A source located at position $\vec{x}^o(t) = \left(\vec{x}_x^o(t), \vec{x}_y^o(t)\right)$ creates a propagating signal that is observed at the array. The direction of propagation appears to arise from an azimuth θ with a range r.

$$= \int_0^{t-T} e^{\widetilde{\mathbf{A}}(t-\tau)}\widetilde{\mathbf{u}}(\tau)\,d\tau + \int_{t-T}^{t} e^{\widetilde{\mathbf{A}}(t-\tau)}\widetilde{\mathbf{u}}(\tau)\,d\tau$$

$$= \underbrace{e^{\widetilde{\mathbf{A}}T}}_{\mathbf{A}} \underbrace{\int_0^{t-T} e^{\widetilde{\mathbf{A}}(t-T-\tau)}\widetilde{\mathbf{u}}(\tau)\,d\tau}_{\mathbf{v}(t-T)} + \underbrace{\int_{t-T}^{t} e^{\widetilde{\mathbf{A}}(t-\tau)}\widetilde{\mathbf{u}}(\tau)\,d\tau}_{\mathbf{u}(t)}$$

Defining a discrete-time state vector $\mathbf{v}(l)$ to express the source's relative position and velocity at $t = lT$, we find it to be governed by the discrete-time state equation

$$\mathbf{v}(l) = \underbrace{\begin{bmatrix} 1 & T & 0 & 0 \\ 0 & 1 & 0 & 0 \\ 0 & 0 & 1 & T \\ 0 & 0 & 0 & 1 \end{bmatrix}}_{\mathbf{A}} \mathbf{v}(l-1) + \underbrace{\int_{(l-1)T}^{lT} \begin{bmatrix} (lT-\tau)a_x(\tau) \\ a_x(\tau) \\ (lT-\tau)a_y(\tau) \\ a_y(\tau) \end{bmatrix} d\tau}_{\mathbf{u}(l)} \tag{8.1}$$

having initial condition equal to the source's position and velocity relative to the array at the initial snapshot $l = 0$.

An array's estimate of location usually comes in the form of an azimuth θ and a range r, each of which is contaminated by measurement noise (see Fig. 8.1). We form these estimates into the observation vector $\mathbf{z} = \text{col}[\theta\ r]$ according to the relationship†

$$\mathbf{z} = \begin{bmatrix} \tan^{-1}\left(\dfrac{\vec{x}_x^o - x_x}{\vec{x}_y^o - x_y}\right) \\ \sqrt{(\vec{x}_x^o - x_x)^2 + (\vec{x}_y^o - x_y)^2} \end{bmatrix} + \mathbf{n} = \begin{bmatrix} \tan^{-1}\left(\dfrac{v_1}{v_3}\right) \\ \sqrt{v_1^2 + v_3^2} \end{bmatrix} + \mathbf{n} \tag{8.2}$$

Note that none of these measurements is related to the source's velocity; as we shall learn, the tracker infers from sequences of measurements what the source velocity must be. When the source is located in the far field, range measurements become impractical, leaving us with a single-component observation vector that expresses only the azimuth measurement. Components of the additive noise $\mathbf{n}$ represent the errors in measuring azimuth and range as described in previous chapters. According to that discussion, these components have statistically independent components, but their variances depend on range and azimuth according to the array's resolution capability. The state equations

†For the moment we shall ignore the effects caused by the finite speed of propagation. In a real application, by the time the signal has arrived at the sensor array, the target has moved to a new position. See Prob. 8.11 for an example.

governing a source's movement relative to the array and observation of its location are succinctly expressed by

$$\begin{array}{|ll|}\hline \mathbf{v}(l) = \mathbf{A}\mathbf{v}(l-1) + \mathbf{u}(l) & \text{Source dynamics equation} \\ \mathbf{z}(l) = \mathcal{C}[\mathbf{v}(l)] + \mathbf{n}(l) & \text{Observation equation} \\ \hline \end{array} \tag{8.3}$$

where Eq. 8.1 expresses the state matrix **A** and Eq. 8.2 the nonlinear transformation $\mathcal{C}[\cdot]$ relating array location measurements to source location.

Example

Assume we have a stationary array located at $(0, 0)$ and a source moving from initial location $(-50, 75)$ at an initial uniform velocity having magnitude three in the x direction (see Fig. 8.2). Thus, the initial condition for the dynamic equation is $\mathbf{v}(0) = \text{col}[-50, 3, 75, 0]$ and the snapshot interval equals 1. The source experiences stochastic accelerations each having variance 0.1 in both directions. The array processing algorithm is assumed to yield only estimates of the source's azimuth, which equals the negative of the direction of propagation. In realistic situations, the received signal diminishes in amplitude as the source's range increases. In most algorithms, the estimated azimuth's variance increases as signal power decreases. In this way, azimuth variance increases with increasing range r in some fashion. Because Cramér-Rao bounds for azimuth estimates are usually proportional to the noise-to-signal ratio of each sensor's observations (see the example {290}), spherical propagation from the source means the azimuth variance is proportional to r^2. Consequently, characteristics of the observation noise vector **n** change with source location in a complicated way. Furthermore, variation of the array's angular resolution with angle means that the measurement noise variance depends on the source's angular location. These effects are discussed in more detail in §8.2.1. Fig. 8.2 displays the motion of our example source and the simulated azimuth estimate produced by the array processing algorithm. In this example, the azimuth measurement process does *not* take into account the dynamic equation. Thus, the lower panel of Fig. 8.2 represents the most primitive kind of tracking algorithm: Provide source locations once each snapshot is obtained without regard to previous measurements or to the source's predicted location.

8.2 Single-Array Location Estimate Properties

Conventional and adaptive array processing algorithms attempt to provide two basic kinds of information about the sources of propagating energy: What signals they are producing and where they are located. Before we develop tracking algorithms, we must quantify the accuracy of location measurements that serve as the tracker's primary inputs. We categorize the location results as range and direction of propagation estimates, assessing their accuracies according to the algorithm's range and direction of propagation resolution. Two phenomena limit a single array's ability to produce high-quality estimates: the array's limited resolution owing to its finite aperture and signal processing algorithms, and its inability to account for source motion.

8.2.1 Resolution Effects

In active systems, the time difference between pulse transmission and reception at the array's spatial origin is directly related to range (see Prob. 2.5 {56}). In such cases,

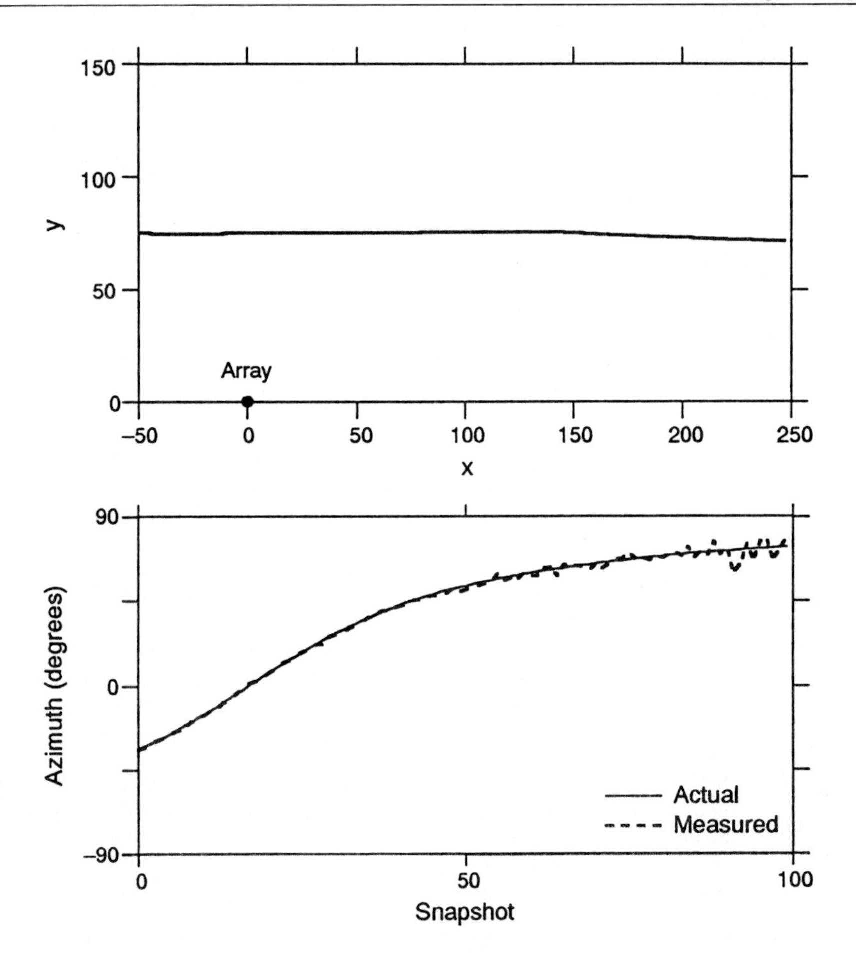

Figure 8.2 The upper panel displays the example motion of a source relative to a fixed linear array while the lower shows the array's estimate of azimuth and the true azimuth as a function of snapshot. The standard deviation of the azimuth estimate equals 1° for the first snapshot, changing proportionally with range to model spherical radiation from the source and inversely with the cosine of azimuth to model a linear array's angular resolution.

range resolution depends on the signal-to-noise ratio, which is inversely proportional to the fourth power of range, and on the signal's time-bandwidth product.† In passive (listening) situations, range information becomes more difficult to extract with a single array. The ability to measure range from wavefront curvature depends on how much of the wavefront is subtended by the array: For a linear array, maximal sensitivity occurs at broadside, decreasing as $\cos\theta$ for off-broadside directions. Furthermore, as the source's location relative to the array passes from the near field to the far field,

†A signal's time-bandwidth product is its duration times its bandwidth. This quantity shares the name, but not the meaning, of the time-bandwidth product for a spatial correlation matrix estimate.

range estimates become poorer and poorer: Range resolution varies inversely with the *square* of increasing range (see Fig. 4.14 {151} for an example and Prob. 4.21 {196}).† Angular resolution also depends on direction of propagation. For a regular linear array, we found that resolution is proportional to the cosine of azimuth over a large class of algorithms {146}. Compounding the inaccuracies owing to resolution is the decrease in signal-to-noise ratio with increasing range. Assuming an inverse-square-law variation of *SNR* with range (spherical propagation in an ideal medium), variances of range and azimuth estimates for a regular linear array depend on source position as

$$\sigma_r^2 \propto \begin{cases} (r^o)^4 & \text{Active systems} \\ \dfrac{(r^o)^6}{\cos^2\theta^o} & \text{Passive systems} \end{cases} \qquad \sigma_\theta^2 \propto \frac{(r^o)^2}{\cos^2\theta^o} \tag{8.4}$$

In adaptive algorithms, angular resolution often depends not only on source location but also on how many signals are present in the field and their angular spacing. Thus, the accuracy of estimated location components depends highly on algorithm, and, regardless of algorithm, on location.

We defined resolution to reflect the width of the steered response about the direction of propagation. No matter what the resolution may be, the steered response displays a peak at the proper range/propagation direction (if unbiased). Why should resolution affect location accuracy? Because we must estimate the spatial correlation matrix from noisy data, an algorithm's steered response differs from its ideal form. The statistics of the *amplitude* of an estimated steered response have been characterized for many algorithms. However, the statistics of the *peak* locations have eluded researchers to date. For example, the steered responses produced by the conventional and minimum variance algorithms have χ^2 distributions for their amplitudes at any assumed propagation direction (Eq. 6.16 {328} for conventional beamforming and Eq. 7.4 {362} for minimum variance). No one knows, however, what the probability distribution of steered response peaks is. Because of this ignorance, we assume that width of the ideal peak, which is inversely related to resolution, reflects the width of the underlying probability distribution *for a given algorithm*. Just because two algorithms yield the same resolution does *not* mean that their peak locations vary the same amount. Note that the χ^2 distribution's degrees of freedom for conventional and minimum variance algorithms differ, which means the statistical variability of their steered responses differ with processing parameters such as time-bandwidth product. We use resolution as a measure of a peak location's standard deviation only within the context of a given algorithm. As for the peak location's expected value, we have seen that algorithms yield biased peaks when two sources come sufficiently close together (Fig. 7.6 {364}). When two sources are widely separated or when only one source is present, peak location corresponds to direction of propagation or range: The estimate is unbiased. To approximate this complicated behavior, assume that the mean peak location equals the true value.

†Nearer to the array, the range resolution decreases somewhat less dramatically. In succeeding simulations, we assume an inverse relation between range resolution and range to describe this situation. This approximation represents an optimistic view of range estimate quality, which deteriorates much more rapidly as the propagating wave more closely resembles a plane wave.

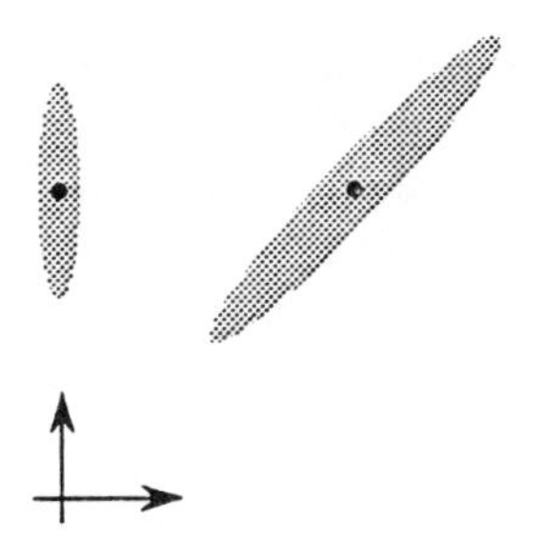

Figure 8.3 The error ellipses of locating two sources by a linear array are diagrammed. Note how ellipse size varies with the source's position relative to the array. The source to the right has the same y coordinate as the other but has an azimuth of 45°. Thus, its error ellipse has twice the diameter in the radial direction and $\sqrt{2}$ the diameter in azimuth.

We further assume that measurements of range and propagation directions from steered response peaks are statistically independent and, when expressed in Cartesian coordinates, are jointly Gaussian.† In a two-dimensional problem, for example, the measured location $\vec{x}^o = (\vec{x}_x^o, \vec{x}_y^o)$ is a Gaussian random vector with expected value $\mathcal{E}[\vec{x}^o]$ equaling the true location and covariance matrix $\mathbf{K}_{\vec{x}^o}$. Statistical independence of range and azimuth estimates means that $\vec{x}_x^o$ and $\vec{x}_y^o$ are *not* statistically independent. To calculate the covariance matrix, we must relate the Cartesian and polar representations of deviation $\delta\vec{x}^o$ of the estimated position from the actual one. Taking into account our convention that azimuth θ is measured as deflection from the y axis (see Fig. 8.1), this relationship is linear for small deviations, resulting in the coordinate transformation

$$\begin{bmatrix} \delta\vec{x}_x^o \\ \delta\vec{x}_y^o \end{bmatrix} = \begin{bmatrix} \sin\theta^o & r^o\cos\theta^o \\ \cos\theta^o & -r^o\sin\theta^o \end{bmatrix} \begin{bmatrix} \delta\vec{x}_r^o \\ \delta\vec{x}_\theta^o \end{bmatrix}$$

Assuming that σ_r^2 and σ_θ^2 describe the variances of range and azimuthal coordinates, respectively, the covariance matrix of position estimates expressed in Cartesian coordinates becomes

$$\boxed{\mathbf{K}_{\vec{x}^o} = \begin{bmatrix} \sigma_r^2\sin^2\theta^o + (r^o)^2\sigma_\theta^2\cos^2\theta^o & \left(\sigma_r^2 - (r^o)^2\sigma_\theta^2\right)\sin\theta^o\cos\theta^o \\ \left(\sigma_r^2 - (r^o)^2\sigma_\theta^2\right)\sin\theta^o\cos\theta^o & \sigma_r^2\cos^2\theta^o + (r^o)^2\sigma_\theta^2\sin^2\theta^o \end{bmatrix}}$$

Recall that range and azimuthal variances are not, in general, constants; they depend on source location in a way that varies with the medium, the array geometry, and the array processing algorithm. Using the expressions for a linear array (Eq. 8.4), the variances of Cartesian position estimates are proportional to the sixth power of range, reinforcing the notion that a single array produces noisy range estimates.

Because of the Gaussian assumption, we quantify the ability of an array and its associated algorithms to locate a given source with an *error ellipse*. As shown in Fig. 8.3, the dependence of Cartesian position estimates manifests itself as an alignment of this ellipse's axes with range and azimuthal directions with axis lengths equal to $2\sigma_r$ and $2r\sigma_\theta$, respectively. By so defining the ellipse to encompass ± 2 standard deviations about the true position, we cover an area corresponding to a probability of $1 - e^{-1}$ (63%) that

†The authors confess little justification for these assumptions. They are widely used in the literature, but preponderance does not make them true.

the source location lies within. We conclude that single-array estimates of source position can be quite noisy. Assuming the source did not move, we could employ multiple arrays to reduce this area by intersecting each array's ellipses. Tracking algorithms employ this effect as well as source motion models to reduce position estimate variance.

8.2.2 Source Motion

Failure to take source or array motion into account can severely hamper direction of propagation estimation algorithms that implicitly assume static sources. Take, for example, methods based on Bartlett-like averages to produce the spatial correlation matrix estimate. We model static far-field signals as $\mathbf{s} = A\,\text{col}\left[e^{j\vec{k}^o\cdot\vec{x}_0}, \ldots, e^{j\vec{k}^o\cdot\vec{x}_{M-1}}\right]$ throughout a sequence of L snapshots. Suppose instead the source moves so that $\vec{k}^o_l = \vec{k}^o_{l-1} + \delta\vec{k}$ while maintaining constant amplitude.† Ignoring for simplicity's sake observation noise, each element of estimated correlation matrix in the *static* case would be

$$\left[\frac{1}{L}\sum_{l=0}^{L-1}\mathbf{s}_l\mathbf{s}_l^t\right]_{m_1 m_2} = A^2 e^{j\vec{k}^o\cdot(\vec{x}_{m_1}-\vec{x}_{m_2})}$$

as expected. In the *dynamic* case,

$$\left[\frac{1}{L}\sum_{l=0}^{L-1}\mathbf{s}_l\mathbf{s}_l^t\right]_{m_1 m_2} = A^2 e^{j\vec{k}^o_{(L-1)/2}\cdot(\vec{x}_{m_1}-\vec{x}_{m_2})}\,\frac{\sin\frac{L}{2}\delta\vec{k}\cdot(\vec{x}_{m_1}-\vec{x}_{m_2})}{L\sin\frac{1}{2}\delta\vec{k}\cdot(\vec{x}_{m_1}-\vec{x}_{m_2})}$$

meaning that spatial correlation matrix elements are weighted by a term that varies throughout the matrix. Because this term depends on the co-array $c(m_1 - m_2) = \vec{x}_{m_1} - \vec{x}_{m_2}$, source motion appears to modify the implicit shading produced by the co-array, with shading reduced more at larger lags than at smaller ones. We would predict, therefore, that the conventional beampattern would have a wider mainlobe, which expresses increased uncertainty in the source's location. Adaptive methods would be affected more adversely because no one signal model applies to the measured correlation matrix (see Fig. 8.4). In the example portrayed there, the minimum variance algorithm hints that two sources are present when in fact only one moving source occupies the field.

8.3 Prediction-Correction Algorithms

A variety of tracking algorithms that incorporate source motion into location estimates can be characterized as a location predictor followed by an observation-dependent corrector. First, they *predict* a source's position state vector $\mathbf{v}$ by incorporating past measurements with the source's dynamic equation, then they *correct* the prediction with current measurements to update the state vector's estimate $\widehat{\mathbf{v}}$.

†Such motion may seem highly artificial (see Prob. 8.3), but the analysis serves to indicate the problems encountered when we use block methods in dynamic environments.

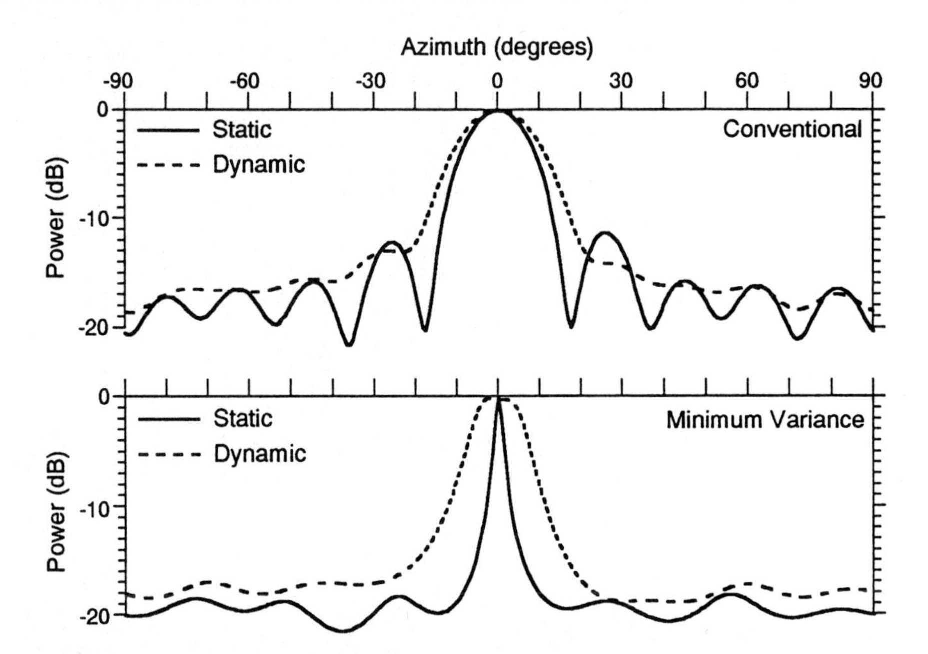

Figure 8.4 A source moved as described in the text [$\vec{\delta k}_x = (2\pi/\lambda)\sin(0.2°)$] relative to a motionless regular linear array. Using a time-bandwidth product of 50, the upper panel displays conventional beamforming's response to the static and dynamic situations, whereas the lower panel displays results produced by the minimum variance technique. In the static case, the source is located at broadside. In the dynamic case, the source's direction of propagation at midaverage coincided with broadside: $\vec{k}^o_{(L-1)/2} \Longleftrightarrow \theta = 0°$.

The source's motion is governed by the dynamic equation

$$\mathbf{v}(l) = \mathcal{A}[\mathbf{v}(l-1)] + \mathbf{u}(l)$$

where $\mathbf{u}$ models accelerations experienced by the moving source. As described by Eq. 8.2 {426} Cartesian coordinate source motion can usually be described by a linear equation, with $\mathcal{A}[\mathbf{v}(l-1)] = \mathbf{A}\mathbf{v}(l-1)$. The source's dynamic equation may be nonlinear when the source's position is expressed in earth coordinates (latitude and longitude, for example) or the source's velocity is better expressed in terms of speed and heading rather than x and y components. Suppose we have any estimate $\widehat{\mathbf{v}}(l-1|l-1)$ of the target's state at snapshot $l-1$ based on the set $\mathbf{Z}_{l-1}$ of observations taken up to and including time $l-1$. If we had constant-velocity motion so that $\mathbf{u} = \mathbf{0}$, we can predict the target's state at snapshot l directly from the dynamic equation with a *one-step* predictor.

$$\widehat{\mathbf{v}}(l|l-1) = \mathcal{A}[\widehat{\mathbf{v}}(l-1|l-1)]$$

This prediction does not take into account an array's observations; we have nothing with which to correct the estimate!

To find the best correction, consider the observation equation expressed by Eq. 8.3 {427}:

$$\mathbf{z}(l) = \mathcal{C}[\mathbf{v}(l)] + \mathbf{n}(l)$$

Given the predicted state $\widehat{\mathbf{v}}(l|l-1)$, we can predict this measurement according to $\widehat{\mathbf{z}}(l|l-1) = \mathcal{C}[\widehat{\mathbf{v}}(l|l-1)]$ and compare it with actual measurements of location $\mathbf{z}(l)$ at snapshot l. The state estimate at snapshot l can be formed by correcting the predicted location by a function of the difference between the observations and their one-step predicted values.

$$\widehat{\mathbf{v}}(l|l) = \widehat{\mathbf{v}}(l|l-1) + \mathcal{G}[\mathbf{z}(l) - \widehat{\mathbf{z}}(l|l-1)] \tag{8.5}$$

The function $\mathcal{G}[\cdot]$ represents a gain function, which we must determine according to some optimality criterion. In many practical cases, this function corresponds to multiplication by a gain matrix $\mathbf{G}$. The difference $\mathbf{z}(l) - \widehat{\mathbf{z}}(l|l-1)$ is known as the *innovations sequence* in tracking and control theory literature. If the predicted measurement equals the actual measurement, this difference equals 0 and the new state estimate should equal the predicted state. We demand, therefore, that the gain function be homogeneous: $\mathcal{G}[\mathbf{0}] = \mathbf{0}$. The performance of a predictor-corrector tracker depends on the choice of the gain function. With small gains, the new state estimate is very close to the predicted state with not much weight given how measurements deviate from their predicted values. Such behavior would be reasonable when the measurement noise $\mathbf{n}$ is relatively large. Conversely, large gains give relatively little credibility to the predicted state and apply when the accelerations $\mathbf{u}$ driving the dynamics are large.

We can find gain function explicitly under (unfortunately) overly ideal circumstances. Suppose we have no observation noise ($\mathbf{n} = \mathbf{0}$) and that the function $\mathcal{C}[\cdot]$ relating the state to the measurement is invertible. This latter assumption means that the array has a geometry with no source location ambiguities {66}. In particular, this assumption would *not* apply when only direction of propagation estimates are available. If valid, however, the state can be predicted directly from the measurement by

$$\mathbf{v}(l) = \mathcal{C}^{-1}[\mathbf{z}(l)]$$

which means that the gain function $\mathcal{G}[\cdot]$ of the state update equation equals $\mathcal{C}^{-1}[\cdot]$.

To derive the optimum gain function $\mathcal{G}[\cdot]$, let's define an error vector as the difference between the actual state $\mathbf{v}$ and the estimated state $\widehat{\mathbf{v}}(l|l)$: $\epsilon(l|l) = \mathbf{v}(l) - \widehat{\mathbf{v}}(l|l)$. Expanding this equation using the state dynamics equation, the state estimate (predictor-corrector) equation, and the measurement equation, we find that

$$\begin{aligned}\epsilon(l|l) = {} & \mathcal{A}[\mathbf{v}(l-1)] - \mathcal{A}[\widehat{\mathbf{v}}(l-1|l-1)] + \mathbf{u}(l) \\ & -\mathcal{G}\left\{\mathcal{C}\left[\mathcal{A}[\mathbf{v}(l-1)] + \mathbf{u}(l)\right] + \mathbf{n}(l) - \mathcal{C}\left[\mathcal{A}[\widehat{\mathbf{v}}(l-1|l-1)]\right]\right\}\end{aligned} \tag{8.6}$$

At this point, the analytically inclined would attempt to form the squared error $\mathcal{E}[\epsilon^2] = \mathcal{E}[\epsilon^t \epsilon]$ and minimize it by varying the gain function $\mathcal{G}[\cdot]$. Because of the nonlinearities, analytic solutions are exceedingly difficult to find. In the following sections, we shall investigate more tractable but practical special cases. The general form of predictor-corrector estimates is

$$
\begin{array}{|rl|}
\hline
\widehat{\mathbf{v}}(l|l-1) = \mathcal{A}[\widehat{\mathbf{v}}(l-1|l-1)] & \text{One-step state predictor} \\
\widehat{\mathbf{z}}(l|l-1) = \mathcal{C}[\widehat{\mathbf{v}}(l|l-1)] & \text{Observation predictor} \\
\widehat{\mathbf{v}}(l|l) = \widehat{\mathbf{v}}(l|l-1) + \mathcal{G}[\mathbf{z}(l) - \widehat{\mathbf{z}}(l|l-1)] & \text{State estimate} \\
\hline
\end{array}
\tag{8.7}
$$

Before proceeding, we should recognize that these equations also apply to the multi-array situation when either one or more than one source is present. Each array's observations can be construed to be statistically independent of the others if their observation noises $\mathbf{n}$ are independent. The observation vector now consists of the concatenation of each array's observations: $\mathbf{z} = \text{col}[\mathbf{z}_1, \ldots, \mathbf{z}_{N_a}]$. When only one source is present, the one-step predictor $\widehat{\mathbf{z}}(l|l-1)$ represents a computation performed locally by each array based on current source location information obtained from the global tracking system. The gain function then combines the location prediction errors from the various arrays to form an updated source location estimate.

In the multisource case, the data association problem arises. The state vector consists of the concatenated position states of the sources in the field: $\mathbf{v} = \text{col}[\mathbf{v}_1, \ldots, \mathbf{v}_{N_s}]$. Assuming source motions are independent of each other, the dynamic operator $\mathcal{A}[\cdot]$ partitions into single-source operators and the acceleration vector $\mathbf{u}$ contains independent components when grouped by source. As in the single-source case, each array produces predicted locations based on global information and measured locations, passing the resulting errors to a global tracking system. The global tracker must associate each array's location estimate errors to a source. This association becomes difficult when a new source enters the field or when tracks cross; more on these difficulties later.

8.4 Tracking Based on Kalman Filtering

Kalman filter-based algorithms make three fundamental assumptions about the tracking problem.

1. The source dynamic equation and the observation equation are *linear*. Under this assumption, the source motion and observation equations Eq. 8.3 {427} become

$$
\begin{aligned}
\mathbf{v}(l) &= \mathbf{A}(l)\mathbf{v}(l-1) + \mathbf{u}(l) \\
\mathbf{z}(l) &= \mathbf{C}(l)\mathbf{v}(l) + \mathbf{n}(l)
\end{aligned}
$$

   This assumption means that the array measurements $\mathbf{z}$ equal linear combinations of the source's position and velocity: In particular, Cartesian coordinates emerge from the array directly.

2. The initial state $\mathbf{v}(-1)$ has a Gaussian probability distribution with mean $\widehat{\mathbf{v}}(-1|-1)$ and covariance matrix $\mathbf{K}_\epsilon(-1|-1)$. Thus, the initial estimation error $\epsilon(-1|-1)$ must have zero mean and some *a priori* covariance.

3. The state input vector and the measurement noise vector are Gaussian random vectors; they are statistically independent from snapshot to snapshot and have covariance matrices $\mathbf{K}_u(l)$ and $\mathbf{K}_n(l)$ respectively. If we assume that the *continuous-time* model is driven by white, Gaussian, component-independent noise accelerations having spectral heights† $\sigma_{a_x}^2 T$ and $\sigma_{a_y}^2 T$, the state-input covariance matrix of the discrete-time state equation (Eq. 8.1 {426}) has the form

$$\mathbf{K}_u = \begin{bmatrix} \frac{1}{3}T^4\sigma_{a_x}^2 & \frac{1}{2}T^3\sigma_{a_x}^2 & 0 & 0 \\ \frac{1}{2}T^3\sigma_{a_x}^2 & T^2\sigma_{a_x}^2 & 0 & 0 \\ 0 & 0 & \frac{1}{3}T^4\sigma_{a_y}^2 & \frac{1}{2}T^3\sigma_{a_y}^2 \\ 0 & 0 & \frac{1}{2}T^3\sigma_{a_y}^2 & T^2\sigma_{a_y}^2 \end{bmatrix}$$

The first assumption amounts to linearizing the relation between measured location coordinates and dynamic equation coordinates. As we have seen, array processing algorithms do not produce Cartesian locations; we have no linear relation in any interesting case. How this relation can be linearized is addressed in §8.4.4. The second assumption means that source accelerations are random and possibly time varying. Such accelerations mean that we model source motion as a kind of random walk. More complex motions, such as maneuvers, are discussed in §8.4.3. Despite doubts concerning the realism of these assumptions, trackers based on linear, Gaussian models represent the simplest possible tracking situations and serve as the springboard for more realistic problems.

8.4.1 Kalman Filter Tracking

We can easily rephrase the Kalman filter equations (Eq. 6.12 {309}) to derive the Kalman tracking algorithm that minimizes the mean-squared estimation error. Repeating those results here,

Kalman Prediction Equations:

$$\begin{aligned} \widehat{\mathbf{v}}(l|l-1) &= \mathbf{A}(l)\widehat{\mathbf{v}}(l-1|l-1) \\ \widehat{\mathbf{z}}(l|l-1) &= \mathbf{C}(l)\widehat{\mathbf{v}}(l|l-1) \\ \boldsymbol{\nu}(l) &= \mathbf{z}(l) - \widehat{\mathbf{z}}(l|l-1) \\ \widehat{\mathbf{v}}(l|l) &= \widehat{\mathbf{v}}(l|l-1) + \mathbf{G}(l)\boldsymbol{\nu}(l) \end{aligned} \tag{8.8a}$$

†These rather curious expressions arise from the vagaries of white noise. Because the covariance function of white noise equals $S\delta(t_1 - t_2)$, "samples" have infinite variance. The quantity S is known as the spectral height and has units of variance/Hertz. Expressing the spectral height as $\sigma^2 T$ allows us to describe continuous-time variances in the context of a sampled-data model.

Kalman Gain Equations:

$$\boxed{\begin{aligned} \mathbf{K}_\nu(l) &= \mathbf{C}(l)\mathbf{K}_\epsilon(l|l-1)\mathbf{C}^t(l) + \mathbf{K}_n(l) \\ \mathbf{K}_\epsilon(l|l-1) &= \mathbf{A}(l)\mathbf{K}_\epsilon(l-1|l-1)\mathbf{A}^t(l) + \mathbf{K}_u(l) \\ \mathbf{G}(l) &= \mathbf{K}_\epsilon(l|l-1)\mathbf{C}^t(l)\mathbf{K}_\nu^{-1}(l) \\ \mathbf{K}_\epsilon(l|l) &= \mathbf{K}_\epsilon(l|l-1) - \mathbf{G}(l)\mathbf{C}(l)\mathbf{K}_\epsilon(l|l-1) \end{aligned}} \tag{8.8b}$$

As promised, the Kalman prediction equations correspond to those given in Eq. 8.7 {434} with the explicit introduction of the innovations sequence $\boldsymbol{\nu}(l)$. The Kalman gain equations govern the computation of the Kalman gain matrix $\mathbf{G}(l)$, which is expressed as a complicated function of the measurement error's covariance matrix $\mathbf{K}_\epsilon$.

As noted before, if we have no measurement noise ($\mathbf{K}_n(l) = \mathbf{0}$) and $\mathbf{C}$ is invertible, then $\mathbf{G}(l) = \mathbf{C}^{-1}(l)$.† Conversely, if the measurement error is large compared with the accelerations, the gain matrix is approximately given by $\mathbf{G}(l) \approx \mathbf{K}_\epsilon(l|l-1)\mathbf{C}^t(l)\mathbf{K}_n^{-1}(l)$, which means that the gain matrix is inversely proportional to the measurement error's covariance matrix. A large amount of measurement noise implies a small gain, forcing the state estimate to rely more heavily on the previous state estimate's prediction; correcting by the current measurement has little impact in this case.

Note that the prediction equations do not influence the Kalman gain equations: The gain matrix $\mathbf{G}$ is seemingly independent of the measurements $\mathbf{Z}_l$. The state estimation error covariance matrix, as it evolves in time, tells us how well we can track the source's motion. This sequence of matrices can then be stored and accessed as needed during an actual tracking operation, making it possible to estimate a track's quality without actually making any measurements or doing any tracking. The reason for this seeming clairvoyance lies in the assumption that we know the measurement noise and acceleration covariance matrices before we track. Unfortunately, this assumption is true in few tracking problems: The variance of an array's location estimate depends on the actual location because resolution depends on location {430}. Consequently, tracking algorithms scale the covariance matrix $\mathbf{K}_n$ dynamically according to tracking estimates, prohibiting precalculation of the error covariance matrix.

Example

Let's simplify the tracker to source motion in the x direction alone. Thus, the position state vector has two components: the source's relative position and velocity in the x direction. The state update matrix equals $\mathbf{A} = \begin{bmatrix} 1 & T \\ 0 & 1 \end{bmatrix}$. The state input vector has a covariance given by $\mathbf{K}_u = \begin{bmatrix} T^4/3 & T^3/2 \\ T^3/2 & T^2 \end{bmatrix} \sigma_a^2$. If we observe only target's position (no velocity data) with some uncertainty because of measurement noise, then $\mathbf{C} = [1 \;\; 0]$, and the measurement noise has a covariance given by $\mathbf{K}_n = [\sigma_n^2]$. Thus, in this simplistic example, measurement noise variance does not vary with source position. All of these assumptions describe a stationary model for the source's motion and the array's measurements. Figs. 8.5 and 8.6 illustrate a sample run of a Kalman tracker for this situation. Note how closely the tracker predicts the source's motion; the tracker's accuracy exceeds that of the raw position measurements.

†Deriving this result implicitly assumes that $\mathbf{K}_\epsilon(l|l-1)$ is invertible. We fulfill this condition when acceleration noise is present.

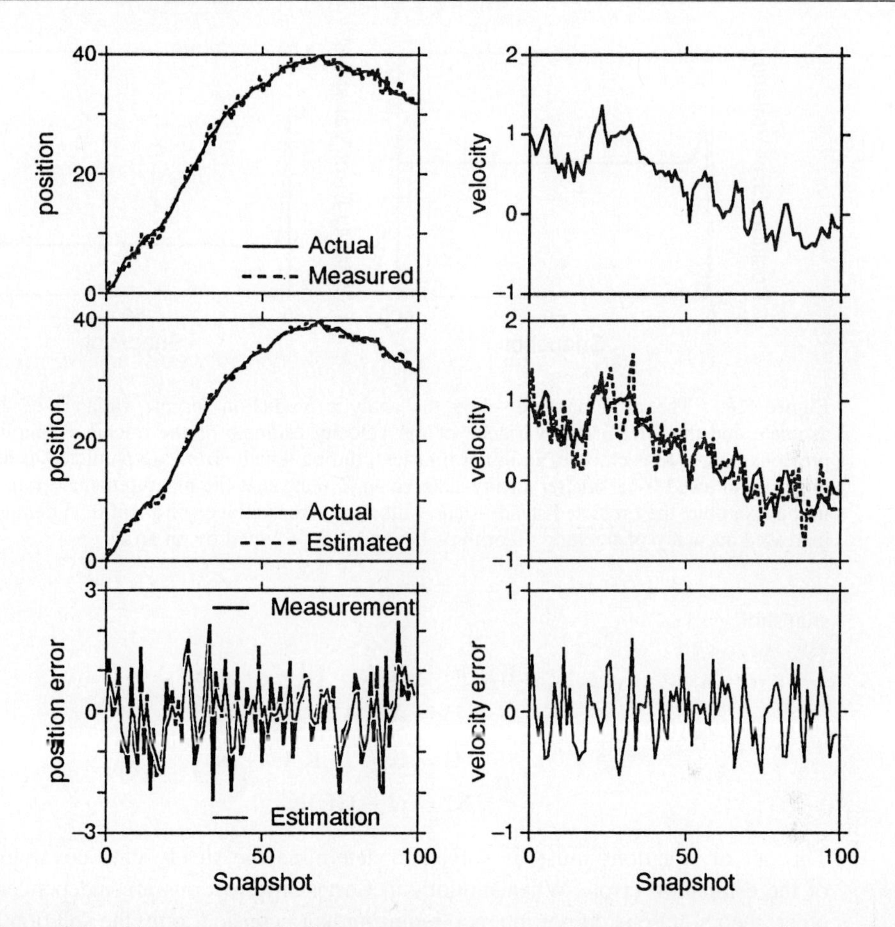

Figure 8.5 The upper panels depict the source's x coordinate position and velocity and the position measurement as a function of snapshot index. No velocity measurement is provided to the tracker. The middle panels display the actual position and velocity, and the Kalman tracker's estimates of them. The bottom panels display the measurement and tracking errors. Note how less noisy the position estimate is compared with the measurements. The parameters of this simulation are $T = 1$, $\sigma_a^2 = 0.1$, and $\sigma_n^2 = 1$. The initial state estimate equaled its actual value of col[0 1] and the initial estimation error covariance matrix equaled identity.

For a tracking algorithm to be viable, it *must* have this property; otherwise, why should we use it?

When the Kalman tracker has converged, the state error covariance matrix $\mathbf{K}_\epsilon(l|l)$ doesn't change from iteration to iteration. In this situation, the predicted state error covariance matrix $\mathbf{K}_\epsilon(l|l-1)$ and the Kalman gain $\mathbf{G}$ also cease to change. As shown in Fig. 8.6, convergence can occur quite rapidly for examples governed by stationary statistics. In steady state, the covariance and Kalman gain matrices must not depend on

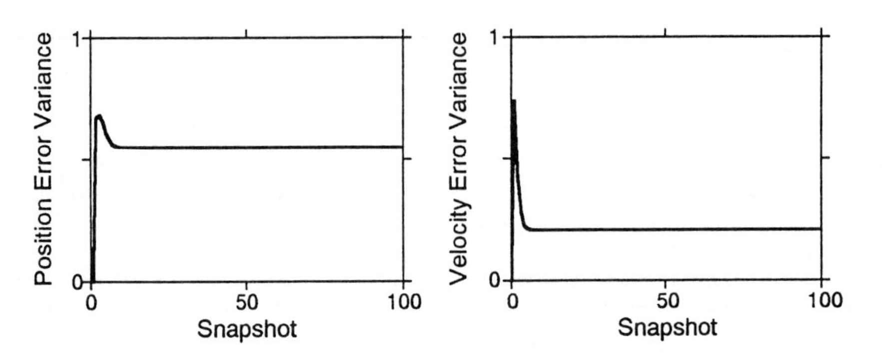

Figure 8.6 The right panel displays the analytic prediction for the variance of the position estimate and the left for the variance of the velocity estimate of the tracker exemplified in the previous figure. In each case, an initial transient, during which errors are predicted to be relatively large, is followed by a smaller steady-state value. Noting that the measurement variance equals 1 in this example, the predicted steady-state estimation error variance of about 0.55 demonstrates the increased accuracy of tracking algorithms beyond that provided by an array.

snapshot.

$$\begin{aligned}
\mathbf{K}_\epsilon(l|l) &= \mathbf{K}_\epsilon(l-1|l-1) \equiv \mathbf{K}_\epsilon \\
\mathbf{K}_\epsilon(l|l-1) &= \mathbf{A}\mathbf{K}_\epsilon\mathbf{A}^t + \mathbf{K}_u \equiv \mathbf{K}_\epsilon^1 \\
\mathbf{G} &= \mathbf{K}_\epsilon^1\mathbf{C}^t\left[\mathbf{C}\mathbf{K}_\epsilon^1\mathbf{C}^t + \mathbf{K}_n\right]^{-1} \\
\mathbf{K}_\epsilon &= (\mathbf{I} - \mathbf{G}\mathbf{C})\mathbf{K}_\epsilon^1
\end{aligned}$$

This set of equations must be solved to determine the steady-state covariance matrix of the estimation error. When motions in Cartesian directions are independent of each other, their solutions do not interact, requiring us to consider only the solution for one of them. Detailing the special one-dimensional tracking problem outlined in the previous example in which we have only position measurements, denote the elements of $\mathbf{K}_\epsilon$ by K_{ij} and those of $\mathbf{K}_\epsilon^1$ by K_{ij}^1.[†] Using the dimensionless parameters $\mu = \sigma_a T^2/\sigma_n$ and $\gamma = \sqrt{\mu^2/3 + 16}$, the steady-state solutions for the one-step error covariance matrix, the estimation error covariance matrix, and the gain matrix are given by [51]

$$\frac{\mathbf{K}_\epsilon^1}{\sigma_n^2} = \begin{bmatrix} \dfrac{\left(\mu + \gamma + \sqrt{\dfrac{4\mu^2}{3} + 2\mu\gamma}\right)^2}{16} - 1 & \dfrac{\mu^2 + \mu\gamma + \mu\sqrt{\dfrac{4\mu^2}{3} + 2\mu\gamma}}{4T} \\ \dfrac{K_{12}^1}{\sigma_n^2} & \dfrac{\mu\left(\sqrt{\dfrac{4\mu^2}{3} + 2\mu\gamma} + 1\right)}{2T^2} \end{bmatrix} \tag{8.9a}$$

[†] Because $\mathbf{K}_\epsilon$ and $\mathbf{K}_\epsilon^1$ are symmetric matrices, use K_{12} (K_{12}^1) to represent K_{21} (K_{21}^1): their respective off-diagonal terms.

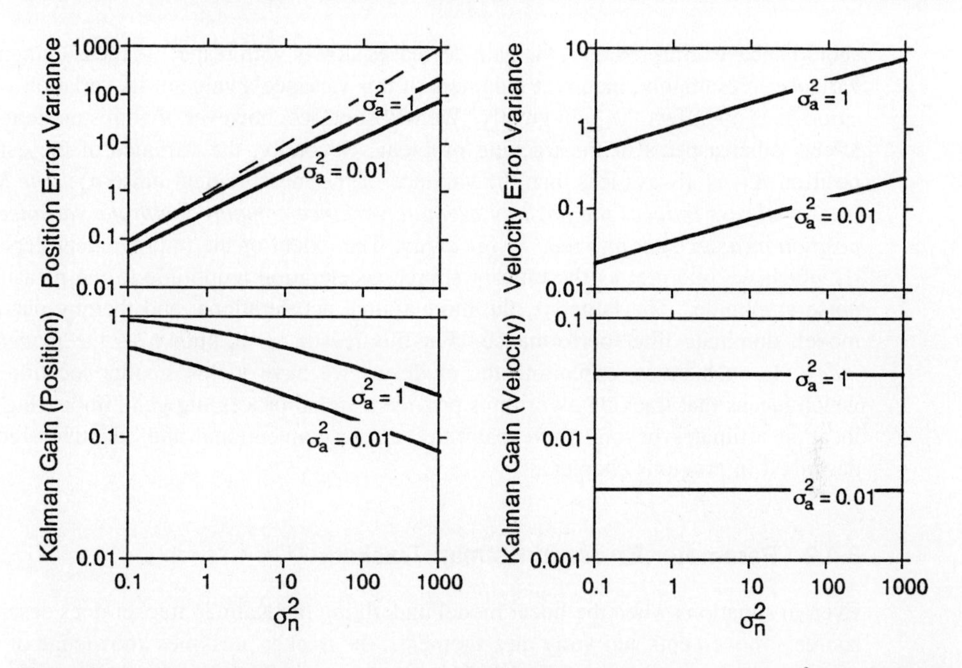

Figure 8.7 The panels show, as a function of the measurement noise variance σ_n^2, the variances of position and velocity estimates obtained from steady-state solutions (Eq. 8.9) to the Kalman filter equations for two values (0.01 and 1.0) of the acceleration variance σ_a^2. The top panels display the position estimate error variance (left) and the velocity estimate error variance produced by the Kalman filter. The dashed line indicates equality between position error variance and measurement error variance σ_n^2. By yielding a variance smaller than σ_n^2, the Kalman filter always improves upon position estimates produced by the measurements. The lower panels display the position and velocity components of the Kalman gain.

$$\frac{\mathbf{K}_\epsilon}{\sigma_n^2} = \begin{bmatrix} \dfrac{K_{11}^1}{\sigma_n^2} - \dfrac{\mu\gamma + \dfrac{2\mu^2}{3}}{2} & \dfrac{\mu^2 + \mu\gamma - \mu\sqrt{\dfrac{4\mu^2}{3} + 2\mu\gamma}}{4T} \\ \dfrac{K_{12}}{\sigma_n^2} & \dfrac{\mu\left(\sqrt{\dfrac{4\mu^2}{3} + 2\mu\gamma} - 1\right)}{2T^2} \end{bmatrix} \tag{8.9b}$$

$$\mathbf{G} = \begin{bmatrix} K_{11}/\sigma_n^2 \\ K_{12}/\sigma_n^2 \end{bmatrix} \tag{8.9c}$$

Despite the complexity of these expressions, plotting them (Fig. 8.7) reveals some important facts. First of all, as the measurement noise variance increases (σ_n^2 increases, which decreases μ and leads to $\gamma \to 4$), the estimation error variance increases and the Kalman gain decreases. This reduced gain puts increasing emphasis on the one-step prediction of the state, which coincides with our interpretation of how predictor-corrector algorithms should behave. Surprisingly, the gain decreases slowly with respect to measurement noise variance; presumably, incorrect judgment of this variance's value or its variation with position does not affect the gain greatly. We will soon see, however, that this modeling error affects other aspects of the tracking problem. Secondly, the variance of the estimated position K_{11} is always less than the variance of the measurement noise σ_n^2: *The Kalman tracker always reduces the position estimate variance compared with the variance of the position measurement provided by the array.* The extent of the improvement depends on μ, which we interpret as the ratio of source acceleration amplitude to the measurement noise amplitude. The larger μ, the more source accelerations, and thereby our motion model, dominate filter performance. For this reason, μ is known as the *maneuvering index*. In such cases, exploiting the model as we have yields smaller location errors, which means that tracking algorithms provide a signal processing gain, improving source location estimates beyond the capabilities of the conventional and adaptive algorithms described in previous chapters.

8.4.2 Parameter Errors in Kalman Trackers

Even in situations when the linear model underlying the Kalman tracker does describe the source's movements and array measurements, the tracker presumes knowledge of critical parameters: The variance of accelerations σ_a^2, the measurement noise variance σ_n^2, and the dynamic equation's initial conditions. If the tracker's values do not coincide with actual ones, the Kalman tracker may *not* compensate for these errors dynamically, producing location and motion estimates more erroneous than analysis would suggest. For example, contrast the results shown in Fig. 8.5 {437}, where parameters coincide, with those in Figs. 8.8 and 8.9 where they don't. Because such parameter errors deteriorate tracker performance, we need either methods of estimating the parameter values or methods to detect model inaccuracy.

Initial conditions. As stated in assumption 2 {434}, our theoretical framework imposes the restriction that the initial conditions of the dynamic equation be Gaussian, having mean equal to the initial state estimate and covariance matrix equal to the estimate's covariance.

$$\mathbf{v}(-1) \sim \mathcal{N}\big(\widehat{\mathbf{v}}(-1|-1), \mathbf{K}_\epsilon(-1|-1)\big)$$

Note that specifying this initial condition requires knowledge of the source's position and velocity, which may not be available from measurements. In particular, velocity may not be produced from an array processing algorithm that does not provide Doppler processing. Initial conditions for position components can usually be obtained from the measurements. In a position-only system, an estimate for the initial velocity can be obtained by differencing two position estimates. As the measurement noise is assumed to be uncorrelated from snapshot to snapshot, some knowledge of the position estimate's variance can be used to form a covariance matrix consistent with this initialization. We thus form initial conditions for the one-dimensional Kalman filter equations when

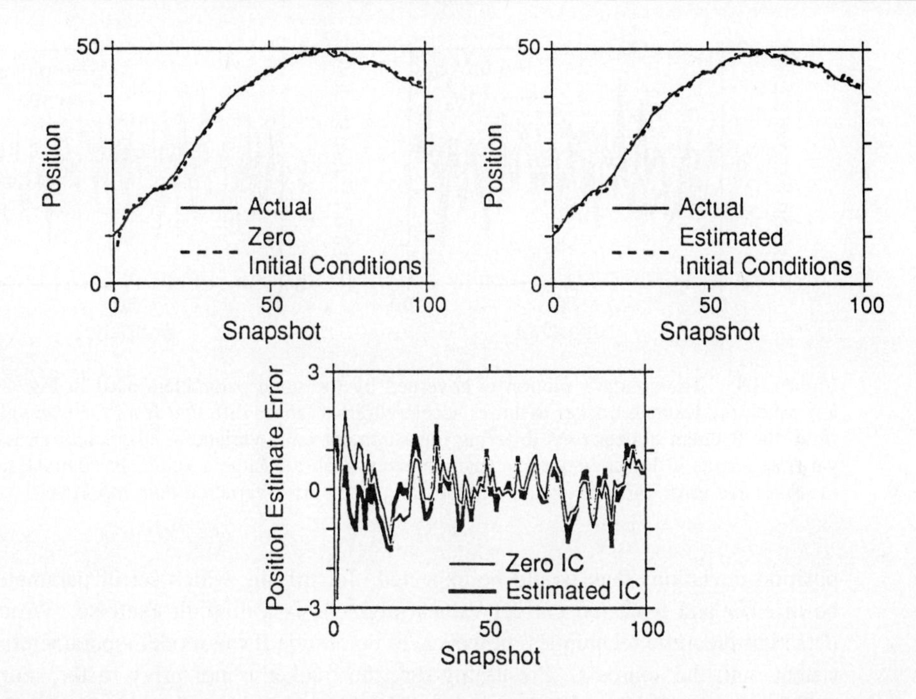

Figure 8.8 The source motion and measurement parameters used in Fig. 8.5 apply here save for the initial conditions (the source's initial location equals 10 instead of 0). In the upper panels, the solid lines indicate the actual x position and the dashed lines the Kalman filter's estimates. In the left panel, the tracker assumes the source's initial position state to be col[0 0] when it actually equals col[10 1] and the initial error covariance matrix to be identity. In the right, the tracker uses two measurements to estimate the initial state. The errors resulting from the two initialization procedures are shown in the bottom panel.

position-only measurements are available according to

$$\boxed{\begin{aligned} v_1(-1) &= z_1(-1) \\ v_2(-1) &= \frac{z_1(-1) - z_1(-2)}{T} \\ \mathbf{K}_\epsilon(-1|-1) &= \begin{bmatrix} \sigma_n^2 & \sigma_n^2/T \\ \sigma_n^2/T & 2\sigma_n^2/T^2 \end{bmatrix} \end{aligned}}$$

Fig. 8.8 demonstrates the improvement in estimate quality made by employing this approach. The computed initial conditions are clearly superior to zero initial conditions at the track's beginning, but they become essentially equal for the remainder of the track.

Model parameter consistency testing. Errors in the Kalman filter model can lead to dramatic differences between estimated and actual source positions. Fig. 8.9 demonstrates errors related to *parameter* errors. Such errors clearly lead to somewhat larger

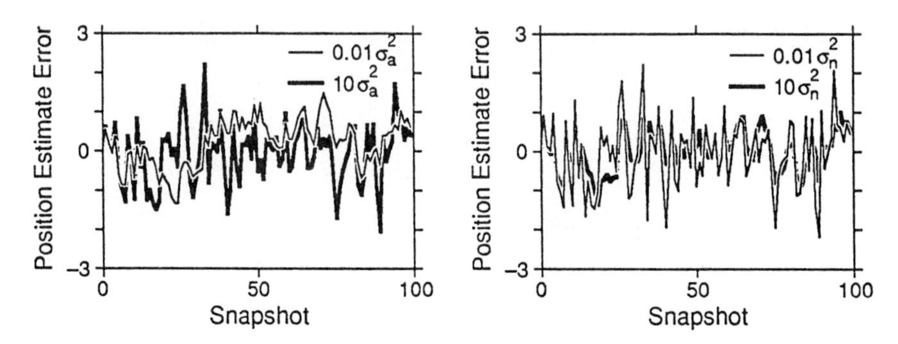

Figure 8.9 The source's motion is governed by the same parameters used in Fig. 8.5. In the left panel, the Kalman tracker assumes acceleration variances different from the true value. In the right, the Kalman tracker uses differing measurement error variances. Too small an acceleration variance results in lower frequency position errors than too large a value. In contrast, too small a measurement error variance yields a higher frequency error variation than too large.

position deviations than would be expected. Identifying which set of parameters might be in error and what the correct values might be is a difficult exercise. We can apply detection-theoretic techniques, however, to determine if the model's parameters are consistent with the source's. Presuming that the tracker is not privy to the source's true position (why would we have a tracker if it did?), we can exploit the statistical structure of the innovations sequence $\boldsymbol{\nu}(l) = \mathbf{z}(l) - \widehat{\mathbf{z}}(l|l-1)$, which depends only on the observations. The innovations should be Gaussian, having zero mean and covariance matrix $\mathbf{K}_\nu(l) = \mathbf{C}(l)\mathbf{K}_\epsilon(l|l-1)\mathbf{C}^t(l) + \mathbf{K}_n(l)$. We can test that the innovations do indeed have these characteristics by noting that the random vector $\mathbf{K}_\nu^{-1/2}(l)\boldsymbol{\nu}(l)$ should have statistically identical (zero mean, unit variance) and uncorrelated components independent of snapshot. Because of this property, this vector's squared norm should have a χ^2 distribution having degrees of freedom equaling the dimension of the observation vector. By summing this norm across L snapshots, a more statistically reliable estimate of it can be obtained. Noting that $\|\mathbf{K}_\nu^{-1/2}\boldsymbol{\nu}\|^2 = \boldsymbol{\nu}^t\mathbf{K}_\nu^{-1}\boldsymbol{\nu}$, we pose our model consistency hypothesis test as

$$\begin{aligned}
\mathcal{H}_0 &: S(L) = \sum_{l=1}^{L} \boldsymbol{\nu}^t(l)\mathbf{K}_\nu^{-1}(l)\boldsymbol{\nu}(l) \sim \chi^2_{L\cdot\dim[\mathbf{z}]} \\
\mathcal{H}_1 &: S(L) = \sum_{l=1}^{L} \boldsymbol{\nu}^t(l)\mathbf{K}_\nu^{-1}(l)\boldsymbol{\nu}(l) \not\sim \chi^2_{L\cdot\dim[\mathbf{z}]}
\end{aligned}$$

where $S(L)$ denotes the consistency statistic summed across L snapshots. This kind of hypothesis test is known as a *null hypothesis test* and is described in §5.1.3 {209}. We determine the decision region $\Re_0$ according to a false-alarm probability specification: $\Pr[S(L) \in \Re_0] = 1 - P_F$. For the χ^2 probability distribution, this region is an interval. Fig. 8.10 demonstrates using this test on our misparameterized examples. Simulations demonstrate that smaller model variances for the accelerations and measurement noise

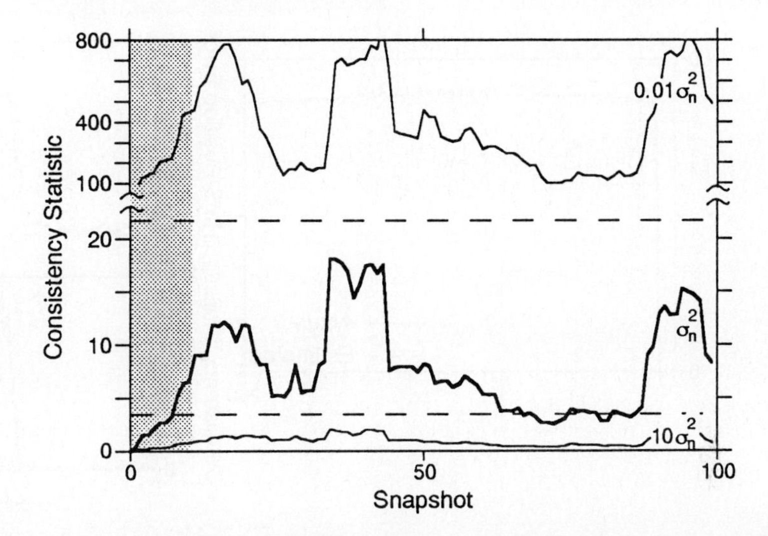

Figure 8.10 The simulations producing Fig. 8.5 and 8.9 are used to test consistency of the Kalman filter's notion of the dynamic and observation equation parameters. The consistency statistic $S(10)$ is computed for the correctly parameterized tracker (heavy line) and the incorrectly parameterized measurement noise variance tracker (lighter lines). The indicated decision boundaries [3.48, 21.85] correspond to a false-alarm probability of 0.05 for a χ^2_{10} distribution. The grayed initial portions indicate the transient portion of the ten-sample running average used to compute $S(10)$.

tend to yield decision statistics that exceed the decision interval's upper limit; larger model variances tend to yield statistics that lie below the decision region's lower limit. Such indications could be used to modify parameters for better agreement between model and reality.

8.4.3 Maneuvering Sources

The dynamic model describing source motion as $\mathbf{v}(l) = \mathbf{A}\mathbf{v}(l-1) + \mathbf{u}(l)$, where $\mathbf{u}(\cdot)$ is a Gaussian random vector uncorrelated from snapshot to snapshot, could be termed the "aimless" model: The source experiences random accelerations that vary randomly with time. Nonzero initial conditions for the source's velocity can be used to describe motion along a particular course. Assuming acceleration variances can be found to well describe a source moving along a particular general course, sudden changes in course—*maneuvers*—are not well modeled by white accelerations having a variance consistent with a fixed course. Because the heart of the Kalman tracking algorithm is the one-step predictor, which essentially means "assume the source is moving exactly like it was before," the position and velocity estimates become suddenly inaccurate postmaneuver. We need to consider methods of tracking sources that tend to change their minds in midtravel.

Checking parameter consistency can be used to detect maneuvers. Fig. 8.11 demonstrates tracking errors during a maneuver and how they can be detected this way. Note

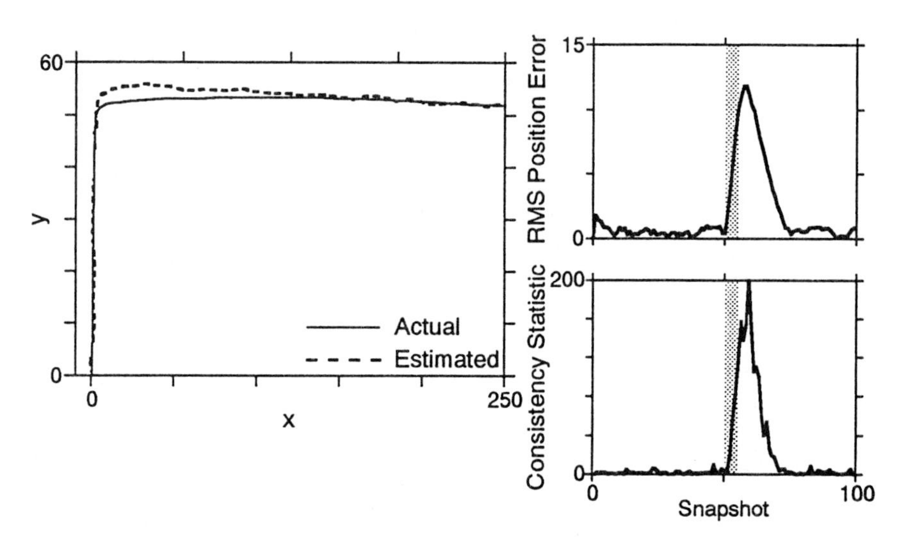

Figure 8.11 A source having an initial heading in the positive y direction makes a sharp right turn at snapshot $l = 49$. The left panel indicates the source's actual path and that estimated by a Kalman tracker employing only position measurements and having the correct source parameters. The top-right panel shows the tracker's root-mean-squared position error and the bottom-right the consistency statistic $S(10)$. The optimum confidence interval for this statistic equals $[9.87, 35.45]$, which corresponds to χ^2_{20} and $P_F = 0.05$. The grayed area indicates the snapshots during which the maneuver occurs. The simulation parameters are $T = 1$, $\sigma^2_{a_x} = 10^{-3}$, $\sigma^2_{a_y} = 10^{-3}$, $\sigma^2_n = 1$, and $\mathbf{v}(-1) = \text{col}[0, 0, 0, 1]$. The maneuver is described by an acceleration lasting five snapshots having constant amplitude equaling $[a_x, a_y] = [1, -0.2]$.

that although the tracker eventually reacquires the source's position, it takes many snapshots to do so. This time is determined by the Kalman gain, which governs how much measurement error contributes to the tracker's position estimate. In the example, measurement noise variance greatly exceeded the source's acceleration noise; as described in §8.4.1 {436}, the gain is roughly proportional to the reciprocal of the measurement noise variance, which increases the reacquisition time. Furthermore, because the error covariances do not depend greatly on this unanticipated maneuver, the tracker's estimate of the position error variances do not indicate the error's dramatic increase during the maneuver. The consistency statistic does, however, dramatically indicate something is wrong in the tracker's assumptions. It is this statistic that can be used to *adapt* the Kalman tracker to maneuvers.

Many techniques have been used to take maneuvers into account; for a summary, see Chap. 4 of Bar-Shalom and Fortmann. One frequently used approach employs multiple Kalman trackers operating in parallel. These trackers' predictions are combined to produce a track (Fig. 8.12). Because maneuvers correspond to unpredictable departures of the dynamic position model from its assumed form, each tracker's assumptions about the measurement process should be the same. They differ in the assumed values for acceleration variances, with trackers that use larger values than the nominal becoming

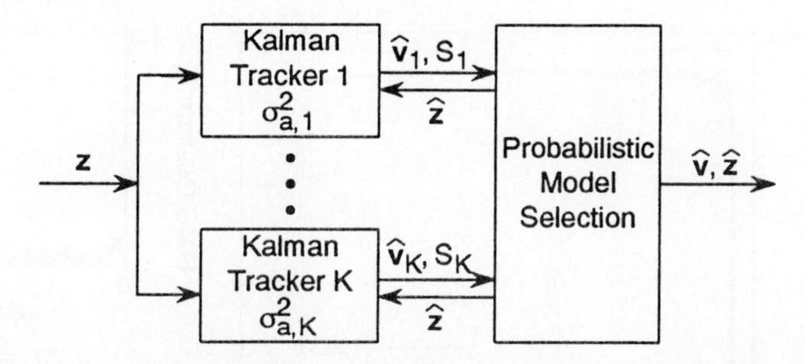

Figure 8.12 Rather than employing a single Kalman tracker, whose position estimate may occasionally deviate significantly from the actual, multiple parallel trackers each produce position estimates. Either the most likely correct estimate is selected or their estimates are linearly combined, weighted by continually updated probabilities of the models' correctness. The predicted measurement associated with the final position state estimate is passed back to the individual trackers.

more appropriate (we hope) during a maneuver. To determine the model most consistent with the observations, we need to compute the consistency statistic from each tracker's innovations sequence and then combine them to produce probabilities that describe each model's correctness during a particular snapshot. To calculate the kth model's correctness probability for the lth snapshot $\Pr[\mathcal{M}_k|\mathbf{Z}_l]$, use Bayes's Rule to create a recursive update formula.

$$\Pr[\mathcal{M}_k|\mathbf{Z}_l] = \frac{\Pr[\mathcal{M}_k|\mathbf{Z}_{l-1}]p_{\boldsymbol{\nu}_k(l)}(\boldsymbol{\nu})}{\sum_k \Pr[\mathcal{M}_k|\mathbf{Z}_{l-1}]p_{\boldsymbol{\nu}_k(l)}(\boldsymbol{\nu})} \tag{8.10}$$

The innovations vector $\boldsymbol{\nu}_k(l)$ for the kth model has a Gaussian probability distribution as described on page 442: $\boldsymbol{\nu}_k(l) \sim \mathcal{N}\big(\mathbf{0}, \mathbf{K}_\nu^{(k)}(l)\big)$, where the covariance matrix is computed according to the kth model's assumptions. To initialize the recursion, *a priori* model probability assignments $\Pr[\mathcal{M}_k|\mathbf{Z}_{-1}]$ must be conjured.

This recursion must be modified to prevent "lock-up": Note that if any model's probability should ever equal 0, it remains so forever afterward whether it becomes most consistent or not. To prevent this effect, we must set a lower bound on each model's probability and adjust all the probabilities so that they sum to 1. Letting $\widetilde{\Pr}[\cdot]$ denote such modified probabilities, this adjustment amounts to

$$\widetilde{\Pr}[\mathcal{M}_k|\mathbf{Z}_l] = \frac{\max\{p_{\text{lower}}, \Pr[\mathcal{M}_k|\mathbf{Z}_l]\}}{\sum_k \max\{p_{\text{lower}}, \Pr[\mathcal{M}_k|\mathbf{Z}_l]\}}$$

Before a maneuver, we should find that the most correct model yields the largest model-correctness probability. Other models intended to describe maneuvers by incorporating larger values for acceleration noise variance should yield smaller probabilities. During a maneuver, these probabilities should change, with one of the alternative models becoming more appropriate. The estimate of the source's position state can be calculated in two ways: We can use that produced by the most likely tracker, or we can simply weight each model's estimate by its probability. In either case, we must employ this consensus state estimate in the calculation of *all* the models' innovation vectors.

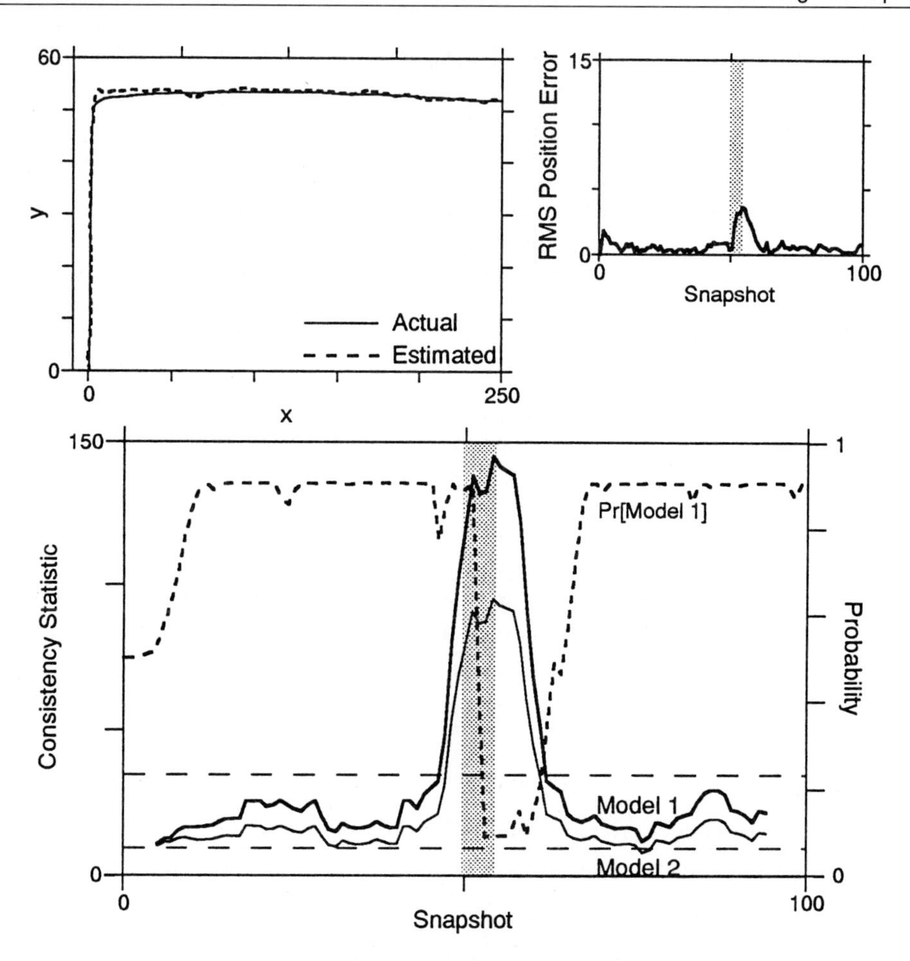

Figure 8.13 The top panel on the left illustrates the multiple-model track for the same maneuvering source shown in Fig. 8.11. Here two models are used: One has the same characteristics as that used in Fig. 8.11, whereas the second assumes an acceleration variance 50 times that used in the first. The top-right panel displays the *RMS* position error for the probabilistically combined tracker. The consistency statistic for both models is shown in the bottom panel and the probability of the first model, the one accurately describing the source except during the maneuver, is overlaid. Note how rapidly model choice changes. The models are initially assumed to be equally likely with a lower probability bound p_{lower} set at 0.1.

Fig. 8.13 illustrates applying two parallel models to the maneuvering source example. This multiple-model tracker exhibits superior abilities to track a maneuvering source, rapidly switching between the models during the turn. Despite neither model passing the consistency test during this critical time, switching to the higher variance, more erratic model greatly reduces the tracker's error from the single-model case.

8.4.4 The Extended Kalman Tracker

Despite the Kalman tracker's importance, as originally derived it is limited to problems in which the dynamic and measurement equations are *linear* functions of the current state. We have seen, however, that using an array to measure azimuth and range to a source yields measurements that depend *nonlinearly* on the source's position and velocity components. The predictor-corrector tracker expressed by Eq. 8.7 {434} provides a starting point for deriving a tracker for such situations. Desiring to apply Kalman tracker ideas despite the nonlinearities, we need to *extend* it to these situations by a careful linearization of the nonlinear equations [7]. As we shall see, the predictor-corrector equations are correct to first-order; higher-order corrections to these equations are needed in the general case.

The dynamic and measurement equations have the form

$$\begin{aligned}\mathbf{v}(l) &= \mathcal{A}[\mathbf{v}(l-1)] + \mathbf{u}(l)\\ \mathbf{z}(l) &= \mathcal{C}[\mathbf{v}(l)] + \mathbf{n}(l)\end{aligned}$$

where $\mathcal{A}[\cdot]$ relates the state to its next value, and $\mathcal{C}[\cdot]$ relates state to measurements, usually in a nonlinear way. The general predictor-corrector formulation for the state estimate is given by

$$\widehat{\mathbf{v}}(l|l) = \widehat{\mathbf{v}}(l|l-1) + \mathcal{G}\big[\mathbf{z}(l) - \mathcal{C}[\widehat{\mathbf{v}}(l|l-1)]\big]$$

Let's linearize the dynamic equation about the most recent estimate of the source's position state using a Taylor series for the dynamic operator $\mathcal{A}[\cdot]$.

$$\begin{aligned}\mathcal{A}[\mathbf{v}(l-1)] = &\underbrace{\mathcal{A}[\widehat{\mathbf{v}}(l-1|l-1)]}_{\text{zeroth order}} + \underbrace{\nabla\mathcal{A}(l-1)\cdot[\mathbf{v}(l-1) - \widehat{\mathbf{v}}(l-1|l-1)]}_{\text{first order}}\\ &\underbrace{+\frac{1}{2}\sum_i \delta_i\big[\mathbf{v}(l-1) - \widehat{\mathbf{v}}(l-1|l-1)\big]^t\,\nabla^2\mathcal{A}_i(l-1)\cdot[\mathbf{v}(l-1) - \widehat{\mathbf{v}}(l-1|l-1)] + \cdots}_{\text{second order}}\end{aligned}$$

Here, $\nabla\mathcal{A}(l-1)$ represents the gradient of the dynamic operator evaluated at the state estimate and $\nabla^2\mathcal{A}_i(l-1)$ the similarly evaluated Hessian of the operator's ith component (see §C.1 {500}).†

$$\nabla\mathcal{A}(l-1) = \left(\nabla_{\mathbf{v}}\mathcal{A}^t[\mathbf{v}]\right)^t\Big|_{\mathbf{v}=\widehat{\mathbf{v}}(l-1|l-1)} \qquad \nabla^2\mathcal{A}_i(l-1) = \nabla^2_{\mathbf{v}}\,\mathcal{A}_i[\mathbf{v}]\Big|_{\mathbf{v}=\widehat{\mathbf{v}}(l-1|l-1)}$$

Because $\mathcal{A}[\cdot]$ is a vector-valued function, its gradient is a square matrix as is the Hessian of each of its components. The quantity δ_i denotes a column matrix containing zero-valued entries save for the ith, which contains unity.

†The notation for the gradient becomes somewhat contorted here because of the seemingly excessive use of transposes. We have taken all vector-valued quantities to be column matrices. To yield the quantity we want consistent with this convention, this contortion accurately describes the required collection of derivatives.

Using second-order Taylor series approximations to the dynamic source motion and measurement equations,

$$\begin{aligned}\mathbf{v}(l) &= \mathcal{A}[\widehat{\mathbf{v}}(l-1|l-1)] + \nabla\mathcal{A}(l-1)\cdot[\mathbf{v}(l-1)-\widehat{\mathbf{v}}(l-1|l-1)] \\ &+\frac{1}{2}\sum_i \delta_i[\mathbf{v}(l-1)-\widehat{\mathbf{v}}(l-1|l-1)]^t\,\nabla^2\mathcal{A}_i(l-1)\cdot[\mathbf{v}(l-1)-\widehat{\mathbf{v}}(l-1|l-1)]+\cdots\end{aligned}$$

$$\begin{aligned}\mathbf{z}(l) &= \mathcal{C}[\widehat{\mathbf{v}}(l|l-1)] + \nabla\mathcal{C}(l)\cdot[\mathbf{v}(l-1)-\widehat{\mathbf{v}}(l|l-1)] \\ &+\frac{1}{2}\sum_i \delta_i[\mathbf{v}(l-1)-\widehat{\mathbf{v}}(l|l-1)]^t\,\nabla^2\mathcal{C}_i(l)\cdot[\mathbf{v}(l-1)-\widehat{\mathbf{v}}(l|l-1)]+\cdots\end{aligned}$$

We find the one-step predictor for the source's position state by evaluating the conditional expected value of our second-order approximation to the dynamic equation with respect to $\mathbf{Z}_{l-1}$.

$$\widehat{\mathbf{v}}(l|l-1) = \mathcal{A}[\widehat{\mathbf{v}}(l-1|l-1)] + \frac{1}{2}\sum_i \delta_i \operatorname{tr}\big[\nabla^2\mathcal{A}_i(l-1)\mathbf{K}_\epsilon(l-1|l-1)\big]$$

Because of the approximation, this predictor can well be biased. Thus, $\mathbf{K}_\epsilon(l-1|l-1)$ represents here a correlation matrix rather than a covariance matrix of the vector $\mathbf{v}(l-1)-\widehat{\mathbf{v}}(l-1|l-1)$. Similar calculations applied to the measurement equation yield its predictor.

$$\widehat{\mathbf{z}}(l|l-1) = \mathcal{C}[\widehat{\mathbf{v}}(l|l-1)] + \frac{1}{2}\sum_i \delta_i \operatorname{tr}\big[\nabla^2\mathcal{C}_i(l)\mathbf{K}_\epsilon(l|l-1)\big]$$

If we eliminated second-order terms from these equations, the predictor-corrector Eq. 8.7 {434} would result. Thus, we see that the predictor-corrector equations represent first-order approximations to some general nonlinear predictor-corrector. Finally, the covariance of the innovations equals

$$\begin{aligned}\mathbf{K}_\nu(l) &= \nabla\mathcal{C}(l)\mathbf{K}_\epsilon(l|l-1)\big(\nabla\mathcal{C}(l)\big)^t + \mathbf{K}_n(l) \\ &+\frac{1}{2}\sum_{i,j}\delta_i\delta_j^t\,\nabla^2\mathcal{C}_i(l)\mathbf{K}_\epsilon(l|l-1)\,\nabla^2\mathcal{C}_j(l)\mathbf{K}_\epsilon(l|l-1)\end{aligned}$$

Making the assumption that the gain operator $\mathcal{G}[\cdot]$ is simply multiplication by the matrix $\mathbf{G}(l)$ as in the linear case, we find that it governed by an equation greatly resembling that of the linear Kalman filter.

$$\mathbf{G}(l) = \mathbf{K}_\epsilon(l|l-1)\big(\nabla\mathcal{C}(l)\big)^t\mathbf{K}_\nu^{-1}(l)$$

We have now derived all of the components of the *extended Kalman filter*. With the identifications

$$\boxed{\begin{aligned}\mathbf{A}(l) &\iff \big(\nabla_{\mathbf{v}}\,\mathcal{A}^t[\mathbf{v}]\big)^t\Big|_{\mathbf{v}=\widehat{\mathbf{v}}(l-1|l-1)} \\ \mathbf{C}(l) &\iff \big(\nabla_{\mathbf{v}}\,\mathcal{C}^t[\mathbf{v}]\big)^t\Big|_{\mathbf{v}=\widehat{\mathbf{v}}(l|l-1)}\end{aligned}} \tag{8.11}$$

the equations governing the first-order extended Kalman tracker are

Kalman Prediction Equations:

$$\begin{aligned}
\widehat{\mathbf{v}}(l|l-1) &= \mathcal{A}[\widehat{\mathbf{v}}(l-1|l-1)] \\
\widehat{\mathbf{z}}(l|l-1) &= \mathcal{C}[\widehat{\mathbf{v}}(l|l-1)] \\
\boldsymbol{\nu}(l) &= \mathbf{z}(l) - \widehat{\mathbf{z}}(l|l-1) \\
\widehat{\mathbf{v}}(l|l) &= \widehat{\mathbf{v}}(l|l-1) + \mathbf{G}(l)\boldsymbol{\nu}(l)
\end{aligned} \tag{8.12a}$$

Kalman Gain Equations:

$$\begin{aligned}
\mathbf{K}_\epsilon(l|l-1) &= \mathbf{A}(l)\mathbf{K}_\epsilon(l-1|l-1)\mathbf{A}^t(l) + \mathbf{K}_u(l) \\
\mathbf{K}_\nu(l) &= \mathbf{C}(l)\mathbf{K}_\epsilon(l|l-1)\mathbf{C}^t(l) + \mathbf{K}_n(l) \\
\mathbf{G}(l) &= \mathbf{K}_\epsilon(l|l-1)\mathbf{C}^t(l)\mathbf{K}_\nu^{-1}(l) \\
\mathbf{K}_\epsilon(l|l) &= \mathbf{K}_\epsilon(l|l-1) - \mathbf{G}(l)\mathbf{C}(l)\mathbf{K}_\epsilon(l|l-1)
\end{aligned} \tag{8.12b}$$

We must remember that the extended Kalman tracker is an approximation; under some circumstances, the approximation can be poor. Using second-order rather than first-order approximations can decrease the error, but at the expense of much more complicated computations.

Kalman Prediction Equations:

$$\begin{aligned}
\widehat{\mathbf{v}}(l|l-1) &= \mathcal{A}[\widehat{\mathbf{v}}(l-1|l-1)] + \frac{1}{2}\sum\nolimits_i \delta_i \operatorname{tr}\left[\nabla^2\mathcal{A}_i(l-1)\mathbf{K}_\epsilon(l-1|l-1)\right] \\
\widehat{\mathbf{z}}(l|l-1) &= \mathcal{C}[\widehat{\mathbf{v}}(l|l-1)] + \frac{1}{2}\sum\nolimits_i \delta_i \operatorname{tr}\left[\nabla^2\mathcal{C}_i(l)\mathbf{K}_\epsilon(l|l-1)\right] \\
\boldsymbol{\nu}(l) &= \mathbf{z}(l) - \widehat{\mathbf{z}}(l|l-1) \\
\widehat{\mathbf{v}}(l|l) &= \widehat{\mathbf{v}}(l|l-1) + \mathbf{G}(l)\boldsymbol{\nu}(l)
\end{aligned} \tag{8.13a}$$

Kalman Gain Equations:

$$\begin{aligned}
\mathbf{K}_\epsilon(l|l-1) &= \mathbf{A}(l)\mathbf{K}_\epsilon(l-1|l-1)\mathbf{A}^t(l) + \mathbf{K}_u(l) \\
&\quad + \frac{1}{2}\sum\nolimits_{i,j} \delta_i\delta_j^t\, \nabla^2\mathcal{A}_i(l-1)\mathbf{K}_\epsilon(l-1|l-1)\, \nabla^2\mathcal{A}_j(l-1)\mathbf{K}_\epsilon(l-1|l-1) \\
\mathbf{K}_\nu(l) &= \mathbf{C}(l)\mathbf{K}_\epsilon(l|l-1)\mathbf{C}^t(l) + \mathbf{K}_n(l) \\
&\quad + \frac{1}{2}\sum\nolimits_{i,j} \delta_i\delta_j^t\, \nabla^2\mathcal{C}_i(l)\mathbf{K}_\epsilon(l|l-1)\, \nabla^2\mathcal{C}_j(l)\mathbf{K}_\epsilon(l|l-1) \\
\mathbf{G}(l) &= \mathbf{K}_\epsilon(l|l-1)\mathbf{C}^t(l)\mathbf{K}_\nu^{-1}(l) \\
\mathbf{K}_\epsilon(l|l) &= \mathbf{K}_\epsilon(l|l-1) - \mathbf{G}(l)\mathbf{C}(l)\mathbf{K}_\epsilon(l|l-1)
\end{aligned} \tag{8.13b}$$

Other techniques, such as artificially increasing the terms in the state dynamics noise covariance matrix $\mathbf{K}_u$ to reflect additional uncertainties in the state estimate related to the nonlinearities, can be used to make the tracker more robust to errors introduced by Taylor series approximations. An important modification is to *iterate* the extended Kalman filter on the same measurements; this method can reduce the tracker's ultimate error caused by the nonlinearities [58: 190–91].

Example

Let's return to the realistic tracking example begun on page 427. Because we have assumed the source dynamic equation to be linear (Eq. 8.3 {427}), the gradient of $\mathcal{A}[\cdot]$ equals the matrix $\mathbf{A}$ and its Hessian equals 0. Assume that the source lies in the array's far field, which means that the array provides *only* azimuth measurements. Using the usual geometrical definitions for a linear array, the operator $\mathcal{C}[\mathbf{v}]$, which specifies how measurements relate to the source's position vector, is given by (see Eq. 8.2 {426})

$$\mathcal{C}[\mathbf{v}] = \tan^{-1}\left(\frac{v_1}{v_3}\right)$$

where the state's first and third components correspond, respectively, to the source's x and y coordinates relative to the array. The matrix $\mathbf{C}$ thus equals

$$\mathbf{C} = \begin{bmatrix} \dfrac{v_3}{v_1^2 + v_3^2} & 0 & -\dfrac{v_1}{v_1^2 + v_3^2} & 0 \end{bmatrix}$$

The measurement-noise covariance matrix equals a scalar, which depends on both azimuth and range (Eq. 8.4 {429}).

$$\mathbf{K}_n(l) = \sigma_n^2(l) \propto \frac{r^2}{\cos^2\theta}$$

Lacking range measurements, not only is the Kalman tracker unable to yield accurate source position estimates, accurate specification of the noise variance becomes impossible and tracker initialization becomes ad hoc. Denoting the initial range estimate as $\hat{r}_0$, the components of the initial position state vector equal

$$v_1(l) = \hat{r}_0 \sin z(l) \qquad v_3(l) = \hat{r}_0 \cos z(l), \; l = -1, -2$$
$$v_2(-1) = \frac{v_1(-1) - v_1(-2)}{T} \qquad v_4(-1) = \frac{v_3(-1) - v_3(-2)}{T}$$

This initial range determines the evolution of the position estimate and, because no range measurements are available, must be chosen arbitrarily. Fig. 8.14 displays the result of tracking the motion described in the previous example. Amazingly, the azimuth estimate remains accurate despite clear errors in estimating the source's position. The tracker's lack of range measurements, however, hampers its ability to initialize itself and monitor measurement noise variance accurately. These effects force the extended Kalman tracker to produce inaccurate position estimates. Fig. 8.15 shows that estimated azimuth varies less than the measured azimuth: Tracking has *improved* our ability to estimate azimuth over that provided by an array calculating individual estimates from snapshot to snapshot. Consequently, employing a tracker even in situations in which it produces inaccurate position estimates can improve measurement quality.

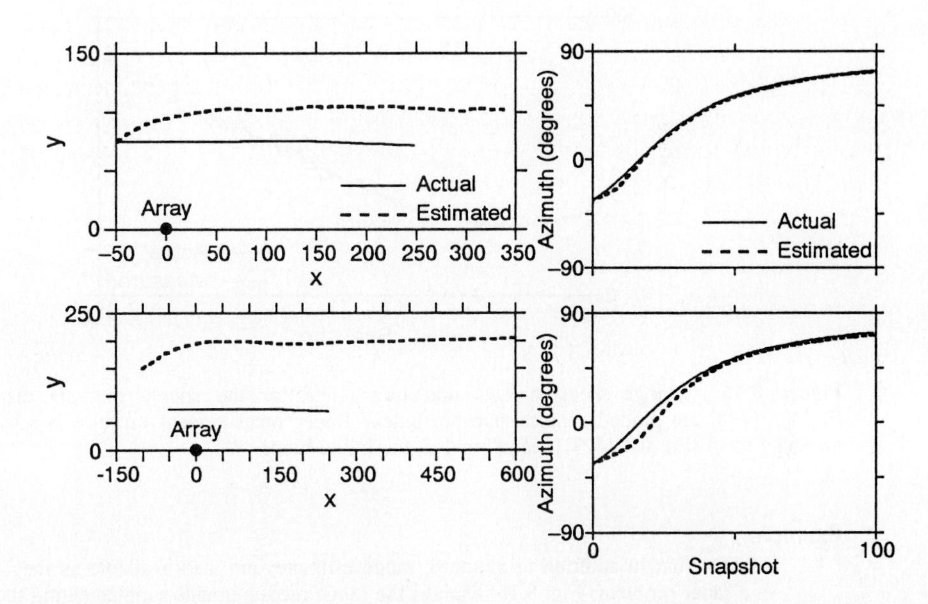

Figure 8.14 The left panels display the true source position in Cartesian coordinates and the extended Kalman tracker's position estimate and the right panels show true azimuth and its estimate. The motion and measurements are identical to those shown in Fig. 8.2 {428}. The tracker uses *only* azimuth measurements from one array in this simulation. In the top row, the initial condition for the source's range clairvoyantly equals the true range, whereas in the bottom row the initial condition is inaccurate (equaling twice the actual initial range of 90.1). Initial conditions for the Cartesian locations and velocities are estimated from the azimuth measurements using the ad hoc range estimate described in the example's text. The initial value of measurement noise variance equals $(1^\circ)^2/\cos^2\theta(-1)$.

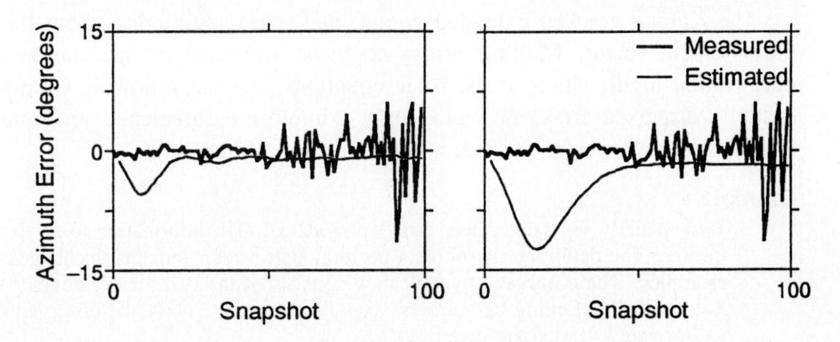

Figure 8.15 The tracker estimates azimuth by calculating its one-step predictor. This prediction along with the measured azimuth for the simulations portrayed in Fig. 8.14 are displayed. The left panel corresponds to the correctly range-initialized simulation, the right to the inaccurately initialized one.

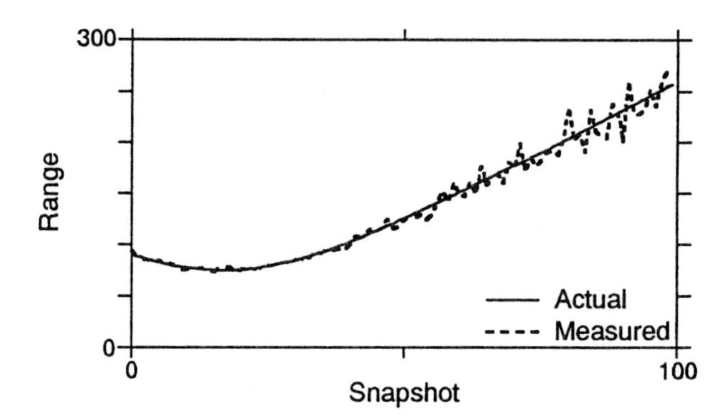

Figure 8.16 Range measurements and true range for the source motions displayed in Fig. 8.2 {428} are plotted against snapshot index. Range measurement variance is assumed proportional to r^4 (Eq. 8.4 {429}), equaling 1 at the initial range.

Example

Assume that in addition to azimuth, range estimates are also available as they would be in a radar problem. Fig. 8.16 displays the range measurements supplementing the azimuth measurements of the previous example. Now observations are related to measurements by Eq. 8.2 {426}. The observation matrix created by linearizing this relationship equals

$$\mathbf{C} = \begin{bmatrix} \dfrac{v_3}{v_1^2+v_3^2} & 0 & -\dfrac{v_1}{v_1^2+v_3^2} & 0 \\ \dfrac{v_1}{\sqrt{v_1^2+v_3^2}} & 0 & \dfrac{v_3}{\sqrt{v_1^2+v_3^2}} & 0 \end{bmatrix}$$

As shown in Fig. 8.17, the extended Kalman tracker now accurately estimates the source's position.

The Kalman tracker, extended or not, makes no assumption about the origin of its measurement vector: Multiple arrays could be providing the measurements as well as one *without* modification of the basic equations. As the following example illustrates, spatially dispersed arrays providing only azimuth measurements can be merged by the Kalman tracker to yield accurate source position estimates.

Example

Two spatially separated linear arrays provide azimuth information to an extended Kalman tracker. The definition of source's position state vector remains unchanged from previous examples. The observation vector now consists of the two array's angular measurements: $\mathbf{z} = \text{col}[\theta_1\ \theta_2]$. Letting both arrays' axes lie along the x axis, the observation operator and its linearized version are given by

$$\mathcal{C}[\mathbf{v}] = \begin{bmatrix} \tan^{-1}\left(\dfrac{v_1}{v_3}\right) \\ \tan^{-1}\left(\dfrac{v_1 - \vec{x}_x}{v_3 - \vec{x}_y}\right) \end{bmatrix}$$

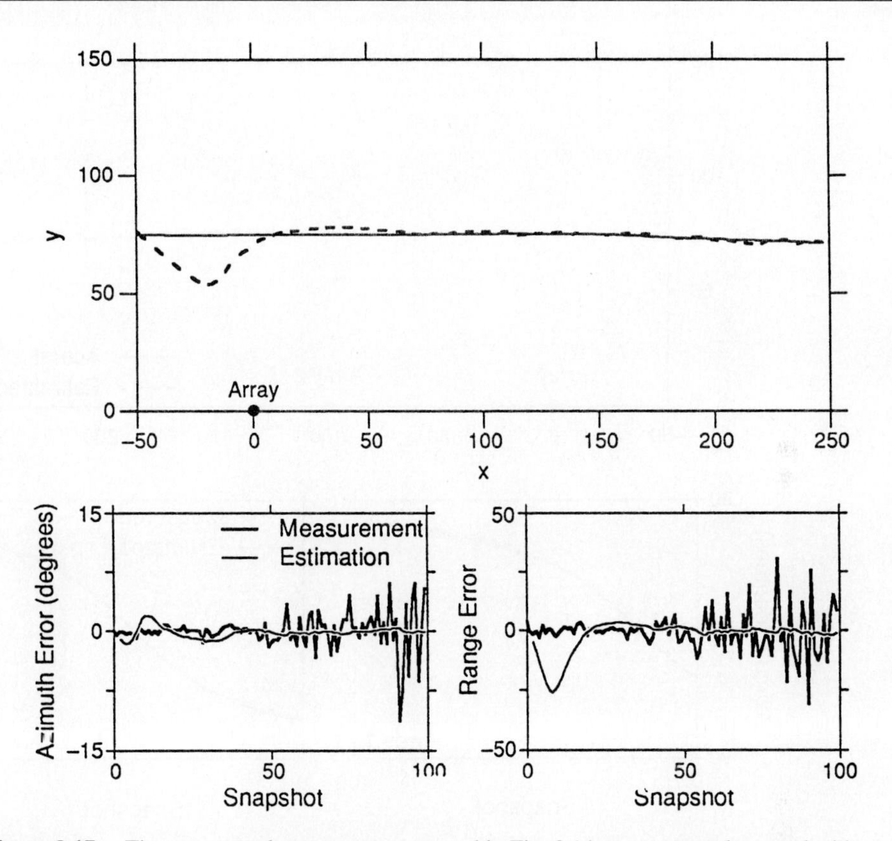

Figure 8.17 The same angular measurements used in Fig. 8.14 are now supplemented with range measurements. The upper panel displays the actual and estimated Cartesian positions when the tracker computed initial position state from the measurements. The lower panels portray how well the extended Kalman tracker improves azimuth (left panel) and range (right) measurements.

$$\mathbf{C} = \begin{bmatrix} \dfrac{v_3}{v_1^2 + v_3^2} & 0 & -\dfrac{v_1}{v_1^2 + v_3^2} & 0 \\ \dfrac{v_3 - \vec{x}_y}{(v_1 - \vec{x}_x)^2 + (v_3 - \vec{x}_y)^2} & 0 & -\dfrac{v_1 - \vec{x}_x}{(v_1 - \vec{x}_x)^2 + (v_3 - \vec{x}_y)^2} & 0 \end{bmatrix}$$

where the first array rests at the coordinate system's origin and the second array at $\vec{x}$. For the simulations, the second array's position is taken to be $(100, 0)$, and the source undergoes motion identical to that described in the example {427}. Fig. 8.18 portrays the arrays' locations and the source's motion as well as the extended Kalman tracker's position estimate. Values for the components of the initial condition $\mathbf{v}(-1)$ are derived from the

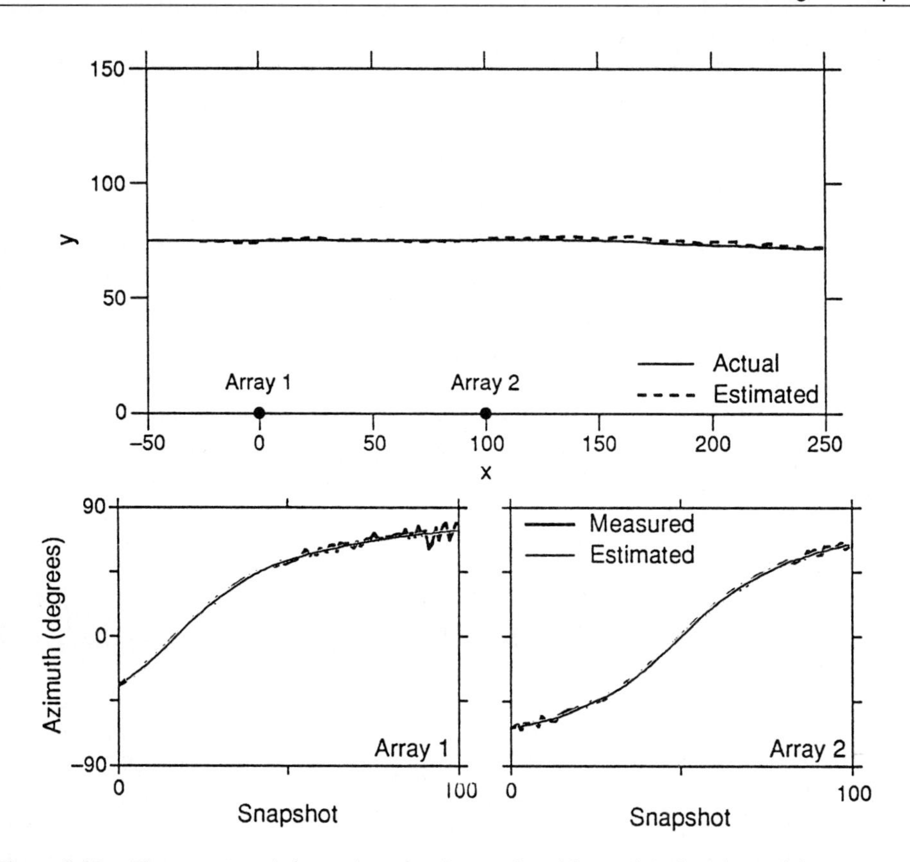

Figure 8.18 The upper panel shows the source's actual position and the locations of the two arrays that provide angular measurements to the extended Kalman tracker estimating source position. The lower panels show the two arrays' angular measurements as well as the tracker's angular estimates.

array's measurements according to†

$$
\begin{aligned}
v_1(l) &= \frac{\vec{x}_x \cos[z_2(l)] \sin[z_1(l)]}{\sin[z_1(l) - z_2(l)]} & v_3(l) &= \frac{\vec{x}_x \cos[z_2(l)] \cos[z_1(l)]}{\sin[z_1(l) - z_2(l)]} \\
v_2(-1) &= \frac{v_1(-1) - v_1(-2)}{T} & v_4(-1) &= \frac{v_3(-1) - v_3(-2)}{T}
\end{aligned}
\tag{8.14}
$$

As opposed to single-array case (Fig. 8.14 {451}), position estimates are much more accurate. The extended Kalman filter essentially supplements triangulation with a source motion model to produce its estimate. Furthermore, the tracker combines the arrays' angular measurements to produce much more accurate angular estimates than could be provided by either array alone.

†This expression holds *only* for the case in which the arrays are separated in the x direction. If the separation vector $\vec{x}$ has components in the y direction, these expressions become much more complicated.

8.5 Multiarray Tracking in Clutter

The Kalman filter–based tracking algorithms discussed in previous sections assume that *all* measurements truly resulted from sources. Reality is much harsher: Because of false alarms and misses, some "measurements" are in reality spurious detections unrelated to sources, and some "measurements" are not provided to the tracker when they fall below the array's detection threshold. Furthermore, ambient sources rather than moving sources can yield measurements. For example, in radar problems, buildings and mountains reflect electromagnetic energy as well as airplanes. Extraneous measurements unrelated to moving sources are termed *clutter*, an annoyance with which viable tracking algorithms must cope. Any "good" clutter removal algorithm must be able to take into account source motion to distinguish source-related measurements from clutter. The algorithms used in previous examples assumed clutter-free measurements, meaning that their applicability to real-world problems is limited. As we shall see, they do lie at the core of viable algorithms.

When we have multiple sources observed by several arrays, another type of clutter arises. As we saw in the example portrayed by Fig. 1.4 {6}, two arrays each producing two source-related measurements can easily lead to ambiguous interpretations. How does the tracker determine the number of sources? Each array indicates two, but in the bearing-only case, triangulation yields *four* possible source locations, two of which are valid and the remaining two are "ghosts."† Why should the tracker believe that all of the measurements are related to sources? Because they could be false alarms, the number of sources could range from zero to four. Thus, clutter-removal, multisource tracking algorithms employing measurements from more than one array must face the well-known *data association* problem: How does a tracking algorithm associate data with sources? Source motion models provide the keys to escaping this quandary: "Ghosts" move in unrealistic ways, and clutter presumably comes and goes, having no physically realizable motion (see Prob. 8.8). Source motion also results in complexity. Sources may cross each other, momentarily occupying the same spatial location as far as the tracker's limited position estimation capability can discern. Once they reseparate, how can the sources be reidentified? This situation represents an additional variant of the data association problem: Associate sequences of source position measurements with reasonable tracks that describe source motion. All of these issues must be addressed by the multiarray tracking algorithm that copes with clutter and multiple sources. Needless to say, these problems are quite difficult; computationally feasible, high-performance trackers remain elusive, with current algorithms making many assumptions (more than a few of which are unreasonable).

Multiarray tracking algorithms capable of dealing with clutter and multiple sources all have a common form. All are based on dynamic source motion and measurement models, which yield Kalman or extended Kalman trackers. These algorithms, however, require addressing the data association problem; the measurements provided to them must in-

†Note how this situation differs greatly from that described in the previous example in which two arrays each produced one measurement. Facing the multiarray, multisource case is much more formidable than the tasks presented in our previous uncluttered example.

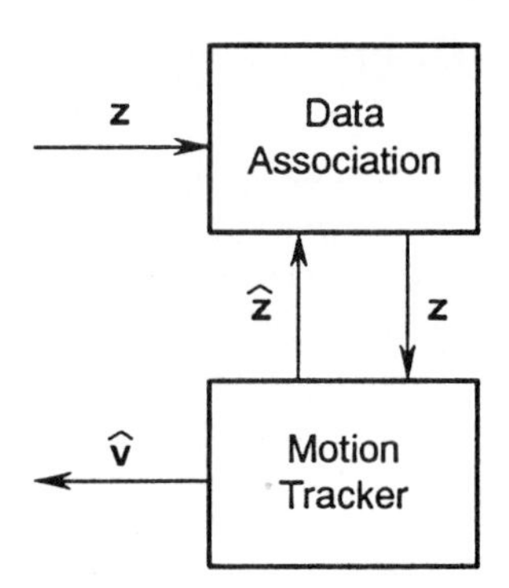

Figure 8.19 Realistic tracking algorithms have two basic components: a high-level component that screens currently available measurements and a dynamic tracker that receives screened measurements and produces updated source position state estimates, initiates new tracks, and deletes irrelevant ones. These components must be designed together because each interacts with the other.

deed be related to sources. Consequently, data association algorithms operate at a higher level, taking raw measurements, considering them in light of previous measurements and tracks, and producing only those measurements that might correspond to viable sources. Measurement evaluation must occur with the cooperation of the tracker: Data association algorithms use model consistency tests to evaluate measurements for assignment to previous tracks. Thus, full-fledged tracking algorithms have the following basic components (Fig. 8.19):

1. High-level *data association* component that sifts through raw measurements, screening out spurious measurements, and produces, with the help of the tracking algorithm, viable measurements
2. *Tracking* algorithm of the Kalman type to evaluate measurements and to produce an updated estimate of source position(s)

As detailed in the next few pages, the tracking algorithm component differs little from that described in previous sections. This section describes the high-level component and how these two components interact. A tracker's high-level component can take two basic forms. The one described here, the *probabilistic data association filter* (*PDAF*) [10: Chaps. 7, 9], is based on statistical models for clutter and source appearance-disappearance. Alternatively, we could use heuristic approaches that try to reason from previous position state estimates what measurements might be viable. Such *artificial intelligence* approaches lead us far afield from the theme of this book; they are described elsewhere [10: §9.5].

When discussing multisource tracking, array measurements other than angular and radial positions become important. Signal characteristics, such as bandwidth, center frequency, and waveform, can all be used to help associate measurements with sources. A tracker can use subtle quantities such as signal amplitude to help determine whether a source is moving toward or away from a given array. Thus, if the array processing algorithms can yield these quantities, they should be included in the measurement vector $\mathbf{z}$, and the relation $\mathcal{C}[\cdot]$ between measurement and state should describe how these nonposition quantities are related to source position and velocity.

The vector of available measurements $\mathbf{z}(l)$ at the lth snapshot consists of a collection of measurement vectors produced by the N_a arrays providing data to the tracker. The nth

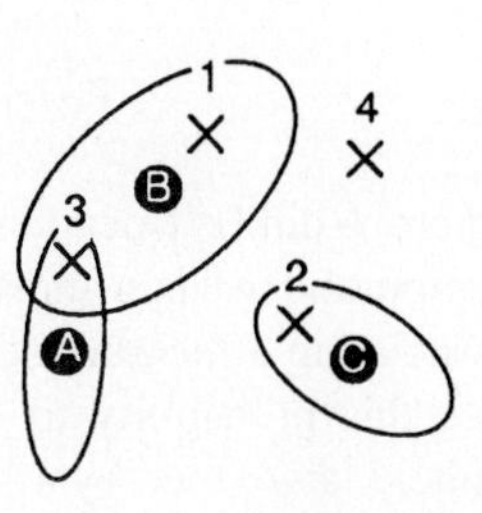

Figure 8.20 Centered about the one-step prediction for the next measurement vector for a given source, labeled here by a capital letter, is an elliptically shaped association region. The region's volume is indirectly defined by the probability P_A for correctly associating source-related measurements. A measurement may be associated with a single source (measurement 1 $\Leftrightarrow$ B, 2 $\Leftrightarrow$ C), more than one source if it lies within the overlap of several association regions (3 $\Leftrightarrow$ A or B), or none (4). In the *PDAF* algorithm, measurement 3 is more closely associated with A than with B: The probability of the former association is greater.

array's measurement vector consists of a basic set of data $\mathbf{z}^b_{n,i_n}$ (bearing, range, etc.) for each of the candidate $N_s^{(n)}(l)$ sources detected by the array during the lth snapshot. Note that these measurements could correspond to clutter as well as sources. The number of measurements observed by the various arrays need not agree even when all observe the same field. The dimension of the measurement vector $\mathbf{z}(l)$ thus equals $\sum_{n=1}^{N_a} N_s^{(n)}(l)$, a potentially large number that varies from one snapshot to another.†

$$\mathbf{z} = \text{col}\big[\underbrace{\mathbf{z}^b_{1,1} \cdots \mathbf{z}^b_{1,N_s^{(1)}}}_{\text{array 1}}\ \underbrace{\mathbf{z}^b_{2,1} \cdots \mathbf{z}^b_{2,N_s^{(2)}}}_{\text{array 2}}\ \cdots\ \underbrace{\mathbf{z}^b_{N_a,1} \cdots \mathbf{z}^b_{N_a,N_s^{(N_a)}}}_{\text{array } N_a}\big]$$

Components selected from each array's measurement vector must be associated with a source, thereby contributing to a track, or must be clutter, spurious measurements, or ghost associations, not contributing to any track.

8.5.1 Association Regions

Because we ultimately feed viable measurements to Kalman trackers, we can use Kalman tracker–based consistency tests as part of the screening procedure to tease out those "sources" that do not seem to be related to a track. Assuming we are in the midst of tracking a source, the statistical characteristics of the innovations, the difference between measurements and their one-step predictions, delineates a region of measurement space most likely related to a source. Taking the one-step predicted measurement vector as the point of focus for this *association* region and noting the Gaussian measurement noise model, source-related measurements $\mathbf{z}(l)$ most likely arise in an elliptically shaped region centered at $\widehat{\mathbf{z}}_j(l|l-1)$, the one-step prediction of measurements related to the jth source (Fig. 8.20). This region is defined explicitly by the quadratic form $\boldsymbol{\nu}_j^t\left(\mathbf{K}_j^\nu\right)^{-1}\boldsymbol{\nu}_j$ appearing in the exponent of the multivariate Gaussian density for the innovations. This quantity equals the test statistic for the consistency test described in §8.4.2 {441}. Consequently, this region's size can be related to a user-specified probability P_A for associating

†Note that this dimension equals the number of basic measurements available during a given snapshot, not the actual dimension of $\mathbf{z}(l)$. As details unfold, this dimension becomes more important than the actual one.

a measurement as source-related according to the statistic's χ^2 distribution.

$$\Pr[\boldsymbol{\nu}_j^t \left(\mathbf{K}_j^\nu\right)^{-1} \boldsymbol{\nu}_j \le \gamma^2] = \Pr[\chi^2_{N\dim[\mathbf{z}^b]} \le \gamma^2] = P_A$$

where $N\dim[\mathbf{z}^b]$ denotes the dimension of collected measurement vectors for an association and γ^2 is a threshold value determined by the user-specified probability P_A for associating a measurement with a source when it really is associated. Ideally, we would like this probability to equal 1; however, this would make the association region infinitely large, thereby associating all measurements with a source (the probability of false associations would also be 1). A more reasonable value for P_A is 0.99.

To cull proposed single-snapshot measurements for a given source, we compute the trial innovations $\mathbf{z}(l) - \widehat{\mathbf{z}}_j(l|l-1)$ for the amalgamated measurements provided by all the arrays and compute the associated consistency statistic. By dismissing those measurements lying outside the ellipse defined by γ^2, we associate measurements likely to be related to the source, but we could associate more than one measurement from an array to a source. As described later, we deal with multiply associated measurements by examining them over several snapshots. All possible combinations of merging the arrays' measurements would need to be formed; the candidate collected measurement vector for an association would be

$$\widetilde{\mathbf{z}}_i = \mathrm{col}\left[\mathbf{z}^b_{1,i_1}\, \mathbf{z}^b_{2,i_2} \cdots \mathbf{z}^b_{N_a,i_{N_a}}\right]$$

where the index i ranges over the possible collections of the tuples $(i_1, i_2, \ldots, i_{N_a})$. Thus, the number of trial associations is huge: $N_A = \prod_{n=1}^{N_a} N_s^{(n)}$. Another approach would be to associate measurements with predicted measurements on an array-by-array basis. In this way, each array's data would be tested individually and compared with the tracker's predictions. Fewer associations would require testing $\left(\sum_{n=1}^{N_a} N_s^{(n)}\right)$, but the associations would be of lower quality: A locally viable measurement may not conform to the arrays' collective wisdom or one locally discarded measurement may not be dismissed from the pooled array perspective. Relying on the rule of thumb "garbage in, garbage out," the more computationally intense algorithm is usually selected.[†] To perform an association algorithm, we need to compute the one-step predictor for each source, defining an association region for each. The sole complication is that a single measurement can lie in more than one association region (Fig. 8.20). At this point in the tracking algorithm, we cannot resolve to which source the associated measurement might belong, but we have eliminated measurements inconsistent with current tracks.

8.5.2 Multiple-Source Tracking in Clutter

The key idea in the *PDAF* algorithm is defining probabilities for viable measurement-to-source associations rather than trying to cull incorrect associations without the tracker's complete support. Generous association regions can be used to discard wildly inconsistent measurements before refining the associations with *PDAF*. Let $P_{i,j}(l)$ denote the

[†]Note that the complexity may not be as daunting as it seems; once trial associations are made, they can be screened in parallel with a consequent reduced computation time.

association probability that the ith collected measurement vector $\widetilde{\mathbf{z}}_i$ is associated with the jth track. Because a measurement may fall within multiple association regions, we cannot impose the restriction that this probability is nonzero for only one combination of i and j. However, for a given track, the probabilities must sum to 1.

$$\sum_{i=0}^{N_A} P_{i,j}(l) = 1$$

In this expression, $P_{0,j}$ denotes the probability that none of the measurements is associated with the jth track.

The estimated position state for the jth track is

$$\widehat{\mathbf{v}}_j(l|l) = \sum_{i=0}^{N_A} \widehat{\mathbf{v}}_{i,j}(l|l) P_{i,j}(l)$$

where $\widehat{\mathbf{v}}_{i,j}(l|l)$ is the estimated position state conditioned on the assignment of the ith collected measurement to the jth track. Clearly, we do not need to compute state estimates for measurement-to-track associations having zero probability. For each specific assignment, the usual form of the Kalman filter predictor equation applies:

$$\widehat{\mathbf{v}}_{i,j}(l|l) = \widehat{\mathbf{v}}_j(l|l-1) + \mathbf{G}_j(l)\boldsymbol{\nu}_{i,j}(l)$$

where the innovations sequence $\boldsymbol{\nu}_{i,j}(l)$ is computed as $\widetilde{\mathbf{z}}_i(l) - \widehat{\mathbf{z}}_j(l|l-1)$, with the one-step measurement predictor equaling $C_j\big[\widehat{\mathbf{v}}_j(l|l-1)\big]$. For the case in which no measurements are associated with a given track, the prediction equation becomes $\widehat{\mathbf{v}}_j(l|l) = \widehat{\mathbf{v}}_j(l|l-1)$, which represents the situation "when you have no data, assume sources keep going in the same direction." The Kalman estimate for the jth track's position state becomes

$$\widehat{\mathbf{v}}_j(l|l) = \widehat{\mathbf{v}}_j(l|l-1) + \mathbf{G}_j(l)\boldsymbol{\nu}_j(l)\,, \quad \boldsymbol{\nu}_j(l) = \sum_{i=1}^{N_A} P_{i,j}(l)\boldsymbol{\nu}_{i,j}(l)$$

with the associated error covariance matrix equaling

$$\mathbf{K}_j^\epsilon(l|l) = \mathbf{K}_j^\epsilon(l|l-1) - [1 - P_{0,j}(l)]\mathbf{G}_j(l)\mathbf{C}_j(l)\mathbf{K}_j^\epsilon(l|l-1) + \widetilde{\mathbf{K}}_j(l)$$

The matrix $\widetilde{\mathbf{K}}_j$ arises because multiple but interacting measurements appear in the tracker.

$$\widetilde{\mathbf{K}}_j(l) = \mathbf{G}_j(l)\left[\left(\sum_{i=1}^{N_A} P_{i,j}(l)\boldsymbol{\nu}_{i,j}(l)\boldsymbol{\nu}_{i,j}^t(l)\right) - \boldsymbol{\nu}_j(l)\boldsymbol{\nu}_j^t(l)\right]\mathbf{G}_j^t(l)$$

Thus, the Kalman filter equations differ little from their structure for ivory tower problems (where no false alarms or multiple associations occur). However, we have assumed that each track is followed independently of others. In terms of the source motion model, we are presuming that intersource acceleration components are uncorrelated. This assumption may be true for widely separated sources, but when positioned close together, airplanes and ships tend to be subjected to related natural phenomena: winds and currents to name a few. Consequently, the commonly used independent-track assumption could use refinement under some conditions.

The key missing algorithmic component is determining the association probabilities $P_{i,j}(l)$. Computing them amounts to the high-level component of the *PDAF* algorithm. Let $\omega(l)$ denote the collection of events (in the sense of probability theory) that measurements have been associated with tracks, clutter, and new tracks at the lth snapshot.

$$\omega(l) = \left\{\omega_{i,j}(l)\right\} \ , \ i = 0, \ldots, N_A(l), \ j = 0, \ldots, T(l)$$

Here, $T(l)$ denotes the number of tracks present at the lth snapshot. The association probability equals the probability of one of these events conditioned on the entire set of observations including the most recent: $P_{i,j}(l) = \Pr[\omega_{i,j}(l)|\mathbf{Z}_l]$. To derive a recursive formula for these probabilities, we must consider the probabilistic structure of how measurements influence track associations and how associations affect measurement models. From the laws of conditional probability,

$$\Pr[\omega(l)|\mathbf{Z}_l] \propto p[\mathbf{z}(l)|\omega(l), \mathbf{Z}_{l-1}] \cdot \Pr[\omega(l)|\mathbf{Z}_{l-1}]$$

where the proportionality constant is inversely related to the unconditional probability density of the measurements. Various tracking algorithms emerge according to the assumptions made for this expression's components.

- $p[\mathbf{z}(l)|\omega(l), \mathbf{Z}_{l-1}]$
  This quantity represents the probability density function of the measurements available at the lth snapshot conditioned on current measurement-to-track associations and on all previous measurements. Usually, each measurement is considered to be statistically independent of the others, leaving

$$p[\mathbf{z}(l)|\omega(l), \mathbf{Z}_{l-1}] = \prod_{i=1}^{N_A} p[\widetilde{\mathbf{z}}_i(l)|\omega(l), \mathbf{Z}_{l-1}]$$

  When measurements arise from different arrays, the intermeasurement independence assumption may well be justified because the noise components of the fields observed at widely separated arrays are (at least) highly uncorrelated. Closely spaced arrays or noiselike energy sources (groups of sound-emitting sea animals in sonar or electromagnetic interference from a city) would mitigate against the interarray independence assumption.

  For measurements arising from the same array, the intermeasurement independence assumption can be seriously questioned. When measurements arise from observations made by an array in disjoint frequency bands or disjoint time intervals within a snapshot, intra-array intermeasurement independence can be safely assumed. If not disjoint in either sense, the observation noise must be uncorrelated spatially for intermeasurement independence. Propagating noise fields fail to have this property (§2.6.3 {48}). Tracking algorithms ignore this correlation, particularly in the difficult (from a tracking perspective) situation when sources approach each other. Such denial of statistical reality must be considered an algorithmic flaw, but the extent to which tracking algorithms suffer as a consequence is not known.

For each basic observation, its conditional density depends on whether it has been associated with a track or not. If associated, we *assume* that this density is Gaussian, having mean equaling its one-step prediction $\widehat{\mathbf{z}}_j(l|l-1)$ and covariance matrix $\mathbf{K}_j^{\nu}(l)$. We have the option of associating measurement with track by assumption or by association regions. If by association regions, the support of this Gaussian density should be restricted to the region and the density scaled by P_A^{-1}. Otherwise, the Gaussian density's support is unrestricted.

If we do not associate a measurement with a track, $\omega_{i,j}(l) = \emptyset$, we need to specify a *spatial clutter model.* Usually, clutter is assumed to be uniformly distributed throughout the measurement space provided by the array. This measurement space is defined by the dim[$\mathbf{z}^b$]-dimensional interval over which the array can provide values. For example, a linear array would provide bearing and range information only over a restricted range. The measurement space would thus be a circle of radius equaling the maximum range. False alarms, the primary source of clutter, can vary in occurrence depending on range, however. If thresholds are lowered in non-*CFAR* detectors as range increases, false-alarm probabilities increase. Furthermore, conventional beamforming algorithms tend to yield false alarms related to sidelobes. In this situation, clutter would tend to occur close to, but not next to, sources and would have a spatial structure. Tracking algorithms have ignored the complexities and used the uniform assumption, assuming $p[\widetilde{\mathbf{z}}_i(l)|\omega_{i,j}(l) = \emptyset, \mathbf{Z}_{l-1}] = 1/V$, where V is the volume of measurement space.

- $\Pr[\omega(l)|\mathbf{Z}_{l-1}]$
  The measurement-to-track association event $\omega(l)$ details the complete tracking situation at the lth snapshot: available tracks, measurement-to-track associations, measurement-to-clutter associations, and measurement-to-new-track associations. A particular tracking situation clearly depends on previous measurements, thereby forcing us to account for all possible paths through the maze of possible tracks and measurement associations. The resultant combinatorial explosion forces tracking algorithms to limit the assumed dependence on the past. The simplest assumption is independence: The probability of a particular tracking situation does not depend on previous measurements. This severe assumption leads to a computationally complex algorithm; making a more reasonable assumption, such as dependence only on the immediate past, leads to extremely complex algorithms that have exceeded computational resources to date.

Under the independence assumption, $\Pr[\omega(l)]$ represents the a priori probability of a tracking situation. Because the event ω expresses details like the existence of a specific track during a snapshot, the number of measurements associated with tracks, and the number of measurements denoted clutter, this probability can be expressed as[†]

$$\begin{aligned}\Pr[\omega] &= \Pr[\omega, \boldsymbol{\delta}(\omega), N_t(\omega), N_c(\omega)] \\ &= \Pr[\omega|\boldsymbol{\delta}(\omega), N_t(\omega), N_c(\omega)] \cdot \Pr[\boldsymbol{\delta}(\omega), N_t(\omega), N_c(\omega)]\end{aligned}$$

where $\boldsymbol{\delta}$ is the vector of indicators that a track exists ($\delta_j = 1$) or not ($\delta_j = 0$), N_t is the number of measurements associated with tracks, and N_c the number associated with clutter. The first term denotes the conditional probability of a tracking situation given these data. This probability amounts to an accounting of the various ways measurements can be associated. *If* each implied situation is equally likely, this probability equals

$$\Pr[\omega|\boldsymbol{\delta}(\omega), N_t(\omega), N_c(\omega)] = \frac{N_t!N_c!}{N_A!}$$

Specification of the last term $\Pr[\boldsymbol{\delta}(\omega), N_t(\omega), N_c(\omega)]$ amounts to detailing the clutter model.

$$\Pr[\boldsymbol{\delta}(\omega), N_t(\omega), N_c(\omega)] = \prod_{j=1}^{T} \left(P_D^j\right)^{\delta_j} \left(1 - P_D^j\right)^{1-\delta_j} \Pr[N_c]\Pr[N_t]$$

P_D^j denotes the detection probability of a track, $\Pr[N_c]$ the probability that N_c clutter measurements are present in the measurement space, and $\Pr[N_t]$ the probability that N_t new tracks arise during a snapshot. Choosing a value for P_D^j amounts to specifying how well the array detects source. As we have seen (the example on page 237, for instance), detection probabilities usually depend on unestimated parameters, making their calculation difficult. A reasonable guess for P_D^j is used in practice. The other probability mass functions $\Pr[N_c]$ and $\Pr[N_t]$ are similarly ad hoc. Usually, Poisson distributions are chosen.

$$\Pr[N_c = n] = \frac{\mu_c^n}{n!}e^{-\mu_c} \qquad \Pr[N_t = n] = \frac{\mu_t^n}{n!}e^{-\mu_t} \tag{8.15}$$

Values for the parameters μ_c and μ_t are as difficult to justify as detection probability values.

With these component probabilities, we can calculate the probability of a particular measurement-to-track association $P_{i,j}(l)$, allowing us to develop a Kalman-like tracker for a clutter-filled environment. Recalling our notation that i indexes basic measurements and j tracks, the equations governing the *PDAF* algorithm coupled with the first-order extended Kalman tracker are

---

[†]The dependence on snapshot has been dropped temporarily from this and succeeding expressions for notational simplicity. Clearly, $\boldsymbol{\delta}$, N_t, and N_c all depend on l.

Kalman Prediction Equations:

$$
\begin{aligned}
\widehat{\mathbf{v}}(l|l-1) &= \mathcal{A}[\widehat{\mathbf{v}}(l-1|l-1)] \\
\widehat{\mathbf{z}}_j(l|l-1) &= \mathcal{C}_j[\widehat{\mathbf{v}}_j(l|l-1)] \\
\widetilde{\mathbf{z}}_i(l) &= \mathrm{col}[\mathbf{z}^b_{1,i_1}(l) \cdots \mathbf{z}^b_{N_a,i_{N_a}}(l)] \\
\boldsymbol{\nu}_{i,j}(l) &= \widetilde{\mathbf{z}}_i(l) - \widehat{\mathbf{z}}_j(l|l-1) \\
\boldsymbol{\nu}_j(l) &= \sum_{i=0}^{N_A} P_{i,j}(l)\boldsymbol{\nu}_{i,j}(l) \\
\widehat{\mathbf{v}}_j(l|l) &= \widehat{\mathbf{v}}_j(l|l-1) + \mathbf{G}_j(l)\boldsymbol{\nu}_j(l)
\end{aligned}
\tag{8.16a}
$$

Kalman Gain Equations:

$$
\begin{aligned}
\mathbf{K}^\epsilon_j(l|l-1) &= \mathbf{A}(l)\mathbf{K}^\epsilon_j(l-1|l-1)\mathbf{A}^t(l) + \mathbf{K}^u_j(l) \\
\mathbf{K}^\nu_j(l) &= \mathbf{C}_j(l)\mathbf{K}^\epsilon_j(l|l-1)\mathbf{C}^t_j(l) + \mathbf{K}^n_j(l) \\
\mathbf{G}_j(l) &= \mathbf{K}^\epsilon_j(l|l-1)\mathbf{C}^t_j(l)\left[\mathbf{K}^\nu_j(l)\right]^{-1} \\
\mathbf{K}^\epsilon_j(l|l) &= \mathbf{K}^\epsilon_j(l|l-1) - [1 - P_{0,j}(l)]\mathbf{G}_j(l)\mathbf{C}_j(l)\mathbf{K}^\epsilon_j(l|l-1) + \widetilde{\mathbf{K}}_j(l) \\
\widetilde{\mathbf{K}}_j(l) &= \mathbf{G}_j(l)\left[\left(\sum_{i=1}^{N_A} P_{i,j}\boldsymbol{\nu}_{i,j}(l)\boldsymbol{\nu}^t_{i,j}(l)\right) - \boldsymbol{\nu}_j(l)\boldsymbol{\nu}^t_j(l)\right]\mathbf{G}^t_j(l)
\end{aligned}
\tag{8.16b}
$$

Data Association Equations:

$$
\begin{aligned}
\Pr[\omega(l)] &\propto p[\mathbf{z}(l)|\omega(l), \mathbf{Z}_{l-1}] \cdot \Pr[\omega(l)|\mathbf{Z}_{l-1}] \\
p[\mathbf{z}(l)|\omega(l), \mathbf{Z}_{l-1}] &= \prod_{i=1}^{N_A} p[\widetilde{\mathbf{z}}_i(l)|\omega(l), \mathbf{Z}_{l-1}] \\
p[\widetilde{\mathbf{z}}_i(l)|\omega(l), \mathbf{Z}_{l-1}] &= \begin{cases} \mathcal{N}\left(\widehat{\mathbf{z}}_j(l|l-1), \mathbf{K}^\nu_j(l)\right), & \omega_{i,j}(l) \neq \emptyset \\ 1/V, & \omega_{i,j}(l) = \emptyset \end{cases} \\
\Pr[\omega(l)|\mathbf{Z}_{l-1}] &= \frac{N_t!N_c!}{N_A!}\prod_{j=1}^{T(l)}\left(P^j_D\right)^{\delta_j}\left(1 - P^j_D\right)^{1-\delta_j}\Pr[N_c]\Pr[N_t] \\
P_{i,j}(l) &\propto \sum_{(i\Leftrightarrow j)\in\omega(l)} \Pr[\omega(l)]
\end{aligned}
\tag{8.16c}
$$

In the next-to-last expression, $\left(P^j_D\right)^{\delta_j}\left(1 - P^j_D\right)^{1-\delta_j}$ means that if the track is present, the probability P^j_D is taken; otherwise, we use the probability $1 - P^j_D$. The mathematical phrase ($i \Leftrightarrow j$) means the association of the ith collected measurement with the jth track. To determine the proportionality constant for the association probabilities, require $\sum_i P_{i,j}(l) = 1$.

Example

When two or more arrays provide measurements to the tracker, the data association problem becomes an important issue. Consider the case of a single source whose angular locations are observed by two arrays. In addition to noisy angular measurements, they also provide clutter: a random number of clutter measurements within each snapshot that are uniformly distributed over the field. The *PDAF* algorithm can be used to sort out which measurements are associated with a source. The arrays are positioned as in the example described on page 452. Fig. 8.18 {454} demonstrates tracking of this source in an uncluttered environment. Because of the clutter, the angular measurements from each array had to be associated. If the two arrays on a given snapshot produced $N^{(1)}$ and $N^{(2)}$ measurements respectively, the number of possible pairs of angles that can be triangulated equals $N^{(1)} \cdot N^{(2)}$. In this example, all possible associations are screened with an acceptance region corresponding to a χ_2^2 random variable and an acceptance probability of $P_A = 0.99$ (these choices yielded $\gamma^2 = 9.12$).

Fig. 8.21 portrays a track produced by the *PDAF* algorithm in a heavily cluttered environment. Clearly, clutter worsens tracking performance, increasing the mean-squared tracking error particularly when the measurement noise increases. Without the data association algorithm, clutter would have completely confused the tracker.

Example

In addition to the moving source described in the previous example, let's add another source that comes close to the first during its movements. Using the same *PDAF* algorithm, the resulting tracks are shown in Fig. 8.22. During the critical time when the sources approach each other, the plot (Fig. 8.23) of the association probability $P_{4,2}(l)$, measurement 4 being the pair of angular measurements that corresponds to source 2, reveals that the tracker tends to "coast" during this period, forcing the Kalman tracker to assume little change in course. Although this example illustrates that a complicated tracking situation can be tackled by the *PDAF* algorithm, incorporating maneuvers as well as clutter into it would probably exceed the tracker's capabilities.

8.5.3 Track Initiation

If a measurement falls outside all of the association regions, more than one measurement falls within a track's association region, or the tracker starts from scratch, we need to initiate a trial track for each unassociated measurement. The initiation procedure amounts to starting a Kalman tracker with the measurement, a straightforward process. What is less straightforward is maintaining a newly initiated track. Because initial conditions for the source's position state must be estimated from measurements, the first few snapshots contributing to a track fail to be consistent with the model hypothesized for the source even when the model is correct (see Fig. 8.18 {454}). Consequently, *if* measurements remain available for the track, it should be maintained over several snapshots until it satisfies the consistency criterion (we thus have a trackable source and away we go!) or it fails (by remaining inconsistent, we must denote the measurements as clutter and stop the new track).† We associate data with a new track by increasing the association

†Dropping a track is always suspect. We drop a track when a source *fails to satisfy our modeling assumptions*. Data termed clutter according to such model-based assumptions may well be an actual source having different

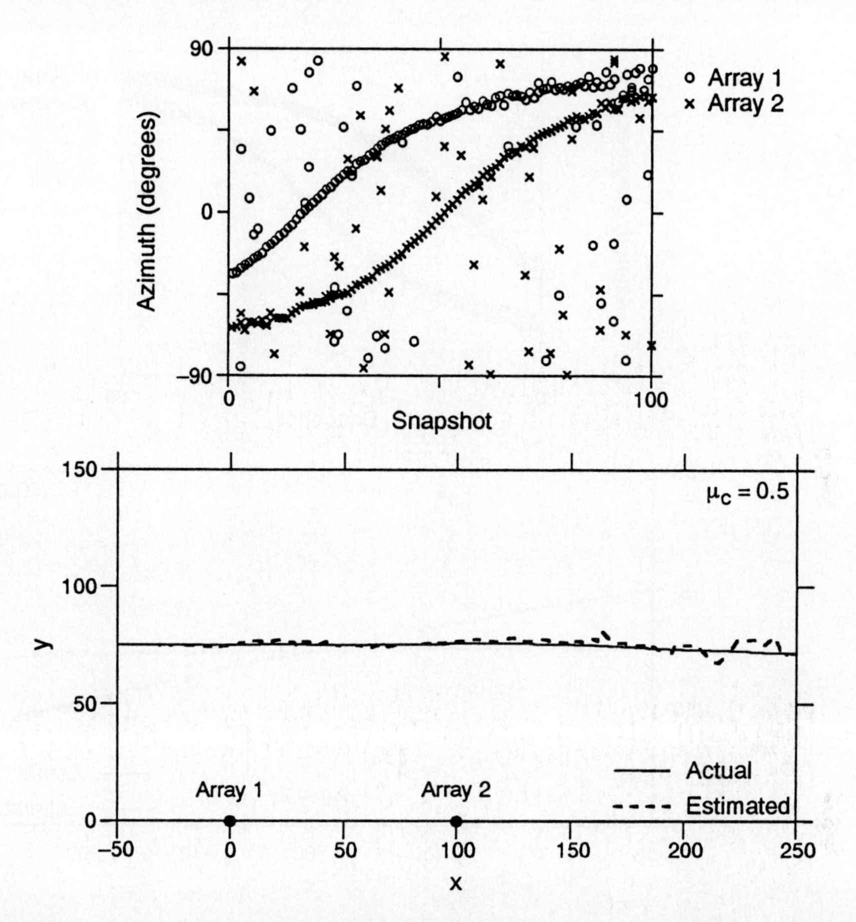

Figure 8.21 A source travels in the field of two linear arrays, one located at the origin and the other at (100, 0). Each array provides only azimuth measurements, but these are contaminated by clutter as shown in the top panel. Employing the Poisson model for clutter (Eq. 8.15 {462}), the parameter μ_c describes the average number of clutter measurements produced by each array. For these simulations, $\mu_c = 0.5$, which corresponds to each array producing, on the average, one clutter datum every two snapshots. The bottom panel shows the track produced by the *PDAF* tracker from these measurements. Fig. 8.18 {454} shows the track produced in the comparable no-clutter situation ($\mu_c = 0$). In these simulations, the clutter produced by each array is independent of that produced by the other array. The tracker presumed only one source in the field and initiated no new tracks ($\mu_t = 0$). The acceptance probability P_A equals 0.99.

probability P_A temporarily until the tracker's transient has subsided.

---

dynamic characteristics. To counteract this phenomenon, we can hypothesize multiple models as we did for maneuvers (see §8.4.3 {443}). Be forewarned that each hypothesized model increases the tracker's complexity.

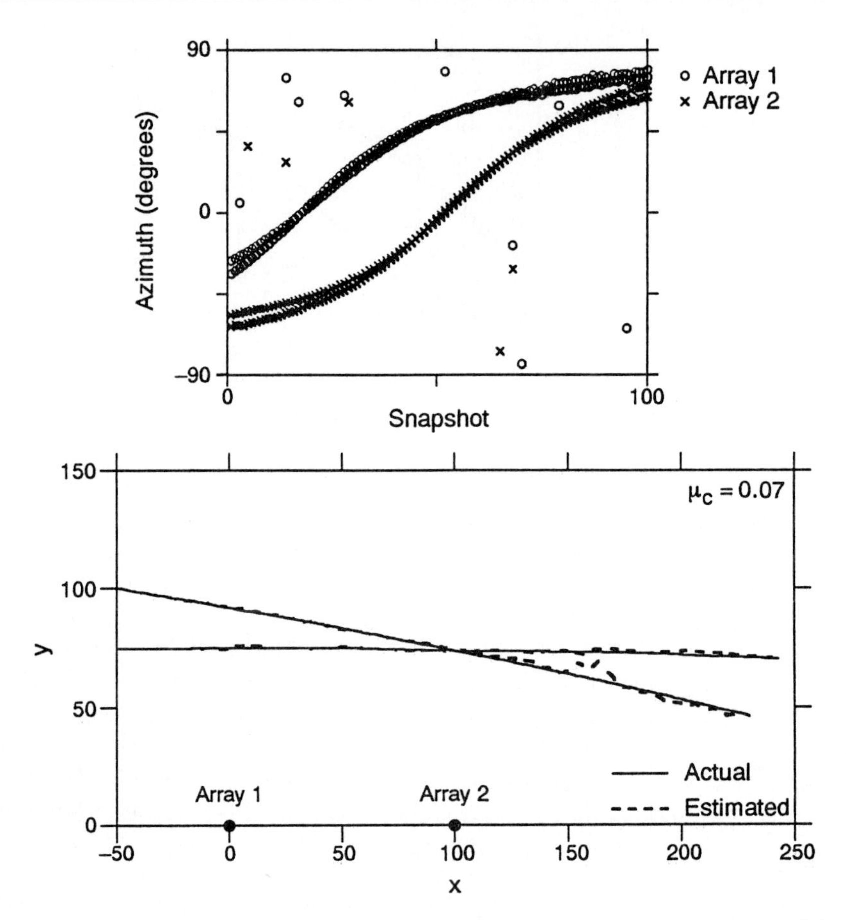

Figure 8.22 Two sources having different courses cross each other near the spatial location (100, 75). The *PDAF* algorithm assimilated angular location measurements (shown in the top panel) provided by the same two arrays used in previous examples. Note how close the measurements become toward the end of the simulation. Clutter is present ($\mu_c = 0.07$) and the algorithm presumed two sources throughout the tracking simulation. The bottom panel displays the paths taken by the sources and the tracker's estimates. Despite light clutter, crossing tracks, and nearby measurements, the *PDAF* algorithm provides reasonably accurate tracks.

Summary

By incorporating source motion models with array measurements, tracking algorithms can provide superior source location estimates than can the array. The nonlinear relation between source position and array measurements causes us to consider Taylor-series-like approximations to the nonlinear dynamic estimation problem. These approximations are ad hoc; how well these approximate trackers perform relative to the optimum ones is not known. Despite these analytic difficulties, the extended Kalman filter underlies virtually all tracking algorithms.

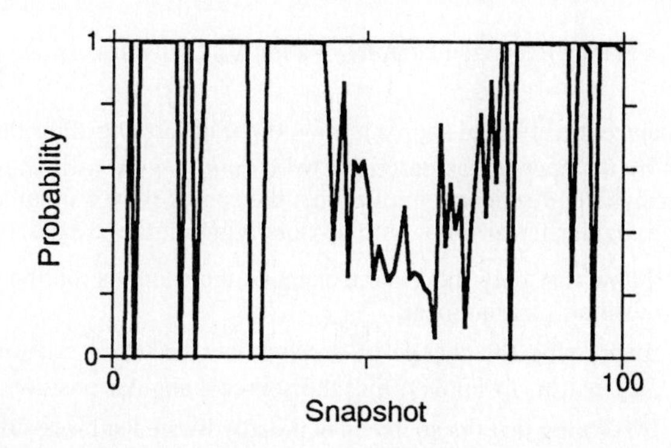

Figure 8.23 The panel plots the tracker's estimate of the probability that measurements are correctly associated with the second source.

The tracker's high-level component must screen the measurements, providing those to the extended Kalman tracker that seems consistent with the tracker's previous notions of source position and velocity. The complexities of maneuvering sources, partial availability of measurements for a source, the data association problem, and physical interactions among source motions makes the screening problem difficult to say the least, leaving the high-level component the weakest. The *PDAF* algorithm described herein is but one approach; many more will emerge as the years go by. It is an example of *centralized tracking*: The tracker merges measurements from all arrays. As mentioned, reasoning systems may help by complementing centralized tracking algorithms. An alternative is *distributed tracking*, in which local source position estimates are refined based on local measurements and occasionally provided estimates from other arrays [148].

Thus, this chapter can only be considered an introduction to evolving field of tracking. Deeper explorations can be pursued by referring to specialized books [10, 17] and to the literature, particularly *IEEE Transactions on Aerospace and Electronic Systems*.

Problems

8.1 We found that constraints among source position and its derivatives and the convolution-like relation governing the evolution of a state equation yielded the discrete-time state equation $\mathbf{v}(l) = \mathbf{A}\mathbf{v}(l-1) + \mathbf{u}(l)$ detailed in Eq. 8.1 {426}.

(a) Show that the state transition matrix $\mathbf{A}$ has the form indicated by the equation.

(b) The continuous-time accelerations $a_x(t)$ and $a_y(t)$ may be statistically independent, whereas the components of $\mathbf{u}(l)$ cannot be. Assuming the accelerations to be independent and white, find the covariance $\mathcal{E}[\mathbf{u}(k)\mathbf{u}'(l)]$.

8.2 In radar applications especially, Doppler measurements supplement the usual static position measurements. From an array's viewpoint, a moving source producing a sinusoidal signal at frequency ω_0 appears to be producing a sinusoid at frequency $\omega_0(1 - \dot{\vec{x}}_\perp/c)$, where $\dot{\vec{x}}_\perp$ is the source's radial velocity relative to the array (Eq. 2.7 {19}).

(a) Find the measurement equation $\mathcal{C}[\mathbf{v}]$ that expresses the relation between Doppler frequency and the source's position state.

(b) Find the linearized Doppler relation $\nabla_{\mathbf{v}} \mathcal{C}[v]$ employed in the extended Kalman tracker.

8.3 To analyze the effect of source motion on beamforming algorithms, we assumed that the path taken by the source was described by $\vec{k}_l^o = \vec{k}_{l-1}^o + \delta\vec{k}$ with $\delta\vec{k}$ a constant. This path, chosen for analytic convenience, implies that the source moves in an artificial manner. Assume we have a regular linear array with sensors lying on the x axis.

(a) Show that only the x component of the source's motion influences the array's source position measurement.

(b) Expressing the change in wavenumber vector in relation to the sine of some angle, $\delta\vec{k}_x = (2\pi/\lambda)\sin(\delta\theta)$, find the source's angular position variation with snapshot.

(c) Assuming that the source is nominally located at broadside, how is the angle $\delta\theta$ related to azimuthal variation?

(d) Translate the angular movement into Cartesian movement by assuming that the source is located at a range r from the array. What kind of accelerations is the source experiencing to cause this motion? Again assume a nominal broadside location.

8.4 The α-β *tracker* is a special-purpose linear tracker that takes the form of the steady-state solution for the Kalman filter. In the special case in which only position measurements are available, $\mathbf{C} = [1 \;\; 0]$ and $\mathbf{K}_n = \sigma_n^2$ in a one-dimensional problem, the Kalman gain matrix becomes a column matrix. In stationary situations, the steady-state Kalman tracker is often expressed by practitioners as[†]

$$\widehat{\mathbf{v}}(l|l) = \widehat{\mathbf{v}}(l|l-1) + \begin{bmatrix} \alpha \\ \beta/T \end{bmatrix} [z(l) - \widehat{z}(l|l-1)]$$

The coefficients α and β are dimensionless, chosen by rules of thumb or from the steady-state Kalman filter equations directly.

(a) Show that the coefficients α and β are related as

$$\mu^2 = \frac{\sigma_a^2 T^4}{\sigma_n^2} = \frac{\beta^2}{1-\alpha}$$

(b) Express β in terms of μ. This expression, in conjunction with the previous result, yields a simplified way of specifying the Kalman gain.

8.5 The expressions for the steady-state variances of position and velocity measurements (Eq. 8.9 {438}) can be used to predict how well Kalman trackers improve location estimates beyond that provided by the array.

(a) Show that the maximum value of K_{11}/σ_n^2, the ratio of the position estimation error variance to the measurement error variance, is one and that this maximum occurs as $\mu \to \infty$.

(b) As the maneuvering index $\mu = \sigma_a T^2/\sigma_n$ decreases, how does this relative variance depend on μ?

(c) Interpret these results: When does using a Kalman tracker help locating a moving source? Can measurement noise swamp a Kalman tracker, rendering attempts to locate a source under extremely noisy conditions pointless? Can a source's motion be so erratic that Kalman trackers cannot predict its location? For what values of the maneuvering index does the tracker reduce the standard deviation of location error by more than a factor of 10?

[†]In those cases in which the position state also includes acceleration as well as position and velocity, a third entry, γ/T^2, sprouts in the gain matrix and we have an α-β-γ tracking filter.

8.6 The multimodel approach to tracking maneuvering sources relies heavily on each component tracker's consistency statistic. Eq. 8.10 {445}, which describes how model probabilities are updated, does not make this dependence explicit.

(a) Show that the probability density function $p_\nu(\nu)$ of each model's innovations vector depends on the data through its consistency statistic at each snapshot and on acceleration variance through $\mathbf{K}_\nu$, the covariance matrix of the innovations.

(b) Assume that one model's innovations probability density is consistently larger than the others. Assuming the models are assumed equally likely at track initiation, show that this model "captures" Eq. 8.10, resulting eventually in its probability equaling 1 and the others equaling 0. Because of this effect, we had to add the concept of p_{lower} to the recursion.

8.7 We can use the Cramér-Rao bound {276ff} to determine how well an unbiased two-array tracking algorithm can perform in a single-source problem. Assume two linear arrays spatially separated along their axes in the x direction produce statistically independent, Gaussian distributed angular measurements.

(a) Find the conditional joint distribution of the measurements given the source's true Cartesian position. Use realistic descriptions of how angular variance depends on source position. Force the standard deviation of each angular estimate to equal 10° when the source lies unit distance from array broadside.

(b) From the Cramér-Rao bound on the mean-squared error produced by any unbiased source position estimator, how does the total mean-squared position error vary with distance from the two arrays? Assume the source's x position is halfway between the arrays and that the array separation is one distance unit

8.8 In the two-array, two-source tracking problem in which only angular measurements are available, measurements must somehow be associated with sources rather than "ghosts." Let's determine how ghosts and sources move in an example. Assume array A is located at the origin of a spatial coordinate system and array B is located at $(\vec{x}_x, \vec{x}_y) = (100, 0)$. Two sources move parallel to each other in the x direction with common speed $s = 20$, which has units of distance/snapshot. At snapshot $l = 0$, source 1 is located at $(10, 100)$ and source 2 at $(10, 150)$.

(a) Derive the triangulation formula given in Eq. 8.14 {454}.

(b) Determine how each array's angular measurements change over the first five snapshots.

(c) Find the positions of the "ghosts" as a function of snapshot.

(d) Over the total observation period of five snapshots for the two sources, how many tracks could be created by an indiscriminate tracker?

(e) Form two reasonable tracks for the ghosts from the position measurements. Do they have constant speed and course?

(f) During the sixth snapshot, source 1 makes a maneuver, turning completely in the positive y direction at the same speed. How does the maneuver of one source affect the ghosts' tracks?

8.9 One interesting aspect of the Kalman filtering equations is that only the dependence of measurements on state is required; the inverse relationship is not explicitly needed. Consequently, auxiliary quantities can be measured and employed to supplement more typical measurements. For example, in tracking applications, Doppler shift could be used to help infer velocity and amplitude to infer range. Let's consider received signal amplitude in some detail.

(a) Assume that the amplitude of the source's radiation is known. How is received signal amplitude related to position state?

(b) Formulate the extended Kalman filter equations that track source position from amplitude *alone*. What kind of source motions would render useless an amplitude-based tracking algorithm?

(c) If amplitude information were used to *supplement* an azimuth-based tracker, what would the extended Kalman filter equations become?

8.10 The *PDAF* algorithm should represent a generalization of the Kalman tracker algorithm that need not worry about associations. In the pristine world in which no clutter or multiple associations occur, the association probabilities $P_{i,j}$ should have rather special values. Define this ideal world more precisely: What probability assignments define the situation in which each source produces a single measurement at each array, no spurious measurements appear, and associations are unique? These choices should yield the result that *PDAF* Eq. 8.16 {463} distills to the extended Kalman equations.

8.11 In problems in which the sources' distance from the tracking array varies considerably and the speed of propagation is relatively small, the source's position may be significantly different by the time the radiation has propagated to the array.

(a) How is measured azimuth at time t related to source position relative to the array and propagation delay $\Delta(t)$?

(b) How is the propagation delay related to source position? Assume a homogeneous medium.

(c) By assuming that source's change of position within Δ is small, linearize these equations using Taylor series approximations. For example, approximate the x component of the position $\vec{x}^o_x$ as

$$\vec{x}^o_x\big(t - \Delta(t)\big) \approx \vec{x}^o_x(t) - \dot{\vec{x}}^o_x(t)\Delta(t)$$

Show that $\Delta(t) = \big(r(t)/c\big)/\big(1 + \dot{r}(t)/c\big)$, where $r(t)$ is the distance between the array and the source at time t.

(d) Formulate the extended Kalman filter equations for this situation.

Appendix A

Probability and Stochastic Processes

A.1 Foundations of Probability Theory

A.1.1 Basic Definitions

The basis of probability theory is a set of events, sample space, and a systematic set of numbers, probabilities, assigned to each event. The key aspect of the theory is the system of assigning probabilities. Formally, a *sample space* is the set Ω of all possible outcomes ω_i of an experiment. An *event* is a collection of sample points ω_i determined by some set-algebraic rules governed by the laws of Boolean algebra. Letting A and B denote events, these laws are

$$
\begin{aligned}
A \cup B &= \{\omega : \omega \in A \text{ or } \omega \in B\} \text{ (union)} \\
A \cap B &= \{\omega : \omega \in A \text{ and } \omega \in B\} \text{ (intersection)} \\
\overline{A} &= \{\omega : \omega \notin A\} \text{ (complement)} \\
\overline{A \cup B} &= \overline{A} \cap \overline{B}
\end{aligned}
$$

The null (or empty) set $\emptyset$ is the complement of Ω. Events are said to be *mutually exclusive* if there is no element common to both events: $A \cap B = \emptyset$.

Associated with each event A_i is a *probability measure* $\Pr[A_i]$ that obeys the *axioms of probability*.

- $\Pr[A_i] \geq 0$
- $\Pr[\Omega] = 1$
- If $A \cap B = \emptyset$, then $\Pr[A \cup B] = \Pr[A] + \Pr[B]$.

The consistent set of probabilities $\Pr[\cdot]$ assigned to events are known as the a priori *probabilities*. From the axioms, probability assignments for Boolean expressions can be computed. For example, simple Boolean manipulations ($A \cup B = A \cup (\overline{A}B)$) lead to

$$\Pr[A \cup B] = \Pr[A] + \Pr[B] - \Pr[A \cap B]$$

Suppose $\Pr[B] \neq 0$. Suppose we know that the event B has occurred; what is the probability that event A also occurred? This calculation is known as the *conditional probability* of A given B and is denoted by $\Pr[A \mid B]$. To evaluate conditional probabilities, consider B to be the sample space rather than Ω. To obtain a probability assignment under these circumstances consistent with the axioms of probability, we must have

$$\Pr[A \mid B] = \frac{\Pr[A \cap B]}{\Pr[B]}$$

The event is said to be *statistically independent* of B if $\Pr[A \mid B] = \Pr[A]$: The occurrence of the event B does not change the probability that A occurred. When independent, the probability of their intersection $\Pr[A \cap B]$ is given by the product of the a priori probabilities $\Pr[A] \cdot \Pr[B]$. This property is necessary and sufficient for the independence of the two events. As $\Pr[A \mid B] = \Pr[A \cap B]/\Pr[B]$ and $\Pr[B \mid A] = \Pr[A \cap B]/\Pr[A]$, we obtain *Bayes's Rule*.

$$\Pr[B \mid A] = \frac{\Pr[A \mid B] \cdot \Pr[B]}{\Pr[A]}$$

A.1.2 Random Variables and Probability Density Functions

A *random variable* X is the assignment of a number, real or complex, to each sample point in sample space. Thus, a random variable can be considered a function whose domain is a set and whose range is, most commonly, the real line. The *probability distribution function* or *cumulative* is defined to be

$$P_X(x) = \Pr[X \leq x]$$

Note that X denotes the random variable and x denotes the argument of the distribution function. Probability distribution functions are increasing functions: If $A = \{\omega : X(\omega) \leq x_1\}$ and $B = \{\omega : x_1 < X(\omega) \leq x_2\}$, $\Pr[A \cup B] = \Pr[A] + \Pr[B] \Longrightarrow P_X(x_2) = P_X(x_1) + \Pr[x_1 < X \leq x_2]$, which means that $P_X(x_2) \geq P_X(x_1)$, $x_1 \leq x_2$.

The *probability density function* $p_X(x)$ is defined to be that function when integrated yields the distribution function.

$$P_X(x) = \int_{-\infty}^{x} p_X(\alpha)\, d\alpha$$

As distribution functions may be discontinuous, we allow density functions to contain impulses. Furthermore, density functions must be nonnegative because their integrals are increasing.

A.1.3 Expected Values

The *expected value* of a function $f(\cdot)$ of a random variable X is defined to be

$$\mathcal{E}[f(X)] = \int_{-\infty}^{\infty} f(x) p_X(x)\, dx$$

Several important quantities are expected values, with specific forms for the function $f(\cdot)$.

- $f(X) = X$
  The *expected value* or *mean* of a random variable is the center of mass of the probability density function. We shall often denote the expected value by m_X or just m when the meaning is clear. Note that the expected value can be a number never assumed by the random variable ($p_X(m)$ can be 0). An important property of the expected value of a random variable is *linearity*: $\mathcal{E}[aX] = a\,\mathcal{E}[X]$, a being a scalar.
- $f(X) = X^2$
  $\mathcal{E}[X^2]$ is known as the *mean-squared value* of X and represents the "power" in the random variable.
- $f(X) = (X - m_X)^2$
  The so-called second central difference of a random variable is its *variance*, usually denoted by σ_X^2. This expression for the variance simplifies to $\sigma_X^2 = \mathcal{E}[X^2] - \mathcal{E}^2[X]$, which expresses the variance operator $\mathcal{V}[\cdot]$. The square root of the variance σ_X is the *standard deviation* and measures the spread of the distribution of X. Among all possible second differences $(X - c)^2$, the minimum value occurs when $c = m_X$ (simply evaluate the derivative with respect to c and equate it to 0).
- $f(X) = X^n$
  $\mathcal{E}[X^n]$ is the nth *moment* of the random variable and $\mathcal{E}\big[(X - m_X)^n\big]$ the nth central moment.
- $f(X) = e^{juX}$
  The *characteristic function* of a random variable is essentially the Fourier Transform of the probability density function.

$$\mathcal{E}\big[e^{juX}\big] \equiv \Phi_X(ju) = \int_{-\infty}^{\infty} p_X(x) e^{+jux}\, dx$$

  The moments of a random variable can be calculated from the derivatives of the characteristic function evaluated at the origin.

$$\mathcal{E}[X^n] = j^{-n} \left.\frac{d^n \Phi_X(ju)}{du^n}\right|_{u=0}$$

A.1.4 Jointly Distributed Random Variables

Two (or more) random variables can be defined over the same sample space. Just as with jointly defined events, the *joint distribution function* is easily defined.

$$P_{X,Y}(x, y) \equiv \Pr[\{X \leq x\} \cap \{Y \leq y\}]$$

The *joint probability density function* $p_{X,Y}(x, y)$ is related to the distribution function via double integration.

$$P_{X,Y}(x, y) = \int_{-\infty}^{x}\int_{-\infty}^{y} p_{X,Y}(\alpha, \beta)\, d\alpha\, d\beta \quad \text{or} \quad p_{X,Y}(x, y) = \frac{\partial^2 P_{X,Y}(x, y)}{\partial x \partial y}$$

Because $\lim_{y\to\infty} P_{X,Y}(x, y) = P_X(x)$, the so-called *marginal density functions* can be related to the joint density function.

$$p_X(x) = \int_{-\infty}^{\infty} p_{X,Y}(x, \beta)\, d\beta \text{ and } p_Y(y) = \int_{-\infty}^{\infty} p_{X,Y}(\alpha, y)\, d\alpha$$

Extending the ideas of conditional probabilities, the *conditional probability density function* $p_{X|Y}(x \mid Y = y)$ is defined (when $p_Y(y) \neq 0$) as

$$p_{X|Y}(x \mid Y = y) = \frac{p_{X,Y}(x, y)}{p_Y(y)}$$

Two random variables are *statistically independent* when $p_{X|Y}(x \mid Y = y) = p_X(x)$, which is equivalent to the condition that the joint density function is separable: $p_{X,Y}(x, y) = p_X(x) \cdot p_Y(y)$.

For jointly defined random variables, expected values are defined similarly as with single random variables. Probably the most important joint moment is the *covariance*:

$$\text{cov}[X, Y] \equiv \mathcal{E}[XY] - \mathcal{E}[X] \cdot \mathcal{E}[Y], \qquad \text{where } \mathcal{E}[XY] = \int_{-\infty}^{\infty}\int_{-\infty}^{\infty} xy p_{X,Y}(x, y)\, dx\, dy$$

Related to the covariance is the (confusingly named) *correlation coefficient*: the covariance normalized by the standard deviations of the component random variables.

$$\rho_{X,Y} = \frac{\text{cov}[X, Y]}{\sigma_X \sigma_Y}$$

When two random variables are *uncorrelated*, their covariance and correlation coefficient equals 0 so that $\mathcal{E}[XY] = \mathcal{E}[X]\mathcal{E}[Y]$. Statistically independent random variables are always uncorrelated, but uncorrelated random variables can be dependent.[†]

A *conditional expected value* is the mean of the conditional density.

$$\mathcal{E}[X \mid Y] = \int_{-\infty}^{\infty} p_{X|Y}(x \mid Y = y)\, dx$$

[†]Let X be uniformly distributed over $[-1, 1]$ and let $Y = X^2$. The two random variables are uncorrelated but are clearly not independent.

Note that the conditional expected value is now a function of Y and is therefore a random variable. Consequently, it too has an expected value, which is easily evaluated to be the expected value of X.

$$\mathcal{E}\big[\mathcal{E}[X \mid Y]\big] = \int_{-\infty}^{\infty} \left[\int_{-\infty}^{\infty} x p_{X|Y}(x \mid Y = y)\, dx \right] p_Y(y)\, dy = \mathcal{E}[X]$$

More generally, the expected value of a function of two random variables can be shown to be the expected value of a conditional expected value: $\mathcal{E}\big[f(X, Y)\big] = \mathcal{E}\big[\mathcal{E}[f(X, Y) \mid Y]\big]$. This kind of calculation is frequently simpler to evaluate than trying to find the expected value of $f(X, Y)$ "all at once." A particularly interesting example of this simplicity is the *random sum of random variables*. Let L be a random variable and $\{X_l\}$ a sequence of random variables. We will find occasion to consider the quantity $\sum_{l=1}^{L} X_l$. Assuming that the each component of the sequence has the same expected value $\mathcal{E}[X]$, the expected value of the sum is found to be

$$\begin{aligned} \mathcal{E}[S_L] &= \mathcal{E}\left[\mathcal{E}\left[\sum_{l=1}^{L} X_l \mid L\right]\right] \\ &= \mathcal{E}\big[L \cdot \mathcal{E}[X]\big] \\ &= \mathcal{E}[L] \cdot \mathcal{E}[X] \end{aligned}$$

A.1.5 Random Vectors

A *random vector* $\mathbf{X}$ is an ordered sequence of random variables $\mathbf{X} = \text{col}[X_1, \ldots, X_L]$. The density function of a random vector is defined in a manner similar to that for pairs of random variables considered previously. The expected value $\mathcal{E}[\mathbf{X}] = \mathbf{m}_X$ of a random vector is the vector of expected values.

$$\mathcal{E}[\mathbf{X}] = \int_{-\infty}^{\infty} \mathbf{x} p_{\mathbf{X}}(\mathbf{x})\, d\mathbf{x} = \text{col}\big[\mathcal{E}[X_1], \ldots, \mathcal{E}[X_L]\big]$$

The *covariance matrix* $\mathbf{K}_X$ is an $L \times L$ matrix consisting of all possible covariances among the random vector's components.

$$K_{ij}^X = \text{cov}[X_i, X_j] = \mathcal{E}[X_i X_j^*] - \mathcal{E}[X_i]\,\mathcal{E}[X_j^*] \quad i, j = 1, \ldots, L$$

Using matrix notation, the covariance matrix can be written as $\mathbf{K}_X = \mathcal{E}\big[(\mathbf{X} - \mathcal{E}[\mathbf{X}])(\mathbf{X} - \mathcal{E}[\mathbf{X}])'\big]$. Using this expression, the covariance matrix is seen to be a symmetric matrix and, when the random vector has no zero-variance component, its covariance matrix is positive definite. Note in particular that when the random variables are real valued, the diagonal elements of a covariance matrix equal the variances of the components: $K_{ii}^X = \sigma_{X_i}^2$. *Circular* random vectors are complex valued with uncorrelated, identically distributed, real and imaginary parts. In this case, $\mathcal{E}\left[|X_i|^2\right] = 2\sigma_{X_i}^2$ and

$\mathcal{E}\left[X_i^2\right] = 0$. By convention, $\sigma_{X_i}^2$ denotes the variance of the real (or imaginary) part. The characteristic function of a real-valued random vector is defined to be

$$\Phi_{\mathbf{X}}(j\boldsymbol{\nu}) = \mathcal{E}\left[e^{j\boldsymbol{\nu}'\mathbf{X}}\right]$$

The maximum of a random vector is a random variable whose probability density is usually quite different than the distributions of the vector's components. The probability that the maximum is less than some number μ equals the probability that *all* of the components are less than μ.

$$\Pr[\max \mathbf{X} < \mu] = P_{\mathbf{X}}(\mu, \ldots, \mu)$$

Assuming that the components of $\mathbf{X}$ are statistically independent, this expression becomes

$$\Pr[\max \mathbf{X} < \mu] = \prod_{i=1}^{\dim \mathbf{X}} P_{X_i}(\mu)$$

A.1.6 Gaussian Random Variables

The random variable X is said to be a *Gaussian random variable*† if its probability density function has the form

$$p_X(x) = \frac{1}{\sqrt{2\pi\sigma^2}} \exp\left\{-\frac{(x-m)^2}{2\sigma^2}\right\}$$

The mean of such a Gaussian random variable is m and its variance σ^2. As a shorthand notation, this information is denoted by $X \sim \mathcal{N}(m, \sigma^2)$, where "$\sim$" means "distributed as." The characteristic function $\Phi_X(\cdot)$ of a Gaussian random variable is given by

$$\Phi_X(ju) = e^{jmu} \cdot e^{-\sigma^2 u^2/2}$$

No closed form expression exists for the probability distribution function of a Gaussian random variable. For a zero-mean, unit-variance, Gaussian random variable $\big(\mathcal{N}(0,1)\big)$, the probability that it *exceeds* the value x is denoted by $Q(x)$.

$$\Pr[X > x] = 1 - P_X(x) = \frac{1}{\sqrt{2\pi}} \int_x^\infty e^{-\alpha^2/2}\, d\alpha \equiv Q(x)$$

A plot of $Q(\cdot)$ is shown in Fig. A.1. When the Gaussian random variable has nonzero mean or nonunit variance, the probability of it exceeding x can also be expressed in terms of $Q(\cdot)$.

$$\Pr[X > x] = Q\left(\frac{x-m}{\sigma}\right), \quad X \sim \mathcal{N}(m, \sigma^2)$$

The inverse function $Q^{-1}(\cdot)$, tabulated at the decade points in Table A.1, is required to define thresholds in Gaussian hypothesis testing and detection problems.

†Gaussian random variables are also known as *normal* random variables.

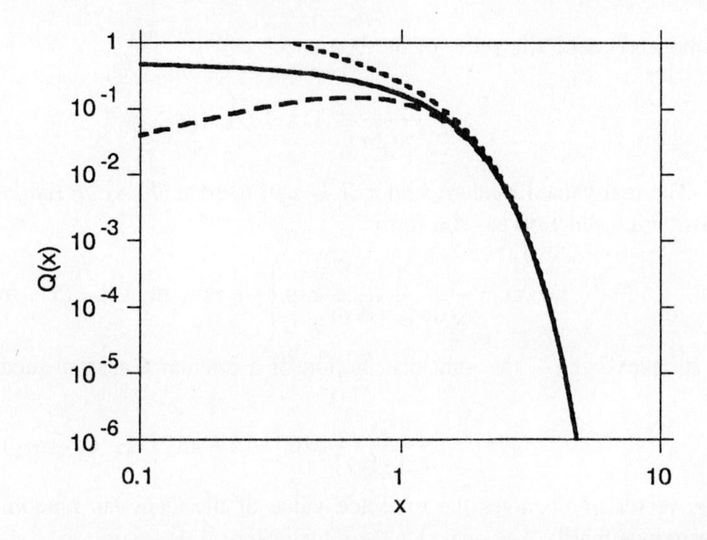

Figure A.1 The function $Q(\cdot)$ is plotted on logarithmic coordinates. Beyond values of about two, this function decreases quite rapidly. Two approximations are also shown that correspond to the upper and lower bounds given by Eq. A.1.

x	$Q^{-1}(x)$
10^{-1}	1.281
10^{-2}	2.396
10^{-3}	3.090
10^{-4}	3.719
10^{-5}	4.265
10^{-6}	4.754

Thresholds in the Neyman-Pearson variant of the likelihood ratio test can be determined from the tabulated values. Note how little the inverse function changes for decade changes in its argument; $Q(\cdot)$ is indeed *very* nonlinear.

TABLE A.1 Values of $Q^{-1}(\cdot)$.

Integrating by parts, $Q(\cdot)$ is bounded (for $x > 0$) by

$$\frac{1}{\sqrt{2\pi}} \cdot \frac{x}{1+x^2} e^{-x^2/2} \leq Q(x) \leq \frac{1}{\sqrt{2\pi}\,x} e^{-x^2/2} \tag{A.1}$$

As x becomes large, these bounds approach each other and either can serve as an approximation to $Q(\cdot)$; the upper bound is usually chosen because of its relative simplicity. The lower bound can be improved; noting that the term $x/(1+x^2)$ decreases for $x < 1$ and that $Q(x)$ increases as x decreases, the term can be replaced by its value at $x = 1$

without affecting the sense of the bound for $x \leq 1$.

$$\frac{1}{2\sqrt{2\pi}} e^{-x^2/2} \leq Q(x), \quad x \leq 1 \tag{A.2}$$

The real-valued random vector $\mathbf{X}$ is said to be a *Gaussian random vector* if its joint distribution function has the form

$$p_{\mathbf{X}}(\mathbf{x}) = \frac{1}{\sqrt{\det[2\pi \mathbf{K}_X]}} \exp\left\{-\frac{1}{2}(\mathbf{x} - \mathbf{m}_X)^t \mathbf{K}_X^{-1}(\mathbf{x} - \mathbf{m}_X)\right\}$$

If complex-valued, the joint distribution of a circular Gaussian random vector is given by

$$p_{\mathbf{X}}(\mathbf{x}) = \frac{1}{\det[\pi \mathbf{K}_X]} \exp\left\{-(\mathbf{x} - \mathbf{m}_X)' \mathbf{K}_X^{-1}(\mathbf{x} - \mathbf{m}_X)\right\} \tag{A.3}$$

The vector $\mathbf{m}_X$ denotes the expected value of the Gaussian random vector and $\mathbf{K}_X$ its covariance matrix.

$$\mathbf{m}_X = \mathcal{E}[\mathbf{X}] \qquad \mathbf{K}_X = \mathcal{E}[\mathbf{X}\mathbf{X}'] - \mathbf{m}_X \mathbf{m}_X'$$

As in the univariate case, the Gaussian distribution of a random vector is denoted by $\mathbf{X} \sim \mathcal{N}(\mathbf{m}_X, \mathbf{K}_X)$. After applying a linear transformation to a Gaussian random vector, such as $\mathbf{Y} = \mathbf{A}\mathbf{X}$, the result is also a Gaussian random vector (a random variable if the matrix is a row vector): $\mathbf{Y} \sim \mathcal{N}(\mathbf{A}\mathbf{m}_X, \mathbf{A}\mathbf{K}_X\mathbf{A}')$.

The characteristic function of a real-valued Gaussian random vector is given by

$$\Phi_{\mathbf{X}}(j\boldsymbol{\nu}) = \exp\left\{+j\boldsymbol{\nu}^t \mathbf{m}_X - \frac{1}{2}\boldsymbol{\nu}^t \mathbf{K}_X \boldsymbol{\nu}\right\}$$

From this formula, the Nth-order moment formula for jointly distributed Gaussian random variables is easily, but tediously, derived.† The general formula for $\mathcal{E}[X_1 \cdots X_N]$ is more easily stated in words than by equations. Let K_{ij} denote the covariance between X_i and X_j and let m_i denote the expected value of X_i. The Nth joint moment consists of the sum of all possible combinations of *distinct* products between covariances and expected values in which the number of indices in each product equals N and each index appears only once. For example, $\mathcal{E}[X_1 X_2] = K_{12} + m_1 m_2$: The covariance appears by itself as it consumes two indices and K_{21} does not appear separately as it equals K_{12} (it's not a distinct term). The first four joint moments illustrate the pattern.

$$\begin{aligned}
\mathcal{E}[X_1] &= m_1 \\
\mathcal{E}[X_1 X_2] &= K_{12} + m_1 m_2 \\
\mathcal{E}[X_1 X_2 X_3] &= m_1 K_{23} + m_2 K_{13} + m_3 K_{12} + m_1 m_2 m_3 \\
\mathcal{E}[X_1 X_2 X_3 X_4] &= K_{12}K_{34} + K_{13}K_{24} + K_{14}K_{23} + m_1 m_2 K_{34} + m_1 m_3 K_{24} \\
&\quad + m_1 m_4 K_{23} + m_2 m_3 K_{14} + m_2 m_4 K_{13} + m_3 m_4 K_{12} + m_1 m_2 m_3 m_4
\end{aligned}$$

† $\mathcal{E}[X_1 \cdots X_N] = (-j)^N \left. \frac{\partial^N}{\partial \nu_1 \cdots \partial \nu_N} \Phi_{\mathbf{X}}(j\boldsymbol{\nu}) \right|_{\boldsymbol{\nu}=\mathbf{0}}$

These expressions apply when some or all of the random variables are equal; the "distinctiveness" criterion only applies to the symbolic formula, not to specific cases in which terms might be equal. Consequently, in the zero-mean case, all odd moments are zero, $\mathcal{E}[X_1X_2X_3X_4] = \mathcal{E}[X_1X_2]\,\mathcal{E}[X_3X_4] + \mathcal{E}[X_1X_3]\,\mathcal{E}[X_2X_4] + \mathcal{E}[X_1X_4]\,\mathcal{E}[X_2X_3]$, and $\mathcal{E}[X_1^2X_2^2] = \sigma_1^2\sigma_2^2 + 2K_{12}^2$.

A.1.7 Distributions Related to the Gaussian

Detection and estimation algorithms important to array signal processing are nonlinear functions of the noise contaminated observations. Because of the Gaussian noise assumptions, the distributions of these quantities are derived from the Gaussian. For example, if $\mathbf{X} \sim \mathcal{N}(\mathbf{m}, \mathbf{K})$, the quadratic form $\mathbf{X}^t\mathbf{A}\mathbf{X}$ ($\mathbf{A}$ a symmetric matrix of rank l) has a noncentral chi-squared distribution with l degrees of freedom and centrality parameter $\mathbf{m}^t\mathbf{A}\mathbf{m}$ ($\mathbf{X}^t\mathbf{A}\mathbf{X} \sim \chi_l'^2(\mathbf{m}^t\mathbf{A}\mathbf{m})$) if and only if $\mathbf{AKA} = \mathbf{A}$ [110: 31]. Estimates of the covariance matrix of a Gaussian random vector are governed by the Wishart distribution. Despite its complexity, this multivariate density has an extensive literature (see Muirhead, pp. 85–108, for a summary). Fortunately, all of these distributions have been known to statisticians for decades [82].

Chi-Squared: $\chi_n^2 = \sum_{i=1}^n X_i^2,\ X_i \sim \mathcal{N}(0,1)$

$$p_{\chi_n^2}(x) = \frac{1}{2^{n/2}\Gamma(n/2)} x^{n/2-1} e^{-x/2}$$

$$\mathcal{E}[\chi_n^2] = n; \quad \mathcal{V}[\chi_n^2] = 2n$$

Noncentral Chi-Squared: $\chi_n'^2(\lambda) = \sum_{i=1}^n X_i^2,\ X_i \sim \mathcal{N}(m_i,1),\ \lambda = \sum_{i=1}^n m_i^2$

$$p_{\chi_n'^2}(x) = \sum_{k=0}^{\infty} \frac{(\lambda/2)^k}{k!} e^{-\lambda/2} p_{\chi^2_{m+2k,n}}(x)$$

$$= \frac{1}{2}\left(\frac{x}{\lambda}\right)^{(n-2)/4} I_{\frac{n-2}{2}}(\sqrt{\lambda}x)e^{-(\lambda+x)/2}$$

$$\mathcal{E}[\chi_n'^2(\lambda)] = n + \lambda; \quad \mathcal{V}[\chi_n'^2(\lambda)] = 2n + 4\lambda$$

The quantity $I_q(\cdot)$ appearing in the formula for the density is the modified Bessel function of first kind, order q [1].

F distribution: $F_{m,n} = \dfrac{\chi_m^2/m}{\chi_n^2/n}$

$$p_{F_{m,n}}(x) = \frac{\Gamma\left(\frac{m+n}{2}\right)}{\Gamma\left(\frac{m}{2}\right)\Gamma\left(\frac{n}{2}\right)} \left(\frac{m}{n}\right)^{m/2} \frac{x^{(m-2)/2}}{\left(1 + \frac{m}{n}x\right)^{(m+n)/2}}$$

$$\mathcal{E}[F_{m,n}] = \frac{n}{n-2},\ n > 2; \quad \mathcal{V}[F_{m,n}] = \frac{2n^2(m+n-2)}{m(n-2)^2(n-4)},\ n > 4$$

Noncentral F: $F'_{m,n}(\lambda) = \dfrac{\chi'^2_m(\lambda)/m}{\chi^2_n/n}$

$$p_{F'_{m,n}}(x) = \sum_{k=0}^{\infty} \frac{\left(\frac{\lambda}{2}\right)^k}{k!} e^{-\lambda/2} p_{\beta_{\frac{m}{2}+k,\frac{n}{2}}}\left(\frac{mx}{mx+n}\right)$$

$$\mathcal{E}[F'_{m,n}] = \frac{n}{n-2}, n > 2$$

$$\mathcal{V}[F'_{m,n}] = 2\left(\frac{n}{m}\right)^2 \frac{(m+\lambda)^2 + (m+2\lambda)(n-2)}{(n-2)^2(n-4)}, n > 4$$

Beta: $\beta_{m,n} = \frac{\chi^2_m}{\chi^2_m + \chi^2_n}$

$$p_{\beta_{m,n}}(x) = \beta(x, m/2, n/2)$$

$$= \frac{\Gamma\left(\frac{m+n}{2}\right)}{\Gamma\left(\frac{m}{2}\right)\Gamma\left(\frac{n}{2}\right)} x^{m/2-1}(1-x)^{n/2-1}$$

$$\mathcal{E}[\beta_{m,n}] = \frac{m}{m+n}; \quad \mathcal{V}[\beta_{m,n}] = \frac{2mn}{(m+n)^2(m+n+2)}$$

Wishart:[†] $\mathbf{W}_M(L, \mathbf{K}) = \sum_{l=1}^{L} \mathbf{X}_l \mathbf{X}'_l$, $\mathbf{X}_l \sim \mathcal{N}(\mathbf{0}, \mathbf{K})$, $\dim[\mathbf{K}] = M$

$$p_{\mathbf{W}_M(L,\mathbf{K})}(\mathbf{w}) = \frac{(\det[\mathbf{w}])^{\frac{L-M-1}{2}}}{2^{LM/2}\Gamma_M(\frac{L}{2})\,(\det[\mathbf{K}])^{\frac{L}{2}}} \exp\left\{-\frac{1}{2}\operatorname{tr}[\mathbf{K}^{-1}\mathbf{w}]\right\}$$

$$\mathcal{E}[\mathbf{W}_M(L, \mathbf{K})] = L\mathbf{K}; \quad \operatorname{cov}[\mathbf{W}_{ij}\mathbf{W}_{kl}] = L(\mathbf{K}_{ik}\mathbf{K}_{jl} + \mathbf{K}_{il}\mathbf{K}_{jk})$$

$\Gamma_M(L/2) = \pi^{M(M-1)/4} \prod_{m=0}^{M-1} \Gamma\left(\frac{L}{2} - \frac{m}{2}\right)$, where $\Gamma(n+1) = n\Gamma(n)$ is the usual Gamma function.

A.1.8 Central Limit Theorem

Let $\{X_l\}$ denote a sequence of independent, identically distributed, random variables. Assuming they have zero means and finite variances (equaling σ^2), the Central Limit Theorem states that the sum $\sum_{l=1}^{L} X_l/\sqrt{L}$ converges in distribution to a Gaussian random variable.

$$\frac{1}{\sqrt{L}} \sum_{l=1}^{L} X_l \overset{L\to\infty}{\longrightarrow} \mathcal{N}(0, \sigma^2)$$

Because of its generality, this theorem is often used to simplify calculations involving *finite* sums of non-Gaussian random variables. Attention is seldom paid, however, to the *convergence rate* of the Central Limit Theorem. Kolmogorov, the famous 20th century mathematician, is reputed to have said, "The Central Limit Theorem is a dangerous tool in the hands of amateurs." Let's see what he meant.

[†]Note that the Wishart distribution is multivariate, governing the elements of a sample covariance matrix.

Taking $\sigma^2 = 1$, the key result is that the magnitude of the difference between $P(x)$, defined to be the probability that the sum given earlier exceeds x, and $Q(x)$, the probability that a unit-variance Gaussian random variable exceeds x, is bounded by a quantity inversely related to the square root of L [42: Theorem 24].

$$|P(x) - Q(x)| \leq \frac{c}{\sqrt{L}} \cdot \mathcal{E}\left[|X|^3\right]$$

The constant of proportionality c is a number known to be about 0.8 [65: 6]. The ratio of absolute third moment of X_l to the cube of its standard deviation, known as the skew and denoted by γ_X, depends only on the distribution of X_l and is independent of scale. This bound on the absolute error has been shown to be tight [42: 79ff]. Using our lower bound for $Q(\cdot)$ (Eq. A.2 {478}), we find that the relative error in the Central Limit Theorem approximation to the distribution of finite sums is bounded for $x > 0$ as

$$\boxed{\frac{|P(x) - Q(x)|}{Q(x)} \leq c\gamma_X \sqrt{\frac{2\pi}{L}} e^{+x^2/2} \cdot \begin{cases} 2, & x \leq 1 \\ \frac{1+x^2}{x}, & x > 1 \end{cases}}$$

Suppose we require that the relative error not exceed some specified value ϵ. The normalized† boundary x at which the approximation is evaluated must not violate

$$\frac{L\epsilon^2}{2\pi c^2 \gamma_X^2} \geq e^{x^2} \cdot \begin{cases} 4 & x \leq 1 \\ \left(\frac{1+x^2}{x}\right)^2 & x > 1 \end{cases}$$

As shown in Fig. A.2, the right side of this equation is a monotonically increasing function.

Example

For example, if $\epsilon = 0.1$ and taking $c\gamma_X$ arbitrarily to be unity (a reasonable value), the upper limit of the preceding equation becomes $1.6 \times 10^{-3} L$. Examining Fig. A.2, we find that for $L = 10,000$, x must not exceed 1.17. Because we have normalized to unit variance, this example suggests that the Gaussian approximates the distribution of a 10,000-term sum only over a range corresponding to an 76% area about the mean. Consequently, the Central Limit Theorem, as a finite-sample distributional approximation, is only guaranteed to hold near the mode of the Gaussian, with *huge* numbers of observations needed to specify the tail behavior. Realizing this fact will keep us from being ignorant amateurs.

A.2 Stochastic Processes

A.2.1 Basic Definitions

A *random* or *stochastic* process is the assignment of a function of a real variable to each sample point ω in sample space. Thus, the process $X(\omega, t)$ can be considered a function of two variables. For each ω, the time function must be well behaved, and may or may

† Taking $\sigma^2 = 1$ has essentially normalized the random variable by the standard deviation. To obtain the unnormalized result, replace x by x/σ.

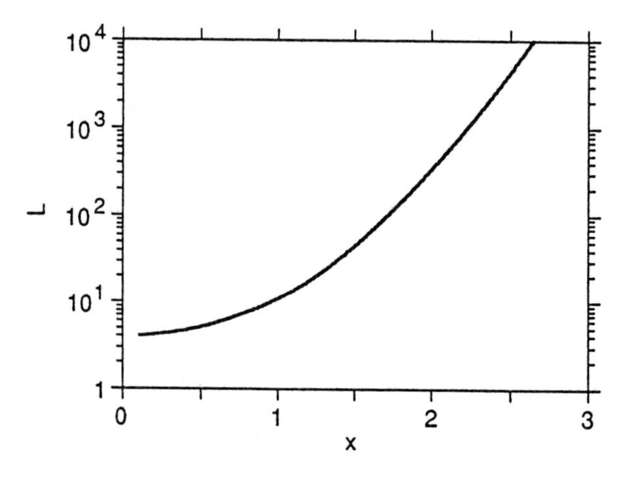

Figure A.2 The quantity that governs the limits of validity for numerically applying the Central Limit Theorem on finite numbers of data is shown over a portion of its range. To judge these limits, we must compute the quantity $L\epsilon^2/2\pi c^2\gamma_X^2$, where ϵ denotes the desired percentage error in the Central Limit Theorem approximation and L the number of observations. By determining the value of x that corresponds to this quantity with the depicted curve, we find the normalized ($x = 1$ implies unit standard deviation) upper limit on an L-term sum to which the Central Limit Theorem is guaranteed to apply. Note how rapidly the curve increases, suggesting that large amounts of data are needed for accurate approximation few standard deviations from the mean.

not look random to the eye. Each time function of the process is called a *sample function* and must be defined over the entire domain of interest. For each t, we have a function of ω, which is precisely the definition of a random variable. Hence the *amplitude* of a random process is a random variable. The *amplitude distribution* of a process refers to the probability density function of the amplitude: $p_{X(t)}(x)$. By examining the process's amplitude at several instants, the joint amplitude distribution can also be defined. For the purposes of this book, a process is said to be *stationary* when the joint amplitude distribution depends on the differences between the selected time instants.

The *expected value* or *mean* of a process is the expected value of the amplitude at each t.

$$\mathcal{E}[X(t)] = m_X(t) = \int_{-\infty}^{\infty} x p_{X(t)}(x)\, dx$$

For the most part, we take the mean to be 0. The *correlation function* is the first-order joint moment between the process's amplitudes at two times.

$$R_X(t_1, t_2) = \int_{-\infty}^{\infty}\int_{-\infty}^{\infty} x_1 x_2 p_{X(t_1),X(t_2)}(x_1, x_2)\, dx_1\, dx_2$$

Because the joint distribution for stationary processes depends only on the time difference, correlation functions of stationary processes depend only on $|t_1 - t_2|$. In this case, correlation functions are really functions of a single variable (the time difference) and are usually written as $R_X(\tau)$ where $\tau = t_1 - t_2$. Related to the correlation function is the *covariance function* $K_X(\tau)$, which equals the correlation function minus the square of the mean.

$$K_X(\tau) = R_X(\tau) - m_X^2$$

The variance of the process equals the covariance function evaluated as the origin. The *power spectrum* of a stationary process is the Fourier Transform of the correlation function.

$$\mathcal{S}_X(\omega) = \int_{-\infty}^{\infty} R_X(\tau) e^{-j\omega\tau}\, d\tau$$

A particularly important example of a random process is *white noise*. The process $X(t)$ is said to be white if it has zero mean and a correlation function proportional to an impulse.

$$\mathcal{E}\big[X(t)\big] = 0 \quad R_X(\tau) = \frac{N_0}{2}\delta(\tau)$$

The power spectrum of white noise is constant for all frequencies, equaling πN_0, which is known as the *spectral height.*

When a stationary process $X(t)$ is passed through a stable linear, time-invariant filter, the resulting output $Y(t)$ is also a stationary process having power density spectrum

$$\mathcal{S}_Y(\omega) = |H(j\omega)|^2 \mathcal{S}_X(\omega)$$

where $H(j\omega)$ is the filter's transfer function.

A.2.2 Gaussian Process

A random process $X(t)$ is Gaussian if the joint density of any N amplitudes $X(t_1), \ldots, X(t_N)$ comprise a Gaussian random vector. The elements of the required covariance matrix equal the covariance between the appropriate amplitudes: $K_{ij} = K_X(t_i, t_j)$. Assuming the mean is known, the entire structure of the Gaussian random process is specified once the correlation function or, equivalently, the power spectrum are known. As linear transformations of Gaussian random processes yield another Gaussian process, linear operations such as differentiation, integration, linear filtering, sampling, and summation with other Gaussian processes result in a Gaussian process.

A.2.3 Sampling and Random Sequences

The usual Sampling Theorem applies to random processes, with the spectrum of interest being the power spectrum. If a stationary process $X(t)$ is bandlimited, $\mathcal{S}_X(\omega) = 0$, $|\omega| > W$, as long as the sampling interval T satisfies the classic constraint $T < \pi/W$ the sequence $X(lT)$ represents the original process. A sampled process is itself a random process defined over discrete time. Hence, all of the random process notions introduced in the previous section apply to the random sequence $\widetilde{X}(l) \equiv X(lT)$. The correlation functions of these two processes are related as

$$R_{\widetilde{X}}(k) = \mathcal{E}\big[\widetilde{X}(l)\widetilde{X}(l+k)\big] = R_X(kT)$$

We note especially that for distinct samples of a random process to be uncorrelated, the correlation function $R_X(kT)$ must equal zero for all nonzero k. This requirement places severe restrictions on the correlation function (hence the power spectrum) of the original process. One correlation function satisfying this property is derived from the random process that has a bandlimited, constant-valued power spectrum over precisely the frequency region needed to satisfy the sampling criterion. *No other power spectrum satisfying the sampling criterion has this property.* Hence, sampling does not normally yield uncorrelated amplitudes, meaning that *discrete-time white noise* is a rarity. White noise has a correlation function given by $R_{\widetilde{X}}(k) = \sigma^2\delta(k)$, where $\delta(\cdot)$ is the unit sample. The power spectrum of white noise is a constant: $\mathcal{S}_{\widetilde{X}}(\omega) = \sigma^2$.

Appendix B

Matrix Theory

B.1 Basic Definitions

An $m \times n$ *matrix* $\mathbf{A}$ is a rectangular (square if $n = m$) arrangement of scalar elements A_{ij} (ith row, jth column).

$$\mathbf{A} = \begin{bmatrix} A_{11} & A_{12} & \cdots & A_{1n} \\ A_{21} & A_{22} & & A_{2n} \\ \vdots & & \ddots & \vdots \\ A_{m1} & A_{m2} & \cdots & A_{mn} \end{bmatrix}$$

When the matrix is square, the *main diagonal* extends from the upper left corner to the lower right and consists of the elements A_{ii}. The *cross diagonal* is the opposite diagonal and consists of the elements $A_{n-i+1,i}$. A *vector* is common terminology for a column ($m \times 1$) matrix. The *dimension* of the vector equals m. Rectangular matrices are (usually) denoted by boldface uppercase letters ($\mathbf{A}, \mathbf{B}, \ldots$) and vectors by boldface lowercase letters ($\mathbf{a}, \mathbf{b}, \ldots$). The dimension of a vector is expressed as $\dim[\mathbf{a}]$.

To add matrices, the number of rows in each matrix as well as the number of columns must agree. The sum of $m \times n$ matrices $\mathbf{A}$ and $\mathbf{B}$ is defined to be an $m \times n$ matrix $\mathbf{C}$ whose elements are

$$C_{ij} = A_{ij} + B_{ij}$$

Matrix addition is commutative ($\mathbf{A} + \mathbf{B} = \mathbf{B} + \mathbf{A}$). The product $\mathbf{AB}$ of two matrices $\mathbf{A}$ and $\mathbf{B}$ is only defined if the number of rows of $\mathbf{A}$ equals the number of columns of $\mathbf{B}$. Thus, if $\mathbf{A}$ represents an $l \times m$ matrix and $\mathbf{B}$ an $m \times n$ matrix, the result is an $l \times n$ matrix $\mathbf{C}$, each term of which is defined to be

$$C_{ij} = \sum_{k=1}^{m} A_{ik} B_{kj}, \quad i = 1, \ldots, l;\ j = 1, \ldots, n$$

Clearly, the matrix product is not necessarily commutative ($\mathbf{AB} \neq \mathbf{BA}$), but is distributive [$\mathbf{A}(\mathbf{B} + \mathbf{C}) = \mathbf{AB} + \mathbf{AC}$] and associative [$\mathbf{A}(\mathbf{BC}) = (\mathbf{AB})\mathbf{C}$].

Several methods of rearranging the elements of a matrix are frequently used. The *complex conjugate* of the matrix **A** is denoted by $\mathbf{A}^*$ and consists of the complex conjugates of the elements of **A**.

$$[\mathbf{A}^*]_{ij} = A_{ij}^*$$

The *transpose* of an $m \times n$ matrix **A** is an $n \times m$ matrix $\mathbf{A}^t$ whose rows are the columns of **A**.

$$[\mathbf{A}^t]_{ij} = A_{ji}$$

The transpose of a product of two matrices equals the product of their transposes but in reversed order.

$$(\mathbf{AB})^t = \mathbf{B}^t\mathbf{A}^t$$

This property applies whether the matrices are square or not. The *conjugate transpose* (sometimes known as the Hermitian transpose) of **A** is denoted by $\mathbf{A}'$ and equals $(\mathbf{A}^*)^t$.

$$[\mathbf{A}']_{ij} = A_{ji}^*$$

B.2 Basic Matrix Forms

The relations between the values of the elements of a matrix define special matrix cases. Matrices having special internal structure are important to recognize in manipulating matrix expressions.

- A *diagonal* matrix, denoted by $\text{diag}[A_{11}, A_{22}, \ldots, A_{nn}]$, has nonzero entries only along the main diagonal ($i = j$) of the matrix.

$$\begin{bmatrix} A_{11} & 0 & \cdots & 0 \\ 0 & A_{22} & 0 & \vdots \\ \vdots & & \ddots & 0 \\ 0 & \cdots & 0 & A_{nn} \end{bmatrix}$$

- The *identity* matrix, denoted by **I**, is a special diagonal matrix having unity on the main diagonal.

$$\mathbf{I} = \begin{bmatrix} 1 & 0 & \cdots & 0 \\ 0 & 1 & 0 & \vdots \\ \vdots & & \ddots & \\ 0 & \cdots & 0 & 1 \end{bmatrix}$$

The identity matrix is so named because it is the multiplicative identity for square matrices ($\mathbf{IA} = \mathbf{AI} = \mathbf{A}$).

- The ordering of the rows or columns of a square matrix can be reversed by premultiplication or postmultiplication by the *exchange* matrix **J**.

$$\mathbf{J} = \begin{bmatrix} 0 & \cdots & 0 & 1 \\ \vdots & \iddots & 1 & 0 \\ 0 & \iddots & \iddots & \vdots \\ 1 & 0 & \cdots & 0 \end{bmatrix}$$

- A *lower triangular* matrix **L** has nonzero values on and "below" its main diagonal.

$$\mathbf{L} = \begin{bmatrix} L_{11} & 0 & \cdots & 0 \\ L_{21} & L_{22} & 0 & \vdots \\ \vdots & \vdots & \ddots & \\ L_{n1} & L_{n2} & \cdots & L_{nn} \end{bmatrix}$$

An upper triangular matrix can be similarly defined.

- A *Vandermonde* matrix consists of columns containing geometric sequences.

$$\mathbf{A} = \begin{bmatrix} 1 & \cdots & 1 \\ a_0 & \cdots & a_{n-1} \\ a_0^2 & \cdots & a_{n-1}^2 \\ \vdots & & \vdots \\ a_0^{m-1} & \cdots & a_{n-1}^{m-1} \end{bmatrix}$$

One special case of a square Vandermonde matrix is the *DFT* matrix **W** in which the elements are $a_k = \exp\{-j2\pi k/m\}$. The discrete Fourier transform of a vector **x** can be expressed as

$$\text{DFT}[\mathbf{x}] = \mathbf{W}\mathbf{x}$$

- A *symmetric* matrix equals its transpose ($\mathbf{A}^t = \mathbf{A}$). A *conjugate symmetric* (Hermitian) matrix equals its own conjugate transpose ($\mathbf{A}' = \mathbf{A}$). Correlation matrices, which have the form $\mathbf{A} = \mathcal{E}[\mathbf{x}\mathbf{x}']$, are Hermitian.
- A *Toeplitz* matrix has constant values along each of its diagonals ($A_{ij} = a_{i-j}$).

$$\begin{bmatrix} a_0 & a_1 & \cdots & a_{n-1} \\ a_{-1} & a_0 & a_1 & \\ \vdots & \ddots & \ddots & \\ a_{-(n-1)} & a_{-(n-2)} & \cdots & a_0 \end{bmatrix}$$

In time series, correlation matrices are not only Hermitian, they are also Toeplitz. Hermitian Toeplitz matrices have no more than n unique elements in the matrix. Because of this extensive redundancy of values, many efficient techniques for computing expressions involving them have been developed.

- Constant values along the cross-diagonals of a square matrix define a *Hankel* matrix.

$$A_{ij} = a_{i+j-2} \qquad \mathbf{A} = \begin{bmatrix} a_0 & a_1 & a_2 & \cdots & a_{n-1} \\ a_1 & a_2 & \cdots & \cdot^{\cdot^{\cdot}} & a_n \\ a_2 & \cdot^{\cdot^{\cdot}} & \cdot^{\cdot^{\cdot}} & \cdot^{\cdot^{\cdot}} & \vdots \\ \vdots & \cdot^{\cdot^{\cdot}} & \cdot^{\cdot^{\cdot}} & & a_{2n-3} \\ a_{n-1} & a_n & \cdots & a_{2n-3} & a_{2n-2} \end{bmatrix}$$

All Hankel matrices are symmetric. If $\mathbf{A}$ is a Hankel matrix, then both $\mathbf{JA}$ and $\mathbf{AJ}$ are Toeplitz matrices.

- A square matrix is said to be *circulant* if each row is a circular rotation of the previous one.

$$\begin{bmatrix} a_0 & a_1 & \cdots & a_{n-1} \\ a_{n-1} & a_0 & a_1 & \cdots \\ a_{n-2} & a_{n-1} & a_0 & \cdots \\ \vdots & \vdots & & \\ a_1 & a_2 & \cdots & a_0 \end{bmatrix}$$

All circulant matrices are Toeplitz; circulant matrices are symmetric only when the sequence $a_1, a_2, \ldots$ is even: $a_i = a_{n-i}, i = 1, \ldots, n/2$ (a_0 is arbitrary).

- The square matrix $\mathbf{A}$ is said to be *orthogonal* if it satisfies $\mathbf{A}^t\mathbf{A} = \mathbf{I}$. $\mathbf{A}$ is *unitary* if it satisfies $\mathbf{A}'\mathbf{A} = \mathbf{I}$.

- The square matrix $\mathbf{A}$ is said to be a *projection* matrix if $\mathbf{A}^2 = \mathbf{A}$. Because of this property, $\mathbf{A}^m = \mathbf{A}^n$ for all choices of positive integers n, m. The reason for this name will be given once eigenvalues are defined (§B.6 {497}).

- A matrix is said to have a *null space* if a set of nonzero vectors $\mathbf{x}$ exist that satisfy $\mathbf{Ax} = \mathbf{0}$. These vectors would thereby be orthogonal to the rows of $\mathbf{A}$. The matrix could be square or, more interestingly, it could be rectangular, having more columns than rows. The rows would then define a collection of vectors that represent all vectors for which $\mathbf{Ax} \neq 0$.

- The vector $\mathbf{y}$ is said to be in the *range* of the matrix $\mathbf{A}$ if it can be expressed as $\mathbf{Ax} = \mathbf{y}$ for some $\mathbf{x}$.

B.3 Operations on Matrices

Several operations on matrices and between them are so commonly used that specific terminology has evolved.

- The *inner product* $\mathbf{x}'\mathbf{y}$ between two vectors $\mathbf{x}$ and $\mathbf{y}$ is defined to be the scalar expressed by†

$$\mathbf{x}'\mathbf{y} = \sum_{i=1}^{n} x_i^* y_i$$

Because an inner product is a scalar, expressions involving inner products are frequently rearranged by noting that the transpose of a scalar is the same scalar. Therefore, $(\mathbf{x}'\mathbf{y})^t = \mathbf{y}^t\mathbf{x}^* = \mathbf{x}'\mathbf{y}$. Two vectors are said to be *orthogonal* if their inner product is 0.

- The *outer product* $\mathbf{x}\mathbf{y}'$ between two vectors x (dimension m) and y (dimension n) is a $m \times n$ matrix whose elements are

$$[\mathbf{x}\mathbf{y}']_{ij} = x_i y_j^*$$

- The *Kronecker product* $\mathbf{A} \otimes \mathbf{B}$ between two matrices $\mathbf{A}$ ($m_\mathbf{A} \times n_\mathbf{A}$) and $\mathbf{B}$ ($m_\mathbf{B} \times n_\mathbf{B}$) is the $m_\mathbf{A} m_\mathbf{B} \times n_\mathbf{A} n_\mathbf{B}$ matrix given by

$$\mathbf{A} \otimes \mathbf{B} = \begin{bmatrix} A_{11}\mathbf{B} & A_{12}\mathbf{B} & \cdots & A_{1n_\mathbf{A}}\mathbf{B} \\ A_{21}\mathbf{B} & A_{22}\mathbf{B} & & A_{2n_\mathbf{A}}\mathbf{B} \\ \vdots & & \ddots & \vdots \\ A_{m_\mathbf{A}1}\mathbf{B} & A_{m_\mathbf{A}2}\mathbf{B} & \cdots & A_{m_\mathbf{A}n_\mathbf{A}}\mathbf{B} \end{bmatrix}$$

The matrix $\mathbf{B}$ multiplied by the scalars A_{ij} is thus repeated throughout the matrix. The Kronecker product of two positive-definite matrices is also positive definite.

- The *inverse* of a matrix $\mathbf{A}$ is denoted by $\mathbf{A}^{-1}$ that satisfies $\mathbf{A}\mathbf{A}^{-1} = \mathbf{A}^{-1}\mathbf{A} = \mathbf{I}$. The inverse of a matrix is not guaranteed to exist. Numerous conditions on the inverse's existence are detailed in the following. When it does exist, the following properties hold.

  1. If $\mathbf{A}$, $\mathbf{B}$ are invertible matrices, $(\mathbf{A}\mathbf{B})^{-1} = \mathbf{B}^{-1}\mathbf{A}^{-1}$.
  2. If $\mathbf{A}$, $\mathbf{B}$ are invertible matrices, $(\mathbf{A} \otimes \mathbf{B})^{-1} = \mathbf{A}^{-1} \otimes \mathbf{B}^{-1}$.
  3. Assuming that all inverses exist where used, the inverse of a sum of matrices can be written several useful ways.

$$\begin{aligned} (\mathbf{A} + \mathbf{BCD})^{-1} &= \mathbf{A}^{-1} - \mathbf{A}^{-1}\mathbf{B}\left(\mathbf{D}\mathbf{A}^{-1}\mathbf{B} + \mathbf{C}^{-1}\right)^{-1}\mathbf{D}\mathbf{A}^{-1} \\ &= \mathbf{A}^{-1} - \mathbf{A}^{-1}\mathbf{BC}\left(\mathbf{I} + \mathbf{D}\mathbf{A}^{-1}\mathbf{BC}\right)^{-1}\mathbf{D}\mathbf{A}^{-1} \\ &= \mathbf{A}^{-1} - \mathbf{A}^{-1}\mathbf{B}\left(\mathbf{I} + \mathbf{CD}\mathbf{A}^{-1}\mathbf{B}\right)^{-1}\mathbf{CD}\mathbf{A}^{-1} \end{aligned}$$

---

†In more abstract settings, the inner product between two vectors x and y is denoted by $\langle x, y \rangle$. There, the inner product becomes any mapping of two vectors to a scalar that satisfies three properties.

1. $\langle y, x \rangle = \langle x, y \rangle^*$
2. $\langle ax + by, z \rangle = a^* \langle x, z \rangle + b^* \langle y, z \rangle$, a, b scalars
3. $\langle x, x \rangle > 0, x \neq 0$

Inner products obey the important Schwarz Inequality, $|\langle x, y \rangle| \leq \|x\| \, \|y\|$, with equality occurring only when $x \propto y$.

Note that the matrices **B** and **D** need not be invertible for these relationships to be valid. In the latter two, the matrix **C** need not be invertible. Notable special cases of this result are

$$(\mathbf{A}+\mu\mathbf{x}\mathbf{y}')^{-1} = \mathbf{A}^{-1}\left(\mathbf{I} - \frac{\mathbf{x}\mathbf{y}'}{\frac{1}{\mu}+\mathbf{y}'\mathbf{A}^{-1}\mathbf{x}}\mathbf{A}^{-1}\right)$$

$$(\mathbf{A}+\mathbf{B})^{-1} = \mathbf{A}^{-1} - \mathbf{A}^{-1}\left(\mathbf{A}^{-1}+\mathbf{B}^{-1}\right)^{-1}\mathbf{A}^{-1}$$

4. If the matrix **A** is symmetric, Hermitian, circulant, or triangular, its inverse has the same structure. The inverse of a Toeplitz matrix is *not* necessarily Toeplitz. The inverse of a Hankel matrix is symmetric.

- A *Givens rotation* is the multiplication of a matrix **A** by a unitary matrix **Q** that zeros a specific element of **A** [60: 43–47]. This simple operation lies at the core of many numerical algorithms for computing matrix inverses and eigensystems. The rotation matrix has the form of an identity matrix augmented by a square submatrix of sine and cosine rotation values. For example, the matrix that zeros the second element of the fourth column of a 5×5 matrix has the form

$$\mathbf{Q} = \begin{bmatrix} 1 & 0 & 0 & 0 & 0 \\ 0 & c & 0 & s & 0 \\ 0 & 0 & 1 & 0 & 0 \\ 0 & -s & 0 & c & 0 \\ 0 & 0 & 0 & 0 & 1 \end{bmatrix}$$

where $c = A_{44}/\left(A_{24}^2 + A_{44}^2\right)^{1/2}$ and $s = A_{24}/\left(A_{24}^2 + A_{44}^2\right)^{1/2}$. The product **QA** performs the rotation by modifying only the second and fourth rows of **A**. Column modification is possible by postmultiplying by the rotation matrix and suitably defining the rotation.

- The *determinant* of a square matrix **A** is denoted by det[**A**] and is given by

$$\det[\mathbf{A}] = A_{11}\widetilde{A}_{11} + A_{12}\widetilde{A}_{12} + \cdots + A_{1n}\widetilde{A}_{1n}$$

where $\widetilde{A}_{ij}$, the *cofactor* of **A**, equals $(-1)^{j+1}$ times the determinant of the $(n-1)\times(n-1)$ matrix formed by deleting the ith row and jth column of **A**. A nonzero determinant is a necessary and sufficient condition for the existence of the inverse of a matrix. The determinant of **A**$'$ equals the conjugate of the determinant of **A**: $\det[\mathbf{A}'] = \det[\mathbf{A}]^*$. The determinant of a unitary matrix has unity magnitude. The determinant of the product of two square matrices is the product of their determinants.

$$\det[\mathbf{AB}] = \det[\mathbf{A}]\det[\mathbf{B}]$$

The determinant of a sum of matrices is *not* equal to the sum of their determinants. The determinant of a Kronecker product of square matrices equals $\det[\mathbf{A}\otimes\mathbf{B}] = \det[\mathbf{A}]^n\det[\mathbf{B}]^m$, where $\dim[\mathbf{A}] = m$ and $\dim[\mathbf{B}] = n$.

- The *trace* $\text{tr}[\mathbf{A}]$ of the square matrix $\mathbf{A}$ equals the sum of its elements along the main diagonal.

$$\text{tr}[\mathbf{A}] = \sum_{i=1}^{n} A_{ii}$$

The trace of a sum of matrices equals the sum of their traces.

$$\text{tr}[\mathbf{A} + \mathbf{B}] = \text{tr}[\mathbf{A}] + \text{tr}[\mathbf{B}]$$

The trace of a product of two square matrices does *not* necessarily equal the product of the traces of the individual matrices; the product of two matrix's traces does equal the trace of their Kronecker product: $\text{tr}[\mathbf{A} \otimes \mathbf{B}] = \text{tr}[\mathbf{A}]\,\text{tr}[\mathbf{B}]$. One of the more interesting properties of the trace is

$$\text{tr}[\mathbf{AB}] = \text{tr}[\mathbf{BA}]$$

Proving this property is straightforward. Using the full expression for the product of two matrices given previously, $\text{tr}[\mathbf{AB}] = \sum_i \sum_k A_{ik} B_{ki}$. Writing a similar expression for $\mathbf{BA}$ easily demonstrates this property. With this result, computation of the trace can be simplified in circumstances in which the component matrices are not square; whichever product yields the smaller matrix can be used to compute the trace.

- The *gradient* with respect to a vector of a scalar-valued function $f(\mathbf{x})$ equals a column matrix of the partial derivatives.

$$\nabla_{\mathbf{x}} f(\mathbf{x}) = \text{col}\left[\frac{\partial f}{\partial x_1}, \ldots, \frac{\partial f}{\partial x_N}\right]$$

Examples of gradient calculation derived from this definition are:

$$\begin{aligned}
&\nabla_{\mathbf{x}}\, \mathbf{x}^t\mathbf{y} = \nabla_{\mathbf{x}}\, \mathbf{y}^t\mathbf{x} = \mathbf{y}\\
&\nabla_{\mathbf{x}}\, \mathbf{x}^t\mathbf{A}\mathbf{x} = 2\mathbf{A}\mathbf{x}\\
&\nabla_{\mathbf{x}}\, \text{tr}[\mathbf{x}\mathbf{y}^t] = \nabla_{\mathbf{x}}\, \text{tr}[\mathbf{y}\mathbf{x}^t] = \mathbf{y}
\end{aligned}$$

The gradient with respect to a matrix of a scalar-valued function can be defined similarly:

$$[\nabla_{\mathbf{A}} f(\mathbf{A})]_{ij} = \frac{\partial f}{\partial A_{ij}}$$

Examples are:

$$\begin{aligned}
&\nabla_{\mathbf{A}}\, \text{tr}[\mathbf{A}] = \mathbf{I} && \nabla_{\mathbf{A}}\, \text{tr}[\mathbf{AB}] = \mathbf{B}\\
&\nabla_{\mathbf{A}}\, \text{tr}[\mathbf{BA}] = \mathbf{B}^t && \nabla_{\mathbf{A}}\, \text{tr}[\mathbf{B}\mathbf{A}^{-1}] = -[\mathbf{A}^{-1}\mathbf{B}\mathbf{A}^{-1}]^t\\
&\nabla_{\mathbf{A}}\, \exp\{\mathbf{x}^t\mathbf{A}\mathbf{x}\} = \mathbf{x}\mathbf{x}^t \exp\{\mathbf{x}^t\mathbf{A}\mathbf{x}\} &&\\
&\nabla_{\mathbf{A}}\, \det[\mathbf{A}] = \det[\mathbf{A}]\left(\mathbf{A}^{-1}\right)^t && \nabla_{\mathbf{A}}\, \ln\det[\mathbf{A}] = \left(\mathbf{A}^{-1}\right)^t
\end{aligned}$$

B.4 Quadratic Forms

Quadratic forms are key quantities of study in statistical signal processing. They are comprised by a Hermitian matrix $\mathbf{A}$ and a vector $\mathbf{x}$ to produce the real-valued scalar $\mathbf{x}'\mathbf{A}\mathbf{x}$. The matrix is termed the *kernel* of the quadratic form. In the special case that $\mathbf{A} = \mathbf{I}$, the quadratic form reduces to the inner product of $\mathbf{x}$ with itself; quadratic forms are thus considered generalizations of the inner product. If $\mathbf{x}'\mathbf{A}\mathbf{x} > 0$ for all nonzero choices of $\mathbf{x}$, the matrix $\mathbf{A}$ is said to be *positive definite*. If $\mathbf{x}'\mathbf{A}\mathbf{x} \geq 0$ under the same conditions, $\mathbf{A}$ is *nonnegative definite*. The structure of Hermitian matrices, even if they are Toeplitz, is not sufficient to guarantee that they are nonnegative definite. In contrast, Hankel matrices having no negative elements are always nonnegative definite.

The argument of the exponential in the probability density function of a Gaussian random vector is a quadratic form. The kernel of that quadratic form is the inverse of the vector's covariance matrix. Assuming zero mean for algebraic simplicity, the covariance matrix is the expected value of the outer product of the random vector $\mathbf{z}$ with itself: $\mathbf{A} = \mathcal{E}[\mathbf{z}\mathbf{z}']$. Such matrices are *always* nonnegative definite. To show this result, simply consider an expanded version of the quadratic form: $\mathbf{x}'\mathbf{A}\mathbf{x} = \mathcal{E}\left[|\mathbf{x}'\mathbf{z}|^2\right]$. Because of the squared magnitude, this quantity can never be negative.

Quadratic forms can be rewritten using the trace operation. The trace of a quadratic form simply equals its value because a quadratic form is a scalar. Using the properties of the trace operation,

$$\operatorname{tr}\left[\mathbf{x}'\mathbf{A}\mathbf{x}\right] = \operatorname{tr}\left[\mathbf{A}\mathbf{x}\mathbf{x}'\right]$$

Thus, a quadratic form equals the trace of the product between the kernel and the outer product of the vector with itself. This seemingly more complicated expression has utility in signal processing. See §A.1.6 {476} on the multivariate Gaussian distribution for the most well-known application of this formula.

Positive-definite Hermitian matrices can be expanded using the *Cholesky factorization*.

$$\mathbf{A} = \mathbf{L}\mathbf{D}\mathbf{L}'$$

where $\mathbf{D}$ is a diagonal matrix and $\mathbf{L}$ is lower-triangular with all of the diagonal elements equal to unity.

$$\mathbf{L} = \begin{bmatrix} 1 & 0 & 0 & \cdots \\ L_{21} & 1 & 0 & \cdots \\ L_{31} & L_{32} & \ddots & \ddots \\ \vdots & \vdots & & \ddots \end{bmatrix}, \quad \mathbf{D} = \begin{bmatrix} D_{11} & 0 & 0 & \cdots \\ 0 & D_{22} & 0 & \cdots \\ \vdots & 0 & D_{33} & \ddots \\ \vdots & \vdots & & \ddots \end{bmatrix}$$

The inverse of a lower-triangular matrix is also lower-triangular; the inverse of the matrix $\mathbf{A}$ is written as $\mathbf{A}^{-1} = \mathbf{L}'^{-1}\mathbf{D}^{-1}\mathbf{L}^{-1}$.

In optimization problems, the quadratic form frequently represents a squared error cost function: The kernel imposes a shape to the range of values of $\mathbf{x}$ and the quadratic form grows quadratically as the vector's length increases. Analytic solutions to such optimization problems are found by evaluating the *gradient* of the cost function with respect to the vector $\mathbf{x}$. As discussed in appendix C {499}, the gradient with respect to $\mathbf{x}$ is evaluated by treating the vector's conjugate as a constant and *vice versa*. In this way, we find that

$$\nabla_{\mathbf{x}}\, \mathbf{x}'\mathbf{A}\mathbf{x} = \mathbf{A}^t\mathbf{x}^*$$
$$\nabla_{\mathbf{x}^*}\, \mathbf{x}'\mathbf{A}\mathbf{x} = \mathbf{A}\mathbf{x}$$

B.5 Matrix Eigenanalysis

One of the most powerful concepts in matrix algebra is eigenanalysis. Letting the matrix $\mathbf{A}$ be square, the nonzero vector $\mathbf{v}$ is said to be an *eigenvector* of $\mathbf{A}$ if it satisfies

$$\mathbf{A}\mathbf{v} = \lambda\mathbf{v}$$

where λ is the scalar termed the *eigenvalue* associated with $\mathbf{v}$. A scaled eigenvector is also an eigenvector. Because of this property, we define an eigenvector to *always* have unit inner product ($\mathbf{v}'\mathbf{v} = 1$). The *generalized eigenvector* $\mathbf{v}$ of the matrix pair $(\mathbf{A}, \mathbf{B})$ satisfies

$$\mathbf{A}\mathbf{v} = \lambda\mathbf{B}\mathbf{v}$$

Expressed in a slightly different form, the defining equation of eigenanalysis becomes $(\mathbf{A} - \lambda\mathbf{I})\,\mathbf{v} = \mathbf{0}$; for generalized eigenanalysis, $(\mathbf{A} - \lambda\mathbf{B})\,\mathbf{v} = \mathbf{0}$. If the matrices within the parentheses had inverses, the *only* solutions to these equations would be $\mathbf{v} = \mathbf{0}$, a trivial and uninteresting result. To find any nonzero solutions, the eigenvectors, these matrices must not have inverses, which implies that their determinants must be zero.

$$\det[\mathbf{A} - \lambda\mathbf{I}] = 0 \quad \det[\mathbf{A} - \lambda\mathbf{B}] = 0$$

For an $n\times n$ matrix, each equation becomes an nth order polynomial. The eigenvalues of a matrix are the roots of this polynomial. Because no closed form expression exists for the roots of polynomials of greater than fourth order, the eigenvalues of a matrix must be found numerically in most cases of interest in signal processing. The n eigenvalues are conventionally labeled in decreasing order: $\lambda_1 \geq \lambda_2 \geq \cdots \geq \lambda_n$. Eigenvalues are not necessarily unique; the number of times a value is repeated is termed its multiplicity. For example, every eigenvalue of the identity matrix equals unity and thus has multiplicity n. The eigenvalues of a sum or product of matrices are not easily expressed in terms of the eigenvalues of the component matrices. One remarkably simple result concerns Kronecker products: The eigenvalues of $\mathbf{A} \otimes \mathbf{B}$ equal $\lambda_i(\mathbf{A})\lambda_j(\mathbf{B})$, $i = 1, \ldots, m$, $j = 1, \ldots, n$. We term the set $\{\lambda_i, \mathbf{v}_i\}_{\mathbf{A}}$ of eigenvectors and associated eigenvalues the *eigensystem* of the matrix $\mathbf{A}$. For generalized systems to have n elements, the matrix $\mathbf{B}$ must be invertible. If not, the number of eigenvectors can be 0, less than n, or infinite [60: 252]. When invertible, the generalized eigensystem $\{\lambda_i, \mathbf{v}_i\}_{(\mathbf{A},\mathbf{B})}$ equals the eigensystem $\{\lambda_i, \mathbf{v}_i\}_{\mathbf{B}^{-1}\mathbf{A}}$.

Hermitian matrices. In the special case in which a matrix has Hermitian symmetry, several interesting properties arise. Because of the prevalence of the correlation matrix, which is Hermitian, this situation occurs often in practice.

- *The eigenvalues of a Hermitian matrix are real.* The inner product $\mathbf{v}'\mathbf{v}$ can be expressed as $\mathbf{v}'\mathbf{A}\mathbf{v} = \lambda\mathbf{v}'\mathbf{v}$. The left side of this equation equals its conjugate transpose; because the inner product $\mathbf{v}'\mathbf{v}$ is always real, we have $\lambda^* = \lambda$ and thus the eigenvalues are real.
- *If* $\mathbf{A}$ *is positive definite, all of its eigenvalues are positive.* The quadratic form $\mathbf{v}'\mathbf{A}\mathbf{v}$ is positive; as inner products are positive, $\mathbf{v}'\mathbf{A}\mathbf{v} = \lambda\mathbf{v}'\mathbf{v}$ implies that λ must also be positive.
- *The eigenvectors associated with distinct eigenvalues of a Hermitian matrix are orthogonal.* Consider $\mathbf{A}\mathbf{v}_1 = \lambda_1\mathbf{v}_1$ and $\mathbf{A}\mathbf{v}_2 = \lambda_2\mathbf{v}_2$ for $\lambda_1 \neq \lambda_2$. Noting that $\mathbf{v}_2'\mathbf{A}\mathbf{v}_1 = \lambda_1\mathbf{v}_2'\mathbf{v}_1$ and $\mathbf{v}_1'\mathbf{A}\mathbf{v}_2 = \lambda_2\mathbf{v}_1'\mathbf{v}_2$, these expressions differ on the left side only in that they are conjugate transposes of each other. Consequently $(\lambda_2 - \lambda_1)\mathbf{v}_1'\mathbf{v}_2 = 0$, thereby indicating that $\mathbf{v}_1$ and $\mathbf{v}_2$ are orthogonal.

Define a matrix $\mathbf{V}$ having its columns to be the eigenvectors of the matrix $\mathbf{A}$.

$$\mathbf{V} = \mathrm{col}[\mathbf{v}_1, \mathbf{v}_2, \ldots, \mathbf{v}_n]$$

If $\mathbf{A}$ is Hermitian and has distinct eigenvalues, its eigenvectors are orthogonal, implying that $\mathbf{V}$ must satisfy $\mathbf{V}'\mathbf{V} = \mathbf{I}$: $\mathbf{V}$ is unitary. Furthermore, the product $\mathbf{V}'\mathbf{A}\mathbf{V}$ is a diagonal matrix with the eigenvalues lying along the diagonal.

$$\mathbf{V}'\mathbf{A}\mathbf{V} = \begin{bmatrix} \lambda_1 & 0 & \cdots & 0 \\ 0 & \lambda_2 & 0 & \vdots \\ \vdots & & \ddots & \\ 0 & \cdots & 0 & \lambda_n \end{bmatrix}$$

Because $\mathbf{V}$ is unitary, we find that Hermitian matrices having distinct eigenvalues can be expressed

$$\mathbf{A} = \mathbf{V} \begin{bmatrix} \lambda_1 & 0 & \cdots & 0 \\ 0 & \lambda_2 & 0 & \vdots \\ \vdots & & \ddots & \\ 0 & \cdots & 0 & \lambda_n \end{bmatrix} \mathbf{V}'$$

This equation defines the matrix's *diagonal form*. From this diagonal form, the determinant and trace of $\mathbf{A}$ are easily related to the eigenvalues of $\mathbf{A}$ as

$$\det[\mathbf{A}] = \prod_{i=1}^{n} \lambda_i$$

$$\mathrm{tr}[\mathbf{A}] = \sum_{i=1}^{n} \lambda_i$$

Expressing the diagonal form of $\mathbf{A}$ less concisely, an important conclusion can be drawn: *Any Hermitian matrix* $\mathbf{A}$ *can be expressed with the expansion* $\mathbf{A} = \sum_i \lambda_i \mathbf{v}_i' \mathbf{v}_i$. This result is used frequently when eigenanalysis is applied to signal processing. In particular, quadratic forms can be expressed as

$$\mathbf{x}'\mathbf{A}\mathbf{x} = \sum_{i=1}^{n} \lambda_i \left|\mathbf{x}'\mathbf{v}_i\right|^2$$

In addition, because $\mathbf{v} = \lambda \mathbf{A}^{-1}\mathbf{v}$, the eigenvectors of a matrix and its inverse are identical with the corresponding eigenvalues being reciprocals of each other. Therefore,

$$\mathbf{A}^{-1} = \sum_{i=1}^{n} \frac{1}{\lambda_i} \mathbf{v}_i \mathbf{v}_i'$$

For generalized eigenanalysis, orthogonality of eigenvectors is defined with respect to the matrix $\mathbf{B}$: $\mathbf{v}_i'\mathbf{B}\mathbf{v}_j = \delta_{ij}$. Then,†

$$\mathbf{A} = \sum_{i=1}^{n} \lambda_i \mathbf{B}\mathbf{v}_i \mathbf{v}_i' \mathbf{B}'$$

$$\mathbf{A}^{-1} = \sum_{i=1}^{n} \frac{1}{\lambda_i} \mathbf{v}_i \mathbf{v}_i'$$

Rayleigh quotient. An important quantity in matrix algebra easily manipulated in terms of eigenvectors and eigenvalues is the ratio of two quadratic forms.

$$R = \frac{\mathbf{x}'\mathbf{A}\mathbf{x}}{\mathbf{x}'\mathbf{B}\mathbf{x}}$$

The matrices $\mathbf{A}$ and $\mathbf{B}$ are Hermitian and $\mathbf{B}$ must be positive definite. The key question asked is what vector $\mathbf{x}$ maximizes (or minimizes) this *Rayleigh quotient*. The simplest, and most prevalent case, occurs when $\mathbf{B} = \mathbf{I}$. The Rayleigh quotient is expressed in terms of the eigenvectors and eigenvalues of $\mathbf{A}$.

$$R = \frac{\sum_{i=1}^{n} \lambda_i |\mathbf{x}'\mathbf{v}_i|^2}{\mathbf{x}'\mathbf{x}}$$

The denominator is now seen to normalize the quotient: The quotient's value does not change when $\mathbf{x}$ is multiplied by a scalar. We can thus take the norm of $\mathbf{x}$ to be 1 and the denominator is no longer a concern. To maximize the numerator, the vector must be proportional to the eigenvector $\mathbf{v}_i$ corresponding to the maximum eigenvalue of $\mathbf{A}$. If not chosen this way, some of $\mathbf{x}$'s components would project onto other eigenvectors that

†Despite appearances, the second equation is correct. Test its veracity.

necessarily have smaller eigenvalues and thus provide less weight to the sum. A similar argument applies if the minimizing vector is sought.

$$R_{\min} = \lambda_{\min}, \mathbf{x}_{\min} = \mathbf{v}_{\min} \qquad R_{\max} = \lambda_{\max}, \mathbf{x}_{\max} = \mathbf{v}_{\max}$$

Furthermore, the Rayleigh quotient is bounded by the minimum and maximum eigenvalues.

$$\lambda_{\min} \leq R \leq \lambda_{\max}$$

When $\mathbf{B}$ does not equal the identity matrix, we can use the generalized eigensystem of $(\mathbf{A}, \mathbf{B})$ to find the extrema of the Rayleigh quotient. When $\mathbf{x}$ is a generalized eigenvector, $\mathbf{Ax} = \lambda\mathbf{Bx}$, which implies that the Rayleigh quotient equals λ. Because $\mathbf{A}$ and $\mathbf{B}$ are Hermitian and positive definite, the generalized eigenvectors form a basis and the Rayleigh quotient obeys the preceding inequality for the generalized eigenvalues. We can express the Rayleigh quotient in other ways. First of all, we know that the generalized eigensystem equals the eigensystem of $\mathbf{B}^{-1}\mathbf{A}$. In addition, define $\tilde{\mathbf{x}} = \mathbf{B}^{1/2}\mathbf{x}$ where $\mathbf{B}^{1/2}$ is the square root of $\mathbf{B}$. Computationally, square roots may seem difficult to find; in eigenanalysis terms, a straightforward definition emerges.

$$\mathbf{B}^{1/2} \equiv \sum_{i=1}^{n} \lambda_i^{1/2} \mathbf{v}_i \mathbf{v}_i'$$

With this definition, the square root of a matrix satisfies the intuitive property $\mathbf{B}^{1/2}\mathbf{B}^{1/2} = \mathbf{B}$. The Rayleigh quotient thus becomes

$$R = \frac{\tilde{\mathbf{x}}' \left(\mathbf{B}^{-1/2}\right)' \mathbf{A} \left(\mathbf{B}^{-1/2}\right) \tilde{\mathbf{x}}}{\tilde{\mathbf{x}}'\tilde{\mathbf{x}}}$$

Thus, the vector $\tilde{\mathbf{x}}$ maximizing or minimizing the Rayleigh quotient corresponds to the eigenvector having maximum or minimum eigenvalues of the matrix product $\left(\mathbf{B}^{-1/2}\right)' \mathbf{A} \left(\mathbf{B}^{-1/2}\right)$.

Singular value decomposition. Related to eigenanalysis is the *singular value decomposition* (*SVD*) technique. Letting $\mathbf{A}$ be an $m \times n$ matrix consisting of complex entries, it can be expressed by

$$\mathbf{A} = \mathbf{UDV}'$$

where $\mathbf{D}$ is the $k \times k$ diagonal matrix $\text{diag}[\sigma_1, \ldots, \sigma_k]$ and where $\mathbf{U}$ $(m \times k)$ and $\mathbf{V}$ $(k \times n)$ are matrices satisfying $\mathbf{U}'\mathbf{U} = \mathbf{I}_k$ and $\mathbf{V}'\mathbf{V} = \mathbf{I}_k$. Letting $\mathbf{u}_i$ and $\mathbf{v}_i$ denote the ith columns of $\mathbf{U}$ and $\mathbf{V}$ respectively, then

$$\mathbf{A}\mathbf{v}_i = \sigma_i \mathbf{u}_i \qquad \mathbf{u}_i'\mathbf{A} = \sigma_i \mathbf{v}_i'$$

The scalars σ_i are termed the singular values of the matrix $\mathbf{A}$ whereas $\mathbf{u}_i$ and $\mathbf{v}_i$ are termed the left and right, respectively, singular vectors of $\mathbf{A}$. The number of nonzero singular values equals k, which must not exceed $\min(m, n)$. The number equals this upper limit when the side of the matrix having the smallest length (either the rows or the columns) has linearly independent components. From these expressions, the eigenvectors of $\mathbf{A}'\mathbf{A}$ and $\mathbf{A}\mathbf{A}'$ are found to be

$$\mathbf{A}'\mathbf{A}\mathbf{v}_i = \sigma_i^2 \mathbf{v}_i \quad \mathbf{A}\mathbf{A}'\mathbf{u}_i = \sigma_i^2 \mathbf{u}_i$$

One of these products will not have full rank when the matrix $\mathbf{A}$ is not square. For example, if $\mathbf{A}$ is a column vector, $\mathbf{A}'\mathbf{A}$ is a scalar and thus is invertible, whereas $\mathbf{A}\mathbf{A}'$ is a $m \times m$ matrix having only one nonzero singular value. These matrices are Hermitian; they share the same nonzero eigenvalues, and these equal the *squares* of the singular values. Furthermore, the collections of vectors $\{\mathbf{u}_i\}$ and $\{\mathbf{v}_i\}$ are each orthonormal sets. This property means that $\mathbf{U}\mathbf{U}' = \mathbf{I}_m$ and $\mathbf{V}\mathbf{V}' = \mathbf{I}_n$. The singular value decomposition suggests that all matrices have an orthonormal-like expansion of the form

$$\boxed{\mathbf{A} = \sum_{i=1}^{k} \sigma_i \mathbf{u}_i \mathbf{v}_i'}$$

No nonsquare matrix has an inverse. Singular value decomposition can be used to define the *pseudoinverse* of a rectangular matrix. Assuming that all of the singular values are nonzero, the pseudoinverse $\mathbf{A}^{\sim 1}$ satisfies either $\mathbf{A}\mathbf{A}^{\sim 1} = \mathbf{I}_m$ or $\mathbf{A}^{\sim 1}\mathbf{A} = \mathbf{I}_n$ according to which dimension of the matrix is the smallest. By analogy to the eigenvalue-eigenvector expansion of an invertible matrix, the singular value decomposition of a matrix's pseudoinverse would be

$$\mathbf{A}^{\sim 1} = \sum_{i=1}^{k} \frac{1}{\sigma_i} \mathbf{v}_i \mathbf{u}_i'$$

or in matrix terms $\mathbf{A}^{\sim 1} = \mathbf{V}\mathbf{D}^{-1}\mathbf{U}'$. The pseudoinverses can be defined more directly by either $\mathbf{A}^{\sim 1} = \mathbf{A}'(\mathbf{A}\mathbf{A}')^{-1}$ or $\mathbf{A}^{\sim 1} = (\mathbf{A}'\mathbf{A})^{-1}\mathbf{A}'$. To be more concrete, suppose $\mathbf{A}$ has linearly independent rows. Its pseudoinverse is given by $\mathbf{A}'(\mathbf{A}\mathbf{A}')^{-1}$ so that $\mathbf{A}\mathbf{A}^{\sim 1} = \mathbf{I}$: Only the right inverse exists.

B.6 Projection Matrices

Eigenanalysis and singular value analysis are often used because they express a matrix's properties "naturally." One case in point is the projection matrix. A projection matrix $\mathbf{P}$ has the property $\mathbf{P}^m = \mathbf{P}^n$ for $m, n \geq 0$. The eigenvalues of a projection must therefore obey the relationship $\lambda^m = \lambda^n$. Two solutions are possible: $\lambda = 1$ and $\lambda = 0$. Thus, the eigenvectors of a projection matrix either have zero eigenvalues, in which case any vector that can be expressed by them is annihilated by the matrix ($\mathbf{P}\mathbf{x} = \mathbf{0}$), or the eigenvectors have unity eigenvalues and vectors comprised of them are unaffected by the matrix ($\mathbf{P}\mathbf{x} = \mathbf{x}$). If a vector has components belonging to both sets of eigenvectors, the part of the vector that "survives" the matrix is that portion represented by the eigenvectors having unity eigenvalues. Hence, the origin of the name "projection matrix": Matrices having only unity and zero eigenvalues can be used to find those components of any vector that correspond to the eigenvectors having unity eigenvalues.

For any $m \times n$ matrix $\mathbf{A}$ $(m < n)$ that is used as a linear transformation, it defines a subspace for the *domain* of the transformation. In more exact terms, when the m-dimensional vectors $\mathbf{y} = \mathbf{A}\mathbf{x}$ define a vector space, what is the image of this space in the original, higher dimensional space containing the vectors $\mathbf{x}$? The most elegant response is in terms of the singular values and singular vectors of $\mathbf{A}$. The vector $\mathbf{y}$ is thereby expressed

$$\mathbf{y} = \sum_{i=1}^{m} \sigma_i (\mathbf{v}_i' \mathbf{x}) \mathbf{u}_i$$

The result of the linear transformation is always a linear combination of the left singular vectors of $\mathbf{A}$.

The projection matrix $\mathbf{P_A}$ associated with a linear transformation $\mathbf{A}$ would project all n-dimensional vectors onto a subspace defined by the linear transformation. $\mathbf{P_A}$ would have m unity eigenvalues for the left singular vectors of $\mathbf{A}$ and zero eigenvalues for all remaining $m - n$ eigenvectors. Defining $\mathbf{U}^{\perp}$ to be the matrix of eigenvectors corresponding to these zero eigenvalues, the projection matrix can be expressed as

$$\mathbf{P_A} = \begin{bmatrix} \mathbf{U} & \mathbf{U}^{\perp} \end{bmatrix} \begin{bmatrix} \mathbf{I}_m & \mathbf{0} \\ \mathbf{0} & \mathbf{0} \end{bmatrix} \begin{bmatrix} \mathbf{U} & \mathbf{U}^{\perp} \end{bmatrix}'$$

This projection matrix can be expressed more directly in terms of the matrix $\mathbf{A}$ as $\mathbf{A}'(\mathbf{A}\mathbf{A}')^{-1}\mathbf{A}$. Note than this expression can be written in terms of the transformation's pseudoinverse: While $\mathbf{A}\mathbf{A}^{\sim 1} = \mathbf{I}_m$, $\mathbf{P_A} = \mathbf{A}^{\sim 1}\mathbf{A}$.

Appendix C

Optimization Theory

Optimization theory is the study of the *extremal* values of a function: its minima and maxima. Topics in this theory range from conditions for the existence of a unique extremal value to methods, both analytic and numeric, for finding the extremal values and for what values of the independent variables the function attains its extremes. In this book, minimizing an error criterion is an essential step toward deriving optimal signal processing algorithms.

C.1 Unconstrained Optimization

The simplest optimization problem is to find the minimum of a scalar-valued function of a scalar variable $f(x)$, the so-called *objective function*, and where that minimum is located. Assuming the function is differentiable, the well-known conditions for finding the minima, both local and global, are[†]

$$\frac{df(x)}{dx} = 0$$

$$\frac{d^2 f(x)}{dx^2} > 0$$

All values of the independent variable x satisfying these relations are locations of local minima.

Without the second condition, solutions to the first could be maxima, minima, or inflection points. Solutions to the first equation are termed the *stationary points* of the objective function. To find the *global* minimum, that value (or values) where the function achieves its smallest value, each candidate extremum must be tested: The objective function must be evaluated at each stationary point and the smallest selected. If, however, the objective function can be shown to be strictly convex, then only one solution of $df/dx = 0$ exists and that solution corresponds to the global minimum. The function

[†] The maximum of a function is found by finding the minimum of its negative.

$f(x)$ is *strictly convex* if, for any choice of x_1, x_2, and the scalar a, $f(ax_1 + (1-a)x_2) < af(x_1) + (1-a)f(x_2)$. Convex objective functions occur often in practice and are more easily minimized because of this property.

When the objective function $f(\cdot)$ depends on a complex variable z, subtleties enter the picture. *If* the function $f(z)$ is differentiable, its extremes can be found in the obvious way: Find the derivative, set it equal to 0, and solve for the locations of the extrema. Of particular interest in array signal processing are situations in which this function is *not* differentiable. In contrast to functions of a real variable, nondifferentiable functions of a complex variable occur frequently. The simplest example is $f(z) = |z|^2$. The minimum value of this function obviously occurs at the origin. To calculate this obvious answer, a complication arises: The function $f(z) = z^*$ is not analytic with respect to z and hence not differentiable. More generally, the derivative of a function with respect to a complex-valued variable cannot be evaluated directly when the function depends on the variable's conjugate.

This complication can be resolved with either of two methods tailored for optimization problems. The first is to express the objective function in terms of the real and imaginary parts of z and find the function's minimum with respect to these two variables.[†] This approach is unnecessarily tedious but will yield the solution. The second, more elegant, approach relies on two results from complex variable theory. First, the quantities z and z^* can be treated as independent variables, each considered a constant with respect to the other. A variable and its conjugate are viewed as the result of applying an invertible linear transformation to the variable's real and imaginary parts. Thus, if the real and imaginary parts can be considered as independent variables, so can the variable and its conjugate with the advantage that the mathematics is far simpler. In this way, $\partial|z|^2/\partial z = z^*$ and $\partial|z|^2/\partial z^* = z$. Seemingly, the next step to minimizing the objective function is to set the derivatives with respect to each quantity to 0 and then solve the resulting pair of equations. As the following theorem suggests, that solution is overly complicated.

Theorem. *If the function $f(z, z^*)$ is real valued and analytic with respect to z and z^*, all stationary points can be found by setting the derivative (in the sense just given) with respect to* either *z or z^* to 0* [18].

Thus, to find the minimum of $|z|^2$, compute the derivative with respect to either z or z^*. In most cases, the derivative with respect to z^* is the most convenient choice.[‡] Thus, $\partial(|z|^2)/\partial z^* = z$ and the stationary point is $z = 0$. As this objective function is strictly convex, the objective function's sole stationary point is its global minimum.

When the objective function depends on a vector-valued quantity $\mathbf{x}$, the evaluation of the function's stationary points is a simple extension of the scalar-variable case. Testing stationary points as possible locations for minima is more complicated, however. The *gradient* of the scalar-valued function $f(\mathbf{x})$ of a vector $\mathbf{x}$ (dimension N) equals an N-dimensional vector in which each component is the partial derivative of $f(\cdot)$ with respect

[†] The multivariate minimization problem is discussed in a few paragraphs.

[‡] Why should this be? In the next few examples, try both and see which you think is "easier."

to each component of **x**.

$$\nabla_{\mathbf{x}} f(\mathbf{x}) = \text{col}\left[\frac{\partial f(\mathbf{x})}{\partial x_1} \cdots \frac{\partial f(\mathbf{x})}{\partial x_N}\right]$$

For example, the gradient of $\mathbf{x}^t\mathbf{A}\mathbf{x}$ is $\mathbf{A}\mathbf{x}+\mathbf{A}^t\mathbf{x}$. This result is easily derived by expressing the quadratic form as a double sum ($\sum_{ij} A_{ij}x_i x_j$) and evaluating the partials directly. When $\mathbf{A}$ is symmetric, which is often the case, this gradient becomes $2\mathbf{A}\mathbf{x}$.

The gradient "points" in the direction of the maximum rate of increase of the function $f(\cdot)$. This fact is often used in numerical optimization algorithms. The *method of steepest descent* is an iterative algorithm in which a candidate minimum is augmented by a quantity proportional to the negative of the objective function's gradient to yield the next candidate.

$$\mathbf{x}_k = \mathbf{x}_{k-1} - \alpha \nabla_{\mathbf{x}} f(\mathbf{x}), \quad \alpha > 0$$

If the objective function is "smooth enough" (there aren't too many minima and maxima), this approach will yield the global minimum. Strictly convex functions are certainly smooth enough for this method to work.

The gradient of the gradient of $f(\mathbf{x})$, denoted by $\nabla_{\mathbf{x}}^2 f(\mathbf{x})$, is a matrix in which the jth column is the gradient of the jth component of f's gradient. This quantity is known as the *Hessian*, defined to be the matrix of all the second partials of $f(\cdot)$.

$$\left[\nabla_{\mathbf{x}}^2 f(\mathbf{x})\right]_{ij} = \frac{\partial^2 f(\mathbf{x})}{\partial x_i \partial x_j}$$

The Hessian is always a symmetric matrix.

The minima of the objective function $f(\mathbf{x})$ occur when

$$\boxed{\nabla_{\mathbf{x}} f(\mathbf{x}) = \mathbf{0} \quad \text{and} \quad \nabla_{\mathbf{x}}^2 f(\mathbf{x}) > 0}$$

Thus, for a stationary point to be a minimum, the Hessian evaluated at that point must be a positive-definite matrix. When the objective function is strictly convex, this test need not be performed. For example, the objective function $f(\mathbf{x}) = \mathbf{x}^t\mathbf{A}\mathbf{x}$ is convex whenever $\mathbf{A}$ is positive definite and symmetric.[†]

When the independent vector is complex-valued, the issues discussed in the scalar case also arise. Because of the complex-valued quantities involved, how to evaluate the gradient becomes an issue: Is $\nabla_{\mathbf{z}}$ or $\nabla_{\mathbf{z}^*}$ more appropriate? In contrast to the case of complex scalars, the choice in the case of complex vectors is unique.

Theorem. *Let $f(\mathbf{z}, \mathbf{z}^*)$ be a real-valued function of the vector-valued complex variable $\mathbf{z}$ in which the dependence on the variable and its conjugate is explicit. By treating $\mathbf{z}$ and $\mathbf{z}^*$ as independent variables, the quantity pointing in the direction of the maximum rate of change of $f(\mathbf{z}, \mathbf{z}^*)$ is $\nabla_{\mathbf{z}^*} f(\mathbf{z})$* [18].

[†]Note that the Hessian of $\mathbf{x}^t\mathbf{A}\mathbf{x}$ is $2\mathbf{A}$.

To show this result, consider the variation of f given by

$$\begin{aligned}\delta f &= \sum_i \left(\frac{\partial f}{\partial z_i} \delta z_i + \frac{\partial f}{\partial z_i^*} \delta z_i^* \right) \\ &= (\nabla_{\mathbf{z}} f)^t \, \delta \mathbf{z} + (\nabla_{\mathbf{z}^*} f)^t \, \delta \mathbf{z}^*\end{aligned}$$

This quantity is concisely expressed as $\delta f = 2\,\mathrm{Re}\left[(\nabla_{\mathbf{z}^*} f)' \, \delta \mathbf{z}\right]$. By the Schwarz Inequality, the maximum value of this variation occurs when $\delta \mathbf{z}$ is in the same direction as $(\nabla_{\mathbf{z}^*} f)$. Thus, the direction corresponding to the largest change in the quantity $f(\mathbf{z}, \mathbf{z}^*)$ is in the direction of its gradient with respect to $\mathbf{z}^*$. To implement the method of steepest descent, for example, the gradient with respect to the conjugate *must* be used.

To find the stationary points of a scalar-valued function of a complex-valued vector, we must solve

$$\nabla_{\mathbf{z}^*} f(\mathbf{z}) = \mathbf{0} \tag{C.1}$$

For solutions of this equation to be minima, the Hessian defined to be the matrix of mixed partials given by $\nabla_{\mathbf{z}} (\nabla_{\mathbf{z}^*} f(\mathbf{z}))$ must be positive definite. For example, the required gradient of the objective function $\mathbf{z}'\mathbf{A}\mathbf{z}$ is given by $\mathbf{A}\mathbf{z}$, implying for positive definite $\mathbf{A}$ that a stationary point is $\mathbf{z} = \mathbf{0}$. The Hessian of the objective function is simply $\mathbf{A}$, confirming that the minimum of a quadratic form is always the origin.

C.2 Constrained Optimization

Constrained optimization is the minimization of an objective function subject to constraints on the possible values of the independent variable. Constraints can be either *equality constraints* or *inequality constraints*. Because the scalar-variable case follows easily from the vector one, only the latter is discussed in detail here.

C.2.1 Equality Constraints

The typical constrained optimization problem has the form

$$\min_{\mathbf{x}} f(\mathbf{x}) \text{ subject to } \mathbf{g}(\mathbf{x}) = \mathbf{0}$$

where $f(\cdot)$ is the scalar-valued objective function and $\mathbf{g}(\cdot)$ is the vector-valued *constraint function*. Strict convexity of the objective function is *not* sufficient to guarantee a unique minimum; in addition, each component of the constraint must be strictly convex to guarantee that the problem has a unique solution. Because of the constraint, stationary points of $f(\cdot)$ alone may not be solutions to the constrained problem: They may not satisfy the constraints. In fact, solutions to the constrained problem are often *not* stationary points of the objective function. Consequently, the *ad hoc* technique of searching for all stationary points of the objective function that also satisfy the constraint do not work.

The classical approach to solving constrained optimization problems is the method of *Lagrange multipliers*. This approach converts the constrained optimization problem into an unconstrained one, thereby allowing use of the techniques described in the previous

Figure C.1 The thick line corresponds to the contour of the values of $\mathbf{x}$ satisfying the constraint equation $\mathbf{g}(\mathbf{x}) = \mathbf{0}$. The thinner lines are contours of constant values of the objective function $f(\mathbf{x})$. The contour corresponding to the smallest value of the objective function just tangent to the constraint contour is the solution to the optimization problem with equality constraints.

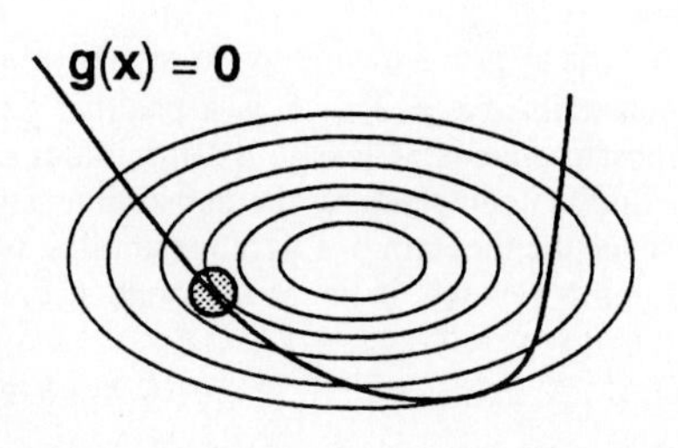

section. The *Lagrangian* of a constrained optimization problem is defined to be the scalar-valued function

$$L(\mathbf{x}, \boldsymbol{\lambda}) = f(\mathbf{x}) + \boldsymbol{\lambda}^t \mathbf{g}(\mathbf{x})$$

Essentially, the following theorem states that stationary points of the Lagrangian are *potential* solutions of the constrained optimization problem: As always, each candidate solution must be tested to determine which minimizes the objective function.

Theorem. *Let* $\mathbf{x}_\diamond$ *denote a local solution to the constrained optimization problem given previously in which the gradients* $\nabla_{\mathbf{x}} g_1(\mathbf{x}), \ldots, \nabla_{\mathbf{x}} g_M(\mathbf{x})$ *of the constraint function's components are linearly independent. There then exists a unique vector* $\boldsymbol{\lambda}_\diamond$ *such that*

$$\nabla_{\mathbf{x}} L(\mathbf{x}_\diamond, \boldsymbol{\lambda}_\diamond) = \mathbf{0}$$

Furthermore, the quadratic form $\mathbf{y}^t \left[\nabla^2_{\mathbf{x}} L(\mathbf{x}_\diamond, \boldsymbol{\lambda}_\diamond)\right] \mathbf{y}$ *is nonnegative for all* $\mathbf{y}$ *satisfying* $\left[\nabla_{\mathbf{x}} \mathbf{g}(\mathbf{x})\right]^t \mathbf{y} = 0$ [13: §1.4].

The latter result in the theorem says that the Hessian of the Lagrangian evaluated at its stationary points is nonnegative definite with respect to all vectors "orthogonal" to the gradient of the constraint. This result generalizes the notion of a positive definite Hessian in unconstrained problems.

The rather abstract result of the preceding theorem has a simple geometrical interpretation. As shown in Fig. C.1, the constraint corresponds to a contour in the $\mathbf{x}$ plane. A "contour map" of the objective function indicates those values of $\mathbf{x}$ for which $f(\mathbf{x}) = c$. In this figure, as c becomes smaller, the contours shrink to a small circle in the center of the figure. The solution to the constrained optimization problem occurs when the smallest value of c is chosen for which the contour just touches the constraint contour. At that point, the gradient of the objective function and of the constraint contour are proportional to each other. This proportionality vector is $\boldsymbol{\lambda}_\diamond$, the so-called *Lagrange multiplier*. The Lagrange multiplier's exact value must be such that the constraint is exactly satisfied. Note that the constraint can be tangent to the objective function's contour map for larger values of c. These potential, but erroneous, solutions can be discarded only by evaluating the objective function.

Example

A typical problem arising in signal processing is to minimize $\mathbf{x}'\mathbf{A}\mathbf{x}$ subject to the linear constraint $\mathbf{c}'\mathbf{x} = 1$.† $\mathbf{A}$ is a positive definite, symmetric matrix (a correlation matrix) in most problems. Clearly, the minimum of the objective function occurs at $\mathbf{x} = \mathbf{0}$, but this solution cannot satisfy the constraint. The constraint $g(\mathbf{x}) = \mathbf{c}'\mathbf{x} - 1$ is a scalar-valued one; hence the theorem of Lagrange applies as there are no multiple components in the constraint forcing a check of linear independence. The Lagrangian is

$$L(\mathbf{x}, \lambda) = \mathbf{x}'\mathbf{A}\mathbf{x} + \lambda(\mathbf{c}'\mathbf{x} - 1)$$

Its gradient is $2\mathbf{A}\mathbf{x} + \lambda\mathbf{c}$ with a solution $\mathbf{x}_\diamond = -\lambda_\diamond \mathbf{A}^{-1}\mathbf{c}/2$. To find the value of the Lagrange multiplier, this solution must satisfy the constraint. Imposing the constraint, $\lambda_\diamond \mathbf{c}'\mathbf{A}^{-1}\mathbf{c} = -2$; thus, $\lambda_\diamond = -2/(\mathbf{c}'\mathbf{A}^{-1}\mathbf{c})$ and the total solution is

$$\mathbf{x}_\diamond = \frac{\mathbf{A}^{-1}\mathbf{c}}{\mathbf{c}'\mathbf{A}^{-1}\mathbf{c}}$$

When the independent variable is complex-valued, the Lagrange multiplier technique can be used *if* care is taken to make the Lagrangian real. If it is not real, we cannot use the theorem {501} that permits computation of stationary points by computing the gradient with respect to $\mathbf{z}^*$ alone. The Lagrangian may not be realvalued even when the constraint is real. Once ensured real, the gradient of the Lagrangian with respect to the conjugate of the independent vector can be evaluated and the minimization procedure remains as before.

Example

Consider slight variations to the previous example: Let the vector $\mathbf{z}$ be complex so that the objective function is $\mathbf{z}'\mathbf{A}\mathbf{z}$ where $\mathbf{A}$ is a positive-definite, Hermitian matrix and let the constraint be linear, but vector-valued ($\mathbf{C}\mathbf{z} = \mathbf{c}$). The Lagrangian is formed from the objective function and the real part of the usual constraint term.

$$L(\mathbf{z}, \boldsymbol{\lambda}) = \mathbf{z}'\mathbf{A}\mathbf{z} + \boldsymbol{\lambda}'(\mathbf{C}\mathbf{z} - \mathbf{c}) + \boldsymbol{\lambda}'(\mathbf{C}^*\mathbf{z}^* - \mathbf{c}^*)$$

For the Lagrange multiplier theorem to hold, the gradients of each component of the constraint must be linearly independent. As these gradients are the columns of $\mathbf{C}$, their mutual linear independence means that each constraint vector must not be expressible as a linear combination of the others. We shall assume this portion of the problem statement true. Evaluating the gradient with respect to $\mathbf{z}^*$, keeping $\mathbf{z}$ a constant, and setting the result equal to 0 yields

$$\mathbf{A}\mathbf{z}_\diamond + \mathbf{C}'\boldsymbol{\lambda}_\diamond = \mathbf{0}$$

The solution is $\mathbf{z}_\diamond = -\mathbf{A}^{-1}\mathbf{C}'\boldsymbol{\lambda}_\diamond$. Applying the constraint, we find that $\mathbf{C}\mathbf{A}^{-1}\mathbf{C}'\boldsymbol{\lambda}_\diamond = -\mathbf{c}$. Solving for the Lagrange multiplier and substituting the result into the solution, we find that the solution to the constrained optimization problem is

$$\mathbf{z}_\diamond = \mathbf{A}^{-1}\mathbf{C}'\left(\mathbf{C}\mathbf{A}^{-1}\mathbf{C}'\right)^{-1}\mathbf{c}$$

The indicated matrix inverses always exist: $\mathbf{A}$ is assumed invertible and $\mathbf{C}\mathbf{A}^{-1}\mathbf{C}'$ is invertible because of the linear independence of the constraints.

---

†See, for example, §7.2.1 {357}.

C.2.2 Inequality Constraints

When some of the constraints are inequalities, the Lagrange multiplier technique can be used, but the solution must be checked carefully in its details. First, however, the optimization problem with equality and inequality constraints is formulated as

$$\min_{\mathbf{x}} f(\mathbf{x}) \text{ subject to } \mathbf{g}(\mathbf{x}) = \mathbf{0} \text{ and } \mathbf{h}(\mathbf{x}) \leq \mathbf{0}$$

As before, $f(\cdot)$ is the scalar-valued objective function and $\mathbf{g}(\cdot)$ is the equality constraint function; $\mathbf{h}(\cdot)$ is the *inequality constraint function.*

The key result that can be used to find the analytic solution to this problem is to first form the Lagrangian in the usual way as $L(\mathbf{x}, \boldsymbol{\lambda}, \boldsymbol{\mu}) = f(\mathbf{x}) + \boldsymbol{\lambda}^t\mathbf{g}(\mathbf{x}) + \boldsymbol{\mu}^t\mathbf{h}(\mathbf{x})$. The following theorem is the general statement of the Lagrange multiplier technique for constrained optimization problems.

Theorem. *Let $\mathbf{x}^\diamond$ be a local minimum for the constrained optimization problem. If the gradients of $\mathbf{g}$'s components and the gradients of those components of $\mathbf{h}(\cdot)$ for which $h_i(\mathbf{x}^\diamond) = 0$ are linearly independent, then*

$$\nabla_{\mathbf{x}} L(\mathbf{x}^\diamond, \boldsymbol{\lambda}^\diamond, \boldsymbol{\mu}^\diamond) = \mathbf{0}$$

where $\mu_i^\diamond \geq 0$ and $\mu_i^\diamond h_i(\mathbf{x}^\diamond) = 0$ [13: §1.4].

The portion of this result dealing with the inequality constraint differs substantially from that concerned with the equality constraint. Either a component of the constraint equals its maximum value (0 in this case), and the corresponding component of its Lagrange multiplier is nonnegative (and is usually positive) *or* a component is less than the constraint and its component of the Lagrange multiplier is 0. This latter result means that some components of the inequality constraint are not as stringent as others and these lax ones do not affect the solution.

The rationale behind this theorem is a technique for converting the inequality constraint into an equality constraint: $h_i(\mathbf{x}) \leq 0$ is equivalent to $h_i(\mathbf{x}) + s_i^2 = 0$. Because the new term, called a *slack variable*, is nonnegative, the constraint must be nonpositive. With the inclusion of slack variables, the equality constraint theorem can be used and the theorem results. To prove the theorem, not only does the gradient with respect to $\mathbf{x}$ need to be considered, but also with respect to the vector $\mathbf{s}$ of slack variables. The ith component of the gradient of the Lagrangian with respect to $\mathbf{s}$ at the stationary point is $2\mu_i^\diamond s_i^\diamond = 0$. If in solving the optimization problem, $s_i^\diamond = 0$, the inequality constraint was in reality an equality constraint and that component of the constraint behaves accordingly. As $s_i = [-h_i(\mathbf{x})]^{1/2}$, $s_i = 0$ implies that that component of the inequality constraint must equal zero. Otherwise, if $s_i \neq 0$, the corresponding Lagrange multiplier must be 0.

Example

Consider the problem of minimizing a quadratic form subject to a linear equality constraint and an inequality constraint on the norm of the linear constraint vector's variation.

$$\min_{\mathbf{x}} \mathbf{x}^t\mathbf{A}\mathbf{x} \text{ subject to } (\mathbf{c} + \boldsymbol{\delta})^t\mathbf{x} = 1 \text{ and } \|\boldsymbol{\delta}\|^2 \leq \epsilon$$

This kind of problem arises in robust estimation. One seeks a solution in which one of the "knowns" of the problem, $\mathbf{c}$ in this case, is, in reality, only approximately specified. The independent variables are $\mathbf{x}$ and $\boldsymbol{\delta}$. The Lagrangian for this problem is

$$L(\{\mathbf{x}, \boldsymbol{\delta}\}, \lambda, \mu) = \mathbf{x}'\mathbf{A}\mathbf{x} + \lambda\left[(\mathbf{c} + \boldsymbol{\delta})'\mathbf{x} - 1\right] + \mu\left[\|\boldsymbol{\delta}\|^2 - \epsilon\right]$$

Evaluating the gradients with respect to the independent variables yields

$$2\mathbf{A}\mathbf{x}^\diamond + \lambda^\diamond(\mathbf{c} + \boldsymbol{\delta}^\diamond) = \mathbf{0}$$
$$\lambda^\diamond\mathbf{x}^\diamond + 2\mu^\diamond\boldsymbol{\delta}^\diamond = \mathbf{0}$$

The latter equation is key. Recall that either $\mu^\diamond = 0$ or the inequality constraint is satisfied with equality. If $\mu^\diamond$ is 0, that implies that $\mathbf{x}^\diamond$ must be 0, which does not allow the equality constraint to be satisfied. The inescapable conclusion is that $\|\boldsymbol{\delta}^\diamond\|^2 = \epsilon$ and that $\boldsymbol{\delta}^\diamond$ is parallel to $\mathbf{x}^\diamond$: $\boldsymbol{\delta}^\diamond = -(\lambda^\diamond/2\mu^\diamond)\mathbf{x}^\diamond$. Using the first equation, $\mathbf{x}^\diamond$ is found to be

$$\mathbf{x}^\diamond = -\frac{\lambda^\diamond}{2}\left(\mathbf{A} - \frac{\lambda^{\diamond 2}}{4\mu^\diamond}\mathbf{I}\right)^{-1}\mathbf{c}$$

Imposing the constraints on this solution results in a pair of equations for the Lagrange multipliers.

$$\frac{1}{4}\left(\frac{\lambda^{\diamond 2}}{\mu^\diamond}\right)^2 \mathbf{c}'\left[\mathbf{A} - \frac{1}{4}\left(\frac{\lambda^{\diamond 2}}{\mu^\diamond}\right)\mathbf{I}\right]^{-2}\mathbf{c} = \epsilon$$
$$\mathbf{c}'\left[\mathbf{A} - \frac{1}{4}\left(\frac{\lambda^{\diamond 2}}{\mu^\diamond}\right)\mathbf{I}\right]^{-1}\mathbf{c} = -\frac{2}{\lambda^\diamond} - 2\epsilon\left(\frac{\mu^\diamond}{\lambda^{\diamond 2}}\right)$$

Multiple solutions are possible and each must be checked. The rather complicated completion of this example is left to the (numerically oriented) reader.

Bibliography

The pages in this book containing references to each bibliographical item are contained within braces.

1. M. Abramowitz and I. A. Stegun, editors. *Handbook of Mathematical Functions*. U.S. Government Printing Office, 1968. {245, 247, 479}
2. H. Akaike. A new look at the statistical model identification problem. *IEEE Trans. Auto. Control*, AC-19:716–723, Dec. 1974. {336}
3. S. T. Alexander. *Adaptive Signal Processing*. Springer-Verlag, New York, 1986. {311, 313, 314, 318}
4. B. D. O. Anderson and J. B. Moore. *Optimal Filtering*. Prentice Hall, Englewood Cliffs, NJ, 1979. {294, 305}
5. S. P. Applebaum. Adaptive arrays. *IEEE Trans. Ant. Prop.*, AP-24:585–598, Sept. 1976. {358}
6. S. P. Applebaum and D. J. Chapman. Adaptive arrays with main beam constraints. *IEEE Trans. Ant. Prop.*, AP-24:650–662, Sept. 1976. {396}
7. M. Athans, R. P. Wishner, and A. Bertolini. Suboptimal state estimation for continuous-time nonlinear systems from discrete noisy measurements. *IEEE Trans. Auto. Control*, AC-13:504–514, Oct. 1968. {447}
8. A. B. Baggeroer. Confidence intervals for regression (MEM) spectral estimates. *IEEE Trans. Info. Th.*, IT-22:534–545, Sept. 1976. {337}
9. A. B. Baggeroer, W. A. Kuperman, and H. Schmidt. Matched field processing: Source localization in correlated noise as an optimum parameter estimation problem. *J. Acoust. Soc. Am.*, 83:571–587, 1988. {39}
10. Y. Bar-Shalom and T. E. Fortmann. *Tracking and Data Association*. Academic Press, San Diego, CA, 1988. {444, 456, 467}
11. M. S. Bartlett. Smoothing periodograms from time series with continuous spectra. *Nature*, 161:686–687, 1948. {327}
12. S. D. Bedrosian. Nonuniform linear arrays: Graph theoretic approach to minimum redundancy. *Proc. IEEE*, 74:1040–1043, July 1986. {100, 101}
13. D. P. Bertsekas. *Constrained Optimization and Lagrange Multiplier Methods*. Academic Press, New York, 1982. {503, 505}
14. G. Bienvenu and L. Kopp. Optimality of high resolution array processing using the eigensystem approach. *IEEE Trans. Acoustics, Speech and Signal Processing*,

ASSP-31:1235–1248, Oct. 1983. {383, 384}

15. G. J. Bierman. *Factorization Methods for Discrete Sequential Estimation.* Academic Press, New York, 1977. {317}
16. R. B. Blackman and J. W. Tukey. *The Measurement of Power Spectra.* Dover Press, New York, 1958. {321}
17. S. S. Blackman. *Multiple Target Tracking with Radar Applications.* Artech House, Dedham, MA, 1986. {467}
18. D. H. Brandwood. A complex gradient operator and its application in adaptive array theory. *IEE Proc., Pts. F and H*, 130:11–16, Feb. 1983. {500, 501}
19. L. Brekhovskikh and Y. Lysanov. *Fundamentals of Ocean Acoustics.* Springer-Verlag, New York, 1982. {34-36}
20. T. P. Bronez and J. A. Cadzow. An algebraic approach to superresolution array processing. *IEEE Trans. Aerosp. Electron. Sys.*, AES-19:123–132, Jan. 1983. {355}
21. H. P. Bucker. Use of calculated sound fields and matched field detection to locate sound sources in shallow water. *J. Acoust. Soc. Am.*, 59:368–373, 1976. {39}
22. K. M. Buckley. Broad-band beamforming and the generalized sidelobe canceller. *IEEE Trans. Acoustics, Speech and Signal Processing*, ASSP-34:1322–1323, Oct. 1986. {369}
23. K. M. Buckley and L. J. Griffiths. An adaptive sidelobe canceller with derivative constraints. *IEEE Trans. Ant. Prop.*, AP-34:311–319, Mar. 1986. {398}
24. J. P. Burg. *Maximum Entropy Spectral Analysis.* PhD thesis, Department of Geophysics, Stanford University, Stanford, CA, 1975. {334}
25. J. P. Burg. The relationship between maximum entropy spectra and maximum likelihood spectra. *Geophysics*, 37:375–376, Apr. 1972. {344}
26. D. Burshtein and E. Weinstein. Confidence intervals for the maximum entropy spectrum. *IEEE Trans. Acoustics, Speech and Signal Processing*, ASSP-35:504–510, Apr. 1987. {337, 367}
27. A. Cantoni and P. Butler. Properties of the eigenvectors of persymmetric matrices with applications to communication theory. *IEEE Trans. Comm.*, COM-24:804–809, Aug. 1976. {372}
28. J. Capon. High-resolution frequency-wavenumber spectrum analysis. *Proc. IEEE*, 57:1408–1418, 1969. {330, 359}
29. J. Capon. Maximum-likelihood spectral estimation. In S. Haykin, editor, *Nonlinear Methods of Spectral Analysis*, pages 155–179. Springer-Verlag, New York, 1979. {330, 362}
30. J. Capon and N. R. Goodman. Probability distributions for estimators of the frequency-wavenumber spectrum. *Proc. IEEE*, 58:1785–1786, Oct. 1970. {333, 362}
31. J. Capon, R. J. Greenfield, and R. J. Kolker. Multidimensional maximum-likelihood processing of a large aperture seismic array. *Proc. IEEE*, 55:192–211, Feb. 1967. {358}
32. B. D. Carlson. Covariance matrix estimation errors and diagonal loading in adaptive arrays. *IEEE Trans. Aerosp. Electron. Sys.*, AES-24:397–401, July 1988. {401}

33. J. W. Carlyle and J. B. Thomas. On nonparametric signal detectors. *IEEE Trans. Info. Th.*, IT-10:146–152, Apr. 1964. {246}
34. D. Chazan, M. Zakai, and J. Ziv. Improved lower bounds on signal parameter estimation. *IEEE Trans. Info. Th.*, IT-21:90–93, Jan. 1975. {290}
35. R. T. Compton, Jr. *Adaptive Antennas*. Prentice Hall, Englewood Cliffs, NJ, 1988. {357, 368, 393, 413, 423}
36. C. F. N. Cowan and P. M. Grant, editors. *Adaptive Filters*. Prentice Hall, Englewood Cliffs, NJ, 1985. {310, 510}
37. H. Cox. Resolving power and sensitivity to mismatch of optimum array processors. *J. Acoust. Soc. Am.*, 54:771–785, Sept. 1973. {137, 364}
38. H. Cox, R. M. Zeskind, and T. Kooij. Practical supergain. *IEEE Trans. Acoustics, Speech and Signal Processing*, ASSP-34:393–398, June 1986. {198}
39. H. Cox, R. M. Zeskind, and M. M. Owen. Effects of amplitude and phase errors on linear predictive array processors. *IEEE Trans. Acoustics, Speech and Signal Processing*, 36:10–19, Jan. 1988. {393}
40. H. Cox, R. M. Zeskind, and M. M. Owen. Robust adaptive beamforming. *IEEE Trans. Acoustics, Speech and Signal Processing*, ASSP-35:1365–1375, Oct. 1987. {399, 411}
41. H. Cramér. *Mathematical Methods of Statistics*. Princeton University Press, Princeton, NJ, 1946. {200, 216, 218, 276, 279, 280}
42. H. Cramér. *Random Variables and Probability Distributions*. Cambridge University Press, third edition, 1970. {481}
43. R. E. Crochiere and L. R. Rabiner. *Multirate Digital Signal Processing*. Prentice Hall, Englewood Cliffs, NJ, 1983. {163, 170}
44. S. R. DeGraaf and D. H. Johnson. Capability of array processing algorithms to estimate source bearings. *IEEE Trans. Acoustics, Speech and Signal Processing*, ASSP-33:1368–1379, Dec. 1985. {364, 365}
45. R. D. DeGroat and R. A. Roberts. Efficient, numerically stabilized rank-one eigenstructure updating. *IEEE Trans. Acoustics, Speech and Signal Processing*, 38:301–316, Feb. 1990. {407}
46. J. E. Dennis, Jr. and R. B. Schnabel. *Numerical Methods for Unconstrained Optimization and Nonlinear Equations*. Prentice Hall, Englewood Cliffs, NJ, 1983. {311}
47. C. L. Dolph. A current distribution of broadside arrays which optimizes the relationship between beamwidth and side-lobe level. *Proc. IRE*, 34:335–356, 1946. {326}
48. D. E. Dudgeon. Fundamentals of digital array processing. *Proc. IEEE*, 65:898–904, June 1977. {172}
49. D. E. Dudgeon and R. M. Mersereau. *Multidimensional Digital Signal Processing*. Prentice Hall, Englewood Cliffs, NJ, 1984. {40, 81, 162, 172}
50. D. J. Edelblute, J. M. Fisk, and G. L. Kinnison. Criteria for optimum-signal-detection theory for arrays. *J. Acoust. Soc. Am.*, 41:199–205, Jan. 1967. {344}
51. B. Ekstrand. Analytical steady state solution for a Kalman tracking filter. *IEEE Trans. Aerosp. Electron. Sys.*, AES-19:815–819, Nov. 1983. {438}

52. M. H. Er and A. Cantoni. Derivative constraints for broad-band element space antenna array processors. *IEEE Trans. Acoustics, Speech and Signal Processing*, ASSP-31:1378–1393, Dec. 1983. {396}
53. M. H. Er and A. Cantoni. A unified approach to the design of robust narrow-band antenna array processors. *IEEE Trans. Ant. Prop.*, 38:17–23, Jan. 1990. {393, 422}
54. J. E. Evans, J. R. Johnson, and D. F. Sun. Application of advanced signal processing techniques to angle of arrival estimation in ATC navigation and surveillance systems. Technical Report 582, MIT Lincoln Laboratory, Lexington, MA, 1982. {380, 390}
55. A. Feuer and E. Weinstein. Convergence analysis of LMS filters with uncorrelated Gaussian data. *IEEE Trans. Acoustics, Speech and Signal Processing*, ASSP-33:222–229, Feb. 1985. {314}
56. B. Friedlander. Adaptive algorithms for finite impulse response filters. In Cowan and Grant [36], pages 29–59. {313, 314, 317}
57. O. L. Frost, III. An algorithm for linearly constrained adaptive array processing. *Proc. IEEE*, 60:926–935, Aug. 1972. {356, 403}
58. A. Gelb, editor. *Applied Optimal Estimation.* MIT Press, Cambridge, MA, 1974. {305, 450}
59. J. D. Gibson and J. L. Melsa. *Introduction to Non-Parametric Detection with Applications.* Academic Press, New York, 1975. {216}
60. G. H. Golub and C. F. van Loan. *Matrix Computations.* Johns Hopkins University Press, Baltimore, 1983. {490, 493}
61. J. W. Goodman. *Introduction to Fourier Optics.* McGraw-Hill Book Company, San Francisco, 1968. {36, 63, 64, 76, 77, 113}
62. U. Grenander and G. Szego. *Toeplitz Forms and Their Applications.* University of California Press, Berkeley, CA, 1958. {376}
63. L. J. Griffiths. A simple adaptive algorithm for real-time processing in antenna arrays. *Proc. IEEE*, 57:1696, Oct. 1969. {423}
64. L. J. Griffiths and C. W. Jim. An alternative approach to linearly constrained adaptive beamforming. *IEEE Trans. Ant. Prop.*, AP-30:27–34, Jan. 1982. {369, 371}
65. P. Hall. *Rates of convergence in the central limit theorem*, volume 62 of *Research Notes in Mathematics.* Pitman Advanced Publishing Program, 1982. {481}
66. F. J. Harris. On the use of windows for harmonic analysis with the discrete Fourier transform. *Proc. IEEE*, 66:51–83, Jan. 1978. {323}
67. R. A. Haubrich. Array design. *Bull. Seismological Soc. Am.*, 58:977–991, 1968. {95}
68. S. Haykin. *Adaptive Filter Theory.* Prentice Hall, Englewood Cliffs, NJ, 1986. {297, 305, 310, 313, 314, 318, 339}
69. S. Haykin, editor. *Advances in Spectrum Analysis and Array Processing*, volume 1. Prentice Hall, Englewood Cliffs, NJ, 1991. {413}
70. S. Haykin, editor. *Advances in Spectrum Analysis and Array Processing*, volume 2. Prentice Hall, Englewood Cliffs, NJ, 1991. {512}
71. S. Haykin and J. Reilly. Mixed autoregressive-moving average modelling of the response of a linear array antenna to incident plane waves. *Proc. IEEE*, 68:622–623,

May 1980. {366}

72. C. W. Helstrom. *Statistical Theory of Signal Detection.* Pergamon Press, Oxford, second edition, 1968. {206, 220, 246}
73. R. T. Hoctor and S. A. Kassam. The unifying role of the coarray in aperture synthesis for coherent and incoherent imaging. *Proc. IEEE*, 78:735–752, April 1990. {72, 94, 186}
74. M. L. Honig and D. G. Messerschmitt. *Adaptive Filters.* Kluwer Academic Publishers, Boston, 1984. {310}
75. P. W. Howells. Explorations in fixed and adaptive resolution at GE and SURC. *IEEE Trans. Ant. Prop.*, AP-24:575–584, Sept. 1976. {358, 368}
76. J. E. Hudson. *Adaptive Array Principles.* Peter Peregrinus, London, 1981. {413}
77. N. K. Jablon. Adaptive beamforming with the generalized sidelobe canceller in the presence of array imperfections. *IEEE Trans. Ant. Prop.*, AP-34:996–1012, Aug. 1986. {393, 410}
78. E. T. Jaynes. On the rationale of the maximum entropy method. *Proc. IEEE*, 70:939–952, Sept. 1982. {334}
79. G. M. Jenkins and D. G. Watts. *Spectral Analysis and its Applications.* Holden-Day, Inc., San Francisco, 1968. {320, 326}
80. D. H. Johnson. The application of spectral estimation methods to bearing estimation problems. *Proc. IEEE*, 70:1018–1028, Sept. 1982. {331, 355, 357, 365, 384, 415}
81. D. H. Johnson and S. DeGraaf. Improving the resolution of bearing in passive sonar arrays by eigenvalue analysis. *IEEE Trans. Acoustics, Speech and Signal Processing*, ASSP-30:638–647, Aug. 1982. {384, 392}
82. N. L. Johnson and S. Kotz. *Continuous Univariate Distributions-2.* Houghton Mifflin, Boston, 1970. {479}
83. R. E. Kalman. A new approach to linear filtering and prediction problems. *J. Basic Engineering* (ASME Trans.), 82D:35–45, 1960. {305}
84. R. L. Kashyap. Inconsistency of the AIC rule for estimating the order of autoregressive models. *IEEE Trans. Auto. Control*, AC-25:996–998, Oct. 1980. {336}
85. S. A. Kassam and H. V. Poor. Robust techniques for signal processing: A survey. *Proc. IEEE*, 73:433–481, 1985. {339}
86. M. Kaveh and A. J. Barabell. The statistical performance of the MUSIC and the minimum-norm algorithms in resolving plane waves in noise. *IEEE Trans. Acoustics, Speech and Signal Processing*, ASSP-34:331–341, Apr. 1986. {385}
87. M. Kaveh and H. Wang. Threshold properties of narrow-band signal-subspace array processing methods. In Haykin [70], pages 173–220. {385}
88. S. Kay and C. Demeure. The high-resolution spectrum estimator—A subjective entity. *Proc. IEEE*, 72:1815–1816, Dec. 1984. {418}
89. S. M. Kay. *Modern Spectral Estimation.* Prentice Hall, Englewood Cliffs, NJ, 1988. {339}
90. E. J. Kelly, I. S. Reed, and W. L. Root. The detection of radar echoes in noise. I. *J. Soc. Indust. Appl. Math.*, 8:309–341, June 1960. {238}
91. E. J. Kelly, I. S. Reed, and W. L. Root. The detection of radar echoes in noise. II. *J. Soc. Indust. Appl. Math.*, 8:481–507, Sept. 1960. {238}

92. L. E. Kinsler et al. *Fundamentals of Acoustics.* John Wiley & Sons, New York, third edition, 1982. {11}

93. L. H. Koopmans. *The Spectral Analysis of Time Series.* Academic Press, New York, 1974. {326}

94. J. Krolik. Focused wide-band array processing for spatial spectral estimation. In Haykin [70], pages 221–261. {350}

95. R. Kumaresan and D. W. Tufts. Estimating the angles of arrival of multiple plane waves. *IEEE Trans. Aerosp. Electron. Sys.*, AES-19:134–139, Jan. 1983. {384}

96. R. T. Lacoss. Data adaptive spectral analysis methods. *Geophysics*, 36:661–675, Aug. 1971. {331, 355, 357, 359}

97. C. L. Lawson and R. J. Hanson. *Solving Least Squares Problems.* Prentice Hall, Englewood Cliffs, NJ, 1974. {401}

98. E. L. Lehmann. *Testing Statistical Hypotheses.* John Wiley & Sons, New York, second edition, 1986. {200, 202, 219}

99. D. A. Linebarger. *Parametric and Non-Parametric Methods of Improving Bearing Estimation in Narrowband Passive Sonar Systems.* PhD thesis, Dept. Electrical and Computer Eng., Rice University, Houston, TX, 1986. {354, 388}

100. L. Ljung, M. Morf, and D. Falconer. Fast calculation of gain matrices for recursive estimation techniques. *Inter. J. Control*, 27:1–19, Jan. 1978. {318}

101. J. Makhoul. Linear prediction: A tutorial review. *Proc. IEEE*, 63:561–580, Apr. 1975. {304, 334, 336}

102. J. Makhoul. On the eigenvectors of symmetric toeplitz matrices. *IEEE Trans. Acoustics, Speech and Signal Processing*, ASSP-29:868–872, Aug. 1981. {372}

103. J. N. Maksym. A robust formulation of an optimum cross-spectral beamformer for line arrays. *J. Acoust. Soc. Am.*, 65:971–975, Apr. 1979. {400, 411}

104. J. D. Markel and A. H. Gray, Jr. *Linear Prediction of Speech.* Springer-Verlag, New York, 1976. {304}

105. S. L. Marple, Jr. *Digital Spectral Analysis.* Prentice Hall, Englewood Cliffs, NJ, 1987. {304, 305, 321, 327, 330, 334, 336, 339, 344}

106. D. P. McGinn and D. H. Johnson. Estimation of all-pole model parameters from noise-corrupted sequences. *IEEE Trans. Acoustics, Speech and Signal Processing*, ASSP-37:433–436, Mar. 1989. {348}

107. A. T. Moffet. Minimum-redundancy linear arrays. *IEEE Trans. Ant. Prop.*, AP-16:172–175, Mar. 1968. {100}

108. R. A. Monzingo and T. W. Miller. *Introduction to Adaptive Arrays.* Wiley Interscience, New York, 1980. {413}

109. P. M. Morse and K. U. Ingard. *Theoretical Acoustics.* Princeton University Press, Princeton, NJ, 1986. {56}

110. R. J. Muirhead. *Aspects of Multivariate Statistical Theory.* John Wiley & Sons, New York, 1982. {391, 479}

111. B. R. Musicus. Fast MLM power spectrum estimation from uniformly spaced correlations. *IEEE Trans. Acoustics, Speech and Signal Processing*, ASSP-33:1333–1335, Oct. 1985. {335}

112. P. J. Napier, A. R. Thompson, and R. D. Ekers. The Very Large Array: Design and performance of a modern synthesis radio telescope. *Proc. IEEE*, 71:1295–1320, Nov. 1983. {109}

113. S. H. Nawab, F. U. Dowla, and R. T. Lacoss. Direction determination for wideband signals. *IEEE Trans. Acoustics, Speech and Signal Processing*, ASSP-33:1114–1122, Oct. 1985. {198}

114. J. Neyman and E. S. Pearson. On the problem of the most efficient tests of statistical hypotheses. *Phil. Trans. Roy. Soc. Ser. A*, 231:289–337, Feb. 1933. {205}

115. E. Nicolau and D. Zaharia. *Adaptive Arrays*, volume 35 of *Studies in Electrical and Electronic Engineering*. Elsevier Science Publishers, New York, 1989. {413}

116. G. Oetken, T. W. Parks, and H. W. Schüssler. New results in the design of digital interpolators. *IEEE Trans. Acoustics, Speech and Signal Processing*, ASSP-23:301–309, June 1975. {162}

117. A. V. Oppenheim and D. H. Johnson. Discrete representation of signals. *Proc. IEEE*, 60:681–691, June 1972. {84}

118. A. V. Oppenheim and R. W. Schafer. *Discrete-Time Signal Processing*. Prentice Hall, Englewood Cliffs, NJ, 1989. {xi, 80, 179, 321}

119. N. Owsley. Sonar array processing. In S. Haykin, editor, *Array Signal Processing*, pages 115–193. Prentice Hall, Englewood Cliffs, NJ, 1985. {359}

120. A. Papoulis. *Probability, Random Variables, and Stochastic Processes*. McGraw-Hill, New York, second edition, 1984. {272}

121. V. F. Pisarenko. The retrieval of harmonics from a covariance function. *Geophys. J. R. Astr. Soc.*, 33:347–366, 1973. {372}

122. H. V. Poor. *An Introduction to Signal Detection and Estimation*. Springer-Verlag, New York, 1988. {200, 220, 339}

123. W. H. Press, B. P. Flannery, S. A. Teukolsky, and W. T. Vetterling. *Numerical Recipes in C*. Cambridge University Press, New York, 1988. {84}

124. R. G. Pridham and R. A. Mucci. A novel approach to digital beamforming. *J. Acoust. Soc. Am.*, 63:425–434, Feb. 1978. {161}

125. R. G. Pridham and R. A. Mucci. Digital interpolation beamforming for lowpass and bandpass signals. *Proc. IEEE*, 67:904–919, June 1979. {161, 163}

126. R. L. Pritchard. Optimum directivity patterns for linear point arrays. *J. Acoust. Soc. Am.*, 25:879–891, 1953. {198}

127. L. R. Rabiner and R. W. Schafer. *Digital Processing of Speech Signals*. Prentice Hall, Englewood Cliffs, NJ, 1978. {182}

128. Lord Rayleigh (J. W. Strutt). *The Theory of Sound*. Dover Publications, New York, 1945. {56}

129. S. S. Reddi. Multiple source location — A digital approach. *IEEE Trans. Aerosp. Electron. Sys.*, AES-15:95–105, Jan. 1979. {383}

130. J. Rissanen. Modeling by shortest data description. *Automatica*, 14:465–471, Sept. 1978. {336}

131. R. Roy and T. Kailath. ESPRIT—Estimation of signal parameters via rotational invariance techniques. *IEEE Trans. Acoustics, Speech and Signal Processing*, ASSP-37:984–995, July 1989. {419}

132. H. Sakai and H. Tokumaru. Statistical analysis of a spectral estimator for ARMA processes. *IEEE Trans. Auto. Control*, AC-25:122–124, Feb. 1980. {347}
133. R. W. Schafer and L. R. Rabiner. A digital signal processing approach to interpolation. *Proc. IEEE*, 61:692–720, June 1973. {162}
134. D. J. Scheibner and T. W. Parks. Slowness aliasing in the discrete Radon transform: A multirate approach to beamforming. *IEEE Trans. Acoustics, Speech and Signal Processing*, ASSP-32:1160–1165, Dec. 1984. {167, 171}
135. R. O. Schmidt. Multiple emitter location and signal parameter estimation. In *Proc. RADC Spectral Estimation Workshop*, pages 243–258, Rome, NY, 1979. {384}
136. R. O. Schmidt. *A Signal Subspace Approach to Multiple Emitter Location and Spectral Estimation.* PhD thesis, Stanford University, Stanford, 1981. {383, 384}
137. R. Schreiber. Implementation of adaptive array algorithms. *IEEE Trans. Acoustics, Speech and Signal Processing*, ASSP-34:1038–1045, Oct. 1986. {408}
138. G. Schwarz. Estimating the dimension of a model. *Ann. Stat.*, 6:461–464, Mar. 1978. {336}
139. F. C. Schweppe. Sensor-array data processing for multiple-signal sources. *IEEE Trans. Info. Th.*, IT-14:294–305, Mar. 1968. {353}
140. C. D. Seligson. Comments on 'High-resolution frequency-wavenumber spectrum analysis.' *Proc. IEEE*, 58:947–949, June 1970. {388}
141. T.-J. Shan, M. Wax, and T. Kailath. On spatial smoothing for direction-of-arrival estimation of coherent signals. *IEEE Trans. Acoustics, Speech and Signal Processing*, ASSP-33:806–811, Aug. 1985. {388}
142. W. M. Siebert. *Circuits, Signals, and Systems.* M.I.T. Press, Cambridge, MA, 1986. {41, 80}
143. B. W. Silverman. *Density Estimation.* Chapman & Hall, London, 1986. {339}
144. M. I. Skolnik. *Radar Handbook.* McGraw-Hill Publishing Comapny, New York, second edition, 1990. {111, 137}
145. A. K. Steele. Comparison of directional and derivative constraints for beamformers subject to multiple linear constraints. *IEE Proc., Pts. F and H*, 130:41–45, Feb. 1983. {394, 421}
146. B. D. Steinberg. *Principles of Aperture and Array System Design.* John Wiley & Sons, New York, 1976. {103}
147. K. Takao, H. Fujita, and T. Nishi. An adaptive array under directional constraint. *IEEE Trans. Ant. Prop.*, AP-24:662–669, Sept. 1976. {394}
148. R. R. Tenney and J. R. Delaney. A distributed aeroacoustic tracking algorithm. In *Proc. Amer. Control Conf.*, pages 1440–1450, 1984. {467}
149. T. J. Ulrych and R. W. Clayton. Time series modeling and maximum entropy. *Phys. Earth Planetary Interiors*, 12:188–200, Aug. 1976. {417}
150. R. J. Urick. *Principles of Underwater Sound.* McGraw-Hill, New York, second edition, 1975. {32}
151. P. P. Vaidyanathan and S. K. Mitra. Polyphase networks, block digital filtering, LPTV systems, and alias-free QMF banks: A unified approach based on pseudocirculants. *IEEE Trans. Acoustics, Speech and Signal Processing*, 36:381–391, March 1988. {163}

152. H. L. van Trees. *Detection, Estimation, and Modulation Theory, Part I.* John Wiley & Sons, New York, 1968. {51, 200, 206, 218-220, 238, 257, 259, 260, 297, 299, 339, 341}
153. R. G. Vaughan, N. L. Scott, and D. R. White. The theory of bandpass sampling. *IEEE Trans. Signal Processing*, 39:1973–1984, Sept. 1991. {157}
154. E. Vertatschitsch and S. Haykin. Nonredundant arrays. *Proc. IEEE*, 74:217, Jan. 1986. {100}
155. A. M. Vural. Effects of perturbations on the performance of optimum/adaptive arrays. *IEEE Trans. Aerosp. Electron. Sys.*, AES-15:76–87, Jan. 1979. {393}
156. M. T. Wasan. *Stochastic Approximation.* Cambridge University Press, 1969. {313}
157. M. Wax and T. Kailath. Detection of signals by information theoretic criteria. *IEEE Trans. Acoustics, Speech and Signal Processing*, ASSP-33:387–392, Apr. 1985. {390, 392}
158. M. Wax and T. Kailath. Optimum localization of multiple sources by passive arrays. *IEEE Trans. Acoustics, Speech and Signal Processing*, ASSP-31:1210–1217, Oct. 1983. {353}
159. A. J. Weiss and E. Weinstein. Fundamental limitations in passive time delay estimation: I. Narrow-band systems. *IEEE Trans. Acoustics, Speech and Signal Processing*, ASSP-31:472–486, Apr. 1983. {290, 292}
160. P. D. Welch. The use of the fast Fourier transform for the estimation of power spectra: A method based on time averaging over short modified periodograms. *IEEE Trans. Audio Electroacoust.*, AU-15:70–73, Mar. 1967. {327}
161. B. Widrow and S. D. Stearns. *Adaptive Signal Processing.* Prentice Hall, Englewood Cliffs, NJ, 1985. {310}
162. N. Wiener. *Extrapolation, Interpolation, and Smoothing of Stationary Time Series.* MIT Press, Cambridge, MA, 1949. {297}
163. R. P. Wishner. Distribution of the normalized periodogram detector. *IRE Trans. Info. Th.*, IT-8:342–349, Sept. 1962. {247}
164. E. Wong and B. Hajek. *Stochastic Processes in Engineering Systems.* Springer-Verlag, New York, 1985. {46}
165. P. M. Woodward. *Probability and Information Theory, with Applications to Radar.* Pergamon Press, Oxford, second edition, 1964. {205}
166. J. Ziv and M. Zakai. Some lower bounds on signal parameter estimation. *IEEE Trans. Info. Th.*, IT-15:386–391, May 1969. {290}

List of Symbols

Operators and Conventions

$\emptyset$ The empty or null set {471}.

$\mathbf{A}, \boldsymbol{\Phi}$ A matrix; a bold-faced, upper-cased Roman or Greek letter {485}.

$\mathbf{a}, \boldsymbol{\phi}$ A vector (column matrix); a bold-faced, lower-cased Roman or Greek letter {485}.

$\mathbf{A}^t$ Transpose of a matrix {486}.

$\mathbf{A}'$ Conjugate transpose of a matrix {486}.

$\mathbf{x}^c$ A concatenated vector: $\mathbf{x}^c = \text{col}[\mathbf{x}_1, \ldots, \mathbf{x}_N]$ {249}.

$\langle \mathbf{a}, \mathbf{b} \rangle$ The inner product between two vectors {489}.

$\|\mathbf{a}\|$ The norm of a vector {273}.

$\lfloor x \rfloor$ The floor of x: the greatest integer less than or equal to x.

$(n)_N$ The integer n evaluated modulo N: the remainder of the division $n \div N$.

$\det[\mathbf{A}]$ The determinant of a matrix {490}.

$\dim[\cdot]$ The dimension of a vector ($\dim[\mathbf{a}]$) or of a square matrix ($\dim[\mathbf{A}]$) {485}.

$(\lambda_i, \mathbf{v}_i)_\mathbf{A}$ The eigensystem of the matrix $\mathbf{A}$ {493}.

$\text{tr}[\mathbf{A}]$ The trace of a matrix {491}.

$\text{col}[\cdot]$ A column matrix formed from the listed values.

$$\text{col}\big[a_1, \ldots, a_N\big] = \begin{bmatrix} a_1 \\ \vdots \\ a_N \end{bmatrix}$$

$\text{diag}[\cdot]$ A diagonal matrix formed from the listed values.

$$\text{diag}\big[a_1, \ldots, a_N\big] = \begin{bmatrix} a_1 & & \\ & \ddots & \\ & & a_N \end{bmatrix}$$

$\mathbf{1}$ A column matrix with each element equaling one: $\mathbf{1} = \text{col}[1, \ldots, 1]$ {54}.

$\mathcal{E}[\cdot]$ The expectation of a random variable: $\mathcal{E}[x] = \int x p_x(x)\, dx$ {473}.

$\mathcal{V}[\cdot]$ Variance of a random variable: $\mathcal{V}[x] = \mathcal{E}[x^2] - (\mathcal{E}[x])^2$ {473}.

$\Pr[A]$ The probability of the event A {471}.

F The symbol denoting the Fourier Transform (continuous- or discrete-time) of the signal f; an uppercase version of the signal's symbol, be it Roman or Greek.

$$F(j\omega) = \int_{-\infty}^{\infty} f(t)e^{-j\omega t}\,dt \qquad F(e^{j\omega}) = \sum_{n=-\infty}^{\infty} f(n)e^{-j\omega n}$$

$\widehat{a}, \widehat{s}(\cdot)$ Estimate of the quantity a {267}; the Hilbert Transform of the signal $s(\cdot)$ {154}.

$\widehat{x}(l_1|l_2)$ Estimate of the signal x at index l_1 based on observations made at times up to and including l_2 {306}.

$s^+(\cdot), s^b(\cdot)$ The former denotes the analytic signal corresponding to $s(\cdot)$, the latter the baseband counterpart of $s(\cdot)$ {154}.

$s^I(\cdot), s^Q(\cdot)$ The in-phase and quadrature components of a signal {155}.

$\mathcal{L}[\cdot]$ A generic linear operator: $\mathcal{L}[ax + by] = a\,\mathcal{L}[x] + b\,\mathcal{L}[y]$ {272}.

$\mathcal{Q}[\cdot]$ The quantization operator {152}.

$\mathcal{R}[x]$ The Radon Transform of the two-dimensional sequence $x(m,n)$ {167}.

$\dot{f}(\cdot)$ The derivative of the function $f(\cdot)$ with respect to its argument.

z^* The complex conjugate of z.

$\mathrm{Re}[\cdot]$ Real part of a complex quantity: $\mathrm{Re}[a + jb] = a$.

$\mathrm{Im}[\cdot]$ Imaginary part of a complex quantity: $\mathrm{Im}[a + jb] = b$.

$\uparrow_I[\cdot]$ Upsample the signal by a factor of I {162}.

$\sim$ Shorthand notation for "distributed as" that applies to a random variable or vector: $X \sim \mathcal{N}(m, \sigma^2)$ means the random variable X has a Gaussian distribution {476}.

$\star$ The convolution operator; applies to both continuous- and discrete-time convolutions.

$$a(t) \star b(t) = \int a(\tau)b(t-\tau)\,d\tau \qquad a(n) \star b(n) = \sum_k a(k)b(n-k)$$

∇ The gradient operator; can be the operator defined over spatial coordinates or over an abstract linear vector space {11}, {500}.

$$\nabla = \frac{\partial}{\partial x}\vec{\imath}_x + \frac{\partial}{\partial y}\vec{\imath}_y + \frac{\partial}{\partial z}\vec{\imath}_z \qquad \nabla_{\mathbf{x}} f(\mathbf{x}) = \mathrm{col}\left[\frac{\partial f}{\partial x_1}, \ldots, \frac{\partial f}{\partial x_N}\right]$$

∇^2 The Laplacian operator defined over spatial coordinates {11}.

$$\nabla^2 f(x,y,z) = \frac{\partial^2 f}{\partial x^2} + \frac{\partial^2 f}{\partial y^2} + \frac{\partial^2 f}{\partial z^2}$$

$\nabla^2_{\mathbf{x}} f(\mathbf{x})$ The Hessian matrix, which comprises all the second partials of $f(\cdot)$ {501}.

$\overset{H_1}{\underset{H_0}{\gtrless}}$ Comparison between two scalars that determines which of two hypotheses to accept {201}.

$$a \overset{H_1}{\underset{H_0}{\gtrless}} b \Longrightarrow a > b, \text{ choose } H_1; a < b, \text{ choose } H_0.$$

$\Re_0, \Re_1$ Decision regions for hypothesis testing and detection problems {200}.

Greek Symbols

$\vec{\alpha}, \alpha, \vec{\alpha}^o$ The generic slowness vector, its magnitude, and the slowness vector of a specific propagating wave {14}.

β The root-mean-squared bandwidth {293}; constraining value on the norm of the shading vector {399}.

$\beta(x, m, n)$ The probability density of a beta random variable having (m, n) degrees of freedom {480}.

γ The threshold for the sufficient statistic in the likelihood ratio test {202}.

$\delta(x - x_0)$ The Dirac delta or impulse function located at x_0.

$\boldsymbol{\delta}_k$ A column matrix consisting of zeros save for the kth position where one appears: $\boldsymbol{\delta}_1 = \text{col}[1, 0, \ldots, 0]$. We take $\boldsymbol{\delta} \equiv \boldsymbol{\delta}_1$ {334}, {365}.

$\delta x, \boldsymbol{\delta}$ A small quantity; a vector of small quantities {399}.

Δ The delay imposed on a signal, be it continuous- or discrete-time: $s(t - \Delta)$ or $s(l - \Delta)$. In the latter case, Δ is always an integer.

ϵ The estimation error {267}.

$\vec{\zeta}^o, \vec{\zeta}$ A unit vector denoting the direction of propagation {14}; the assumed propagation direction of an array processing algorithm {15}.

η The threshold in the likelihood ratio test {201}.

θ, Θ Azimuthal angle and the incidence angle of a propagating wave onto a linear array {10}.

κ The radius of curvature {36}.

λ Wavelength {14}; eigenvalue {493}; Lagrange multiplier {503}.

μ Adaptation parameter in the *LMS* algorithm {311}; maneuvering index {438}.

Λ The likelihood ratio {201}.

$\nu, \boldsymbol{\nu}$ The innovations sequence of a Kalman filter {308}.

$\xi, \boldsymbol{\xi}$ A generic symbol for a parameter (vector) {267}.

$\rho, \boldsymbol{\rho}$ The correlation coefficient between two random variables {474} and a matrix of correlation coefficients {50}.

σ^2 The variance of a random variable {473}.

τ Temporal lag variable {482}; a generic variable for temporal quantities.

Υ The sufficient statistic of a detection or estimation system {202}.

ϕ, Φ Elevation angle {10}; incidence angle of a plane wave onto a linear array {63}.

$\Phi_x(j\nu)$ The characteristic function of the random variable x: $\Phi_x(j\nu) = \int_{-\infty}^{\infty} p_x(x) e^{j\nu x}\, dx$ {473}.

$\chi, \vec{\chi}$ Spatial lag magnitude and vector: $\chi = |\vec{\chi}|$ {46}.

ψ The angle between two quantities.

ω, Ω The temporal frequency variable; the lowercase variable applies to both continuous-time and discrete-time signal spectra, the uppercase only in continuous-time.

Calligraphic Symbols

$\mathcal{A}$ The operator in a source's dynamic motion equation that relates current position to previous position {432}.

$\mathcal{C}[\cdot]$ The observation operator that transforms Cartesian coordinates into polar coordinates {426}.

$\mathcal{E}[\cdot]$ The expectation of a random variable: $\mathcal{E}[x] = \int x p_x(x)\, dx$ {473}.

$\mathcal{H}[\mathcal{S}]$ Entropy of a power spectrum {335}.

$\mathcal{L}[\cdot]$ A generic linear operator: $\mathcal{L}[ax + by] = a\,\mathcal{L}[x] + b\,\mathcal{L}[y]$ {272}.

$\mathcal{M}$ Symbol used to represent source motion models, which include the dynamics and measurement equations as well as their parameters {445}.

$\mathcal{N}(m, \sigma^2)$ The symbol denoting the Gaussian (normal) probability density function having mean m and variance σ^2 {476}.

$\mathcal{P}$ The beam power associated with an array and its signal processing algorithm {135}.

$\mathcal{Q}[\cdot]$ The quantization operator {152}.

$\mathcal{R}[x]$ The Radon Transform of the two-dimensional sequence $x(m, n)$ {167}.

$\mathcal{S}_x(\omega)$ The power spectrum of the stationary random process x {483}.

$\mathcal{V}[\cdot]$ Variance of a random variable: $\mathcal{V}[x] = \mathcal{E}[x^2] - \left(\mathcal{E}[x]\right)^2$ {473}.

$\mathcal{W}$ The array pattern associated with an array and its array processing algorithm {120}.

Roman Symbols

A The amplitude of a signal.

$\mathbf{A}$ The matrix used in state variable characterizations to denote signal dynamics {306}. Symbol used to denote the exemplar matrix when describing matrix calculations and properties.

$b, \mathbf{b}$ The bias (vector) of an estimator {267}.

B The bandwidth of a bandpass signal or system {153}.

B Blocking matrix for the generalized sidelobe canceller {267}.

c The speed of propagation {11}; the co-array associated with a continuous aperture {72} or with an array {94}.

C Cost in a Bayes's decision rule {200}.

c, **C** The intersignal coherence and intersignal coherence matrix {54}.

c, **C** The vector (matrix) of constraints employed in deriving adaptive array processing algorithms using constrained optimization {355}.

C The matrix in state variable representations that relates state to observations {306}.

d The separation between adjacent sensors (spatial samples) in a regular linear or rectangular array {77}.

D The temporal duration of a frame, particularly for short-time Fourier transforms {133}.

D A diagonal matrix {486}.

e The elemental frequency vector, which also models the phase shift of a narrowband signal across an array: $\mathbf{e}(\omega) = \text{col}\left[1, e^{j\omega}, e^{j2\omega}, \ldots, e^{j(M-1)\omega}\right]$ {54}.

E Signal energy {190}.

$f(\vec{x}, t)$, $F(\vec{k}, \omega)$ The field, usually consisting of signal + noise, measured by a sensor {1}; the spatiotemporal Fourier Transform of a field {40}.

F The Fisher information matrix {277}.

g, **G** Generic symbol for gain or a gain matrix.

G Array gain {138}.

$h(\cdot)$, $H(\cdot)$ The impulse response (continuous-time) or unit-sample response (discrete-time) of a linear, time- (shift-) invariant system and its associated frequency response.

$\vec{\imath}_x, \vec{\imath}_y, \vec{\imath}_z$ The unit vectors denoting the three spatial directions {10}.

I The identity matrix {486}.

$I(\cdot)$ The indicator function {281}.

j The square root of -1.

J The exchange matrix {487}.

$\vec{k}, k, \breve{k}$ The symbol denoting a wavenumber vector {13}. The magnitude of the wavenumber vector, denoted by k {14}, equals $2\pi/\lambda$. $\breve{k}$ denotes the spatial frequency derived from spatial samples {78}.

$\vec{k}^o$ The wavenumber vector of a specific source's propagating energy toward an array {40}.

$K(\cdot)$, **K** The covariance function {483} and matrix of a signal.

$$K_x(\tau) = \mathcal{E}[x(t)x(t+\tau)] - m^2 \qquad \mathbf{K}_x = \mathcal{E}[\mathbf{x}\mathbf{x}'] - \mathbf{m}\mathbf{m}'$$

ℓ Arc length along a path {18}.

L A lower triangular matrix {487}.

m, **m** The mean (vector of means) of a (vector) random variable {473}.

m, M The index for and the number of sensors comprising an array {1}.

$M_{1/2}$ "Half" the number of sensors in an array: $M_{1/2} = (M - 1)/2$ {87}.

n Index of refraction: $n = c_0/c$ {31}; temporal index for discrete-time signals.

$n(\cdot)$ A noise process that interferes with all array processing algorithms.

N_a The number of arrays participating in multiarray tracking {434}.

N_A The number of candidate associations considered in data association algorithms {458}.

N_s The number of signals present in the observations; the number of sources present in the field {53}.

p, q The intercept and slope (respectively) associated with the Radon transform {167}.

$P(\vec{\xi}\,)$ The power in the array output when the assumed propagation direction equals $\vec{\xi}$ {355}.

$\mathbf{P_C}$ The projection matrix associated with the matrix **C**: $\mathbf{P_C} = \mathbf{C}'(\mathbf{CC}')^{-1}\mathbf{C}$ {488}.

$p_x(\cdot), P_x(\cdot)$ The probability density (distribution) function of the random variable x {472}.

P_D A detection system's detection probability {205}.

P_e A detection system's probability of error {204}.

P_F A detection system's false-alarm probability {205}.

P_M A detection system's miss probability {205}.

PW The parabolic width of a mainlobe {143}.

r The range or distance between two spatial locations; the distance from the origin in polar or spherical coordinates {10}.

r, $R(\cdot)$, **R** A correlation sequence or function {482} and the associated correlation matrix.

$$R_x(\tau) = \mathcal{E}\big[x(t)x(t+\tau)\big] \qquad \mathbf{R}_x = \mathcal{E}\big[\mathbf{xx}'\big]$$

$s(\cdot)$ A signal, defined with respect to either continuous time or discrete time, whose properties are to be determined. $s_c(t)$ denotes explicitly a continuous-time signal {152}.

SNR Signal-to-noise ratio.

T Sampling interval {152}.

$\mathrm{u}(\cdot)$ The unit step function: $\mathrm{u}(x) = \begin{cases} 1, & x>0 \\ 0, & x<0 \end{cases}$.

u, **u** The input to the dynamic state equation {306}.

u, v The discrete wavenumber and discrete temporal frequency indices {175}.

U The periodicity matrix in multidimensional sampling {81}.

v, $\mathbf{v}$ The symbol denoting a system's state (vector) {306}.

$\mathbf{v}$ Generic symbol for velocity.

$\mathbf{V}$ The matrix formed from the eigenvectors of a matrix {379}; the sampling matrix in multidimensional sampling {81}.

$w, \mathbf{w}, W, \mathbf{W}$ The shading or weight (vector) applied to sensor outputs {87}. The aperture function for continuous apertures {59}. The diagonal matrix formed by the shading sequence ($\mathbf{W} = \text{diag}[w_0, \ldots, w_{M-1}]$) {135} and the *DFT* matrix {487}.

$W(\cdot)$ The aperture smoothing function for a continuous aperture {60} or the array pattern for an array {88}.

$\vec{x}, \vec{x}^o$ Spatial position vector: $\vec{x} = (x, y, z) = (\vec{x}_x, \vec{x}_y, \vec{x}_z)$ {10}; the location of a source, near- or far-field, relative to the array's coordinate system.

$y(\cdot), \mathbf{y}, \mathbf{Y}$ The waveform (vector of waveforms) produced by a sensor; usually equals signal plus noise: $y(l) = s(l) + n(l)$ {3}. $\mathbf{Y}$ denotes the vector comprised of short-time Fourier transforms calculated from sensor outputs {134}.

$z(\cdot)$ The output of a system, particularly an array processor {3}.

$\mathbf{z}, \widetilde{\mathbf{z}}$ A collection of observations used in data association algorithms {458}.

$\mathbf{Z}_l$ The collection of observations made up to and including time l {432}.

Index

The following index not only lists pages referring to the indicated topics, but also highlights definitions with italicized page numbers and examples with boldfaced ones.

A

I

J

K

L

M

N

O

P

Q

T

U

V

W

Y

Z